CHEMISTRY AND BIOLOGY OF HYALURONAN

CHEMISTRY AND BIOLOGY OF HYALURONAN

Edited by

Hari G. Garg and Charles A. Hales

Harvard Medical School
Pulmonary and Critical Care Unit
Massachusetts General Hospital
Boston
MA 02114-2696
U.S.A.

2004

ELSEVIER

Amsterdam – Boston – Heidelberg – London – New York – Oxford
Paris – San Diego – San Francisco – Singapore – Sydney – Tokyo

ELSEVIER B.V.
Sara Burgerhartstraat 25
P.O. Box 211, 1000 AE
Amsterdam, The Netherlands

ELSEVIER Inc.
525 B Street, Suite 1900
San Diego, CA 92101-4495
USA

ELSEVIER Ltd
The Boulevard, Langford Lane
Kidlington, Oxford OX5 1GB
UK

ELSEVIER Ltd
84 Theobalds Road
London WC1X 8RR
UK

First edition 2004

Library of Congress Cataloging in Publication Data
A catalog record is available from the Library of Congress.

British Library Cataloguing in Publication Data
A catalogue record is available from the British Library.

ISBN: 0 08 044382 6

∞ The paper used in this publication meets the requirements of ANSI/NISO Z39.48-1992 (Permanence of Paper).
Printed in The Netherlands.

To my wife, Mithlesh and our children, Ashima and Arvin for their loyal support and generous love

Hari G. Garg

To my wife, Mary Ann and our children, Sam, Chris, and John for their loyal support and generous love

Charles A. Hales

Preface

It was probably the French chemist Portes who first reported in 1880 that the mucin in the vitreous body, which he named hyalomucine, behaved differently from other mucoids in cornea and cartilage. Fifty-four years later, Karl Meyer isolated a new polysaccharide from the vitreous, which he named hyaluronic acid. Today its official name is hyaluronan.

A hyaluronan molecule is generally of high molecular weight and occupies through its random-coil configuration a large hydrodynamic volume in solution. At higher concentrations, hyaluronan molecules entangle and form continuous networks. Such solutions are endowed with interesting rheological properties. They are visco-elastic and the viscosity is strongly shear-dependent. For this reason, hyaluronan can act as a lubricant. It is found in joints and other tissues such as muscles at surfaces which are moving over each other. The human body is a well-oiled machine and hyaluronan seems to be that oil.

However, hyaluronan has also been assigned functions such as space filling, filter effects, promotion of cell migration, regulation of the cell cycle, a regulator in embryonic development and many others. In recent years it has also become apparent that hyaluronan plays a key role in various pathological processes such as inflammatory edema. The hyaluronan molecule may be degraded by enzymes or free radicals so that short fragments are formed. They have unexpected biological properties, e.g. to promote angiogenesis and wound healing. There are thus many interesting aspects to cover in a book on this unusual polysaccharide.

The 27 chapters in this volume present a sequence leading from the chemistry and biochemistry of hyaluronan, followed by its role in various pathological conditions, to modified hyaluronans as potential therapeutic agents and finally to the functional, structural, and biological properties of hyaluronidases.

Chapter 1 focuses on the solution properties of hyaluronan. Chapter 2 describes methods for analysis of hyaluronan and its fragments. Chapter 3 includes the methods for determination of hyaluronan molecular size. About one-third of the total hyaluronan in the human body is metabolically removed and replaced daily. Chapter 4 focuses on the biodegradation of hyaluronan.

The cellular hyaluronan binding proteins, CD44 and RHAMM, contain key basic residues that wrap around and secure the hyaluronan chain. Chapters 5 and 6 describe these hyaluronan binding proteins. Hyaluronan functions in a variety of biological processes. Chapter 7 reviews signal transduction associated with hyaluronan.

Unlike other glycosaminoglycans, hyaluronan is non sulfated. Many diverse biological functions have been attributed to hyaluronan due to its associated proteins. Chapters 8 to 10 address this aspect.

Hyaluronan was first implicated in lung disorders about 60 years ago. Since then it has been found to play a role in other diseases. Chapters 11 to 18 focus on its role in different diseases including cancer, injury caused by the mechanical ventilator, aging, and scarring etc.

Introduction of radioisotopic labeling and microscopic staining techniques prompted the study of hyaluronan present in thin epithelial tissues. Chapter 19 focuses on its role in the epidermis and other epithelial tissues.

Due to unique physiochemical properties and distinctive biological functions of hyaluronan, this polyionic polymer is an attractive material in drug delivery, tissue engineering, and viscosupplementation. Chapters 20 to 24 describe the chemical modification and the potential uses of these biomaterials in tissue engineering and drug delivery.

The enzymes that make hyaluronan were so difficult to solubilize and purify that 65 years elapsed between the identification of hyaluronan and purification of an active synthase. Two chapters focus on hyaluronan synthases. Chapter 25 summarizes structure, function and mechanisms of hyluronan synthases. Chapter 26 deals with the molecular genetic dissection of hyaluronan in the mouse.

This volume on hyaluronan would be incomplete without a chapter on degradative hyaluronan macromolecules present in tissues. Therefore, Chapter 27 focuses on the functional, structural, and biological properties of hyaluronidases.

In summary, this book offers a detailed panoramic review of the chemistry and biology of hyaluronan.

Hari G. Garg
Charles A. Hales

Contributors

Pasquale Aragona, M.D., Ph.D. Assistant Professor of Ophthalmology, Head of Ocular Surface Unit, Department of Surgical Sciences, Section of Ophthalmology, University of Messina, Italy

Akira Asari, Ph.D. Manager, Glyco-Research Project, Central Research Laboratories, Seikagaku Corporation, Tokyo, Japan

Luis Z. Avila, Ph.D. Principal Research Scientist, Genzyme Corporation, Cambridge, Massachusetts, U.S.A.

Endre A. Balazs, M.D. Chairman, Matrix Biology Institute, Edgewater, New Jersey, and Professor Emeritus, College of Physicians and Surgeons, Columbia University, New York, U.S.A.

Charles D. Blundell, M.Biochem., D.Phil. Post-Doctoral Research Assistant, MRC Immunochemistry Unit, Department of Biochemistry, University of Oxford, United Kingdom

Bonnie Anderson Bray, Ph.D. Special Research Scientist, James P. Mara Center for Lung Disease, St. Luke's-Roosevelt Hospital Center, Columbia University College of Physicians and Surgeons, New York, New York, U.S.A.

Andrew Burd, M.D., FRCS, FHKAM (Surgery). Professor, Department of Plastic and Reconstructive Surgery, Department of Surgery, Chinese University of Hong Kong, Prince of Wales Hospital, Shatin, Hong Kong, S.A.R., P.R.C.

Ishan Capila, Ph.D. Post-Doctoral Associate, Division of Biological Engineering, Massachusetts Institute of Technology, Cambridge, Massachusetts, U.S.A.

Gregory E. Conner, Ph.D. Associate Professor of Cell Biology and Medicine, Department of Cell Biology and Anatomy, University of Miami School of Medicine, Miami, Florida, U.S.A.

Mary K. Cowman, Ph.D. Associate Professor of Biochemistry, Department of Chemical and Biological Sciences and Engineering, Polytechnic University, Brooklyn, New York, U.S.A.

Anthony J. Day, M.A., D. Phil. Senior Scientist, MRC Immunochemistry Unit, Department of Biochemistry, University of Oxford, United Kingdom

Regina M. Day, Ph.D. Division of Pulmonary, Critical Care and Sleep Medicine, Department of Medicine and Pharmacology, Georgetown University, Washington, D.C., U.S.A.

Stephen Evanko, Ph.D. Research Scientist II, The Hope Heart Institute, Seattle, Washington, U.S.A.

Andrew Farb, Ph.D. Cardiovascular Pathologist, Armed Forces Institute of Pathology, Washington, D.C., U.S.A.

Christine Fehrer, M.Sc., Ph.D. student, Institute for Biomedical Aging Research, Austrian Academy of Sciences, Innsbruck, Austria

Rosanna Forteza, M.D. Assistant Professor of Medicine, Division of Pulmonary and Critical Care Medicine, University of Miami School of Medicine, Miami, Florida, U.S.A.

Hari G. Garg, Ph.D., D.Sc. Associated Biochemist, Pulmonary and Critical Care Unit, Department of Medicine, Massachusetts General Hospital, and Principal Associate, Harvard Medical School, Boston, Massachusetts, U.S.A.

Charles A. Hales, M.D. Chief, Pulmonary and Critical Care Unit, Department of Medicine, Massachusetts General Hospital, and Professor, Harvard Medical School, Boston, Massachusetts, U.S.A.

Sara R. Hamilton, B.Sc. Department of Biochemistry, London Regional Cancer Center, London, Canada

Tim Hardingham, Ph.D., D.Sc. Professor of Biochemistry, Wellcome School of Biological Sciences, University of Manchester, Manchester, United Kingdom

Naoki Itano, Ph.D. Assistant Professor, Institute for Molecular Science of Medicine, Aichi Medical University, Nagakute, Aichi and CREST, Japan Science and Technology Agency, Kawaguchi, Saitama, Japan

Koji Kimata, Ph.D. Professor, Institute for Molecular Science of Medicine, Aichi Medical University, Nagakute, Aichi, Japan

Warren Knudson, Ph.D. Professor, Department of Biochemistry, Rush Medical College, Rush University Medical Center, Chicago, Illinois, U.S.A.

Frank Kolodgie, Ph.D. Senior Research Scientist, Armed Forces Institute of Pathology, Washington, D.C., U.S.A.

Günther Kreil, Ph.D. Professor Emeritus, Institute of Molecular Biology, Austrian Academy of Sciences, Salzburg, Austria

Günter Lepperdinger, Ph.D. Section Head, Institute for Biomedical Aging Research, Austrian Academy of Sciences, Innsbruck, Austria

María O. Longas, Professor of Chemistry, Department of Chemistry and Physics, Purdue University Calumet, Hammond, Indiana, U.S.A.

Marcella M. Mascarenhas, Ph.D. Fellow, Pulmonary and Critical Care Unit, Department of Medicine, Massachusetts General Hospital, and Research Fellow in Medicine, Harvard Medical School, Boston, Massachusetts, U.S.A.

Raniero Mendichi, Ph.D. Researcher in Molecular Characterization, Instituto per lo Studio delle Macromolecole (Consiglio Nazionale delle Ricerche), Milan, Italy

Robert J. Miller, Ph.D. Senior Scientific Director, Genzyme Corporation, Cambridge, Massachusetts, U.S.A.

Tsunemasa Nonogaki, M.D., Ph.D. Professor, Tokai College of Medical Engineering, Nishi-Kamo, Aichi, Japan

Martin J. Page, Ph.D. Senior Director, UK Cancer Biology, (OSI) Pharamaceuticals, Oxford, United Kingdom

Sonal Patel, Ph.D. Team Leader, Piramed Ltd., Slough, United Kingdom

Richard S. Peterson, Ph.D. Instructor, Department of Biochemistry, Rush Medical College, Rush University Medical Center, Chicago, Illinois, U.S.A.

Glenn D. Prestwich, Ph.D. Presidential Professor, Department of Medicinal Chemistry, University of Utah, Salt Lake City, Utah, U.S.A.

Deborah A. Quinn, M.D. Assistant Physician, Massachusetts General Hospital, and Assistant Professor of Medicine, Harvard Medical School, Boston, Massachusetts, U.S.A.

Stephan Reitinger, M.Sc., Ph.D. student, Institute for Biomedical Aging Research, Austrian Academy of Sciences, Innsbruck, Austria

Matthias Salathe, M.D. Associate Professor of Medicine and Molecular Cellular Pharmacology, Division of Pulmonary and Critical Care Medicine, University of Miami School of Medicine, Miami, Florida, U.S.A.

Ram Sasisekharan, Ph.D. Professor, Division of Biological Engineering, Massachusetts Institute of Technology, Cambridge, Massachusetts, U.S.A.

Nicholas T. Seyfried, B.Sc. MRC Immunochemistry Unit, Department of Biochemistry, University of Oxford, Oxford, United Kingdom

Li Shen, M.D. Researcher, Institute for Molecular Science of Medicine, Aichi Medical University, Nagakute, Aichi, Japan

Xiaozheng Shu, Ph.D. Department of Medicinal Chemistry, University of Utah, Salt Lake City, Utah, U.S.A.

Andrew P. Spicer, Ph.D. Associate Professor, Center for Extracellular Matrix Biology, Institute of Biosciences and Technology, Texas A&M University System Health Science Center, Houston, Texas, U.S.A.

Markku I. Tammi, M.D., Ph.D. Professor, Department of Anatomy, University of Kuopio, Kuopio, Finland

Raija H. Tammi, M.D., Ph.D. Assistant Professor, Department of Anatomy, University of Kuopio, Kuopio, Finland

Susan L. Thibeault, Ph.D., CCC/SLP. Research Assistant Professor, Division of Otolaryngology—Head and Neck Surgery, University of Utah, School of Medicine, Salt Lake City, Utah, U.S.A.

Janet Y. Lee Tien, B.S. Department of Human Anatomy and Cell Biology, University of California Davis, School of Medicine, Davis, California, U.S.A.

Cornelia Tölg, Ph.D. Scientist, Department of Biochemistry, London Regional Cancer Center, London, Canada

Eva A. Turley, Ph.D. Senior Scientist, Department of Biochemistry, London Regional Cancer Center, London, Canada

Renu Virmani, M.D. Chairperson, Department of Cardiovascular Pathology, Armed Forces Institute of Pathology, Washington, D.C., U.S.A.

Hideto Watanabe, M.D., Ph.D. Associate Professor, Institute for Molecular Science of Medicine, Aichi Medical University, Nagakute, Aichi, Japan

Paul H. Weigel, Ph.D. Professor and Chairman, Department of Biochemistry and Molecular Biology and The Oklahoma Center for Medical Glycobiology, University of Oklahoma Health Sciences Center, Oklahoma City, U.S.A.

Thomas N. Wight, Ph.D. Chair, The Hope Heart Institute, Seattle, Washington, U.S.A.

Jiwen Wu, M.D. Foreign Researcher, Institute for Molecular Science of Medicine, Aichi Medical University, Nagakute, Aichi, Japan

Masahiko Yoneda, Ph.D. Associate Professor, Aichi Prefectural College of Nursing and Health Nagoya, Aichi, Japan

Masahiro Zako, M.D. Associate Professor, Department of Ophthalmology, Aichi Medical University, Nagakute, Aichi, Japan

Lisheng Zhuo, Ph.D. Research Associate, Institute for Molecular Science of Medicine, Aichi Medical University, Nagakute, Aichi and CREST, Japan Science and Technology Agency, Kawaguchi, Saitama, Japan

Contents

Preface vii

Contributors xi

1. Solution Properties of Hyaluronan 1
Tim Hardingham
   - I. Introduction 1
   - II. Historical Perspective 1
   - III. Fundamentals of Hyaluronan Properties 4
   - IV. Conformation of Hyaluronan in Solution 5
   - V. Concentration Dependence of Hyaluronan Self-Diffusion 7
   - VI. Effects of Electrolytes on Hyaluronan Solution Properties 8
   - VII. The Effects of Alkali pH on Hyaluronan Self-Diffusion and Tracer Diffusion in Hyaluronan Solutions 11
   - VIII. Temperature Dependence of Hyaluronan Self-Diffusion 14
   - IX. Effects of Urea on Hyaluronan Properties 14
   - X. Effects of Hyaluronan Oligosaccharides on Polymeric Hyaluronan Properties 16
   - XI. Conclusions 16

2. Methods for Analysis of Hyaluronan and Its Fragments 21
Ishan Capila and Ram Sasisekharan
   - I. Introduction 21
   - II. Biological Role of Hyaluronan and Its Fragments 23
   - III. Preparation and Isolation of Hyaluronan Oligomers 24
   - IV. Analysis and Characterization of Hyaluronan Oligomers 29
   - V. Summary and Conclusion 35

3. Methods for Determination of Hyaluronan Molecular Weight 41
Mary K. Cowman and Raniero Mendichi
   - I. Introduction 41
   - II. Viscometry 42
   - III. Electrophoresis 49
   - IV. Light Scattering and Size Exclusion Chromatography 55

4. Biodegradation of Hyaluronan 71
Gunther Lepperdinger, Christine Fehrer and Stephan Reitinger
   - I. Introduction 71
   - II. Turnover in the Organism and Cellular Uptake 71
   - III. Summary and Conclusion 79

5. The Hyaluronan Receptor: CD44 83
Warren Knudson and Richard S. Peterson
   - I. Introduction 83
   - II. Structure of CD44 Primary Transcripts 85

III. Regulation of CD44 Expression ... 89
IV. Protein Structure of CD44 ... 90
V. Cellular Functions of CD44 ... 97
VI. Concluding Remarks ... 110

6. The Role of the Hyaluronan Receptor RHAMM in Wound Repair and Tumorigenesis ... 125
Cornelia Tölg, Sara R. Hamilton and Eva A. Turley
I. Introduction ... 125
II. Hyaluronan in Wound Repair and Cancer ... 126
III. RHAMM is an Atypical Hyaladherin ... 129
IV. Classification of RHAMM Protein Forms by Binding Properties and Structure ... 131
V. RHAMM Expression Influences Wound Repair and Tumor Progression ... 136
VI. Conclusions ... 142

7. Signal Transduction Associated with Hyaluronan ... 153
Regina M. Day and Marcella M. Mascarenhas
I. Introduction ... 153
II. CD44 Receptor ... 154
III. RHAMM Receptor ... 164
IV. Toll-like Receptor 4 ... 168
V. Layilin ... 170
VI. PH-20 Hyaluronidase and HA Receptor ... 171
VII. The LYVE-1 and HARE Receptors for HA Endocytosis and Degradation ... 172
VIII. Summary and Conclusions ... 174

8. Structural and Functional Diversity of Hyaluronan-Binding Proteins ... 189
Charles D. Blundell, Nicholas T. Seyfried and Anthony J. Day
I. Introduction ... 189
II. The Link Module Superfamily ... 191
III. Insights into the Molecular Basis of HA Binding ... 196
IV. Summary ... 200

9. Biological Function of SHAP–Hyaluronan Covalent Complex ... 205
Lisheng Zhuo, Li Shen, Tsunemasa Nonogaki, Jiwen Wu, Naoki Itano, Hideto Watanabe and Koji Kimata
I. Introduction ... 205
II. Formation of the SHAP–Hyaluronan Complex ... 207
III. Physiological Significance of the SHAP–Hyaluronan Complex ... 209
IV. SHAP–Hyaluronan Complex in Disease ... 211
IV. Future Prospects ... 216

10. Hyaluronan and Associated Proteins in the Visual System ... 223
Masahiro Zako and Masahiko Yoneda
I. Introduction ... 223
II. Physiological Role of Hyaluronan and Its Binding Proteins in Ocular Tissues ... 223
III. Ocular Diseases Involving Hyaluronan and Its Binding Proteins ... 232
IV. Summary and Conclusion ... 235

11. Hyaluronan in the Pulmonary Alveolus and Interstitium 247
Bonnie Anderson Bray
I. Introduction 247
II. Lung Hyaluronan (HA) 249
III. Hyaluronan Oligosaccharides 252
IV. Hyaluronan in the Alveolus 255
V. Hyaluronan in the Pulmonary Interstitium 258
VI. Conclusion 261

12. Hyaluronan in Ventilator-Induced Lung Injury 271
Hari G. Garg, Deborah A. Quinn, Marcella M. Mascarenhas and Charles A. Hales
I. Introduction 271
II. Hyaluronan 271
III. Ventilator-Induced Lung Injury 275
IV. Conclusion 279

13. The Role of Hyaluronan in Cancer 285
Sonal Patel and Martin J. Page
I. Introduction 285
II. Clinical Indications of Hyaluronan in Cancer 286
III. Hyaluronan in Cancer Biology: Expression of Hyaluronan Receptors 288
IV. Hyaluronan in Cancer Biology: Role of Hyaluronan in Adhesion, Migration, and Invasion of Cancer 289
V. Interaction of Hyaluronan with the Extracellular Matrix 292
VI. Hyaluronan and Angiogenesis 293
VII. Hyaluronan-Mediated Signalling Mechanisms in Cancer 294
VIII. Manipulation of HA Function as Potential Therapeutic Strategies 296

14. Hyaluronan in Atherosclerosis and Restenosis 307
Thomas N. Wight, Stephen Evanko, Frank Kolodgie, Andrew Farb and Renu Virmani
I. Introduction 307
II. Hyaluronan in Atherosclerosis 308
III. Hyaluronan in Restenosis 310
IV. Hyaluronan in Other Vascular Diseases 314
V. Conclusions 315

15. Hyaluronan in the Airways 323
Rosanna Forteza, Gregory E. Conner and Matthias Salathe
I. Introduction 323
II. Airway Hyaluronan: General Aspects 324
III. Airways Host Defence and Hyaluronan 325
IV. Tissue Kallikrein and Hyaluronan: Interaction and Enzyme Inhibition 326
V. Hyaluronan Serves as an Anchor for Secreted Proteins, Preventing Their Removal by Mucociliary Clearance 327
VI. Receptor for Hyaluronan-Mediated Motility is Expressed at the Apical Border of Epithelial Cells 328
VII. Ciliary Beating and Hyaluronan 328
VIII. Hyaluronan Size and Airway Pathophysiology 329
IX. Concluding Remarks and Outlook 331

16. Hyaluronan Biology in Vocal Fold Morphology and Biomechanics 339
Susan L. Thibeault
I. Introduction 339
II. Hyaluronan Localization in the Vocal Fold 340
III. Hyaluronan Localization in Vocal Fold Pathologies 342
IV. Biomechanical Properties of Hyaluronan 342
V. Therapeutic Uses of Hyaluronan in the Vocal Folds 345
VI. In Vivo Alteration of Hyaluronan Production 347
VII. Future Directions 348

17. Hyaluronan in Aging 351
María O. Longas
I. Introduction 351
II. Hyaluronan Structures in Solution: Relevance to Tissue Biologic Functions and Aging 353
III. Summary and Conclusion 359

18. Hyaluronan and Scarring 367
Andrew Burd
I. Introduction 367
II. Wound Healing 368
III. Hyaluronan and Wound Healing-Hypotheses 370
IV. Perfect Skin 371
V. Properties of Hyaluronan 372
VI. Role of Hyaluronan in Scarring 374
VII. Medical Applications 376
VIII. Hyaluronan Derivatives 377
IX. Hyaluronan, Wound Healing and Scarring: A New Perspective 382

19. Hyaluronan in the Epidermis and Other Epithelial Tissues 395
Raija H. Tammi and Markku I. Tammi
I. Introduction 395
II. Epidermal Hyaluronan 395
III. Hyaluronan in Epidermal Appendices and Other Derivatives of Ectodermal Epithelium 399
IV. Gastrointestinal and Respiratory Epithelium 400
V. Genital and Urinary Tracts and Mesothelial Cells as Hyaluronan Producers 403
VI. Epithelial Barrier Function and Hyaluronan 405
VII. Hyaluronan in Wound Healing 405
VIII. The Role of Hyaluronan in Transformation and Spreading of Malignant Epithelial Cells 406
IX. Summary and Conclusions 407

20. Viscoelastic Properties of Hyaluronan and Its Therapeutic Use 415
Endre A. Balazs
I. Elastoviscosity of Hyaluronan Solutions 415
II. Therapeutic Applications 420
III. Viscosurgery 421
IV. Viscoaugmentation 424
V. Viscoseparation 427

VI. Viscosupplementation ... 429
VII. Viscoprotection ... 439
VIII. Drug Delivery with Hyaluronan ... 442
IX. Hyaluronan in Matrix and Tissue Engineering ... 443

21. Medical Applications of Hyaluronan ... 457
Akira Asari
I. Introduction ... 457
II. Applications of Physicochemical Properties of Hyaluronan ... 458
III. Application of Cell Biological Functions of Hyaluronan ... 460
IV. Applications of Both Physicochemical Properties and Cell Biological Functions of Hyaluronan ... 461
V. Conclusions ... 468

22. Therapeutic Biomaterials from Chemically Modified Hyaluronan ... 475
Xiao Zheng Shu and Glenn D. Prestwich
I. Introduction ... 475
II. Chemical Modifications ... 476
III. Hyaluronan-Based Biomaterials ... 483
VI. Conclusions ... 494

23. Medicinal Uses of Modified Hyaluronate ... 505
Robert J. Miller and Luis Z. Avila
I. Introduction ... 505
II. Surgical Adhesion Reduction ... 507
III. Tissue Augmentation ... 513
IV. Conjugation of Therapeutic Agents of Hyaluronate ... 514
V. Summary and Conclusion ... 518

24. Hyaluronan in the Treatment of Ocular Surface Disorders ... 529
Pasquale Aragona
I. Introduction ... 529
II. Hyaluronan and Ocular Surface ... 534
III. Hyaluronan as a Tear Substitute ... 537
IV. Hyaluronan Administration on the Ocular Surface ... 538
V. Other Uses of Hyaluronan in Ocular Surface Disorders ... 544
VI. Summary and Conclusions ... 545

25. The Hyaluronan Synthases ... 553
Paul H. Weigel
I. Introduction ... 553
II. Molecular Size and Biology of Hyaluronan ... 554
III. Hyaluronan Synthesis by Hyaluronan Synthases ... 555
IV. Topological Organization of the Hyaluronan Synthase Proteins ... 557
V. Molecular Size of Functional Hyaluronan Synthases ... 558
VI. Purification of Hyaluronan Synthases ... 558
VII. The Class I Hyaluronan Synthases Polymerize Hyaluronan by Addition to the Reducing End ... 560
VIII. Are Hyaluronan Synthases Processive? ... 562

IX. Size Distributions of Hyaluronan Products Made by Hyaluronan Synthases 563
X. Summary and Conclusions 564

26. Molecular Genetic Dissection of Hyaluronan Function in the Mouse 569
Andrew P. Spicer and Janet Y. Lee Tien
I. Introduction 569
II. Hyaluronan Synthases 571
III. Hyaluronan Synthases Gene Expression 571
IV. Hyaluronan Synthase 2 Null Allele 573
V. Hyaluronan Synthase 3 Null Allele 576
VI. Hyaluronan Synthase 2 Conditional Gene Knockout 577
VII. Summary and Conclusion 579

27. Functional, Structural and Biological Properties of Hyaluronidases 585
Günter Lepperdinger and Günther Kreil
I. Introduction 585
II. Mammalian-Type Hyaluronidase 585
III. Bacterial Hyaluronidase 591
IV. Summary and Conclusion 593

Index 599

Chemistry and Biology of Hyaluronan
H.G. Garg and C.A. Hales (editors)

Chapter 1

Solution Properties of Hyaluronan

TIM HARDINGHAM

*Wellcome Trust Centre for Cell-Matrix Research,
School of Biological Sciences,
University of Manchester, Manchester, UK*

I. Introduction

Hyaluronan (HA) is a high molecular weight ($10^5 - 10^7$ Da) unbranched glycosaminoglycan, composed of repeating disaccharides (β1-3 D-*N*-acetylglucosamine, β1-4 D-glucuronic acid). It is a widely distributed component of the extracellular matrix of vertebrate tissues (1). It acts as a scaffold for the binding of other matrix molecules including aggrecan and other members of the hyalectan family (2,3). It has an interesting mechanism of synthesis in which chain extension is by monosaccharide addition at the reducing end of the chain (4). This is thus in the opposite direction to other vertebrate glycosaminoglycans. It also appears to be synthesised by a glycosyltransfererase with two catalytic activities, for the glucuronic acid transfer and the *N*-acetylglucosamine transfer (5). The enzyme also appears to be embedded in the plasma membrane of cells with the product being translocated out of the cell as synthesis proceeds and there are three related mammalian HA synthases.

II. Historical Perspective

HA was initially discovered and named hyaluronic acid in a paper published in 1934 by Karl Meyer (6). It was isolated from vitreous of the eye as a polysaccharide containing D-glucuronic acid and D-*N*-acetylglucosamine, but it was not until 20 years later that he completed the determination of its structure and showed that it contained a repeating β1-3, β1-4 linked disaccharide. In the meantime, HA was isolated from many tissue sources, including synovial fluid, cock's comb and umbilical cord. Its extraction from tissues was not easy and HA preparations always retained some protein. The controversy over whether HA was linked to a protein remained in the literature for many years (see Refs. 7–9) and the issue was particularly debated during the 1960s and 1970s as other structurally related glycosaminoglycans were characterised and their covalent attachment to protein as proteoglycans was being explored. The question naturally arose, was HA a proteoglycan? The issue was not easily resolved because of the unusual properties of HA and the difficulty of preparing it free of protein using classical biochemical methods. Exhaustive isolation and fractionation methods resulted in a low but significant content of residual protein (~0.5% w/w). This combined with the high molecular weight of HA left the possibility that each chain was attached to a small protein. At the same time, important discoveries were establishing highly specific interactions of HA with proteins, the first of which identified its role in binding to aggrecan and forming supramolecular aggregates (10). However, the issue of HA's covalent link to protein as a requirement for biosynthesis was finally resolved with the discovery of the mechanism of biosynthesis of HA (4) and the subsequent cloning of the HA synthase enzyme (5), which showed that HA could be made without any protein primer. It is interesting that subsequently novel mechanisms have been discovered by which covalent protein–HA bonds can be formed extracellularly with inter α-trypsin inhibitor and related proteins (11) and this may explain some of the difficulty of removing final traces of protein from HA prepared from some tissue sources. It is now clear that HA is synthesised by the cells of higher organisms without the need for any protein primer. HA has thus evolved from quite a different evolutionary origin from the other structurally related glycosaminoglycans, which are synthesised attached to proteins and whose chains are extended by single sugar addition to the non-reducing end of each chain.

From its initial isolation, the physical properties of HA have been the dominant feature that distinguished it from other components of extracellular matrix. In the early work characterising HA, even including the simple determination of its molecular weight presented great difficulty. The properties of HA provided a challenge to the classical biophysical methods, in which simple analysis was developed for proteins and required that the properties approached those of perfect Newtonian solutes. The behaviour of HA in solutions even at low concentration is far from Newtonian or 'ideal', and it presented a challenge in the 1950s and 1960s that some very distinguished researchers took up, notably Sandy Ogston, Torvard Laurent, Endre (Bandi) Balazs and later Bob Cleland (7,8,12–17). Their work established a theoretical and experimental framework that underpins HA research to this day and the concepts they developed are fundamental to understanding the biophysical properties of HA.

The key elements they identified were:

- It was a high molecular weight unbranched polysaccharide, which behaved as a stiffened random coil in solution (Fig. 1).
- It occupied a large hydrated volume and therefore showed solute–solute interactions at unusually low concentration.
- It showed excluded volume effects, as it restricted access to this domain by other macromolecules.
- These properties were compounded by the fact that HA was a polyelectrolyte and therefore the solution properties were also greatly affected by ionic strength.
- HA was also established to be polydisperse and its properties were therefore the aggregate properties of a population of molecules of varying chain length, rather than those of a unique species.

Much of this early work focussed on relating biophysical measurements of light scattering, osmometry, viscosity and sedimentation to models of behaviour. An important early development in this process was the recognition of the need to extrapolate experimental results to vanishingly low concentrations in order to

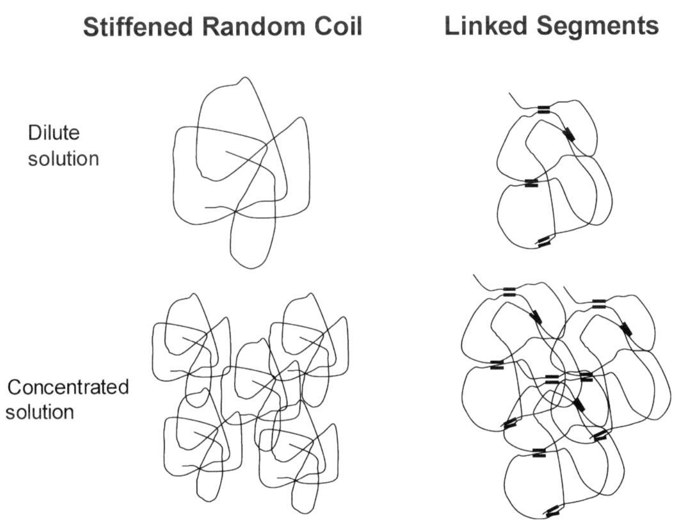

Figure 1 Models of hyaluronan behaviour in solution. In dilute solution, hyaluronan behaves as a stiffened random coil. The presence of linked segments would act in opposition to chain stiffening in determining the hydrated domain. In concentrated solutions, stiffened random coils show entanglement; they form viscoelastic solutions and retain flow and do not become gels. The presence of linked segments would create a network and lead to gel formation.

determine intrinsic properties and free the measured parameters from non-ideal effects caused by interaction between adjacent molecules. With an evolving understanding of non-ideal behaviour of polyelectrolyte biopolymers such as HA it was found possible to obtain consistent results from different biophysical techniques.

There is a theme that runs through the fascinating history of HA research. HA presents complex biophysical properties from an inherently simple chemical structure. This apparent complexity has spawned many novel hypotheses and mechanisms to explain the complexity. Some of these ideas have been 'red herrings' and have not themselves stood the test of time. However, they have helped the progress of research by acting as a challenge to stimulate renewed effort for more experimental measurements and to instil more rigour into the models of behaviour that are consistent with results from a broad range of independent techniques. Time has also led to progress in experimental techniques and many of the questions that were once difficult to study have now become more accessible.

III. Fundamentals of Hyaluronan Properties

HA solutions have pronounced viscoelastic properties and the biophysical basis of its 'non-ideal' behaviour has been the source of much interest and speculation. At neutral pH and physiological ionic strength much of the early work of the groups of Laurent and Balazs led to the conclusion that HA behaved as a stiffened random coil in solution (Fig. 1). Later the stiffening was proposed to be at least in part due to hydrogen bonding between adjacent saccharides, combined with some effect from the mutual electrostatic repulsion between carboxyl groups (18–21) and these proposals have been substantiated by later results using different techniques (22–25). This provided a rational basis for understanding the unusual hydrodynamic and rheological behaviour of HA, which distinguished it from other polymers such as dextran. However, this relatively simple model has in recent years been challenged by a proposal that HA chains self-associate and that this dominates the solution properties (Fig. 1). The core evidence for this was in two strands. The first was that apparent association between HA chains was visualised in EM preparations and it was interpreted as anti-parallel double helices, bundles and ropes (26,27) and the second was that NMR spectra were interpreted to suggest that chain–chain association occurs in solution (28,29). However, the principal driver for this model of self-association was the observation that HA in a 2-fold helix could contain hydrophobic patches and these might provide sites for self-association between chains (27). This provided a possible mechanism for interaction between HA chains and experiments were designed to seek evidence for it. The important question to be asked is: if there is self-association of HA chains in solutions under physiological conditions, is this compatible with the mass of experimental data describing HA properties from the experiments of Ogston, Laurent, Balazs, Cleland and many others?

It is pertinent to examine exactly what properties might result if HA chains were significantly bound to each other in solution.

- First, if the associating chains stacked together this would seem likely to result in insolubility, such as is seen with cellulose, rather than HA's highly soluble properties. (Solubility might only be retained if self-association was weak and the interactions transient.)
- Secondly, if the chains bound to each other, this would most likely be intramolecular, particularly in dilute solution, i.e., between different segments of the same chain, rather than intermolecular, linking different chains. This would have the opposite effect to chain stiffness and create a smaller rather than a larger molecular domain in solution (Fig. 1).
- Thirdly, at high concentration the intermolecular association should create a stable 'gel' structure, i.e., a cross-linked network (Fig. 1).

Therefore if self-association between HA chains occurred in a significant fraction of HA chains all the time, we might expect HA to be small and compact and/or poorly soluble and form gels. Self-association would also be incompatible with the ability to isolate HA fractions of different molecular weight by gel filtration and for these fractions to re-chromatograph consistently and with low polydispersity.

These simple considerations would seem to rule out strong lasting self-association of HA in aqueous solutions under physiological conditions. HA self-association thus carries with it a host of consequences that seem at odds with the properties of HA studied over 50 years.

The abundant data in the literature therefore suggest that HA chains in aqueous solution do not strongly self-associate. However, it might be that HA self-association is very weak and transient, such that it contributes to non-ideal behaviour, but does not cause gel formation. Could it then explain some of HA's unusual properties? How weak would the interaction need to be to allow HA to be highly soluble, with a highly expanded domain in solution, able to be chromatographed, yet have an influence on the rheological properties?

IV. Conformation of Hyaluronan in Solution

With the premise that HA may self-associate, but the interactions may be weak and transient, we set out in our research group to investigate HA properties in concentrated solutions and look for evidence of self-association. For this study we used a newly developed technique, confocal-FRAP, which is uniquely suited for detecting weak associations and is able to study at high concentration, where chain–chain interaction could be expected to be maximal. This new technique could thus provide results beyond those obtained by other methods.

Confocal-FRAP is a powerful method for determining concentrated solution properties of polymers such as HA, as it is also an equilibrium method and is carried out in the absence of flow and shear forces with no concentration gradients (30,31). It is therefore free of many of the artefacts that accompany the study of the biophysical properties of viscous, non-ideal polymers in solution. It provides measurements of lateral self-diffusion. In dilute solution these measurements are of free diffusion of each HA chain, which by Stokes–Einstein equation can be related to the hydrodynamic radius. However, at higher concentrations, each HA will begin to be slowed in diffusion by interaction with its neighbours and it becomes less than free diffusion. The measurements then reveal the extent of these interactions and experiments can be carried out at high concentration when the molecular domains overlap and the interactions become very large. The method thus determines the bulk properties of HA, rather than any local chain motion or flexing and anything that affects the hydrodynamic interactions between HA molecules will affect the long-term lateral translational diffusion determined by this technique. The technique also provides a method to analyse how at high concentration HA chains impede the diffusion of other tracer molecules (24,32). Fluorescent tracer molecules of known size can thus be used to interrogate the random network of chains in a concentrated HA solution and derive apparent pore sizes that are sensitive to the concentration, organisation and mobility of the HA chains.

Initial experiments characterised the system and used it to assess how different factors might contribute to HA properties in solution.

Several factors were considered:

1. electrostatic interaction of the regularly placed carboxyl groups;
2. hydrogen bonding between adjacent saccharides;
3. domain overlap and polymer entanglement;
4. chain–chain association through mechanisms such as interaction of hydrophobic patches.

An important aspect, and indeed the value, of using confocal-FRAP is that it permitted analysis at concentrations of HA up to and far exceeding the critical concentration at which there is predicted molecular domain overlap (23,24,31, 32). This analysis was thus ideally suited to investigations of entanglement and intermolecular chain–chain association, as these would be concentration-dependent and strongly favoured at high concentration (Fig. 1). The concentration dependence helps distinguish these effects from electrostatic interactions and hydrogen bonding of adjacent saccharides, which occurs at all concentrations.

Electrostatic and ionic effects on the HA network and its sensitivity to counter-ion type and valency were determined as these are known to greatly affect rheological and hydrodynamic properties (33). The role of hydrogen bonds was investigated by comparing concentration-dependent solution properties in de-ionised water, 0.5 M NaCl and 0.5 M NaOH. This revealed the effect of electrostatic shielding of the supporting electrolyte and also the profound affect of alkali on HA chain stiffness. The presence of HA chain–chain associations that might involve hydrophobic interactions were investigated under physiological conditions, in solvents of varying polarity and in the presence of chaotropic agents (18,24,25,27–29,34,35). Intermolecular chain–chain associations were also investigated using HA oligosaccharides as low molecular weight competitors of such interactions (19,36,37).

V. Concentration Dependence of Hyaluronan Self-Diffusion

In characterising the general behaviour of HA solutions in 0.2 M NaCl at neutral pH (24) it was shown that the lateral translational self-diffusion coefficients of HA (average MW ~900 kDa) showed a progressive fall with increasing concentration as it approached and exceeded the predicted critical concentration for domain overlap (Fig. 2). Similar smooth transitions between dilute, semi-dilute and concentrated regimes have been observed experimentally in many polymer/solvent systems (38–40) due to polymer entanglement, including HA solutions (20). The lateral self-diffusion coefficients reduced steeply with concentration in a manner consistent with phenomenological descriptions of

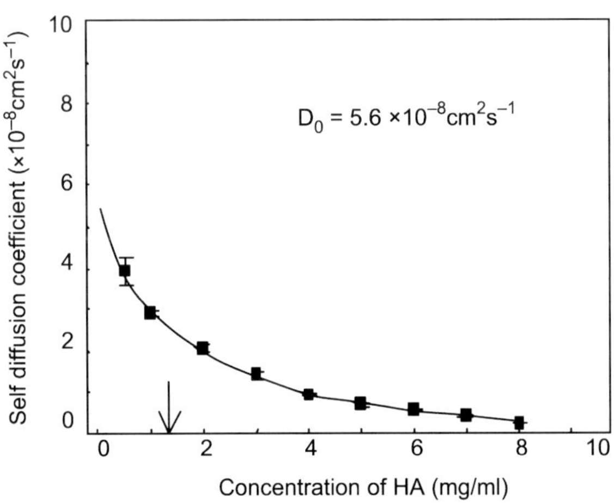

Figure 2 Concentration dependence of the lateral translational self-diffusion coefficients of hyaluronan. The lateral translational self-diffusion was determined by confocal-FRAP for hyaluronan (830 kDa) at 0.5–8.0 mg/mL in PBS at 25 °C. The vertical arrow marks the critical overlap concentration c^*. The solid line shows the data fitted to Eq. 1 and extrapolated to zero concentration to give D_0. (Data from Ref. 24. Confocal-FRAP technique described by Gribbon and Hardingham (31,32).)

polymer self-diffusion in terms of a universal scaling equation (39)

$$D = D_0 \exp^{-\alpha c^{\nu}} \tag{1}$$

where D_0 is the polymer free self-diffusion defined in the limit of zero concentration and α and ν are empirically derived. The parameter α describes the strength of interpolymer hydrodynamic interactions and the deviation of ν from unity arises from chain contraction at high concentrations. Data were fitted to Eq. 1 using a non-linear least squares fit (non-weighted). Analysis of data from results with HA (830 kDa) gave $D_0 = 5.6 \times 10^{-8}$ cm^2/s, $\alpha = 0.63$ mL/mg and $\nu = 0.74$. These measurements were insensitive to pH over a broad range (pH 4–8). The absence of any effect of concentration on HA self-diffusion beyond that predicted by simple polymer entanglement was the first sign that other concentration-dependent effects such as chain–chain association were absent.

VI. Effects of Electrolytes on Hyaluronan Solution Properties

Investigation of the effect of increasing electrolyte concentration on HA solution properties (24) showed that the self-diffusion coefficient of HA was very low in the absence of any supporting electrolyte, but increased dramatically with small increases in NaCl concentration with a 2.8-fold increase in lateral self-diffusion coefficient from 0 to 100 mM (Fig. 3). This is consistent with increased

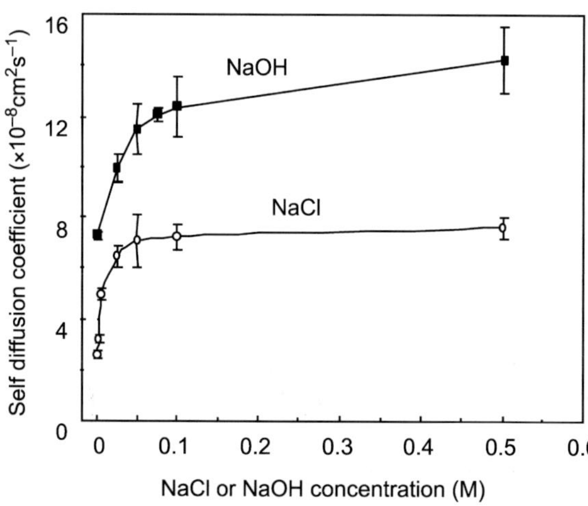

Figure 3 Effect of increasing concentrations of NaCl and NaOH on hyaluronan self-diffusion. The lateral translational self-diffusion was determined for hyaluronan (500 kDa, 0.2 mg/mL) in increasing concentrations of NaCl (○) or NaOH (■). Ionic strength of solutions in NaOH was maintained at 500 mM by the addition of NaCl. All measurements at 25 °C (25).

electrostatic shielding resulting in polyanion coil contraction and as this was largely complete at 100 mM NaCl, the contribution of electrostatic effects to macromolecular stiffness under physiological conditions of ionic strength and pH is suggested to be small. If, at concentrations higher than that required for domain overlap, HA was a network of molecules linked by hydrophobic interactions in solution, increasing the ionic strength might have been expected to favour chain–chain association and cause a decrease in self-diffusion, but there was no evidence for this.

The effects of different counter-ions on the self-diffusion of HA showed that Ca^{2+} caused a significant increase compared with Na^{+}, with less increase with Mn and Mg (25) (Fig. 4). The concentration dependence of HA properties up to 10 mg/mL showed that the self-diffusion coefficients were greater in $CaCl_2$ than in NaCl (Fig. 5), although the difference became smaller at high concentration The intrinsic effects of these counter-ions on the conformation of HA in dilute solution were investigated by gel filtration and multi-angle laser light scattering (MALLS) analyses. With HA (930 kDa) solutions in different electrolytes, but at similar ionic strength (Table 1), the R_g (radius of gyration) in $CaCl_2$ and $MnCl_2$ were significantly lower than in KCl ($p < 0.001$) and NaCl ($p < 0.001$). As the peak concentration of HA was <30 μg/mL, this measurement provided a comparison of the R_g of individual chains and showed that there was a direct effect of Ca^{2+} in contracting the free solution domain of HA. The equivalent Stokes sphere radius (R_H) (see Eq. 2) for HA (830 kDa) was 43 nm in 0.5 M NaCl and 36 nm in 0.5 M $CaCl_2$.

The effect of Ca^{2+} on the properties of HA were also found to be dominant over the effects of Na^{+}, as in the presence of 0.15 M NaCl the addition of

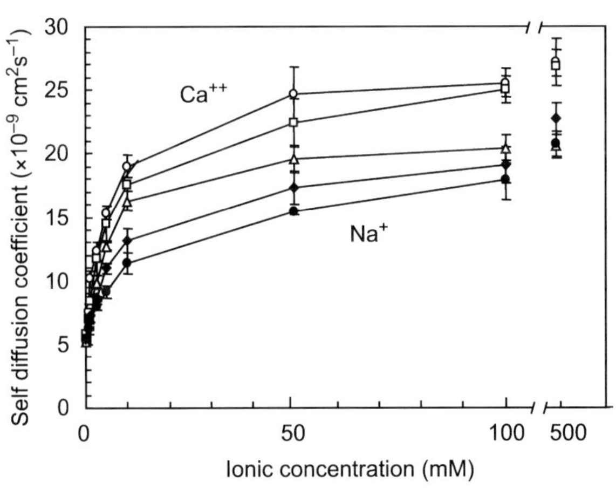

Figure 4 Effect of increasing cation concentration on hyaluronan self-diffusion. Self-diffusion coefficients of HA (830 kDa, 2 mg/mL) were determined as a function of salt concentration for NaCl, KCl, $MgCl_2$, $MnCl_2$ and $CaCl_2$. Hyaluronan was most mobile in $CaCl_2$ and least mobile in NaCl. All measurements at 25 °C (25).

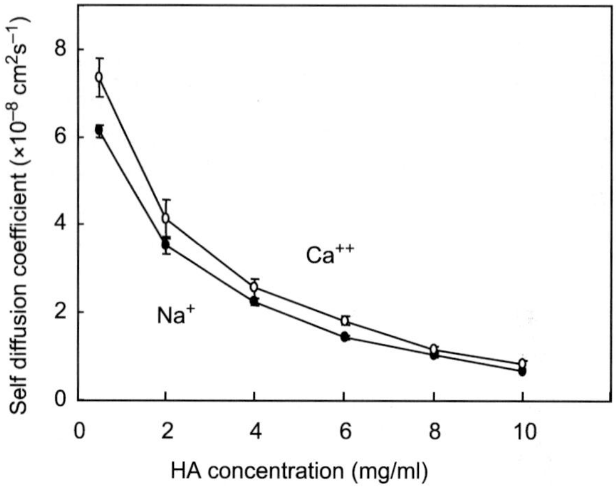

Figure 5 Comparison of hyaluronan self-diffusion in $CaCl_2$ and NaCl. Lateral translational diffusion coefficients of HA (830 kDa) were determined from 0.5 to 10 mg/mL, in $CaCl_2$ (○) and NaCl (●) solutions (both at 0.5 ionic strength) at 25 °C (25).

$CaCl_2$ at low concentration (<10 mM) caused a further significant increase in the self-diffusion of HA (Fig. 6). The differences between the self-diffusion of HA in Ca^{2+} and Na^+ were also accompanied by differences in the tracer diffusion of FITC-dextran (2000 kDa) in solutions of HA at up to 20 mg/mL, and tracer mobility in 150 mM NaCl was also increased by the addition of $CaCl_2$ (25).

The contraction of the HA domain in calcium solutions suggested that Ca^{2+} increased the flexibility of the chain by promoting a greater range of movement at each glycosidic bond. Recently, molecular dynamics simulations of HA have demonstrated that short lengths of HA (five disaccharides) access a range of compact configurations including hairpin loops (41). It might be proposed that individual Ca^{2+} ions may coordinate two carboxyl groups, on the same HA chain, and promote chain contraction. However, if this mechanism occurred Ca^{2+} would

Table 1 Weight-Averaged Radius of Gyration Calculated from Multi-angle Laser Light Scattering (MALLS) Analyses of 930 kDa HA in 100 mM Ionic Strength Electrolyte Solutions Chromatographed on a Sephacryl S-1000 Size Exclusion Column (25)

Electrolyte	Ionic strength	R_g (± SE) (nm) (weight-averaged)
NaCl	0.10	98.9 ± 1.0
KCl	0.10	97.7 ± 2.2
$MgCl_2$	0.10	92.9 ± 2.0
$CaCl_2$	0.10	88.0 ± 1.3
$MnCl_2$	0.10	87.4 ± 3.2

Error values represent the SEM ($N = 3$).

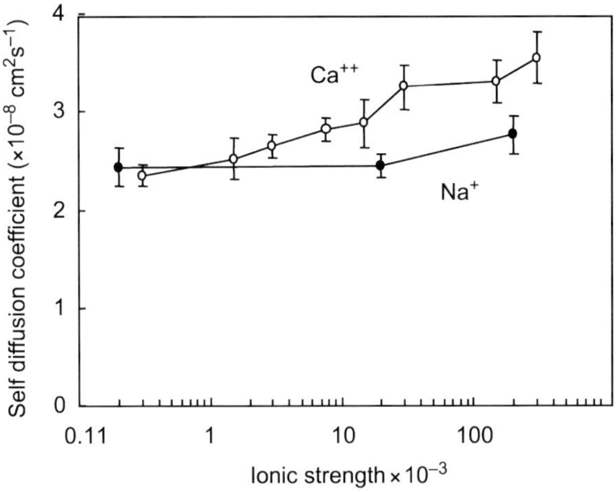

Figure 6 Effect of low concentrations of $CaCl_2$ on hyaluronan self-diffusion in NaCl (0.15 M). The self-diffusion coefficient of HA (830 kDa, 2 mg/mL) was determined in 150 mM NaCl with increasing concentrations of $CaCl_2$ (○) or NaCl (●) at 25 °C (25).

also be able to stabilise interchain associations and the results show that it does not. Alternatively, Ca^{2+} may alter the coordination of water molecules with HA chains, thereby disrupting hydrogen bonds involving water bridges. The presence of Ca^{2+} may therefore cause less stability in the range of hydrogen bonds that bridge adjacent sugars in these linkages (42,43). Molecular dynamics simulations incorporating counter-ions would be required to further substantiate this model. Overall, the changes in HA properties caused by Ca^{2+} are small compared to the effects of strong alkali (see later), which appears to disrupt the hydrogen bonds between adjacent saccharides and causes a major reduction in chain stiffness (24). However, it may be speculated that HA 'destiffening' by Ca^{2+} may have a role in cell-mediated matrix re-modelling processes.

VII. The Effects of Alkali pH on Hyaluronan Self-Diffusion and Tracer Diffusion in Hyaluronan Solutions

The effect of high pH in NaOH also contracted the domain size of HA, but the effect (Fig. 7) was much greater than the reduction found due to electrostatic shielding (Fig. 2). This effect was consistent with previously reported reductions in R_g and intrinsic viscosity (44). Changes in the R_H and hydrodynamic volume of HA with NaCl and NaOH were calculated using the Stoke's Einstein approximation for the behaviour of a sphere

$$D_0 = \frac{\kappa T}{6\pi \eta R_H} \tag{2}$$

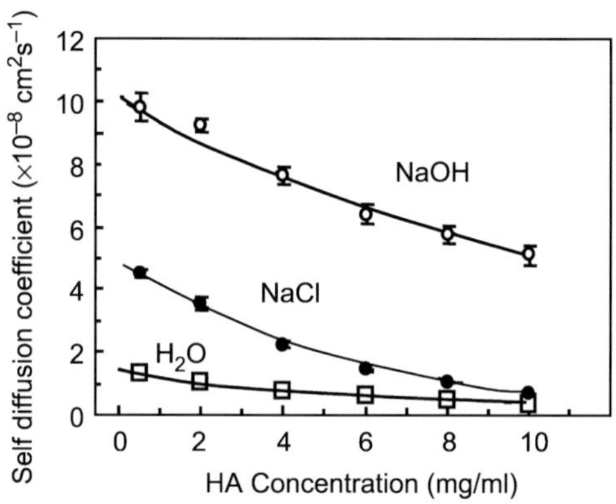

Figure 7 Comparison of the concentration dependence of hyaluronan self-diffusion in NaOH, NaCl and water. The concentration dependence of the lateral translational self-diffusion coefficient of hyaluronan (830 kDa) at 0.5–10 mg/mL was determined in 0.5 M NaOH (○), 0.5 M NaCl (●), and de-ionised water (□). All measurements at 25 °C (24).

where κ is the Boltzmann's constant, T the temperature and η the solvent. If it is assumed that the self-diffusion coefficient at 0.2 mg/mL is approximately equal to the free diffusion coefficient (see Fig. 2), then for HA of 500 kDa, from the Stoke's Einstein equation, R_H contracted from 95 to 33.5 nm, in going from de-ionised water to 0.5 M NaCl, reducing further to 17.5 nm in 0.5 M NaOH (Fig. 8). These results show that in going from 0.5 M NaOH to de-ionised water, the apparent domains of HA chains were increased by more than 100 times and this most likely resulted from increased electrostatic interactions and hydrogen bond

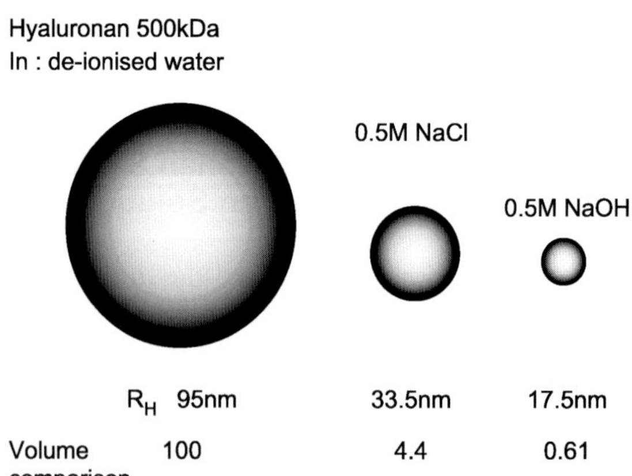

Figure 8 Comparison of the hydrodynamic radius (R_H) of hyaluronan (500 kDa) in de-ionised water, salt and alkaline solutions.

formation (24). In the most compact configuration in alkali, the hydrodynamics of HA (500 kDa) became similar to those of the partly branched dextran (2000 kDa, $R_H = 19$ nm), which is neither charged nor predicted to form comparable hydrogen bonds. For HA (500 kDa) in 0.5 M NaOH (Fig. 4), domain overlap is predicted to occur at 37 mg/mL. This implies that at 2–10 mg/mL solutions are well below c^* and this is entirely consistent with the comparatively greater network mobility observed in self-diffusion experiments, including those with higher molecular weight HA (830 kDa, Fig. 7). These effects in alkali were reversible and caused no significant depolymerisation under the conditions used.

Tracer diffusion results at low HA concentrations (1–4 mg/mL) (Fig. 9) show analogous behaviour to the changes in self-diffusion (Fig. 7). The network is both more permeable and more mobile in 0.5 M NaCl than in de-ionised water and this supports a model involving contraction of the HA chain conformation in the presence of increasing electrolyte. However, as the concentration of HA approached 20 mg/mL (Fig. 9), tracer mobility became progressively independent of salt concentration, indicating that at high concentrations chain density was the major determinant of matrix permeability.

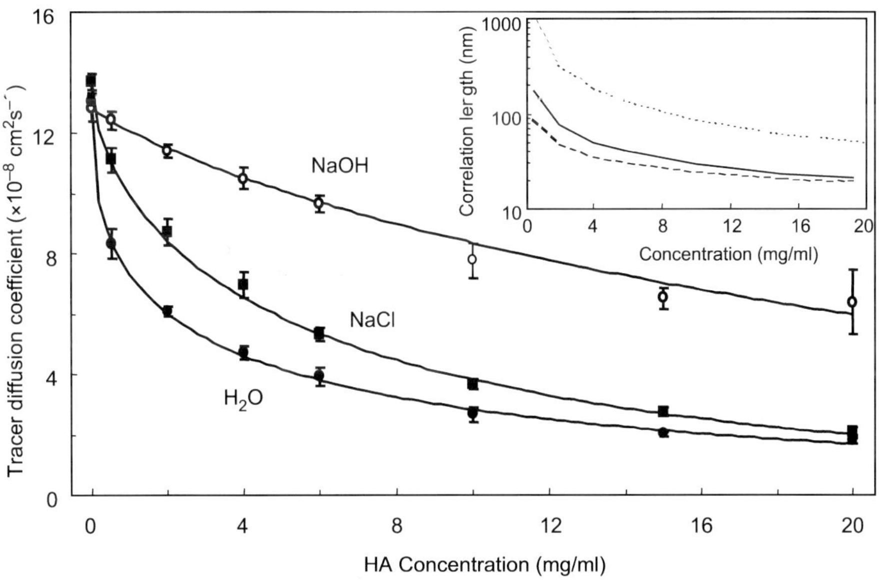

Figure 9 Comparison of tracer diffusion in hyaluronan solutions of different concentrations in NaOH, NaCl and water. Lateral translational diffusion coefficients of FITC-dextran (2000 kDa) were determined in hyaluronan solutions (930 kDa) 0–20 mg/mL in de-ionised water (●), 0.5 M NaCl (■) and 0.5 M NaOH (○). Inset shows the correlation length parameter (ξ) versus hyaluronan concentration for de-ionised water (long dash), 0.5 M NaCl (solid line) and 0.5 M NaOH (short dash). All measurements in PBS at 25 °C (24).

The lack of a salt effect at high HA concentration was most interesting, as it suggested that there was no evidence of hydrophobic interactions between chains, as by analogy with RNA and DNA, high salt would be expected to favour chain–chain association.

The tracer studies also provided a measure of the major changes induced by NaOH. In HA (930 kDa), at 20 mg/mL (Fig. 8), the translational diffusion of FITC-dextran was independent of NaCl concentration, but not of NaOH concentration. This reflected, as noted previously, that for 930 kDa HA, 20 mg/mL was likely to represent a semi-dilute regime in the presence of NaOH, whereas it was clearly a concentrated, entanglement-dominated regime, both in NaCl and in de-ionised water. These results strongly suggested that the solution properties at higher concentration in various solvents were directly related to the hydrodynamic volumes of single chains in the same solvent. Results at high pH, showing high mobility and permeability of HA, clearly revealed the degree to which, at neutral pH, intrachain hydrogen bonds profoundly affected chain stiffness, chain entanglement and interchain hydrodynamic interactions, but the results lacked any evidence of HA self-associating to form a network in solution.

VIII. Temperature Dependence of Hyaluronan Self-Diffusion

As a further test of the possible presence of HA self-association, we investigated the temperature dependence of diffusion under conditions of domain overlap where self-diffusion should be critically dependent on any interaction that linked neighbouring molecules. If we consider the thermodynamics of chain–chain self-association, it would have a certain free energy and a loss of entropy associated with it, which would suggest that the equilibrium of self-association should be temperature dependent. It should therefore be reversibly dissociated by increasing temperature (melted in the way that cDNA oligonucleotides are in a PCR reaction). However, there was no evidence of any transient 'melting' in the translational diffusion coefficient between 20 and 70 °C. The bulk properties and mobility of individual HA molecules were thus unaffected by temperature in this range.

IX. Effects of Urea on Hyaluronan Solution Properties

In concentrated polymer solutions, if the network properties are determined solely by chain entanglements, then they should be independent of agents that disrupt other associative mechanisms. To investigate further for evidence of hydrophobic chain–chain interactions, self-diffusion properties were investigated in the presence of urea, a potent disrupter of hydrophobic association (25). Initially, the effect of urea on individual chain hydrodynamics was investigated by analysing HA self-diffusion at low polymer concentrations. In a dilute solution of HA, if there is an association between segments of chains this will tend to contract its hydrodynamic domain, whereas in concentrated solution it might serve to additionally make linkages between adjacent molecules. However, the self-diffusion of HA (500 kDa) in the presence of 6 M urea was consistently higher than in de-ionised water (Fig. 10). This showed that in urea the polymer domain of HA became smaller and thus showed no evidence for the disruption of intramolecular chain–chain association. The increased free diffusion coefficient of HA in dilute solution in the presence of urea was therefore inconsistent with the presence of intramolecular chain associations. On the contrary, the reduced hydrodynamic domain size shows that urea increases the flexibility of HA chains. If the primary intramolecular chain-stiffening mechanism for HA arises from hydrogen bonding the effect of urea was compatible with it reducing the hydrogen bonding between adjacent saccharides. However, the de-stiffening caused by urea in these experiments was substantially less than that caused by 0.5 M NaOH (24). Therefore, urea may, e.g., disrupt only a sub-fraction of hydrogen bonds, such as those involving a water bridge. Interestingly urea had no detectable effect on HA diffusion in the presence of 0.5 M NaCl as supporting electrolyte, which suggested that it did not affect the chain-stiffening hydrogen bonds present in 0.5 M salt, but could affect those additionally present in de-ionised water. Further investigation would be required to confirm this interpretation. Urea appeared to have little effect on intermolecular interactions between HA molecules, as the concentration dependence of self-diffusion, which

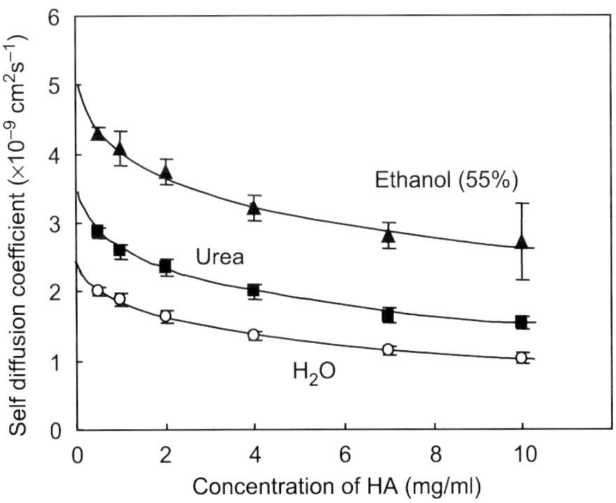

Figure 10 Effects of urea and ethanol/water on hyaluronan self-diffusion. The concentration dependence of self-diffusion of HA (500 kDa) 0.5–10 mg/mL was determined for solutions in 55% v/v ethanol/water (▲), 6 M urea (■) and de-ionised water (○). All data at 25 °C and corrected for solvent viscosity. Solid lines show data fitted to the polymer scaling equation (25).

is a measure of intermolecular interaction, follows a very similar form in urea and de-ionised water (Fig. 9). These results, therefore, suggest that there are no chain–chain associations of HA in aqueous solution that is sensitive to this chaotropic agent.

X. Effects of Hyaluronan Oligosaccharides on Polymeric Hyaluronan Properties

The presence of chain–chain interactions was also investigated using another strategy by competition with HA oligosaccharides. However, the presence of increasing concentrations of HA_{20-26} had no significant effect on the self-diffusion of FA-HA (2 mg/mL), even at an oligosaccharide concentration twice that of the full-length HA. If HA mobility at 2 mg/mL was restricted by associations between chains, then HA_{20-26} should compete to form oligosaccharide chain associations and this would increase the diffusion coefficient of HA. These results therefore also suggest that intermolecular chain–chain associations are not important in determining the concentrated solution properties of HA (25).

XI. Conclusions

The data presented in these studies (24,25) suggest that intramolecular hydrogen bonds and polyanionic properties of HA both contribute to provide a highly expanded macromolecular conformation. However, under physiological conditions of ionic strength the results predict the electrostatic effects to be modest. The most plausible explanation for the large hydrodynamic volume of HA and hence its other important non-Newtonian viscoelastic properties, is the presence of multiple dynamic hydrogen bonds between adjacent saccharides. This restricts rotation and flexion at the glycosidic bonds and creates a stiffened yet mobile polymer chain (Fig. 1). The flexibility and permeability properties of the HA network can then be accounted for in terms of interchain hydrodynamic interactions of this extended structure, with entanglement being especially important at elevated concentrations. However, even at high concentrations, under physiological conditions, individual HA chains remain mobile and at no stage do HA solutions undergo transition to a gel-like state. These observations are incompatible with any significant degree of intermolecular self-association that is stable or cooperative. The results suggest that even at high concentrations the properties can be directly predicted from the behaviour in dilute solution. It seems reasonable to conclude that the major solution properties of HA can be best described by a simple hydrodynamic model with chains stiffened by a dynamic family of hydrogen bonds between adjacent saccharides.

Acknowledgements

The Wellcome Trust and Seikagaku Corporation (Tokyo, Japan) are thanked for their support for these studies.

References

1. Balazs EA, Gibbs DA. The rheological properties and the biological function of hyaluronic acid. In: Balazs EA, ed. Chemistry and Molecular Biology of the Intercellular Matrix. London: Academic press, 1970:1241–1254.
2. Laurent TC. Structure of the extracellular matrix and the biology of hyaluronan. In: Reed RK, McHale NG, Bert JL, Winlove CP, Laine GA, eds. Intersitium, Connective Tissue and Lymphatics. London: Portland Press, 1995:1–12.
3. Lapcik L Jr, Lapcik L, De Smedt S, Demeeste J, Chabrecek P. Hyaluronan: preparation, structure, properties and applications. Chem Rev 1998; 98: 2663–2684.
4. Prehm P. Synthesis of hyaluronic acid in differentiated teratocarcinoma cells—mechanism of chain growth. Biochem J 1983; 220:597–600.
5. De Angelis PL, Papaconstantinou J, Weigel PH. Molecular cloning, identification and sequence of the hyaluronic acid synthase gene from Group A Streptococcus pyogenes. J Biol Chem 1993; 268:19181–19184.
6. Meyer K, Palmer JW. The polysaccharide of the vitreous humour. J Biol Chem 1934; 107:629–634.
7. Ogston AG, Stanier JE. Further observations on the preparation and composition of the hyaluronic acid complex from ox synovial fluid. Biochem J 1952; 52: 149–156.
8. Laurent TC. Studies on hyaluronic acid in the vitreous body. J Biol Chem 1955; 216:263–271.
9. Swann DA. Studies on hyaluronic acid: II. The protein component of rooster comb hyaluronic acid. Biochem Biophys Acta 1968; 160:96–105.
10. Hardingham TE, Muir H. The specific interaction of hyaluronic acid with cartilage proteoglycans. Biochim Biophys Acta 1972; 279:401–405.
11. Huang L, Yoneda M, Kimata K. A serum derived hyaluronan-associated protein (SHAP) is the heavy chain of inter α-trypsin inhibitor. J Biol Chem 1993; 268: 26725–26730.
12. Balazs EA, Laurent TC. Viscosity function of hyaluronic acid as a polyelectrolyte. J Polym Sci 1951; 6:665–667.
13. Balazs EA. Physical chemistry of hyaluronic acid. Fed Proc 1958; 17:1086–1093.
14. Laurent TC, Ryan M, Pietruszkiewicz A. Fractionation of hyaluronic acid. The polydispersity of hyaluronic acid from the vitreous body. Biochem Biophys Acta 1960; 42:476–485.
15. Ogston AG. Some thermodynamic relationships in ternary systems with special reference to the properties of systems containing hyaluronic acid and protein. Arch Biochem Biophys 1962; 1:39–51.
16. Balazs EA. Sediment volume and viscoelastic behaviour of hyaluronic acid solutions. Fed Proc 1966; 25:1817–1822.
17. Cleland RL, Wang JL. Ionic polysaccharides: III. Dilute solution properties of hyaluronic acid fractions. Biopolymers 1970; 9:799–810.

18. Scott JE, Tigwell MJ. Periodate oxidation and the shapes of glycosaminoglycuronans in solution. Biochem J 1978; 173:103–114.
19. Morris ER, Rees DA, Welsh EJ. Conformation and dynamic interactions in HA solutions. J Mol Biol 1980; 138:383–400.
20. Wik KO, Comper WD. Hyaluronate diffusion in semi-dilute solutions. Biopolymers 1982; 21:583–599.
21. Sheehan JK, Atkins EDT. X-ray fiber diffraction study of the conformational changes in hyaluronic acid induced in the presence of sodium, potassium and calcium cations. Int J Biol Macromol 1983; 5:215–221.
22. Almond A, Sheehan JK, Brass A. Molecular dynamics simulations of the disaccharides of hyaluronan in solution. Glycobiology 1997; 7:597–604.
23. Hardingham TE, Gribbon P, Heng BC. New approaches to the investigation of hyaluronan networks. Biochem Soc Trans 1999; 27:124–127.
24. Gribbon P, Heng BC, Hardingham TE. The molecular basis of the solution properties of hyaluronan investigated by confocal fluorescence recovery after photobleaching. Biophys J 1999; 77:2210–2216.
25. Gribbon P, Heng BC, Hardingham TE. The analysis of intermolecular interactions in concentrated hyaluronan solutions suggest no evidence for chain–chain association. Biochem J 2000; 350:329–335.
26. Engel J. Figure 1 and discussion comment. The biology of hyaluronan. Ciba Found Symp 1989; 143:18–19.
27. Scott JE, Cummings C, Brass A, Chen Y. Secondary and tertiary structures of HA in aqueous solution, investigation by rotary shadowing electron microscopy and computer simulation. Biochem J 1991; 274:699–705.
28. Scott JE, Heatley F. Hyaluronan forms specific stable tertiary structures in aqueous solution: a C-13 NMR study. Proc Natl Acad Sci 1999; 96:4850–4855.
29. Scott JE, Heatley F. Biological properties of hyaluronan in aqueous solution are controlled and sequestered by reversible tertiary structures defined by NMR spectroscopy. Biomacromolecules 2002; 3:547–553.
30. Kubitscheck H, Wedekind P, Peters R. Lateral diffusion measurements at high spatial resolution by scanning microphotolysis in a confocal microscope. Biophys J 1994; 67:946–965.
31. Gribbon P, Hardingham TE. Macromolecular diffusion of biological polymers measured by confocal fluorescence recovery after photobleaching. Biophys J 1998; 75:1032–1039.
32. Hardingham TE, Gribbon P. Confocal-FRAP analysis of ECM molecular interactions. In: Strueli C, Grant M, eds. Methods in Molecular Biology, Extracellular Matrix Protocols. Totowa, NJ: Humana Press, 2000; vol. 139:83–93.
33. Sheehan JK, Arundel C, Phelps CF. Effect of the cations sodium, potassium and calcium on the interactions of hyaluronate chains: a light scattering and viscometric study. Int J Biol Macromol 1983; 5:222–228.
34. Cowman MK, Hittner DM, Feder-Davies J. ^{1}C NMR studies of hyaluronan: conformational sensitivity to various environments. Macromolecules 1996; 29: 2894–2902.
35. Geciavo R, Flaibani A, Delben F, Liut G, Urbani R, Cesaro A. Physico-chemical properties of hyaluronan and its hydrophobic derivatives: a calorimetric study. Macromol Chem Phys 1995; 196:2891–2903.

36. Cowman MK, Cozart D, Nakanishi K, Balazs EA. ^{1}H NMR of glycosaminoglycans and hyaluronic acid oligosaccharides in aqueous solutions: the amide proton environment. Arch Biochem Biophys 1984; 230:203–212.
37. Fujii K, Kawate M, Kobayashi Y, Okamoto A. Effects of the addition of hyaluronate segments with different chain lengths on the viscoelasticity of hyaluronic acid solutions. Biopolymers 1996; 38:583–591.
38. Callaghan PT, Pinder DN. Influence of multiple length scales on the behaviour of polymer self-diffusion in the semidilute region. Macromolecules 1984; 17: 431–437.
39. Phillies DJ. The hydrodynamic scaling model for polymer self-diffusion. J Phys Chem 1989; 93:5029–5039.
40. Imhoff A, Van Blaadren A, Maret G, Mallema J, Dhont JKG. A comparison between the long time self diffusion of and low shear viscosity of concentrated dispersions of charged colloidal silica spheres. J Chem Phys 1994; 100:2170–2181.
41. Sheehan JK, Brass A, Almond A. The conformations of hyaluronan in aqueous solution: comparison of theory and experiment. Biochem Soc Trans 1999; 27: 121–124.
42. Almond A, Brass A, Sheehan JK. Deducing polymeric structure from aqueous molecular dynamics simulations of oligosaccharides: predictions from simulations of hyaluronan tetrasaccharides compared with hydrodynamic and X-ray fibre diffraction data. J Mol Biol 1998; 284:1425–1437.
43. Almond A, Brass A, Sheehan JK. Dynamic exchange between stabilized conformations predicted for hyaluronan tetrasaccharides: comparison of molecular dynamics simulations with available NMR data. Glycobiology 1998; 8:973–980.
44. Ghosh S, Khobal I, Zanette D, Reed WF. Conformational contraction and hydrolysis of hyaluronate in sodium hydroxide solutions. Macromolecules 1993; 26:4684–4691.

Chemistry and Biology of Hyaluronan
H.G. Garg and C.A. Hales (editors)

Chapter 2

Methods for Analysis of Hyaluronan and Its Fragments

**ISHAN CAPILA and
RAM SASISEKHARAN**

*Division of Biological Engineering,
Massachusetts Institute of Technology,
Cambridge, MA, USA*

I. Introduction

A. Historical Perspective

Hyaluronan was biochemically purified in 1934 by Karl Meyer and John Palmer from the highly viscous vitreous humor of bovine eyes. It was described as a 'polysaccharide acid of high molecular weight' and believed to consist of 'a uronic acid, an amino sugar, and possibly a pentose' (1). As a result, they proposed the name 'hyaluronic acid' (HA) from hyaloid (=vitreous) and uronic acid. They also reported that the isolated material did not contain any sulfur. Subsequently, in the late 1930s and 1940s, hyaluronan was isolated from a number of sources including synovial fluid, skin, umbilical cord, tumor tissue, and rooster comb (2). It was also isolated from certain strains of bacteria, like hemolytic streptococci (3). In conjunction with the isolation of HA, the discovery of the hyaluronidases as a 'spreading factor' from mammalian testicular extracts was also described (4,5). The purification and characterization of these enzymes established the foundation for further structural insights into this polymer. Initial chemical characterization of carefully purified hyaluronan preparations showed that it was composed of equimolar concentrations of glucuronic acid and *N*-acetylglucosamine (6). Enzymatic hydrolysis of hyaluronan by testicular hyaluronidase further confirmed that it was composed of a uniform structure of alternating acetylglucosamine and glucuronic acid residues (7). In 1951,

Karl Meyer and his co-workers isolated and characterized a crystalline disaccharide, hyalobiuronic acid from HA by enzymatic digestion and mild acid hydrolysis (8). Structural studies on this disaccharide were useful in establishing the β(1 → 3) glucuronidic linkage in the polymer (9). Similar approaches involving the use of hyaluronan degrading enzymes and structural analyses of isolated oligosaccharides led to the determination of the β(1 → 4) glucosaminidic linkage and thereby defined the structure of the basic structural unit in hyaluronan (10–12). Therefore, the chemical structure of hyaluronan was solved almost 20 years after its initial biochemical purification.

Over the course of the next decade, the physicochemical properties of hyaluronan were elucidated. Electron microscopy studies on hyaluronan revealed that it is a linear polymer (13). The polymer was shown to be polydisperse and at higher concentrations exhibited an extremely high viscosity in solution. A better understanding of the physical properties of hyaluronan provided a platform to define its possible physiological functions as a structural component of connective tissue (14). In the 1970s, Hardingham and Muir found that cartilage proteoglycans interact specifically with hyaluronan (15), and this discovery set the stage for studying hyaluronan–protein interactions. Over the course of the last 30 years more insight into the biosynthesis, catabolism and turnover of hyaluronan has been provided (16,17). The effects of hyaluronan in various aspects of biology including cell migration and differentiation, growth and metastases of tumors, inflammation and wound healing have also been reported (18). Therefore, hyaluronan is currently recognized as versatile polysaccharide with unique physical properties, having multiple structural and physiological functions.

B. Structural Overview

Hyaluronan (HA) is a high molecular mass homogeneous polysaccharide widely distributed in mammalian cells and tissue. It is classified as a member of the glycosaminoglycan (GAG) family of polymers. The repeating disaccharide unit in hyaluronan consists of D-glucuronic acid attached β(1 → 3) to *N*-acetylglucosamine (Fig. 1). Adjacent disaccharide units are linked by a β(1 → 4) linkage and there can be anywhere between 200–20,000 disaccharide units per chain. Therefore, the molecular weight of this polymer can range

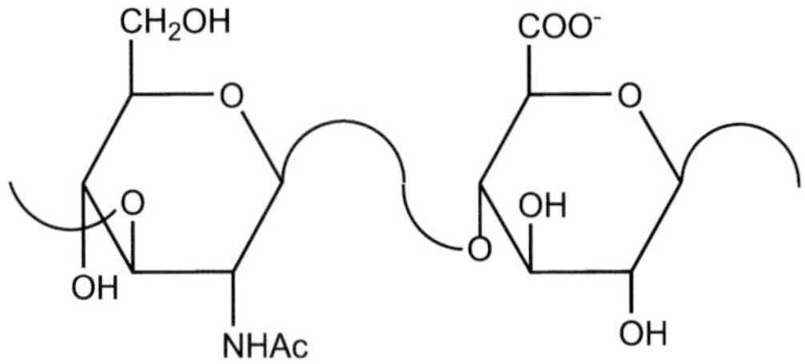

Figure 1 Structure of the disaccharide repeating unit found in hyaluronan (Ac, acetate).

from ~200 to 10,000 kDa. While other members of the GAG family, like heparin and chondroitin sulfate, have varying degrees of sulfation, no such modifications have been reported for hyaluronan. Hence in terms of chemical structure it is a simple molecule, however, its high molecular mass and rheological properties have led to a growing interest in its physiological and biological roles.

Based on bulk solution properties, hyaluronan is best described as a random coil with considerable intrinsic stiffness (19). This inherent stiffness is attributed to direct or water-mediated intramolecular hydrogen-bonding across the two glycosidic linkages. However, it is also suggested that solvent access and chain length play an important role in the ordering of this hydrogen-bonding pattern across the glycosidic linkages (20). The conformation of the individual saccharides in hyaluronan is also an important factor contributing to the overall shape of the molecule. The glucuronic acid and the *N*-acetylglucosamine moiety in the molecule appear to exist predominantly in the chair forms. This limits the saccharide-based flexibility that is a common feature in other GAGs like heparin that contain highly flexible 2-*O*-sulfated iduronic acid residues. Depending on these constraints, hyaluronan in solution could have an overall stiff random coil structure but may also have highly flexible and dynamic local regions (21). The conformation of the hyaluronan molecule can, however, vary due to the binding of proteins. Crystal structure studies have shown that the reducing end *N*-acetylglucosamine residue complexed to hyaluronidase from bee venom, adopts a boat form instead of the usual $^4C_1$ chain conformation (22).

II. Biological Role of Hyaluronan and Its Fragments

Numerous functions have been ascribed to hyaluronan. Its ability to regulate water balance and fill space, and interact with a variety of extracellular molecules makes it an essential structural component in the organization of the extracellular matrix (ECM). It can also interact with different cell surface receptors thereby activating intracellular signaling pathways and inducing proliferative and migratory responses. The majority of extracellular HA-binding proteins belong to the Link protein superfamily. Proteins in this family contain a conserved Link module that consists of a disulfide-linked domain of about 100 amino acid residues (23).

The interaction of hyaluronan with the cell surface receptor, CD44, is one of the most widely known and studied interactions. The signaling processes activated by the HA–CD44 interaction are varied and the specific intermediates involved are beginning to be elucidated (24,25). Other putative HA receptors which are homologous to CD44 and members of the Link protein superfamily have also been described (26,27). While many studies have focused on the extracellular functions of hyaluronan there is also a growing interest in putative intracellular roles of hyaluronan (28). It is suggested that HA may play a role in cell cycle and may also modulate the trafficking of specific kinases within the cell, thereby regulating cell behavior (25).

The physical properties of hyaluronan as a high molecular weight polymer account for its role as an essential structural component of the ECM. However, in order to better understand its role in interaction with different proteins, low molecular weight hyaluronan oligosaccharides have been very useful. Using hyaluronan oligomers of defined sizes, Lesley et al. were able to clearly demonstrate the role of cooperativity in the binding of HA by cell surface CD44 (29). Furthermore, hyaluronan decasaccharides were the minimum size oligosaccharides required to displace hyaluronan from CD44 on keratinocytes (30). This ability of hyaluronan oligomers has also been reported to play a role in the inhibition of tumor growth (31). The sequestration of *Plasmodium falciparum*-infected erythrocytes in the placenta was inhibited by structurally defined hyaluronan dodecasaccharides (32). Structurally defined and pure hyaluronan hexasaccharides and tetrasaccharides have also been instrumental in understanding the mechanism of action of the *Streptococcus pneumoniae* hyaluronate lyase (33).

There have also been reports suggesting that the biological effects of hyaluronan may vary depending upon its average mass (34). In many instances low molecular weight fragments of hyaluronan have exhibited effects not associated with native HA. Hyaluronan fragments have been used to establish a novel signal transduction cascade downstream from CD44 activation that involves Ras and protein kinase C in T-24 carcinoma cells (24). During inflammation, there is an increased degradation of components of the ECM and as expected, hyaluronan is broken down to lower molecular weight forms (35). These small hyaluronan fragments have been reported to interact with dendritic cells and induce their maturation (36) and also induce the expression of macrophage genes that are important in the development and maintenance of the inflammatory response (37).

The above examples highlight the importance of hyaluronan oligomers as essential tools for studying protein–HA interactions as well as determining new biological roles for HA. Therefore, it is essential that the hyaluronan oligomers used in similar studies be well characterized and highly pure so as to avoid any misinterpretation of data generated from the use of these fragments.

III. Preparation and Isolation of Hyaluronan Oligomers

A. Degradation of Hyaluronan

Hyaluronan can be degraded to smaller fragments by chemical methods (acidic and alkaline conditions), physical stress (high-speed stirring or critical shearing), sonication (38), free-radical-based cleavage (39), or enzymatic methods. Free-radical-based cleavage of hyaluronan in the connective tissue has physiological implications in arthritis and aging. The hydroxyl radical has been shown to be a primary factor in the initiation of hyaluronan degradation by causing non-specific scission of the glycosidic linkage. The higher the concentration of the free radical, greater the decrease in the molecular mass of hyaluronan (40). The use of high-energy ultrasound or sonication is another established method for the cleavage of hyaluronan chains. Preliminary studies have defined the relationship between sonication time, intensity, and reduction in chain length (41,42). After prolonged sonication at a fixed intensity, the molecular size of depolymerized hyaluronan does not change. The fragments produced after this procedure predominantly have *N*-acetylglucosamine (86%) at their reducing end and glucuronic acid (98%) at the non-reducing end. This suggests that there is some level of specificity during sonication and certain weak linkages related to *N*-acetylglucosamine are extremely susceptible to sonication (38).

While both the methods described above have been used in various studies for the generation of low molecular weight hyaluronan, the most popular method for the generation of hyaluronan oligomers involves the digestion of hyaluronan by the hyaluronidase enzymes. There are three different types of hyaluronidases known, and they degrade hyaluronan by different mechanisms (5). The first group includes the mammalian-type hyaluronidases (EC 3.2.1.35), which are endo-β-*N*-acetyl-D-hexosaminidases that degrade hyaluronan to tetrasaccharides and hexasaccharides as the major end products. The testicular hyaluronidase, which belongs to this group, has been shown to have the ability to catalyze transglycosylation reactions also, in addition to its hydrolytic activity (43,44). The second group consists of hyaluronidases (EC 3.2.1.36) that are endo-β-glucuronidases, from leeches and other parasites. Bacterial hyaluronidases (EC 4.2.2.1) form the third group, and they act on hyaluronan via a β-elimination reaction. Based on the bacterial source these enzymes can either yield tetrasaccharides and hexasaccharides, or disaccharides as the final products. The elimination reaction generates a modified uronic acid having a C-4, 5 double bond at the non-reducing end. The formation of this double bond enables detection of the products of digestion by monitoring absorbance at 232 nm (Fig. 2). Therefore, this ultraviolet chromophore represents a good internal 'tag' for following the digestion products during isolation. However, since the conformation adopted by the unsaturated uronic acid is different from the internal glucuronic acid residues it could significantly affect the overall conformation of shorter hyaluronan oligosaccharides. This may eventually affect how they interact with proteins in subsequent studies.

Enzymatic digestion of hyaluronan is usually carried out in sodium acetate buffer adjusted with acetic acid to an acidic pH range (4.8–6.0) where most of these enzymes are active. The temperature for the digestion is usually 37 °C for

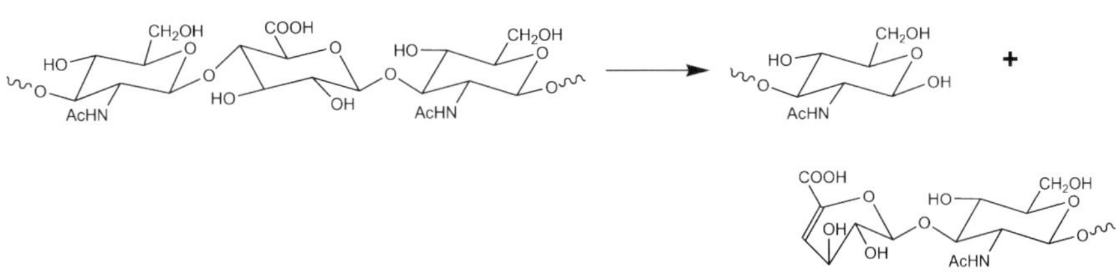

Figure 2 The action of bacterial hyaluronidases on hyaluronan (Ac, acetate).

the mammalian enzymes; however, the bacterial enzymes have been shown to be active from room temperature to 60 °C (45). By varying the time of enzymatic digestion the average size of the resulting oligosaccharide pool also varies. Increasing digestion time leads to a larger proportion of low molecular weight species. After the hyaluronan polymer has been treated with the enzyme for the desired time, further reaction is stopped by boiling the digestion mixture for around 5 min. The mixture can then be analyzed and hyaluronan oligomers can be purified using different column chromatography techniques.

B. Separation and Purification of Hyaluronan Oligomers

Various methods have been described for the separation of hyaluronan oligomers. Most widely used among these include size-exclusion chromatography (SEC), ion-exchange chromatography and reversed-phase ion pair (RPIP) high-performance liquid chromatography (HPLC).

SEC separates molecules according to their hydrodynamic volume, i.e., the space a particular polymer molecule takes up when in solution. This results in a separation according to decreasing molecular mass for hyaluronan digestion mixtures (29). This technique is also useful for determining the relative molecular mass of hyaluronan. Based on the enzyme used for digestion, the products eluting off the column can be monitored at either 232 nm (bacterial hyaluronidase) or 206–210 nm (mammalian hyaluronidase). In cases where the buffer contains acetate or citrate, which have strong background UV absorption at these wavelengths (206–210 nm), the uronic acid assay and its modifications are used (46). The smaller size oligomers (4–16mer) are well resolved using this technique and there is also less cross-contamination among peaks (30). However, larger oligosaccharides (>18mer) show clusters of 3–8 sizes within a peak obtained off the column and may need further analysis to determine the individual components (Fig. 3) (29). Therefore, SEC is useful for obtaining pure low molecular weight hyaluronan oligomers and fragments thus obtained are useful in biochemical and crystallography studies.

The HPLC methods for separation and purification of oligosaccharides produced by enzymatic or chemical hydrolysis of hyaluronan include normal phase partition (47), weak anion-exchange (48,49), size-exclusion (50), and RPIP (51). Detection can be based on pulsed amperometric detection (PAD), UV absorbance or fluorescence. The reversed-phase method has been used for the quantification of hyaluronan in biological tissues and samples (51). Hyaluronidase from *Streptomyces hyalurolyticus* is specific for hyaluronan and quantitatively yields a tetrasaccharide and a hexasaccharide as the final products. These two products were resolved by RPIP HPLC on a C-18 column in the presence of the ion-pairing agent, tetrabutylammonium hydroxide, at pH 7.6 in an acetonitrile gradient. The products were detected and quantified by their absorbance at 232 nm. Based on this quantification, the starting concentration of hyaluronan was estimated to within 93%. A modification of this RPIP method has been applied to study the degradation kinetics of purified hyaluronan oligomers

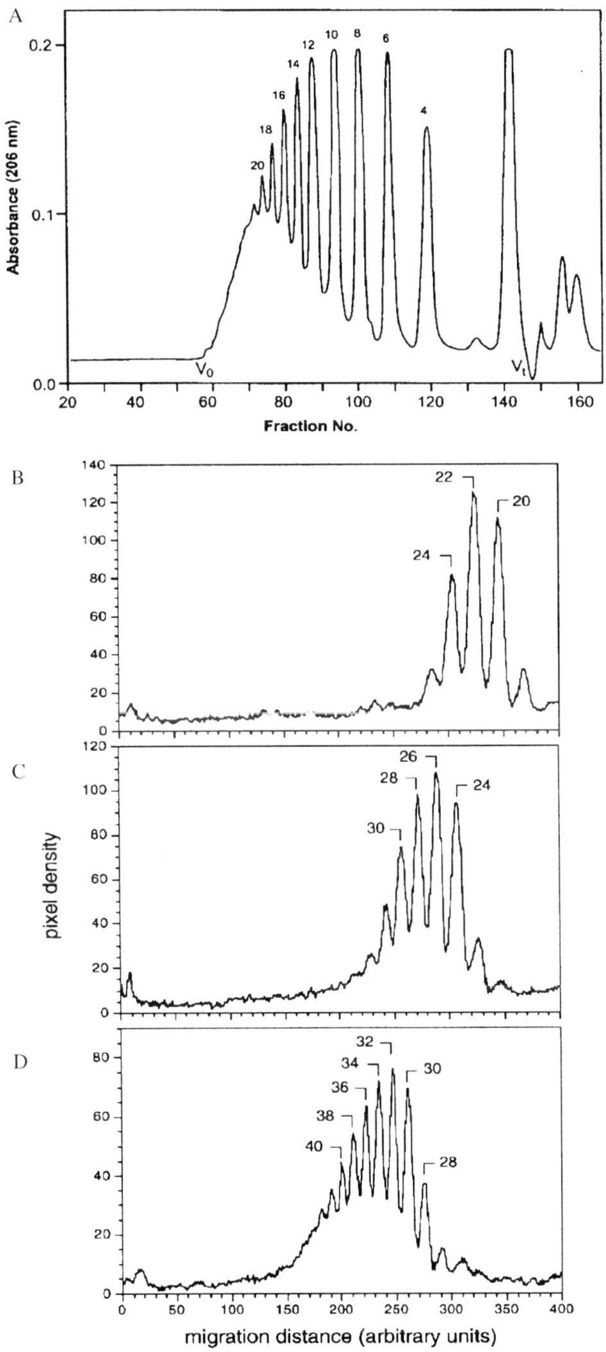
A
Absorbance (206 nm)
Fraction No.
V0
Vt
B
C
D
pixel density
migration distance (arbitrary units)

by bovine testicular hyaluronidase (52). An isocratic elution at pH 9.0 was used and this was consistent with a post-column derivatization procedure using 2-cyanoacetamide. The 2-cyanoacetamide reacted with the reducing end *N*-acetylglucosamine of hyaluronan oligomers eluting off the column, to yield products that were monitored at 276 nm. This labeling agent offers a variety of detection modes including fluorescence and PAD, and therefore may be compatible with different systems.

Weak-anion exchange HPLC methods utilize an amine-modified stationary phase that becomes protonated under acidic conditions to an extent proportional to the pH of the mobile phase. Modifying the solvent composition and pH of the mobile phase has enabled optimization of the separation of weakly acidic hyaluronan species (49). With most of the earlier HPLC methods the largest hyaluronan fragment that was separated was a dodecasaccharide (52). The development of high-performance anion-exchange chromatography (HPAEC) for the separation of neutral and acidic oligosaccharides (53,54) facilitated the resolution of larger hyaluronan fragments. The initial studies on neutral carbohydrates were performed at high pH to ensure deprotonation of the ring hydroxyls, which could then interact with a pellicular anion-exchange resin. However, since hyaluronan is highly susceptible to degradation at high pH, from 'alkali-peeling' reactions (55), the separation was done in the pH range of 6.3–5.0 by utilizing the carboxylate group. Using this method, hyaluronan oligomers of between 2 and 20 disaccharide units have been resolved (56).

Another approach to address the issue of hyaluronan degradation at the high alkaline pH used in HPAEC, involves the reduction of the hyaluronan oligomers to their alditol forms using borohydride, thereby making them stable to alkali (57). This procedure allows use of the carbohydrate separation abilities of a CarboPac PA1 column when run under alkaline conditions. The chromatographic conditions used afforded a high-resolution and highly sensitive method for the compositional analysis of hyaluronan, and chondroitin sulphates in minute quantities of biological samples. In addition, these conditions were also shown to be ideal for the separation of hyaluronan oligosaccharide alditols in the range of hexasaccharide to dodecasaccharide (12mer). This method has been modified subsequently, by altering the elution conditions, so as to separate hyaluronan

---

Figure 3 Fractionation of oligosaccharides from a testicular hyaluronidase digest of hyaluronan. A, elution profile of digest. Fractions under the UV absorbing peaks were pooled, and several additional pools were made from fractions 58–77, containing clusters of higher oligomers. V_0 and V_t indicate the approximate void and total volumes of the column, respectively. B–D, examples of size distribution of the fluorotagged higher oligomer pools on 20% polyacrylamide minigels. The densitogram in B was material collected under peak 20° in A, while C and D represent material closer to V_0, with a nominal size of $HA_{\sim 26}$ and $HA_{\sim 34}$, respectively. The number of monosaccharide units in each oligosaccharide peak is indicated *above* each *peak*. Reproduced from Ref. 29 with permission from the publisher.

oligosaccharides up to a hexadecasaccharide and chondroitin and dermatan sulfate oligosaccharides up to a hexasaccharide in size (58). A combination of SEC and HPAEC has been used in a recent study for the separation of hyaluronan oligomers ranging from tetrasaccharides to 34mers (59). SEC was primarily used for the separation and purification of low molecular weight hyaluronan oligomers (4–12mer). High molecular weight hyaluronan oligomers were generated by reducing the enzymatic digestion time. An initial gel filtration step was used to select an oligosaccharide pool corresponding to the larger fragments, and these were subsequently purified by anion-exchange HPLC. As noted previously, the larger purified fragments (22–34mer) were mixtures of hyaluronan oligomers of different lengths. It is suggested that these mixtures can be further purified by re-running the samples on the ion-exchange column.

IV. Analysis and Characterization of Hyaluronan Oligomers

A. Capillary Electrophoresis

The application of capillary electrophoresis (CE) for the analysis of acidic oligosaccharides has been an important development in the GAG field (60). Its high-resolution separation, rapid analysis, quantitation of analytes, and low amount of sample usage makes it a very valuable analytical technique (61). Procedures for the separation and identification of hyaluronan and its fragments by CE are well established (62,63). In normal polarity mode (sample application at anode and detection at cathode; buffer pH = neutral or basic), a migration buffer of phosphate and borate at pH = 9.0, in the presence of sodium dodecyl sulfate (SDS) is used. The complete depolymerization of hyaluronan by testicular hyaluronidase yields a tetrasaccharide as the major product. This tetrasaccharide migrates as a sharp peak in the normal polarity mode and the peak area can be used to quantify the polymer (63). The detection of hyaluronan depolymerization by CE has also been used in an assay for determination of hyaluronidase activity in bee and snake venom (64).

In reversed polarity mode sample is applied at the cathode and detected at the anode. An acidic buffer ($pH \leq 3$) is used that effectively reduces the electroosmotic flow, therefore, the migration of an acidic oligosaccharide is a function of its charge and size. This method works well for resolution of oligosaccharides of heparin, heparan sulfate and chondroitin sulfates, however, since hyaluronan oligomers have a uniform charge density (one carboxylate group per disaccharide repeat), there is poor separation between oligosaccharides of differing size (60). This problem has been addressed by introduction of a single dominant charge group at the reducing end of each oligosaccharide chain. In a study that investigated the action pattern of hyaluronate lyase from *S. hyalurolyticus*, enzymatic digestion products were derivatized at the reducing end by reductive amination with AGA (7-amino-1,3-naphthalene disulfonic acid) (45). At low pH, only the sulfonate groups on the AGA tag remain charged thus giving each oligosaccharide a fixed charge. Under these conditions the oligosaccharide size becomes the dominating factor that influences migration through the capillary. This technique was successfully used to follow the depolymerization of hyaluronan kinetically.

The application of CE to larger saccharides has been made possible by using polymeric sieving media to coat the inner surface of separation capillaries, thereby inducing size-dependent migration of the saccharides. This approach, known as capillary gel electrophoresis (CGE), has been coupled with laser-induced fluorescence (LIF) detection to enable resolution of large hyaluronan saccharides of up to 190 disaccharide units (Fig. 4) (65). In this case, the inner surface of the capillary was coated with polyacrylamide and the hyaluronan oligomers were derivatized with APTS (1-aminopyrene-3,6,8-trisulfonic acid) to facilitate LIF detection. Although this technique led to a substantial improvement in resolution compared to previous methods (66,67), it showed additional peaks in the oligomer

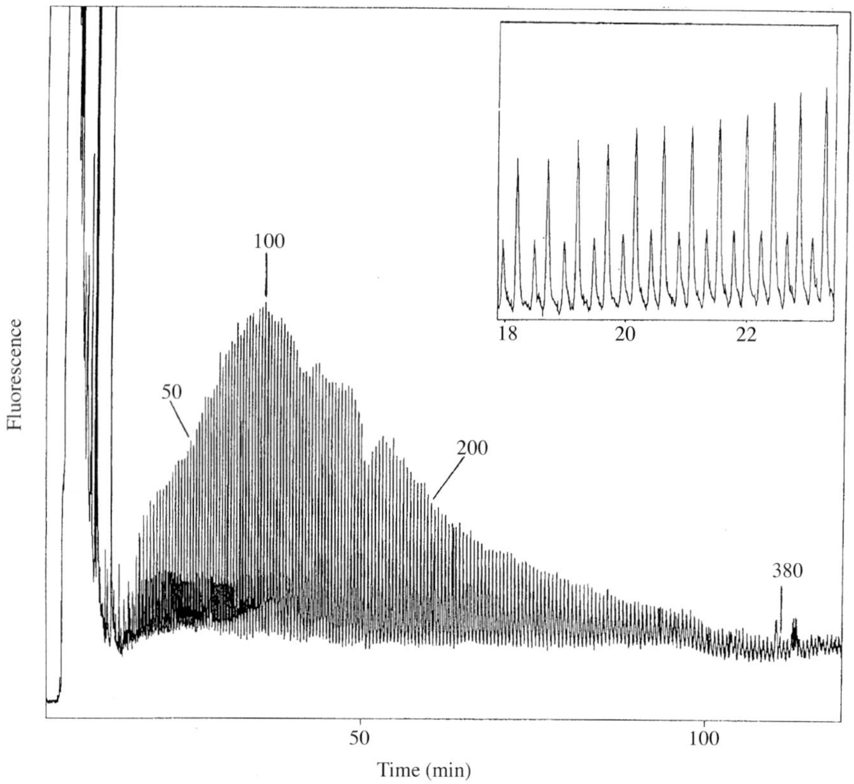

Figure 4 Electropherogram of HA in an entangled polymer solution. Conditions: −403 V/cm (10 μA) using 25 mM citric acid and 12.5 mM Tris buffer as the electrolyte (pH 3.0); 5% LPAA. The effective length of the separation capillary was 50 cm. The inset in the upper corner corresponds to a detail of the electropherogram. The numbers indicate the degree of polymerization. Reproduced from Ref. 65 with permission from the publisher.

profile in-between the regular oligomer peaks. The authors assume that these are the result of hyaluronan fragments existing in two different conformations in solution, which migrate slightly differently through the polyacrylamide matrix. Another issue that has been reported with hyaluronan fragment movement through a polymer matrix is the anomalous migration of smaller oligosaccharides, which migrate in the reverse order of their molecular masses, whereas larger oligomers migrate in order of their molecular masses (68,69).

B. Fluorophore-Assisted Carbohydrate Electrophoresis of Hyaluronan Oligomers

Fluorophore-assisted carbohydrate electrophoresis (FACE) represents a simple, rapid and highly sensitive technique for quantification and determination of chain length and fine structure of different GAGs. The technique basically involves generation of oligosaccharides by either chemical or enzymatic methods ensuring that the free reducing end is preserved. The free reducing end is fluorescently labeled and the labeled oligosaccharides are separated on high percent polyacrylamide gels (20–40%) at a high voltage and then imaged. The fluorescent labeling is via reductive amination chemistry and the labeling efficiency has been optimized to yield 95–100% labeling independent of the GAG structure or composition. FACE has been applied to the analysis of enzyme digests of hyaluronan and other GAGs (70,71).

For fragments of hyaluronan there are two procedures that have been used. In the first case, the hyaluronan oligomers are derivatized with 2-aminoacridone (AMAC) and run on polyacrylamide gels in the presence of borate (both in the gel and the running buffer). This has enabled the resolution of hyaluronan oligomers from disaccharides up to 50mers (Fig. 5) (71). While the migration of each oligomer in the gel is mainly influenced by its size (since AMAC is an uncharged label), an anomalous migration is observed for disaccharides up to hexasaccharides. These fragments show an inversion of mobility, with the disaccharide and tetrasaccharide migrating slower than the hexasaccharide. The hexasaccharide migration overlaps with that of the octasaccharide, and for higher fragments the order of migration is as expected. Since the borate in this system is involved in forming a complex with the sugars and influencing separation, the authors suggest that this anomalous mobility is possibly the result of smaller oligosaccharides interacting with borate in the electrophoresis buffer differently. Due to its high sensitivity, this method has been used for quantification and analysis of hyaluronan from cartilage (72) and also for the analysis of purified hyaluronan oligosaccharides (59). For the low molecular weight hyaluronan oligosaccharides (4–20mer) size can be determined by running sugar standards as markers. However, for longer oligosaccharides in addition to the sugar standards it is useful to run a previously purified and characterized long hyaluronan oligomer of known size (59).

Another procedure that has been applied to the resolution of hyaluronan oligomers by FACE involves derivatization with a charged ANTS

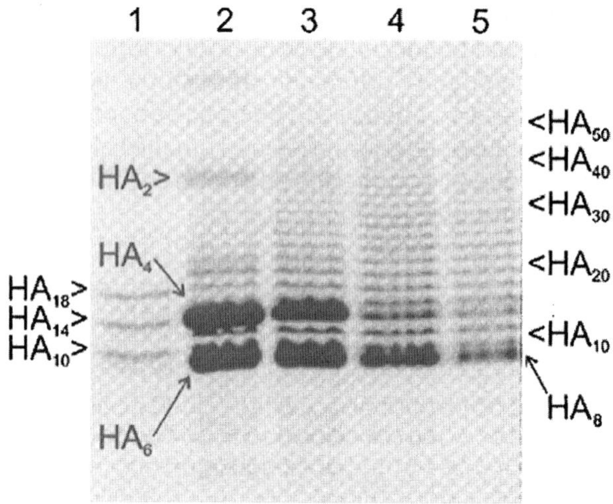

Figure 5 FACE analyses of AMAC-derivatized products from partial digestion of a constant amount of hyaluronan (100 μg) for 4 h at 37 °C with 1:3 serial dilutions of testicular hyaluronidase starting at 1000 U/mL (lanes 2–5). The relative positions of the saturated hyaluronan oligomers containing 1 (HA_2), 2 (HA_4), 3 (HA_6), 4 (HA_8), 5 (HA_{10}), 10 (HA_{20}), 15 (HA_{30}), 20 (HA_{40}), and 25 (HA_{50}) disaccharides are indicated. Lane 1 contains a standard mixture of three purified, AMAC-derivatized hyaluronan oligomers (HA_{10}, HA_{14}, and HA_{18}) used to index the ladder. Reproduced from Ref. 71 with permission from the publisher.

(8-aminonaphthalene-1,3,6,-trisulphate) label and separation using a non-borate containing electrophoresis system (73). In this system, the additional charge on each fluoro-tagged HA oligosaccharide is a major factor, in addition to the mass, that influences migration; therefore samples migrate further into the gel than with an uncharged AMAC label. There is also no observed inversion in mobility for the smaller oligosaccharides using this method and this may be attributed to the absence of borate in this system. This method was used for the characterization of purity and size of hyaluronan oligosaccharides ranging from 4mer to 20mer (73).

C. Electrospray-Ionization Mass Spectrometry of Hyaluronan Oligomers

Electrospray-ionization mass spectrometry (ESI MS) is a powerful analytical technique for carbohydrate analysis. Studies on anionic sugars using ESI MS in the negative-ion mode have demonstrated its usefulness in the area of GAG analysis. Hyaluronan fragments generated by enzymatic digestion with hyaluronate lyase from *S. hyalurolyticus* were characterized using HPAEC and ESI MS (74). The observed m/z values for hyaluronan oligomers ranging from a 2mer to a 16mer, were compared with the predicted possible charge states to get an idea of the extent of polyanionization as a function of oligomer length. It was observed that the charge distribution tends to cluster in a small range of charge states, with the smaller oligomers existing predominantly in the lower charge states, while the larger ones exhibit a distribution maximum at higher charge states. It is also important to note that the charge distributions reported here were observed at a cone voltage of 25 V. If the cone voltage is increased the charge distribution for any given oligomer is shifted to a some what lower charge state.

The control of the cone voltage is essential for proper interpretation of results obtained from ESI MS studies on hyaluronan oligomers (75). The cone voltage plays a part in accelerating the ions into the mass analyzer. Therefore, it cannot be set too low since this will lead to insufficient ions being channeled into the mass analyzer, resulting in an unusable mass spectrum. However, increasing cone voltage beyond a certain limit imparts more internal energy into the ions through collision and results in fragmentation of the oligomer, thereby giving rise to an additional series of negatively charged species. The susceptibility of an oligomer to undergo fragmentation at a certain cone voltage is dependent on the length of the oligomer. Thus, for higher oligomers (14–16mer) there is a greater possibility for observing an odd-numbered oligomer (15mer) at a lower cone voltage. This is important to keep in mind since there have been reports of odd-numbered saccharides in enzymatic digests of hyaluronan using testicular hyaluronidase (59) and hyaluronate lyase (74). This is contrary to the known specificity of these enzymes, which should generate only even-numbered saccharides. However, in these cases the presence of the odd-numbered oligomers was confirmed independently by techniques other than ESI MS. Hence the authors conclude that the odd-numbered oligomers are possibly the result of some contaminating hydrolase or glucuronidase activity in the enzyme preparation.

While fragmentation definitely leads to a more complicated mass spectrum, it can also yield important sequence information. Fragmentation down to the next lower oligomer was useful in confirming the identity of the odd-numbered oligomers (5 and 7mer) observed in the enzymatic digests of hyaluronan (59). Therefore, ESI MS represents a good technique for the characterization of moderately sized hyaluronan oligomers; provided proper care is taken in the control and choice of the cone voltage and the results are corroborated by other analytical techniques. Since it is tough to avoid fragmentation of larger oligomers even at lower cone voltages in ESI MS, it is advisable to use other techniques like MALDI MS (described below) for their characterization.

D. Matrix-Assisted Laser Desorption Ionization (MALDI) Mass Spectrometry of Hyaluronan Oligomers

MALDI time-of-flight (TOF) mass spectrometry is a popular tool for the analysis of complex carbohydrate compounds. Its low sample requirement, simplicity, reasonable tolerance towards salts and buffers and high sensitivity make it an attractive choice as an analytical technique. One of the main issues that limited the use of MALDI for GAG analysis earlier on was the low extent of desorption of highly acidic polysaccharides. This problem has been addressed by the addition of a positively charged peptide that binds to the acidic sugar and the resulting neutral peptide–saccharide complex can desorb easily and be analyzed (76). This has enabled development of structure analysis approaches for highly acidic polysaccharides like heparin with MALDI-TOF MS as the analytical platform (77). In the case of less acidic hyaluronan, procedures have been developed that do not need any addition to or derivatization of the sample prior to analysis.

MALDI-TOF MS was used to analyze hyaluronan oligomers resulting from the degradation of hyaluronan by hyaluronate lyase (78). Since hyaluronan is less acidic, both positive and negative mode spectra can be collected for hyaluronan oligomers. However, in the positive mode only tetrasaccharide fragment peaks are observed, whereas in the negative mode for the same sample, there are intense peaks corresponding to a hexasaccharide and an octasaccharide as well (Fig. 6). This indicated that the negative mode is more sensitive for detection of hyaluronan oligomers. For these studies a 2,5-dihydroxybenzoic acid (DHB) solution in water containing trifluoroacetic acid was used as a matrix. Due to the limitations of ESI MS for the characterization of larger hyaluronan oligomers, the use of MALDI-TOF MS for these samples has been very useful since it does not result in sample fragmentation or generation of multiple ion species. Therefore, it has been possible to use MALDI to characterize a wide range of oligomer sizes (8–34mer) (59). In this study, the DHB matrix which had been used previously did not give good results so an alternate co-matrix of 2,4,6-trihydroxyacetophenone and triammonium citrate was used.

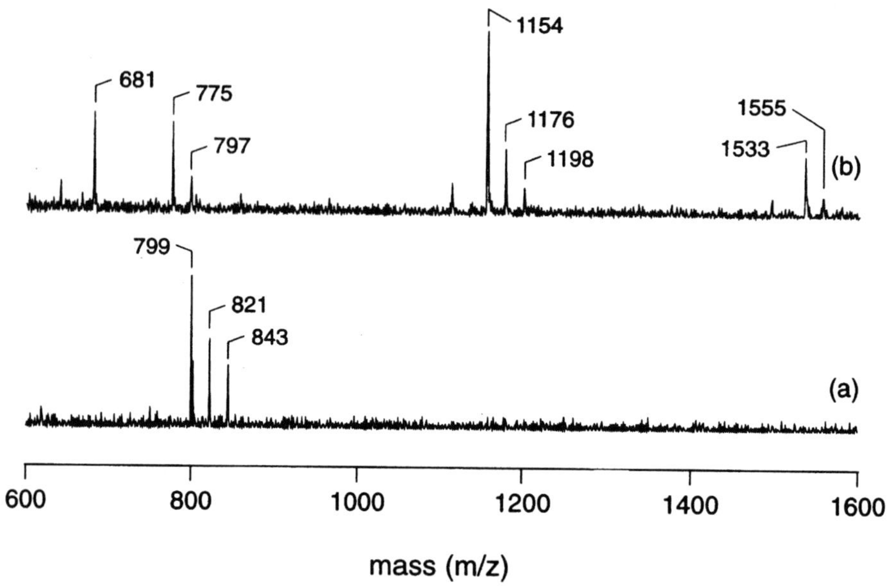

Figure 6 Positive (a) and negative ion (b) MALDI-TOF mass spectra of a 10 mg/mL hyaluronic acid solution digested with 2 U hyaluronidase for 9 h. The spectrum was recorded in 2,5-dihydroxybenzoic acid matrix after a 1:10 dilution of the carbohydrate solution. Reproduced from Ref. 78 with permission from the publisher.

MALDI MS has also been used for the molecular weight characterization of polydisperse mixtures of larger hyaluronan oligomers (79). Therefore, MALDI-TOF MS is useful for providing qualitative information on pure hyaluronan oligosaccharides and polydisperse mixtures of hyaluronan, however, cannot be used as a quantitative method. This is because the desorption efficiency of different size hyaluronan oligomers varies considerably, with shorter oligomers undergoing preferential desorption from the matrix.

E. Other Methods for Analysis of Hyaluronan Oligosaccharides

Nuclear magnetic resonance (NMR) spectroscopy has proved useful in determining the purity of isolated hyaluronan oligomers, before they are used in biological assays (32). Although one of the limitations with NMR is the large amount of pure sample required to acquire a spectrum, it has provided a lot of insight into the possible conformations adopted by hyaluronan oligomers. NMR studies on hyaluronan oligomers in conjunction with molecular dynamics have helped elucidate the arrangement of intramolecular hydrogen bonding around the glycosidic linkage and how this plays a role in affecting the overall conformation of the oligomer (21).

New approaches to analyze hyaluronan oligosaccharides obtained by enzymatic digestion include a combination of techniques like online CE/ESI MS (80) and LC/ESI MS (81). The coupling of CE to ESI MS has been achieved through a sheath–liquid interface that compensates for the low CE flow rate and also incorporates organic solvent to facilitate a stable electrospray (80). The use of an ion trap mass analyzer yields further structural information through MS/MS and MS^n experiments. Using this method the separation and online identification of hyaluronan oligomers up to 16mer was possible. The analysis of hyaluronan fragments in pharmaceutical formulations has recently been performed by liquid chromatography–electrospray tandem mass spectrometry (LC/MS/MS). This technique has enabled the quantification of hyaluronan oligomers ranging from a 4mer to 10mer in different pharmaceutical formulations (81). Therefore, these online techniques represent important advances in the rapid characterization of hyaluronan oligomers, which are now being used more frequently in the pharmaceutical industry.

V. Summary and Conclusion

In recent years, a growing number of physiological and biological roles have been ascribed to hyaluronan in addition to is role as an essential structural component of the ECM. Its rheological properties have led to an increasing number of applications for hyaluronan in the pharmaceutical industry. Hyaluronan preparations from different biological sources are finding applications in ophthalmology, cosmetics and drug delivery. The chemical modification of hyaluronan to alter its physical properties has also led to important applications for this molecule. There have also been reports suggesting that the biological effects of hyaluronan vary depending on its average mass. Hyaluronan oligomers have demonstrated activities that are not observed with the full-length molecule and have proved to be important for studying various protein-hyaluronan interactions. Recent advances in the chemoenzymatic synthesis of hyaluronan oligosaccharides have opened up possibilities of using these fragments as therapeutics (82). The application of hyaluronan fragments in various pharmaceutical applications is also under investigation. Therefore, it is essential from an analytical point of view to define the composition of mixtures of low molecular weight hyaluronan fragments. The purity and structure of individual hyaluronan oligomers is also important to establish before they are used in biological systems. This will enable a better understanding of the biological responses these molecules elicit and will help unravel the various physiological roles played by hyaluronan.

Acknowledgements

The authors would like to acknowledge support from NIH Grants GM57073 and CA90940.

References

1. Meyer K, Palmer J. The polysaccharide of the vitreous humor. J Biol Chem 1934; 107:629–634.
2. Meyer K. The biological significance of hyaluronic acid and hyaluronidase. Physiol Rev 1947; 27:335–359.
3. Kendall FE, Heidelberger M, Dawson MH. A serologically inactive polysaccharide elaborated by mucoid strains of group A hemolytic streptococcus. J Biol Chem 1937; 118:61–69.
4. Duran-Reynals F. Tissue permeability and the spreading factors in infection. Bacteriol Rev 1942; 6:197–252.
5. Kreil K. Hyaluronidases—a group of neglected enzymes. Protein Sci 1999; 4: 1666–1669.
6. Meyer K, Palmer JW. On glycoproteins. II. The polysaccharides of vitreous humor and of umbilical cord. J Biol Chem 1936; 114:689–703.
7. Rapport MM, Meyer K, Linker A. Analysis of the products formed on hydrolysis of hyaluronic acid by testicular hyaluronidase. J Am Chem Soc 1951; 73:2416–2420.
8. Rapport MM, Weissmann B, Linker A, Meyer K. Isolation of a crystalline disaccharide, hyalobiuronic acid, from hyaluronic acid. Nature 1951; 168:996–997.
9. Weissmann B, Meyer K. Structure of hyaluronic acid. The glucuronidic linkage. J Am Chem Soc 1952; 74:4729.
10. Weissmann B, Meyer K. The structure of hyalobiuronic acid and of hyaluronic acid from umbilical cord. J Am Chem Soc 1954; 76:1753–1757.
11. Linker A, Meyer K. Production of unsaturated uronides by bacterial hyaluronidases. Nature 1954; 174:1192–1193.
12. Weissmann B, Meyer K, Sampson P, Linker A. Isolation of oligosaccharides enzymatically produced from hyaluronic acid. J Biol Chem 1954; 208:417–429.

13. Fessler JH, Fessler LI. Electron microscopic visualization of the polysaccharide hyaluronic acid. Proc Natl Acad Sci USA 1966; 56:141–147.
14. Comper WD, Laurent TC. Physiological function of connective tissue polysaccharides. Physiol Rev 1978; 58:255–315.
15. Hardingham TE, Muir H. The specific interaction of hyaluronic acid with cartilage proteoglycans. Biochim Biophys Acta 1972; 279:401–405.
16. McCourt PAG. How does the hyaluronan scrap-yard operate? Matrix Biol 1999; 18: 427–432.
17. Tammi MI, Day AJ, Turley EA. Hyaluronan and homeostasis: a balancing act. J Biol Chem 2002; 277:4581–4584.
18. Laurent TC, Fraser JRE. Hyaluronan. FASEB J 1992; 6:2397–2404.
19. Scott JE. Secondary structures in hyaluronan solutions: chemical and biological implications, The Biology of Hyaluronan, Ciba Foundation Symposium. Chichester, England: Wiley, 1989; vol. 143: 6–20.
20. Cowman MK, Feder-Davis J, Hittner DM. ^{13}CNMR studies of hyaluronan. 2. Dependence of conformational dynamics on chain length and solvent. Macromolecules 2001; 34:110–115.
21. Day AJ, Sheehan JK. Hyaluronan: polysaccharide chaos to protein organisation. Curr Opin Struct Biol 2001; 11:617–622.
22. Markovic-Housley Z, Miglierini G, Soldatova L, Rizkallah PJ, Muller U, Schirmer T. Crystal structure of hyaluronidase, a major allergen of bee venom. Structure 2000; 8:1025–1035.
23. Kohda D, Morton CJ, Parkar AA, Hatanaka H, Inagaki FM, Campbell ID, Day AJ. Solution structure of the link module: a hyaluronan-binding domain involved in extracellular matrix stability and cell migration. Cell 1996; 86:767–775.
24. Fitzgerald KA, Bowie AG, Skeffington BS, O'Neill LAJ. Ras, protein kinase Cξ, and IκB kinases 1 and 2 are downstream effectors of CD44 during the activation of NF-κB by hyaluronic acid fragments in T-24 carcinoma cells. J Immunol 2000; 164: 2053–2063.
25. Lee JY, Spicer AP. Hyaluronan: a multifunctional, megadalton, stealth molecule. Curr Opin Cell Biol 2000; 12:581–586.
26. Banerji S, Ni J, Wang SX, Clasper S, Su J, Tammi R, Jones M, Jackson DJ. LYVE-1, a new homologue of the CD44 glycoprotein, is a lymph-specific receptor for hyaluronan. J Cell Biol 1999; 144:789–801.
27. Tsifrina E, Ananyeva NM, Hastings G, Liau G. Identification and characterization of three cDNAs that encode putative novel hyaluronan-binding proteins, including an endothelial cell-specific hyaluronan receptor. Am J Path 1999; 155:1625–1633.
28. Evanko SP, Wight TN. Intracellular localization of hyaluronan in proliferating cells. J Histochem Cytochem 1999; 47:1331–1342.
29. Lesley J, Hascall VC, Tammi M, Hyman R. Hyaluronan binding by cell surface CD44. J Biol Chem 2000; 275:26967–26975.
30. Tammi R, MacCallum D, Hascall VC, Pienimaki J-P, Hyttinen M, Tammi M. Hyaluronan bound to CD44 on keratinocytes is displaced by hyaluronan decasaccharides and not hexasaccharides. J Biol Chem 1998; 273:28878–28888.
31. Zeng C, Toole BP, Kinney SD, Kuo J-W, Stamenkovic I. Inhibition of tumor growth *in vivo* by hyaluronan oligomers. Int J Cancer 1998; 77:396–401.
32. Chai W, Beeson JG, Kogelberg H, Brown GV, Lawson AM. Inhibition of adhesion of *Plasmodium falciparum*-infected erythrocytes by structurally defined hyaluronic acid dodecasaccharides. Infect Immun 2001; 69:420–425.

33. Jedrzejas MJ, Mello LV, deGroot BL, Li S. Mechanism of hyaluronan degradation by *Streptococcus pneumoniae* hyaluronate lyase—structures of complexes with the substrate. J Biol Chem 2002; 277:28287–28297.
34. Camenisch TD, McDonald JA. Hyaluronan—is bigger better? Am J Respir Cell Mol Biol 2000; 23:431–433.
35. Saari H. Oxygen derived free radicals and synovial fluid hyaluronate. Ann Rheum Dis 1991; 50:389–392.
36. Termeer CC, Hennies J, Voith U, Ahrens T, Weiss JM, Prehm P, Simon JC. Oligosaccharides of hyaluronan are potent activators of dendritic cells. J Immunol 2000; 165:1863–1870.
37. McKee CM, Penno MB, Cowman M, Burdick MD, Strieter RM, Bao C, Noble PW. Hyaluronan (HA) fragments induce chemokine gene expression in alveolar macrophages—the role of HA size and CD44. J Clin Investig 1996; 98:2403–2413.
38. Kubo K, Nakamura T, Takagaki K, Yoshida Y, Endo M. Depolymerization of hyaluronan by sonication. Glycoconj J 1993; 10:453–459.
39. Phillips GO. Degradation of hyaluronan systems by free radicals. In: Laurent TC, ed. The Chemistry, Biology and Medical Applications of Hyaluronan and Its Derivatives. London: Portland Press, 1998:93–111.
40. Hokputsa S, Jumel K, Alexander C, Harding SE. A comparison of molecular mass determination of hyaluronic acid using SEC/MALLS and sedimentation equilibrium. Eur Biophys J 2003; 32:450–456.
41. Orviský E, Šoltés L, Chabreček P, Novák I, Stančíková M. Size exclusion chromatographic characterization of sodium hyaluronate fractions prepared by high energetic sonication. Chromatographia 1993; 37:20–22.
42. Chabreček P, Šoltés L, Kállay Z, Novák I. Gel permeation chromatographic characterization of sodium hyauronate and its fractions prepared by ultrasonic degradation. Chromatographia 1990; 30:201–204.
43. Takagaki K, Nakamura T, Izumi J, Saitoh H, Endo M, Kojima K, Kato I, Majima M. Characterization of hydrolysis and transglycosylation by testicular hyaluronidase using ion-spray mass spectrometry. Biochemistry 1994; 33:6503–6507.
44. Cramer JA, Bailey LC, Bailey CA, Miller RT. Kinetic and mechanistic studies with bovine testicular hyaluronidase. Biochim Biophys Acta 1994; 1200:315–321.
45. Park Y, Cho S, Linhardt RJ. Exploration of the action pattern of *Streptomyces* hyaluronate lyase using high-resolution capillary electrophoresis. Biochim Biophys Acta 1997; 1337:217–226.
46. Bitter T, Muir HM. A modified uronic acid carbazole reaction. Anal Biochem 1962; 4:330–334.
47. Takazono I, Tanaka Y. Quantitative analysis of hyaluronic acid by high-performance liquid chromatography of streptomyces hyaluronidase digests. J Chromatogr 1984; 288:167–176.
48. Nebinger P, Koel M, Franz A, Werries E. High-performance liquid chromatographic analysis of even- and odd-numbered hyaluronate oligosaccahrides. J Chromatogr 1983; 265:19–25.
49. Gherezghiher T, Koss MC, Nordquist RE, Wilkinson CP. Rapid and sensitive method for measurement of hyaluronic acid and isomeric chondroitin sulphates using high-performance liquid chromatography. J Chromatogr 1987; 413:9–15.
50. Knudsen PJ, Eriksen PB, Fenger M, Florentz K. High-performance liquid chromatography of hyaluronic acid and oligosaccharides produced by bovine testes hyaluronidase. J Chromatogr 1980; 187:373–379.

51. Chun LE, Koob TJ, Eyre DF. Quantitation of hyaluronic acid in tissues by ion-pair reverse-phase high-performance liquid chromatography of oligosaccharide cleavage products. Anal Biochem 1988; 171:197–206.
52. Cramer JA, Bailey LC. A reversed-phase ion-pair high-performance liquid chromatography method for bovine testicular hyaluronidase digests using postcolumn derivatization with 2-cyanoacetamide and ultraviolet detection. Anal Biochem 1991; 196:183–191.
53. Townsend RR, Hardy MR, Hindsgaul O, Lee YC. High-performance anion-exchange chromatography of oligosaccharides using pellicular resins and pulsed amperometric detection. Anal Biochem 1988; 174:459–470.
54. Hotchkiss AT Jr, Hicks KB. Analysis of oligogalacturonic acids with 50 or fewer residues by high-performance anion-exchange chromatography and pulsed amperometric detection. Anal Biochem 1990; 184:200–206.
55. Whistler RL, BeMiller JN. Alkaline degradation of polysaccharides. Adv Carbohydr Chem 1958; 13:289–329.
56. Holmbeck S, Lerner L. Separation of hyaluronan oligosaccharides by the use of anion-exchange HPLC. Carbohydr Res 1993; 239:239–244.
57. Midura RJ, Salustri A, Calabro A, Yanagishita M, Hascall VC. High-resolution separation of disaccharide and oligosaccharide alditols from chondroitin sulfate, dermatan sulfate and hyaluronan using CarboPac PA1 chromatography. Glycobiology 1994; 4:333–342.
58. Lauder RM, Huckerby TN, Nieduszynski IA. A fingerprinting method for chondroitin/dermatan sulfate and hyaluronan oligosaccharides. Glycobiology 2000; 10:393–401.
59. Mahoney DJ, Aplin RT, Calabro A, Hascall VC, Day AJ. Novel methods for the preparation and characterization of hyaluronan oligosaccharides of defined length. Glycobiology 2001; 11:1025–1033.
60. Linhardt RJ, Pervin A. Separation of negatively charged carbohydrates by capillary electrophoresis. J Chromatogr A 1996; 720:323–335.
61. Grimshaw J. Analysis of glycosaminoglycans and their oligosaccharide fragments by capillary electrophoresis. Electrophoresis 1997; 18:2408–2414.
62. Grimshaw J, Kane A, Trocha-Grimshaw J, Douglas A, Chakravarthy U, Archer D. Quantitative analysis of hyaluronan in vitreous humor using capillary electrophoresis. Electrophoresis 1994; 15:936–940.
63. Grimshaw J, Trocha-Grimshaw J, Fisher W, Rice A, Smith S, Spedding P, Duffy J, Mollan R. Quantitative analysis of hyaluronan in human synovial fluid using capillary electrophoresis. Electrophoresis 1996; 17:396–400.
64. Pattanaargson S, Roboz J. Determination of hyaluronidase activity in venoms using capillary electrophoresis. Toxicon 1996; 34:1107–1117.
65. Hong M, Sudor J, Stefansson M, Novotny MV. High-resolution studies of hyaluronic acid mixtures through capillary gel electrophoresis. Anal Chem 1998; 70:568–573.
66. Liu J, Dolnik V, Hsieh YZ, Novotny M. Experimental evaluation of the separation efficiency in capillary electrophoresis using open tubular and gel-filled columns. Anal Chem 1992; 64:1328–1336.
67. Hayase S, Oda Y, Honda S, Kakehi K. High-performance capillary electrophoresis of hyaluronic acid: determination of its amount and molecular mass. J Chromatogr A 1997; 768:295–305.

68. Kakehi K, Kinoshita M, Hayase S, Oda Y. Capillary electrophoresis of *N*-acetylneuraminic acid polymers and hyaluronic acid: correlation between migration order reversal and biological functions. Anal Chem 1999; 71:1592–1596.
69. Kinoshita M, Okino A, Oda Y, Kakehi K. Anomalous migration of hyaluronic acid oligomers in capillary electrophoresis: correlation to susceptibility to hyaluronidase. Electrophoresis 2001; 22:3458–3465.
70. Calabro A, Midura R, Wang A, West L, Plaas A, Hascall VC. Fluorophore-assisted carbohydrate electrophoresis (FACE) of glycosaminoglycans. Osteoarthritis Cartilage 2001; 9(Suppl A):S16–S22.
71. Calabro A, Benavides M, Tammi M, Hascall VC, Midura RJ. Microanalysis of enzyme digests of hyaluronan and chondroitin/dermatan sulfate by fluorophore-assisted carbohydrate electrophoresis (FACE). Glycobiology 2000; 10:273–281.
72. Calabro A, Hascall VC, Midura RJ. Adaptation of FACE methodology for microanalysis of total hyaluronan and chondroitin sulfate composition from cartilage. Glycobiology 2000; 10:283–293.
73. Tawada A, Masa T, Oonuki Y, Watanabe A, Matsuzaki Y, Asari A. Large-scale preparation, purification, and characterization of hyaluronan oligosaccharides from 4-mers to 52-mers. Glycobiology 2002; 12:421–426.
74. Price KN, Tuinman A, Baker DC, Chisena C, Cysyk RL. Isolation and characterization by electrospray–ionization mass spectrometry and high-performance anion-exchange chromatography of oligosaccharides derived from hyaluronic acid by hyaluronate lyase digestion: observation of some heretofore unobserved oligosaccharides that contain an odd number of units. Carbohydr Res 1997; 303: 303–311.
75. Prebyl BS, Kaczmarek C, Tuinman AA, Baker DC. Characterizing the electrospray–ionization mass spectral fragmentation pattern of enzymatically derived hyaluronic acid oligomers. Carbohydr Res 2003; 338:1381–1387.
76. Juhasz P, Biemann K. Utility of non-covalent complexes in the matrix-assisted laser desorption ionization mass spectrometry of heparin-derived oligosaccharides. Carbohydr Res 1995; 270:131–147.
77. Venkataraman G, Shriver Z, Raman R, Sasisekharan R. Sequencing complex polysaccharides. Science 1999; 286:537–542.
78. Schiller J, Arnhold J, Benard S, Reichl S, Arnold K. Cartilage degradation by hyaluronate lyase and chondroitin ABC lyase: a MALDI-TOF mass spectrometric study. Carbohydr Res 1999; 318:116–122.
79. Yeung B, Marecak D. Molecular weight determination of hyaluronic acid by gel filtration chromatography coupled to matrix-assisted laser desorption ionization mass spectrometry. J Chromatogr A 1999; 852:573–581.
80. Kühn AV, Rüttinger HH, Neubert RHH, Raith K. Identification of hyaluronic acid oligosaccharides by direct coupling of capillary electrophoresis with electrospray ion trap mass spectrometry. Rapid Commun Mass Spectrom 2003; 17:576–582.
81. Kühn AV, Raith K, Sauerland V, Neubert RHH. Quantification of hyaluronic acid fragments in pharmaceutical formulations using LC-ESI-MS. J Pharm Biomed Anal 2003; 30:1531–1537.
82. DeAngelis PI, Oatman IC, Guy DF. Rapid chemoenzymatic synthesis of monodisperse hyaluronan oligosaccharides with immobilized enzyme reactors. J Biol Chem 2003; 278:35199–35203.

Chemistry and Biology of Hyaluronan
H.G. Garg and C.A. Hales (editors)

Chapter 3

Methods for Determination of Hyaluronan Molecular Weight

MARY K. COWMAN

Department of Chemical and Biological Sciences and Engineering, Polytechnic University, Brooklyn, New York USA

RANIERO MENDICHI

Istituto per lo Studio delle Macromolecole Consiglio Nazionale delle Ricerche (CNR), Milan, Italy

I. Introduction

Hyaluronan samples are not generally monodisperse in molecular weight. When we characterize the molecular weight, we either determine an average molecular weight, or we characterize the distribution of molecular weights present. The type of average measured depends on the method used. A method such as light scattering (LS), the signal for which depends on the molecular weight of each species present, determines a weight-average molecular weight. A method such as osmometry or end group analysis, for which the number of molecules of each size is determined, gives a number-average molecular weight. The ratio of the weight-average to the number-average molecular weights is the polydispersity index, which is equal to one for a monodisperse sample, but greater than one for a polydisperse sample. Consider a sample containing just two different molecular weights (m_i) of a polymer, 1×10^6 and 2×10^6. If the two species are present in equal weights (w_i), 1 g each, then the weight average molecular weight is

$$\overline{M}_w = \frac{\sum w_i m_i}{\sum w_i} = \frac{1(1 \times 10^6) + 1(2 \times 10^6)}{1+1} = \frac{(1 \times 10^6) + (2 \times 10^6)}{2}$$
$$= 1.5 \times 10^6 \qquad (1)$$

The number average is calculated by recognizing that the number of moles (n_i) of a species present is equal to its weight divided by its molecular weight (g/(g/mol) = mol). The previously described sample contains 1×10^{-6} mole of the species with a molecular weight of 1×10^{6}, and 0.5×10^{-6} mole of the species with a molecular weight of 2×10^{6}. The number average molecular weight is thus

$$\overline{M}_n = \frac{\sum n_i m_i}{\sum n_i} = \frac{(1\times10^{-6})(1\times10^{6}) + (0.5\times10^{-6})(2\times10^{6})}{(1\times10^{-6}) + (0.5\times10^{-6})}$$

$$= \frac{1+1}{1.5\times10^{-6}} = 1.33\times10^{6} \tag{2}$$

and the polydispersity is $M_w/M_n = 1.13$.

We shall discuss in detail only three types of methods for the determination of hyaluronan molecular weight, because these are the most widely used methods today. Viscometry allows the routine determination of viscosity-average (close to weight-average) molecular weight for hyaluronan over a wide range of molecular weights. Electrophoretic techniques may be used for characterization of hyaluronan molecular weight distributions (MWDs) ranging from oligosaccharides to polymers with molecular weights up to about 6×10^{6}. LS gives a weight-average molecular weight and when used in conjunction with separation by size exclusion chromatography (SEC) the complete MWD may be determined.

II. Viscometry

A. Theory

Measurement of the viscosity of a solution containing hyaluronan allows the determination of the polymer viscosity-average (close to the weight-average) molecular weight. The theoretical basis for this effect is conceptually simple (1). The Stokes–Einstein relation for the specific viscosity, η_{sp}, of a dilute suspension of n spherical particles per unit volume of suspension, each with volume V, is

$$\eta_{sp} = \frac{\eta - \eta_0}{\eta_0} = 2.5nV \tag{3}$$

where η and η_0 are the respective viscosities of the suspension and pure fluid. The product, nV, is the volume fraction (sometimes denoted by φ) of the suspension occupied by the particles. Thus this equation states that the extent to which the suspension viscosity is greater than the pure fluid is determined by the fraction of the volume which is filled with particles.

For a polymer solution, the corresponding relation is

$$\eta_{sp} = 2.5\left(\frac{cN_A}{M}\right)v_h = 2.5c\left(\frac{N_A v_h}{M}\right) = 2.5\,cV_s = c[\eta] \tag{4}$$

where c is the polymer concentration in g/cm^3, N_A Avagadro's number, M the polymer molecular weight, v_h the hydrodynamic volume in cm^3 of a single polymer molecule, V_s the specific hydrodynamic volume of the polymer in cm^3/g, and $[\eta]$ the intrinsic viscosity also in cm^3/g. In Eq. 4, the volume fraction of solution that is filled with polymer molecules is equal to cV_s or $c[\eta]/2.5$. We see that measurement of the specific viscosity of a hyaluronan solution at a known concentration should allow the intrinsic viscosity to be determined and that the intrinsic viscosity is related to the specific volume of the polymer, which is in turn related to the ratio of the hydrodynamic volume of a hyaluronan chain to its molecular weight. For a solid sphere like a globular protein, the hydrodynamic volume increases directly with an increase in molecular weight and the specific volume (or intrinsic viscosity) would not depend on molecular weight. But for a random coil molecule or a rod-like molecule, the effective volume of solution occupied by the polymer chain increases faster than the molecular weight and the intrinsic viscosity retains a dependence on molecular weight. Once the intrinsic viscosity is determined, the molecular weight of the polymer can be determined using the Mark-Houwink-Sakurada equation

$$[\eta] = KM^a \qquad (5)$$

where K and a are constants for a given polymer–solvent–temperature condition. The values of K and a are determined from the dependence of the intrinsic viscosity on molecular weight for well-characterized samples of low polydispersity (Fig. 1). For high molecular weight hyaluronan in neutral aqueous salt solution (0.15 M NaCl) at 25 °C, values of K and a of 0.029 and

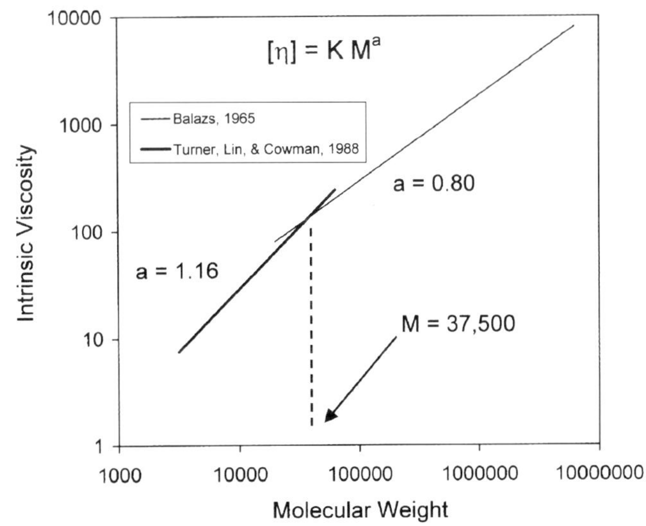

Figure 1 Experimental dependence of intrinsic viscosity determined by capillary viscometry on molecular weight for hyaluronan in 0.15 M NaCl solution. Note the different behaviors of low- and high molecular weight samples. Data from Refs. 2 and 11. Figure from Ref. 1.

0.80, respectively (with units of cm^3/g for $[\eta]$) were proposed by Balazs (2), after measuring the intrinsic viscosity of hyaluronan samples for which the molecular weights were determined by primary methods like LS. Similar values were determined by Laurent et al. (3), Cleland and Wang (4), Shimada and Matsumura (5), Bothner et al. (6), Fouissac et al. (7), Gamini et al. (8), Yanaki and Yamaguchi (9), and Takahashi et al. (10). These values are typical for a semi-flexible polymer in a good solvent. For low molecular weight hyaluronan, the intrinsic viscosity depends on molecular weight slightly differently; a K value of 6.54×10^{-4} and a value of 1.16 were found by Turner et al. (11). Similar results were reported by Shimada and Matsumura (5) and Cleland (12,13). Hayashi et al. (14) found an intermediate dependence on molecular weight in fitting data that spanned the two molecular weight ranges. The differences in the constants arise because short hyaluronan chains are effectively semi-stiff rods whereas high molecular weight hyaluronan has greater evident coiling, with a time-average domain shape of a sphere. The molecular weight at which hyaluronan begins to behave hydrodynamically like high molecular weight polymer is approximately 40,000. The transition from low- to high molecular weight behavior is not as sharp as it appears in the plot of limiting cases in Fig. 1, but can be well fit as a smooth transition using the worm-like chain model. One advantage of the use of limiting cases is the ability to identify molecular weight regions in which the intrinsic viscosity of hyaluronan can be reasonably converted to molecular weight using one of the two simple equations.

Recently Mendichi et al. (15) used coupled SEC and viscometry to analyze the $[\eta]-M$ relation over a wide range of M at 37 °C of temperature and fit the dependence to three equations as follows

a. $[\eta] = 1.29 \times 10^{-3} M^{1.056}$ for $M < 10^5$
b. $[\eta] = 3.39 \times 10^{-2} M^{0.778}$ for $10^5 < M < 10^6$
c. $[\eta] = 3.95 \times 10^{-1} M^{0.604}$ for $M > 10^6$

where the polydispersity of hyaluronan fractions used to create the equations was extremely low.

Eqs. 3 and 4 are applicable only when the solutions are extremely dilute and the polymer chains do not interact with each other. This is a condition never reached in practice. At finite but still dilute concentrations, the chains begin to approach each other. Transient hydrodynamic coupling occurs between chains. This is known to increase the solution viscosity. Frisch and Simha (16) proposed the form of the dependence to be a power series in the volume fraction occupied by polymer chains. Their equation was limited to volume fractions less than one. Matsuoka and Cowman (17,18) point out that the term $k'c[\eta]$, where k' is explicitly $0.4 (= 1/2.5)$ based on the Stokes–Einstein equation, represents a probability of interaction or coil overlap of polymer chains, each of which physically occupies only a tiny fraction of its effective hydrodynamic volume. Thus coil overlap is not restricted to values less than one and overlap simply results in greater intermolecular contact as concentration and molecular size increase. They derive a polynomial in the coil overlap term with no adjustable coefficients and retain only the first four terms

$$\eta_{sp} = c[\eta]\left(1 + k'c[\eta] + \frac{(k'c[\eta])^2}{2!} + \frac{(k'c[\eta])^3}{3!}\right) \tag{6}$$

where $k' = 0.4$.

The choice of four terms was made on the basis of comparison with high quality experimental zero shear viscosity data (19) for hyaluronan as a function of concentration and intrinsic viscosity. The first two terms are identical to the Huggins equation, which is commonly used in the experimental determination of intrinsic viscosity for polymers, with k' (usually referred to as the Huggins' constant) considered to be an adjustable parameter

$$\eta_{sp} = c[\eta] + k'(c[\eta])^2 \tag{7}$$

If Eq. 6 were to include all possible terms in the power series, the exponential equation described by Martin (20), again with an adjustable k' value, would be obtained

$$\eta_{sp} = c[\eta]\exp(k'c[\eta]) \tag{8}$$

Comparison of the experimental data of Berriaud et al. (19) with the three equations, assuming $k' = 0.4$ in each case, is shown in Fig. 2. The four-term equation is an excellent fit, whereas the Huggins equation underestimates the viscosity and the Martin equation overestimates it.

The value of the Huggins constant, usually considered to be an adjustable parameter, has been measured for high molecular weight hyaluronan by several investigators and found to be close to the theoretical value of 0.4. Reported values include 0.35–0.45 (5), 0.37–0.43 (8), 0.33–0.57 (21), 0.396–0.427 (19), 0.35 (9), 0.34–0.43 (14), 0.37–0.45 (22), and 0.4 (23). Anomalously high and low values have been reported for low molecular weight hyaluronan (5,8,11,14). This reflects the tendency for the more extended short chains to aggregate. Generally, significant deviation from the predicted k' value of 0.4 may be taken as an indication of association or other non-ideal behavior of a polymer.

B. Experimental Procedures and Data Analysis

The hyaluronan sample should be dissolved in physiological salt solution, generally 0.15–0.20 M NaCl. This is necessary to eliminate contributions to the viscosity from intramolecular and intermolecular electrostatic repulsion. After dissolution, the sample should be dialyzed in cold against a large volume of the solvent to establish osmotic equilibrium between the hyaluronan solution and the dialysate to be used in making dilutions of the sample. The sample should be diluted with the dialysate to a concentration below the coil overlap point, $c = 2.5/[\eta]$, then clarified to remove dust, lint, and any other particulate material.

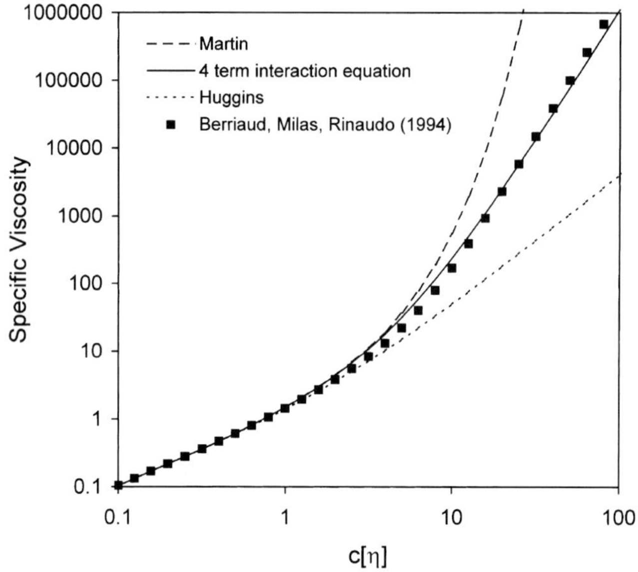

Figure 2 Dependence of the specific viscosity on the coil overlap parameter. The experimental data of Berriaud et al. (19) are compared with equations proposed by Martin, Huggins, and Matsuoka and Cowman.

Both the sample and dialysate may be filtered through a membrane with an effective pore size of 0.20–0.45 μm (depending on the hyaluronan molecular weight). Alternatively, the sample may be centrifuged. Since the hyaluronan concentration may be reduced by losses during filtration, the concentration should be determined after this procedure and adjusted to an appropriate value (see later).

The intrinsic viscosity is determined by measuring the specific viscosity for hyaluronan solutions at several (at least four) low concentrations and extrapolating the data to zero concentration. The usual data analysis technique employs the Huggins equation, for which the reduced viscosity, $\eta_{red} = \eta_{sp}/c$, is plotted versus c (Fig. 3). The intercept yields $[\eta]$, and the slope yields $k'[\eta]^2$. A limitation of this method is the fact that the Huggins equation includes only the first two terms of the expression for viscosity dependence on coil overlap. The error in specific viscosity relative to the experimentally validated four-term equation is shown in Fig. 4. By a $c[\eta]$ value of about 1, the error is already about 5%. This would require hyaluronan with a molecular weight of 2×10^6, where $[\eta] \cong 3200\ \mathrm{cm}^3/\mathrm{g}$, to be analyzed at concentrations at or below about 300 μg/mL. In contrast, the Martin equation remains close to the four-term equation, and a $c[\eta]$ value of perhaps 3.5 is required to reach a 5% error in specific viscosity. The Martin equation may be used to determine $[\eta]$ by plotting $\ln(\eta_{sp}/c)$ versus c, yielding $\ln[\eta]$ as the intercept and $k'[\eta]$ as the slope. It may be a preferable data analysis procedure where it is advantageous to analyze data over a more broad concentration range.

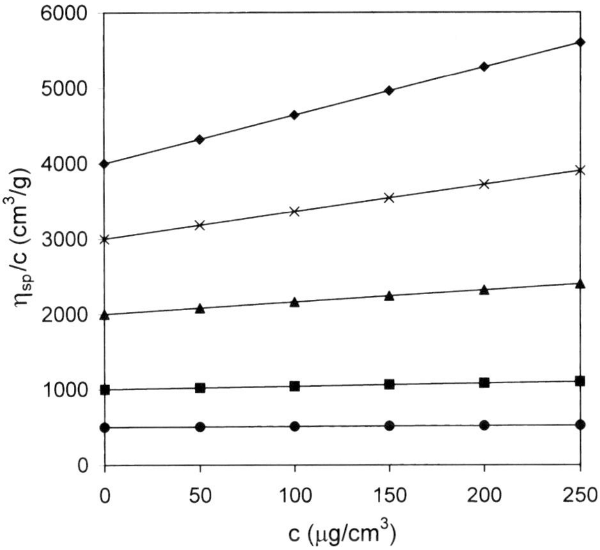

Figure 3 The dependence of reduced viscosity on the concentration, for samples of differing intrinsic viscosity, assuming Huggins equation validity. The intercept yields the intrinsic viscosity.

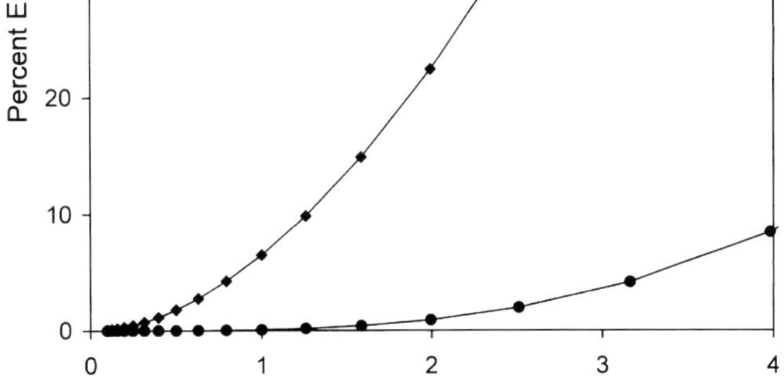

Figure 4 The error involved in the analysis of specific viscosity assuming either the Huggins equation or the Martin equation to be valid.

The experimental apparatus required includes a constant temperature bath capable of maintaining a temperature of 25.0 °C, and a glass capillary viscometer with suspended level outflow allowing the dilution of the sample directly in the viscometer (24,25). This eliminates the need for sample removal followed by viscometer cleaning and drying between samples differing only in concentration, thus ensuring the relatedness of the concentrations according to the dilution factor. The viscometer should be chosen to provide the lowest shear rate possible. For high molecular weight hyaluronan, a good choice for the viscometer is the Cannon-Ubbelohde semi-micro dilution viscometer, type 75 (26). This viscometer has a viscometer constant of 0.008 mm^2/s^2, a kinematic viscosity range of 1.6–8 mm^2/s, an internal diameter of tube *R* (the capillary) of 0.30 mm, a volume of the bulb *C* of 0.30 cm^3, and a funnel-shaped lower capillary end. The viscometer is designed so that no correction for kinetic energy effects is required for flow times of approximately 200 s or more (26). For flow times of about 200–360 s, corresponding to relative viscosities (η/η_0) of about 1.6–3, the mean shear rate ranges from about 375 s^{-1} (for the lowest sample concentration) to about 210 s^{-1} (for the highest sample concentration). The mean shear rate for solvent would be about 600 s^{-1}. The hyaluronan concentration should be chosen to give relative viscosities in the above range. For high molecular weight hyaluronan, an approximate concentration range is 75–250 $\mu g/cm^3$. This range would allow reasonable use of the Huggins equation for data analysis.

If the shear rate significantly exceeds the values above, the viscosity will be artifactually reduced by perturbation of the hyaluronan hydrodynamic volume. In such a case, the intrinsic viscosity relationship to molecular weight will be altered. For high molecular weight hyaluronan analyzed in a capillary viscometer with a shear rate of about 1000 s^{-1}, Bothner et al. (6) propose the use of the Mark-Houwink-Sakurada equation (Eq. 5) with the constants $K = 0.397$ and $a = 0.601$.

C. Applications

Viscometry has been widely used to analyze hyaluronan molecular weight. For impure preparations of hyaluronan and soluble extracts from vitreous, synovial fluid, or other tissues the viscometric method can give a reasonable estimate of hyaluronan molecular weight, because the solution viscosity is mainly determined by the high molecular weight flexible polymer and not by the much smaller soluble globular proteins (27). For pure hyaluronan, the method has been widely applied as a standard characterization method and to follow degradation reactions caused by hydroxyl free radicals, peroxynitrite, enzymatic degradation, sonication, etc. (see, for example, references (28–32)). The equipment required is inexpensive and the procedure is quite simple and not excessively time-consuming. This method is expected to remain a common method for hyaluronan molecular weight determination.

III. Electrophoresis

Hyaluronan is a polyanion with a constant charge to mass ratio, regardless of molecular weight. In order to obtain a molecular weight-dependent separation of hyaluronan using electrophoresis, a gel matrix is commonly employed as a sieving and separation-stabilizing medium. Under appropriate conditions, hyaluronan molecules may thus be separated according to molecular weight, with the smallest species having the greatest mobility. The size range over which separation can be achieved depends on the size of the gel pores relative to the size of the hyaluronan molecules. The porosity of the gel matrix, in turn, depends on the concentration of the gel-forming polymer and its degree of crosslinking. Gels with a small average pore size, such as crosslinked polyacrylamide gels are suitable for the separation of low molecular weight oligosaccharides and fragments of hyaluronan. Gels with a large average pore size such as agarose gels are used for the separation of high molecular weight hyaluronan.

A. Polyacrylamide Gel Electrophoresis (PAGE) of Hyaluronan Oligosaccharides and Fragments

Early electrophoretic analyses of hyaluronan and other glycosaminoglycans were designed to separate polysaccharides primarily according to charge density (33–35). Size separation had not been established by analysis of well-characterized samples differing solely in molecular weight. In 1984, three separate groups (36–38) reported high-resolution separation of glycosaminoglycan oligosaccharides according to size. In each case, a polyacrylamide gel was used to separate oligosaccharides into discrete bands, with adjacent bands in the final pattern differing in size by one disaccharide repeat. The gel composition varied from 10 to 25% acrylamide and the continuous buffer systems used were 50–100 mM Tris–borate, pH 8.3, with 1–2.4 mM EDTA or 100 mM Tris–glycine, pH 8.9, with 1.25 mM EDTA. If the glycosaminoglycan fragments were radiolabeled, the separation patterns were visualized by fluorography of the gel after incorporation of a fluor and subsequent drying of the gel. Separation patterns for non-labeled oligosaccharides were visualized by staining with alcian blue, which effectively precipitated the fragments in the gel. The separation of up to 30–40 different hyaluronan oligosaccharides into distinguishable bands could be observed (37). Turner and Cowman (39) showed that the bands could be identified by co-electrophoresis of purified hyaluronan oligosaccharides. They also showed that short oligosaccharides (less than about eight disaccharides in length) were not visualized using alcian blue in water and oligosaccharides shorter than about 12 disaccharides were underrepresented in the stained pattern for digests as a result of the difficulty in immobilizing the separated fragments in the gel by dye binding. A number of useful modifications were subsequently made to the electrophoretic methods. Min and Cowman (40) developed an improved procedure using long thin gels and a two-step staining process: precipitation of fragments in the gel with alcian blue followed by silver staining.

The sensitivity of the improved alcian blue/silver stain procedure was approximately 50 ng per band, or 2–5 μg for a complex mixture. Multiple loadings of samples on the gel after different delay times, a technique employed by Hampson and Gallagher (37) for the separation of dermatan sulfate oligosaccharides, was used to separate HA fragments containing up to 250 disaccharides into discrete bands. Lyon and Gallagher (41) found even more sensitive detection (as low as 1–2 ng per band for sulfated glycosaminoglycan fragments) using azure A and an ammoniacal silver stain, but data for hyaluronan were not detailed. Separation techniques using gradient gels (12–25% or 20–30% acrylamide) and discontinuous buffer systems were developed primarily for sulfated glycosaminoglycans with significant charge density heterogeneity (42,43), but were also reported to work for hyaluronan. Electrotransfer to positively charged nylon membranes allowed better fluorography of labeled fragments, and/or isolation of separated fragments (43,44). The trend toward mini-gel systems was recently exploited by Ikegami-Kawai and Takahashi (45) in the development of a rapid method for hyaluronan fragment analysis, using a 15% polyacrylamide gel with a Tris–borate–EDTA continuous buffer system, and staining using alcian blue/silver. Excellent separation was achieved in a 45 min run for fragments containing up to approximately 50 disaccharides. As previously observed, short fragments are poorly retained in the gel during staining. Samples containing species shorter than about 11 disaccharides should be analyzed with this restriction in mind.

Some of the applications of PAGE to the molecular weight characterization of hyaluronan samples include [1] calibration of gel permeation chromatography columns by direct analysis of the MWD in each fraction (46); [2] determination of the weight-average and number-average molecular weights of isolated hyaluronan subfractions for the purpose of comparison with light-scattering data showing self-association of hyaluronan fragments (Fig. 5) (11); [3] assays of extremely high sensitivity for hyaluronidase activity (45); and [4] characterization of oligosaccharides with both odd and even numbers of sugars prepared by chemoenzymatic synthesis of hyaluronan fragments (47).

B. Fluorophore-Assisted Carbohydrate Electrophoresis (FACE) of Short Hyaluronan Oligosaccharides

The PAGE techniques described above are not optimal for the quantitative analysis of polydisperse hyaluronan samples containing species with fewer than about 11 disaccharides, because the current detection methods do not retain such species in the gel during the staining processes. An alternative gel electrophoretic procedure known as FACE is more suited to these cases. In FACE techniques, the sample is derivatized with a fluorescent group at the reducing end prior to electrophoresis. In a procedure developed for monosaccharide compositional analysis (but useful for oligosaccharides as well), the label may be 2-aminoacridone, denoted AMAC. The samples are subjected to electrophoresis on 20% PAGE gels using a discontinuous borate-containing buffer system

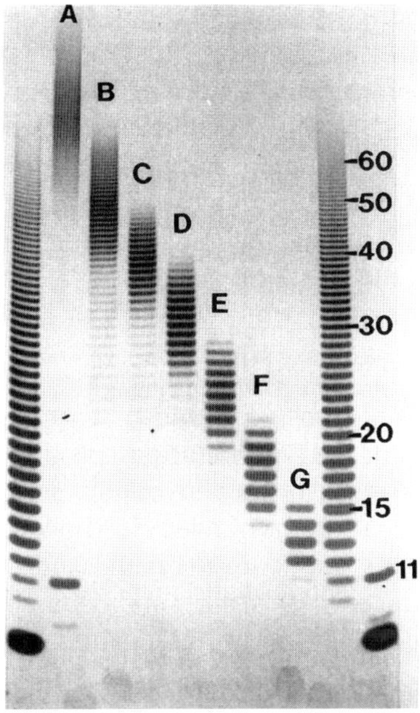

Figure 5 Polyacrylamide gel electrophoresis of hyaluronan fragment samples. Samples A–G were prepared by enzymatic digestion of hyaluronan, and subsequent fractionation by gel filtration chromatography. The unfractionated digest is electrophoresed in the outer lanes, and the number of disaccharide repeats in each band is given. The smallest fragments migrate most rapidly in the gel. (From Ref. 11).

(Tris–borate; Tris–glycine–borate). Borate plays an important role in the separation by complexing with the sugars. A second procedure developed for oligosaccharide profiling employs derivatization at the reducing end with disodium 8-amino-1,3,6-naphthalene trisulfonate, denoted ANTS. This label adds three negative charges to each oligosaccharide. Samples are electrophoresed on 20% PAGE gels using a discontinuous non-borate buffer system (Tris–HCl; Tris–glycine). For both types of systems, mini-gels are run at high voltages in the cold for short run times. The fluorescent labels allow immediate visualization of the separation by UV light illumination and fluorescence photography, so that even small mono- or oligosaccharides remain trapped in the gel. The original systems were proprietary, but a detailed description of substitute gel and buffer descriptions has recently been published (48).

The borate-containing system has been employed by Calabro et al. (49) and Mahoney et al. (50) in the analysis of short hyaluronan oligosaccharides. AMAC-labeled hyaluronan fragments ranging in size from the unsaturated disaccharide to 25 disaccharides in length were separated and visualized in the gel. Although the shortest oligosaccharides (disaccharide through hexasaccharide) moved more slowly than expected, other oligosaccharides showed decreasing mobility with increasing molecular weight, as seen in normal PAGE. The FACE method was used to establish the identity of enzymatic digestion products and the purity of isolated oligosaccharides of hyaluronan.

The non-borate system was employed by Tawada et al. (51) for the characterization of purified hyaluronan oligosaccharides. ANTS-labeled oligosaccharides containing two to approximately 10 disaccharide repeats were separated into discrete bands. The migration distance of each band decreased with increase in molecular weight. There was no unexpectedly slow migration of short oligosaccharides in this system.

C. Capillary Electrophoresis of HA

Recent studies have applied the technique of capillary electrophoresis to the analysis of hyaluronan or its degradation products (52,53). In capillary electrophoresis, electrophoresis is carried out in narrow (50–100 μm internal diameter) silica capillaries at high voltages. In 'normal polarity' capillary zone electrophoresis, the capillary walls are uncoated and the pH is neutral or basic. This leads to deprotonation of silanol sites and causes the capillary wall to have an excess negative charge. The loosely associated cationic counterions can be induced to move toward the cathode under the influence of an electric field. In a narrow capillary, this leads to bulk flow of the aqueous medium in the capillary, and co-transport of solute molecules with the solvent (electro-osmotic flow). Even uncharged molecules migrate, but charged solutes move faster or slower than the bulk solvent, depending on their charges and frictional coefficients. This method has been employed to separate unsaturated disaccharides of glycosaminoglycans (detected by absorbance at 232 nm) or HA oligosaccharides containing 3–7 disaccharides (detected by absorbance at 200 nm) in borated buffers (54,55). In 'reverse polarity' capillary zone electrophoresis, the buffer pH is low, suppressing ionization of the capillary walls. For hyaluronan, the low pH reduces the net negative charge by protonation of the carboxyl groups, but derivatization of the reducing end with a small charged group can be employed to drive migration under an applied electric field. Park et al. (56) used this technique to obtain high-resolution separation of HA oligosaccharides containing up to 25 disaccharide repeat units.

Capillary gel electrophoresis applies the concept of hindered motion through a gel or entangled polymer matrix to the capillary environment, further minimizing diffusion and enhancing separation by size. Hayase et al. (57) used pullulan as the entangled polymer medium and a weakly acidic buffer (reverse polarity conditions) to separate hyaluronan and obtained some low-resolution separation on the basis of molecular weight. They also found this method useful in quantitation, such that concentrations of high molecular weight hyaluronan as low as 10 μg/mL were accurately analyzed in electropherograms. Hong et al. (58) used linear polyacrylamide, covalently tethered to the capillary wall, to aid hyaluronan separation at low pH. The hyaluronan was also derivatized with a fluorescent dye to aid detection. Excellent resolution of individual oligosaccharides over a size range of 2 to approximately 190 disaccharides was obtained, but unexplained shadow peaks complicated interpretation of the profile. Kakehi et al. (59) and Kinoshita et al. (60) employed polysiloxane-coated capillaries (to eliminate electro-osmotic flow) and a polyethylene glycol entangled polymer matrix in a Tris–borate buffer to obtain excellent separation of hyaluronan oligosaccharides up to 100 disaccharides in length. Detection was by absorbance at 200 nm. Such separations have the potential to replace PAGE for hyaluronan oligosaccharide and fragment analysis, but it is important to note that quantitative correlation of the electropherograms with independent methods for analysis of MWD has not yet been established.

D. Agarose Gel Electrophoresis of High Molecular Weight Hyaluronan

Lee and Cowman (61) adapted methods used in the electrophoretic separation of high molecular weight nucleic acids for the separation of high molecular weight hyaluronan. They proposed the use of agarose gel at 0.5% in a continuous Tris–acetate–EDTA buffer for the separation of hyaluronan (Fig. 6). Sample loads of approximately 4–7 μg were required for polydisperse samples and the separated pattern was visualized by staining with the dye Stains-All (3,3′-dimethyl-9-methyl-4,5,4′,5′-dibenzothiacarbocyanine). For hyaluronan standards of known average molecular weight, the electrophoretic mobility was found to be approximately linearly related to the logarithm of the hyaluronan molecular weight over the range of $0.2 \times 10^6 - 6 \times 10^6$. Larger molecules may be separable by this method, but no suitable standards have been available. The method was shown to be useable preparatively, but yields are low and some degradation occurs during extraction of the hyaluronan from the gel. Impure hyaluronan samples containing high levels of contaminating protein (e.g., synovial fluid) were found to require predigestion with a proteolytic enzyme. Contaminating sulfated glycosaminoglycans were readily identifiable by their faster mobility and different color upon staining. The hyaluronan could be transferred from the gel by semi-dry electroblotting onto positively charged nylon. Hyaluronan could be subsequently detected by staining the membrane with alcian blue, or it could be specifically stained using biotin-labeled HA binding protein and streptavidin-gold followed by silver staining for sensitive visualization.

AGE proved to be a facile method for the determination of the MWD of hyaluronan. It was applied to the characterization of low molecular weight hyaluronan samples implicated in the induction of inflammatory gene expression in macrophages (62,63). It was used to demonstrate the degradation of hyaluronan by peroxynitrite, which may be generated during inflammation by the reaction of nitric oxide with superoxide anion (31). The pattern of degradation in the presence of different scavengers was similar to that caused by hydroxyl radicals. The average molecular weight determined by electrophoresis for hyaluronan as a function of degradation by peroxynitrite was in good agreement with the results obtained by viscometric analysis.

Recently an improved blotting and detection procedure was developed by Armstrong and Bell (64). A number of nylon-based membranes were tested for

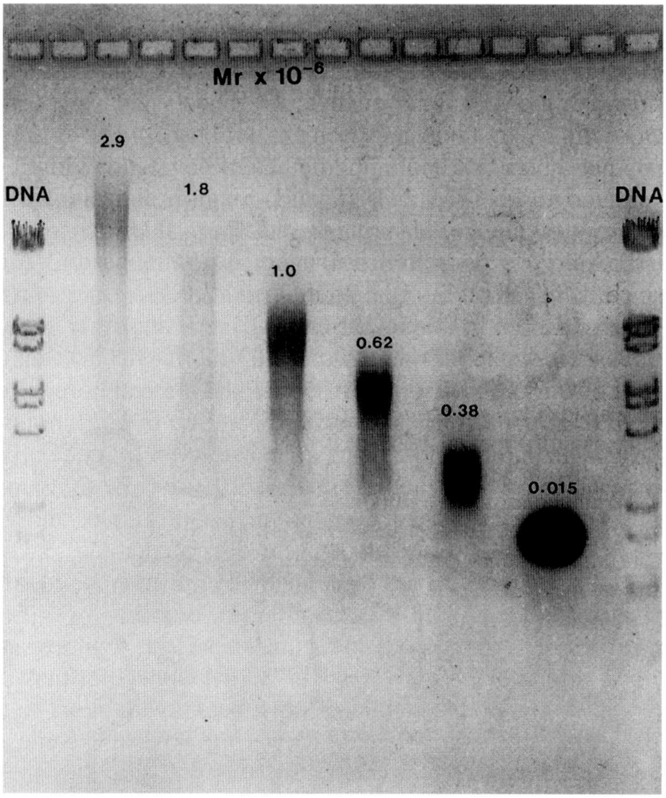

Figure 6 Agarose gel electrophoresis of low polydispersity hyaluronan samples obtained by electrophoretic fractionation and subsequent recovery from the gel. Low molecular weight hyaluronan migrates more rapidly than high molecular weight hyaluronan. From Ref. 61.

hyaluronan binding, and the two best were found to be Gene Screen plus and Hybond-N$^+$. Hyaluronan was transferred from the electrophoretic gel by capillary blotting, and the separation pattern detected by binding of ^{125}I-labeled HA binding protein and autoradiography. The authors employ their procedure to analyze hyaluronan from solid tissues after protease digestion in the presence of desferoxamine. Although the exact molecular weight was impossible to determine because it was higher than the standards employed, the fraction of very high molecular weight hyaluronan ($>4 \times 10^6$) was approximately 58% for hyaluronan from tissues as diverse as skin, skeletal muscle, heart, lung, small intestine, and large intestine. This surprising result indicated that normal isolation procedures generally degrade hyaluronan and that in the tissue the average molecular weight is probably on the order of 6×10^6, as it was found to be in human knee joint synovial fluid and owl monkey eye vitreous (61).

Slightly higher agarose concentrations were used by Pummill and DeAngelis (65) to optimize the separation of hyaluronan in the 0.2×10^6–1×10^6 range and study the molecular weight of hyaluronan produced by single amino acid mutated forms of a vertebrate hyaluronan synthase. Radio-labeled hyaluronan was separated on 1.35% agarose gel and detected by fluorography. The mutated forms were shown to produce hyaluronan of larger or smaller size, determined by the nature of the single site mutation.

A major handicap in the use of AGE to determine hyaluronan MWD has been the difficulty in obtaining suitable molecular weight standards. Recently, nearly monodisperse hyaluronan standards have been produced by Hyalose LLC. The pmHAS enzyme, the HA synthase from the Gram-negative bacterium *Pasteurella multocida*, catalyzes the synthesis of HA polymer utilizing monosaccharides from UDP-sugar precursors. The recombinant pmHAS will also elongate exogenously supplied HA oligosaccharide acceptors *in vitro* (66). HA oligosaccharides substantially boost the overall incorporation rate in comparison to *de novo* synthesis of HA polymer chains because chain initiation is slower than chain elongation. The chemoenzymatic synthesis of HA polymers of any desired molecular weight ($\sim 5 \times 10^3$ to $\sim 1.5 \times 10^6$) with very narrow size distributions using pmHAS has been developed ('selectHA', Jing and DeAngelis, in preparation). HA polymers of a desired size are produced by controlling the reaction stoichiometry (i.e., ratio of UDP-sugar precursors and acceptor molecules). The total amount of precursors determines the final mass of HA polymer that can be synthesized. If a small number of acceptor molecules (e.g., HA tetrasaccharide) are present in the reaction mixture, then a few long chains will be made. Conversely, if a large number of acceptor molecules are present, then many short chains will result. The polymerization process is synchronized in the presence of acceptor (i.e., bypassing slow *de novo* initiation step) thus all polymer products are very similar. In contrast, reactions without acceptor produce HA polymers with a wider size distribution. Each specific size class of selectHA had a polydispersity value in the range of 1.01–1.2 (1 is the ideal monodisperse size distribution) as assessed by SEC/multi-angle laser LS analysis. The selectHA preparations migrate on electrophoretic gels (agarose or polyacrylamide) as very tight bands facilitating their use as size standards (Fig. 7). Furthermore, these HA preparations should be of great utility for elucidating the relationship between HA size and its biological activities.

IV. Light Scattering and Size Exclusion Chromatography

Fundamental properties of hyaluronan, such as viscoelasticity and flow behavior primarily depend on the MWD, size, and conformation of the macromolecules. A primary method in estimating the molecular weight and the size of macromolecules is LS. LS and a few other methods such as osmometry, sedimentation, and mass spectrometry are absolute techniques. However, only the LS technique can be used online to a SEC system in obtaining the whole

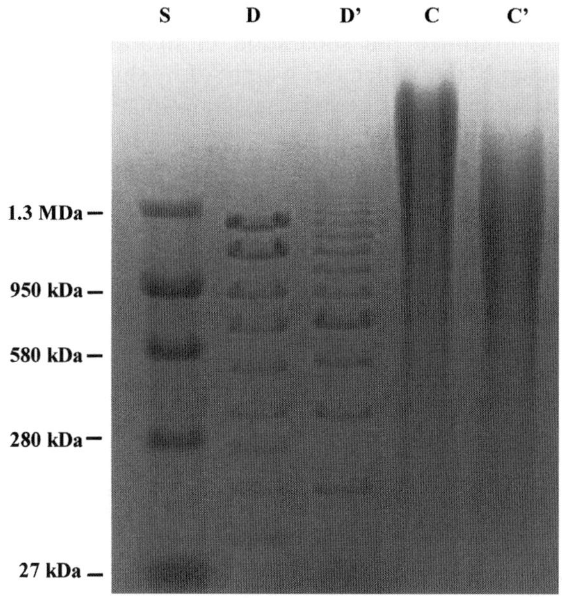

Figure 7 Agarose gel electrophoresis of nearly monodisperse hyaluronan standards and commercial hyaluronan. Gel was 0.7% agarose in Tris–acetate–EDTA (minigel format), stained with Stains-All by the method of Lee and Cowman (61). S: a mixture of 5 different monodisperse SelectHA preparations with indicated M_w determined by SEC–MALS; C and C′: commercial hyaluronan samples; D: DNA standards, Bioline Hyperladder 1, containing DNA of 10, 8, 6, 5, 4, 3, 2.5, 2, 1.5 kb; D′: DNA standards, BioRad 1 Kilobase Ruler, containing DNA of 12, 11, 10, 9, 8, 7, 6, 5, 4, 3, 2 kb. Figure kindly provided by P DeAngelis and W Jing.

MWD. As a consequence, LS is a fundamental method for the characterization of hyaluronan.

LS concerns the interaction of light with matter in the specific case with macromolecules in solution. The interaction of light with matter is a very complex topic. Depending on the type of scattering analyzed (elastic, quasi-elastic, Raman, etc.) different information may be obtained. For the characterization of macromolecules (molecular weight and size) only elastic and quasi-elastic LS are of interest. In an elastic LS experiment (also known as static or total intensity or Rayleigh scattering) we measure the intensity of the scattering. In this case, we assume that the scattered light has the same wavelength and polarization of the incident light. On the contrary, in a quasi-elastic LS experiment (also known as dynamic or photon correlation spectroscopy) we measure the fluctuations of the intensity of the scattering due to the Brownian movement of the macromolecules.

Following Zimm (67) the intensity of the scattering of a solution of macromolecules is related to the molecular weight M of the sample by the following equation

$$\frac{Kc}{\Delta R(\theta)} = \frac{1}{MP(\theta)} + 2A_2c + \cdots \tag{9}$$

where $\Delta R(\theta)$ is the scattering excess (Rayleigh factor) at angle θ of the solution with regard to the pure solvent, θ the angle between the incident light and the detector, c the concentration, A_2 the second virial coefficient, $P(\theta)$ the form factor, $K = (4\pi^2 n_0^2(\mathrm{d}n/\mathrm{d}c)^2)/(N_\mathrm{a}\lambda_0^4)$ an optical constant, n_0 the refractive index of the solvent, $\mathrm{d}n/\mathrm{d}c$ the refractive index increment of the polymer, λ_0 the wavelength of light in a vacuum, N_A the Avagadro's number.

Modern LS photometer uses coherent light, that is a laser (often a He–Ne laser with wavelength $\lambda = 632.8$ nm), and vertical polarization. The constant K puts together all the physical and optical parameters. All the parameters of the K optical constant are known, only the $\mathrm{d}n/\mathrm{d}c$ of the polymer is unknown. Often the $\mathrm{d}n/\mathrm{d}c$ of the polymer may be found in the literature, otherwise the value must be measured, generally by an offline refractometer at the same wavelength, solvent and temperature of the LS experiment. The $\mathrm{d}n/\mathrm{d}c$ value for hyaluronan is well known: 0.15 mL/g, in 0.15 M NaCl solvent, at 25 °C and $\lambda = 632.8$ nm. However, other than λ, $\mathrm{d}n/\mathrm{d}c$ also depends a little on the solvent (salt, buffer) used and in general it is better to measure it. The fundamental parameter of interest in obtaining the molecular weight and the size of macromolecules is the intensity of the scattering $R(\theta)$ that depends on the angle θ and on the concentration c. Specifically, we need $R(\theta)$ at zero angle, $\theta = 0°$, and infinite dilution, $c = 0$. The condition of infinite dilution, that is an isolated macromolecule, could be obtained quite easily by measuring $R(\theta)$ at decreasing finite concentrations (3–5) followed by an extrapolation to zero concentration. In this way, other than the molecular weight it is also possible to estimate the fundamental thermodynamic parameter of macromolecules in solution A_2. Very complex is the estimation of $R(\theta)$ at zero angle (68,69). In fact, $R(\theta)$ at zero angle is not experimentally measurable as a consequence of the interference with the intense primary incident light. In measuring $R(\theta)$ at zero angle we can use two different strategies corresponding with two different LS instrument: low-angle LS (LALS) and multi-angle LS (MALS). A LALS photometer measures $R(\theta)$ at a scattering angle as low as possible and assumes that this value corresponds to $R(\theta)$ at zero angle. Considering the experimental physical limit this means θ about 4–6° for LALS. On the contrary, a MALS photometer measures $R(\theta)$ in a wide range of angles, by means of an array of photodiodes, and $R(\theta)$ at zero angle is calculated by an extrapolation. Both LALS and MALS photometers are commercially available. In the case of MALS there are several instruments with 2, 3, 18, and recently also 7 angles. All the LS photometers could be used both offline, batch mode, and online to a SEC/HPLC system.

Quite complex is the definition of the form factor $P(\theta)$. A macromolecule could not be considered as a single point of scattering. Hence, the light scattered from two different points of the same macromolecules will be not in phase and the total intensity of the scattering for large molecules is lower as a consequence of the destructive interference. The interference depends on the angle of measure of the intensity of the scattering. The interference is absent at 0° angle, highest at 180°. The interference depends on the shape and on the dimension of the molecules. Therefore a form factor $P(\theta)$ has been introduced that quantifies the interference. $P(\theta)$ is defined as the ratio between $R(\theta)$ in the presence of interference, $\theta > 0°$, and $R(\theta)$ in the absence of interference, $\theta = 0°$. Thus, by definition

$$P(\theta) \equiv \frac{R(\theta)}{R(\theta = 0°)} \tag{10}$$

A direct consequence of the previous equation is that $P(\theta) = 1$ for $\theta = 0°$ independent of the size of the molecules, $P(\theta) < 1$ for $\theta > 0°$ when the size of the molecules is comparable with the wavelength λ. Debye (70) found that $P(\theta)$ could be expressed independently of the shape and of the conformation of the macromolecules. Considering the reciprocal of $P(\theta)$, that is $P(\theta)^{-1}$, Debye found the following equation

$$P(\theta)^{-1} = 1 + \frac{1}{3}\mu^2\langle s^2\rangle \tag{11}$$

where $\mu = 4\pi/\lambda \sin(\theta/2)$ and $\lambda = \lambda_0/n_0$ is the wavelength of the light in the solvent. Fortunately, the presence of the destructive interference is not only a problem, because from $P(\theta)$ it is possible to measure the size of the macromolecules. Indeed, it is evident that combining Eqs. 9 and 11 from the initial slope of $P(\theta)$ versus $\sin^2(\theta/2)$ plot we can estimate the radius of gyration $\langle s^2\rangle^{1/2}$ of the macromolecules. Obviously, this fact is true only for the MALS photometer in which $R(\theta)$ is measured at different angles. In a LALS photometer where $R(\theta)$ is measured only at low angle, we assume $P(\theta) = 1$ and the information on the size of the macromolecules is completely lost.

Regarding the size of the macromolecules from elastic LS two additional considerations are of interest. First, the size of the macromolecules is expressed in terms of the radius of gyration. R_g is defined as the mass average of the distance r_i from the center of gravity of the repeating units (segment) of mass m_i (Eq. 12). In other word, R_g is an equilibrium parameter, distribution of masses with regard to the center of gravity of the molecule and it is different from the hydrodynamic radius R_H obtained in a quasi-elastic LS experiment. Secondly, R_g is obtained from the angular variation of the scattering. If the macromolecules are smaller in size, the angular variation of the scattering is not experimentally measurable and the size is not obtained with accuracy. Practically, using an elastic MALS photometer the lower measurable R_g is about 10 nm. This is an other important difference between elastic and quasi-elastic LS. In fact, the lower measurable

R_H by quasi-elastic LS is about 1–2 nm

$$R_g = \left(\frac{\langle \sum m_i r_i^2 \rangle}{\sum m_i} \right)^{1/2} \tag{12}$$

A. Elastic Offline Light Scattering (Batch Mode)

Average values of the molecular weight and of the size could be obtained by elastic offline LS, that is in batch mode. In batch LS mode a number of concentrations, usually from 3 to 5, is prepared and the intensity of the scattering is measured. Obviously in a MALS instrument the intensity of the scattering at different angles is also simultaneously measured by an array of photodiodes in each single scan. The concentration of the sample solutions has to be chosen on the basis of the signal-to-noise ratio. Substantially, the intensity of the scattering (Eq. 9), depends on M, c and $(\mathrm{d}n/\mathrm{d}c)^2$ and the concentration must be as low as possible to obtain a good signal-to-noise ratio for each solution. LS is particularly sensitive to the presence of dust and in general to any extraneous particle. In a LS characterization an accurate clarification of the solutions is absolutely important especially with water solvent. The clarification could be obtained by using very clean solvent and an accurate filtration of each solution before each test. The dimension of the filter (0.2, 0.45 μm or more) depends on the molecular weight (size) of the hyaluronan sample. In addition, we need to consider that hyaluronan in solution is an anionic polyelectrolyte. It is well known that LS needs dilute solutions in thermodynamic equilibrium. Consequently, before the experiment all the hyaluronan solutions have to be exhaustively dialyzed against the proper solvent both for $\mathrm{d}n/\mathrm{d}c$ and LS measurements.

In batch mode, the experimental LS data (in MALS there are three variables $R(\theta)$, θ and c) are analyzed using the classical double extrapolation ($c \longrightarrow 0$ and $\theta \longrightarrow 0$) generally known as the Zimm-Plot. Using the Zimm procedure (67) three average values are obtained: the weight-average molecular weight M_w, the z-average root mean square radius $\langle s^2 \rangle_z^{1/2}$ (generally known as radius of gyration R_g) and the second virial coefficient A_2. Fig. 8 shows the Zimm-Plot of an ultra-high molecular weight hyaluronan. In this specific case we have used the Zimm formalism with a 2° order polynomial for the angular variation, considering the marked curvature, and a 1° order polynomial (linear) for the concentration. The curvature of the angular pattern of this hyaluronan sample is an exception due to the ultra-high molecular weight usually a linear extrapolation could also be used for the angular extrapolation. The results for the hyaluronan sample are: $M_w = 7.4 \times 10^6$ g/mol, $R_g = 385.0$ nm, $A_2 = 1.66 \times 10^{-3}$ mol mL/g^2. If the hyaluronan characterization is performed by a LALS instrument, only the extrapolation to zero concentration is possible. In this case, only M_w and A_2 are obtained and the information on the size of the macromolecules is lost. However, it is important to note that in some particular cases M_w value from LALS may be more accurate than M_w from MALS because the angular pattern of the scattering could be very complex and the extrapolation to zero angle not simple.

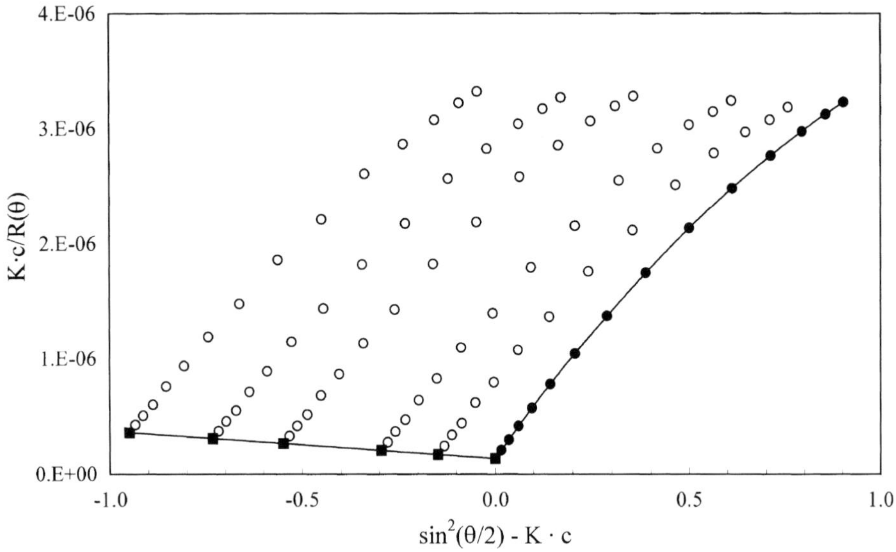

Figure 8 Zimm-Plot of a ultra-high molecular weight hyaluronan in 0.15 M NaCl: $M_w = 7400$ kg/mol, $R_g = 385.0$ nm, $A_2 = 1.66 \times 10^{-3}$ mol mL/g^2.

B. Elastic LS Online to a SEC System

In general, average values of M_w and R_g are not sufficient to characterize the complex properties of hyaluronan. Indeed, the biological functions of hyaluronan are closely related to the viscoelastic properties and consequently to the whole distributions of the molecular weight and of the size of the macromolecules. Hence, the determination of the whole distribution, MWD and RGD, is a main concern for hyaluronan. It is well known that in obtaining the MWD of a broad polymer an online fractionation method is more convenient. In fact, all the offline fractionation methods of broad MWD polymers are time-consuming and, more importantly, often the final results are not adequate. Theoretically, three online fractionation methods are effective for hyaluronan broad MWD polymers with medium-high molecular weight: SEC, hydrodynamic chromatography (HD), and flow-field flow fractionation (F-FFF). Except the particular case of ultra-high molecular weight samples, surely SEC is the more successful fractionation method for hyaluronan. In the following, only SEC fractionation will be considered although some interesting results were obtained by F-FFF (71). Many articles concerning the SEC fractionation of broad MWD hyaluronan can be found in the literature (32,72–75).

In general, the use of a conventional SEC system using a concentration detector, typically a differential refractometer (DRI), and calibration with narrow or broad MWD standards is not accurate for hyaluronan. In fact, it is quite difficult to find adequate hyaluronan standards for the calibration of the SEC system and mainly the concentration effect is meaningful. As a consequence, the difference between experimental MWD and true MWD for hyaluronan may be dramatic. Usually, a LS detector online to a SEC system (SEC–LS) is the most accurate method in estimating MWD and RGD for hyaluronan.

Usually, a SEC–LS system is composed of a LS detector (LALS or MALS) and a concentration detector. Both the DRI and the UV spectrophotometer are effective for hyaluronan as concentration detectors. The more usual set-up of the SEC–LS system is serial in the following order: pump–injector–columns–LS–DRI. The LS signal for a specific polymer (known dn/dc) depends on the molecular weight and on the concentration (Eq. 9). In a batch LS experiment the concentration of the solution is known. In SEC–LS, we need the concentration of each polymeric fraction eluting from the SEC columns. In this case the concentration is measured slice by slice from the concentration detector. In addition, for each elution volume (slice) only one concentration of the polymer is known. In other words, in a SEC–LS experiment the extrapolation to infinite dilution is not possible. Usually in SEC–LS the extrapolation to infinite dilution is neglected, that is, the $2A_2c$ term of Eq. 9) is ignored. As a first approximation, this term is neglected because the concentration of each slice is extremely dilute and A_2 is low (approximately ranging from 10^{-4} to 10^{-5}). In the characterization of high molecular weight hyaluronan the influence of the $2A_2c$ term is not negligible (A_2 for hyaluronan is about 2×10^{-3}). However, using the A_2 value obtained by offline LS it is possible to overcome this problem. It is well known that A_2 is a function of the molecular weight. Mendichi et al. (76) published the A_2 molecular weight dependence for hyaluronan in 0.15 M NaCl.

SEC fractionation of hyaluronan is not simple. Typical SEC experimental conditions applied to the fractionation of hyaluronan present several drawbacks such as shear degradation, concentration effects, anomalous elution (viscous fingering), and in general poor resolution. With hyaluronan each detail of the SEC experimental protocol (mobile phase, flow rate, sample concentration, temperature, and injection volume) have to be optimized methodically for reliable results. As mobile phase, 0.1–0.2 M NaCl ionic strength is sufficient in screening out the anionic charge of the hyaluronan chain. Many other salts or buffers have been used for SEC of hyaluronan and in general the choice of the mobile phase is relatively simple. Flow rate and sample concentration should be as low as possible. Usually, as flow rate we use 0.8, 0.4 or 0.2 mL/min, depending on the molecular weight of the hyaluronan sample. A flow rate value of 0.2 mL/min is used only for ultra-high molecular weight hyaluronan. It is important to note that SEC fractionation of hyaluronan is not unlimited. Using a SEC system optimized for ultra-high molecular weight hyaluronan samples a successful fractionation was obtained up to about $M_w = 3 \times 10^6$ g/mol (75). In addition, the concentration of the sample depends on the molecular weight of the sample and in general ranges from 0.01 to 0.5 mg/mL. The minimal effective concentration depends on the signal-to-noise ratio of the concentration detector. Obviously, the crucial point in the SEC fractionation of hyaluronan is the column set. The sizes of hyaluronan macromolecules are very large and one needs to use aqueous SEC columns specifically suitable for high molecular weight polymers. Only SEC columns with larger particle size and larger pore size are able to successfully fractionate hyaluronan. Specifically, polymeric aqueous columns with particle size of 17 μm (TosoHaas, TSKgel PW), 15 μm (Polymer Laboratories, PL aquagel-OH), 13 μm (Shodex, OHpak), and 10 μm (Waters, Ultrahydrogel) are commercially available for high molecular weight polysaccharides. On the contrary, silica columns are not appropriate for the fractionation of hyaluronan. In fact, hyaluronan macromolecules are adsorbed on the silica packing and practically do not elute at all. In general, a column set composed of two columns with larger particle size and larger pore size are suitable for the hyaluronan SEC fractionation. However, the exact type of the columns depends on the molecular weight range of the hyaluronan samples.

When the SEC–LS system is optimized a number of important pieces of information about hyaluronan MWD, RGD, and conformation can be obtained. Fig. 9 shows the extrapolation to zero angle for a hyaluronan fraction (slice) from the SEC–MALS system. Slice by slice three parameters c_i (from the concentration detector), M_i and R_{gi} (from MALS) are calculated. As a result, two experimental functions, $M = f(V)$ and $R_g = f(V)$, where V is the elution volume, are obtained. Fig. 10 shows the two experimental functions $M = f(V)$ and $R_g = f(V)$ obtained from the SEC–MALS system applied to a hyaluronan sample with $M_w = 1017$ kg/mol. The plots were obtained by using two polymeric TSK-Gel PW columns (G6000 and G5000) from TosoHaas, 0.15 M NaCl mobile phase and 0.4 mL/min flow rate. By using the $M = f(V)$

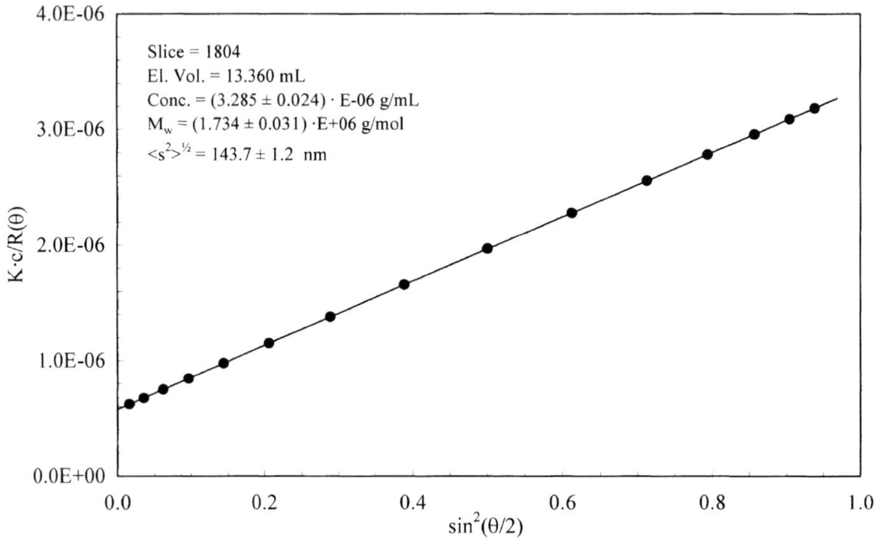

Figure 9 Debye plot, $Kc/R(\theta)$ versus $\sin^2(\theta/2)$, for a fraction (slice) of a hyaluronan sample with $M_w = 656$ kg/mol.

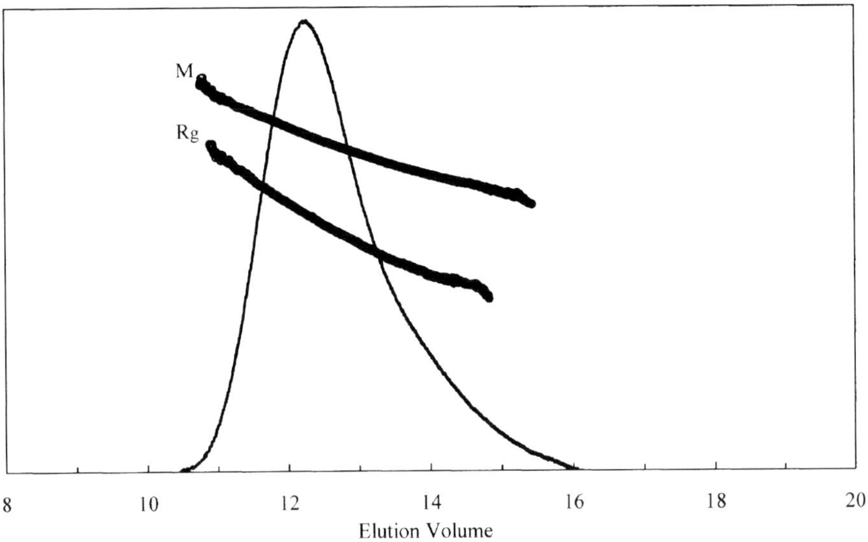

Figure 10 $M = f(V)$ and $R_g = f(V)$ experimental functions for a hyaluronan sample with $M_w = 1017$.

experimental function (i.e., the classical SEC calibration curve) and the concentration curve (chromatogram), the calculation of the MWD is immediate. The MWDs, both differential and cumulative, of the previous hyaluronan sample are shown in Fig. 11. Starting from the previous MWDs all the molecular weight

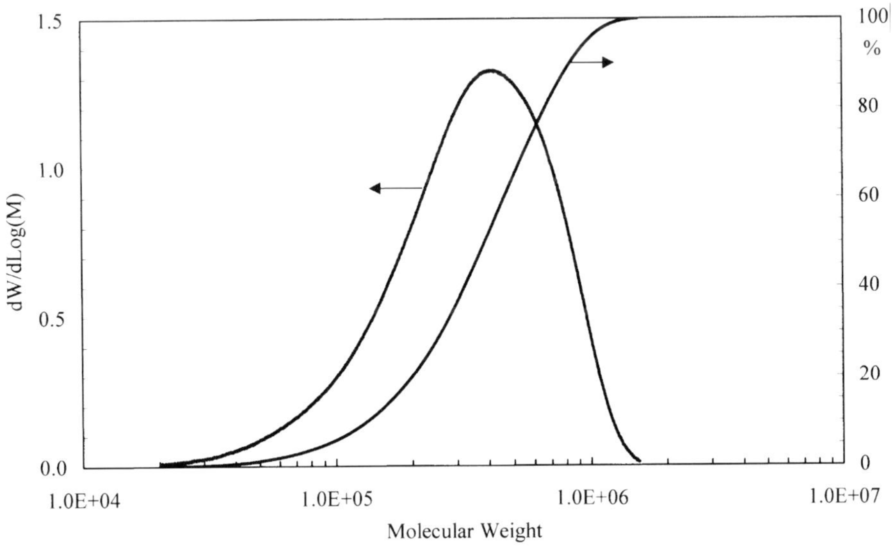

Figure 11 Molecular weight distributions, differential and cumulative, of a hyaluronan sample.

averages (numeric-, weight-, z-) could be calculated easily using the proper definitions. In addition, by using the $R_g = f(V)$ experimental function, the distribution of the radius of gyration RGD, both differential and cumulative, could be obtained. Such R_g distributions are shown in Fig. 12. Finally, in a similar way to the molecular weight starting from the RGD, the different averages of R_g could also be calculated.

Because for each fraction of the sample the online MALS detector measures both M_i and R_{gi} from a single sample, it is possible to obtain the $R_g = f(M)$ scaling law (generally known as conformation plot) of the polymer. The $R_g = f(M)$ scaling law is a very important function in understanding the conformation (flexible coil, compact sphere, rigid rod) of the polymers. In fact, if the molar mass distribution of the sample is sufficiently broad, it is possible, from a linear regression of Log(R_g) versus Log(M), to estimate the coefficients, K and α, of the $R_g = KM^{\alpha}$ scaling law. The $R_g = f(M)$ experimental function for hyaluronan is shown in Fig. 13. The plot was obtained by applying the SEC–MALS system to a hyaluronan sample with $M_w = 1017$ kg/mol and the polydispersity index $D = 1.7$. The slope of the plot is about 0.6, that is the typical value of semi-stiff polymers such as hyaluronan (15).

Finally, all the experimental functions $M = f(V)$, $R_g = f(V)$ and $R_g = f(M)$ are powerful tools in checking the chemical modification (derivatization) of hyaluronan. Indeed, SEC fractionates the macromolecules on the basis of size (specifically on the basis of the hydrodynamic volume $V_H \propto M[\eta]$). At constant V_H (i.e., specific elution volume) a native and a derivatizated hyaluronan have different M or R_g. As a consequence, a simple comparison of the previously

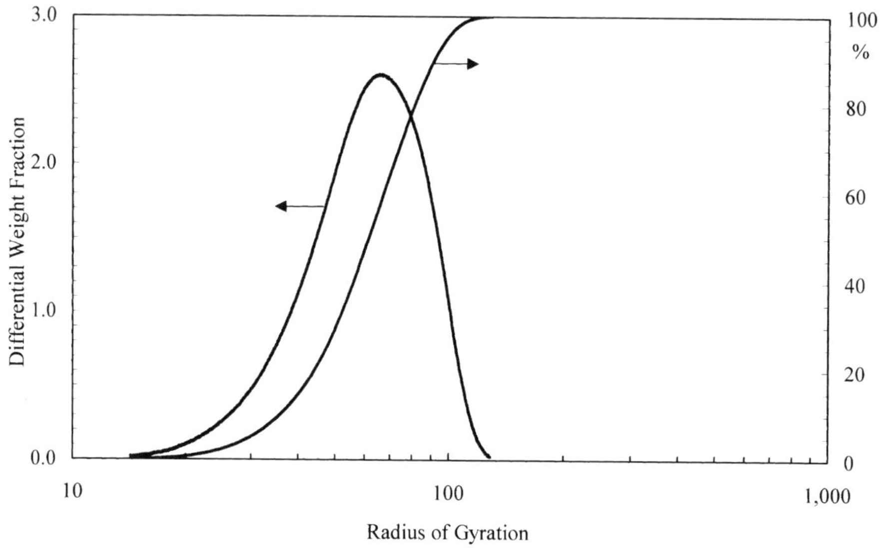

Figure 12 Radius of gyration distributions, differential and cumulative, of a hyaluronan sample.

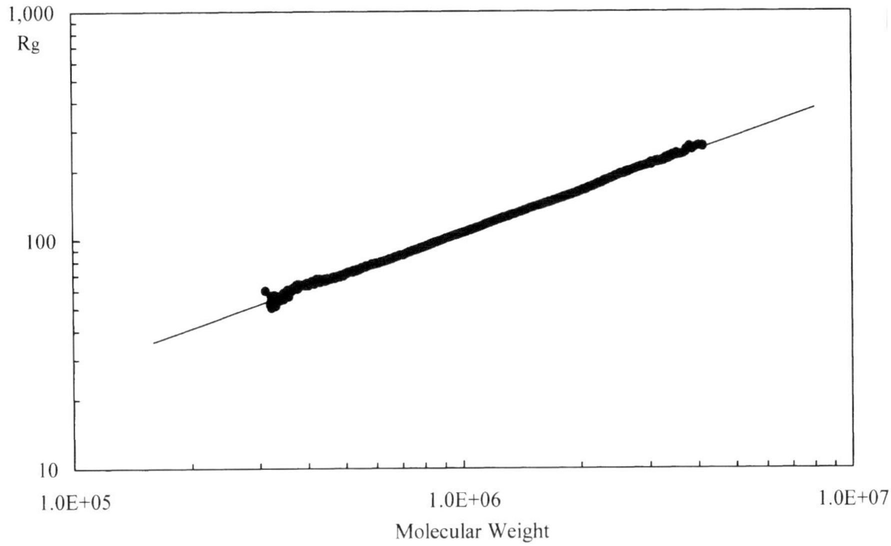

Figure 13 $R_g = f(M)$ power law of a hyaluronan sample with $M_w = 1017$ kg/mol in 0.15 M NaCl solvent.

mentioned plots immediately allows a check of the extent of the derivatization. The method was applied successfully by Soltes et al. (77) to hyaluronan derivatized with β-cyclodextrin and *N*-acylurea.

Acknowledgements

We are grateful to Paul DeAngelis and Wei Jing for providing information and a figure describing the synthesis and properties of hyaluronan standards of nearly monodisperse molecular weight.

References

1. Cowman MK, Matsuoka S. The intrinsic viscosity of hyaluronan. In: Kennedy JF, Phillips GO, Williams PA, eds. Hyaluronan. Cambridge: Woodhead Publishing, 2002:75–78.
2. Balazs EA. Amino sugar-containing macromolecules in the tissues of the eye and the ear. In: Balazs EA, Jeanloz RW, eds. The Amino Sugars: The Chemistry and Biology of Compounds Containing Amino Sugars. New York: Academic Press, 1965; vol. 2A:401–460.
3. Laurent TC, Ryan M, Pietruszkiewicz A. Fractionation of hyaluronic acid. The polydispersity of hyaluronic acid from the bovine vitreous body. Biochim Biophys Acta 1960; 42:476–485.
4. Cleland RL, Wang JL. Ionic polysaccharides. III. Dilute solution properties of hyaluronic acid fractions. Biopolymers 1970; 9:799–810.

5. Shimada E, Matsumura G. Viscosity and molecular weight of hyaluronic acids. J Biochem 1975; 78:513–517.
6. Bothner H, Waaler T, Wik O. Limiting viscosity number and weight average molecular weight of hyaluronate samples produced by heat degradation. Int J Biol Macromol 1988; 10:287–291.
7. Fouissac E, Milas M, Rinaudo M, Borsali R. Influence of the ionic strength on the dimensions of sodium hyaluronate. Macromolecules 1992; 25:5613–5617.
8. Gamini A, Paoletti S, Zanetti F. Chain rigidity of polyuronates: static light scattering of aqueous solutions of hyaluronate and alginate. In: Harding SE, Satelle DB, Bloomfield VA, eds. Laser Light Scattering in Biochemistry. Cambridge: Royal Society of Chemistry, 1992:294–311.
9. Yanaki T, Yamaguchi M. Shear-rate dependence of the intrinsic viscosity of sodium hyaluronate in 0.2 M sodium chloride solution. Chem Pharm Bull 1994; 42: 1651–1654.
10. Takahashi R, Kubota K, Kawada M, Okamoto A. Effect of molecular weight distribution on the solution properties of sodium hyaluronate in 0.2 M NaCl solution. Biopolymers 1999; 50:87–98.
11. Turner RE, Lin P, Cowman MK. Self-association of hyaluronate segments in aqueous NaCl solution. Arch Biochem Biophys 1988; 265:484–495.
12. Cleland RL. Molecular weight distribution in hyaluronic acid. In: Balazs EA, ed. Chemistry and Molecular Biology of the Intercellular Matrix. New York: Academic Press, 1970:733–742.
13. Cleland RL. Viscometry and sedimentation equilibrium of partially hydrolyzed hyaluronate: comparison with theoretical models of wormlike chains. Biopolymers 1984; 23:647–666.
14. Hayashi K, Tsutsumi K, Nakajima F, Norisuye T, Teramoto A. Chain-stiffness and excluded-volume effects in solutions of sodium hyaluronate at high ionic strength. Macromolecules 1995; 28:3824–3830.
15. Mendichi R, Šoltés L, Giacometti Schieroni A. Evaluation of radius of gyration and intrinsic viscosity molar mass dependence and stiffness of hyaluronan. Biomacromolecules 2003; 4:1805–1810.
16. Frisch HL, Simha R. In: Eirich FR, ed. Rheology—Theory and Practice. New York: Academic Press, 1956:525–614.
17. Matsuoka S, Cowman MK. Viscosity of polymer solutions revisited. In: Kennedy JF, Phillips GO, Williams PA, eds. Hyaluronan. Cambridge: Woodhead Publishing, 2002:79–88.
18. Matsuoka S, Cowman MK. Equation of state for polymer solution. Polymer 2002; 43:3447–3453.
19. Berriaud N, Milas M, Rinaudo M. Rheological study on mixtures of different molecular weight hyaluronates. Int J Biol Macromol 1994; 16:137–142.
20. Martin AF. ACS Meeting, Memphis, Tennessee, 1942.
21. Fouissac E, Milas M, Rinaudo M. Shear-rate, concentration, molecular weight, and temperature viscosity dependences of hyaluronate, a wormlike polyelectrolyte. Macromolecules 1993; 26:6945–6951.
22. Mo Y, Takaya T, Nishinari K, Kubota K, Okamoto A. Effects of sodium chloride, guanidine hydrochloride, and sucrose on the viscoelastic properties of sodium hyaluronate solutions. Biopolymers 1999; 50:23–34.
23. Milas M, Rinaudo M, Roure I, Al Assaf S, Phillips GO, Williams PA. Comparative rheological behavior of hyaluronan from bacterial and animal sources

with cross-linked hyaluronan (hylan) in aqueous solution. Biopolymers 2001; 59:191–204.
24. Onyon PF. Viscometry. In: Allen PW, ed. Techniques of Polymer Characterization. London: Butterworths Scientific Publications, 1959:171–206.
25. Billmeyer FW Jr. Textbook of Polymer Science. New York: Interscience Publishers, 1962; 81.
26. ASTM Standard Method D 446-97^{e1}, Standard specifications and operating instructions for glass capillary kinematic viscometers. ASTM Standard Method D 2857–95. Standard practice for dilute solution viscosity of polymers, section 10.3 and NOTE 13.
27. Balazs EA, Sundblad L. Viscosity of hyaluronic acid solution containing proteins. Acta Soc Med Upsaliensis 1959; 64:137–146.
28. Wedlock DJ, Phillips GO, Davies A, Gormally J, Wyn-Jones E. Depolymerization of sodium hyaluronate during freeze drying. Int J Biol Macromol 1983; 5:186–188.
29. Lowry KM, Beavers EM. Thermal stability of sodium hyaluronate in aqueous solution. J Biomed Mater Res 1994; 28:1239–1244.
30. Vercruysse KP, Lauwers AR, Demeester JM. Absolute and empirical determination of the enzymic activity and kinetic investigation of the action of hyaluronidase on hyaluronan using viscosimetry. Biochem J 1995; 306:153–160.
31. Li M, Rosenfeld L, Vilar RE, Cowman MK. Degradation of hyaluronan by peroxynitrite. Arch Biochem Biophys 1997; 341:245–250.
32. Šoltés L, Mendichi R, Lath D, Mach M, Bakoš D. Molecular characteristics of some commercial high-molecular-weight hyaluronans. Biomed Chromatogr 2002; 16: 459–462.
33. Horner AA. Electrophoresis of acidic mucopolysaccharides in agarose gel. Can J Biochem 1967; 45:1009–1013.
34. Hilborn JC, Anastassiadis PA. Acrylamide gel electrophoresis of acidic mucopolysaccharides. Anal Biochem 1969; 31:51–58.
35. Bader JP, Ray DA, Steck TL. Electrophoretic determinations of hyaluronate produced by cells in culture. Biochim Biophys Acta 1972; 264:73–84.
36. Knudson W, Gundlach MW, Schmid TM, Conrad HE. Selective hydrolysis of chondroitin sulfates by hyaluronidase. Biochemistry 1984; 23:368–375.
37. Hampson IN, Gallagher JT. Separation of radiolabelled glycosaminoglycan oligosaccharides by polyacrylamide-gel electrophoresis. Biochem J 1984; 221:697–705.
38. Cowman MK, Slahetka MF, Hittner DM, Kim J, Forino M, Gadelrab G. Polyacrylamide-gel electrophoresis and alcian blue staining of sulphated glycosaminoglycan oligosaccharides. Biochem J 1984; 221:707–716.
39. Turner RE, Cowman MK. Cationic dye binding by hyaluronate fragments: dependence on hyaluronate chain length. Arch Biochem Biophys 1985; 237:253–260.
40. Min H, Cowman MK. Combined alcian blue and silver staining of glycosaminoglycans in polyacrylamide gels: application to electrophoretic analysis of molecular weight distribution. Anal Biochem 1986; 155:275–285.
41. Lyon M, Gallagher JT. A general method for the detection and mapping of submicrogram quantities of glycosaminoglycan oligosaccharides on polyacrylamide gels by sequential staining with azure A and ammoniacal silver. Anal Biochem 1990; 185:63–70.
42. Rice KG, Rottink MK, Linhardt RJ. Fractionation of heparin-derived oligosaccharides by gradient polyacrylamide-gel electrophoresis. Biochem J 1987; 244: 515–522.

43. Turnbull JE, Gallagher JT. Oligosaccharide mapping of heparan sulphate by polyacrylamide-gradient-gel electrophoresis and electrotransfer to nylon membrane. Biochem J 1988; 251:597–608.
44. Al-Hakim A, Linhardt RJ. Isolation and recovery of acidic oligosaccharides from polyacrylamide gels by semi-dry electrotransfer. Electrophoresis 1990; 11:23–28.
45. Ikegami-Kawai M, Takahashi T. Microanalysis of hyaluronan oligosaccharides by polyacrylamide gel electrophoresis and its application to assay of hyaluronidase activity. Anal Biochem 2002; 311:157–165.
46. Hittner DM, Cowman MK. High-performance gel permeation chromatography of glycosaminoglycans. Column calibration by gel electrophoresis. J Chromatogr 1987; 402:149–158.
47. DeAngelis PL, Oatman LC, Gay DF. Rapid chemoenzymatic synthesis of monodisperse hyaluronan oligosaccharides with immobilized enzyme reactors. J Biol Chem 2003; 278:35199–35203.
48. Gao N, Lehrman MA. Alternative sources of reagents and supplies for fluorophore-assisted carbohydrate electrophoresis (FACE). Glycobiology 2003; 13:1G–3G.
49. Calabro A, Benavides M, Tammi M, Hascall VC, Midura RJ. Microanalysis of enzyme digests of hyaluronan and chondroitin/dermatan sulfate by fluorophore-assisted carbohydrate electrophoresis (FACE). Glycobiology 2000; 10:273–281.
50. Mahoney DJ, Aplin RT, Calabro A, Hascall VC, Day AJ. Novel methods for the preparation and characterization of hyaluronan oligosaccharides of defined length. Glycobiology 2001; 11:1025–1033.
51. Tawada A, Masa T, Oonuki Y, Watanabe A, Matsuzaki Y, Asari A. Large scale preparation, purification, and characterization of hyaluronan oligosaccharides from 4-mers to 52-mers. Glycobiology 2002; 12:421–426.
52. Grimshaw J. Analysis of glycosaminoglycans and their oligosaccharide fragments by capillary electrophoresis. Electrophoresis 1997; 18:2408–2414.
53. Koketsu M, Linhardt RJ. Electrophoresis for the analysis of acidic oligosaccharides. Anal Biochem 2000; 283:136–145.
54. Al-Hakim A, Linhardt RJ. Capillary electrophoresis for the analysis of chondroitin sulfate- and dermatan sulfate-derived disaccharides. Anal Biochem 1991; 195:68–73.
55. Carney SL, Osborne DJ. The separation of chondroitin sulfate disaccharides and hyaluronan oligosaccharides by capillary zone electrophoresis. Anal Biochem 1991; 195:132–140.
56. Park Y, Cho S, Linhardt RJ. Exploration of the action pattern of Streptomyces hyaluronate lyase using high-resolution capillary electrophoresis. Biochim Biophys Acta 1997; 1337:217–226.
57. Hayase S, Oda Y, Honda S, Kakehi K. High-performance capillary electrophoresis of hyaluronic acid: determination of its amount and molecular mass. J Chromatogr A 1997; 768:295–305.
58. Hong M, Sudoe J, Stefansson M, Novotny MV. High-resolution studies of hyaluronic acid mixtures through capillary gel electrophoresis. Anal Chem 1998; 70:568–573.
59. Kakehi K, Kinoshita M, Hayase S, Oda Y. Capillary electrophoresis of *N*-acetylneuraminic acid polymers and hyaluronic acid: correlation between migration order reversal and biological functions. Anal Chem 1999; 71:1592–1596.
60. Kinoshita M, Okino A, Oda Y, Kakehi K. Anomalous migration of hyaluronic acid oligomers in capillary electrophoresis: correlation to susceptibility to hyaluronidase. Electrophoresis 2001; 22:3458–3465.

61. Lee HG, Cowman MK. An agarose gel electrophoretic method for analysis of hyaluronan molecular weight distribution. Anal Biochem 1994; 219:278–287.
62. Noble PW, McKee CM, Cowman M, Shin HS. Hyaluronan fragments activate an NF-κB/I-κBα autoregulatory loop in murine macrophages. J Exp Med 1996; 183: 2373–2378.
63. McKee CM, Penno MB, Cowman M, Burdick MD, Strieter RM, Bao C, Noble PW. Hyaluronan (HA) fragments induce chemokine gene expression in alveolar macrophages. J Clin Investig 1996; 98:2403–2413.
64. Armstrong SE, Bell DR. Measurement of high-molecular-weight hyaluronan in solid tissue using agarose gel electrophoresis. Anal Biochem 2002; 308:255–264.
65. Pummill PE, DeAngelis PL. Alteration of polysaccharide size distribution of a vertebrate hyaluronan synthase by mutation. J Biol Chem 2003; 278:19808–19814.
66. DeAngelis PL. Molecular directionality of polysaccharide polymerization by the Pasteurella multocida hyaluronan synthase. J Biol Chem 1999; 274:26557–26562.
67. Zimm BH. Apparatus and methods for measurement and interpretation of the angular variation of light scattering; preliminary results on polystyrene solutions. J Chem Phys 1948; 16(12):1099–1117.
68. Kratochvil P. Classical Light Scattering from Polymer Solutions. Amsterdam: Elsevier, 1987.
69. Wyatt PJ. Light scattering and the absolute characterization of macromolecules. Anal Chim Acta 1993; 272:1–40.
70. Debye P. Molecular weight determination by light scattering. J Phys Colloid Chem 1947; 51:18–31.
71. Takahashi R, Al-Assaf S, Williams PA, Kubota K, Okamoto A, Nishinari K. Asymmetrical-flow field-flow fractionation with online multiangle light scattering detection. 1. application to wormlike chain analysis of weakly stiff polymer chains. Biomacromolecules 2003; 4(2):404–409.
72. Rinaudo M, Roure I, Milas M. Use of steric exclusion chromatography to characterize hyaluronan a semirigid polysaccharide. Int J Polym Anal Charact 1999; 5:277–287.
73. Milas M, Rinaudo M, Roure I, Al-Assaf S, Phillips GO, Williams PA. Comparative rheological behavior of hyaluronan from bacterial and animal sources with cross-linked hyaluronan (hylan) in aqueous solution. Biopolymers 2001; 59(4):191–204.
74. Adam N, Ghosh P. Hyaluronan molecular weight and polydispersity in some commercial intra-articular injectable preparations and in synovial fluid. Inflamm Res 2001; 50:294–299.
75. Mendichi R, Giacometti Schieroni A. Fractionation and characterization of ultra-high molar mass hyaluronan: 2. On-line size exclusion chromatography methods. Polymer 2002; 43(23):6115–6121.
76. Mendichi R, Giacometti Schieroni A, Grassi C, Re A. Characterization of ultra-high molar mass hyaluronan: 1. Off-line static methods. Polymer 1998; 39(25): 6611–6620.
77. Soltes L, Mendichi R. Molecular characterization of two host-guest associating hyaluronan derivatives. Biomed Chromatogr 2003; 37(6):376–384.

Chemistry and Biology of Hyaluronan
H.G. Garg and C.A. Hales (editors)

Chapter 4

Biodegradation of Hyaluronan

GÜNTER LEPPERDINGER,
CHRISTINE FEHRER and
STEPHAN REITINGER

Institute for Biomedical Aging Research, Austrian Academy of Sciences, Rennweg 10 A-6020, Innsbruck, Austria

I. Introduction

Biodegradation of hyaluronan (HA) is a step-wise process (1). In the extracellular matrix (ECM) of most mature tissues, HA is of high molecular weight. Together with other structural macromolecules, HA contributes to the mechanical properties of the meshwork. Hence, in order to be released from this firm network, the polymer has to be at least partially degraded (2). Subsequently, the cells can take up intermediate-sized chains either by receptor-mediated mechanisms or by endocytosis. In consequence, HA is sorted into the lysosomal compartment, where it becomes hydrolyzed to small oligosaccharides by intracellular hyaluronidases. These fragments are eventually degraded to monosaccharides by exoglucosidases present in lysosomes (3). This chapter describes the molecular pathways as well as the cellular mechanisms of HA catabolism.

II. Turnover in the Organism and Cellular Uptake

A. General Consideration

The concentration of HA in the human body varies greatly from about 4 g/kg in umbilical cord, 2–4 g/L in synovial fluid, 0.2 g/kg in dermis, about 10 mg/L in thoracic lymph to as low as 0.1–0.01 mg/L in normal serum. The total body content of HA in a 70 kg human is approximately 15 g. The largest amount is found in the intercellular matrix of skin and musculoskeletal tissues. In the normal body, depending on its location, most of the HA is catabolized within days. Besides direct enzymatic degradation in the extracellular space and non-enzymatic depolymerization, there are two pathways in the body that are engaged in HA catabolism: local turnover (internalization and degradation within tissues) (4), and secondly, release from the matrix, drainage into the vasculature and clearance in lymph nodes, the liver and the kidneys (Fig. 1) (5).

B. Direct Uptake

In densely structured tissues such as bone or cartilage with no or little lymphatic drainage, HA is degraded in situ together with other ECM molecules (collagens

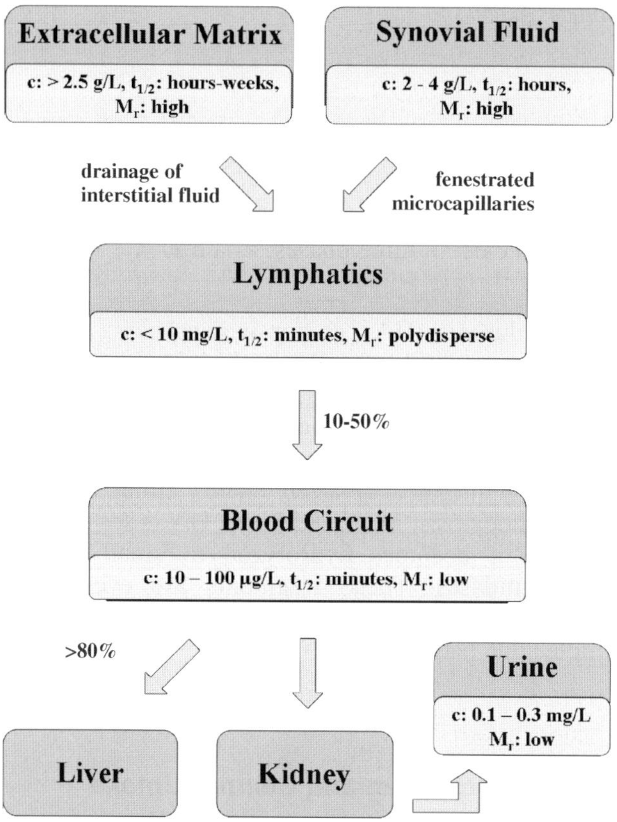

Figure 1 Catabolic pathway of HA from ECM and synovial fluid to liver, kidney and urine. Arrows indicate the flow of HA. Concentration (*c*), half-life time ($t_{1/2}$) and molecular weight (M_r) of the polymer within the organ systems are shown in light gray boxes.

and proteoglycans). In skin and joints, only a minor fraction (approximately 20–30%) of HA is internalized and degraded within the tissue. About 50% of HA in the body resides in skin tissue. In the dermis, concentrations of about 0.5 mg/g wet tissue and in the epidermis about 0.1 mg/g wet tissue have been observed. As HA is restricted to the small intercellular space, its concentration around the cells can be estimated with a concentration of about 2.5 g/L. The metabolic half-life is about 1.5 days (6). HA synthesized in the epidermis has a short half-life time of 2–3 h. Rat epidermal keratinocytes (REK) internalize a large proportion of their newly synthesized HA into non-clathrin-coated endosomes in a receptor-mediated way and rapidly transport it for slower degradation in the endosomal/lysosomal system (7). In chondrocytes and liver endothelial cells, sulfated glycosaminoglycans (8) and HA_6 and HA_8 oligosaccharides (9) compete for HA uptake. This is in contrast to the receptor-mediated fashion in REK, which is not sensitive to sulfated GAGs and requires HA decasaccharides to compete. HA oligosaccharides are mainly endocytosed through bulk phase pinocytosis, whereas uptake of longer HA molecules involves multivalent interactions with several cell surface receptors (10). This is also consistent with the patchy distribution of HA and CD44 in keratinocytes (11).

Keratinocytes that are tightly arranged in layers express CD44 on their surfaces (12). CD44 is thought to be the principal cell surface receptor for HA. It is abundant on plasma membrane domains, facing open intercellular spaces, rich in HA. CD44 is synthesized as a single-pass transmembrane glycoprotein. However, it appears to exist in three phases: as a transmembrane receptor, as an integral component of the matrix and as a soluble protein found in body fluids (13). The extracellular domain is the primary site of alternative splicing variation among CD44 isoforms. Many cell types, e.g., stromal cells such as fibroblasts and smooth muscle cells, epithelial cells and immune cells such as neutrophils, macrophages and lymphocytes express CD44. The cellular functions of CD44 appear to be manifold. For instance, it can mediate migration in a variety of cell types, or it can communicate cell–matrix interactions into the cell via 'outside-in signaling'. Although glycosaminoglycan side chains associated with some CD44 isoforms mediate binding of heparin-binding growth factors, cytokines and ECM proteins such as fibronectin, most of the functions ascribed to CD44 can be attributed to its ability to bind and internalize HA (14). Mice lacking detectable CD44 expression in skin keratinocytes have HA accumulation in the superficial dermis and the epidermis (15). Therefore, CD44 is an important key player in maintaining HA homeostasis in the skin. Generally, CD44-deficient mice appear to be healthy suggesting that the CD44 deficiency could be partially compensated for by other HA receptors. However, to our knowledge, HA turnover has not yet been evaluated in detail in CD44-deficient mice.

C. Turnover in the Circulation

The HA in ECM of relatively dense tissues can also be removed through the lymphatics. Large native HA molecules, 10^7 Da in size, are thought to be partially degraded (~10^6 Da) before they are released from the matrix. The polymer then enters the lymphatic system. Metabolic degradation mainly occurs in lymphatic tissues as HA passes from peripheral tissues to the bloodstream (2). The primary function of the lymphatics is to collect leaked plasma and interstitial fluid and return it into the blood vasculature. After injection of radio-labeled HA into afferent lymph vessels, most of it is taken up or degraded by lymphatic tissue (16). The lymph nodes can extract 50–90% from the peripheral lymph. Remaining HA, which was found to be smaller in size (~10^5 Da) (re-) enters the blood stream, from which it is rapidly eliminated by sinusoidal endothelial cells (SEC) in the liver. There is a high rate of turnover in the circulation (1 mg/kg a day), which in other words means that the fractional turnover ranges as high as 70% per minute (17). This obviously leads to low concentrations of HA in these fluids (10–100 ng/mL). In the bloodstream, liver removes more than 80% and kidneys contribute for another 10%. Under normal circumstances, only 0.1–0.3 mg/L are being excreted in the urine. Altogether, the daily turnover of HA is in the order of one-third of the total body content. As described in more detail later, cellular uptake is primarily effected by specialized HA receptors of endothelial cells.

1. HARE, the HA Receptor for Endocytosis

SECs in liver and in lymph nodes show a high rate of endocytosis that removes ECM-derived fragments of HA and chondroitin sulfate (18). These cell linings express HARE (19), which has been purified to raise a monoclonal antibody (20). In perfused liver, this antibody could block receptor-mediated HA uptake (21). Other groups independently characterized the receptor and called it stabilin-2 (22) or FEEL (fasciclin EGF-like, laminin-type EGF-like and link domain-containing scavenger receptor) (23). Presently, two isoforms have been described in humans with the sizes of 190 and 315 kDa, respectively. The smaller HARE, which is derived from the larger precursor by proteolysis is glycosylated and exhibits high affinity for HA. It mediates HA endocytosis through the coated pit pathway in the absence of the large HARE (24). Therefore, the small and the large HARE appear to be independent iso-receptors for HA, though it is possible that both HARE species form a complex *in vivo*. Interestingly, the overall ratio of the small and large HARE species varies in different tissues. In liver, more of the small HARE is present; in spleen and lymph nodes, more of the large species could be detected (25). The large HARE appears to have higher affinity for high molecular weight HA, whereas the small HARE may interact more efficiently with smaller HA. Thus, endothelial cells of liver, spleen and lymph node exhibit some preference for different sized HA. Hence, it appears likely that both HARE species are necessary for efficient uptake and degradation of polydisperse HA, which is present in tissues throughout the body (5). Furthermore, it has been shown that recombinant HARE functions as a recycling endocytotic receptor. Internalized HA is delivered to lysosomes and degraded.

2. *LYVE-1, the Lymphatic Vessel Endothelial HA Receptor*

LYVE-1 is a close relative of CD44 with an overall similarity of 43%. It is preferentially expressed on lymphatic endothelial cells (26), which do not express CD44. Furthermore, LYVE-1 can also be found in the sinusoidal endothelium of liver and spleen and is expressed in discrete populations of activated macrophages (27,28). It has been shown that LYVE-1 mediates HA endocytosis in transfected fibroblasts (29); hence, it is highly likely that LYVE-1 functions primarily as an HA transporter. However, in LYVE-1 transfected cells HA uptake is low compared with the rate of HA uptake by HARE. Therefore, it is reasonable to think that LYVE-1 acts as a co-receptor rather than the primary receptor for HA uptake in the lymphatic vasculature. Indeed, LYVE-1 is co-expressed with HARE in liver and spleen sinusoids. Being a member of the link module superfamily, its extracellular domain is comprised of an HA-binding structure resembling a C-type lectin fold (30,31). Similar to CD44, this receptor also exhibits an enlarged specialized link module, most likely containing elements that enable modulation of HA binding in response to extracellular factors. This suggests an important regulatory role for LYVE-1 in the catabolism of HA in the aforementioned organs as well as a function in HA signaling or cellular migration and differentiation (28).

D. Hyaluronidase-Mediated Degradation

Hyaluronidases are considered responsible for much of the HA catabolism taking place in the organisms (see Chapter 17 for details).

1. *Testicular Hyaluronidase, PH-20*

Successful fertilization in mammals is dependent on the action of hyaluronidase, when the sperm traverses the outermost HA-rich cumulus layer surrounding the egg (32). Therefore, all mammalian spermatozoa bear hyaluronidase activity. A protein, first of all characterized with the aid of a monoclonal antibody against posterior head determinants and thus named PH-20 (33), could later be shown to exhibit this particular activity (34). Unfortunately, PH-20 was also named sperm adhesion molecule 1 (SPAM1) (35). The PH-20 gene has been deleted in mice by homologous recombination. Homozygous male offspring, however, produce sperm, which still can penetrate the cumulus layer, though at a reduced rate (36). This suggests that sperm is synthesizing another enzyme with features comparable to those found with PH-20. Since antibodies raised against PH-20 completely block cumulus sperm penetration, it is highly likely that the complementary enzyme is a close relative to PH-20. Expressed sequence tags encoding HYAL3 and HYAL5 have been found in cDNA libraries derived from mouse spermatids. It is therefore highly likely that one of them is another hyaluronidase capable of degrading extracellular HA at physiological conditions.

2. Cell Surface-Bound Somatic Hyaluronidase

HYAL2 is almost ubiquitously expressed, which suggests a prime role in HA catabolism. However, the gene was found to be silent in adult brain. In line with this, brain is known to be rich in HA. As an interesting detail, HYAL2 expressed by means of a recombinant vaccinia virus hydrolyzed high molecular weight HA to products comprising of approximately 100 sugar moieties in length indicating that in solution HA exhibits heterogeneous structural elements, some of those more prone to degradation by HYAL2 (37). Human HYAL2 activity exerts its optimal activity at pH 4. A somewhat different result was obtained with HYAL2 from *Xenopus laevis* expressed in frog oocytes. Using a sensitive assay (38), it could be shown that frog HYAL2 is also active under physiological conditions. Ectopic expression of this enzyme in *Xenopus* embryos indeed leads to extensive degradation of HA in ECM (39). The cellular localization of the HYAL2 protein is complex: the wide distribution, acidic pH-optimum is typical for a lysosomal enzyme. However, upon expression of HYAL2 in different cells, part of the enzyme was always found in the membrane fraction as a GPI-anchored protein (40). This indicates that the enzyme may be transported to lysosomes via secretion and re-uptake at the plasma membrane. In consequence, HYAL2 can be considered a good candidate for an extracellular hyaluronidase needed by somatic cells, which is actively involved in hydrolysis of the HA meshwork prior to uptake and thus might have an important role for tissue remodeling and cellular migration.

A putative PH-20-like enzyme was isolated from *X. laevis*. In the adult frog, this mRNA was only found to be expressed in the kidney and therefore named XKH1 (41). It solely exhibits enzymatic activity at neutral pH, at physiologic ionic strength and in weakly acidic solutions. To date, it is the only known hyaluronidase exclusively active above pH 5.4. In addition to HA hydrolysis, the enzyme also degrades chondroitin sulfate. The enzyme is sorted to the outer surface of the cell membrane. From there, it cannot be removed by phospholipase C. Hence, XKH1 represents a membrane-bound HA-degrading enzyme exclusively expressed in cells of the adult frog kidney. Most probably, it is involved in the reorganization of the extracellular architecture or in supporting physiological demands for proper renal function in the frog. In mammals, functional homologues have not yet been characterized.

3. Serum and Lysosomal Hyaluronidase, HYAL1

HA-degrading activity can be observed in acidified human plasma (42). Although the specific activity of hyaluronidases is easily detected by zymography (43), purification of the human serum hyaluronidase, HYAL1 from plasma has been exceptionally difficult, in fact, due to its low concentration, 60 ng/mL, and instability in the absence of detergents (44). Furthermore, HYAL1-specific activity has also been detected in human urine. Using a monoclonal antibody against human serum hyaluronidase, unlike in plasma, two HYAL1 isozymes (57 and 45 kDa) could be purified. The lower molecular weight isozyme is made up of two chains (45). Hyaluronidase partially purified from urine of bladder cancer patients showed a slightly different pH activity profile than HYAL1 isolated from serum or urine of healthy people (46). Suggestively, this is due to some posttranslational modification of the HYAL1 polypeptide chain. In line with this, hyaluronidases secreted from various human carcinoma cells were found to differ from serum hyaluronidase with regard to glycosylation. Supposedly, glycosylation can modulate the enzymatic properties of hyaluronidase as well (47).

HYAL1 is widely expressed. Albeit highly similar to PH-20, it is enzymatically active only below pH 5.5 (48). HYAL1 was first characterized from serum as a soluble enzyme. None the less, HYAL1 is a lysosomal enzyme that cleaves HA to di- and tetrasaccharides. The latter view is supported by the finding that the lysosomal storage disease (49), mucopolysaccharidosis IX is caused by mutated forms of HYAL1 (50). It is not clear why a presumably lysosomal enzyme is found in the circulation, or what its possible function there could be. Yet, it appears most reasonable to think that an acid-active hyaluronidase, HYAL1 degrades internalized HA, when it reaches the lysosomal compartment (51).

4. pH Matters!

Partially degraded HA is taken up at a faster rate than high molecular mass HA (52). Obviously, in the case of high molecular weight HA, or even larger, if other macromolecules such as proteoglycans are bound to the HA, internalization is sterically inhibited. Furthermore, high molecular HA can also be tethered to the cell surface by HA receptors. Hence, internalization can only be initiated when the matrix is first eroded. For instance, in cartilage, matrix metalloproteinases participate in this process and consequently allow HA internalization. Alternatively, HYAL2 may cleave HA to a smaller size permissive to uptake (39). This takes place at physiological pH. The resulting fragments are delivered to endosomes and finally to lysosomes, where HYAL1 activity degrades the fragments to small oligosaccharides, which only occurs at an efficient rate at very low pH. Concomitantly, lysosomal exoglycosidases, such as β-glucuronidase and β-*N*-acetyl-glucosaminidase, participate to complete degradation at low pH (1).

E. Degradation by Non-enzymatic Means

In contrast to being a chemical compound stable under normal physiological conditions, chain scission of the HA polymer can be induced in an unspecific manner by chemical reactions other than enzyme-catalyzed degradation as well as by physical stresses such as freeze-drying, shearing or stirring at critical conditions. Moreover, irradiation can cause depolymerization. In line with this, it was found that free radicals can interact with HA and cause degradation of the polymer (53).

Cells form reactive oxygen species (ROS) as a consequence of normal aerobic respiration. Several candidates have been described to cause HA chain scission such as the superoxide anion, the hydroxyl radical (54) or hypochloride and species derived from peroxynitrite (55). ROS are suggested to be involved in several biodegenerative and inflammatory processes such as arthritis, but evidence for them participating in disease, still remains circumstantial. However, depolymerization of HA in synovial fluids during the early onset of inflammatory arthritis is believed to be caused by ROS and not by hyaluronidases (56).

Normal synovial fluid is viscoelastic, which is largely determined by the presence of high molecular weight HA. The normal high levels (2–4 g/L) of HA (average molecular weight 7×10^6 Da) in healthy joints are essential for functional joint articulation. Synoviocytes continuously secrete macromolecules into the synovial fluid including HA. Fluid is pushed out of the joint cavity into microcapillaries embedded in the synovium every time the pressure is raised. In this way, HA molecules escape through the interstitial drainage pores in the synovial lining (diameter 30–90 nm), but with decreased motility compared to smaller molecules (Fig. 2). It is believed that high molecular weight HA forms a layer at the tissue–fluid interface, not to be absorbed by the microcapillaries (57). The normal intra-articular turnover time for HA is <40 h. However, in arthritis

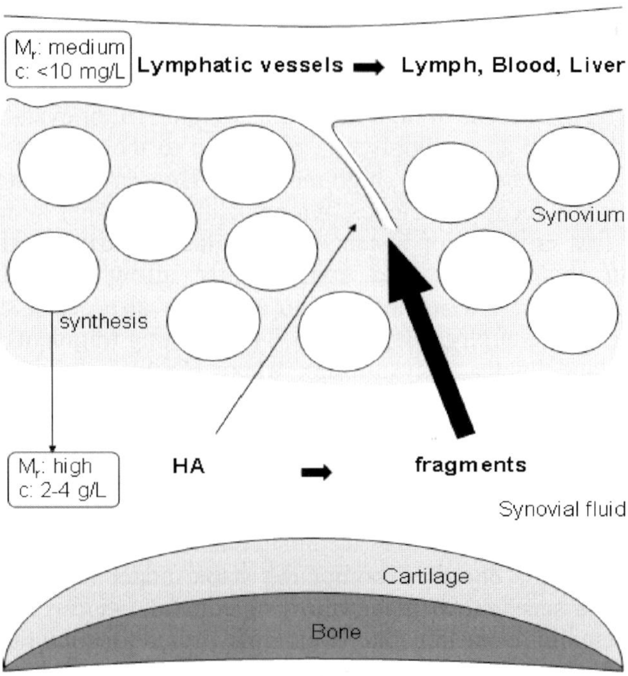

Figure 2 Drainage of HA from the synovial fluid to the lymphatic system. HA is primarily synthesized by fibroblast-like cells. High molecular weight HA can be taken up into the lymphatics at a low rate (slim arrow). Fragmented chains can pass the interstitial drainage pores embedded in the matrix at an elevated rate indicated by the bold arrow. Concentration (*c*) and molecular weight (M_r) of HA are indicated.

patients both HA chain length and HA concentration are decreased, consequently leading to significantly reduced lubricant viscosity. It is widely believed that this is being evoked by ROS, because treatment with radical scavengers inhibits degradation of HA.

III. Summary and Conclusion

The metabolism of HA is very dynamic. Some cells, such as chondrocytes in cartilage, actively synthesize and catabolize HA in a balanced fashion throughout life thereby maintaining a constant concentration in the tissue. Other cells, e.g., dermal cells synthesize more HA than they catabolize. HA can only leave the tissue of origin, when the ECM is at least partially disintegrated. Extracellular hyaluronidases or ROS can render HA short enough to be released from the matrix. Then, it is either immediately internalized by cells and degraded in lysosomes or first transferred to the circulation, from where it is cleared at special sites in the liver, lymph nodes or the kidneys. The end products of degradation, glucuronic acid and *N*-acetylglucosamine can thus be re-used for polysaccharide biosynthesis. In the kidneys, only trace amounts are lost into the urinary system. In this manner, about one-third of the total HA in the human body can be metabolically removed and replaced daily.

Acknowledgements

GL is an APART fellow of the Austrian Academy of Sciences and supported by the Jubilee Fund of the Austrian National Bank (Project 10451). StR is a DOC fellow of the Austrian Academy of Sciences. The Austrian Science Foundation (FWF) has furthermore supported work in the authors' institute.

References

1. Roden L, Campbell P, Fraser JR, Laurent TC, Pertoft H, Thompson JN. Enzymic pathways of hyaluronan catabolism. Ciba Found Symp 1989; 143:60–76.
2. Fraser JR, Brown TJ, Laurent TC. Catabolism of Hyaluronan. The Chemistry, Biology and Medical Applications of Hyaluronan and Its Derivatives. London: Portland Press, 1998; 92–95.
3. Laurent TC, Fraser JR. Hyaluronan. Faseb J 1992; 6:2397–2404.
4. Knudson W, Chow G, Knudson CB. CD44-mediated uptake and degradation of hyaluronan. Matrix Biol 2002; 21:15–23.
5. Weigel PH, Yik JH. Glycans as endocytosis signals: the cases of the asialoglycoprotein and hyaluronan/chondroitin sulfate receptors. Biochim Biophys Acta 2002; 1572:341–363.
6. Tammi R, Saamanen AM, Maibach HI, Tammi M. Degradation of newly synthesized high molecular mass hyaluronan in the epidermal and dermal compartments of human skin in organ culture. J Investig Dermatol 1991; 97: 126–130.

7. Tammi R, Rilla K, Pienimaki JP, MacCallum DK, Hogg M, Luukkonen M, Hascall VC, Tammi M. Hyaluronan enters keratinocytes by a novel endocytic route for catabolism. J Biol Chem 2001; 276:35111–35122.
8. Knudson CB. Hyaluronan receptor-directed assembly of chondrocyte pericellular matrix. J Cell Biol 1993; 120:825–834.
9. Hua Q, Knudson CB, Knudson W. Internalization of hyaluronan by chondrocytes occurs via receptor-mediated endocytosis. J Cell Sci 1993; 106:365–375.
10. Lesley J, English N, Charles C, Hyman R. The role of the CD44 cytoplasmic and transmembrane domains in constitutive and inducible hyaluronan binding. Eur J Immunol 2000; 30:245–253.
11. Tammi R, MacCallum D, Hascall VC, Pienimäki J, Hyttinen M, Tammi M. Hyaluronan bound to CD44 on keratinocytes is displaced by hyaluronan decasaccharides and not hexasaccharides. J Biol Chem 1998; 273:28878–28888.
12. Tuhkanen AL, Agren UM, Tammi MI, Tammi RH. CD44 expression marks the onset of keratinocyte stratification and mesenchymal maturation into fibrous dermis in fetal human skin. J Histochem Cytochem 1999; 47:1617–1624.
13. Cichy J, Pure E. The liberation of CD44. J Cell Biol 2003; 161:839–843.
14. Sherman L, Sleeman J, Herrlich P, Ponta H. Hyaluronate receptors: key players in growth, differentiation, migration and tumor progression. Curr Opin Cell Biol 1994; 6:726–733.
15. Kaya G, Rodriguez I, Jorcano JL, Vassalli P, Stamenkovic I. Selective suppression of CD44 in keratinocytes of mice bearing an antisense CD44 transgene driven by a tissue-specific promoter disrupts hyaluronate metabolism in the skin and impairs keratinocyte proliferation. Genes Dev 1997; 11:996–1007.
16. Fraser JR, Kimpton WG, Laurent TC, Cahill RN, Vakakis N. Uptake and degradation of hyaluronan in lymphatic tissue. Biochem J 1988; 256:153–158.
17. Fraser JR, Dahl LB, Kimpton WG, Cahill RN, Brown TJ, Vakakis N. Elimination and subsequent metabolism of circulating hyaluronic acid in the fetus. J Dev Physiol 1989; 11:235–242.
18. Eriksson S, Fraser JR, Laurent TC, Pertoft H, Smedsrod B. Endothelial cells are a site of uptake and degradation of hyaluronic acid in the liver. Exp Cell Res 1983; 144:223–228.
19. Yannariello Brown J, Frost SJ, Weigel PH. Identification of the Ca(2 +)-independent endocytic hyaluronan receptor in rat liver sinusoidal endothelial cells using a photoaffinity cross-linking reagent. J Biol Chem 1992; 267: 20451–20456.
20. Zhou B, Oka JA, Singh A, Weigel PH. Purification and subunit characterization of the rat liver endocytic hyaluronan receptor. J Biol Chem 1999; 274:33831–33834.
21. Weigel JA, Raymond RC, McGary C, Singh A, Weigel PH. A blocking antibody to the hyaluronan receptor for endocytosis (HARE) inhibits hyaluronan clearance by perfused liver. J Biol Chem 2003; 278:9808–9812.
22. Politz O, Gratchev A, McCourt PA, Schledzewski K, Guillot P, Johansson S, Svineng G, Franke P, Kannicht C, Kzhyshkowska J, Longati P, Velten FW, Goerdt S. Stabilin-1 and -2 constitute a novel family of fasciclin-like hyaluronan receptor homologues. Biochem J 2002; 362:155–164.
23. Tamura Y, Adachi H, Osuga J, Ohashi K, Yahagi N, Sekiya M, Okazaki H, Tomita S, Iizuka Y, Shimano H, Nagai R, Kimura S, Tsujimoto M, Ishibashi S. FEEL-1 and FEEL-2 are endocytic receptors for advanced glycation end products. J Biol Chem 2003; 278:12613–12617.

24. Zhou B, Weigel JA, Saxena A, Weigel PH. Molecular cloning and functional expression of the rat 175-kDa hyaluronan receptor for endocytosis. Mol Biol Cell 2002; 13:2853–2868.
25. Zhou B, Weigel JA, Fauss L, Weigel PH. Identification of the hyaluronan receptor for endocytosis (HARE). J Biol Chem 2000; 275:37733–37741.
26. Banerji S, Ni J, Wang SX, Clasper S, Su J, Tammi R, Jones M, Jackson DG. LYVE-1, a new homologue of the CD44 glycoprotein, is a lymph-specific receptor for hyaluronan. J Cell Biol 1999; 144:789–801.
27. Mouta Carreira C, Nasser SM, di Tomaso E, Padera TP, Boucher Y, Tomarev SI, Jain RK. LYVE-1 is not restricted to the lymph vessels: expression in normal liver blood sinusoids and down-regulation in human liver cancer and cirrhosis. Cancer Res 2001; 61:8079–8084.
28. Jackson DG. The lymphatics revisited: new perspectives from the hyaluronan receptor LYVE-1. Trends Cardiovasc Med 2003; 13:1–7.
29. Prevo R, Banerji S, Ferguson DJ, Clasper S, Jackson DG. Mouse LYVE-1 is an endocytic receptor for hyaluronan in lymphatic endothelium. J Biol Chem 2001; 276:19420–19430.
30. Kohda D, Morton CJ, Parkar AA, Hatanaka H, Inagaki FM, Campbell ID, Day AJ. Solution structure of the link module: a hyaluronan-binding domain involved in extracellular matrix stability and cell migration. Cell 1996; 86:767–775.
31. Day AJ. The structure and regulation of hyaluronan-binding proteins. Biochem Soc Trans 1999; 27:115–121.
32. Salustri A, Camaioni A, Di Giacomo M, Fulop C, Hascall VC. Hyaluronan and proteoglycans in ovarian follicles. Hum Reprod Update 1999; 5:293–301.
33. Primakoff P, Hyatt H, Myles DG. A role for the migrating sperm surface antigen PH-20 in guinea pig sperm binding to the egg zona pellucida. J Cell Biol 1985; 101: 2239–2244.
34. Gmachl M, Sagan S, Ketter S, Kreil G. The human sperm protein PH-20 has hyaluronidase activity. FEBS Lett 1993; 336:545–548.
35. Jones MH, Davey PM, Aplin H, Affara NA. Expression analysis, genomic structure, and mapping to 7q31 of the human sperm adhesion molecule gene SPAM1. Genomics 1995; 29:796–800.
36. Baba D, Kashiwabara S, Honda A, Yamagata K, Wu Q, Ikawa M, Okabe M, Baba T. Mouse sperm lacking cell surface hyaluronidase PH-20 can pass through the layer of cumulus cells and fertilize the egg. J Biol Chem 2002; 277:30310–30314.
37. Lepperdinger G, Strobl B, Kreil G. HYAL2, a human gene expressed in many cells, encodes a lysosomal hyaluronidase with a novel type of specificity. J Biol Chem 1998; 273:22466–22470.
38. Müllegger J, Reitinger S, Lepperdinger G. Hapten-labeled hyaluronan, a substrate to monitor hyaluronidase activity by enhanced chemiluminescence-assisted detection on filter blots. Anal Biochem 2001; 293:291–293.
39. Müllegger J, Lepperdinger G. Degradation of hyaluronan by a Hyal2-type hyaluronidase affects pattern formation of vitelline vessels during embryogenesis of *Xenopus laevis*. Mech Dev 2002; 111:25–35.
40. Rai SK, Duh FM, Vigdorovich V, Danilkovitch-Miagkova A, Lerman MI, Miller AD. Candidate tumor suppressor HYAL2 is a glycosylphosphatidylinositol (GPI)-anchored cell-surface receptor for jaagsiekte sheep retrovirus, the envelope protein of which mediates oncogenic transformation. Proc Natl Acad Sci USA 2001; 98: 4443–4448.

41. Reitinger S, Mullegger J, Lepperdinger G. *Xenopus* kidney hyaluronidase-1 (XKH1), a novel type of membrane-bound hyaluronidase solely degrades hyaluronan at neutral pH. FEBS Lett 2001; 505:213–216.
42. Bonner WM Jr, Cantey EY. Colorimetric method for determination of serum hyaluronidase activity. Clin Chim Acta 1966; 13:746–752.
43. Fiszer-Szafarz B, Litynska A, Zou L. Human hyaluronidases: electrophoretic multiple forms in somatic tissues and body fluids. Evidence for conserved hyaluronidase potential N-glycosylation sites in different mammalian species. J Biochem Biophys Methods 2000; 45:103–116.
44. Frost GI, Csoka TB, Wong T, Stern R. Purification, cloning, and expression of human plasma hyaluronidase. Biochem Biophys Res Commun 1997; 236:10–15.
45. Csoka TB, Frost GI, Wong T, Stern R. Purification and microsequencing of hyaluronidase isozymes from human urine. FEBS Lett 1997; 417:307–310.
46. Lokeshwar VB, Young MJ, Goudarzi G, Iida N, Yudin AI, Cherr GN, Selzer MG. Identification of bladder tumor-derived hyaluronidase: its similarity to HYAL1. Cancer Res 1999; 59:4464–4470.
47. Podyma KA, Yamagata S, Sakata K, Yamagata T. Difference of hyaluronidase produced by human tumor cell lines with hyaluronidase present in human serum as revealed by zymography. Biochem Biophys Res Commun 1997; 241:446–452.
48. Afify AM, Stern M, Guntenhoner M, Stern R. Purification and characterization of human serum hyaluronidase. Arch Biochem Biophys 1993; 305:434–441.
49. Natowicz MR, Short MP, Wang Y, Dickersin GR, Gebhardt MC, Rosenthal DI, Sims KB, Rosenberg AE. Clinical and biochemical manifestations of hyaluronidase deficiency. N Engl J Med 1996; 335:1029–1033.
50. Triggs-Raine B, Salo TJ, Zhang H, Wicklow BA, Natowicz MR. Mutations in HYAL1, a member of a tandemly distributed multigene family encoding disparate hyaluronidase activities, cause a newly described lysosomal disorder, mucopolysaccharidosis IX. Proc Natl Acad Sci USA 1999; 96:6296–6300.
51. Stern R. Devising a pathway for hyaluronan catabolism: are we there yet? Glycobiology 2003; 13:105R–115R.
52. McGuire PG, Castellot JJ Jr, Orkin RW. Size-dependent hyaluronate degradation by cultured cells. J Cell Physiol 1987; 133:267–276.
53. Phillips GO. Degradation of hyaluronansystems by free radicals. In: Laurent TC, ed. The Chemistry, Biology and Medical Applications of Hyaluronan and Its Derivatives. London: Portland Press, 1998:93–112.
54. Al-Assaf S, Meadows J, Phillips GO, Williams PA, Parsons BJ. The effect of hydroxyl radicals on the rheological performance of hylan and hyaluronan. Int J Biol Macromol 2000; 27:337–348.
55. Al-Assaf S, Navaratnam S, Parsons BJ, Phillips GO. Chain scission of hyaluronan by peroxynitrite. Arch Biochem Biophys 2003; 411:73–82.
56. Flugge LA, Miller-Deist LA, Petillo PA. Towards a molecular understanding of arthritis. Chem Biol 1999; 6:R157–R166.
57. Coleman PJ, Scott D, Mason RM, Levick JR. Role of hyaluronan chain length in buffering interstitial flow across synovium in rabbits. J Physiol (Lond) 2000; 526: 425–434.

Chemistry and Biology of Hyaluronan
H.G. Garg and C.A. Hales (editors)

Chapter 5

The Hyaluronan Receptor: CD44

WARREN KNUDSON and RICHARD S. PETERSON

Department of Biochemistry, Rush Medical College, Rush University Medical Center, 1653 W. Congress Parkway, Chicago, IL 60612, USA

I. Introduction

One reason hyaluronan attracts considerable attention is its apparent ability to influence cell behavior. This capacity is due in part to the unique property of hyaluronan to establish a highly hydrated extracellular matrix and second, the capacity of extracellular hyaluronan to directly interact with cells. In fact, hyaluronan was the first extracellular matrix macromolecule documented to bind to a protein present on the surface of cells (1). For example, early studies documented the role of hyaluronan in cell–cell aggregation—a cross-bridging phenomenon that could be inhibited by the presence of small hyaluronan oligosaccharides (2–4). Somewhat later studies documented that the binding of ^{3}H-labeled hyaluronan to the surface of cells was a saturable event, displayed competition with excess unlabeled ligand and, was sensitive to the pretreatment of cells with trypsin—all common criteria used to define the cell–hyaluronan interaction as likely a true receptor–ligand interaction (5–7). An example of such saturable binding of ^{3}H-labeled hyaluronan to human bladder carcinoma cells is shown in Fig. 1 (8). The receptor in question was identified as a protein already under investigation as a lymphocyte marker and, which had been given such names as Hermes-1, Pgp-1 (phagocytic glycoprotein-1), In[Lu]-related p80, HUTCH-1, gp90, gp85, H-CAM, brain granulocyte T lymphocyte antigen and ECMRIII (9–11). At the Third International Workshop and Conference on

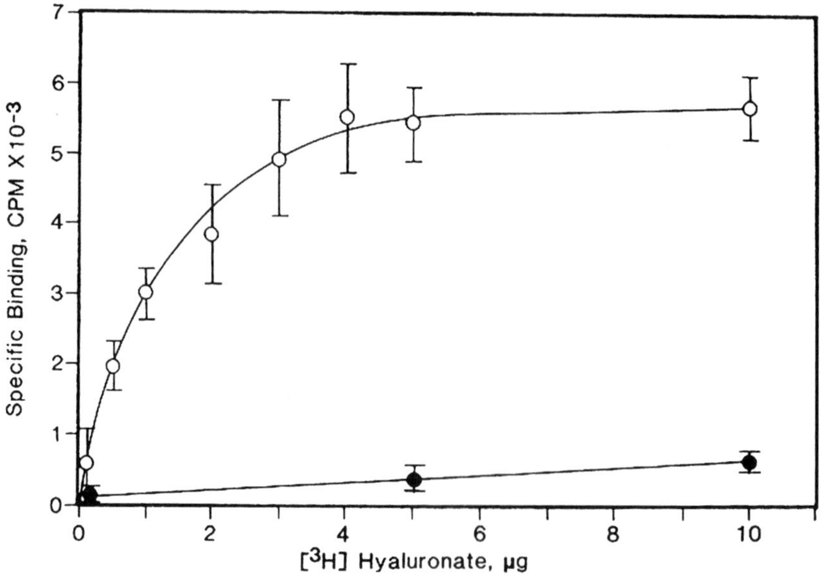

Figure 1 Effect of increasing concentrations of ^{3}H-hyaluronan on the specific binding of hyaluronan to HCV-29T and normal fibroblasts. Various concentrations of ^{3}H-hyaluronan were incubated with 1×10^{6} HCV-29T (–O–) cells or normal human skin fibroblasts (–●–) for 60 min at room temperature with gentle shaking in the presence or absence of 500 µg/mL unlabeled hyaluronan (to determine background binding). Following incubation the cells were pelleted, washed, resuspended in 0.5% SDS and analyzed by scintillation counting. Each point represents specific binding determined by the subtraction of background binding. Error bars represent the SEM of five experiments, each assayed in duplicate (Reproduced with permission from Ref. 8).

Human Leucocyte Differentiation Antigens held at Oxford in 1986, the protein was given the formal name of CD44, becoming a member of the cluster of common leukocyte antigens (Clusters of Differentiation (CD) (12)). Since that time research concerning CD44 has expanded rapidly. CD44 serves as an example of how evolution allows utilization of one protein for a wide variety of tasks with small deviations to its structure or context of expression. CD44 is at times a hyaluronan receptor—at other times an integral membrane proteoglycan. CD44 serves as a docking protein for matrix metalloproteinases, an activator of c-Met and, can function by itself as a nuclear transcription factor. CD44 participates in signal transduction and rearrangements of the actin cytoskeleton. CD44 has been implicated in cell migration during morphogenesis, angiogenesis, and tumor invasion and metastasis; retention, organization and endocytosis of hyaluronan-rich extracellular matrix; mediating the adhesion and rolling of lymphocytes and; coordinating matrix signals that balance cell survival and cell death pathways. It was thus somewhat of a shock to the hyaluronan research community that the CD44 null mouse exhibited no apparent changes in phenotype (13), however, more recent studies are challenging this conclusion (14–17). Nonetheless, a comprehensive discussion of all these topics related to CD44 is beyond the scope of a single book chapter and readers are directed to three recent excellent reviews that cover many aspects of CD44 biology in depth (18–21). This chapter will outline the basic features of CD44, primarily concerning the role of CD44 as a hyaluronan receptor in addition to functions of CD44 that are independent of hyaluronan interactions.

II. Structure of CD44 Primary Transcripts

The CD44 gene is typically thought of as consisting of 20 exons (Fig. 2). Although a single gene is located on the short arm of human chromosome 11, multiple mRNA transcripts arise from the alternative splicing of 12 of the 20

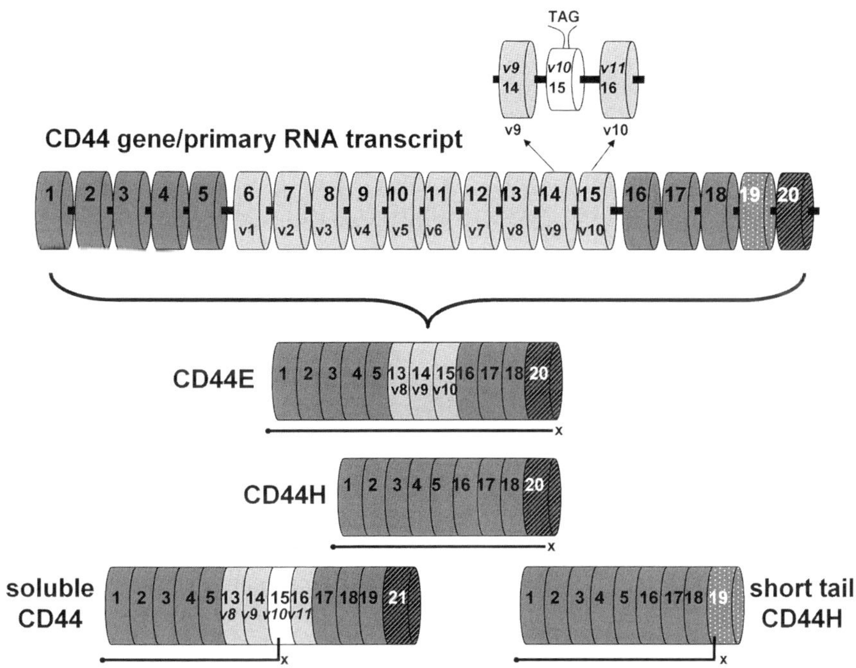

Figure 2 The intron/exon structure of the CD44 gene and pre-mRNA. Depicted are the 20 exons (horizontal tube sections) and introns (intervening lines) of the CD44 gene that give rise to a CD44 pre-mRNA and alternatively spliced CD44 mRNA. Depicted is the exon numbering of the 20 exons as well as the associated numbering of the variant exons. The exon compositions of mRNAs of four representative CD44 isoforms are shown including CD44E, CD44H, soluble CD44 and short-tail CD44H. The line beneath each of these structures denotes the approximate start (●) and stop (x) site of the coding sequence. At the top of the figure is the revised numbering nomenclature (suggested in Ref. 27) associated with the generation of soluble CD44 isoforms.

exons (22,23). The most prevalent form of CD44 is a protein encoded by exons 1–5, 16–18 and 20 (Fig. 2) and is sometimes denoted as the standard form of CD44 or CD44s. Since CD44s is also the predominant isoform expressed by hematopoetic cells it is also designated by the more common name CD44H (Fig. 3). Exons 1–16 all code for extracellular domains of the protein; exon 18 encodes the highly conserved transmembrane domain; and exon 20 encodes the intracellular cytoplasmic domain (Fig. 2). The 1482 bp of open reading frame mRNA for human CD44s results in the translation of a polypeptide chain of ~37 kDa (22). Subsequent post-translational addition of N-linked and O-linked oligosaccharides gives rise to a protein of ~85–90 kDa (24).

Numerous isoforms of CD44 arise from alternative splicing of the remaining exons and result in CD44 proteins that range in size from ~75 to

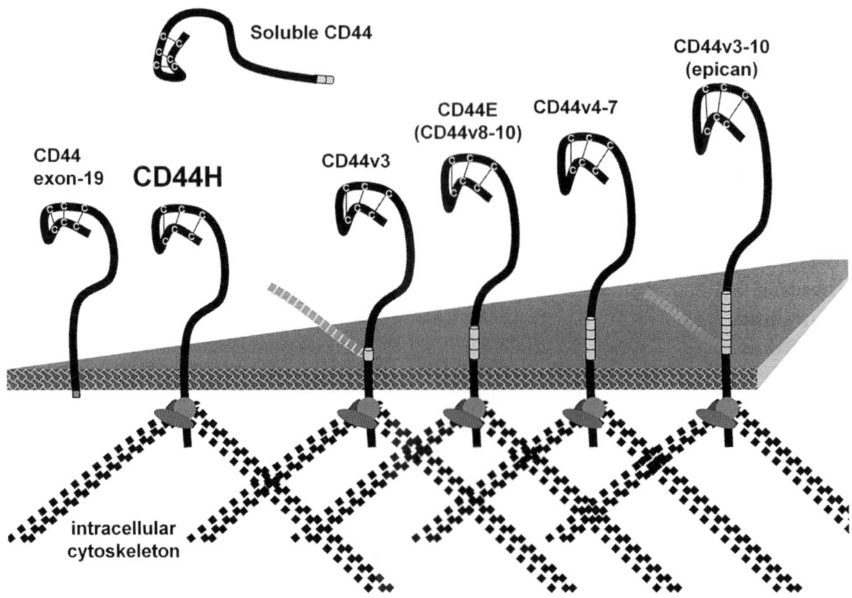

Figure 3 The protein structure of the several representative CD44 isoforms. The idealized protein structure of several CD44 isoforms is depicted. The gray triangle represents a cut-away of the plasma membrane. Each CD44 is composed of a distal extracellular link-protein-homology domain, a membrane-proximal stalk domain into which are inserted peptide segments due to alternative splicing of variant exons (light gray tube sections). If the CD44 expresses the v3 exon, the domain may become substituted with a heparan sulfate or chondroitin sulfate glycosaminoglycan chain. With the exception of soluble CD44, all other CD44 proteins exhibit a highly conserved transmembrane domain as well as a cytoplasmic tail domain. If exon 19 is utilized, the cytoplasmic tail domain is short (five amino acids). If exon 20 is used as in the more common isoforms of CD44, a longer 72 amino acid cytoplasmic tail domain is present that can participate in interactions with actin adapter proteins (ovals) and the actin cytoskeleton.

250 kDa (22,25,26). The variant isoforms fall into three groupings: [1] isoforms with extensions within the extracellular domain; [2] one isoform with a truncated cytoplasmic tail (tail-less CD44); and [3] a CD44 isoform truncated within the extracellular domain resulting in its secretion from the cell (soluble CD44) (27). The extension CD44 isoform group arises from alternative splicing of exons 6–15 in a wide assortment of combinations (28,29). These internal 10 exons (exons 6–15) are also denoted as 'variable' exons, v1–v10. Alternative splicing of exons 6–15 increases the length of the exposed receptor, possibly modulating some of its functions. For example, transfectants expressing CD44v6+7 gained the capacity to bind chondroitin sulfate as well as hyaluronan (30,31). Isoforms such as CD44v4-7 are characteristic of cells undergoing metastasis and as such, serve as important diagnostic markers (32–36). The nomenclature of these variant isoforms is not yet standardized. Thus in specific areas of study the variant isoforms have user-friendly names such as epican, descriptive names such as gp116 (CD44v10 found in endothelial cells), and variant number designation names, CD44vn, with the 'n' corresponding to the variant exons inserted, e.g., CD44v7-9. Some of the variant isoforms of CD44, such as CD44v3 and CD44v3-10 (epican) contain Ser/Gly sites substituted with chondroitin sulfate or heparan sulfate glycosaminoglycan chains, converting CD44 to a proteoglycan (Fig. 3) (37). Although other potential Ser/Gly sites for the addition of glycosaminoglycan chains are present within CD44H and CD44v10, only the site within exon 8 (variant exon 3) has been found to contain chains.

Another example of an extension isoform is a CD44 originally isolated from epithelial cells termed CD44E (38,39). This isoform arises from the inclusion due to alternative splicing of three additional exons—exons 13–15 into the CD44 mRNA. Thus, CD44E is also denoted as CD44v8-10. With the exception of variant exon 10, which is slightly larger, each variant exon is approximately 100 bp in length, coding for approximately 33 additional amino acids per exon (22). These additional amino acids are inserted between amino acids 202 and 203 of human CD44 giving rise to CD44 isoforms with variant extension lengths of the extracellular domain (Fig. 3). Although we have gained considerable insight into some of the physiological functions of the variant isoforms, especially the variant isoforms that provide for glycosaminoglycan attachment, the complete picture of the expression and function of such a diversity of structure remains an active area of investigation.

Alternative splicing also occurs for exons encoding the intracellular domain of CD44. Either exon 19 or 20 is differentially expressed due to alternative splicing and represents two variations of the intracellular 'tail' portion of the molecule (Fig. 2) (22,40,41). Transcripts containing exon 20 are prominent in most cells encoding a ~7 kDa, 72 amino acid, cytoplasmic domain. Interestingly, the 5′ end of exon 19 differs from exon 20 by only one base pair. However, this difference results in a chain termination stop codon and the expression of a truncated cytoplasmic domain of only six amino acids (Fig. 3). Thus the alternatively spliced message containing exon 19 in lieu of exon 20 generates a 'short-tailed' or essentially 'tail-less' form of the CD44 protein. The lack of intracellular signaling motifs and protein domains necessary for interaction with cytoskeletal components has fueled intense speculation over the role of this CD44 isoform. Although the expression of exon 19 is uncommon in most cells, we have identified CD44 exon 19 mRNA transcripts in human chondrocytes—the same cells that also express transcripts for CD44s (CD44 exon 20) (26). Expression of exon 19 containing transcripts varies from human donor to donor cartilage samples ranging from 5 to 40% of the total CD44 mRNA. Evidence that the exon 19 containing transcripts are translated into protein is indirect. Antisense phosphorothioate oligonucleotides directed against unique sequence within the 3′ untranslated region of exon 19 resulted in the selective inhibition of all exon 19 containing mRNA and a 'sharpening' of the typically broad bands observed for CD44 on western blots (i.e., lower molecular mass region of the bands was eliminated) (26). Interestingly, transient transfection and expression of human recombinant CD44HΔ67 (equivalent to expected protein translated from exon 19 containing mRNA) acts as a dominant negative receptor, interfering with the function of the endogenous (exon 20 containing) wild type CD44 isoforms. Thus, one might speculate that exon 19 containing transcripts, if expressed in sufficient amounts as protein, might serve as a natural regulator of CD44 function.

Yu and Toole have described another group of alternatively spliced CD44 isofoms termed 'soluble' CD44. This new series of variant CD44 mRNAs were observed in RNA isolated from murine embryonic muscle and cartilage tissues. In addition to CD44v8, CD44v8+9 and CD44v8-10 isoforms, a new variant of CD44v8-10 was observed containing an additional 93 bp of sequence between the v9 and v10 exons (27). The new exon arose from the usage of intron sequence between variant exons 9 and 10 (Fig. 2). The authors suggested that what had been thought of as variant exon 9 (i.e., exon 14) represents only the 5′ half of a larger potential exon. In subsequent experiments, three additional minor transcripts containing intron sequence between variant exons 9 and 10 were documented. Yu and Toole (27) suggested renaming the new exon as variant exon 10 (or exon 15) resulting in a register shift of exon numbering of the CD44 gene. When this nomenclature is utilized, v10 would become v11, and the CD44 would include a total of 21 exons instead of 20 (Fig. 2, top). A unique feature of inclusion of these minor exons between the original v9–v10 is that all four contain stop codons. Thus, translation of any of these mRNA transcripts results in a truncated CD44—truncated within the membrane-proximal extracellular domain. Upon cloning and expression of these mRNAs, the authors found that the proteins were synthesized and processed correctly but, having no transmembrane or cytoplasmic domains, the proteins were secreted as 'soluble CD44'. These constructs were put to practical use in subsequent studies. Overexpression of these C-terminal truncation isoforms of CD44 in various tumor cells resulted in the secretion of soluble CD44 that in turn served as a decoy receptor binding to all available sites on endogenous hyaluronan, interfering with binding and signaling by the wild type CD44 (42). Whether such isoforms have a similar control function when expressed naturally in cells remains to be determined.

Recently, synovial cells from the joints of rheumatoid arthritis (RA) patients have been shown to express an mRNA transcript, encoding an additional trinucleotide between variant exons 4 and 5 of a CD44v3-10 heparan sulfate proteoglycan isoform (43). The authors suggested the name of CD44vRA for the RA-specific CD44v3-10 variant. Like the soluble CD44, transcripts for this sequence arise from the alternative use of intronic sequence between v4 and v5 exons. These RA-specific CD44v3-10 variants were detected in 23 of 30 RA patients analyzed. This splicing event resulted in the inclusion of an additional alanine residue without interference in the reading frame. Upon further examination, the investigators determined that cells transfected with recombinant CD44vRA (or primary RA synovial cells) bound more soluble FGF receptor-1 ectodomain subsequent to FGF-2 binding to the heparan sulfate chains than control cells expressing CD44v3-10.

III. Regulation of CD44 Expression

Increases in CD44 protein expression due to transcriptional activation are often observed following the exposure of cells to proinflammatory cytokines such as IL-1α, IL-1β, TNFα and growth factors such as TGF-β, BMP-7 (aka osteogenic protein-1) and EGF (44–49). Although many classes of cellular mediators have been shown to enhance CD44 expression, we have found that the addition of exogenous high molecular mass hyaluronan or small hyaluronan oligosaccharides does not enhance or diminish the expression of CD44 (50). Information concerning the molecular regulation of the CD44 gene continues to expand. The upstream CD44 promoter/enhancer region does not contain typical TATA or CCAAT boxes (51). Nonetheless, the presence of a 150 bp DNA sequence immediately upstream of the transcriptional start site is sufficient to confer basal transcriptional activity (51). A variety of *cis* elements within the CD44 promoter have been characterized with respect to IL-1 induction of CD44 including an Egr-1 motif at −301 bp in the human gene (45) and a conserved AP-1 site at −110 to −104 bp in the rat gene (46). In the latter, the AP-1 proteins Fos and c-Jun are involved and the activation event is potentiated by the expression of HMG-I(Y)—a high mobility group architectural transcription factor protein, the expression of which is also upregulated by IL-1. Zhang et al. (44) have described an EGF regulatory element at −583 to −604 bp of the mouse CD44 promoter that is critical to the upregulation of CD44 by EGF.

In some cancers, such as prostatic and colorectal carcinomas, tumor progression (increase in tumor grade and stage) is accompanied by a complete loss of CD44H and CD44v6 expression. Several groups have determined that this loss of CD44 expression is due to hypermethylation of CpG islands within the CD44 promoter region (52–54). Treatment of CD44-negative PC346C prostate cancer cells with the demethylating agent, 5-azacytidine, resulted in the re-establishment of CD44 expression (52). More recent studies have documented similar inhibition of CD44 expression and coordinate hypermethylation of the

CD44 promoter in human clinical cancer samples. For example, in the study by Stallmach et al. (54), the frequency of CD44 promoter methylation correlated with advancement of colorectal cancer stage. Hypermethylation is not unique to CD44. E-cadherin is also silenced in human breast and prostate carcinomas by DNA hypermethylation (55). Nonetheless, the silencing of CD44 expression is likely more complex than hypermethylation alone. For example, Hyman found that the CD44-negative T-cell lymphoma cell line AKR1 did exhibit a more heavily methylated upstream CD44 promoter region as compared to CD44-positive cells. However, re-activation of CD44 expression in AKR1 cells by transient transfection of c-jun, or treatment with sodium butyrate, occurred without significant demethylation of the CD44 promoter (56).

IV. Protein Structure of CD44

CD44 is a type I single-pass transmembrane glycoprotein consisting of four functional domains. The distal extracellular domain, representing the amino-terminal portion of the protein, is the region primarily responsible for the binding of hyaluronan (Fig. 3). It consists of ~100 amino acids brought together by three critical cysteine disulfide cross-links. This region bears strong homology to the hyaluronan binding domain of cartilage link protein and thus is often termed the 'link module'. The next domain is the membrane-proximal, extracellular domain or 'stalk' region of CD44. It is within this region that amino acid extensions due to alternative splicing are incorporated. In other words, extensions within this region give rise to many of the variant isoforms of CD44. For example, when v3-containing CD44 mRNAs are expressed, such as CD44v3 or CD44v3-10, the additional amino acids within the stalk region provide new serine/glycine attachment sites for glycosaminoglycans (Fig. 3) (37,57). Depending on the cell type, these attachment sites are often substituted with either chondroitin sulfate or heparan sulfate side chains and thus CD44 is often considered a member of the proteoglycan family.

The CD44 transmembrane domain consists of 22 amino acids and is fairly typical of most single-pass membrane glycoproteins (58). This domain is highly conserved among species as, e.g., seen in the homology between chick and human CD44 wherein there is 90% identity between the transmembrane domains as compared with a 46% homology for the remaining coding region (59). Site-directed mutagenesis studies as well as work with detergent-solubilization of CD44 suggest that associated lipids or accessory membrane proteins interacting with this domain may modulate hyaluronan binding as well as CD44 interaction with the cytoskeleton (60,61). One suggestion is that the clustering of CD44 is, in part, regulated by post-translational modifications within the transmembrane domain (62). For example, evidence is accumulating that CD44 exhibits the capacity to partition into cholesterol/sphingolipid-rich lipid rafts and, in some cases, is responsible for cytoskeletal reorganization and recruiting of other proteins, kinases, etc., into the lipid raft microdomain (63–66).

In most CD44 isoforms (except exon 19 containing transcripts), the carboxy-terminal end of the protein consists of an ~72 amino acid cytoplasmic domain or 'tail' domain (Fig. 3) (22). As shown in Fig. 4, the cytoplasmic domain exhibits several protein motifs that indicate a capacity for interaction with cytoskeletal linker proteins such as ERM proteins (ezrin, radixin, moesin and merlin) and a second site for interaction with ankyrin (67–69). There are also at least three potential sites for serine phosphorylation (70) and a putative SH3 domain. All or parts of these sites are likely involved in outside–in signaling mediated via hyaluronan–CD44 interactions, CD44 clustering and CD44-mediated internalization of hyaluronan. Notably absent from this domain is an AP-2 adaptor binding site (i.e., YQRL or LL) indicative of interactions with clathrin (71–73) or caveolin-binding domains such as ΦXΦXXXXΦ, ΦXXXΦXXΦ, ΦXΦXXXXΦXXΦ where X is any amino acid and Φ, is an aromatic residue (74,75). When CD44 exon 19 containing transcripts are

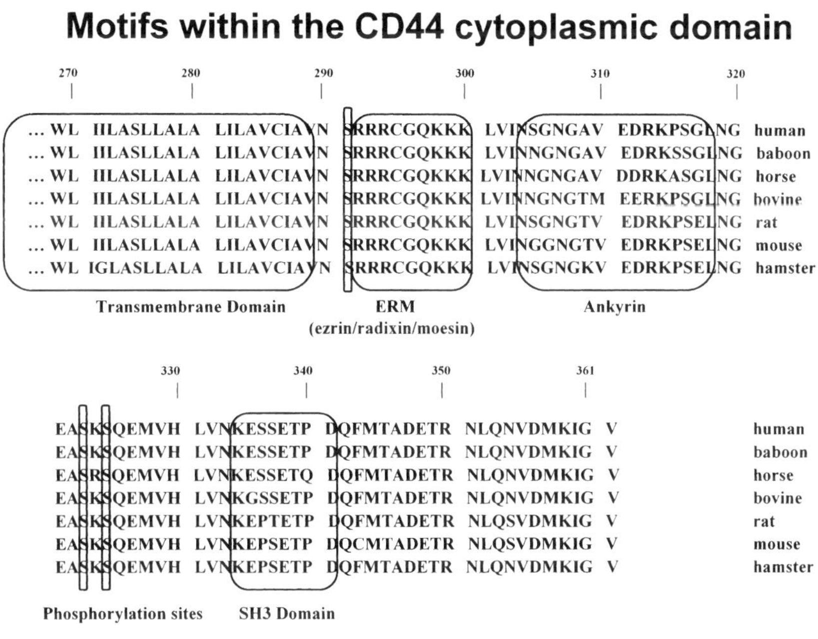

Figure 4 The amino acids and binding motifs present within the transmembrane and cytoplasmic domain of CD44. Depicted is an alignment of sequences denoted for the transmembrane and cytoplasmic domains of CD44 from seven mammalian species. The putative ERM (ezrin, radixin, moesin) and ankyrin-binding domains are enclosed in rectangles as well as a potential SH3 domain. Sites for serine phosphorylation are shown by vertical enclosing rectangles. Of note, there are no Leu–Leu repeats in the cytoplasmic domain and only one aromatic amino acid, Phe-343—critical features of consensus sites for interaction with clatherin or caveolin. Exon 19 containing CD44 isoforms (and CD44Δ67) terminate at Arg-294.

expressed, the cytoplasmic tail is terminated following Arg-294 and thus exhibits no capacity for interaction with cytoskeletal components but retains the potential serine phosphorylation site, Ser-291 (26,76).

A. Post-translational Modifications

As with other transmembrane glycoprotein receptors, CD44 undergoes extensive post-translational modifications including extensive glycosylation, phosphorylation, protein palmitoylation and, in some CD44 isoforms, addition of heparan sulfate or chondroitin sulfate glycosaminoglycan chains. All these modifications are being actively explored as potential modulators of hyaluronan binding, CD44 signaling, CD44 clustering and other CD44 functions. The open reading frame of CD44H mRNA codes for a protein of ~37 kDa (22). Thus, the 85–90 kDa CD44H isolated from the cell surface represents a protein that has undergone extensive glycosylation, in particular, the addition of N-linked carbohydrate. Some investigators maintain that CD44 glycosylation is a necessary prerequisite for hyaluronan binding (77,78), while others find that recombinant non-glycosylated CD44 still binds hyaluronan (79). What may be of more importance is the structure and composition of the carbohydrate residues attached to CD44. For example, recent data suggest that enzymatic removal of sialic acid from N-linked oligosaccharide structures of CD44 activates the binding activity of hyaluronan to CD44. Treatment of Chinese hamster ovary cells with exogenous neuraminidase activated the hyaluronan-binding capacity of CD44 (80). Neuraminidase treatment also enhanced the hyaluronan binding capacity of CD44–immunoglobulin fusion proteins. Equally effective were reagents that removed all complex sugars from CD44 including tunicamycin during biosynthesis or endoglycosidase F together with *N*-glycanase F treatment of cell surface CD44. Subsequently, investigators have determined that agents known to induce/activate CD44 binding, such as LPS, TNF-α or phorbol 12-myristate 13-acetate, function by enhancing the expression of an endogenous cell surface bound neuraminidase (81,82). Additionally, the neuraminidase inhibitor, 2-deoxy-2,3-dehydro-*N*-acetyl-neuraminic acid diminishes the LPS-induced activation of monocyte CD44 (81). The LPS-initiated cascade has been subsequently documented as a TNF-α-mediated, p38 MAPK-induced activation of endogenous neuraminidase (83). Interestingly, expanded neuroblastoma cell lines that continue to express CD44 but lose the ability to bind to hyaluronan are characterized by increased sialylation (84). The inference from all these studies is that the presence of sialic acid residues in some way interferes with the binding of hyaluronan to CD44. These results may explain in part, the observation that in some cells, particularly nucleated blood cells, CD44 is expressed but requires activation before it gains the capacity to bind hyaluronan (inducible CD44) whereas in other cells CD44 is constitutively active or totally inactive. These three different states of CD44 binding activity could be explained by cellular control of endogenous sialytransferase and/or neuraminidase activities (85). Nonetheless, most investigators concede that the presence or absence of sialic acid residues on CD44 is but one component participating in the overall regulation of hyaluronan binding activity. Most conclude that other events such as receptor clustering are also likely required to obtain optimal CD44 binding of hyaluronan (81,82). In turn, CD44 clustering is dependent on other post-translational events such as associations with the actin cytoskeleton, protein palmitoylation (86) and even cross-linking of adjacent receptors. Association of CD44 with the cytoskeleton is controlled in part by phosphorylation of the intercellular domain of CD44. Phosphorylation/dephosphorylation of serine residues of the cytoplasmic tail of CD44 represents dynamic modifications that suggest a role for 'outside–in' signal transduction (70,87). Phosphorylation has also been shown to influence linking and unlinking the tethering of CD44 to the actin cytoskeleton, and one might speculate the transit of CD44 into and out of lipid rafts. Nonetheless, it still remains unclear how these post-translational modification events contribute to the activation of hyaluronan binding capacity and/or functional usage of CD44 as will be covered in later sections.

B. Molecular Interactions Responsible for Hyaluronan Binding to CD44

As discussed earlier, CD44 exhibits a high degree of homology to the B-loop structure of link protein and aggrecan. However, unlike link protein and aggrecan, CD44 exhibits only one half of the tandem repeat structure. CD44 is thus a member of a family of hyaluronan-binding proteins termed hyaladherins and a subgroup of the hyaladherins that express a 'link protein' module the module responsible for binding to hyaluronan (88). Other members of the 'link protein-like' hyaladherins include link protein, aggrecan, versican, neurocan, brevican, TSG-6 (tumor necrosis factor-α-stimulated glycoprotein-6) and LYVE-1 (88). Peach et al. (89) made the first careful molecular analysis of hyaluronan–CD44 binding and identified two clustered regions of basic amino acids, one within the distal portion of the region of homology to the link protein and another closer to the proximal domain. One residue in particular, Arg-41, was determined to be critical for binding. However, their data also demonstrated that the other, more proximal cluster of basic amino acids was also necessary. Mutations in either domain significantly reduced CD44 binding to hyaluronan. CD44–immunoglobulin fusion proteins, truncated to contain only the distal link protein-homologous domain (i.e., no proximal domain), displayed a low capacity to bind to hyaluronan-coated plates. In addition, point mutation of any of the four clustered basic arginine residues in the proximal domain also reduced hyaluronan-binding activity. The investigators suggested that the two domain regions may work cooperatively, in some fashion, to bind hyaluronan. Their studies also suggested that the anti-CD44 monoclonal antibody IM7.8.1 recognized an epitope near the proximal domain site necessary for hyaluronan binding. This may explain why, in some experimental models, IM7.8.1 displays no capacity as a blocking antibody whereas in others, blocking of hyaluronan binding is observed (90). Subsequently, a three-dimensional structural model of CD44 has been generated based on homology of CD44 to TSG-6 (91,92). TSG-6, like CD44, is another member of the link protein homologous superfamily. However, in the case of TSG-6, the three-dimensional solution structure of the link protein domain of the protein has been determined and was found to be similar to a calcium-dependent lectin fold such as that found in a rat mannose-binding protein. Using the site of this putative fold domain for direction and Arg-41 as a starting point, Bajorath et al. (91) generated and tested additional conservative and non-conservative amino acid point mutations. These studies revealed a large surface region of the link protein-homologous domain of CD44 likely to be responsible for hyaluronan binding. A particular cluster of basic amino acids namely, Arg-41, Arg-78, as well as the hydrophobic residues Tyr-42 and Tyr-79 were found to be essential for hyaluronan binding to CD44. Mutations in this region inhibit the capacity of CD44 to bind to hyaluronan, and inhibit the binding of anti-CD44 monoclonal antibodies that block hyaluronan binding within the distal link module. In the predicted model structure, all four of these amino acids run along a ridge on the surface of the CD44 protein. Additional amino acids, namely, Asn-100, Asn-101, Lys-38, Lys-68 and Tyr-105 also contribute to binding, and together make up a coherent patch on the surface of the CD44 (92).

C. Interaction of CD44 and the Actin Cytoskeleton and Lipid Rafts

When detergents such as Triton X-100 or NP-40 are used to extract CD44 from the plasma membrane a significant pool of CD44 remains, still associated with the residual membrane fraction (93,94). In order to solubilize the remaining CD44 on mouse 3T3 cells, Underhill et al. found that cytoskeleton-disrupting reagents such as phalloidin, cytochalasin B or DNase, in additional to detergent, were necessary (93). The NP-40 resistant pool found in resident macrophages (42% of the total CD44) was solubilized using 1% zwitterionic detergent, Empigen-BB, or inclusion of 1 mg/mL DNase-I. An example of such pools of CD44 can also be seen in Fig. 5 (lanes 1 and 4) depicting the differential extraction of transiently transfected and overexpressed recombinant CD44H in COS-7 cells. This NP-40 resistant membrane-residual pool (e.g., Fig. 5, lane 4) has been viewed as a pool of CD44 in tight association with the underlying actin cytoskeleton (93–95). For example, in bovine articular chondrocytes, approximately 50% of the total CD44 is in each pool (95,96). However, upon pretreatment of the cells with hyaluronidase, a substantially higher proportion of CD44 is extractable with NP-40 alone. In addition, pretreatment of the chondrocytes with dihydrocytochalasin B or latrunculin A, like the earlier studies in 3T3 cells, increased the percentage of CD44 extractable with mild NP-40 detergent. These data suggest that the CD44 residing in the plasma membrane is present in two pools, one tightly tethered to the actin cytoskeleton and another pool presumably free of such anchoring associations and easily solubilized. Furthermore, the proportion of CD44 in each pool can be altered (an intracellular rearrangement) by events such as a change in the extracellular occupancy of the receptor. Observations such as these have spurred investigators

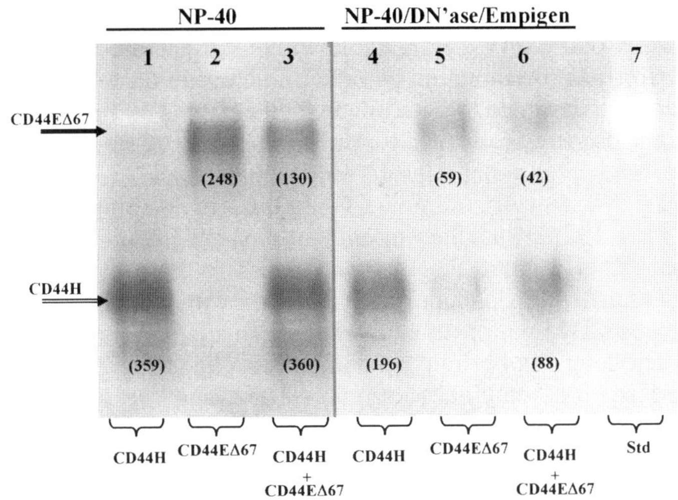

Figure 5 Differential extraction of co-transfected CD44H and CD44EΔ67 isoforms. Seventy-two hours following transfection of COS-7 cells with pCD44H, CD44EΔ67 or co-transfection with pCD44H and CD44EΔ67, the cells were extracted with Nonidet P-40 (NP-40) followed by a subsequent solubilization of the residual pellet with NP-40 lysis buffer containing DNase and Empigen. Equivalent aliquots of protein were electrophoresed, electroblot transferred to nitrocellulose and the various CD44 isoforms detected by western blotting. Arrows depict migration of the CD44EΔ67 and CD44H isoforms. Lanes 1 and 2 represent NP-40 lysates from COS-7 cells transfected with 2.0 μg/mL of pCD44H and CD44EΔ67, respectively. Lane 3 represents NP-40 lysates from cells co-transfected with 2.0 μg/mL each of pCD44H and CD44EΔ67. Lanes 4 and 5 represent NP-40 plus DNase and Empigen lysates from COS-7 cells transfected with 2.0 μg/mL of pCD44H and CD44EΔ67, respectively. Lane 6 represents NP-40, DNase and Empigen lysates from cells co-transfected with 2.0 μg/mL each of pCD44H and CD44EΔ67. Values shown in parentheses represent the digital intensity values for each band obtained by image analysis. These results suggest that CD44 may exist in three pools within the plasma membrane.

to investigate whether the on and off binding of CD44 to the actin cytoskeleton was responsible for outside–in signaling or, the reverse, 'inside–out' signals whereby intracellular rearrangements could modulate extracellular hyaluronan binding. As discussed previously, two regions within the cytoplasmic domain of CD44 have been documented as sites for attachment of cytoskeleton adapter proteins. Using specifically constructed truncation mutants expressed in COS-7 cells, a region between Asp-304 and Leu-318 of the human cytoplasmic tail domain (Fig. 4) was identified as a binding site for ankyrin (97–99). A second, nine amino acid membrane-proximal binding motif for ezrin/radixin/moesin (ERM) cytoskeletal linker proteins (Arg-292 to Lys-301 in human CD44, Fig. 4) was also documented in the cytoplasmic tail domain of CD44 (68,100). When both domains are truncated such that the CD44 contains a cytoplasmic tail domain with only five amino acids (pCD44HΔ67, Fig. 5, lanes 2 and 5), greater than 80% of the total CD44 is extractable with NP-40 alone. This also implicates that CD44 exon 19 protein would be predominantly in the NP-40 soluble pool.

Experiments designed to determine whether sites within the cytoplasmic or transmembrane domain are critical for the ability of the extracellular link domain of CD44 to bind hyaluronan (inside–out signaling) have been controversial. First, soluble CD44 as well as soluble CD44–IgG1 fusion proteins readily bind to hyaluronan (101,102). Thus, the presence of the transmembrane or intracellular domains of CD44 is not an inherent necessity for hyaluronan binding. Nonetheless, transfection of mouse AKR1 cells with a pCD44HΔ66 construct (i.e., a truncated CD44 with extracellular and transmembrane domains as well as a six amino acid long cytoplasmic tail domain) resulted in severely reduced CD44 binding capacity for soluble hyaluronan (103), and we have found similar reductions in COS-7 cells transfected with pCD44HΔ67 (76). However, in subsequent work, Lesley et al. (104) demonstrated that the CD44HΔ66 construct could be induced to bind hyaluronan by pretreatment of transfected XJ(3)/CD44 cells with the anti-mouse CD44 monoclonal antibody IRAWB14. This suggested that the cytoplasmic tail is not required if CD44 can be otherwise aggregated. The anti-human CD44 monoclonal antibody F10-44-2, shown to enhance hyaluronan binding to CD44 in transfected Jurkat cells (105), was not effective in rescuing binding in COS-7 cells transfected with pCD44HΔ67 (76).

In some systems, serine phosphorylation appears to control the attachment of CD44 to the actin cytoskeleton and be critical for hyaluronan binding to CD44. Pure et al. (106) found that constitutively phosphorylated CD44 serine residues Ser-323 and Ser-325 within the cytoplasmic domain were critical for efficient CD44 binding of hyaluronan when mutated serine constructs were expressed in T lymphoma cells. However, Uff et al. (87), performing similar mutations of cytoplasmic tail serines and expression in AKR1 cells, found no changes in the capacity of CD44 to bind hyaluronan. Similar conflicting results have been reported concerning the ankyrin-binding motif within the cytoplasmic tail as well as clusters of basic residues that constitute the ezrin-binding motif. Some studies suggested that these sites were critical for efficient HA binding (98,107), only to be confronted by results demonstrating the lack of importance of these sites (41,68,87). As discussed earlier, Liu et al. (61,62) have demonstrated that sites within the transmembrane domain of CD44 appear critical for efficient HA binding namely, Cys-286. The interaction of Cys-286 residues was suggested to participate in the dimerization of CD44 molecules. Using a domain swapping approach, a chimera construct consisting of CD44HΔ66 containing the transmembrane of CD3ξ (a transmembrane domain that readily forms disulfide cross-bridges within the membrane) also promoted hyaluronan binding (41). Again, this suggests that dimerization of CD44, mediated by interactions within the transmembrane domain, is more critical to hyaluronan binding than interaction with the underlying cytoskeleton. However, the CD44HΔ66-CD3ξ transmembrane chimera could not rescue hyaluronan binding in 'inducible' cells suggesting that additional signals are needed to promote inducible binding (104).

In the same study, Lesley et al. also tested a CD44 chimera containing the extracellular domain of CD44 linked to the transmembrane and cytoplasmic domain of β5 integrin (and expressed in AKR1 cells). The chimera bound hyaluronan with the same capacity as wild-type CD44 (104). Their conclusion was that the exact sequence or motifs within the transmembrane and/or cytoplasmic domain are not critical for constitutive CD44-mediated hyaluronan binding. What is critical may be merely a compatible transmembrane domain together with a minimum length of cytoplasmic domain (i.e., >6 amino acids). In sum, it is difficult to make generalized conclusions concerning inside–out control over hyaluronan binding to CD44. In some cell systems, disruption of the actin cytoskeleton or the deletion of CD44 motifs responsible for linking CD44 to the cytoskeleton significantly reduces or blocks hyaluronan binding to the cell surface. In other systems, there is no disruption of binding. Why CD44 does not bind hyaluronan at all times, like soluble CD44, is unclear. Why there are two membrane pools of CD44, differential phosphorylation of CD44, ERM as well as ankyrin binding motifs, transmembrane palmitoylation, etc., if these are of no regulatory importance, is also unclear. One suggestion is that these regulatory events are more important for outside–in signaling as opposed to an inside–out control mechanism.

As discussed previously, evidence is accumulating that CD44, in part through its interaction with the actin cytoskeleton, partitions into and out of spingolipid/cholesterol-rich lipid rafts (62–65). Furthermore, some have suggested that the detergent resistant membrane pool of CD44 may reflect not only CD44 in association with the actin cytoskeleton, but also CD44 present in a lipid raft—a microdomain that is also detergent resistant under these conditions. As discussed earlier, COS-7 cells were transfected with pCD44H, pCD44EΔ67 or co-transfected with both constructs. Following expression at the cell surface the CD44 was extracted using a sequential detergent extraction protocol. Interestingly, as shown in Fig. 5 (lanes 2 and 5) approximately 20% of the CD44EΔ67 remains resistant to extraction even following treatment with NP-40, DNase and Empigen-BB combined. Given that this CD44 isoform has little physical capacity for interaction with the cytoskeleton, perhaps this 20% pool represents CD44 present within a lipid raft environment. When CD44HΔ67 is co-transfected with an equivalent proportion of pCD44H (wild type CD44), significantly less pCD44H (15% as compared to 35% for pCD44H alone) is now resistant to extraction with NP-40 followed by NP-40 plus DNase and Empigen-BB. This may imply that the two isoforms are competing for lipid raft sites within the membrane. Future studies using cholesterol depletion experiments should define this potentially third pool of CD44 present within the plasma membrane.

V. Cellular Functions of CD44

The wide range of cellular functions purportedly mediated by CD44 can be separated into two general categories. The first are functions related to CD44 as a hyaluronan receptor and second, functions that are seemingly independent of receptor–ligand interactions. However, these distinctions are seldom truly independent. The most well-characterized functions relating to CD44 as a hyaluronan receptor include mediating: [1] the locomotion or adherence of a cell to a hyaluronan-rich extracellular matrix; [2] the retention of a hyaluronan-rich matrix to the cell surface; [3] the endocytosis of hyaluronan and associated bound molecules; and [4] signal transduction due to changes in hyaluronan occupancy including the coordination of hyaluronan-mediated events associated with apoptosis. Functions that may or may not be dependent on ligand binding include: [1] functions as a membrane-intercalated small proteoglycan including the retention, activation or presentation of basic growth factors or cytokines and [2] functions as a docking protein (e.g., for matrix metalloproteinases) or as a co-receptor/co-activator protein. The latter roles make use of the capacity of CD44 to exhibit on/off interactions with the actin cytoskeleton and undergo lipid raft localization and thus acting as a useful transmembrane adaptor protein. The participation of CD44 during tumor invasion is likely similar to events that occur during angiogenesis, lymphocyte rolling and neural crest cell migration.

A. CD44 Mediates the Retention of Hyaluronan-Enriched Pericellular Matrices

In the presence of low concentrations of hyaluronan, cells that express hyaluronan receptors such as CD44 undergo cell–cell aggregation (4,108–110). However, when the hyaluronan levels reach the point of saturation of available CD44 receptor sites, the cells disaggregate. The now independent cells remain autonomous within their own cell-associated or 'pericellular' matrix. These cell-associated pericellular matrices, often extending to a distance equal to the cell diameter, can be visualized around living cells *in vitro* by the use of a particle exclusion assay (111–114). The particle exclusion assay, first developed by Clarris and Fraser (115), consists of the addition of a suspension of small particles onto cells in monolayer culture. Upon settling the particles are excluded from an otherwise transparent region surrounding the cells. The particles outline a gel-like halo structure or 'cell coat' that is representative of the pericellular matrix. This assay represents the most useful method of visualizing the full extent of a pericellular matrix because conventional staining of the matrix by histochemical or immunohistochemical techniques often leads to significant collapse of this hydrated structure (90). In addition, the assay can be performed readily on either living or non-living cells. Early studies demonstrated that pericellular matrices could be removed by the treatment of cells with a dilute solution of Streptomyces hyaluronidase (6,113,115), demonstrating that hyaluronan was an essential component of the cell coat. On some cells such as chondrocytes, the pericellular matrix can also be removed by treatment of the cells with hyaluronan hexasaccharides—small oligosaccharides that compete with hyaluronan–receptor interactions (111–113). Thus, not only are pericellular matrices dependent on hyaluronan, but also receptor-mediated association of hyaluronan with the plasma membrane serves to retain or 'anchor' the cell coat. When the hyaluronidase or oligosaccharides are removed from cells such as chondrocytes, a native cell-associated matrix can be synthesized and reassembled within 4–12 h (90,113). However, reassembly can be prevented by incubation of the recovering cells in the presence of blocking anti-CD44 antibodies, dilute hyaluronidase or hyaluronan hexasaccharides (but not chondroitin sulfate hexasaccharides) (90). These results demonstrate that CD44 is the primary binding protein responsible for retention and organization of pericellular matrix coats surrounding many cell types. Inhibition of CD44 expression with antisense phosphorothioate oligonucleotides also leads to a reduction in matrix retention (116,117). Results such as these have led to the hypothesis that the hyaluronan/proteoglycan-rich cell-associated matrix is tethered to the plasma membrane of chondrocytes (and many other cell types) via the interaction of hyaluronan with CD44.

It should be noted that the retention of hyaluronan-rich pericellular matrices does not occur exclusively by interactions with receptors. It is clear that on a variety of cell types, including fibroblasts and mesothelioma cells (118,119), hyaluronan is also bound to the plasma membrane through a 'non-displaceable' interaction. This interaction most likely represents continued association of hyaluronan with the plasma membrane localized hyaluronan synthase (120,121). In fact, the capacity to establish synthase-bound pericellular coats was the output measure used to isolate the hyaluronan synthase genes, both in eukaryotic cells (122) and Streptococci (123). Prior to chondrogenesis, chick limb bud mesenchymal cells also display prominent pericellular matrices that are Streptomyces hyaluronidase sensitive but not displaceable with hyaluronan oligosaccharides (112). However after chondrogenesis, the newly differentiated chondrocytes again exhibit a prominent matrix, but one that is now displaceable by oligosaccharides. In the latter cells, greater than 70% of the cell-associated ^{35}S-labeled aggrecan is also released into the culture medium in the presence of hyaluronan oligosaccharides. In subsequent studies, the pericellular matrix of ^{3}H-acetate-labeled chick chondrocytes was removed using Streptomyces hyaluronidase, followed by re-growth in the absence of the enzyme. The kinetics of re-growth of cell surface hyaluronan showed that after 2 or 4 h of synthesis, only 2–10% of hyaluronan was bound to receptors, but by 6 h, 80% of the hyaluronan was displaced by hyaluronan oligosaccharides (124). Thus, newly synthesized hyaluronan is at first non-displaceable but with time, becomes transferred to receptor sites.

One of the interesting properties of CD44 is that the receptor maintains its capacity to bind hyaluronan following mild fixation with glutaraldehyde (5). In some protocols, fixed CD44+ cells have been used as an affinity-column resin to purify hyaluronan or hyaluronan oligosaccharides (8,125). In cells such as chondrocytes, once their endogenous pericellular matrix is removed using oligosaccharides or Streptomyces hyaluronidase treatment, the 'matrix-free' cells can then be fixed with glutaraldehyde, eliminating the contribution of endogenous synthesis of matrix macromolecules during binding studies (113). Chondrocytes treated in such a fashion bind saturable levels of radiolabeled hyaluronan, yet this binding does not result in the assembly of a cell-associated matrix (113). However, if purified aggrecan is added to the cells in addition to the hyaluronan, a pericellular matrix assembles within 2 h of incubation. Other hyaluronan-binding proteoglycans such as versican will also support hyaluronan-mediated matrix assembly. The assembly of a pericellular matrix utilizing only purified exogenous macromolecules can be blocked by the addition of hyaluronan hexasaccharides or anti-CD44 blocking monoclonal antibodies. This led to the hypothesis that three components were critical to the establishment of a pericellular matrix namely, hyaluronan, a hyaluronan-binding proteoglycan and a means for retention to the cell surface. In subsequent studies, we determined that any cell type expressing constitutively active CD44, including chondrocytes, tumor cells or capillary endothelial cells, can assemble prominent pericellular matrices in the presence of exogenously added hyaluronan and proteoglycan (126). As with the chondrocytes, these matrices will assemble on live or fixed cells and can be displaced by agents such as hyaluronan oligosaccharides or Streptomyces hyaluronidase. COS-7 cells are a monkey kidney epithelial cell line that does not express CD44. When exogenous hyaluronan and aggrecan are added to these cells, no pericellular matrices assemble (127). However, when the cells are transiently transfected with pCD44H or pCD44E, the cells gain the capacity to assemble pericellular matrices in the presence of the exogenous macromolecules (Fig. 6) (76,127). This is the most direct proof that CD44 has the capacity to anchor and retain these pericellular matrices. Although CD44 likely participates in retaining displaceable coats on many cell types, the results do not exclude the possibility that other hyaluronan receptors may have a similar capacity.

Most of the studies concerning hyaluronan-dependent matrix retention by anchorage to CD44 have been performed using *in vitro* cell cultures. This always leaves open the question as to whether such interactions or structures occur *in vivo*. This is difficult to address because most methods used to liberate cells from tissues, by definition, destroy the modes of interaction of cells with their extracellular matrix. As one approach to address this question, embryonic chick tibia were treated with a highly purified collagenase in the presence of 20% fetal bovine serum, a rich source of inhibitors of potential proteases (128). The chondrocytes liberated under these conditions were cytospin-plated onto culture plates and immediately analyzed using the particle exclusion assay. The majority of the cells exhibited large intact cell-associated matrices—matrices that could be removed competitively by the addition of hyaluronan hexasaccharides. Similarly, preincubation of the embryonic tibiae with oligosaccharides prior to isolation of the cells with collagenase also resulted in recovered cells devoid of a pericellular matrix thus implicating the presence of chondrocyte pericellular matrices *in vivo*.

Another question is how many CD44 receptors must become occupied by hyaluronan for matrix assembly to occur. Obtaining an exact titer or

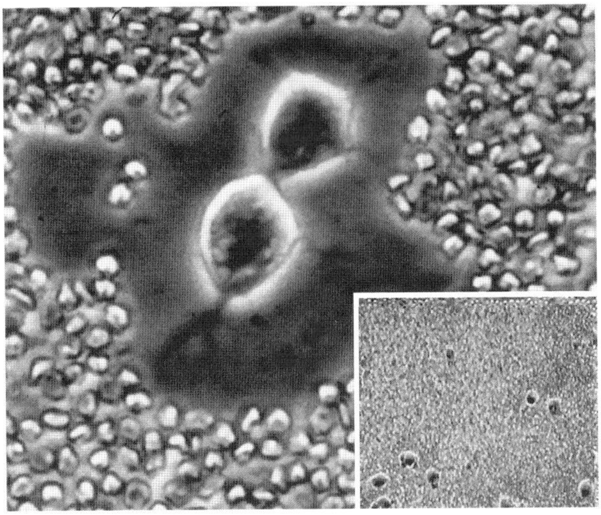

Figure 6 Matrix assembly on live COS-7 cells transfected with CD44 expression constructs. Seventy-two hours post-transfection, COS-7 cell transfectants were trypsinized and allowed to re-attach overnight. Transfectants were then incubated in fresh medium containing exogenous hyaluronan and aggregating proteoglycan for 3 h. Matrices were visualized by the particle exclusion assay. Shown in the main panel are prominent pericellular matrices surrounding cells transfected with pCD44H. COS-7 cells transfected with control, pCDM8 empty vector, did not have the capacity to assemble pericellular matrices (inset panel). Similar results were obtained using non-transfected COS-7 cells (data not shown). (This figure was reproduced with permission from Ref. 76).

density of CD44 at the cell surface is difficult to achieve. One simple method is to remove all cell surface CD44 by treatment of cells with trypsin and then allow new CD44 to be re-synthesized and replenished on the plasma membrane. When cells are fixed at various time points after trypsinization and analyzed by flow cytometry, repopulation of CD44 occurs in a linear, temporal fashion, reaching pretrypsinization levels following 24 h of re-growth (124). When CD44 expression on chondrocytes reached ~25% of the pretrypsinization level, the cells gained the capacity to establish a pericellular matrix in the presence of exogenously added hyaluronan and aggrecan. Additionally, the size of the pericellular matrix did not increase gradually with time indicating that there is a minimum threshold density of CD44 at the cell surface that is required for assembly of pericellular matrices. It remains to be determined whether additional CD44 receptors become occupied with time. However, if not, the data suggest that there is a significant pool of CD44 present at the surface of cells that is always unoccupied. This pool may represent CD44 undergoing some form of activation or serving an alternative function.

B. CD44 Mediates the Internalization of Hyaluronan

The turnover of hyaluronan from the extracellular matrix occurs by essentially two mechanisms, local cell-mediated catabolism and/or drainage into the lymphatic system for catabolism in regional lymph nodes, the liver and spleen (129,130). Although the amount of hyaluronan turnover in some tissues such as the epidermis is high, there is little evidence to support the concept that hyaluronan is extensively degraded within the extracellular matrix. While there may be fragmentation of extracellular hyaluronan via cell surface bound hyaluronidases (131) as well as free-radical-mediated events (132), local turnover of hyaluronan occurs intracellularly within lysosomes by the action low pH hydrolases. It is now clear that the primary mechanism for the transport of hyaluronan into cells for delivery to lysosomes, as well as other intracellular sites, is CD44-mediated internalization (130,133–137). Several cells exhibit the capacity for CD44-mediated internalization and catabolism of hyaluronan including various tumor and transformed cells (138), macrophages (136), chondrocytes (116,135), smooth muscle cells (134) and keratinocytes (137) to list but a few.

The principle evidence that hyaluronan internalization is a receptor-mediated event is that the binding of hyaluronan to the plasma membrane is a prerequisite. When binding is blocked, no internalization is observed. For example, when exogenous fluorescein-hyaluronan was added to chondrocytes in the presence of excess unlabeled hyaluronan, hyaluronan oligosaccharides or incubation with anti-CD44 antibodies, no internalization was observed (135). Further, addition of fluorescein-labeled dextran, a polysaccharide of similar molecular mass as hyaluronan, did not bind or become internalized under similar conditions indicating that internalization of hyaluronan in all these studies does not occur via simple fluid phase pinocytosis (135,137). In more recent studies we demonstrated that CD44-negative COS-7 cells exhibited no capacity to internalize fluorescein-hyaluronan. However, upon transient transfection and expression of recombinant CD44 proteins at the cell surface, the cells gained the capacity to internalize bound hyaluronan (76). Again, this does not exclude the possibility that other hyaluronan receptors may also participate in hyaluronan internalization but rather, that CD44 itself does, in fact, have the capacity to mediate hyaluronan internalization. Interestingly, one of the notable phenotypes observed following the selective antisense transgene suppression of CD44 in mouse keratinocytes was the abnormal accumulation of hyaluronan in the superficial dermis as well as in the corneal stroma (16). Thus, in many tissues, CD44-mediated internalization of hyaluronan may be the primary mechanism for the turnover and catabolism of hyaluronan.

Using ^{3}H-labeled hyaluronan as a probe, chondrocytes were shown to internalize ~5% of the labeled hyaluronan probe bound on the extracellular cell surface within 24 h of incubation (135). Analyses of the intracellular pool displayed two size classes of label, one that eluted in the void volume of a

Sepharose CL-2B column (i.e., $>1 \times 10^6$ Da) and one that eluted in the total volume of the column (i.e., degradation products of <50 kDa). In rat keratinocytes, the size of intracellular hyaluronan was predominately <90 kDa (137). The generation of these small, extensively degraded products was inhibited by the presence of the lysomotropic agent chloroquine, NH_4Cl or the hyaluronidase inhibitor apigenin (135,137). Therefore, the intracellular degradation of hyaluronan occurs within a low pH environment, such as that of the lysosome.

How CD44-mediated internalization is regulated remains an active area of investigation. As discussed earlier, the intracellular domain of CD44 does not exhibit an AP-2 adaptor binding site (i.e., YQRL or LL) indicative of interactions with clathrin (71–73) or the aromatic amino acid rich caveolin-binding domains (74,75) commonly used by classical internalization receptors. Further, Tammi et al. (137) have demonstrated that neither chlorpromazine, an inhibitor of clathrin-mediated uptake, nor filipin III or nystatin, inhibitors of caveolae-mediated uptake, had any effect on the internalization of hyaluronan by rat epidermal keratinocytes. Thus, it is likely that certain, selective interactions of CD44 with the actin cytoskeleton modulate the extent to which hyaluronan is taken up by the cell. However, it remains a question as to how CD44 can participate in two seemingly opposing functions namely, hyaluronan matrix retention and hyaluronan internalization. For example, we have found that treatment of chondrocytes with BMP-7, a growth factor that promotes matrix biosynthesis leading to the generation of large-sized pericellular matrices, includes a substantial upregulation of CD44 expression (48,49). Under these conditions, the presence of increased CD44 is viewed as a means to facilitate enhanced matrix retention. However, when the same cells are treated with the catabolic inflammatory cytokine IL-1α, there is 3 to 6-fold upregulation of CD44 mRNA as well as CD44 protein (47,139). Under these conditions, there is an enhancement of CD44-mediated internalization of hyaluronan. Together with the inhibition of matrix biosynthesis, the pericellular matrices on IL-1 treated cells are significantly reduced. This suggests that transcriptional control of CD44 itself is not responsible for regulating the function of CD44. This makes the interpretation of CD44 expression in pathological tissues ambiguous. For example, by in situ hybridization we have observed enhanced CD44 mRNA expression in samples of human osteoarthritic cartilage (unpublished results). Does this expression represent CD44 participating in attempted repair and retention of matrix following damage or enhanced turnover of hyaluronan within the pericellular matrix?

A second question is whether CD44 is internalized along with the hyaluronan and, if so, is the CD44 recycled back to the cell surface. When bovine articular chondrocytes were incubated with a phycoerythrin-anti-CD44 antibody (non-blocking antibody) together with fluorescein-labeled hyaluronan, both fluorescent conjugates were found co-localized within intracellular organelles as observed by conventional fluorescence and confocal microscopy (140). To insure the visualization of intracellular localization, the cells were trypsinized extensively before viewing by Z-scan optical sectioning. The internalized cell surface-tagged CD44 represented ~20% of the total antibody-tagged CD44 before trypsin treatment. Tammi et al. (137) were also able to observe intracellular hyaluronan–CD44 co-localization in rat epidermal keratinocytes but only after retarding endosomal trafficking through the use of monensin or lowered temperatures. These results led the investigators to suggest that under normal conditions in keratinocytes, CD44 is rapidly recycled back to the cell surface following internalization. Such recycling would explain why only small pools of intracellular CD44 are observed.

C. Role of CD44 in Hyaluronan-Cell Signaling

The direct participation of CD44 in signal transduction has always been controversial. Certainly, as a transmembrane matrix receptor with a cytoplasmic tail domain, CD44 has the basic ingredients necessary to theoretically function in signal transduction. However, the CD44 cytoplasmic tail domain exhibits no inherent receptor kinase or phosphatase activity. The simplest view of CD44-mediated signaling is similar to that of integrins, indirect transfer of information via associated signaling proteins that become linked to CD44 primarily through formation of cytoskeletal complexes. One of the difficulties has been the definition of what constitutes the 'activation' or inducing signaling event related to CD44. In nucleated blood borne cells as well as migrating embryonic, endothelial or malignant cells, unoccupied CD44 receptors are 'activated' by the binding of hyaluronan polysaccharide, initiating extracellular clustering of CD44. The extracellular clustering of CD44 results in intracellular events including the activation of tyrosine kinases such as Src kinases (141) and Rho kinases (141) leading to enhanced association of CD44 into actin cytoskeleton complexes and the recruitment and activation of additional signaling partners (142). That extracellular clustering of CD44 is the inductive event is also evidenced by the use of monoclonal antibody cross-linking to initiate CD44-mediated signaling such as the activation of $p56^{lck}$ and resultant tyrosine phosphorylation of ZAP-70 in T lymphocytes (143). In other tissues where hyaluronan is more ubiquitous and in abundance, the quiescent state of the cells is represented by clustered CD44 receptors occupied in a multivalent fashion with high molecular mass hyaluronan. For these cells, CD44 signaling is initiated by disruption of hyaluronan–CD44 interactions. This disruption may occur by degradation of the hyaluronan (144), the presence of soluble CD44 acting as a competitor for hyaluronan (42), cleavage of the ectodomain of CD44 (145,146) or the presence and competition by small hyaluronan oligosaccharides (50,147,148). In these instances the release of extracellular constraints or clustering imposed by bound hyaluronan polysaccharide is the likely signal induction. As discussed previously, hyaluronidase pretreatment of chondrocytes resulted in a higher proportion of CD44 extractable with NP-40 alone (95) suggesting that release from extracellular constraints does result in the dissociation of CD44–cytoskeletal complexes. Thus, outside–in CD44 signaling can be viewed as extracellular matrix-directed receptor organization that generates changes in CD44 interaction with the actin cytoskeleton and associated signaling protein partners.

The association of CD44 with cytoskeletal complexes is regulated by dynamic interactions with ERM and/or ankyrin adapter proteins. Ponta et al. have proposed a mechanism whereby dynamic associations of CD44 with cytoskeletal complexes could be regulated by changes in the phosphorylation state of the adapter protein, merlin (19). Merlin (also known as the tumor suppressor gene, NF2) is a member of the band 4.1 superfamily. Merlin is highly homologous to ERM proteins and in the non-phosphorylated state, binds to CD44 at the ERM binding site. Unlike ERM, however, merlin has little capacity to engage with the actin cytoskeleton and as such functions as a competitive inhibitor of ERM-mediated CD44 tethering to the actin cytoskeleton. Thus, the regulated phosphorylation/dephosphorylation of merlin could function as a regulatory 'switch'. Other investigators have emphasized the importance of ankyrin–CD44 interactions in regulating the intracellular complex events (141,149). As discussed earlier, CD44 may also differentially localize into or be excluded from lipid raft microdomains providing another mechanism for dynamic association with intracellular signaling kinases such as c-Src (142).

CD44 can also participate in auxiliary functions that are closely related to its role mediating hyaluronan signal transduction. First, CD44 isoforms, such as v3-containing CD44 bearing a heparan sulfate glycosaminoglycan chain, can serve as a support stage for the binding of basic growth factors such as heparin-binding EGF (HB-EGF) and FGF-2 (150–152) as well as a docking protein for matrix metalloproteinases (MMP-7, MMP-9 and MT1-MMP) (152–154). In addition, CD44 can serve as a co-receptor, physically linked to other classical signaling receptors such as c-Met (155), c-ErbB4 (152,156), c-ErbB2, also known as $p185^{HER2}$ (156), RANTES (157) and TGFβR1 (158), and in the process, facilitate the association of intracellular mediators of signal transduction. For example, while c-Met/CD44 activation and complex formation requires the physical interaction of the v6 exon domain of CD44, downstream signaling of c-Met (MEK and Erk phosphorylation) requires the presence of an intact CD44 cytoplasmic tail domain (155). But perhaps the best example of CD44's role as a 'signaling facilitator' is the case of CD44v3 proteoglycan expressed in the lactating mammary gland (152). As a heparan sulfate proteoglycan, CD44v3 facilitates the binding of pro-HB-EGF precursor protein. Subsequently, this same CD44 facilitates the binding of active MMP-7 that functions in part to cleave pro-HB-EGF generating the active form of the growth factor. The activated HB-EGF is then presented to the ErbB4 that is in a stable association complex with CD44.

Recent reports have documented a new mechanism for direct CD44-mediated signaling. The cytoplasmic tail domain of CD44 is enzymatically cleaved and released into cytoplasm where it subsequently translocates to the nucleus and functions as a transcription factor (20). The release is a two-step process initiated first by the cleavage of the ectodomain of CD44 by an MMP

(146,159,160). It is likely that this step is facilitated by the docking of MMPs directly to CD44. The next step is an intramembranous cleavage within the transmembrane domain of CD44 by an Alzheimer's disease-associated, presenilin-dependent γ-secretase activity (146,161,162). This cleavage releases a fragment of CD44 containing the residual portion of the transmembrane domain and the cytoplasmic tail domain, which together have been termed the CD44 'intracellular domain' or CD44-ICD. The CD44-ICD has been shown to promote transcription of various genes with TPA-responsive elements and potentiate transactivation mediated by CBP/p300 (146). One of the potential target genes identified was CD44 itself. The two step MMP/γ-secretase cleavage of CD44 would also be expected to modulate CD44's role as a docking protein and a co-receptor. Interestingly, additional substrates for this dual cleavage pathway besides ameloid β-precursor protein and CD44 include Notch, E-cadherin and ErbB4, the latter a tyrosine kinase receptor that forms a complex with CD44. It is also of interest that treatment of human pancreatic cells with hyaluronan oligosaccharides, conditions that result in cleavage of the ectodomain of CD44, presumably occurs via the action of MT1-MMP (145) and may thus represent one example of how CD44-ICD generation may be initiated extracellularly.

D. CD44/Hyaluronan Interactions in Tumor Invasion and Metastasis

There is long standing evidence that many solid tumors are enriched in hyaluronan (163). As far back as the beginning of the 20th century there was the description of a 'mucinous substance' associated with malignant breast carcinoma, analogous in nature to that found in umbilical cord (164). Higher levels of hyaluronan are associated with poor prognoses in many cancers including human ovarian, breast and prostate carcinomas (165–168). Coincident with this is the finding that CD44 is often upregulated in several of the same tumor tissues (36,169,170). Given the close association of extracellular matrix receptors participating in adhesion and migration, a predicted facilitatory role for CD44 during invasion and metastasis is well warranted. A necessary question is whether binding to hyaluronan is a necessary component of CD44's positive function in invasion and/or metastasis. Bartolazzi et al. (171) demonstrated that stable transfectants of CD44H in human melanoma cell line MC acquired the capacity to form subcutaneous tumors in SCID mice. Tumor weight increased rapidly from 17 to 1765 mg between days 25 and 35 following injection. No measurable tumors were observed in control MC cells or in stable MC cell transfectants expressing CD44H containing an Arg-41 to Ala mutation. As discussed earlier, Arg-41 is one of the critical amino acids within the link protein homology domain of CD44 that is necessary for hyaluronan binding (89,91,92). In the same study, the authors showed that the metastasis of B16F10 melanoma cells could be completely blocked by subcutaneous infusion of soluble IgG–CD44H fusion proteins which compete for cellular CD44–hyaluronan interactions (171). No inhibition of metastasis was observed following the infusion of IgG control protein or soluble IgG-CD44H containing the Arg-41 to Ala mutation. These results demonstrated that the expression of CD44 clearly facilitates tumor growth/metastasis and further, that CD44–hyaluronan interactions are a critical component of the mechanism. *In vitro* studies demonstrated that the CD44H expressing human melanoma cells had enhanced motility on hyaluronan-coated surfaces as compared to cells expressing hyaluronan-binding-defective CD44 mutants (172). Subsequent studies by Sleeman et al. (173) shed doubt on this conclusion by experiments in which highly metastatic, CD44-positive rat pancreatic cells were transfected with a surface-bound form of hyaluronidase. These cells acquired the capacity to degrade all local accumulation of hyaluronan at neutral pH, yet there was no inhibition of tumor growth or metastasis. Their conclusion was that while the expression of CD44, particularly CD44 variants, promote metastasis, hyaluronan–CD44 interactions are not rate limiting for the process. In more recent studies, CD44-positive murine mammary carcinoma cells (TA3/St) readily metastasize to the lung (102,148). However, upon transfecting the TA3/St cells with cDNA encoding the soluble isoform of CD44, no metastases were observed. Like the rat pancreatic tumor cells expressing hyaluronidase directly at the cell surface, endogenous secretion of the soluble CD44 isoform would be expected to interfere with CD44–hyaluronan in the immediate pericellular environment. The debate on this issue continues. Nonetheless, investigators have presented compelling evidence concerning several mechanisms whereby enhancement of CD44 as well as hyaluronan–CD44 interactions may promote tumor growth and/or metastasis. These include providing for: [1] haptotactic migration of neoplastic cells through hyaluronan-enriched extracellular matrices; [2] formation of pericellular matrices that act as protective cocoons enhancing cell survival; [3] enhancing the chemokinetic activity of invasive cells; [4] avoidance of immune surveillance; [5] participation in angiogenesis of capillary endothelial cells; [6] enhanced capacity for hyaluronan endocytosis and degradation; [7] enhanced co-receptor activities with receptors such as c-Met and $p185^{HER2}$; and [8] enhancing cell survival by providing anti-apoptosis signaling.

Evidence that CD44–hyaluronan interactions are critical features in human cancer is more limited and less direct than the experimental model systems. In most carcinomas, the elevated levels of hyaluronan are primarily localized in the adjacent tumor stromal connective tissue with the tumor parenchyma exhibiting little hyaluronan accumulation (163). However, upon close examination some of the neoplastic cells display prominent positive staining for hyaluronan in breast and gastric cancers; the number of positive staining tumor cells is associated with poor cellular differentiation, axillary lymph node positivity and poorer overall survival rate (167,174). The suggestion from these studies is that these hyaluronan-positive parenchyma cells have acquired an enhanced capacity for binding hyaluronan supplied by the stromal cells. However, such results may also reflect changes in hyaluronan synthase activity or in other hyaluronan receptors such as RHAMM. Nonetheless, regardless of the mechanism of hyaluronan retention, the presence of bound hyaluronan is predictive of more aggressive behavior in these tumors and may reflect positive growth selection due to hyaluronan-mediated cocooning, anti-apoptotic signaling, enhancement of chemokinesis, enhanced internalization capacity, etc. However, these conclusions are not applicable to all human cancers. As discussed previously, tumor progression in human colorectal and prostate cancer is associated with a reduction in CD44 expression principally due to promoter methylation (52–54, 175,176). Thus, generalizations concerning the expression and participation of CD44 during tumor invasion and metastasis will depend on the tissue of origin of the malignant cells.

Over the past decade there has been intense interest concerning the expression of particular CD44 variant and tumor progression/metastasis. During differential screening for epitopes present on a metastatic rat pancreatic carcinoma cell line versus a non-metastatic cell line, Gunthert et al. identified two alternatively spliced isoforms of CD44, CD44v4-7 and CD44v6,7 (termed pMeta-1 and pMeta-2, respectively) (32,33). As discussed earlier, these variant CD44s contained additional amino acid sequences within the extracellular domain of the molecule (see Fig. 3). The expression of a v6 exon (i.e., exon 11, Fig. 2) is rare in most normal cell types. One exception is the transient expression by B and T-cells upon antigen activation (35). When an antibody specific to the variant CD44 isoform (CD44v6) was co-injected with the metastasizing cells, metastatic growth of the pancreatic carcinoma was inhibited, and host survival was prolonged. Interest in this CD44 variant peaked when it was found that transfection of non-metastatic tumor cells with CD44v6 enhanced the cell's efficiency for metastasis to regional lymph nodes. Since these initial observations, numerous studies have documented the prevalence as well as diagnostic/prognostic value of CD44 variant isoforms in human cancers, including the expression of alternatively spliced combinations of the v3, v6, v9 and v10 isoforms of CD44 (35,36,169,177–181). For example, upon screening different stages of human colon cancer it was noted that CD44v6 expression begins in the adenomatous polyp stage, becomes more positive in dysplastic polyps, and in carcinomas and metastasis samples, 100% of the cells are positive for CD44v6 (182). As expected there are also reports to the contrary. For example, abundant CD44v6 expression was seen in normal human squamous epithelial cells such as keratinocytes and endothelial cells as well as benign squamous epithelial neoplasms (183). In squamous carcinomas, neither primary tumors nor metastasis express v6 isoforms of CD44 while the same cells do exhibit CD44H (Hermes-3-positive) (183). Thus, generalizations concerning the expression of CD44 variants such as CD44v6 as a marker of malignancy cannot be made. Further, even in cancers in which CD44 variants are expressed preferentially (as compared to surrounding normal tissues), there is disagreement over the prognostic/diagnostic value of CD44 variant expression. In some cancers, CD44 variant expression is well correlated with increased tumor aggressiveness and poor prognosis (180,181,184,185) whereas in others CD44 variants are present but not of prognostic value (36,169,179,186).

E. The Participation of CD44 in Cellular Functions Continues to Expand

As discussed earlier, mouse mammary carcinoma cells expressing soluble CD44 (TA3sCD44) exhibit fewer lung metastases following intravenous tail vein injection into syngeneic mice (102). Interestingly however, both wild type (TA3/St) and TA3sCD44 cells initially adhered to the pulmonary endothelium and invaded into the interstitial layers. The failure of the TA3sCD44 cells to survive and establish metastatic foci was traced to the induction of apoptosis in these cells. Similar inhibition of tumor growth and colony formation could be achieved by the use of ALZET osmotic pump infused hyaluronan oligosaccharides to interfere with hyaluronan–CD44 interactions. Oligosaccharide-induced apoptosis was observed in the mouse mammary carcinoma cell line (TA3/St) as well as rat and human glioma cells and human lung carcinoma cells (148,187). This suggested that the loss of hyaluronan–CD44 interactions resulted in the induction of apoptosis, particularly in anchorage-independent-growth cell types. Subsequent studies revealed that the induction of apoptosis due to oligosaccharides resulted from an inhibition of phosphoinositide 3-kinase activity (PI 3-kinase) leading to an inhibition of Akt phosphorylation, BAD and FKHR phosphorylation all of which impinge on Bcl-2 and the activation of caspase-3 (148). The inhibition of PI 3-kinase was further amplified by hyaluronan oligosaccharide stimulation of PTEN—a phosphatase that dephosphorylates the PI 3-kinase product, phosphatidylinositol-3-phosphate. In addition to hyaluronan oligosaccharides, a similar inhibition of PI 3-kinase activity was also observed by the use of anti-CD44 antibodies. These studies suggest that hyaluronan–CD44 interactions are required for the maintenance of cell survival pathways, a disruption of which may lead to the activation of the apoptotic pathway.

Cell survival in other cells types, besides transformed cells, may also be affected by a disruption in hyaluronan–CD44 interactions. For example, in a reverse experiment to the one described earlier, the addition of hyaluronan to CD44+ T lymphocytes inhibited apoptosis initiated by anti-CD3 monoclonal antibodies (188). The addition of hyaluronan to chondrocytes also inhibited the activation of apoptosis initiated by anti-Fas antibodies (189). The anti-apoptotic effects of added hyaluronan were inhibited by pretreatment of the chondrocytes with blocking anti-CD44 monoclonal antibodies. The addition of hyaluronan oligosaccharides to many cells, including chondrocytes, results in the release of nitric oxide (190,191). This is of interest in that nitric oxide itself has been shown to downregulate PI 3-kinase in chondrocytes, a process that is reversed by the addition of the cell survival cytokine, IGF-1 (192). In sum, the participation of CD44 in the maintenance of cell survival may be restricted to anchorage-independent cell types but remains an exciting new area of study concerning CD44 function.

Another function of CD44 that is under active investigation is the participation of CD44 in lymphocyte rolling and the infiltration of lymphocytes into sites of inflammation. Prior to forming high strength, integrin-mediated, firm adhesions to the endothelial vessel wall, lymphocytes exhibit a weaker adhesion phase characterized by cell rolling. In RA, hyaluronan is elevated in the synovial tissue of joints as well as the endothelial cell lining of synovial blood vessels. In a proteoglycan-induced animal model of RA, Mikecz et al. (193) were able to abrogate all swelling and leukocyte infiltration by systemic treatment of the mice with anti-CD44 monoclonal antibody. Antibody treatment resulted in the rapid shedding of CD44 from circulating lymphocytes and synovial cells. This added to the suggestions in the literature that CD44 homing to areas of inflammation was due, in part, to hyaluronan–CD44 interactions. Studies by Clark et al. (105) demonstrated that the rolling of tonsillar lymphocytes under shear stress on the surface of cultured human tonsillar stromal cells was dependent on hyaluronan–CD44 interactions. Lymphocyte rolling was inhibited by the addition of anti-CD44 monoclonal antibodies, soluble hyaluronan and hyaluronidase treatment. In the same study, the investigators demonstrated that lymphocytes could exhibit rolling under shear to plastic surfaces coated with hyaluronan—rolling that was also inhibited by the use of soluble hyaluronan or anti-CD44 antibodies. More recently, stable-transfectant subclones of a mouse T-cell lymphoma, expressing variable densities of cell surface CD44, were characterized (194). Subclones expressing the highest levels of CD44 bound the most soluble fluorescein-hyaluronan in a proportional fashion. Similarly, cells with the greatest amount of cell surface CD44 continued to exhibit adherence and rolling on hyaluronan-coated surfaces at higher shear forces and with lower rolling velocities than lymphoma cells expressing less CD44. Additional new data demonstrate that CD44-deficient mice exhibit a reduction in incidence and severity of collagen-induced arthritis as compared to wild type mice (195,196). Thus, CD44–hyaluronan interactions play a key role in the initial interaction of lymphocytes with the endothelium, lymphocyte rolling.

VI. Concluding Remarks

CD44 serves a critical role linking one of the most ubiquitous extracellular macromolecule, hyaluronan, with the cell surface and intracellular network. CD44 participates in the binding of hyaluronan, pericellular matrix assembly, hyaluronan internalization, cell migration and adhesion, tumor invasion and metastasis, signal transduction, regulating apoptosis, serving as a co-receptor and docking protein for matrix metalloproteinases, and lymphocyte homing. There is, as yet, no clear understanding as to why the CD44 $-/-$ mouse does not exhibit apparent phenotypic changes (13). The most likely explanation is that another hyaluronan receptor becomes upregulated in the knockout mice and compensates for the absence of CD44. Subsequent studies have demonstrated that adult

CD44 −/− mice do exhibit profoundly different phenotypes when challenged. As described previously, upon challenge in a collagen-induced arthritis model, there is delayed onset and lessened severity of joint swelling in the CD44 −/− mice (195,196). Lymphocyte infiltration into the joint cavity is clearly altered. When CD44 −/− mouse lungs are exposed to bleomycin, a model of lung injury and inflammation, there is no resolution of inflammatory cell infiltrates in the lung and the mice die of respiratory failure after 14 days (14). Interestingly, bleomycin-induced inflammation in wild-type mice only rarely exhibits apoptotic cells. However, in the CD44 −/− mice there was a 17-fold increase in the number of apoptotic cells (14). In a study concerning the inhibition of select CD44 variants, CD44v7 and CD44v6,7 null mice, there was also a 3 to 4-fold increase in apoptotic cells in early inflammation lesion in the colon (15). Again, this suggests that CD44–hyaluronan interactions are important for the maintenance of cell survival.

Another reason for the suggestion that there must be compensation in the CD44 −/− mouse is that earlier studies using a targeted antisense transgene approach, involving a CD44 antisense gene driven by keratin-5 promoter, resulted in a severe disruption in hyaluronan catabolism in the superficial dermis and corneal stroma (16). The increase in hyaluronan was due to inhibition of local turnover rather than increased biosynthesis. In addition, these targeted CD44 null mice exhibited defective keratinocyte proliferation in response to growth factors, morphological alterations of basal keratinocytes, delayed hair growth, and impaired local inflammatory responses. Many of these features are what would be expected given the many functions of CD44 described in Section V. Subsequent studies documented that the changes in the targeted CD44 antisense mouse were similar to a form of skin dermatosis lesions found in humans, termed Lichen sclerosus et atrophicus (17). Lichen sclerosus et atrophicus is characterized in part by an increase in glycosaminoglycan in the dermis, hyaluronan in particular. Upon closer examination of these patients, the investigators noted a substantially elevated accumulation of hyaluronan in epidermal and dermal skin sections and the near absence of CD44 expression as measured by in situ hybridization and immunohistochemistry. Thus, CD44 expression is clearly necessary for select functions in adult tissues and the adult mice do not appear capable of adopting another mechanism to compensate for its absence. Is CD44 important during embryonic development? To address this question, Zoller et al. (197) used an alternative approach to the development of CD44 *null* mice and examined the effects of transient interference of CD44 by way of intravenous injection of anti-CD44 antibody into pregnant rats. In the antibody-injected rats they observed delayed delivery, frequent abortions, smaller fetuses and delayed formation of lung alveoli, the kidney tubular system, and villi of the gut. They reported that the degradation of hyaluronan in the developing kidney was also delayed. Again, all these data suggest the critical importance of CD44 in a wide variety of cellular functions, and it is likely because of the presence of CD44 that hyaluronan plays such a prominent role in cell behavior.

Acknowledgements

The authors thank Cheryl B. Knudson, Ph.D., for helpful discussions, critical review of this manuscript, and the use of preliminary data. Supported in part by NIH grants RO1-AR43384 and P50-AR39239 (SCOR), and a grant from the National Arthritis Foundation.

References

1. Truppe W, Basner R, von Figura K, Kresse H. Uptake of hyaluronate by cultured cells. Biochem Biophys Res Commun 1977; 78:713–719.
2. Underhill CB, Dorfman A. The role of hyaluronic acid in intercellular adhesion of cultured mouse cells. Exp Cell Res 1978; 117:155–164.
3. Underhill CB, Toole BP. Binding of hyaluronate to the surface of cultured cells. J Cell Biol 1979; 82:475–484.
4. Love SW, Shannon BT, Myrvik QN, Lynn WS. Characterization of macrophage agglutinating factor as a hyaluronic acid–protein complex. J Reticuloendothelial Soc 1979; 25:269–282.
5. Underhill CB, Toole BP. Physical characteristics of hyaluronate binding to the surface of simian virus 40-transformed 3T3 cells. J Biol Chem 1980; 255: 4544–4549.
6. Underhill CB, Toole BP. Receptors for hyaluronate on the surface of parent and virus-transformed cell lines. Exp Cell Res 1981; 131:419–423.
7. Underhill CB. Interaction of hyaluronate with the surface of simian virus 40-transformed 3T3 cells: aggregation and binding studies. J Cell Sci 1982; 56:177–189.
8. Nemec RE, Toole BP, Knudson W. The cell surface hyaluronate binding sites of invasive human bladder carcinoma cells. Biochem Biophys Res Commun 1987; 149:249–257.
9. Underhill CB. CD44: the hyaluronan receptor. J Cell Sci 1992; 103:293–298.
10. Knudson CB, Knudson W. Hyaluronan-binding proteins in development, tissue homeostasis and disease. FASEB J 1993; 7:1233–1241.
11. Dalchau R, Kirkley J, Fabre JW. Monoclonal antibody to a human brain-granulocyte-T lymphocyte antigen probably homologous to the W 3/13 antigen of the rat. Eur J Immunol 1980; 10:745–749.
12. Cobbold S, Hale G, Waldmann H. Non-lineage, LFA-1 family, and leucocyte common antigens: new and previously defined clusters. In: McMichael AJ, ed. Leucocyte Typing III: White Cell Differentiation Antigens. New York: Oxford University Press, 1987:788–803.
13. Schmits R, Filmus J, Gerwin N, Senaldi G, Kiefer F, Kundig T, Wakeham A, Shahinian A, Catzavelos C, Rak J, Furlonger C, Zakarian A, Simard JJ, Ohashi PS, Paige CJ, Gutierrez-Ramos JC, Mak TW. CD44 regulates hematopoietic progenitor distribution, granuloma formation, and tumorigenicity. Blood 1997; 90:2217–2233.
14. Teder P, Vandivier RW, Jiang D, Liang J, Cohn L, Pure E, Henson PM, Noble PW. Resolution of lung inflammation by CD44. Science 2002; 296:155–158.
15. Wittig BM, Johansson B, Zoller M, Schwarzler C, Gunthert U. Abrogation of experimental colitis correlates with increased apoptosis in mice deficient for CD44 variant exon 7 (CD44v7). J Exp Med 2000; 191:2053–2064.

16. Kaya G, Rodriguez I, Jorcano JL, Vassalli P, Stamenkovic I. Selective suppression of CD44 in keratinocytes of mice bearing an antisense CD44 transgene driven by a tissue-specific promoter disrupts hyaluronate metabolism in the skin and impairs keratinocyte proliferation. Genes Dev 1997; 11: 996–1007.
17. Kaya G, Augsburger E, Stamenkovic I, Saurat J-H. Decrease in epidermal CD44 expression as a potential mechanism for abnormal hyaluronate accumulation in superficial dermis in Lichen Sclerosus et Atrophicus. J Investig Dermatol 2000; 115:1054–1058.
18. Naor D, Sionov RV, Ish-Shalom D. CD44: structure, function, and the association with the malignant process. Adv Cancer Res 1987; 71:241–319.
19. Ponta H, Sherman L, Herrlich P. CD44: from adhesion molecules to signaling regulators. Nat Rev Mol Cell Biol 2003; 4:33–45.
20. Cichy J, Pure E. The liberation of CD44. J Cell Biol 2003; 161:839–843.
21. Lesley J, Hascall VC, Tammi M, Hyman R. Hyaluronan binding by cell surface CD44. J Biol Chem 2000; 275:26967–26975.
22. Screaton GR, Bell MV, Jackson DG, Cornelis FB, Gerth U, Bell JI. Genomic structure of DNA encoding the lymphocyte homing receptor CD44 reveals at least 12 alternatively spliced exons. Proc Natl Acad Sci USA 1992; 89:12160–12164.
23. Goodfellow PN, Banting G, Wiles MV, Tunnacliffe A, Parkar M, Solomon E, Dalchau R, Fabre JW. The gene, MIC4, which controls expression of the antigen defined by monoclonal antibody F10.44.2, is on human chromosome 11. Eur J Immunol 1982; 12:659–663.
24. Underhill CB, Green SJ, Cologlio PM, Tarone G. The hyaluronate receptor is identical to a glycoprotein of 85,000 Mr (gp 85) as shown by a monoclonal antibody that interferes with binding activity. J Biol Chem 1987; 262:13142–13146.
25. Gunthert U. CD44. A multitude of isoforms with diverse functions. Curr Top Microbiol Immunol 1993; 184:47.
26. Jiang H, Knudson CB, Knudson W. Antisense inhibition of alternatively spliced CD44 variant in human articular chondrocytes promotes hyaluronan internalization. Arthritis Rheum 2001; 44:2599–2619.
27. Yu Q, Toole BP. A new alternatively spliced exon between v9 and v10 provides a molecular basis for synthesis of soluble CD44. J Biol Chem 1996; 271: 20603–20607.
28. Ruiz P, Schwarzler C, Gunthert U. CD44 isoforms during differentiation and development. Bioessays 1995; 17:17–24.
29. Lesley J, Hyman R, Kincade PW. CD44 and its interaction with extracellular matrix. Adv Immunol 1993; 54:271–335.
30. Sleeman JP, Kondo K, Moll J, Ponta H, Herrlich P. Variant exons v6 and v7 together expand the repertoire of glycosaminoglycans bound by CD44. J Biol Chem 1997; 272:31837–31844.
31. Moll J, Khaldoyanidi S, Sleeman JP, Achtnich M, Preuss I, Ponta H, Herrlich P. Two different functions for CD44 proteins in human myelopoiesis. J Clin Investig 1998; 102:1024–1034.
32. Gunthert U, Hofmann M, Rudy W, Reber S, Zoller M, Haussmann I, Matzku S, Wenzel A, Ponta H, Herrlich P. A new variant of glycoprotein CD44 confers metastatic potential to rat carcinoma cells. Cell 1991; 65:13–24.
33. Herrlich P, Rudy W, Hofmann M, Arch R, Zoller M, Zawadzki V, Tolg C, Hekele A, Koopman G, Pals S, Heider K-H, Sleeman J, Ponta H. CD44 and splice variants

of CD44 in normal differentiation and tumor progression. In: Hemler ME, Mihich E, eds. Cell Adhesion Molecules. New York: Plenum Press, 1993:265–278.
34. Tarin D, Matsumura Y. Deranged activity of the CD44 gene and other loci as biomarkers for progression to metastatic malignancy. J Cell Biochem 1993; 17G: 173–185.
35. Arch R, Wirth K, Hofmann M, Ponta H, Matzku S, Herrlich P, Zoller M. Participation in normal immune responses of a metastasis-inducing splice variant of CD44. Science 1992; 257:682–685.
36. Charpin C, Garcia S, Bouvier C, Devictor B, Andrac L, Choux R, Lavaut MN, Allasia C. Automated and quantitative immunocytochemical assays of CD44v6 in breast carcinomas. Hum Pathol 1997; 28:289–296.
37. Milstone LM, Hough-Monroe L, Kugelman LC, Bender JR, Haggerty JG. Epican, a heparan/chondroitin sulfate proteoglycan form of CD44, mediates cell–cell adhesion. J Cell Sci 1994; 107:3183–3190.
38. Stamenkovic I, Aruffo A, Amiot M, Seed B. The hematopoietic and epithelial forms of CD44 are distinct polypeptides with different adhesion potentials for hyaluronate-bearing cells. EMBO J 1991; 10:348–353.
39. Jackson DG, Buckley J, Bell JI. Multiple variants of the human lymphocyte homing receptor CD44 generated by insertions at a single site in the extracellular domain. J Biol Chem 1992; 267:4732–4739.
40. Goldstein LA, Butcher EC. Identification of mRNA that encodes an alternative form of H-CAM (CD-44) in lymphoid and nonlymphoid tissues. Immunogenetics 1990; 32:389–397.
41. Perschl A, Lesley J, English N, Trowbridge I, Hyman R. Role of CD44 cytoplasmic domain in hyaluronan binding. Eur J Immunol 1995; 25:495–501.
42. Peterson RM, Yu Q, Stamenkovic I, Toole BP. Perturbation of hyaluronan interactions by soluble CD44 inhibits growth of murine mammary carcinoma cells in ascites. Am J Pathol 2000; 156:2159–2167.
43. Nedvetzki S, Golan I, Assayag N, Gonen E, Caspi D, Gladnikoff M, Yayon A, Naor D. A mutation in a CD44 variant of inflammatory cells enhances the mitogenic interaction of FGF with its receptor. J Clin Investig 2003; 111: 1211–1220.
44. Zhang M, Wang MH, Singh RK, Wells A, Siegal GP. Epidermal growth factor induces CD44 gene expression through a novel regulatory element in mouse fibroblasts. J Biol Chem 1997; 272:14139–14146.
45. Fitzgerald KA, O'Neill LA. Characterization of CD44 induction by IL-1: a critical role for Egr-1. J Immunol 1999; 162:4920–4927.
46. Foster LC, Wiesel P, Huggins GS, Panares R, Chin MT, Pellacani A, Perrella MA. Role of activating protein-1 and high mobility group-I(Y) protein in the induction of CD44 gene expression by interleukin-1beta in vascular smooth muscle cells. FASEB J 2000; 14:368–378.
47. Nishida Y, D'Souza AL, Thonar JMA, Knudson W. IL-1a stimulates hyaluronan metabolism in human articular cartilage. Arthritis Rheum 2000; 43:1315–1326.
48. Nishida Y, Knudson CB, Kuettner KE, Knudson W. Osteogenic protein-1 promotes the synthesis and retention of extracellular matrix within bovine articular cartilage and chondrocyte cultures. Osteoarthritis Cartilage 2000; 8:127–136.
49. Nishida Y, Knudson CB, Eger W, Kuettner KE, Knudson W. Osteogenic protein-1 stimulates cell-associated matrix assembly by normal human articular chondrocytes. Arthritis Rheum 2000; 43:206–214.

50. Knudson W, Casey B, Nishida Y, Eger W, Kuettner KE, Knudson CB. Hyaluronan oligosaccharides perturb cartilage matrix homeostasis and induce chondrocytic chondrolysis. Arthritis Rheum 2000; 43:1165–1174.
51. Shtivelman E, Bishop JM. Expression of CD44 is repressed in neuroblastoma cells. Mol Cell Biol 1991; 11:5446–5453.
52. Verkaik NS, Trapman J, Romijn JC, Van der Kwast TH, Van Steenbrugge GJ. Down-regulation of CD44 expression in human prostatic carcinoma cell lines is correlated with DNA hypermethylation. Int J Cancer 1999; 80:439–443.
53. Verkaik NS, van Steenbrugge GJ, van Weerden WM, Bussemakers MJ, van der Kwast TH. Silencing of CD44 expression in prostate cancer by hypermethylation of the CD44 promoter region. Lab Investig 2000; 80:1291–1298.
54. Stallmach A, Wittig BM, Kremp K, Goebel R, Santourlidis S, Zeitz M, Menges M, Raedle J, Zeuzem S, Schulz WA. Downregulation of CD44v6 in colorectal carcinomas is associated with hypermethylation of the CD44 promoter region. Exp Mol Pathol 2003; 74:262–266.
55. Graff JR, Herman JG, Lapidus RG, Chopra H, Xu R, Jarrard DF, Isaacs WB, Pitha PM, Davidson NE, Baylin SB. E-cadherin expression is silenced by DNA hypermethylation in human breast and prostate carcinomas. Cancer Res 1995; 55:5195–5199.
56. Hyman R. Lack of a consistent relationship between demethylation of the CD44 promoter and CD44 expression. Immunogenetics 2002; 53:914–924.
57. Jackson DG, Bell JI, Dickinson R, Timans J, Shields J, Whittle N. Proteoglycan forms of the lymphocyte homing receptor CD44 are alternatively spliced variants containing the v3 exon. J Cell Biol 1995; 128:673–685.
58. Liu D, Liu T, Li R, Sy MS. Mechanisms regulating the binding activity of CD44 to hyaluronic acid. Front Biosci 1998; 3:D631–D636.
59. Rousche KT, Knudson CB. Temporal expression of CD44 during embryonic chick limb development and modulation of its expression with retinoic acid. Matrix Biol 2002; 21:53–62.
60. Guo YJ, Lin SC, Wang JH, Bigby M, Sy MS. Palmitoylation of CD44 interferes with CD3-mediated signaling in human T lymphocytes. Int Immunol 1994; 6: 213–221.
61. Liu D, Sy MS. A cysteine residue located in the transmembrane domain of CD44 is important in binding of CD44 to hyaluronic acid. J Exp Med 1996; 183:1987–1994.
62. Liu D, Sy MS. Phorbol myristate acetate stimulates the dimerization of CD44 involving a cysteine in the transmembrane domain. J Immunol 1997; 159: 2702–2711.
63. Oliferenko S, Paiha K, Harder T, Gerke V, Schwarzler C, Schwarz H, Beug H, Gunthert U, Huber LA. Analysis of CD44-containing lipid rafts: recruitment of annexin II and stabilization by the actin cytoskeleton. J Cell Biol 1999; 146: 843–854.
64. Ilangumaran S, Borisch B, Hoessli DC. Signal transduction via CD44: role of plasma membrane microdomains. Leuk Lymphoma 1999; 35:455–469.
65. Foger N, Marhaba R, Zoller M. Involvement of CD44 in cytoskeleton rearrangement and raft reorganization in T cells. J Cell Sci 2001; 114:1169–1178.
66. Lafont F, Tran Van Nhieu G, Hanada K, Sansonetti P, van der Goot FG. Initial steps of Shigella infection depend on the cholesterol/sphingolipid raft-mediated CD44–IpaB interaction. EMBO J 2002; 21:4449–4457.

67. Lokeshwar VB, Bourguignon LYW. Post-translational protein modification and expression of ankyrin-binding site(s) in GP85 (Pgp-1/CD44) and its biosynthetic precursors during T-lymphoma membrane biosynthesis. J Biol Chem 1991; 266: 17983–17989.
68. Legg JW, Isacke CM. Identification and functional analysis of the ezrin-binding site in the hyaluronan receptor, CD44. Curr Biol 1998; 8:705–708.
69. Morrison H, Sherman LS, Legg J, Banine F, Isacke C, Haipek CA, Gutmann DH, Ponta H, Herrlich P. The NF2 tumor suppressor gene product, merlin, mediates contact inhibition of growth through interactions with CD44. Genes Dev 2001; 15:968–980.
70. Peck D, Isacke CM. Hyaluronan-dependent cell migration can be blocked by a CD44 cytoplasmic domain peptide containing a phosphoserine at position 325. J Cell Sci 1998; 111:1595–1601.
71. Kirchhausen T. Adaptors for clathrin-mediated traffic. Annu Rev Cell Dev Biol 1999; 15:705–732.
72. Haucke V, Krauss M. Tyrosine-based endocytic motifs stimulate oligomerization of AP-2 adaptor complexes. Eur J Cell Biol 2002; 81:647–653.
73. Brett TJ, Traub LM, Fremont DH. Accessory protein recruitment motifs in clathrin-mediated endocytosis. Structure (Camb) 2002; 10:797–809.
74. Gaillard S, Bartoli M, Castets F, Monneron A. Striatin, a calmodulin-dependent scaffolding protein, directly binds caveolin-1. FEBS Lett 2001; 508:49–52.
75. Leclerc PC, Auger-Messier M, Lanctot PM, Escher E, Leduc R, Guillemette G. A polyaromatic caveolin-binding-like motif in the cytoplasmic tail of the type 1 receptor for angiotensin II plays an important role in receptor trafficking and signaling. Endocrinology 2002; 143:4702–4710.
76. Jiang H, Peterson RS, Wang W, Bartnik E, Knudson CB, Knudson W. A requirement for the CD44 cytoplasmic domain for hyaluronan binding, pericellular matrix assembly and receptor mediated endocytosis in COS-7 cells. J Biol Chem 2002; 277:10531–10538.
77. Bartolazzi A, Nocks A, Aruffo A, Spring F, Stamenkovic I. Glycosylation of CD44 is implicated in CD44-mediated cell adhesion to hyaluronan. J Cell Biol 1996; 132:1199–1208.
78. Skelton TP, Zeng C, Nocks A, Stamenkovic I. Glycosylation provides both stimulatory and inhibitory effects on cell surface and soluble CD44 binding to hyaluronan. J Cell Biol 1998; 140:431–446.
79. Banerji S, Day AJ, Kahmann JD, Jackson DG. Characterization of a functional hyaluronan-binding domain from the human CD44 molecule expressed in *Escherichia coli*. Protein Expr Purif 1998; 14:371–381.
80. Katoh S, Zheng Z, Oritani K, Shimozato T, Kincade PW. Glycosylation of CD44 negatively regulates its recognition of hyaluronan. J Exp Med 1995; 182:419–429.
81. Katoh S, Miyagi T, Taniguchi H, Matsubara Y, Kadota J, Tominaga A, Kincade PW, Matsukura S, Kohno S. Cutting edge: an inducible sialidase regulates the hyaluronic acid binding ability of CD44-bearing human monocytes. J Immunol 1999; 162:5058–5061.
82. Rochman M, Moll J, Herrlich P, Wallach SB, Nedvetzki S, Sionov RV, Golan I, Ish-Shalom D, Naor D. The CD44 receptor of lymphoma cells: structure–function relationships and mechanism of activation. Cell Adhes Commun 2000; 7:331–347.
83. Gee K, Kozlowski M, Kumar A. Tumor necrosis factor-alpha induces functionally active hyaluronan-adhesive CD44 by activating sialidase through

p38 mitogen-activated protein kinase in lipopolysaccharide-stimulated human monocytic cells. J Biol Chem 2003; 278:37275–37287.

84. Gross N, Balmas K, Beretta Brognara C. Role of CD44H carbohydrate structure in neuroblastoma adhesive properties. Med Pediatr Oncol 2001; 36:139–141.
85. English NM, Lesley JF, Hyman R. Site-specific de-N-glycosylation of CD44 can activate hyaluronan binding, and CD44 activation states show distinct threshold densities for hyaluronan binding. Cancer Res 1998; 58:3736–3742.
86. Bourguignon LYW, Kalomiris EL, Lokeshwar VB. Acylation of the lymphoma transmembrane glycoprotein, GP85, may be required for GP85–ankyrin interaction. J Biol Chem 1991; 266:11761–11765.
87. Uff CR, Neame SJ, Isacke CM. Hyaluronan binding by CD44 is regulated by a phosphorylation-independent mechanism. Eur J Immunol 1995; 25:1883–1887.
88. Day AJ, Prestwich GD. Hyaluronan-binding proteins: tying up the giant. J Biol Chem 2002; 277:4585–4588.
89. Peach RJ, Hollenbaugh D, Stamenkovic I, Aruffo A. Identification of hyaluronic acid binding sites in the extracellular domain of CD44. J Cell Biol 1993; 122: 257–264.
90. Knudson W, Aguiar DJ, Hua Q, Knudson CB. CD44-anchored hyaluronan-rich pericellular matrices: an ultrastructural and biochemical analysis. Exp Cell Res 1996; 228:216–228.
91. Bajorath J, Greenfield B, Munro SB, Day AJ, Aruffo A. Identification of CD44 residues important for hyaluronan binding and delineation of the binding site. J Biol Chem 1998; 273:338–343.
92. Mahoney DJ, Blundell CD, Day AJ. Mapping the hyaluronan-binding site on the link module from human tumor necrosis factor-stimulated gene-6 by site-directed mutagenesis. J Biol Chem 2001; 276:22764–22771.
93. Lacy BE, Underhill CB. The hyaluronate receptor is associated with actin filaments. J Cell Biol 1987; 105:1395–1404.
94. Camp RL, Kraus TA, Pure E. Variations in the cytoskeletal interaction and posttranslational modification of the CD44 homing receptor in macrophages. J Cell Biol 1991; 115:1283–1292.
95. Nofal GA, Knudson CB. Chondrocyte CD44 interacts with the cytoskeleton via ankyrin. Trans Orthop Res Soc 1999; 24:130.
96. Nofal GA, Knudson CB. Latrunculin and cytochalasin decrease chondrocyte matrix retention. J Histochem Cytochem 2002; 50:1313–1324.
97. Bourguignon LY, Walker G, Suchard SJ, Balazovich K. A lymphoma plasma membrane-associated protein with ankyrin-like properties. J Cell Biol 1986; 102: 2115–2124.
98. Lokeshwar VB, Fregien N, Bourguignon LYW. Ankyrin-binding domain of CD44(GP85) is required for the expression of hyaluronic acid-mediated adhesion function. J Cell Biol 1994; 126:1099–1109.
99. Zhu D, Bourguignon LY. The ankyrin-binding domain of CD44s is involved in regulating hyaluronic acid-mediated functions and prostate tumor cell transformation. Cell Motil Cytoskeleton 1998; 39:209–222.
100. Tsukita S, Oishi K, Sato N, Sagara J, Kawai A. ERM family members as molecular linkers between the cell surface glycoprotein CD44 and actin-based cytoskeletons. J Cell Biol 1994; 126:391–401.
101. Aruffo A, Stamenkovic I, Melnick M, Underhill CB, Seed B. CD44 is the principal cell surface receptor for hyaluronate. Cell 1990; 61:1303–1313.

102. Yu BQ, Toole BO, Stamenkovic I. Induction of apoptosis of metastatic mammary carcinoma cells *in vivo* by disruption of tumor cell surface CD44 function. J Exp Med 1997; 186:1985–1996.
103. Lesley J, He Q, Miyake K, Hamann A, Hyman R, Kincade PW. Requirements for hyaluronic acid binding by CD44: a role for the cytoplasmic domain and activation by antibody. J Exp Med 1992; 175:257–266.
104. Lesley J, English N, Charles C, Hyman R. The role of CD44 cytoplasmic domain and transmembrane domains in constitutive and inducible binding. Eur J Immunol 2000; 30:245–253.
105. Clark RA, Alon R, Springer TA. CD44 and hyaluronan-dependent rolling interactions of lymphocytes on tonsillar stroma. J Cell Biol 1996; 134:1075–1087.
106. Pure E, Camp RL, Peritt D, Panettieri RA, Lazaar AL, Nayak S. Defective phosphorylation and hyaluronate binding of CD44 with point mutations in the cytoplasmic domain. J Exp Med 1995; 181:55–62.
107. Liu D, Liu T, Sy MS. Identification of two regions in the cytoplasmic domain of CD44 through which PMA, calcium, and foskolin differentially regulate the binding of CD44 to hyaluronic acid. Cell Immunol 1998; 190:132–140.
108. Green SJ, Tarone G, Underhill CB. Aggregation of macrophages and fibroblasts is inhibited by a monoclonal antibody to the hyaluronate receptor. Exp Cell Res 1988; 178:224–232.
109. Maleski MP, Knudson CB. Hyaluronan meditated aggregation of limb bud mesoderm and mesenchymal condensation during chondrogenesis. Exp Cell Res 1996; 225:55–66.
110. Toole BP. Proteoglycans and hyaluronan in morphogenesis and differentiation. In: Hay ED, ed. Cell Biology of Extracellular Matrix. New York: Plenum Press, 1991:305–339.
111. Goldberg RL, Toole BP. Pericellular coat of chick embryo chondrocytes: structural role of hyaluronate. J Cell Biol 1984; 99:2114–2122.
112. Knudson CB, Toole BP. Changes in the pericellular matrix during differentiation of limb bud mesoderm. Dev Biol 1985; 112:308–318.
113. Knudson CB. Hyaluronan receptor-directed assembly of chondrocyte pericellular matrix. J Cell Biol 1993; 120:825–834.
114. Underhill CB, Toole BP. Transformation-dependent loss of the hyaluronate-containing coats of cultured cells. J Cell Phys 1982; 110:123–128.
115. Clarris BJ, Fraser JRE. On the pericellular zone of some mammalian cells *in vitro*. Exp Cell Res 1968; 49:181–193.
116. Chow G, Nietfeld J, Knudson CB, Knudson W. Antisense inhibition of chondrocyte CD44 expression results in cartilage chondrolysis. Arthritis Rheum 1998; 41:1411–1419.
117. Knudson CB, Nofal GA, Chow G, Peterson RS. CD44: the link between hyaluronan and the cytoskeleton. In: Kennedy JF, Phillips GO, Williams PA, Hascall VC, eds. Hyaluronan. Cambridge: Woodhead Publishing, 2002:331–340.
118. Asplund T, Versnel MA, Laurent TC, Heldin P. Human mesothelioma cells produce factors that stimulate the production of hyaluronan by mesothelial cells and fibroblasts. Cancer Res 1993; 53:388–392.
119. Heldin P, Pertoft H. Synthesis and assembly of the hyaluronan-containing coats around normal human mesothelial cells. Exp Cell Res 1993; 208:422–429.
120. Prehm P. Hyaluronate is synthesized at plasma membranes. Biochem J 1984; 220:597–600.

121. Weigel PH, Hascall VC, Tammi M. Hyaluronan synthases. J Biol Chem 1997; 272:13997–14000.
122. Itano N, Kimata K. Expression cloning and molecular characterization of HAS protein, a eukaryotic hyaluronan synthase. J Biol Chem 1996; 271: 9875–9878.
123. DeAngelis PL, Papaconstantinou J, Weigel PH. Isolation of a Streptococcus pyogenes gene locus that directs hyaluronan biosynthesis in acapsular mutants and in heterologous bacteria. J Biol Chem 1993; 268:14568–14571.
124. Knudson CB, Nofal GA, Pamintuan L, Aguiar DJ. The chondrocyte pericellular matrix: a model for hyaluronan-mediated cell–matrix interactions. Biochem Soc Trans 1999; 27:142–147.
125. Goldberg RL, Underhill CB, Toole BP. Affinity chromatography of hyaluronate glutaraldehyde-fixed SV-3T3 cells. Anal Biochem 1982; 125:59–65.
126. Knudson W, Knudson CB. Assembly of a chondrocyte-like pericellular matrix on non-chondrogenic cells. J Cell Sci 1991; 99:227–235.
127. Knudson W, Bartnik E, Knudson CB. Assembly of pericellular matrices by COS-7 cells transfected with CD44 homing receptor genes. Proc Natl Acad Sci USA 1993; 90:4003–4007.
128. Maleski MP, Knudson CB. Matrix accumulation and retention in embryonic cartilage and *in vitro* chondrogenesis. Connect Tissue Res 1996; 34:75–86.
129. Fraser JRE, Brown TJ, Laurent TC. Catabolism of hyaluronan. In: Laurent TC, ed. The Chemistry, Biology and Medical Applications of Hyaluronan and Its Derivatives. London: Portland Press, 1998:85–92.
130. Knudson W, Chow G, Knudson CB. CD44-mediated uptake and degradation of hyaluronan. Matrix Biol 2002; 21:15–23.
131. Stern R. Devising a pathway for hyaluronan catabolism: are we there yet? Glycobiology 2003; 13:105–115.
132. Al-Assaf S, Navaratnam S, Parsons BJ, Phillips GO. Chain scission of hyaluronan by peroxynitrite. Arch Biochem Biophys 2003; 411:73–82.
133. Culty M, Nguyen HA, Underhill CB. The hyaluronan receptor (CD44) participates in the uptake and degradation of hyaluronan. J Cell Biol 1992; 116:1055–1062.
134. Evanko SP, Wight TN. Intracellular localization of hyaluronan in proliferating cells. J Histochem Cytochem 1999; 43:1331–1341.
135. Hua Q, Knudson CB, Knudson W. Internalization of hyaluronan by chondrocytes occurs via receptor-mediated endocytosis. J Cell Sci 1993; 106:365–375.
136. Underhill CB, Nguyen HA, Shizari M, Culty M. CD44 positive macrophages take up hyaluronan during lung development. Dev Biol 1993; 155:324–336.
137. Tammi R, Rilla K, Pienimaki J-P, MacCallum DK, Hogg M, Luukkonen M, Hascall VC, Tammi M. Hyaluronan enters keratinocytes by a novel endocytic route for catabolism. J Biol Chem 2001; 276:35111–35122.
138. Culty M, Shizari M, Thompson EW, Underhill CB. Binding and degradation of hyaluronan by human breast cancer cell lines expressing different isoforms of CD44: correlation with invasive potential. J Cell Physiol 1994; 160:275–286.
139. Chow G, Knudson CB, Homandberg G, Knudson W. Increased CD44 expression in bovine articular chondrocytes by catabolic cellular mediators. J Biol Chem 1995; 270:27734–27741.
140. Aguiar DJ, Knudson W, Knudson CB. Internalization of the hyaluronan receptor CD44 by chondrocytes. Exp Cell Res 1999; 252:292–302.

141. Bourguignon LY, Zhu H, Shao L, Zhu D, Chen YW. Rho-kinase (ROK) promotes CD44v(3,8-10)-ankyrin interaction and tumor cell migration in metastatic breast cancer cells. Cell Motil Cytoskeleton 1999; 43:269–287.
142. Turley EA, Noble PW, Bourguignon LY. Signaling properties of hyaluronan receptors. J Biol Chem 2002; 277:4589–4592.
143. Taher TEI, Smit L, Griffioen AW, Schilder-Tol EJM, Borst J, Pals ST. Signaling through CD44 is mediated by tyrosine kinases: association with p56lck in T lymphocytes. J Biol Chem 1996; 271:2863–2867.
144. Nishida N, Knudson CB, Knudson W. Extracellular matrix recovery by human articular chondrocytes after treatment with hyaluronan hexasaccharides or Streptomyces hyaluronidase. Mod Rheum 2003; 13:62–68.
145. Sugahara KN, Murai T, Nishinakamura H, Kawashima H, Saya H, Miyasaka M. Hyaluronan oligosaccharides induce CD44 cleavage and promote cell migration in CD44-expressing tumor cells. J Biol Chem 2003; 278:32259–32265.
146. Okamoto I, Kawano Y, Murakami D, Sasayama T, Araki N, Miki T, Wong AJ, Saya H. Proteolytic release of CD44 intracellular domain and its role in the CD44 signaling pathway. J Cell Biol 2001; 155:755–762.
147. Noble PW, McKee CM, Cowman M, Shin HS. Hyaluronan fragments activate an NF-kappa B/I-kappa B alpha autoregulatory loop in murine macrophages. J Exp Med 1996; 183:2373–2378.
148. Ghatak S, Misra S, Toole BP. Hyaluronan oligosaccharides inhibit anchorage-independent growth of tumor cells by suppressing the phosphoinositide 3-kinase/Akt cell survival pathway. J Biol Chem 2002; 277:38013–38020.
149. Bourguignon LY, Jin H. Identification of the ankyrin-binding domain of the mouse T-lymphoma cell inositol 1,4,5-trisphosphate (IP3) receptor and its role in the regulation of IP3-mediated internal Ca2+ release. J Biol Chem 1995; 270: 7257–7260.
150. Bennett KL, Jackson DG, Simon JC, Tanczos E, Peach R, Modrell B, Stamenkovic I, Plowman G, Aruffo A. CD44 isoforms containing exon V3 are responsible for the presentation of heparin-binding growth factor. J Cell Biol 1995; 128:687–698.
151. Sherman L, Wainwright D, Ponta H, Herrlich P. A splice variant of CD44 expressed in the apical ectodermal ridge presents fibroblast growth factors to limb mesenchyme and is required for limb outgrowth. Genes Dev 1998; 12:1058–1071.
152. Yu WH, Woessner JF Jr, McNeish JD, Stamenkovic I. CD44 anchors the assembly of matrilysin/MMP-7 with heparin-binding epidermal growth factor precursor and ErbB4 and regulates female reproductive organ remodeling. Genes Dev 2002; 16:307–323.
153. Yu Q, Stamenkovic I. Localization of matrix metalloproteinase 9 to the cell surface provides a mechanism for CD44-mediated tumor invasion. Genes Dev 1999; 13:35–48.
154. Mori H, Tomari T, Koshikawa N, Kajita M, Itoh Y, Sato H, Tojo H, Yana I, Seiki M. CD44 directs membrane-type 1 matrix metalloproteinase to lamellipodia by associating with its hemopexin-like domain. EMBO J 2002; 21:3949–3959.
155. Orian-Rousseau V, Chen L, Sleeman JP, Herrlich P, Ponta H. CD44 is required for two consecutive steps in HGF/c-Met signaling. Genes Dev 2002; 16:3074–3086.
156. Bourguignon LY, Zhu H, Zhou B, Diedrich F, Singleton PA, Hung MC. Hyaluronan promotes CD44v3-Vav2 interaction with Grb2-p185(HER2) and induces Rac1 and Ras signaling during ovarian tumor cell migration and growth. J Biol Chem 2001; 276:48679–48692.

157. Roscic-Mrkic B, Fischer M, Leemann C, Manrique A, Gordon CJ, Moore JP, Proudfoot AE, Trkola A. RANTES (CCL5) uses the proteoglycan CD44 as an auxiliary receptor to mediate cellular activation signals and HIV-1 enhancement. Blood 2003; 102:1169–1177.
158. Bourguignon LY, Singleton PA, Zhu H, Zhou B. Hyaluronan promotes signaling interaction between CD44 and the transforming growth factor beta receptor I in metastatic breast tumor cells. J Biol Chem 2002; 277:39703–39712.
159. Okamoto I, Kawano Y, Tsuiki H, Sasaki J, Nakao M, Matsumoto M, Suga M, Ando M, Nakajima M, Saya H. CD44 cleavage induced by a membrane-associated metalloprotease plays a critical role in tumor cell migration. Oncogene 1999; 18: 1435–1446.
160. Okamoto I, Kawano Y, Matsumoto M, Suga M, Kaibuchi K, Ando M, Saya H. Regulated CD44 cleavage under the control of protein kinase C, calcium influx, and the Rho family of small G proteins. J Biol Chem 1999; 274: 25525–25534.
161. Lammich S, Okochi M, Takeda M, Kaether C, Capell A, Zimmer AK, Edbauer D, Walter J, Steiner H, Haass C. Presenilin-dependent intramembrane proteolysis of CD44 leads to the liberation of its intracellular domain and the secretion of an Abeta-like peptide. J Biol Chem 2002; 277:44754–44759.
162. Murakami D, Okamoto I, Nagano O, Kawano Y, Tomita T, Iwatsubo T, De Strooper B, Yumoto E, Saya H. Presenilin-dependent gamma-secretase activity mediates the intramembranous cleavage of CD44. Oncogene 2003; 22: 1511–1516.
163. Knudson W. Tumor-associated hyaluronan. Providing an extracellular matrix that facilitates invasion. Am J Pathol 1996; 148:1721–1725.
164. Wells HG. The chemistry of tumors. In: Chemical Pathology. Philadelphia: W.B. Saunders, 1907; 411–430.
165. Anttila MA, Tammi RH, Tammi MI, Syrjanen KJ, Saarikoski SV, Kosma VM. High levels of stromal hyaluronan predict poor disease outcome in epithelial ovarian cancer. Cancer Res 2000; 60:150–155.
166. Auvinen PK, Parkkinen JJ, Johansson RT, Agren UM, Tammi RH, Eskelinen MJ, Kosma VM. Expression of hyaluronan in benign and malignant breast lesions. Int J Cancer 1997; 74:477–481.
167. Auvinen P, Tammi R, Parkkinen J, Tammi M, Agren UM, Johansson R, Hirvikoski P, Eskelinen M, Kosma VM. Hyaluronan in peritumoral stroma and malignant cells associates with breast cancer spreading and predicts survival. Am J Pathol 2000; 156:529–536.
168. Ropponen K, Tammi M, Parkkinen J, Eskelinen M, Tammi R, Lipponen P, Agren UM, Alhava E, Kosma VM. Tumor cell-associated hyaluronan as an unfavorable prognostic factor in colorectal cancer. Cancer Res 1998; 58:342–347.
169. Muller W, Schneiders A, Heider KH, Meier S, Hommel G, Habbert HE. Expression and prognostic value of the CD44 splicing variants v5 and v6 in gastric cancer. J Pathol 1997; 183:222–227.
170. Streit M, Schmidt R, Hilgenfeld RU, Thiel E, Kreuser ED. Adhesion receptors in malignant transformation and dissemination of gastrointestinal tumors. J Mol Med 1996; 74:253–268.
171. Bartolazzi A, Peach R, Aruffo A, Stamenkovic I. Interaction between CD44 and hyaluronate is directly implicated in the regulation of tumor development. J Exp Med 1994; 180:53–66.

172. Thomas L, Byers HR, Vink J, Stamenkovic I. CD44H regulates tumor cell migration on hyaluronate-coated substrate. J Cell Biol 1992; 118:971–977.
173. Sleeman JP, Arming S, Moll JF, Hekele A, Rudy W, Sherman LS, Kreil G, Ponta H, Herrlich P. Hyaluronate-independent metastatic behavior of CD44 variant-expressing pancreatic carcinoma cells. Cancer Res 1996; 56:134–141.
174. Setala LP, Tammi MI, Tammi RH, Eskelinen MJ, Lipponen PK, Agren UM, Parkkinen J, Alhava EM, Kosma VM. Hyaluronan expression in gastric cancer cells is associated with local and nodal spread and reduced survival rate. Br J Cancer 1999; 79:1133–1138.
175. Lipponen P, Aaltomaa S, Tammi R, Tammi M, Agren UM, Kosma VM. High stromal hyaluronan level is associated with poor differentiation and metastasis in prostate cancer. Eur J Cancer 2001; 37:849–856.
176. McCarthy JB. Hyaluronan in prostate cancer. 2003; glycoforum http://www.glycoforum.gr.jp/science/hyaluronan/HA26/HA26E.html.
177. Bourguignon LYW, Zhu D, Zhu H. CD44 isoform–cytoskeleton interaction in oncogenic signaling and tumor progression. Front Biosci 1998; 3:637–649.
178. Frenz DA, Jaikaria NS, Newman SA. The mechanism of precartilage mesenchymal condensation: a major role for interaction of the cell surface with the amino-terminal heparin-binding domain of fibronectin. Dev Biol 1989; 136: 97–103.
179. Friedrichs K, Franke F, Lisboa BW, Kugler G, Gille I, Terpe HJ, Holzel F, Maass H, Gunthert U. CD44 isoforms correlate with cellular differentiation but not with prognosis in human breast cancer. Cancer Res 1995; 55:5424–5433.
180. Mayer B, Jauch KW, Gunthert U, Figdor CG, Schildberg FW, Funke I, Johnson JP. *De novo* expression of CD44 and survival in gastric cancer. Lancet 1993; 342: 1019–1022.
181. Streit M, Schmidt R, Hilgenfeld RU, Thiel E, Kreuser ED. Adhesion receptors in malignant transformation and dissemination of gastrointestinal tumors. Recent Results Cancer Res 1996; 142:19–50.
182. Heider K-H, Hofmann M, Hors E, van den Berg F, Ponta H, Herrlich P, Pals S. A human homologue of the rat metastasis-associated variant of CD44 is expressed in colorectal carcinomas and adenomatous polyps. J Cell Biol 1993; 120:227–233.
183. Salmi M, Gron-Virta K, Sointu P, Grenman R, Kalimo H, Jalkanen S. Regulated expression of exon v6 containing isoforms of CD44 in man: downregulation during malignant transformation of tumors of squamocellular origin. J Cell Biol 1993; 122:431–442.
184. Kaufmann M, Heider KH, Sinn HP, von Minckwitz G, Ponta H, Herrlich P. CD44 variant exon epitopes in primary breast cancer and length of survival. Lancet 1995; 345:615–619.
185. Guriec N, Gairard B, Marcellin L, Wilk A, Calderoli H, Renaud R, Bergerat JP, Oberling F. CD44 isoforms with exon v6 and metastasis of primary N0M0 breast carcinomas. Breast Cancer Res Treat 1997; 44:261–268.
186. Hong RL, Lee WJ, Shun CT, Chu JS, Chen YC. Expression of CD44 and its clinical implication in diffuse-type and intestinal-type gastric adenocarcinomas. Oncology 1995; 52:334–339.
187. Ward JA, Huang L, Guo H, Ghatak S, Toole BP. Perturbation of hyaluronan interactions inhibits malignant properties of glioma cells. Am J Pathol 2003; 162: 1403–1409.

188. Ayroldi E, Cannarile L, Migliorati G, Bartoli A, Nicoletti I, Riccardi C. CD44 (Pgp-1) inhibits CD3 and dexamethasone-induced apoptosis. Blood 1995; 86:2672–2678.
189. Lisignoli G, Grassi F, Zini N, Toneguzzi S, Piacentini A, Guidolin D, Bevilacqua C, Facchini A. Anti-Fas-induced apoptosis in chondrocytes reduced by hyaluronan: evidence for CD44 and CD54 (intercellular adhesion molecule 1) invovement. Arthritis Rheum 2001; 44:1800–1807.
190. Iacob S, Knudson C. Hyaluronan fragments activate nitric oxide synthase and the release of nitric oxide in articular chondrocytes. Orthop Res Soc Trans 2003; 28:584.
191. Rockey DC, Chung JJ, McKee CM, Noble PW. Stimulation of inducible nitric oxide synthase in rat liver by hyaluronan fragments. Hepatology 1998; 27:86–92.
192. Oh C-D, Chun J-S. Signaling mechanisms leading to the regulation of differentiation and apoptosis of articular chondrocytes by insulin-like growth factor-1. J Biol Chem 2003; 278:36563–36571.
193. Mikecz K, Brennan FR, Kim JH, Glant TT. Anti-CD44 treatment abrogates tissue edema and leukocyte infiltration in murine arthritis. Nat Med 1995; 1:558–564.
194. Gal I, Lesley J, Ko W, Gonda A, Stoop R, Hyman R, Mikecz K. Role of the extracellular and cytoplasmic domains of CD44 in the rolling interaction of lymphoid cells with hyaluronan under physiologic flow. J Biol Chem 2003; 278: 11150–11158.
195. Stoop R, Kotani H, McNeish JD, Otterness IG, Mikecz K. Increased resistance to collagen-induced arthritis in CD44-deficient DBA/1 mice. Arthritis Rheum 2001; 44:2922–2931.
196. Stoop R, Gal I, Glant TT, McNeish JD, Mikecz K. Trafficking of CD44-deficient murine lymphocytes under normal and inflammatory conditions. J Immunol 2002; 32:2532–2542.
197. Zoller M, Herrmann K, Buchner S, Seiter S, Claas C, Underhill CB, Moller P. Transient absence of CD44 expression and delay in development by anti-CD44 treatment during ontogeny: a surrogate of an inducible knockout? Cell Growth Differ 1997; 8:1211–1223.

Chemistry and Biology of Hyaluronan
H.G. Garg and C.A. Hales (editors)

Chapter 6

The Role of the Hyaluronan Receptor RHAMM in Wound Repair and Tumorigenesis

CORNELIA TÖLG

London Regional Cancer Centre, London, Ont., Canada

SARA R. HAMILTON and EVA A. TURLEY

Department of Biochemistry/London Regional Cancer Centre, University of Western Ontario, London, Ont., Canada

I. Introduction

Hyaluronan (HA) is a negatively charged polysaccharide belonging to the glycosaminoglycan family and is characterized by the presence of repeated disaccharides composed of amino sugars and β-glucuronic acid residues. HA is unique in this class of polysaccharides since it is not sulfated and is rarely covalently linked to a protein core as is typical for most sulfated glycosaminoglycans, for example, link protein, aggrecan and syndecan (1–5). HA is also unique in its size, reaching up to several million Daltons, and is synthesized at the plasma membrane rather than in the golgi, where sulfated glycosaminoglycans are added to protein cores (6,7). Three transmembrane HA synthases (HAS 1–3) responsible for the production of HA have been identified. UDP-sugar residues bind to the cytoplasmic face of HAS enzymes and are added on to a growing HA chain that is thought to be extruded through a pore created by oligomers of HAS proteins. The functional interrelatedness of the HAS enzymes has not yet been extensively investigated, but they are differentially promoted and expressed during embryogenesis, response-to-injury processes and neoplastic transformation of tissues, and can produce HA chains of different lengths (1,6–8).

A number of studies have linked wound repair processes to susceptibility for neoplastic transformation (9,10). These studies in particular have stressed the dual role of stromal factors in wound repair and cancer initiation/progression, one of which is HA (11–16). The dual role of HA as a regulator of wound repair and neoplastic initiation/progression requires cell HA receptors or hyaladherins such as CD44, RHAMM, LYVE-1 (CSRSBP-1) and layilin, as well as various intracellular and extracellular HA-binding proteins (HABPs) (Fig. 1) (16–22). A brief summary of the known functions of HA during wound repair and cancer are detailed below followed by a review of the role of the cellular hyaladherin, RHAMM, in these processes.

II. Hyaluronan in Wound Repair and Cancer

HA synthesis is upregulated at sites of tissue injury, for example, in the dermis following incisional or excisional wound repair (13,23–26). HA accumulation is enhanced immediately following injury and remains elevated during the inflammatory and early granulation/re-epithelialization stages of wound repair. HA synthesis ceases later in the granulation phase and accumulated HA is de-polymerized by host hyaluronidases into smaller fragments. In adult organisms, healing of excisional wounds almost always involves fibrosis that is associated with extracellular matrix (ECM) remodeling by fibroblasts to a 'reactive stroma' (9,27–29). A reactive stroma is characterized by enhanced inflammatory cell infiltration, deposition of tenascin, extensive neo-angiogenesis and enhanced collagen deposition and fibrillogenesis (30,31). The appearance of a reactive stroma is usually associated with the disappearance of high-molecular weight HA (13,23–26) and the accumulation of HA fragments that are likely to participate in angiogenesis (11,32,33). A prolonged accumulation of high-molecular weight HA, such as occurs during fetal wound repair, is associated

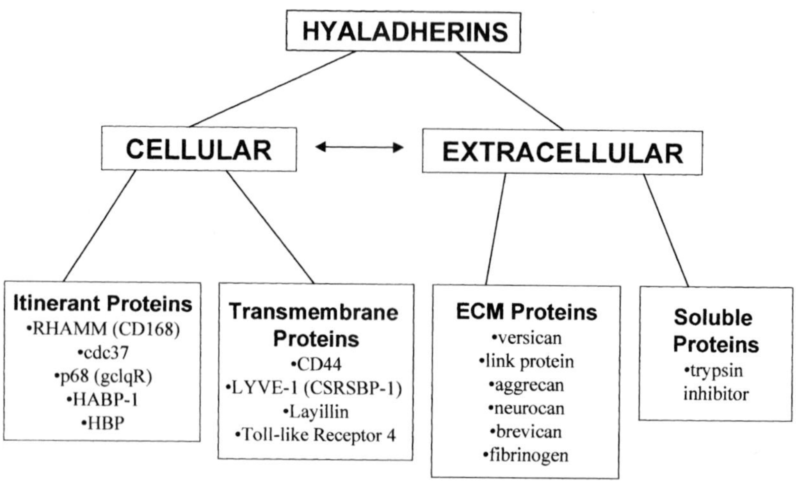

Figure 1 Classification of hyaladherins.

with reduced inflammatory cell infiltration and a regeneration type of healing that does not involve extensive collagen fibril deposition or scarring (23,34,35). In fact, the application of high-molecular weight HA to skin wounds reduces fibrotic repair as detected by reduced collagen deposition and fibrillogenesis (36–39).

HA performs multiple functions during skin wound repair, particularly in the inflammatory and early granulation stages. HA interacts with fibrin clots and initially modulates host inflammatory cell infiltration into the inflamed site. It also induces production of growth factors and cytokines in inflammatory cells, fibroblasts and keratinocytes (23,34,35), and protects and presents growth factors involved in skin wound repair, such as VEGF and PDGF, to their cognate receptors (40–43). Some of these growth factors promote production of HA synthesis in other cell types at the wound site, for example, endothelial cells (25). In addition to regulating gene expression in inflammatory cells, HA promotes their migration, adherence to inflamed tissue, as well as phagocytosis and killing of wound-site pathogens, and can directly inhibit pathogen proliferation (12,13, 16,25). Conversely, HA acts as an antioxidant by scavenging ROS from inflammatory cells, and thus functions to both stimulate and limit inflammation at the wound site (44–49). During the formation of granulation tissue, HA promotes migration of both keratinocytes and fibroblasts (50–54), regulates cell proliferation, possibly progression through G_2M of the cell cycle (55) and contributes to the structure of the provisional matrix of granulation tissue, particularly by restricting collagen deposition and fibril organization (25). Later in the granulation phase, high-molecular weight HA is largely degraded into fragments and these contribute to enhanced angiogenesis that is associated with tissue fibrosis (11,25,32,33). The functional roles of HA have been deduced by noting effects of exogenously administered HA to skin wounds and by studying the consequences of blocking the function of HA receptors, administering hyaluronidases and modifying HAS enzyme expression. These studies are reviewed more extensively in other chapters of this book. A summary of the roles of HA in wound repair is shown in Fig. 2.

The formation of remodeled or reactive stroma that is associated with fibrotic repair is also typical of the connective tissue surrounding many aggressive neoplasms and this type of ECM microenvironment can predispose tissues to neoplastic transformation, increased tumor colonization and enhanced metastasis (56–59). Although a role for HA in stroma-regulated neoplastic growth has not been directly demonstrated, enhanced HA accumulation in the stroma surrounding tumors, e.g. breast cancer, is significantly related to poor differentiation of the tumors, auxiliary lymph node positivity and short overall survival of patients (11,60). HA is normally produced in the stroma, but neoplastic transformation often results in the synthesis of HA by transformed epithelial cells (11,60), and in breast cancer enhanced accumulation of HA in the tumor cells is also an indicator of poor prognosis. Interestingly, in breast cancer enhanced accumulation of stromal or transformed ductal epithelial-associated HA are independent prognostic parameters and the power of the association between HA accumulation and poor outcome is enhanced when these two

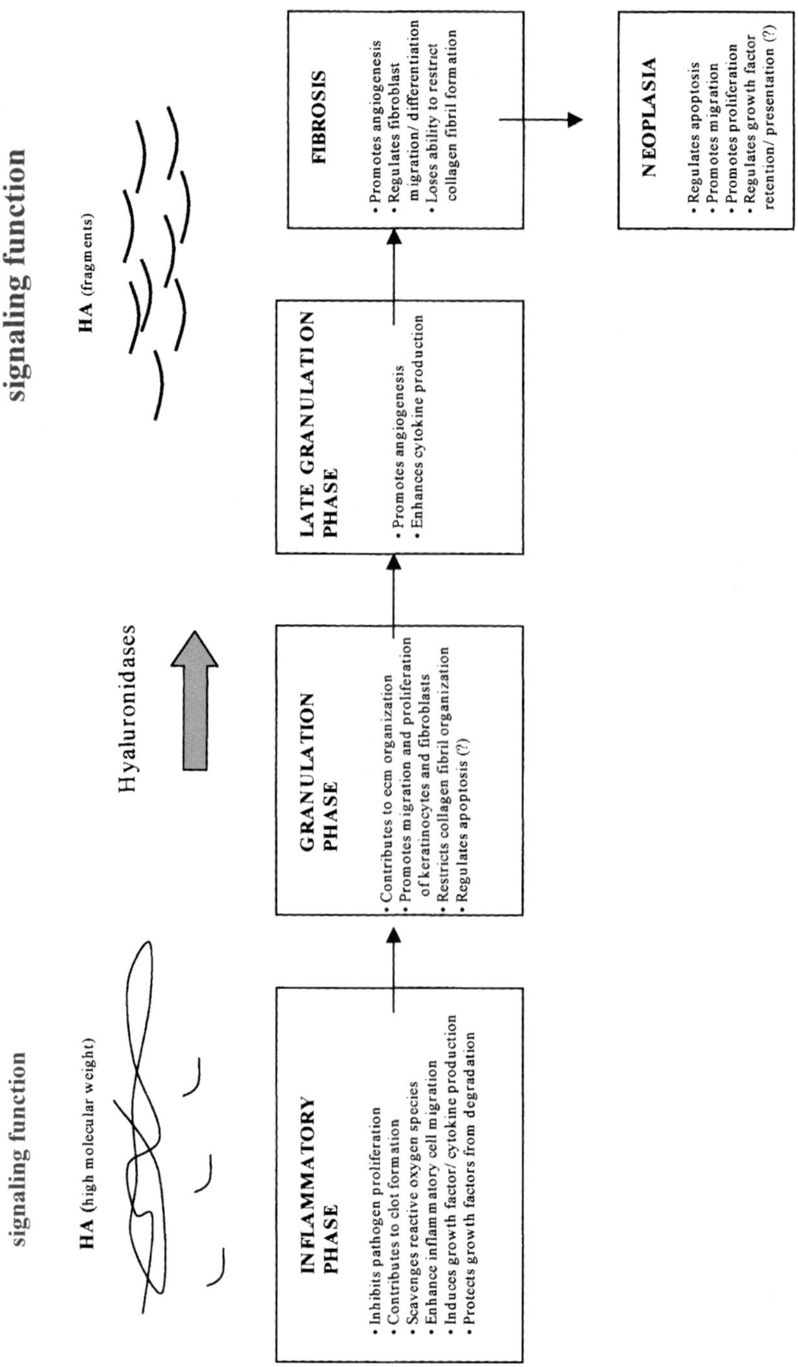

Figure 2 Role of hyaluronan in wound repair and tumor progression.

parameters are combined (60). These results suggest that stromal HA contributes to breast cancer progression by a mechanism that is distinct from cancer cell HA. The molecular significance of these findings has not yet been dissected, but it is likely that stromal HA affects tumor progression, at least in part, by promoting angiogenesis (11,33). A variety of other studies suggest that HA production by tumor cells themselves also has important functional consequences that might promote tumor progression. For example, stable expression of HAS-2 in a rat colon carcinoma cell line (PROb) resulted in higher growth rates and also in a more rapid development of transplantable tumors (61). Furthermore, expression of one HA receptor, CD44, significantly correlates with the survival of human breast tumor xenografts in immune-compromised mice (62) and modification of HA–tumor cell interactions propels breast tumor cells into apoptosis (63). These results suggest that HA-rich environments promote cell survival, and this may be the one key function of HA that is common to normal cells responding to injury and neoplastic cells. In addition to this function, and similar to the multiple functions of wound-site HA, production of HA by tumor cells also promotes migration and invasion, and regulates expression of gene sets that allow tumor cells to remodel their microenvironment and preferentially proliferate (11). In particular, stable expression of HAS2 in breast ductal epithelial cell lines is required for and promotes a conversion to a mesenchymal phenotype (EMT), a process that is clinically associated with increased tumor cell autonomy and aggressiveness (64). The functions of HA in neoplastic processes are also shown in Fig. 2.

HA exerts its effects on transformed cells and cells responding to injury via unique physiochemical properties and an ability to activate signaling cascades (see Chapter 7). The mechanisms by which the highly viscoelastic and hydrating properties of HA contribute to skin wound repair or neoplastic progression have not been well dissected from its signaling properties. However, the ability of HA to swell ECM has been proposed to facilitate cell invasion and the inherent elastic properties of HA may alter the rigidity or stretchability of ECM (65). The latter effect can indirectly activate signaling cascades such as MAP kinases through stretch-sensitive integrin receptors (66,67). HA was first demonstrated to activate protein tyrosine phosphorylation cascades in 1988 (68) and since then has been shown to regulate signaling through src, ras, FAK, PI3 kinase/ATK kinase to control remodeling of the actin cytoskeleton, cell motility, proliferation and apoptosis (11). These HA-mediated signaling events have largely been studied from the perspective of either CD44 or RHAMM acting as the receptor transducing a signal. Signaling properties of CD44 have been well and recently reviewed (see Chapter 7). Here, we examine in detail the dual roles of RHAMM in wound repair and cancer as a signal transducer for HA.

III. RHAMM is an Atypical Hyaladherin

HABPs or hyaladherins are conveniently divided into extracellular and cellular proteins (Fig. 1). Although with increasing characterization, this distinction is often blurred. For example, the cellular HABPs, CD44 and RHAMM can both be shed and occur as ECM proteins (69,70). Cellular hyaladherins have been further classified according to their sequence homology (e.g., CD44 and LYVE-1/CSRSBP-1), the mechanism for their association with the cell surface and the mechanism by which they bind to HA (71). For example, CD44 is an example of a transmembrane HABP that binds to HA through a link module, common to aggrecan, versican and LYVE-1. In contrast to CD44, RHAMM is present in multiple cellular compartments, associates with the cell surface, but is not a transmembrane protein and binds to HA via a region that is rich in basic amino acids, but is not homologous to the link module. The RHAMM–HA binding region is composed of two coiled-coil regions that contain key basic residues that wrap around and secure the HA chain (71). A number of cellular hyaladherins resemble RHAMM in this distinctive binding mechanism and these also occur in multiple cellular compartments. Collectively, these RHAMM-like proteins have been designated itinerant hyaladherins (16). In addition to binding to HA with high affinity (71), RHAMM binds to sulfated glycosaminoglycans such as heparin (72) and, therefore, has the potential to connect signaling pathways regulated by HS proteoglycans to those regulated by HA, as has also been demonstrated for CD44, which binds to growth factors via HS modification of alternatively spliced exons (73–75).

Like an increasing number of proteins that include epimorphin/syntaxin-2, phosphohexose isomerase/autocrine motility factor, galectin-1, HMGB1/amaphoterin, tissue transglutaminase and thioredoxin/ADF (76), RHAMM occurs on the cell surface yet resembles an intracellular protein, lacking a signal peptide that would permit its export through the golgi–ER. Cell surface RHAMM, which has been given the cluster designation of CD168 (77), has been detected on sub-confluent adherent cells such as fibroblasts and endothelial cells, non-adherent cells such as B and T cells, and on many types of tumor cells as quantified by FACS analysis, confocal microscopy, SEM, and as inferred by the ability of anti-RHAMM antibodies to block HA-mediated functions (16,78). Intracellular RHAMM forms are also expressed in multiple sub-cellular compartments of many cell types (42,79–82). To date, the cell surface form of RHAMM is required for the following functions: PDGF and HA-mediated activation of signaling cascades including phospho-tyrosine kinases such as src (54), FAK (83, 84) and other kinases such as PKC (53) and erk (42); random motility in response to serum, scratch wounding, HA and PDGF; progression through G_2M of the cell cycle (55) and tubule formation during angiogenesis (85). Intracellular RHAMM forms that associate with the mitotic spindle and centrosomes have been proposed to be required for spindle stability since strong over-expression of RHAMM or injection of RHAMM antibodies results in the formation of multiple spindles in HeLa cells (82). Both cell surface RHAMM and intracellular RHAMM forms are required for the transformation of fibroblasts (86), repair of excisional wounds (Tolg and Turley, unpublished data) and rapid growth at sub-confluence (86). Whether or not cell surface RHAMM forms coordinate with intracellular RHAMM forms to regulate these functions remains to be investigated.

IV. Classification of RHAMM Protein Forms by Binding Properties and Structure

RHAMM is an extensively coiled-coil protein that has a basic amino-terminal globular head (70) and a $B(X)_7B$ HA-binding domain in its C-terminus (87). It occurs as multiple proteins characterized by multiple molecular weights, some of which are generated by alternative splicing of a single full-length transcript (81,88,89). Since over-expression of an N-terminal truncated RHAMM form is transforming in fibroblasts (89), the shorter RHAMM protein forms appear to represent activated forms. These active shorter RHAMM protein forms are expressed in human tumors (81,90,91) and following tissue injury (92). RHAMM has been shown to localize at the cell surface and in various intracellular compartments including the cytosol, the nucleus (81,93), the cytoskeleton (where it can associate with both the actin and microtubule cytoskeletons (81,93)), centrosomes (82) and cell lamellae. Furthermore, intracellular RHAMM, which contains multiple putative protein kinase recognition sites as well as known sites for protein–protein interactions including SH2 and SH3 binding sites has also been shown to associate with signaling proteins such as erk1 and mek1 (Fig. 4) (42,89).

RHAMM has recently been proposed to belong to several protein families in addition to its classification as a hyaladherin based on its different functions, sub-cellular localization and association with specific proteins or polysaccharides. Its classification as an HABP is based on its ability to bind HA (70,94), as a microtubule-associated protein (MAP) because of its association with the microtubule cytoskeleton (81) and a transforming acidic coiled-coil (TACC) protein because of its localization at the centrosome, as well as putative phylogenetic similarity to other TACC proteins (82). However, these characterizations do not take into account the apparent functional and structural complexity of RHAMM protein forms. The criteria for including RHAMM as a hyaladherin were discussed in detail earlier and will not be addressed further.

A. RHAMM as MAP Proteins

The term MAP refers to a large family of proteins that share the capacity to associate directly and reversibly with microtubules, probably with a regulatory role, co-polymerizing with them during cycles of assembly and disassembly (95–97). These proteins may not have sequence homology as a group, but are able to bind directly to the acidic COOH-terminus of tubulin and have a widespread distribution among cells, though certain MAPs (e.g., tau) are limited to specific cell types (98–101). Therefore, the MAP family of proteins is a functional classification defined as proteins that bind directly to microtubules and that modulate microtubule stability.

Assmann et al. (81) reported an association of intracellular RHAMM, referred to in their publication as IHABP, with microtubules in interphase and mitotic cells, but also with actin filaments. Furthermore, they reported that a region within the basic N-terminal globular domain of RHAMM is responsible for this association (81). Based upon its association with interphase and mitotic spindle microtubules, RHAMM was proposed to be a new member of the MAP family of proteins, although a functional consequence of RHAMM expression on microtubule stability was not assessed (81). A primary function of intracellular RHAMM as a linker protein that modulated the interactions between the microtubule and actin cytoskeletons was proposed (81). This last property and the association of RHAMM with the actin cytoskeleton (81) are not typical of MAPs. Further, in the absence of any evidence for a direct binding to tubulin and functional consequences to microtubule stability, the classification of intracellular RHAMM forms as MAP proteins is premature.

B. RHAMM as a TACC

The TACC family of proteins include human TACC3, murine TACC3, human TACC2/AZU-1/ECTACC, human TACC1, D-TACC and murine AINT (102). Members of this family of proteins are defined by the presence of the so-called TACC domain, a predicted coiled-coil region in their carboxyl terminus (102). In fact, the TACC proteins show the most similarity in this region as they share very little sequence identity throughout the rest of their protein sequence (Fig. 3A) (102). In addition to the TACC domain, all members of this family have an acidic isoelectric point and a proline-rich sequence outside the TACC domain (102,103). Members of this closely related family of proteins are concentrated at the centrosome and have been implicated in processes such as microtubule stabilization, acentrosomal spindle assembly, translational regulation, hematopoietic development and cancer progression (102,104,105). Another common characteristic of TACC genes is their evolutionary conserved relationship with the FGFR genes (106,107). TACC genes may have evolved from a common ancestor as a result of two successive duplications of the chromosomal region accommodating the FGFR and TACC genes. However, this co-evolution of the TACC and FGFR genes was believed to be incomplete because while the human genome contains four FGFR genes, there appeared to be only three TACC genes (human TACC1, 2 and 3 genes map proximal to the FGFR1, 2 and 3 genes. No identified TACC genes map proximal to FGFR4) (82,103). Intracellular RHAMM, which has also been shown to associate with microtubules and is involved in cancer progression (see Section V. C below) also localizes to the centrosome through a basic leucine zipper located in the C-terminal region of the protein, similar to TACC proteins (82). Furthermore, Maxwell et al. (82) noted that the RHAMM gene, like that of TACCS, maps proximal to the FGFR4 gene and largely based on this evidence proposed that RHAMM represents TACC4. However, RHAMM, which has an isoelectric point of 5.8, shares little sequence identity with the centrosomal targeting TACC domain that is highly homologous to the TACC family of proteins and any homology between the two proteins are limited primarily to the coiled-coil amino acids (Fig. 3A). Furthermore, RHAMM associates with dynein and dynactin (82), while TACC proteins do not and

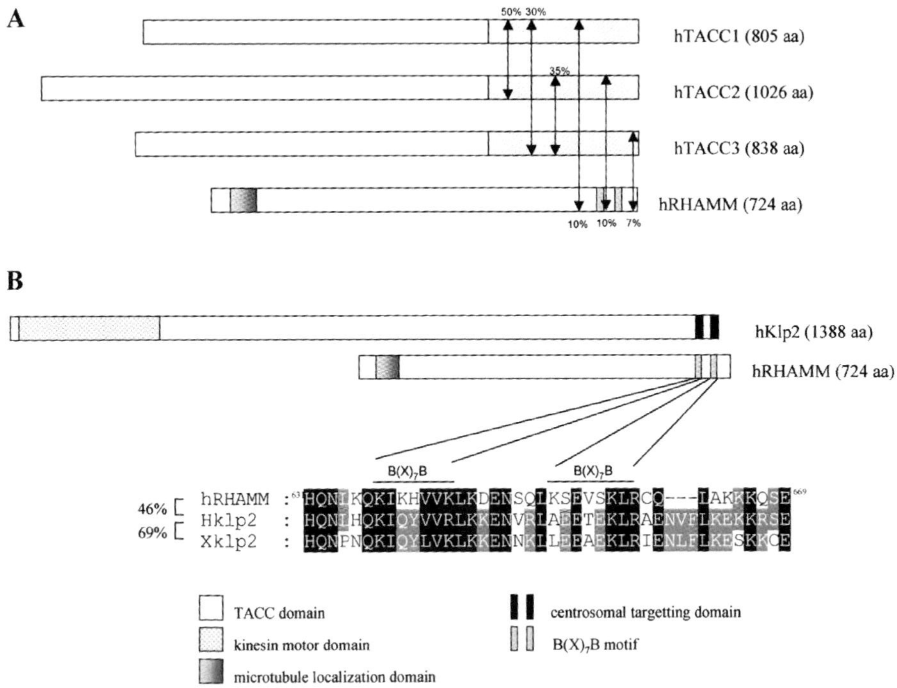

Figure 3 Comparison of RHAMM with the TACC and Klp family of proteins. (A) Schematic diagram of putative structure of RHAMM with the TACC family of proteins. The TACC proteins all contain significant sequence identity in the TACC domain located at their extreme carboxyl-terminus, though they contain little sequence identity in other regions of the proteins. RHAMM, which has been reported to be a member of the TACC family does not contain a carboxyl terminal TACC domain and otherwise has little sequence identity (7–10%) with TACC family members. Any observed identity is located primarily within the amino acids responsible for the coiled-coil structure. (B) The region of RHAMM reported to be required for its centrosomal targeting overlaps with the two $B(X)_7B$ motifs required for hyaluronan binding. This region more closely resembles the centrosomal targeting domain of the Klp (kinesin) family of proteins than that of TACC proteins. However, RHAMM does not contain the highly conserved kinesin motor domain and so is not included in this family of microtubule binding proteins.

RHAMM has no known functional consequence on microtubule stability although high over-expression or microinjection anti-RHAMM antibodies affects mitotic spindle integrity (82). Other more likely candidate genes have been proposed for the role of TACC4. For example, Steadman et al. (103) isolated a centrosomal protein with significant sequence identity to TACC proteins using an A Kinase Anchoring Protein (AKAP350) as bait in a yeast two-hybrid screen of a rabbit parietal cell library. The AKAP350 binding protein was found to have an isoelectric point of 4.6, a coiled-coil motif encompassing the carboxyl-terminal 200 amino acids (TACC domain) required for its association with centrosomes and structural predictions indicated a proline-rich region in its N-terminal domain (103). These two studies raise the important question of how to classify proteins into functional families, in this case, TACC proteins. If sequence homology is considered to be an important common feature for classification to a protein family, the only significant sequence identity between any of the known family members is the C-terminal coiled-coil TACC domain and its presence is therefore crucial for inclusion into this family (82,102,105,108). Since RHAMM does not contain this domain (Fig. 3A) its classification as a TACC protein rests upon its association with centrosomes and its chromosomal localization near FGFR4 (82). In fact, the region of RHAMM found to be important for targeting the centrosome more closely resembles the centrosomal targeting motif of the kinesin-like protein (Klp) family, though the absence of the highly conserved molecular motor domain precludes RHAMM from being a member of the Klp family of proteins (Fig. 3B) (82,109,110). Maxwell et al. raised the important possibility that RHAMM may affect mitotic spindle polarity, which resembles a TACC function (82,108). However, this property must be reconciled with other established functions of RHAMM including its HA-binding ability, requirement for cell motility (70) and progression through G_2M of the cell cycle (55), as well as its association with, and role in activation of protein tyrosine kinases (54), PKC (53) and erk (42,111) in order to assign a classification that illuminates, rather than obscures, the biological roles of RHAMM.

C. RHAMM is a Multi-functional Adapter/Targeting Protein

Murine and human RHAMM genes do not contain significant sequence homology with any single gene from lower organisms for which extensive genomic sequence is available (e.g., *Drosophila* or *C. elegans*), and therefore is admittedly a difficult gene to assign to a family of proteins using standard criteria. Reviewing all of the established functional properties and binding associations of RHAMM, even if collectively these do not fit neatly into currently accepted classification paradigms, and combining these known functional properties with a non-biased analysis of RHAMM sequences is a legitimate alternative approach to establish relationships of RHAMM to specific protein families. This attempt may still be difficult since RHAMM, like other proteins that occur in and outside of the cell, most likely performs distinct functions when it is present on the cell surface vs. when it is present in multiple intracellular compartments. For example, the HA-binding domain of RHAMM overlaps with sequence required for its association with erk as well as with dynein and dynactin (42,82). Very likely, the cell surface form of RHAMM utilizes this domain to bind to HA while intracellular RHAMM forms utilize it for their associations with erk, dynein and dynactin. RHAMM's localization to many sub-cellular compartments, its demonstrated association with at least two kinases, src (54) and erk (42) and its potential for interaction with additional regulatory proteins and kinases suggest that intracellular RHAMM forms function as adapter proteins much like AKAPs, which are docking/targeting proteins for PKA. AKAPs were originally named for their common association with protein kinase A (112–114) although they are otherwise a diverse family of proteins and contain multiple putative docking sites for other regulatory proteins and kinases. Studies have shown that anchored PKA is important in multiple cellular functions including gene transcription, hormone-mediated insulin secretion and ion-channel modulation (113). Furthermore, AKAPs coordinate multiple components of signal transduction pathways to the PKA pathway through their association with additional signaling molecules (113–115). Predicted structural analysis of RHAMM shows that in addition to being a coiled-coil protein (a secondary structure that facilitates protein–protein interactions), RHAMM also has a number of putative SH2 and SH3 protein–protein interaction sites, as well as docking sites for other signaling molecules, including the p85 regulatory subunit of PI 3-kinase and erk1/2 and kinase phospho-acceptor sites (Fig. 4). Indeed, RHAMM is required for signaling through the ras transformation pathway (89) and for activation of erk kinase through PDGF (42), both of which are consistent with an accessory function of RHAMM similar to that performed by AKAPs for PKA. RHAMM's association with the cytoskeleton (81) and centrosome (82) may therefore be indirect and mediated, as an example, through its binding to erk kinase. Erk is present in the same sub-cellular compartments as RHAMM and its association with microtubules mediates stability of this cytoskeletal network (116–118). For example, erk is present in centrosomes and Ahn et al. have identified the centrosomal protein, RanBP1 as an erk substrate (119–122). Furthermore, erk activity has

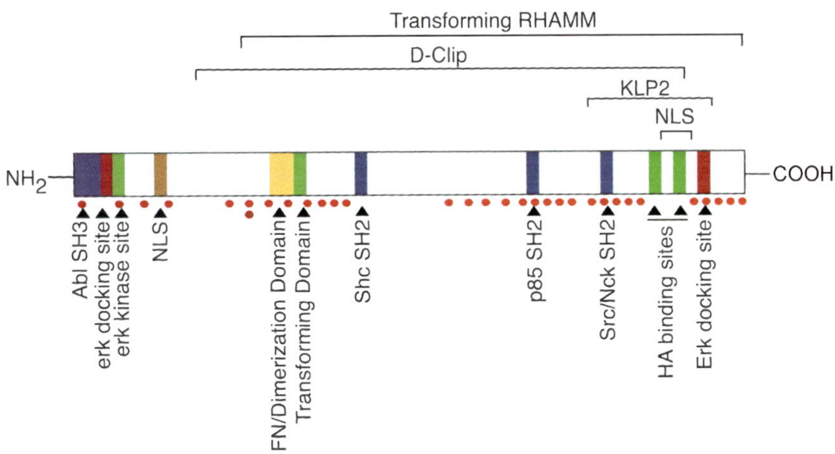

Figure 4 Schematic diagram of RHAMM as an adaptor protein. The identification of multiple protein kinase recognitions sites, its established association with kinases such as src and erk, as well as putative sites for protein–protein interactions including SH2 and SH3 bindings sites, together with its localization to multiple sub-cellular compartments, is consistent with a proposed function of intracellular RHAMM as an adaptor/accessory protein that may link and target multiple signaling cascades in a manner resembling the AKAP group of adaptor proteins.

been linked to genomic stability (116–118) and therefore, the proposed effect of RHAMM in genomic stability and mitotic figure integrity may also be a function of its association with erk.

The difficulty in classifying RHAMM to any single protein family based on functional characteristics is not a unique problem as many proteins are multi-functional and could potentially be included in a number of different protein families. A relevant example is MAP2, which in addition to being a functional MAP was also the first protein found to co-purify and interact with the RII subunit of the PKA haloenzyme, and is in fact a functional AKAP (115, 123–125). Based on the above analysis, and similar to the classification of MAP2, we propose that the intracellular RHAMM is an adapter protein for the erk family of MAP kinases. Cell surface RHAMM may similarly act as an adaptor protein linking growth factor receptors, integrins and possibly other hyaladherins to one another (Fig. 5).

V. RHAMM Expression Influences Wound Repair and Tumor Progression

A. Alternative Use of CD44 and RHAMM in Tissue Response to Injury Processes

The importance of expression profiles of HABPs for cell response to HA was recently demonstrated by a study performed by Nedvetzky et al. (126). In this study, the role of CD44 in collagen II-induced arthritis was analyzed. Surprisingly, whereas CD44 − /− mice developed arthritis 25 days after a single injection of collagen II, a second injection was necessary to induce arthritis in wild-type (wt) mice. This difference was coupled to invading inflammatory cells because wt mice injected with spleenocytes from CD44 − /− mice respond like CD44 − /− mice and vice versa. The degree of arthritis in CD44 − /− and wt mice was reduced by injection of hyaluronidase, suggesting a HABP compensated for the function of CD44 in CD44 − /− mice. Collagen-induced arthritis induced the expression of a smaller, possibly activated, RHAMM isoform not seen in non-inflamed tissue, suggesting RHAMM might be the HABP compensating for the absence of CD44. Anti-RHAMM antibodies had a stronger blocking effect on *in vitro* migration of CD44 − /− spleenocytes through HA or fibronectin-coated boyden chamber filter than on migration of wt spleenocytes (126). Furthermore, injection of anti-RHAMM antibodies reduced the degree of arthritis in CD44 − /− mice, but not in wt mice, confirming that RHAMM was compensating for the absence of CD44. In the presence of CD44, RHAMM had only minor effects on the inflammation process, whereas in the absence of CD44, RHAMM not only compensated for the absence of CD44 but also enhanced the inflammation process resulting in a more complete joint destruction than that observed in wt joints. Although the precise mechanism is unknown, it can be speculated that in the absence of CD44, increased binding of HA to RHAMM and, therefore, a predominance of signaling via RHAMM occurred. These results

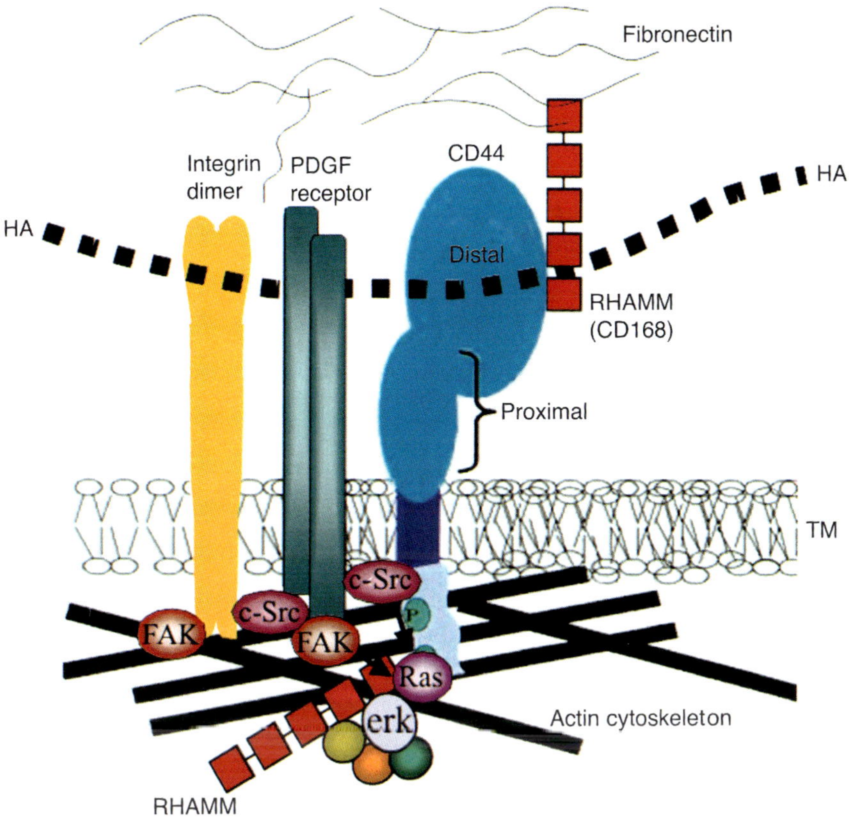

Figure 5 A model for proposed protein–protein and protein–HA interactions mediated by RHAMM. Cell surface RHAMM (CD168) binds to ECM constituents, HA, fibronectin and heparin. RHAMM is also proposed to associate directly or indirectly with other cell surface hyaladherins such as CD44, growth factor receptors such as the PDGFR and integrins such as the fibronectin receptor to create large signaling complexes that are at least in part also mediated by HA–protein interactions. Inside the cell, RHAMM is proposed to associate with kinases such as erk, linking them to the HA-mediated CD44/growth factor receptor/integrin complexes and to the cytoskeleton. Intracellular RHAMM forms are proposed to function primarily as adaptor proteins that regulate the activation, targeting and complexing of kinases with multiple signaling pathways.

predict that the function of RHAMM might be suppressed by signaling via CD44 so that in the absence of CD44 uncontrolled inflammation occurs. From this and other studies, a paradigm is emerging whereupon CD44 and RHAMM are each required, but perform separate functions in a process, e.g., migration or tubule formation (85). In each other's absence, the functions associated with the one receptor are amplified. For example, antibody-blocking studies have established that during tubulogenesis associated with angiogenesis, RHAMM is required for migration of endothelial cells while CD44 is required for regulated proliferation.

We would predict, however, that interplay between RHAMM and CD44 regulates different aspects of processes. For example, CD44 is required for attachment to HA, and RHAMM is required for migration of these cells on HA, possibly mediating detachment (53,85,127,128). Our model would predict that if CD44 is required for attachment and RHAMM is required for detachment during the cycles of attachment/detachment associated with migration, RHAMM − /− cells would have predominantly CD44-mediated signaling resulting in flattened, well-attached cells that migrate poorly in response to HA. Whatever its specific role in processes such as migration or proliferation might be, RHAMM hyper-expression appears to be a hallmark of aggressive tumors, and we will now review in detail the association of RHAMM with wound repair and clinical tumor progression.

B. RHAMM and Wound Repair

During tissue homeostasis, expression and therefore signaling via RHAMM is suppressed whereas during tissue repair and remodeling following wounding RHAMM expression is upregulated (129). For example, using incisional and excisional wounds of fetal skin transplanted sub-cutaneously on immune-suppressed mice, Lovvorn et al. analyzed expression of the HABPs RHAMM and CD44 as well as HA content. Between 1 and 7 days after wounding, RHAMM and CD44 expression were upregulated at the edges of excisional, but not incisional wounds. This expression correlated with decreased HA concentration in excisional wounds compared to incisional wounds and is in agreement with a function of both CD44 and RHAMM in the uptake and subsequent degradation of HA. Because decreased concentration of high-molecular weight HA is thought to be required for fibroplasia and scar formation (39,130), Lovvorn et al. speculated that strategies limiting the expression of CD44 and RHAMM during wound repair might be useful for controlling scar formation. The upregulation of RHAMM during wound repair is not restricted to skin wounds. Capolicchio et al. (131) demonstrated an increase in RHAMM expression in a model of acute stretch injury of bladder where highest expression occurred 5–10 h after stretch injury. As in collagen-induced arthritis (126), the size of RHAMM isoforms changed, shifting from 55 to 120 kDa as a result of bladder stretch injury. The importance of RHAMM isoforms in wound repair was also demonstrated *in vitro* by analyzing the re-surfacing of scratch wounds in smooth muscle cell monolayers (92). Expression of a short (70 kDa) RHAMM isoform was upregulated only 1 h after injury and this correlated with the appearance of cell surface RHAMM at the wound edge. Cell migration was blocked by anti-RHAMM antibodies, demonstrating the importance of cell surface RHAMM (CD168) in directed cell migration (92). These results suggest RHAMM might play an important role in wound repair although the precise functions it regulates during this process have not yet been dissected.

C. RHAMM and Tumor Progression

As we have noted earlier, the tissue remodeling that occurs with wound repair resembles that occurring during neoplastic progression (56–59). In particular, the presence of chronic tissue fibrosis, associated with hyper-expression of both CD44 and RHAMM appears to provide favorable conditions for sustaining neoplastic conversion and progression. The similarity between the two processes is reflected by the fact that RHAMM mRNA expression is specifically upregulated in tumors and the appearance of short RHAMM isoforms, particularly those that appear on the cell surface also correlates with tumor progression.

Serological identification of antigens by recombinant expression cloning, SEREX, identified antibodies against RHAMM in acute myeloid leukemia (42%), chronic myeloid leukemia (31%), melanoma (83%), renal cell carcinoma (40%), breast cancer (67%) as well as ovarian carcinoma (50%) (132), and real-time PCR revealed a 1–13.6 fold increase of RHAMM expression in 97% of colon cancer (133). Furthermore, because of the presence of mononucleotide repeat sequences in the coding region, RHAMM is frequently mutated in subsets of colorectal cancers with defects in mismatch repair genes, suggesting a selective pressure for the accumulation of RHAMM mutations (134). Intriguingly, several reports suggest that small tumor subsets are responsible for hyper-expression of both CD44 and RHAMM in breast cancers (90,135). Further, subsets of tumor cells from breast primary tumors that hyper-express CD44 also express stem/progenitor markers and are several hundred times more tumorigenic upon transplantation into immune-compromised mice than tumor cells that do not highly express these markers (62). Similarly, immuno-staining of breast cancer samples identified subsets of cells with high RHAMM expression (90,135) (Fig. 6) and their presence in a primary tumor was significantly associated with poor clinical outcome and with an occurrence of lymphatic metastasis (90). Possible hyper-expression of either CD44 or RHAMM confers a selective advantage to tumor cells that permits colonization, and part of this ability may be related to suppression of pro-apoptotic pathways. For example, RHAMM hyper-expression strongly correlates with erk kinase hyper-expression, a MAP kinase that has been shown to act on the HA-regulated signaling pathway that confers resistance to anchorage dependent apoptosis or anoikis (136–139). Elevated RHAMM has been reported for other tumor types as well. For example in endometrial carcinoma, 100% of tumors in patients with lymph node involvement were positive for RHAMM expression whereas RHAMM expression was found in 50.7% of tumors in patients without lymph node involvement and 13% of normal control tissue (140). As for wound repair, smaller, possibly activated isoforms of RHAMM are predominantly found in tumor tissue. RT-PCR analysis identified full length RHAMM in 49% of random c-DNA clones isolated from multiple myeloma patients (88). A smaller RHAMM isoform missing exon 4, $RHAMM^{-48}$ was found in 47% of c-DNA clones whereas only one out of eight normal donors expressed $RHAMM^{-48}$ (88).

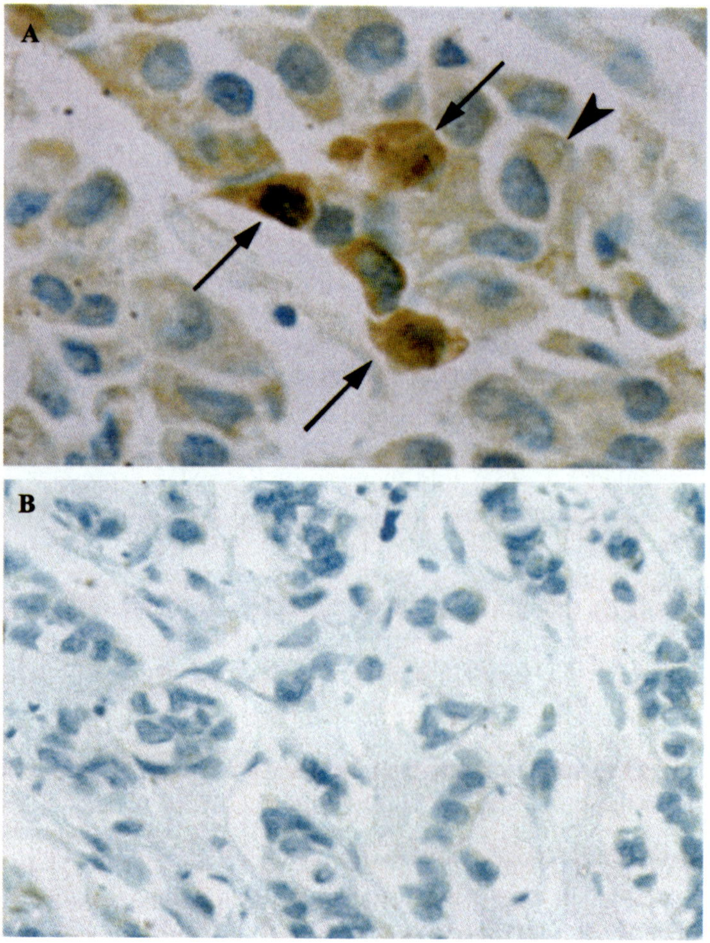

Figure 6 RHAMM localization in breast tumor sections. Paraffin sections of breast tumor samples were stained with anti-RHAMM antibodies. Staining intensity varied between tumors isolated from different patients, but the presence of foci of RHAMM over-expressing tumor cells is prognostic of poor outcome. Foci shown in (A) (arrows) contain nuclear (arrow) and cytoplasmic (arrowhead) staining. Image of another tumor shown in (B) has little RHAMM staining and no foci of RHAMM-hyperexpressing tumor cells for comparison with A. Magnification: (A) 580 ×, (B) 360 ×.

Furthermore, a study by Greiner et al. (132) demonstrated RHAMM expression in 100% of acute myeloid leukemia, 83% of chronic myeloid leukemia and 100% of renal cell carcinoma patients and expression of RHAMM^{-48} was found in all RHAMM-positive samples. Comparison of RHAMM expression between astrocytoma cell lines and tissues with normal astrocytes and brain tissue revealed a 70 kDa isoform in addition to the 86 kDa full length RHAMM in cancerous cells or tissues (141). Furthermore, a splice variant missing exon 4 was predominantly expressed in colon tumors compared to normal tissue (142). These results link RHAMM mRNA and protein hyper-expression to human neoplasia and suggest that small protein forms of RHAMM, previously shown to be transforming in murine cells *in vitro* (89), predominate during neoplastic progression, thus probably representing activated forms of RHAMM.

The clinical correlation between RHAMM expression and tumors suggests that this gene plays key functions in either neoplastic conversion and/or progression. The possibility that RHAMM might play a key role in both processes was first suggested by the demonstration that overexpression of short RHAMM forms transformed 10T1/2 and 3T3 fibroblasts to tumors that were metastatic in tail vein assays. Conversely, expression of either a dominant inhibitory form of RHAMM that cannot bind to HA or soluble recombinant RHAMM forms that compete with cell surface RHAMM (CD168) for HA-binding, blocked ras-mediated transformation (89). The relevance of these *in vitro* studies to tumor formation originating *in vitro* has recently been confirmed by studies using mice in which the RHAMM gene has been deleted by homologous recombination. RHAMM $-/-$ mice were crossed with mice heterozygous for a mutation in the Adenomatous Polyposis Coli (APC) tumor suppressor gene that blocks the tumor suppressor function of this scaffold protein and which results in elevated levels of beta-catenin protein and, as a consequence, development of aggressive fibromatosis (desmoid) tumors. The APC mutation in this transgenic mouse line differs from that of the 'min' mouse, which is predisposed to aggressive intestinal tumors. Although upper intestinal tract pre-neoplastic polyps are formed in the APC transgenic line, these do not proceed to frank neoplasms over the life span of the mice. Instead, the mice die prematurely from desmoid tumors (143). Loss of RHAMM significantly reduced both the number of desmoid tumors and the size of the tumors, the latter providing a measure of the invasiveness of the desmoid tumor (86). This defect was associated with reduced proliferation of RHAMM $-/-$ tumor cells in response to serum supplements and to PDGF, when the cells were sub-confluent. At high culture confluence a difference in proliferation was not observed. Interestingly, the absence of RHAMM had no effect on the number of pre-neoplastic polyps of the upper intestinal tract (86). While gastrointestinal tumors are derived from epithelial cells, desmoid tumors consist of mesenchymal fibroblastic cells resembling cells that predominate in the granulation phase of wound healing, a stage in wound repair where HA plays a key role (26,144). Collectively, these results indicate a role for RHAMM in the neoplastic transformation of mesenchymal cells and for a predominance of RHAMM function in sparse culture conditions such as those most likely found during wound repair and during tumor invasion and metastasis. These results also raise the intriguing possibility that RHAMM function may predominate following EMT of parenchymal cells, a process that is associated with aggressive tumors and poor clinical outcome.

Several studies have addressed how RHAMM might contribute to cell proliferation. For example, Mohapatra et al. (55) showed that soluble recombinant RHAMM protein, down-regulation of RHAMM function by expression of a dominant negative mutant or antisense, results in G_2M arrest of H-ras transformed fibroblasts and suppression of tumor formation as a result of decreased expression of Cdc2/CyclinB1. Maxwell et al. (82) showed that over-expression of full length RHAMM or RHAMM containing a deletion of a putative Cdc2 phosphorylation site in HeLa or Jurkat cells leads to cell cycle arrest and accumulation of cells in prometaphase/metaphase or prophase and abnormal mitotic spindle formation. Although it cannot be excluded that the observed chromosome spindle breakdown is the result of an artificially high RHAMM expression, these results suggest that either increased or decreased RHAMM expression/function blocks cell proliferation. Furthermore, full-length RHAMM may perform a dual inhibitory/stimulatory role in proliferation, unlike the shorter RHAMM forms, which seem to predominantly stimulate. This would be consistent with constitutive expression of shorter RHAMM forms in cancer. The study of Maxwell et al. (82) suggested that RHAMM plays a role in genomic stability although this possibility would be better analyzed by expressing activated RHAMM forms at lower levels than were used in this study. Hyper-expression of several proteins, e.g., beta-catenin, has been shown to promote genomic instability via artifactual processes, i.e., high protein levels result in abnormal accumulation in atypical sub-cellular compartments (145,146). In any event, the accumulation of additional mutations or changes in the cell's stromal microenvironment might allow tumor cells to increasingly rely upon RHAMM as an ECM receptor. This would result in a progressively aggressive tumor, similar to the aggressiveness of CD44 − /− spleenocytes that rely upon RHAMM for functions associated with inflammation and which, in the absence of CD44, result in a strongly enhanced destruction of normal tissues.

VI. Conclusions

The hyaladherin RHAMM belongs to a new group of proteins whose protein structure and cellular localization challenges our current understanding of the potential for polymorphic protein structure and function (76). Using current information, perhaps the best way to describe RHAMM is as an adapter protein connecting multiple signaling pathways with each other, both at the cell surface and inside the cell, and with structures such as the cytoskeleton or chromatin in the nucleus. This adapter function is manifested by direct interactions between RHAMM and receptors such as CD44 and PDGF on the cell surface, and with kinases, microtubule or actin binding proteins inside the cell. RHAMM may also complex a diversity of proteins through mutual binding to HA. A model diagramming such complexes is shown in Fig. 5. These interactions are predicted to contribute to the cells' ability to respond to changes in microenvironment, particularly those involved in wound repair and neoplastic conversion. Wound repair and tumor progression are very dynamic processes characterized by complex changes in cell populations, ECM and cell–ECM interactions. Many reports suggest that increased accumulation of HA and the presence of HA fragments together with upregulated RHAMM expression and the appearance of smaller activated RHAMM isoforms are hallmarks of both processes. Recent *in vitro* and *in vivo* studies suggest important and central functions of RHAMM in both processes. The challenges created by the unique cellular localization, regulation and protein structure of RHAMM forms could contribute to paradigm shifts in our understanding of dynamic processes and to new tumor therapies as well as improvements in wound repair.

Acknowledgements

This work was funded by a CIHR grant (MOP 57694) and Cancer Research Society to ET (MOP 57694), and a Breast Cancer Society of Canada fellowship and CIHR fellowship (UST-63811) to SH.

References

1. Spicer AP, Tien JL, Joo A, Bowling RA Jr. Investigation of hyaluronan function in the mouse through targeted mutagenesis. Glycoconj J 2002; 19: 341–345.
2. McDonald JA, Camenisch TD. Hyaluronan: Genetic insights into the complex biology of a simple polysaccharide. Glycoconj J 2002; 19:331–339.
3. DeAngelis PL. Evolution of glycosaminoglycans and their glycosyltransferases: implications for the extracellular matrices of animals and the capsules of pathogenic bacteria. Anat Rec 2002; 268:317–326.
4. Tammi MI, Day AJ, Turley EA. Hyaluronan and homeostasis: a balancing act. J Biol Chem 2002; 277:4581–4584.
5. McCourt PA. How does the hyaluronan scrap-yard operate? Matrix Biol 1999; 18:427–432.
6. Itano N, Kimata K. Mammalian hyaluronan synthases. IUBMB Life 2002; 54: 195–199.
7. Weigel PH, Hascall VC, Tammi M. Hyaluronan synthases. J Biol Chem 1997; 272: 13997–14000.
8. DeAngelis PL. Hyaluronan synthases: fascinating glycosyltransferases from vertebrates, bacterial pathogens, and algal viruses. Cell Mol Life Sci 1999; 56:670–682.
9. Bissell MJ, Radisky D. Putting tumours in context. Nat Rev Cancer 2001; 1:46–54.
10. Cox R. The multi-step nature of carcinogenesis and the implications for risk analysis. Int J Radiat Biol 1998; 73:373–376.
11. Toole BP, Wight TN, Tammi MI. Hyaluronan–cell interactions in cancer and vascular disease. J Biol Chem 2002; 277:4593–4596.
12. Turino GM, Cantor JO. Hyaluronan in respiratory injury and repair. Am J Respir Crit Care Med 2003; 167:1169–1175.
13. Juhlin L. Hyaluronan in skin. J Intern Med 1997; 242:61–66.
14. Casey G. Wound repair: advanced dressing materials. Nurs Stand 2002; 17:49–53 (quiz 54, 56).
15. Saika S, Ohnishi Y, Ooshima A, Liu CY, Kao WW. Epithelial repair: roles of extracellular matrix. Cornea 2002; 21:S23–S29.

16. Turley EA, Noble PW, Bourguignon LY. Signaling properties of hyaluronan receptors. J Biol Chem 2002; 277:4589–4592.
17. Cichy J, Pure E. The liberation of CD44. J Cell Biol 2003; 161:839–843.
18. Christofori G. Changing neighbours, changing behaviour: cell adhesion molecule-mediated signalling during tumour progression. EMBO J 2003; 22:2318–2323.
19. Jothy S. CD44 and its partners in metastasis. Clin Exp Metastasis 2003; 20: 195–201.
20. Jackson DG. The lymphatics revisited: new perspectives from the hyaluronan receptor LYVE-1. Trends Cardiovasc Med 2003; 13:1–7.
21. Ponta H, Sherman L, Herrlich PA. CD44: from adhesion molecules to signalling regulators. Nat Rev Mol Cell Biol 2003; 4:33–45.
22. Naor D, Nedvetzki S, Golan I, Melnik L, Faitelson Y. CD44 in cancer. Crit Rev Clin Lab Sci 2002; 39:527–579.
23. Bullard KM, Longaker MT, Lorenz HP. Fetal wound healing: current biology. World J Surg 2003; 27:54–61.
24. Anderson I. The properties of hyaluronan and its role in wound healing. Prof Nurs 2001; 17:232–235.
25. Moseley R, Waddington RJ, Embery G. Hyaluronan and its potential role in periodontal healing. Dent Update 2002; 29:144–148.
26. Chen WY, Abatangelo G. Functions of hyaluronan in wound repair. Wound Repair Regen 1999; 7:79–89.
27. Tuxhorn JA, Ayala GE, Rowley DR. Reactive stroma in prostate cancer progression. J Urol 2001; 166:2472–2483.
28. Wiesen JF, Werb Z. The role of stromelysin-1 in stromal–epithelial interactions and cancer. Enzyme Protein 1996; 49:174–181.
29. Triffitt JT, Oreffo RO, Virdi AS, Xia Z. Osteogenic stem-cell characterization and development: potentials for cytotherapy. Cytotherapy 2001; 3:413–416.
30. Huber MA, Kraut N, Park JE, Schubert RD, Rettig WJ, Peter RU, Garin-Chesa P. Fibroblast activation protein: differential expression and serine protease activity in reactive stromal fibroblasts of melanocytic skin tumors. J Investig Dermatol 2003; 120:182–188.
31. Thomasset N, Lochter A, Sympson CJ, Lund LR, Williams DR, Behrendtsen O, Werb Z, Bissell MJ. Expression of autoactivated stromelysin-1 in mammary glands of transgenic mice leads to a reactive stroma during early development. Am J Pathol 1998; 153:457–467.
32. Rossler A, Hinghofer-Szalkay H. Hyaluronan fragments: an information-carrying system? Horm Metab Res 2003; 35:67–68.
33. Rooney P, Kumar S, Ponting J, Wang M. The role of hyaluronan in tumour neovascularization (review). Int J Cancer 1995; 60:632–636.
34. Colwell AS, Longaker MT, Lorenz HP. Fetal wound healing. Front Biosci 2003; 8: S1240–S1248.
35. Croce MA, Dyne K, Boraldi F, Quaglino D Jr, Cetta G, Tiozzo R, Pasquali Ronchetti I. Hyaluronan affects protein and collagen synthesis by *in vitro* human skin fibroblasts. Tissue Cell 2001; 33:326–331.
36. Iocono JA, Ehrlich HP, Keefer KA, Krummel TM. Hyaluronan induces scarless repair in mouse limb organ culture. J Pediatr Surg 1998; 33:564–567.
37. Iocono JA, Krummel TM, Keefer KA, Allison GM, Paul H. Repeated additions of hyaluronan alters granulation tissue deposition in sponge implants in mice. Wound Repair Regen 1998; 6:442–448.

38. Jia C, Chen B, Arnold F. The effect of ultrapure hyaluronic acid with different molecular weights on the healing of porcine full thickness skin wound. Zhongguo Xiu Fu Chong Jian Wai Ke Za Zhi 1998; 12:197–200.
39. West DC, Shaw DM, Lorenz P, Adzick NS, Longaker MT. Fibrotic healing of adult and late gestation fetal wounds correlates with increased hyaluronidase activity and removal of hyaluronan. Int J Biochem Cell Biol 1997; 29:201–210.
40. Bourguignon LY, Singleton PA, Zhu H, Zhou B. Hyaluronan promotes signaling interaction between CD44 and the transforming growth factor beta receptor I in metastatic breast tumor cells. J Biol Chem 2002; 277:39703–39712.
41. Huang SS, Tang FM, Huang YH, Liu IH, Hsu SC, Chen ST, Huang JS. Cloning, expression, characterization and role in autocrine cell growth of cell surface retention sequence binding protein-1. J Biol Chem 2003; 278:43855–43869.
42. Zhang S, Chang MC, Zylka D, Turley S, Harrison R, Turley EA. The hyaluronan receptor RHAMM regulates extracellular-regulated kinase. J Biol Chem 1998; 273:11342–11348.
43. Vincent T, Molina L, Espert L, Mechti N. Hyaluronan, a major non-protein glycosaminoglycan component of the extracellular matrix in human bone marrow, mediates dexamethasone resistance in multiple myeloma. Br J Haematol 2003; 121:259–269.
44. Sato H, Takahashi T, Ide H, Fukushima T, Tabata M, Sekine F, Kobayashi K, Negishi M, Niwa Y. Antioxidant activity of synovial fluid, hyaluronic acid, and two subcomponents of hyaluronic acid. Synovial fluid scavenging effect is enhanced in rheumatoid arthritis patients. Arthritis Rheum 1988; 31:63–71.
45. Moseley R, Walker M, Waddington RJ, Chen WY. Comparison of the antioxidant properties of wound dressing materials—carboxymethylcellulose, hyaluronan benzyl ester and hyaluronan, towards polymorphonuclear leukocyte-derived reactive oxygen species. Biomaterials 2003; 24:1549–1557.
46. Moseley R, Leaver M, Walker M, Waddington RJ, Parsons D, Chen WY, Embery G. Comparison of the antioxidant properties of HYAFF-11p75, AQUACEL and hyaluronan towards reactive oxygen species *in vitro*. Biomaterials 2002; 23: 2255–2264.
47. Campo GM, Avenoso A, Campo S, Ferlazzo A, Altavilla D, Micali C, Calatroni A. Aromatic trap analysis of free radicals production in experimental collagen-induced arthritis in the rat: protective effect of glycosaminoglycans treatment. Free Radic Res 2003; 37:257–268.
48. Balogh GT, Illes J, Szekely Z, Forrai E, Gere A. Effect of different metal ions on the oxidative damage and antioxidant capacity of hyaluronic acid. Arch Biochem Biophys 2003; 410:76–82.
49. Debbasch C, De La Salle SB, Brignole F, Rat P, Warnet JM, Baudouin C. Cytoprotective effects of hyaluronic acid and Carbomer 934P in ocular surface epithelial cells. Investig Ophthalmol Vis Sci 2002; 43:3409–3415.
50. Itano N, Atsumi F, Sawai T, Yamada Y, Miyaishi O, Senga T, Hamaguchi M, Kimata K. Abnormal accumulation of hyaluronan matrix diminishes contact inhibition of cell growth and promotes cell migration. Proc Natl Acad Sci USA 2002; 99:3609–3614.
51. Karvinen S, Pasonen-Seppanen S, Hyttinen JM, Pienimaki JP, Torronen K, Jokela TA, Tammi MI, Tammi R. Keratinocyte growth factor stimulates migration and hyaluronan synthesis in the epidermis by activation of keratinocyte hyaluronan synthases 2 and 3. J Biol Chem 2003; 278:49495–49504.

52. Rilla K, Lammi MJ, Sironen R, Torronen K, Luukkonen M, Hascall VC, Midura RJ, Hyttinen M, Pelkonen J, Tammi M, Tammi R. Changed lamellipodial extension, adhesion plaques and migration in epidermal keratinocytes containing constitutively expressed sense and antisense hyaluronan synthase 2 (Has2) genes. J Cell Sci 2002; 115:3633–3643.
53. Hall CL, Collis LA, Bo AJ, Lange L, McNicol A, Gerrard JM, Turley EA. Fibroblasts require protein kinase C activation to respond to hyaluronan with increased locomotion. Matrix Biol 2001; 20:183–192.
54. Hall CL, Lange LA, Prober DA, Zhang S, Turley EA. pp60(c-src) is required for cell locomotion regulated by the hyaluronanreceptor RHAMM. Oncogene 1996; 13:2213–2224.
55. Mohapatra S, Yang X, Wright JA, Turley EA, Greenberg AH. Soluble hyaluronan receptor RHAMM induces mitotic arrest by suppressing Cdc2 and cyclin B1 expression. J Exp Med 1996; 183:1663–1668.
56. De Wever O, Mareel M. Role of tissue stroma in cancer cell invasion. J Pathol 2003; 200:429–447.
57. Wiseman BS, Werb Z. Stromal effects on mammary gland development and breast cancer. Science 2002; 296:1046–1049.
58. Stamenkovic I. Extracellular matrix remodelling: the role of matrix metalloproteinases. J Pathol 2003; 200:448–464.
59. Sternlicht MD, Bissell MJ, Werb Z. The matrix metalloproteinase stromelysin-1 acts as a natural mammary tumor promoter. Oncogene 2000; 19:1102–1113.
60. Auvinen P, Tammi R, Parkkinen J, Tammi M, Agren U, Johansson R, Hirvikoski P, Eskelinen M, Kosma VM. Hyaluronan in peritumoral stroma and malignant cells associates with breast cancer spreading and predicts survival. Am J Pathol 2000; 156:529–536.
61. Jacobson A, Rahmanian M, Rubin K, Heldin P. Expression of hyaluronan synthase 2 or hyaluronidase 1 differentially affect the growth rate of transplantable colon carcinoma cell tumors. Int J Cancer 2002; 102:212–219.
62. Al-Hajj M, Wicha MS, Benito-Hernandez A, Morrison SJ, Clarke MF. Prospective identification of tumorigenic breast cancer cells. Proc Natl Acad Sci USA 2003; 100:3983–3988.
63. Ghatak S, Misra S, Toole BP. Hyaluronan oligosaccharides inhibit anchorage-independent growth of tumor cells by suppressing the phosphoinositide 3-kinase/Akt cell survival pathway. J Biol Chem 2002; 277:38013–38020.
64. Zoltan-Jones A, Huang L, Ghatak S, Toole BP. Elevated hyaluronan production induces mesenchymal and transformed properties in epithelial cells. J Biol Chem 2003; 278:45801–45810.
65. Nehls V, Hayen W. Are hyaluronan receptors involved in three-dimensional cell migration? Histol Histopathol 2000; 15:629–636.
66. Yamada KM, Pankov R, Cukierman E. Dimensions and dynamics in integrin function. Braz J Med Biol Res 2003; 36:959–966.
67. Force T, Michael A, Kilter H, Haq S. Stretch-activated pathways and left ventricular remodeling. J Card Fail 2002; 8:S351–S358.
68. Turley EA. Hyaluronic acid stimulates protein kinase activity in intact cells and in an isolated protein complex. J Biol Chem 1989; 264:8951–8955.
69. Okamoto I, Kawano Y, Murakami D, Sasayama T, Araki N, Miki T, Wong AJ, Saya H. Proteolytic release of CD44 intracellular domain and its role in the CD44 signaling pathway. J Cell Biol 2001; 155:755–762.

70. Hardwick C, Hoare K, Owens R, Hohn HP, Hook M, Moore D, Cripps V, Austen L, Nance DM, Turley EA. Molecular cloning of a novel hyaluronan receptor that mediates tumor cell motility. J Cell Biol 1992; 117:1343–1350.
71. Day AJ, Prestwich GD. Hyaluronan-binding proteins: tying up the giant. J Biol Chem 2002; 277:4585–4588.
72. Yang B, Hall CL, Yang BL, Savani RC, Turley EA. Identification of a novel heparin binding domain in RHAMM and evidence that it modifies HA mediated locomotion of ras-transformed cells. J Cell Biochem 1994; 56:455–468.
73. Yu WH, Woessner JF Jr, McNeish JD, Stamenkovic I. CD44 anchors the assembly of matrilysin/MMP-7 with heparin-binding epidermal growth factor precursor and ErbB4 and regulates female reproductive organ remodeling. Genes Dev 2002; 16: 307–323.
74. Orian-Rousseau V, Chen L, Sleeman JP, Herrlich P, Ponta H. CD44 is required for two consecutive steps in HGF/c-Met signaling. Genes Dev 2002; 16: 3074–3086.
75. Kamikura DM, Khoury H, Maroun C, Naujokas MA, Park M. Enhanced transformation by a plasma membrane-associated met oncoprotein: activation of a phosphoinositide 3′-kinase-dependent autocrine loop involving hyaluronic acid and CD44. Mol Cell Biol 2000; 20:3482–3496.
76. Radisky DC, Hirai Y, Bissell MJ. Delivering the message: epimorphin and mammary epithelial morphogenesis. Trends Cell Biol 2003; 13:426–434.
77. Regeczy N, Kormos L, Szigetvari CM, Torbagyi E, Hajdu M, Gopcsa L, Banyai A, Paloczi K. Reactivity of new adhesion molecules on lymphocytes from patients with chronic graft versus host disease. Acta Microbiol Immunol Hung 2003; 50: 55–65.
78. Gares SL, Pilarski LM. Balancing thymocyte adhesion and motility: a functional linkage between beta1 integrins and the motility receptor RHAMM. Dev Immunol 2000; 7:209–225.
79. Lynn BD, Turley EA, Nagy JI. Subcellular distribution, calmodulin interaction, and mitochondrial association of the hyaluronan-binding protein RHAMM in rat brain. J Neurosci Res 2001; 65:6–16.
80. Pilarski LM, Masellis-Smith A, Belch AR, Yang B, Savani RC, Turley EA. RHAMM, a receptor for hyaluronan-mediated motility, on normal human lymphocytes, thymocytes and malignant B cells: a mediator in B cell malignancy? Leuk Lymphoma 1994; 14:363–374.
81. Assmann V, Jenkinson D, Marshall JF, Hart IR. The intracellular hyaluronan receptor RHAMM/IHABP interacts with microtubules and actin filaments. J Cell Sci 1999; 112:3943–3954.
82. Maxwell CA, Keats JJ, Crainie M, Sun X, Yen T, Shibuya E, Hendzel M, Chan G, Pilarski LM. RHAMM is a centrosomal protein that interacts with dynein and maintains spindle pole stability. Mol Biol Cell 2003; 14:2262–2276.
83. Lokeshwar VB, Selzer MG. Differences in hyaluronic acid-mediated functions and signaling in arterial, microvessel, and vein-derived human endothelial cells. J Biol Chem 2000; 275:27641–27649.
84. Hall CL, Wang C, Lange LA, Turley EA. Hyaluronan and the hyaluronan receptor RHAMM promote focal adhesion turnover and transient tyrosine kinase activity. J Cell Biol 1994; 126:575–588.
85. Savani RC, Cao G, Pooler PM, Zaman A, Zhou Z, DeLisser HM. Differential involvement of the hyaluronan (HA) receptors CD44 and receptor for HA-mediated

motility in endothelial cell function and angiogenesis. J Biol Chem 2001; 276: 36770–36778.
86. Tölg C, Poon R, Fodde R, Turley EA, Alman BA. Genetic deletion of receptor for hyaluronan-mediated motility (Rhamm) attenuates the formation of aggressive fibromatosis (desmoid tumor). Oncogene 2003; 22:6873–6882.
87. Yang B, Yang BL, Savani RC, Turley EA. Identification of a common hyaluronan binding motif in the hyaluronan binding proteins RHAMM, CD44 and link protein. EMBO J 1994; 13:286–296.
88. Crainie M, Belch AR, Mant MJ, Pilarski LM. Overexpression of the receptor for hyaluronan-mediated motility (RHAMM) characterizes the malignant clone in multiple myeloma: identification of three distinct RHAMM variants. Blood 1999; 93:1684–1696.
89. Hall CL, Yang B, Yang X, Zhang X, Turley M, Samuel S, Lange LA, Wang C, Curpen GD, Savani RC, Greenberg AH, Turley EA. Overexpression of the hyaluronan receptor RHAMM is transforming and is also required for H-ras transformation. Cell 1995; 82:19–26.
90. Wang C, Thor AD, Moore DH 2nd, Zhao Y, Kerschmann R, Stern R, Watson PH, Turley EA. The overexpression of RHAMM, a hyaluronan-binding protein that regulates ras signaling, correlates with overexpression of mitogen-activated protein kinase and is a significant parameter in breast cancer progression. Clin Cancer Res 1998; 4:567–576.
91. Li H, Guo L, Li JW, Liu N, Qi R, Liu J. Expression of hyaluronan receptors CD44 and RHAMM in stomach cancers: relevance with tumor progression. Int J Oncol 2000; 17:927–932.
92. Savani RC, Wang C, Yang B, Zhang S, Kinsella MG, Wight TN, Stern R, Nance DM, Turley EA. Migration of bovine aortic smooth muscle cells after wounding injury. The role of hyaluronan and RHAMM. J Clin Investig 1995; 95:1158–1168.
93. Entwistle J, Hall CL, Turley EA. HA receptors: regulators of signalling to the cytoskeleton. J Cell Biochem 1996; 61:569–577.
94. Turley EA, Moore D, Hayden LJ. Characterization of hyaluronate binding proteins isolated from 3T3 and murine sarcoma virus transformed 3T3 cells. Biochemistry 1987; 26:2997–3005.
95. Cassimeris L, Spittle C. Regulation of microtubule-associated proteins. Int Rev Cytol 2001; 210:163–226.
96. Maccioni RB, Cambiazo V. Role of microtubule-associated proteins in the control of microtubule assembly. Physiol Rev 1995; 75:835–864.
97. MacRae TH. Microtubule organization by cross-linking and bundling proteins. Biochim Biophys Acta 1992; 1160:145–155.
98. Mandelkow EM, Schweers O, Drewes G, Biernat J, Gustke N, Trinczek B, Mandelkow E. Structure, microtubule interactions, and phosphorylation of tau protein. Ann N Y Acad Sci 1996; 777:96–106.
99. Trinczek B, Biernat J, Baumann K, Mandelkow EM, Mandelkow E. Domains of tau protein, differential phosphorylation, and dynamic instability of microtubules. Mol Biol Cell 1995; 6:1887–1902.
100. Gustke N, Trinczek B, Biernat J, Mandelkow EM, Mandelkow E. Domains of tau protein and interactions with microtubules. Biochemistry 1994; 33:9511–9522.
101. Littauer UZ, Giveon D, Thierauf M, Ginzburg I, Ponstingl H. Common and distinct tubulin binding sites for microtubule-associated proteins. Proc Natl Acad Sci USA 1986; 83:7162–7166.

102. Lappin TR, Mullan RN, Stewart JP, Morgan NA, Thompson A, Maxwell AP. AINT/ERIC/TACC: an expanding family of proteins with C-terminal coiled coil domains. Leuk Lymphoma 2002; 43:1455–1459.
103. Steadman BT, Schmidt PH, Shanks RA, Lapierre LA, Goldenring JR. Transforming acidic coiled-coil-containing protein 4 interacts with centrosomal AKAP350 and the mitotic spindle apparatus. J Biol Chem 2002; 277:30165–30176.
104. Gergely F, Karlsson C, Still I, Cowell J, Kilmartin J, Raff JW. The TACC domain identifies a family of centrosomal proteins that can interact with microtubules. Proc Natl Acad Sci USA 2000; 97:14352–14357.
105. Gergely F. Centrosomal TACCtics. Bioessays 2002; 24:915–925.
106. Still IH, Vince P, Cowell JK. The third member of the transforming acidic coiled coil-containing gene family, TACC3, maps in 4p16, close to translocation breakpoints in multiple myeloma, and is upregulated in various cancer cell lines. Genomics 1999; 58:165–170.
107. Still IH, Hamilton M, Vince P, Wolfman A, Cowell JK. Cloning of TACC1, an embryonically expressed, potentially transforming coiled coil containing gene, from the 8p11 breast cancer amplicon. Oncogene 1999; 18:4032–4038.
108. Raff JW. Centrosomes and cancer: lessons from a TACC. Trends Cell Biol 2002; 12:222–225.
109. Kull FJ. Motor proteins of the kinesin superfamily: structure and mechanism. Essays Biochem 2000; 35:61–73.
110. Verhey KJ, Rapoport TA. Kinesin carries the signal. Trends Biochem Sci 2001; 26:545–550.
111. Aitken K, Bagli DJ. Stretch-induced bladder smooth muscle cell (SMC) proliferation is mediated by RHAMM-dependent extracellular-regulated kinase (erk) signaling. Urology 2001; 57:109.
112. Dell'Acqua ML, Scott JD. Protein kinase A anchoring. J Biol Chem 1997; 272: 12881–12884.
113. Michel JJ, Scott JD. AKAP mediated signal transduction. Annu Rev Pharmacol Toxicol 2002; 42:235–257.
114. Feliciello A, Gottesman ME, Avvedimento EV. The biological functions of A-kinase anchor proteins. J Mol Biol 2001; 308:99–114.
115. Dodge K, Scott JD. AKAP79 and the evolution of the AKAP model. FEBS Lett 2000; 476:58–61.
116. Jorquera R, Tanguay RM. Fumarylacetoacetate, the metabolite accumulating in hereditary tyrosinemia, activates the ERK pathway and induces mitotic abnormalities and genomic instability. Hum Mol Genet 2001; 10:1741–1752.
117. Willard FS, Crouch MF. MEK, ERK, and p90RSK are present on mitotic tubulin in Swiss 3T3 cells: a role for the MAP kinase pathway in regulating mitotic exit. Cell Signal 2001; 13:653–664.
118. Shapiro PS, Vaisberg E, Hunt AJ, Tolwinski NS, Whalen AM, McIntosh JR, Ahn NG. Activation of the MKK/ERK pathway during somatic cell mitosis: direct interactions of active ERK with kinetochores and regulation of the mitotic 3F3/2 phosphoantigen. J Cell Biol 1998; 142:1533–1545.
119. Pennisi E. Nuclear transport protein does double duty in mitosis. Science 1999; 284: 1260–1261.
120. Guarguaglini G, Renzi L, D'Ottavio F, Di Fiore B, Casenghi M, Cundari E, Lavia P. Regulated Ran-binding protein 1 activity is required for organization and function of the mitotic spindle in mammalian cells *in vivo*. Cell Growth Differ 2000; 11:455–465.

121. Lewis TS, Hunt JB, Aveline LD, Johscher KR, Louie DF, Yeh JM, Nahreini TS, Resing KA, Ahn NG. Identification of novel MAP kinase pathway signaling targets by functional proteomics and mass spectrometry. Mol Cell 2000; 6: 1343–1354.
122. Di Fiore B, Ciciarello M, Mangiacasale R, Palena A, Tassin AM, Cundari E, Lavia P. Mammalian RanBP1 regulates centrosome cohesion during mitosis. J Cell Sci 2003; 116:3399–3411.
123. Theurkauf WE, Vallee RB. Molecular characterization of the cAMP-dependent protein kinase bound to microtubule-associated protein 2. J Biol Chem 1982; 257: 3284–3290.
124. Sarkar D, Erlichman J, Rubin CS. Identification of a calmodulin-binding protein that co-purifies with the regulatory subunit of brain protein kinase II. J Biol Chem 1984; 259:9840–9846.
125. Harrison A, Hyams JS. Mapping MAP-2. J Cell Sci 1990; 96(Pt 3):347–349.
126. Nedvetzki S, Gonen E, Assayag N, Reich R, Williams RO, Neudecker BA, Stern R, Wang FS, Turley EA, Naor D. RHAMM, a motility-mediated receptor, compensates for CD44 in inflamed CD44-knockout mice: evidence for molecular redundancy. Submitted.
127. Enegd B, King JA, Stylli S, Paradiso L, Kaye AH, Novak U. Overexpression of hyaluronan synthase-2 reduces the tumorigenic potential of glioma cells lacking hyaluronidase activity. Neurosurgery 2002; 50:1311–1318.
128. Akiyama Y, Jung S, Salhia B, Lee S, Hubbard S, Taylor M, Mainprize T, Akaishi K, van Furth W, Rutka JT. Hyaluronate receptors mediating glioma cell migration and proliferation. J Neurooncol 2001; 53:115–127.
129. Lovvorn HN 3rd, Cass DL, Sylvester KG, Yang EY, Crombleholme TM, Adzick NS, Savani RC. Hyaluronan receptor expression increases in fetal excisional skin wounds and correlates with fibroplasia. J Pediatr Surg 1998; 33:1062–1069 (discussion 1069-1070).
130. Meyer LJ, Russell SB, Russell JD, Trupin JS, Egbert BM, Shuster S, Stern R. Reduced hyaluronan in keloid tissue and cultured keloid fibroblasts. J Investig Dermatol 2000; 114:953–959.
131. Capolicchio G, Aitken KJ, Gu JX, Reddy P, Bagli DJ. Extracellular matrix gene responses in a novel ex vivo model of bladder stretch injury. J Urol 2001; 165: 2235–2240.
132. Greiner J, Ringhoffer M, Taniguchi M, Schmitt A, Kirchner D, Krahn G, Heilmann V, Gschwend J, Bergmann L, Dohner H, Schmitt M. Receptor for hyaluronan acid-mediated motility (RHAMM) is a new immunogenic leukemia-associated antigen in acute and chronic myeloid leukemia. Exp Hematol 2002; 30:1029–1035.
133. Yamada Y, Itano N, Narimatsu H, Kudo T, Hirohashi S, Ochiai A, Niimi A, Ueda M, Kimata K. Receptor for hyaluronan-mediated motility and CD44 expressions in colon cancer assessed by quantitative analysis using real-time reverse transcriptase-polymerase chain reaction. Jpn J Cancer Res 1999; 90:987–992.
134. Duval A, Rolland S, Compoint A, Tubacher E, Iacopetta B, Thomas G, Hamelin R. Evolution of instability at coding and non-coding repeat sequences in human MSI-H colorectal cancers. Hum Mol Genet 2001; 10:513–518.
135. Assmann V, Gillett CE, Poulsom R, Ryder K, Hart IR, Hanby AM. The pattern of expression of the microtubule-binding protein RHAMM/IHABP in mammary carcinoma suggests a role in the invasive behaviour of tumour cells. J Pathol 2001; 195:191–196.

136. Chang F, Steelman LS, Shelton JG, Lee JT, Navolanic PM, Blalock WL, Franklin R, McCubrey JA. Regulation of cell cycle progression and apoptosis by the Ras/Raf/MEK/ERK pathway (Review). Int J Oncol 2003; 22:469–480.
137. Mazzoni E, Adam A, Bal de Kier Joffe E, Aguirre-Ghiso JA. Immortalized mammary epithelial cells overexpressing protein kinase C gamma acquire a malignant phenotype and become tumorigenic *in vivo*. Mol Cancer Res 2003; 1:776–787.
138. Reginato MJ, Mills KR, Paulus JK, Lynch DK, Sgroi DC, Debnath J, Muthuswamy SK, Brugge JS. Integrins and EGFR coordinately regulate the pro-apoptotic protein Bim to prevent anoikis. Nat Cell Biol 2003; 5:733–740.
139. Howe AK, Aplin AE, Juliano RL. Anchorage-dependent ERK signaling—mechanisms and consequences. Curr Opin Genet Dev 2002; 12:30–35.
140. Rein DT, Roehrig K, Schondorf T, Lazar A, Fleisch M, Niederacher D, Bender HG, Dall P. Expression of the hyaluronan receptor RHAMM in endometrial carcinomas suggests a role in tumour progression and metastasis. J Cancer Res Clin Oncol 2003; 129:161–164.
141. Zhou R, Wu X, Skalli O. The hyaluronan receptor RHAMM/IHABP in astrocytoma cells: expression of a tumor-specific variant and association with microtubules. J Neurooncol 2002; 59:15–26.
142. Line A, Slucka Z, Stengrevics A, Silina K, Li G, Rees RC. Characterisation of tumour-associated antigens in colon cancer. Cancer Immunol Immunother 2002; 51:574–582.
143. Smits R, van der Houven van Oordt W, Luz A, Zurcher C, Jagmohan-Changur S, Breukel C, Khan PM, Fodde R. Apc1638N: a mouse model for familial adenomatous polyposis-associated desmoid tumors and cutaneous cysts. Gastroenterology 1998; 114:275–283.
144. Murashita T, Nakayama Y, Hirano T, Ohashi S. Acceleration of granulation tissue ingrowth by hyaluronic acid in artificial skin. Br J Plast Surg 1996; 49:58–63.
145. Blaker H, Scholten M, Sutter C, Otto HF, Penzel R. Somatic mutations in familial adenomatous polyps. Nuclear translocation of beta-catenin requires more than biallelic APC inactivation. Am J Clin Pathol 2003; 120:418–423.
146. Kim K, Hay ED. New evidence that nuclear import of endogenous beta-catenin is LEF-1 dependent, while LEF-1 independent import of exogenous beta-catenin leads to nuclear abnormalities. Cell Biol Int 2001; 25:1149–1161.

Chemistry and Biology of Hyaluronan
H.G. Garg and C.A. Hales (editors)

Chapter 7

Signal Transduction Associated with Hyaluronan

REGINA M. DAY

*New England Medical Center,
Tufts University School of Medicine,
Boston, Massachusetts, USA*

MARCELLA M. MASCARENHAS

*Massachusetts General Hospital,
Harvard Medical School,
Boston, Massachusetts, USA*

I. Introduction

Hyaluronan is a non-sulfated glycosaminoglycan (GAG), consisting of repeating units of (β,1-4)-D-glucuronic acid–(β,1-3)-*N*-acetyl-D-glucosamine. HA occurs normally as a part of the ECM of almost all tissues in a high molecular weight (HMW) polymer (>10^6 kDa), and the highest concentrations are found in brain, skin and the central nervous system. Originally, it was believed that the physiological function of HA was only structural (1). However, HA is now recognized as a pharmacological signaling molecule. HA functions in a variety of biological processes, including embryonic development (2–4), inflammation, especially with regard to white blood cell function (5–7), angiogenesis (8–11), mammalian fertilization (12,13) and tissue repair/wound healing (1,14–17). HA is also critical for the maintenance of normal tissue elasticity and hydration (5,7) and normal joint function (18). Abnormal expression of HA fragments and/or HA receptors has been shown to play roles in metastasis and survival in several cancer cell types (19–24), tumor vascularization (21,24,25), complications associated with acute lung injury (26–29) and immunological dysfunctions, including asthma and rheumatoid arthritis (5,30–32). At the cellular level HA can induce migration and adhesion (29,33,34), growth and survival (34,35), endocytosis (36,37) and maintenance of endothelial barrier (35).

There is evidence that HA has different functions based on its molecular weight (38,39). In some cases, low molecular weight (LMW) and HMW HA bind the same receptor, but elicit different cellular processes (7,38,39). LMW and HMW HA have also been shown to bind different receptors. HA interacts with a variety of different cell types as well as with many different extracellular molecules collectively known as hyaladherins (40–42). Hyaladherins are involved in cell–cell interactions, cell–matrix interactions, and clearance of HA from blood or tissues. Some of these molecules are found in the ECM and are important in matrix organization. Others are cell surface receptors, including CD44, the receptor for HA-mediated motility (RHAMM), Toll-like receptor 4, layilin, PH-20 protein, LYVE-1, and the HA-associated receptor for endocytosis (HARE) (Table 1).

II. CD44 Receptor

CD44 proteins are class I transmembrane glycoproteins that bind HA and GAGs. CD44 has been shown to be expressed in most tissues and play a role in inflammation and immune function, organogenesis and development, malignancy and tissue homeostasis (7). At the cellular level, CD44 can induce growth and survival (20,43,44), endocytosis (37,45), differentiation and maturation (46,47), anchoring (48), and motility (49,50). CD44 is also important for resolution of inflammation in the lung following non-infectious injury (51).

A. HA Binding and Activation of CD44

CD44 proteins can vary in size from 80 to 200 kDa, depending on splice variations, and the splice variations themselves can influence the downstream signaling mechanism (52). The number of potential variants of CD44 is not known, but 10 variant exons (v exons) are present in the CD44 gene, located between five non-variant $5'$ exons and five non-variant $3'$ exons. Splicing occurs differently according to cell type and cell activation state (7). The first five non-variable exons (exons 1–5) and the 10 variant exons are extracellular; the final $3'$ non-variant exons (exons 15–19) are believed to be partially extracellular and also encode a single transmembrane domain and a cytoplasmic tail (7,53). The HA-binding region has been mapped to the amino terminal globular domain of CD44 encoded by the first five non-variable exons (7). This region is homologous with the 'link module', also known as the proteoglycan tandem repeat, consisting of two α-helices and two triple-stranded anti-parallel β-sheets (54). The link module serves as the proteoglycan-binding domain of the link protein superfamily, most of which form the protein-stabilized HA structures that function as the load-bearing complexes in cartilage (54).

The splice variants of CD44, together with post-translational modifications, determine the affinity of the receptor for HA and other GAGs (7). Both HA fragments and HA polymers (components of the extracellular matrix) bind CD44, but these different sizes of HA may have different biological outcomes (39). Whereas HA fragments play a role in angiogenesis and inflammation (9,55),

Table 1 Cellular Receptors for Hyaluronan and Their Interacting Proteins

HA receptor	Direct and indirect interactions				Biological activities
	Adapter proteins	Cytoskeletal proteins	Signal transduction proteins	Other	
CD44	GAB1	Actin, ankyrin, ezrin, merlin	Ras, Rho, RhoGEF, ROK, FAK, Pyk2, PI3K, Akt, p42/p44 MAPK, pp60 c-src family, Tiam 1, Rac1, Cas, PKC-ζ	Ca^{2+} channel	Growth, survival, motility
RHAMM		Actin, paxillin, vinculin	FAK, pp60 c-src, p44/p42 MAPK, MEK, H-Ras		Growth, survival, motility
Toll-like receptor 4	MyD88, TIRAP, A20		IRAK, p44/p42 MAPK, p38 MAPK, JNK, Btk, PKR	Extracellular MD-2	Activation of monocytes, macrophages and dendritic cells
Layilin		Actin, talin, vinculin	Unknown		Motility??
PH20			Tyrosine kinases (unidentified)	Ca^{2+} channel?	Hyaluronidase, sperm maturation, cumulus penetration, sperm–egg recognition, oolemmal fusion
LYVE-1			Unknown		Endocytosis, HA degradation
HARE		Clatherin	Unknown	Ca^{2+} channel?	Endocytosis, HA degradation

HMW HA polymers are associated primarily with cell adhesion (56). Additionally, LMW HA is associated with cell growth and metastasis (7,57), while HMW HA prevents growth under some circumstances (58).

The mechanism of HA binding to CD44 has been the subject of some controversy. CD44 has been proposed to exist in three potential states in cells: inactive and unable to bind HA; inducible; and constitutively active. The ability of CD44 to bind HA is believed to depend upon the cell type and/or the activation state of the cell (especially with regard to immune cells), and may require post-translational modifications (especially glycosylations) for changes in the binding (38,59). Using murine pre-B lymphoma and leukocytes, several laboratories reported that HA binding to the CD44 extracellular domain is independent of the cytoplasmic and transmembrane domains of CD44, but occurs at least in part as a function of the oligomerization of CD44 (38,59–63).

In contrast, studies with the GP85 splice variant of CD44 and phorbol ester-activated cells show that CD44 binding to HA requires several key domains in the cytoplasmic portion of the protein. COS cells transfected with a mutant CD44(GP85), in which the actin cytoskeleton-binding domain was deleted, had reduced HA binding compared with COS cells transfected with wild type CD44(GP85) (64). Phorbol ester-induced CD44 binding to HA in human leukemia Jurkat cells was shown to involve CD44 dimerization (65), and required the transmembrane domain and two basic amino acid clusters in the cytoplasmic domain (65,66). This induced binding of HA to CD44 was blocked by cytochalasin D, an inhibitor of actin polymerization, and taxol and colchicine, inhibitors of microtubule function (67). These results were taken to support the hypothesis that, under certain circumstances, activation of the cytoskeleton results in CD44 clustering which in turn activates HA binding (68).

A study by Pure et al. (69) in murine lymphoma cells showed that HA binding to CD44 required two phosphorylated serines (325 and 327) in the cytoplasmic tail. The lack of HA binding in the CD44 serine mutants could be partially overcome by extracellular antibody-induced oligomerization, suggesting that the serines may be involved in an intracellular mechanism for CD44 oligomerization (69). Interestingly, increased affinity of cellular binding of HA was also observed with increasing HA oligomer size, especially increases from 22 to 38 sugars which corresponded to divalent CD44 receptor binding and with an affinity similar to that of the induced form of CD44. However, shorter HA oligomers bound cells in a monovalent manner, with an affinity corresponding to the uninduced form of CD44 (38). The increased affinity of the longer HA oligomers with divalent CD44 binding also suggested a cooperative mechanism for binding.

B. Cell Growth and Survival Signaling by CD44

CD44 activated by HA fragments and the over-expression of certain CD44 variants have been shown to promote cell survival, prevent drug-induced apoptosis and induce proliferation in a variety of untransformed cells as well as cancer cells (70–74). Signal transduction by CD44 has been associated with the activation of signaling molecules which support cell growth, including: members of the Ras family of small GTP-binding proteins; an intracellular membrane-associated tyrosine kinase, focal adhesion kinase (FAK); phosphtidylinositol 3-kinase (PI3K) and its downstream target, the Akt kinase; the cytoplasmic serine/threonine kinase, p42/p44 mitogen-activated protein kinase (MAPK); and members of the src family of cytoplasmic tyrosine kinases (Fig. 1A). These proteins have been independently shown to support cell growth and survival in many cell types (75–78). The pathways activated by CD44, and the mechanism of activation, appear to depend on both the cell type examined and the splice variant of CD44 that is expressed.

In T-24 carcinoma cells, HeLa, MCF7 and J774 cell lines, CD44 binding to HA fragments, but not dimeric or native polymer HA, led to the activation of nuclear factor kappa B (NF-κB) (71), a transcription factor which regulates anti-apoptotic and cell survival gene expression (79). HA activation of NF-κB was blocked by an anti-CD44 antibody, showing that the CD44 receptor was required for the downstream signaling. In T-24 cells, HA fragments binding to CD44 caused the downstream activation of protein kinase C- (PKC-) ζ and the small GTP-binding protein Ras (71). Transfection of T-24 cells with dominant negative (DN) Ras or inhibition of PKC blocked HA fragment-induced kappa B-linked reporter gene expression (71). These results were used to establish a signal transduction cascade emanating from CD44 to Ras, PKC-ζ and NF-κB (71).

In a human small cell lung cancer cell line, expression of CD44 conferred significant resistance to etoposide-induced apoptosis (72). The binding of HA fragments to CD44 induced rapid phosphorylation and activation of FAK, PI3K and subsequently p42/p44 MAPK (72). In this study, CD44-stimulated FAK phosphorylation was inhibited by the over-expression of DN Rho, a member of the Ras family of small GTP-binding proteins, indicating that, as in the case of NF-κB activation (71), a member of the Ras family lies upstream in the signaling cascade. The anti-apoptotic effect of CD44 expression was cancelled by the inhibition of either Rho, FAK or PI3K.

A recent report by Bourguignon et al. (80) showed that CD44 was directly associated with a RhoA-specific guanine nucleotide exchange factor (p115Rho-GEF) and activated the small GTP-binding protein Rho A and its downstream effector Rho-kinase (ROK) in a metastatic human breast cancer cell line. The pathway was further elucidated to show that the adaptor protein GAB1 was phosphorylated downstream of ROK, followed by the activation of PI3K/AKT (80). In agreement with previous findings, CD44-induced cell survival and growth were abrogated by the over-expression of DN Rho, or by a pharmaceutical inhibitor of PI3K (80). Thus, activation of Rho is upstream of PI3K in this system, as was found for CD44 signaling in small cell lung cancer cells.

In colon cancer cells, neutrophils and lymphoid B- and T-cells, CD44 was associated with and activated lck, hck and lyn, members of the pp60 c-src family of cytoplasmic tyrosine kinases (70,81,82). Activation of a CD44 splice variant in

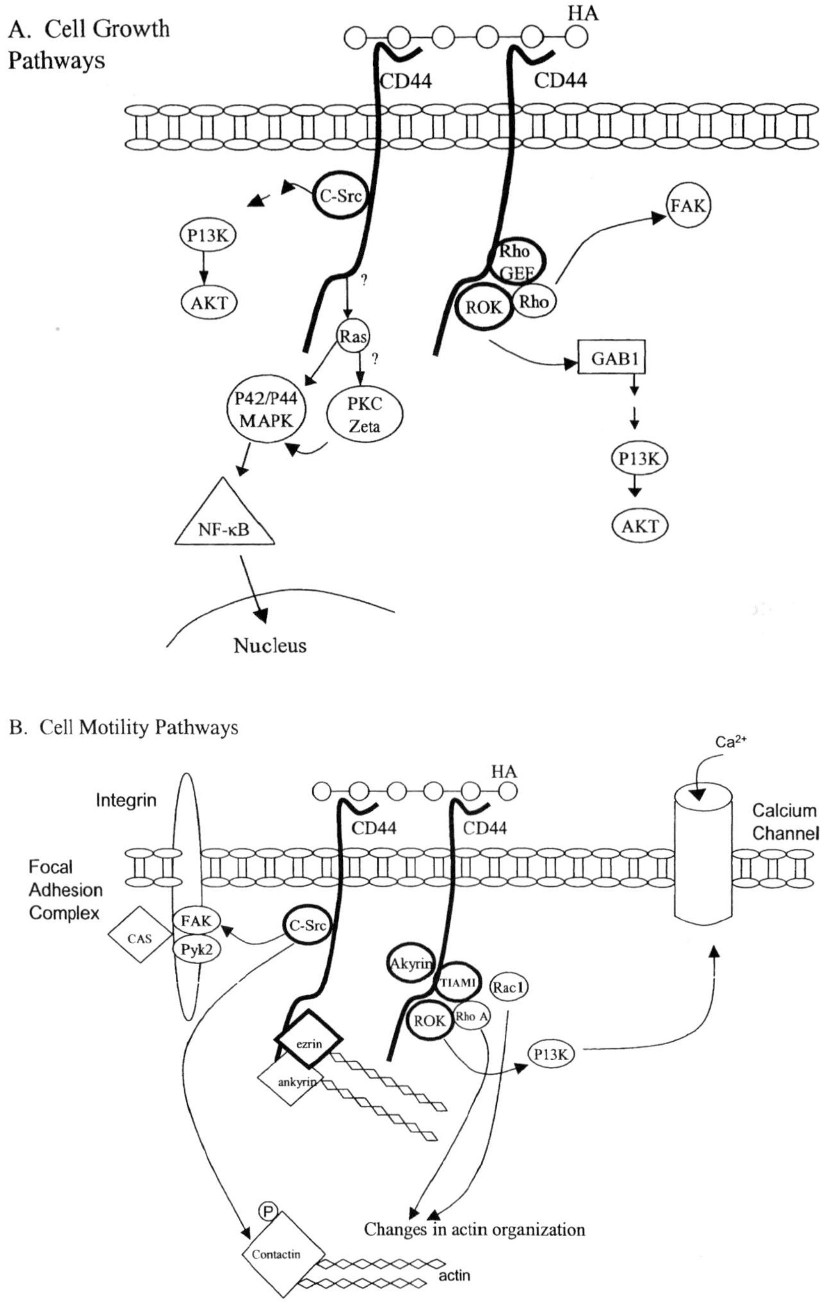

Figure 1 Signal transduction pathways activated downstream of CD44. For individual references, please refer to the text. Not all pathways are activated in all cells. Thick black borders indicate proteins shown to directly interact with CD44. Proteins are shaped according to function: circles, signal transduction; squares, adaptor proteins; triangles,

colon carcinoma cells caused resistance to 1,3-bis(2-chloroethyl)-1-nitrosurea-induced apoptosis. In these cells, lyn activation by CD44 led to the activation of PI3K and its downstream target AKT, a kinase with anti-apoptotic functions (70). Stable CD44-lck and CD44-lyn complexes were observed in non-stimulated lymphoid T- and B-cells, and these two kinases accounted for much of the tyrosine kinase activity associated with CD44 upon stimulation by HA (82). A synthetic peptide (ILAVCIAVNSRRR), corresponding to a sequence of murine CD44 at the plasma membrane-cytoplasmic interface, exhibited affinity for lck and lyn (82). Modification of the cysteine residue completely abolished the binding, while deletion of the three tandem arginines decreased it significantly (82). Because the members of the src kinase family directly bind the CD44 cytoplasmic domain, it is possible that they are activated directly, and do not require upstream activation by members of the Ras family.

While most studies have shown that HA fragments binding to CD44 signaling causes cell growth, HMW HA polymers can have growth inhibitory effects. CD44, together with merlin, the product of the neurofibromatosis-2 gene, an ezrin–radixin–moesin- (ERM-) binding protein, is proposed to form a molecular switch, inducing contact-inhibition associated growth arrest (58). In growing cells at low density, merlin is phosphorylated and complexed with ERM proteins, which in turn bind CD44; this state is growth permissive. At high cell density, merlin and the ERM proteins become hypo-phosphorylated in response to HA, and the ERMs no longer bind merlin. Merlin's growth-inhibitory activity requires direct interaction with the CD44 cytoplasmic tail, disrupting the link with the actin cytoskeleton, and blocking Ras activation (58).

C. CD44 Signaling for Cellular Motility and Adhesion

CD44 activation is associated with cell motility, especially in metastasizing cancer cells (50,83). In some cancer cells, CD44 activity is specifically associated with cell motility, but not with growth (84,85). The cytoplasmic domain of CD44 contains binding sites for several cytoskeleton-associated proteins including ezrin (86) and ankyrin (87,88) (Fig. 1B). Ezrin is a member of the ERM family of structural proteins, which functions in both the organization of the actin cytoskeleton and the regulation of Rho and Rac signal transduction. Ezrin is ordinarily found in cells in a dormant state, which must be activated in order for it to associate with filamentous actin. Both dormant and activated ezrins were found to be associated with CD44, although the interaction was believed to undergo a conformational change upon activation (89,90). Ankyrin links a variety of transmembrane proteins to the actin network through interactions with spectrin and fodrin, and mediates adhesion, endocytosis and migration (37,91,92).

---

transcription factors; diamonds, structural proteins; ovals or cylinders, others. (A) Signalling pathways associated with cell growth and anti-apoptosis. (B) Signalling pathways associated with cell motility.

In murine T lymphoma cells, HA binding to CD44 caused increased concentrations of intracellular Ca^{2+} within seconds (87). Following the rise of intracellular Ca^{2+}, CD44 receptors formed patched/capped structures, and cell adhesion occurred on HA-coated plates (87). HA-induced receptor redistribution and adhesion was inhibited by EGTA (a Ca^{2+} chelator), nefedipine/bepridil (Ca^{2+} channel blockers), W-7 (a calmodulin antagonist) and cytochalasin D (a microfilament inhibitor), but not colchicine (a microtubule disrupting agent) (87). In lymphoma cells as well as several other metastatic cancer cell types, the HA-induced capped structures preferentially accumulated ankyrin, and ankyrin binding in this structure was required for HA-mediated adhesion and migration (49,64,88). Selective expression of some CD44 isoforms are unique for certain metastatic carcinomas and their interaction with the cytoskeleton, especially ankyrin, suggests that CD44 may be involved in regulating tumor development and metastasis (49,74,88,93,94).

In SP1 metastatic breast tumor cells, the binding of HA to the splice variant 3 of CD44 (CD44v3) stimulated T lymphoma invasion and metastasis-inducing protein 1 (Tiam1), a guanine nucleotide exchange factor for the small GTP-binding proteins Rac1 and RhoA, proteins known to function in actin cytoskeleton regulation (89,90). Tiam1 was shown to catalyze Rac1 and RhoA signaling and to subsequently co-localize with ankyrin in membrane projections of migrating cells following treatment with HA (50). Experiments showed that the NH(2)-terminal pleckstrin homology (PHn) domain of Tiam1 interacted directly with the CD44 cytoplasmic domain, and inhibition of Tiam1 signaling by the over-expression of a DN Tiam1 blocked both Rac1 activation and migration in response to HA (50). Similar results were found in aortic endothelial cells and Met-1 metastatic breast cells in which the CD44v3,8-10 and CD44v10 isoforms were shown to require Rho kinase, ROK, activation for migration (95,96). CD44v10 was demonstrated to directly bind ROK, which in turn leads to PI3K-induced Ca^{2+} influx and cell migration (96). The significance of Ca^{2+} influx in this case may also have implications for the regulation of actin–myosin interactions.

As described above, the CD44 cytoplasmic domain contains a binding site for the non-receptor tyrosine kinases from the pp60 c-src family (70,81,82, 97–99). Src was found to be recruited to CD44 following HA binding, and subsequently phosphorylated the cytoskeleton-associated protein cortactin (97). Phosphorylated cortactin has reduced ability to cross-link filamentous actin, thus allowing changes in the actin cytoskeletal organization (97). Activated c-src can be found complexed to CD44 in the membrane projects of migrating tumor cells, and inhibition of src activity, by the overexpression of a DN src kinase or by pharmaceutical agents, blocks migration (84,97). Src activation was also shown to induce downstream phosphorylation and activation of the focal adhesion family kinases Pyk2 and FAK as well as Cas, a protein associated with integrin signaling (100). Regulation of these proteins suggests that the regulation of focal adhesion complexes and integrin complexes may also be required for or associated with CD44-induced cell motility. Together, the findings suggest that activation of src may occur independently of the activation of Tiam1, Rho and ROK, although these pathways may converge in the process of cell migration and tumor metastasis.

D. CD44 Interactions with Receptor Kinases

In human peripheral blood T lymphocytes and endothelial cells, a significant proportion of CD44 was found to be associated with specialized plasma membrane domains containing low-density plasma membrane fractions enriched in glycosphingolipids and glycosylphosphatidylinositol (GPI)-anchored proteins (99). Studies on T cell receptor and IgE receptor mediated signaling in lymphocytes and mast cells have helped develop the hypothesis that microdomains exist in activated cells, consisting of signaling platforms where components of multiple signaling pathways are assembled. Co-isolation of CD44 with microdomains, especially in association with members of the c-src family, strongly suggests that CD44 generates cellular activation signals utilizing the machinery present in the plasma membrane microdomains (57,101).

Additionally, the modification of some splice variants of CD44 to include heparan sulfate side chains is believed to allow CD44 to directly associate with heparin-binding growth factors. In this way, CD44 may serve as a low affinity binding site for factors which can then be presented to their high affinity receptors (102). Thus the signaling platforms along with the heparan sulfate modifications are believed to provide the environment in which some CD44 isoforms function as co-receptors for specific tyrosine kinase receptors, which include: c-Met, the receptor for hepatocyte growth factor (HGF) (53,102); the ERBB receptor tyrosine kinase family (also known as EGFR/HER receptors), which bind several of the epidermal growth factor (EGF)-related factors (103–105); fibroblast growth factor receptors (FGFRs) (106); transforming growth factor beta 1 receptor (22); and components of the T-cell-receptor complex (107). In these complexes, it is believed that CD44 serves to facilitate activation of the receptor or association of intracellular-signaling proteins, and in most cases is not believed to directly transmit signals induced by the growth factors (7).

1. *Co-activation of c-Met Receptor*

c-Met, the transmembrane tyrosine kinase receptor for HGF, is involved in embryonic development, organogenesis and tissue repair (108). Abnormal expression of HGF or its receptor is often associated with increased tumor invasiveness and metastasis (109,110). c-Met activation leads to cellular proliferation, migration, differentiation and/or cell death, depending upon the cell type affected (108,111). Scatchard analysis and competitive binding with HGF showed that two classes of binding sites are present on cells (112–114), in which one class, mediated by c-Met, is high affinity, low capacity (2–25 pM; 200–5000 per cell), and the other class is low affinity, high capacity (0.2–5 nM; ~1,000,000 per cell), which is believed to be mediated by the binding of HGF to heparin sulfate proteoglycans (HSPGs). HSPGs have been shown to modulate

HGF signaling, and the absence of HSPGs, on cells which lack their expression, blunts the activation of c-Met by low doses of HGF (78,115–117). In Namalwa lymphoma cells, expression of CD44v3-10 caused increased responsiveness of cells to HGF. The v3 exon, which contains a site for heparan sulfate modification, was believed to interact with and oligomerize HGF, increasing its affinity for and activation of c-Met (102). Treatment of cells with heparitinase, but not chondroitinase, abrogated this effect. Further, lymphoma cells in which the variant 3 (v3) exon was not expressed (CD44v8-10 or CD44 with no variable exons) responded only weakly to HGF (102).

Based on previous findings that expression of CD44v4-v7 or CD44v6,7 was sufficient to induce the metastatic phenotype in non-metastatic cell lines (118,119), and that antibodies directed against the CD44v6 exon abolished metastatic outgrowth of some tumor cells (7), Orian-Rousseau and colleagues (53) investigated the role of the v6 in c-Met activation. In BSp73ASML and HT29, CD44 co-activation of c-Met required exon 6-containing CD44 variants, including a CD44 isoform containing the v6 exon alone (53). Antibodies directed against the v6- and non-variable exon 15-encoded epitopes, but not the v3-encoded epitope, blocked the ability of HGF to induce c-Met autophosphorylation in two metastatic tumor cell lines (53). The v6 and exon 15 regions of CD44 are believed to comprise a membrane-proximal structure hypothesized to interact with the c-Met receptor. V6-containing CD44 isoforms co-immunoprecipitated with c-Met in a multimeric complex. Treatment of cells with heparinase II had no influence on the ability of CD44v6 variants to co-activate c-Met, consistent with the lack of a heparan sulfate glycosylation site on the v6 epitope (53). Finally, HGF-induced c-Met autophosphorylation and downstream phosphorylation of the Gab1 adapter protein and phospholipase C-γ did not require the cytoplasmic domain of CD44. However, activation of MEK and p42/p44 MAPK by HGF/c-Met was not observed in the absence of the CD44 tail or when the actin-binding protein ezrin was sequestered (53).

Together, these results suggest that CD44 splice variant requirements (v3 versus v6 exon expression) for c-Met signaling may depend in part on the cell type examined (53,102). CD44 appears to support a two-step function in the co-activation of c-Met: in the first step, the extracellular region of CD44 contributes to HGF binding, induction of c-Met autophosphorylation and the formation of a multimeric complex; in the second step, the cytoplasmic domain of CD44 potentiates downstream signaling by the activated c-Met, in part through organization of cytoskeletal binding proteins (53).

2. *Co-activation of ERBB Receptor Tyrosine Kinase Family*

The ERBB family (ERBB1-4) of transmembrane tyrosine kinase receptors is involved in wound healing, proliferation and cell survival, migration, blastocyst implantation and progression of some types of cancer (120). The factors that bind and activate the ERBB family include EGF and EGF-like proteins including heparin-binding EGF (HB-EGF), transforming growth factor alpha, amphiregulin, betacellulin, epiregulin, and neuregulin.

Many of the EGF-related factors are produced as integral membrane precursor proteins which require proteolytic processing to remove the membrane-binding peptide (120). The proteases responsible for activation are mostly unknown, with the exception of HB-EGF which appears to be processed by matrix metalloproteinases (MMPs), in particular MMPs 3 and 7, and MDC9 (120–123). Early evidence of CD44 involvement in ERBB signaling came from the observation that CD44-deficient murine keratinocytes failed to proliferate in response to HB-EGF (105). Based on previous observations that heparan sulfate-modified CD44 variants co-localized with MMP7 (60), Yu and co-workers (123) investigated the association of CD44v3-8, MMP7, pro-HB-EGF and ERBB4 in Namalwa lymphoma cells. Co-localization of CD44, MMP7 and pro-HB-EGF required expression of CD44 containing the v3 exon; the association of MMP and CD44 appeared to be mediated by heparan sulfate (123). CD44v3,8-10 and CD44v3-10, but not CD44 variants lacking v3, allowed activation of pro-HB-EGF and subsequent phosphorylation and activation of ERBB4 (123). CD44−/− mice showed altered MMP7 distribution, decreased pro-HB-EGF processing, decreased ERBB4 activation and abnormal epithelial function in post-partum uterus and lactating mammary glands, thus revealing an *in vivo* function for the associations observed in cell culture (123).

In separate studies in ovarian carcinoma cells, it was found that CD44s (an isoform of CD44 which contains no variable domains, and therefore no heparan sulfate modifications) covalently bound ERBB2 (also known as HER2) through interchain disulfide bonds (104). Stimulation of these cells with HA lead to ERBB2 phosphorylation and activation, resulting in cell growth (104). In a later study with the same cells, CD44v3 was found to associate with Vav2, a guanine nucleotide exchange factor (124). HA binding promoted recruitment of Grb2 and ERBB2 to the CD44v3-Vav2 complex, leading to Ras and Rac1 activation and ovarian tumor cell growth; this association with ERBB2 did not require the addition of an EGF-like factor, but activation of ERBB2 was not addressed (124). CD44 was also found constitutively associated with ERBB2 and ERBB3 in Schwann cells (103). In these cells, CD44 potentiated neuregulin-induced ERBB2 phosphorylation and ERBB2–ERBB3 heterodimerization (103). Reduction of CD44 expression induced Schwann cell apoptosis, suggesting that CD44 expression in these cells was integrally linked to cell survival, possibly in part through its co-receptor functions (103). In these cases, it appeared that the ERBB receptors may be acting as co-activators of downstream signaling from CD44.

The findings of CD44 co-activation of ERBB receptor kinases implies that two mechanisms may be present: one mechanism which requires interaction of the growth factor with the heparan sulfate modification of v3-containing CD44 splice variants; and a separate mechanism which requires covalent binding of CD44 to ERBB receptor(s). This parallels the findings of mechanisms for co-activation of the c-Met receptor tyrosine kinase (see earlier), with the exception that CD44 may also utilize the ERBB receptors as co-activators of its own signaling in response to HA, without the addition of ERBB-binding factors. In the case of ERBB receptor co-activation, the mechanism(s) involved may differ with ERBB receptor subtype, cell type and/or factor.

III. RHAMM Receptor

The receptor for HA-mediated motility (RHAMM) is a membrane-associated protein, ranging in size from ~59 to 80 kDa, depending upon cell type and splice variations (125,126). RHAMM is expressed on most cell types, and its primary function in response to HA binding is the mediation of adhesion and cell motility (57). RHAMM may additionally play a role in joint formation, but this function is not well defined (127). In some cases, RHAMM can participate in cell growth (126), and overexpression of RHAMM causes cellular transformation (128); however, at least one isoform of RHAMM can suppress growth (129).

A. HA Binding and Activation of RHAMM

At least five naturally occurring splice variants of RHAMM have been detected in cells (130,131). RHAMM lacks a traditional leader sequence to direct its secretion, and it does not contain a transmembrane domain (131,132). RHAMM proteins have been found to be distributed throughout the cell, in locations such as the cell surface (133,134), cytoskeleton (135,136), mitochondria (134) and nucleus (137). Because of its localization to multiple compartments in the cell, different functions have been proposed for intracellular RHAMM (131,135,136).

Truncation mutagenesis was used to show that RHAMM binds HA through two clusters of basic amino acids within a 35 amino acid sequence near the carboxy terminus of the protein (138). The two HA-binding sequences were found to contain similar motifs, in which two basic amino acids are spaced seven amino acids apart, termed ‘B(x7)B domains’; the seven amino acid linker regions were characterized by mutagenesis as requiring at least one basic residue and no acidic residues (139). RHAMM binds both HMW HA polymers and LMW HA fragments (126). HA binding occurs on the cell surface, and can be blocked by extracellular addition of anti-RHAMM antibodies (126). Binding of HA to RHAMM may lead to its internalization and down-regulation from the cell surface, although the mechanism(s) for this are unknown (24,126). Different signals are transduced by RHAMM following the binding of different sizes of HA; in some cell types, large HA polymers induce motility, whereas LMW HA is better at promoting cell proliferation through RHAMM (126).

B. RHAMM Signaling in Cellular Motility

RHAMM is expressed on most cell types and has been shown to be involved in HA-mediated motility of lymphocytes (140), hematopoietic cells (141,142), sperm (143), glial cells (129), fibroblasts (130) and human umbilical vein endothelial cells (126,144), as well as in tumor progression and metastasis (128, 145–149). In some cell types, RHAMM is required for motility even though CD44 is also expressed (126,144,150). Peptides corresponding to the HA-binding domains of RHAMM, or antibodies directed against RHAMM, inhibit HA-induced motility, indicating that RHAMM-mediated motility is initiated by HA-binding at the cell surface (125,140,151–153).

RHAMM activation by HA fragments induces the rapid formation and dissociation of focal adhesions, which is required for and precedes motility (154) (Fig. 2). RHAMM induced the transient phosphorylation and activation of FAK and the reorganization of the actin cytoskeleton (126,154). The phosphorylation of FAK was also accompanied by the association of vinculin in focal adhesions (154). In endothelial cells, tyrosine phosphorylation of paxillin was observed following FAK activation (126); when phosphorylated by FAK, paxillin associates with vinculin in focal adhesions, participating in the formation of a complex to coordinate actin filament binding to transmembrane integrin proteins (155,156). Gares and colleagues have hypothesized that focal adhesion changes induced by HA-bound RHAMM regulate adhesion and migration through the modulation of integrin affinity for the extracellular matrix (156).

A key role for pp60 c-src in RHAMM-mediated cell motility was shown by Hall and co-workers (128,130), who showed a physical association between pp60 c-src and RHAMM in Ras-transformed fibroblasts. In fibroblasts derived from

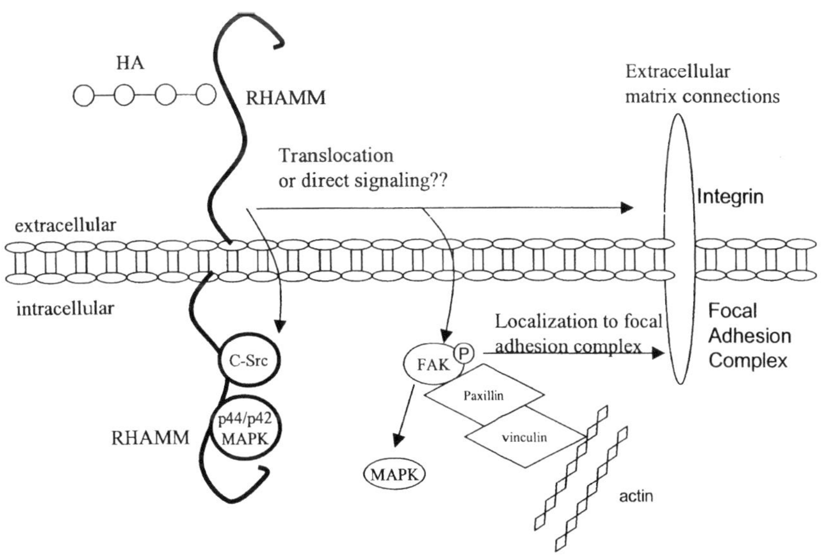

Figure 2 Signal transduction pathways activated downstream of RHAMM. For individual references, please refer to the text. Not all pathways are activated in all cells. Thick black borders indicate proteins shown to directly interact with RHAMM. Proteins are shaped according to function: circle, signal transduction; diamond, structural protein; oval, other.

mice lacking src (src (−/−)) motility was found to be significantly slower than the corresponding wild-type fibroblasts; motility could be restored by the expression of wild type c-src but not by the expression of kinase-deficient src or a truncated src containing only SH2 and SH3 domains. RHAMM was required for the restoration of src (−/−) cell locomotion, and the motility of cells expressing c-src was reduced to src (−/−) levels by RHAMM-blocking antibodies (130). Interestingly, the expression of an activated viral src mutant (v-src) enhanced cell motility in a RHAMM-independent manner, but focal adhesions did not turn over, suggesting that regulation of focal adhesions by RHAMM may occur via an src-independent mechanism (130).

C. RHAMM Signaling in Cell Growth

Although RHAMM was first described as an HA receptor inducing cellular migration, it has become clear that in specific cell types RHAMM can also induce proliferation. RHAMM activation by HA can cause growth in some types of endothelial cells. HA stimulated growth and tyrosine phosphorylation of proteins in cultured primary human pulmonary artery and lung microvascular endothelial cells in a RHAMM-dependent manner (126); although CD44 was expressed on the surfaces of these cells, antibodies directed against CD44 did not block HA-induced tyrosine kinase activity. Phosphorylation of FAK, paxillin and p42/p44 MAPK were detected in primary endothelial cells downstream of RHAMM (126) (Fig. 2); both FAK and MAPK were activated within 1–2 min of HA treatment. In general, tyrosine phosphorylation and cell growth occurred more strongly in response to HA fragments than by polymer HA. While FAK and paxillin are more frequently associated with cell adhesion and migration, both FAK and MAPK activations have been shown to participate in proliferation and cell survival (75,157). RHAMM contains a binding site for p42/p44 MAPK and RHAMM can co-immunoprecipitate with p42/p44 MAPK (133). Furthermore, over-expression of an intracellularly localized RHAMM isoform (RHAMMv4) constitutively associates with MEK and p42/p44 MAPK, and activates p42/p44 MAPK (133). The mechanism(s) by which RHAMM activates either MAPK or FAK are unknown; RHAMM is not believed to function as a scaffold protein, such as yeast STE5, but may involve membrane recruitment, or receptor recruitment, of the proteins (133).

RHAMM was also shown to have a fundamental role in the regulation of proliferation downstream of activated Ras. Over-expression of RHAMM in a fibroblast cell line caused increased motility, anchorage-independent growth and full transformation into a metastatic fibrosarcoma (128). A RHAMM mutant, in which the HA-binding sites were altered, lacked the ability to transform cells; over-expression of this mutant RHAMM (a dominant negative mutant) blocked H-Ras-induced transformation, indicating that RHAMM activity is essential in the growth pathway downstream of activated Ras (128). In experiments with the CIRCAS-3 fibroblast cell line expressing H-Ras, soluble RHAMM protein added to the culture medium blocked the ability of serum to induce cellular proliferation and caused growth arrest at G_2/M phase (129). The inhibition of growth correlated with the reduced expression of two cell cycle proteins, Cdc2 and Cyclin B (129). The reduction in the total amount of cellular Cdc2 protein was associated with a decreased half-life of cdc2 mRNA, but not with a decreased rate of cdc2 gene transcription (129). The mechanism(s) of cdc2 mRNA regulation by RHAMM are unknown, although it is hypothesized that Cdc2 protein levels may be coordinated with cell detachment which is required for cells to enter mitosis (129). HA synthesis and synthase activities increase during mitosis, when cells round up and become loosely adherent (158); thus, soluble HA is believed to be required for detachment from the supporting matrix, and the mutant DN RHAMM protein may interfere with this event (129,158).

D. The Role of RHAMM in Signaling by Receptor Kinases

The discovery of RHAMM as a key regulator of cellular migration and proliferation led to the hypothesis that RHAMM could also function downstream of receptor kinase signaling. Studies showed that the expression of RHAMM is required for tumor cell migration (125). Samuel et al. (152) investigated the role of RHAMM in motogenesis downstream of the serine/threonine kinase receptor for transforming growth factor β-1 (TGFβ-1). TGFβ-1 regulates the expression of a variety of extracellular matrix proteins, suggesting that cell migration induced by TGFβ-1 could potentially proceed through other proteoglycans. In Ras-transformed fibroblasts, TGFβ-1 upregulated expression of RHAMM with a time course which corresponded with increased cell motility; inclusion of anti-RHAMM monoclonal antibodies severely blocked both TGFβ-1-induced motility as well as basal levels of cell migration (152). Interestingly, while a polyclonal antibody directed against RHAMM peptide 268–288 also blocked cell migration, a polyclonal antibody directed against RHAMM peptide 124–145 had no effect on migration (152).

A number of studies with transformed cell lines indicated that RHAMM was required for H-Ras and serum downstream activation of p42/p44 MAPK (see earlier) (128,129). The requirement of RHAMM in these systems led Zhang et al. (133) to study the role of RHAMM in platelet-derived growth factor (PDGF) signal transduction. The ability of PDGF to signal through its tyrosine kinase receptor and activate p42/p44 MAPK, as well as other phosphoproteins, depended upon the level of RHAMM expression on the surface of 10T1/2 murine fibroblasts (133). Decreased expression of RHAMM, due to the expression of RHAMM anti-sense, caused reduced responsiveness to PDGF. Downstream signaling by PDGF was also blocked by a RHAMM antibody, as measured by the level of phosphoproteins present in cellular extracts following PDGF treatment. This finding indicated that *extracellular* RHAMM is essential for PDGF signaling (133). In contrast, results by Zhang et al. (133) also showed that an *intracellular* form of RHAMM was required for activation of p42/p44 MAPK by activated H-Ras (133). The function of RHAMM in signal transduction was later examined in rat pheochromocytoma PC12 cells, a neural cell line (159).

RHAMM was detected in association with the cytoskeleton, neurites and growth cones of these cells. When the PC12 cells were stimulated with nerve growth factor (NGF) or fibroblast growth factor-2 (FGF-2), two growth factors with cell surface tyrosine kinase receptors, RHAMM was found to co-immunoprecipitate with activated p42/p44 MAPK (159).

Thus, RHAMM appears to participate importantly in the signaling of both tyrosine kinase receptors and serine/threonine kinase receptors involved in cellular locomotion, especially related to the activation of p42/p44 MAPK. As stated earlier, it is not believed that RHAMM acts as a scaffold protein, but may function in some way to recruit intracellular signaling targets to the activated receptor, in a manner similar to the function of adapter proteins (57,133). Due to its dual function on the cell surface as well as the cytoplasm, it has been proposed that RHAMM may contribute to a version of 'inside–outside' signaling, similar to that of integrin proteins (57). The ability of extracellular antibodies to block the function of RHAMM and the downstream signaling of both types of receptors indicates that RHAMM on the cellular surface is necessary, but the mechanism by which extracellular RHAMM is then associated with, or induces intracellularly localized RHAMM to associate with, p42/p44 MAPK, an intracellular kinase, is not clear.

IV. Toll-Like Receptor 4

Toll-like receptors (TLRs), are a relatively new group of receptors belonging to the interleukin-1 receptor family, and are the mammalian equivalent of the *Drosophila* Toll protein (160–163). TLR-4 is broadly expressed in tissues, including heart, brain, liver and kidney (164,165). TLRs are responsible for activation of monocytes, macrophages and dendritic cells, and hence participate in the innate defense against bacterial infection (162,163,166). Activation of cells in response to TLR-4 requires the regulation of specific genes, especially cytokines and chemokines, through the actions of transcription factors including NF-κB and AP1 (167,168). LMW (tetra and hexasaccharides) HA, but not intermediate or HMW (80,000–600,000 kDa) HA, binds to TLR-4 and participates in dendritic cell maturation (169). Dendritic cell maturation in response to HA was not due to CD44, as dendritic cells derived from CD44-null mice responded equally well (169).

TLR-4 binds HA as well as the bacterial toxin lipopolysaccharide (LPS) (170). Much of the signal transduction research on TLR-4 has been performed using LPS as a ligand, however, it is likely that HA stimulation of TLR-4 proceeds through similar pathways since the biological outcomes are similar. TLRs contain an extracellular domain with leucine-rich repeats, a single transmembrane domain and a cytoplasmic domain with signaling domains homologous to the interleukin-1 (IL-1) receptor (171). The extracellular domain of TLR-4 interacts with extracellular protein MD-2, a secreted protein on the cell surface which is tethered through its association with proteins containing

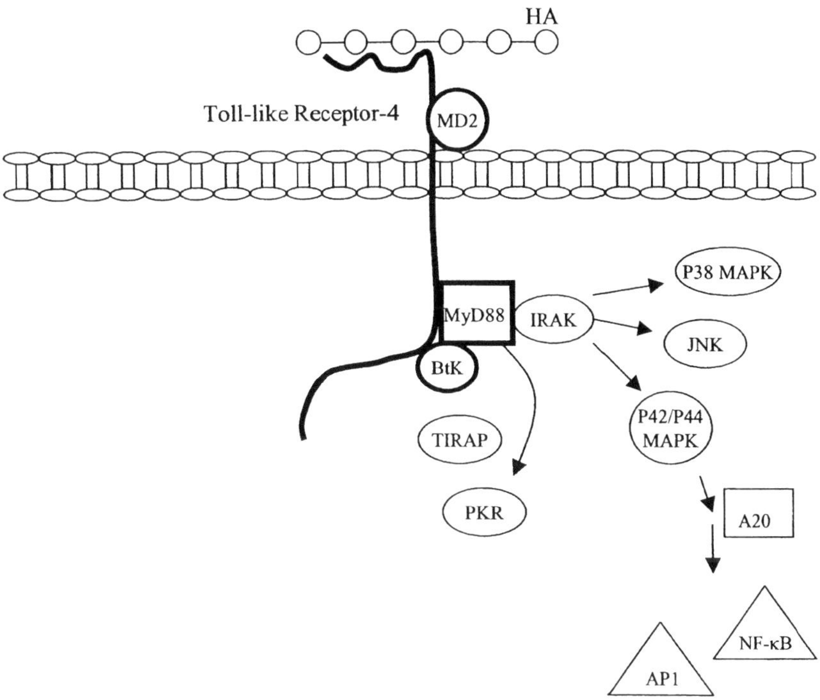

Figure 3 Signal transduction pathways activated downstream of Toll-like receptor 4. For individual references, please refer to the text. Thick black borders indicate proteins shown to directly interact with Toll-like receptor 4. Proteins are shaped according to function: circle, signal transduction; square, adaptor protein; triangle, transcription factor; oval, other.

leucine-rich repeats (171) (Fig. 3). Interactions with MD-2 are believed to be required for the activity and correct subcellular localization of TLR-4. In embryonic fibroblast-derived MD-2 null mice, TLR-4 failed to reach the plasma membrane and was found primarily in the Golgi (167,171,172).

The TLR-4 intracellular and transmembrane domains interact with a number of adaptor proteins; extensive mutagenesis experiments have been performed on the cytoplasmic region of TLR-4 to identify specific regions of the protein required for these interactions (168). The Toll-like receptors contain three conserved sequences in their cytoplasmic regions, termed ‘boxes’ (173). In macrophages, two structural surfaces of the TLR-4 cytoplasmic domain were required for downstream activation of the pro-inflammatory IL-12 p40 and anti-inflammatory IL-10 promoters, as well as for downstream activation of minimal promoters dependent on individual transcription factors (168). The same regions were required for activation of all promoters tested, suggesting that the signaling pathways diverge downstream of the adaptors (168). One of the major cellular adaptor protein identified for TLR-4 signaling is myeloid differentiation factor

88 (MyD88); interaction with MyD88 is required for the formation of a complex with the cytoplasmic serine kinase, IL-1 receptor-associated kinase (IRAK) (162, 174,175). A number of MAPKs are activated in response to LPS treatment of TLR-4 including p38 MAPK, jun kinase (JNK), and p42/p44 MAPK (176) (Fig. 3); activation of IRAK is believed to be required for downstream activation of all these MAPKs and for subsequent translocation of the NF-κB transcription factor to the nucleus (169). The Box 2 and 3 motifs of TLR-4 interact with another intracellular protein tyrosine kinase, Bruton's tyrosine kinase (Btk), a protein shown to be involved in immune function (173). Inhibition of Btk also blocked the activation of NF-κB by TLR-4 (173).

Gene regulation by most TLRs is dependent upon complex formation with MyD88 and activation of the double-stranded RNA-dependent protein kinase (PKR), with the exception of TLR-4 induction of type I interferons (177). TLR-4 can function as either a monomer or a homodimer; the downstream signaling from each of these appears to require different adaptor proteins (178). However, another recently identified adaptor molecule, Toll-IL receptor domain-containing adaptor protein (TIRAP)/MyD88 adaptor-like protein, may be involved in the MyD88-independent pathway (177,179,180). Recently, the zinc finger protein A20 has also been shown to modulate the activation of NF-κB and AP1 by TLR-4 (181); A20 was shown to bind both NF-κB and AP1, and to interrupt TLR-4 signaling at the level of MEKK, an upstream kinase of MAPK (181).

V. Layilin

Layilin is a recently cloned 55 kDa type I transmembrane protein with sequence homology to C-selectin; layilin is expressed in most tissues including ovaries, heart, lung, kidney, brain, liver and mammary tissue (182). Using microtiter plate-binding assays, co-precipitation experiments, and staining of sections pre-digested with different GAG-degrading enzymes and cell adhesion assays, Bono et al. (183) showed that HA, and no other glycosaminoglycan, is an essential ligand for layilin. Interestingly, layilin's extracellular HA-binding domain has no homology with other HA-binding proteins such as CD44, RHAMM or LYVE-1.

Layilin was identified in a yeast-two-hybrid experiment designed to identify proteins interacting with the head domain of talin, a protein involved in the membrane association of the actin cytoskeleton (182). The carboxy domain of talins were known to play a role in integrin-mediated adhesion (focal adhesions) by binding to integrin cytoplasmic domains, Fak, actin and vinculin (184,185). Layilin co-localizes with talin in membrane ruffles and the leading edge of migrating cells, but not in focal adhesions (182). The interaction of layilin with talin may provide a link between the membrane and the actin cytoskeleton in cell migration (182) (Fig. 4). These findings suggest a role for layilin in cell migration, but further work is required to determine the biological function of layilin.

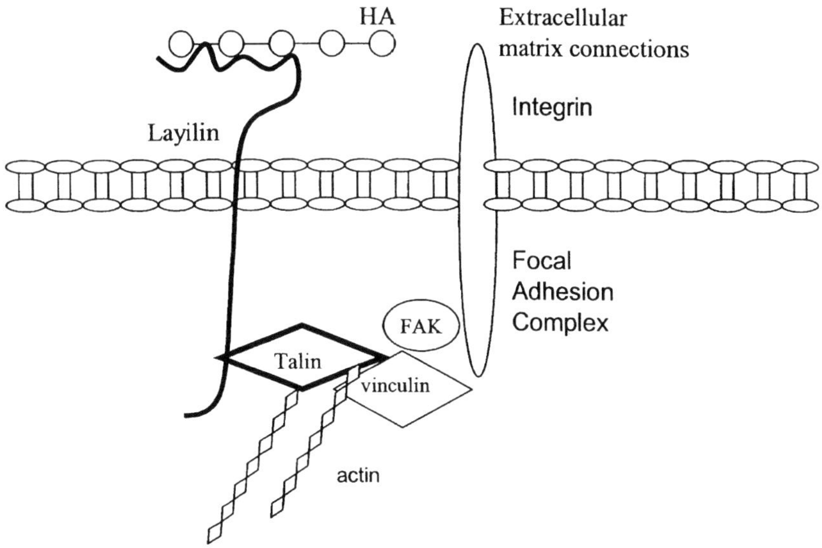

Figure 4 Signal transduction pathways activated downstream of layilin. For individual references, please refer to the text. Thick black borders indicate proteins shown to directly interact with layilin. Proteins are shaped according to function: circle, signal transduction; diamond, structural proteins oval, other.

VI. PH-20 Hyaluronidase and HA Receptor

PH-20 was first identified as a glycoprotein on the plasma membrane of the posterior head of guinea pig sperm involved in the binding of the sperm to the egg zona pellicula (186,187). Based on its similarity with other known hyaluronidases, it was shown that PH-20 bound and catalyzed HA, although PH-20 also recognizes the zona pellicula surrounding the oocyte (186,188). Recent reports show that PH-20 is also produced by other reproductive organs and has been studied in some fibroblasts (12,188–190). Also known as murine sperm adhesion molecule-1 (SPAM-1), PH-20 is produced by the epidiymal epithelium, from where it is released along with a lipid anchor, and can attach to the sperm (12). PH-20 is also expressed in a region-dependent manner in the female reproductive tract and breast tissue (190,191); here, PH-20 expression is 1.5–3 fold lower than that found in sperm, and the levels of expression oscillate with the estrous cycle (191). It is thought that the primary function of PH-20 in these tissues is the metabolism of HA (191). As has also been found for a number of other HA receptors and hyaluronidases, abnormal expression of PH-20 has been associated with some tumor cell types, possibly due to its role as a hyaluronidase and the production of HA fragments with mitogenic function (190,192).

PH-20 has been thought to play roles in sperm maturation, cumulus penetration, sperm–egg recognition and oolemmal fusion (12). PH-20 is glycerolphosphatidylinositol-linked, not transmembrane, and has been found expressed on both the plasma membrane and the inner acrosomal membrane (188). Recognition of the zona pellucida is ascribed to the inner acrosomal membrane PH-20, which appears to differ biochemically from the PH-20 on the sperm surface (188). Antibodies directed against PH-20 inhibited fertilization both *in vitro* and *in vivo* (186,187); reduced PH-20 levels were also associated with lower levels of fertility in some mice (193). However, a more recent study showed that sperms from PH-20 null mice were capable of cumulus penetration, although cells had compromised motility. The authors suggested that residual hyaluronidase activity by other proteins present on the sperm could potentially substitute for PH-20 activity (194).

In immature sperm in the caput, PH-20 exists as a polypeptide with a molecular weight of ~64 kDa and functions primarily in the maturation of sperm (187). During maturation occurring in epididymal transit from the caput to the cauda, PH-20 is cleaved into two domains linked by a disulfide bridge, an amino terminal fragment of ~41–48 kDa and a carboxy terminal fragment of ~27 kDa (13,187). Interestingly, both forms of the protein function as hyaluronidases; this activity is not affected by treatment with trypsin or by O-deglycosylation but was inhibited by N-deglycosylation (12). In caput epididymal sperm, PH-20 is distributed over the entire sperm head while cauda epididymal sperm is only found in the post-acrosomal region of the sperm head (13). When exposed to trypsin, the distribution of PH-20 in caput sperm changes from its more widespread form to the more restricted distribution of cauda sperm, suggesting that maturation is dependent on a trypsin-like mechanism with possible influence from complementary membrane-associated factors (13).

PH-20 acts as both a hyaluronidase and an HA receptor (188), and signal transduction downstream of PH-20 activation has mostly been reported in relation to sperm maturation and function. Research into the mechanism of binding of PH-20 to HA revealed that a region of the PH-20 molecule, termed Peptide 2 (aa 205–235), has an amino acid charge conformity with other HA-binding proteins and is responsible for the affinity of PH-20 for HA (195). Interactions between HA and PH-20, occurring in the Peptide 2 domain, were shown to increase intracellular Ca^{2+}, which causes an aggregation of receptors (195); increased intracellular Ca^{2+} is associated with penetration of the cumulus by the sperm. Studies by Meyers (12) also showed that maturation of PH-20 was associated with increased intracellular protein tyrosine phosphorylation, although the identity of the tyrosine phosphorylated proteins are not yet known. Although other glycerolphosphatidylinositol-linked proteins have been shown to activate pp60 c-src, this activity has not yet been shown for PH-20 (188).

VII. The LYVE-1 and HARE Receptors for HA Endocytosis and Degradation

Throughout the body, HA is continuously synthesized and degraded, and the turnover of HA in mammalian tissues has been extensively studied (5,196–198).

HA levels in the extracellular milieu are carefully modulated, as the level of free HA influences a number of hydrodynamic systems in the organism, such as lung fluid balance (199). High molecular mass HA is synthesized in the plasma membrane of fibroblasts and other cells by the addition of sugars to the reducing end of the growing polymer chain, with the non-reducing end protruding outside the cell (200,201). The rate of synthesis of HA by cells can be affected by cell density (202), mechanical stretch (26), and by growth factors, including transforming growth factor β, EGF and insulin-like growth factor 1 (203,204).

The turnover rate of HA is comparatively rapid for a connective tissue matrix component ($t_{1/2}$ 0.5 to a few days) (5). The lymphatic system accounts for 85% of the HA degradation in the body and the remaining 15% is degraded via the liver (5,196–198,205–209). Degradation of HA in the ECM occurs in two phases. Large native HA molecules (~10,000 kDa in the ECM) are partially degraded to fragments of ~1000 kDa (5). The final HA degradation process involves uptake of free HA released from the matrix into lymphatic vessels, followed by delivery to lymph nodes and finally to liver sinusoids for terminal hydrolysis (5,206,209). Six hyaluronidase-like genes have been identified in the human genome which takes part in HA degradation (210). Hyaluronidase-1 (Hyal-1) is found in mammalian plasma and urine, and also in major organs such as liver, kidney, spleen and heart (210). Data suggest that Hyal-1 and Hyal-2 are the major mammalian hyaluronidases in somatic tissues, acting in concert extracellularly to degrade HMW HA to the tetrasaccharide (210,211). In liver endothelial cells and Kupffer cells, experiments have determined that substantial levels of enzymes are present to fully degrade HA. Nine enzymes, including β-D-glucuronidase, β-*N*-acetyl-D-hexosaminidase and *N*-acetylglucosamine-6-phosphate deacetylase were shown to be present in endothelial cells and Kupffer cells, but were completely absent from hepatocytes, supporting the hypothesis that endothelial and Kupffer cells are primarily responsible for the final degradation of HA in the liver (211).

The lymphatic vessel endothelial HA receptor (LYVE-1) was first identified as the primary protein responsible for HA uptake in the lymph endothelium, and was found based on its cDNA homology (43%) with the HA receptor CD44 (212). Based on its sequence and structure, LYVE-1 was placed along with CD44 in the Link protein superfamily (36,212). LYVE-1 is also expressed in the sinusoidal endothelium of liver and spleen, the sites where uptake and degradation of HMW HA is known to occur (213). LYVE-1 is a 322 residue type I transmembrane glycoprotein, and can bind both immobilized and soluble HA (36,212). Despite similarities between the LYVE-1 and CD44-binding regions for HA, it is likely that LYVE-1 preferentially binds larger HA fragments than those recognized by CD44, probably $>HA_6$ (36). LYVE-1 does not contain sequence homology with the CD44 transmembrane or cytoplasmic domains (36); these domains are responsible for the interaction of CD44 with the cytoskeleton and other signaling molecules. Thus far, similar interactions for LYVE-1 have not been identified.

In addition to LYVE-1, the HARE protein is also believed to endocytose HA for terminal degradation (214,215). HARE was first identified in liver endothelial cells, and is distributed in liver sinusoids, venous sinuses of the red pulp in the spleen and the medullary sinuses of lymph nodes (215). Although the sinusoidal liver and lymph endothelial cells also express CD44, antibodies directed against CD44 did not block HA internalization and degradation (216).

Like LYVE-1, HARE is a HA-binding type I membrane protein. HARE was identified as a member of the protein family of fasciclin, a primarily alpha-helical, lipid-linked cell-surface glycoprotein that can act as a homophilic adhesion molecule in tissue culture (214,217,218). Two forms of HARE have been identified, a 175 and a 300 kDa forms; the 175 kDa form is a monomer, but the 300 kDa form is a trimer, made up of alpha, beta and gamma subunits, at 260, 230 and 97 kDa, respectively (214,215). Based on monoclonal antibody cross-reactivity, it was determined that the 175 HARE monomer was related to the 260 and 230 kDa subunits of the 300 kDa form (215,219,220). Complete inhibition of HA internalization and subsequent degradation requires the simultaneous blocking of both the 300 and 175 kDa HARE proteins, although the two receptors appear to function independently (217,221)

Internalization of the HA–HARE complex occurs via clathrin-coated pit-mediated endocytosis in a Ca^{2+}-independent manner (219,222,223). HARE was substantially co-localized with clathrin in cells, but not with internalized HA that was delivered to lysosomes (217). In a mechanism which corresponds to many other cell surface receptors, once HARE has delivered HA to pre-lysosomal compartments for transport to lysosomes, HARE itself is recycled to the cell surface to be used for further endocytosis cycles (224,225). Internalization of HA by HARE was not affected by treatment of cells with the cytokines tumor nectrosis factor α, interferon γ or interleukin 1, or by treatment with *Escherichia coli* endotoxin (226). The direct interactions of HARE with cytoplasmic protein have not been determined; this issue, along with the signaling mechanisms directing the cellular uptake of HA, remains as topic for further research.

VIII. Summary and Conclusions

HA has been shown to be involved in a wide variety of biological events, ranging from developmental and repair processes to the maintenance of tissue homeostasis and immune cell regulation. HA has also been shown to play roles in abnormal processes including tumorigenesis and abnormal immune function. The variety of HA biological functions is reflected by the variety of cellular receptors, which bind HA. Although some HA receptors contain homologous regions, the majority are completely divergent, including in the HA-binding domains. The further understanding of these receptors and the signaling pathways that they regulate, will provide greater insight for the treatment of diseases and dysfunctions involving HA.

Acknowledgements

We would like to thank Drs EA Turley and YJ Suzuki, and Rubén D Chévere and Jill M Angelosanto for help in proofreading this manuscript.

References

1. Weigel PH, Fuller GM, LeBoeuf RD. A model for the role of hyaluronic acid and fibrin in the early events during the inflammatory response and wound healing. J Theor Biol 1986; 119:219–234.
2. Tammi M, Day AJ, Turley EA. Hyaluronan and homeostasis: a balancing act. J Biol Chem 2002; 277:4581–4584.
3. Toole BP. Hyaluronan in morphogenesis. J Intern Med 1997; 242:35–40.
4. Gakunga P, Frost G, Shuster S, Cunha G, Formby B, Stern R. Hyaluronan is a prerequisite for ductal branching morphogenesis. Development 1997; 124: 3987–3997.
5. Laurent TC, Fraser JR. Hyaluronan. FASEB J 1992; 6:2397–2404.
6. Ohkawara Y, Tamura G, Iwasaki T, Tanaka A, Kikuchi T, Shirato K. Activation and transforming growth factor-beta production in eosinophils by hyaluronan. Am J Respir Cell Mol Biol 2000; 23:444–451.
7. Ponta H, Sherman L, Herrlich PA. CD44: from adhesion molecules to signalling regulators. Nat Mol Cell Biol 2003; 4:33–45.
8. Sattar A, Rooney P, Kumar S, Pye D, West DC, Scott I, Ledger P. Application of angiogenic oligosaccharides of hyaluronan increases blood vessel numbers in rat skin. J Investig Dermatol 1994; 103:576–579.
9. West DC, Hampson IN, Arnold F, Kumar S. Angiogenesis induced by degradation products of hyaluronic acid. Science 1985; 228:1324–1326.
10. Deed R, Rooney P, Kumar P, Norton JD, Smith J, Freemont AJ, Kumar S. Early-response gene signalling is induced by angiogenic oligosaccharides of hyaluronan in endothelial cells. Inhibition by non-angiogenic, high-molecular-weight hyaluronan. Int J Cancer 1997; 71:251–256.
11. Rahmanian M, Pertoft H, Kanda S, Christofferson R, Claesson-Welsh L, Helden P. Hyaluronan oligosaccharides induce tube formation of a brain endotelial cell line *in vitro*. Exp Cell Res 1997; 237:223–230.
12. Meyers SA. Equine sperm–oocyte interaction: the role of sperm surface hyaluronidase. Anim Reprod Sci 2001; 68:291–303.
13. Rutllant J, Meyers SA. Posttranslational processing of PH-20 during epididymal sperm maturation in the horse. Biol Reprod 2001; 65:1324–1331.
14. Chajara A, Levesque H, Courel MN, Chauzy C, Maingonnat C, Bertrand P, Delpech B. Hyaluronan and hyaluronectin production in injured rat thoracic aorta. Atherosclerosis 1996; 125:193–207.
15. Meyers SA, Seaber AV, Glisson RR, Nunley JA. Effect of hyaluronic acid/ chondroitin sulfate on healing of full-thickness tendon lacerations in rabbits. J Orthop Res 1989; 7:683–689.
16. Chen W, Abatangelo G. Functions of hyaluronan in wound repair. Wound Repair Regen 1999; 7:79–89.
17. Burd DAR, Greco RM, Regauer S, Longaker MT, Siebert JW, Garg HG. Hyaluronan and wound healing: a new perspective. Br J Plast Surg 1991; 44:579–584.

18. Scott JE. Extracellular matrix, supramolecular organisation and shape. J Anat 1995; 187:259–269.
19. Ward JA, Huang L, Guo H, Ghatak S, Toole BP. Perturbation of hyaluronan interactions inhibits malignant properties of glioma cells. Am J Pathol 2003; 162: 1403–1409.
20. Yasuda M, Nakano K, Yasumoto K, Tanaka Y. CD44: functional relevance to inflammation and malignancy. Histol Histopathol 2002; 17:945–950.
21. West DC, Kumar S. Tumour-associated hyaluronan: a potential regulator of tumour angiogenesis. Int J Radiat Biol 1991; 60:55–60.
22. Bourguignon LY, Singleton PA, Zhu H, Zhou B. Hyaluronan promotes signaling interaction between CD44 and the transforming growth factor beta receptor I in metastatic breast tumor cells. J Biol Chem 2002; 277:39703–39712.
23. Csoka T, Frost G, Stern R. Hyaluronidase in tissue invasion. Invasion Metastasis 1997; 17:297–311.
24. Delpech B, Girard N, Bertrand P, Courel MN, Chauzy C, Delpech A. Hyaluronan: fundamental principles and applications in cancer. J Intern Med 1997; 242:41–48.
25. Rooney P, Kumar S, Ponting J, Wang M. The role of hyaluronan in tumour neovascularization (review). Int J Cancer 1995; 60:632–636.
26. Mascarenhas MM, Day RM, Ochoa CD, Choi WI, Yu L, Ouyang B, Garg HG, Hales CA, Quinn DA. Low molecular weight hyaluronan from stretched lung enhances IL-8 expression. Am J Respir Cell Mol Biol 2004; 30:51–60.
27. Turino GM, Cantor JO. Hyaluronan in respiratory injury and repair. Am J Respir Crit Care Med 2003; 167:1169–1175.
28. Horton MR, Olman MA, Noble PW. Hyaluronan fragments induce plasminogen activator inhibitor-1 and inhibit urokinase activity in mouse alveolar macrophages: a potential mechanism for impaired fibrinolytic activity in acute lung injury. Chest 1999; 116:17S.
29. Svee K, White J, Vaillant P, Jessurun J, Roongta U, Krumwiede M, Johnson D, Henke C. Acute lung injury fibroblast migration and invasion of a fibrin matrix is mediated by CD44. J Clin Investig 1996; 98:1713–1727.
30. Rothenberg M. CD44—a sticky target for asthma. J Clin Investig 2003; 111: 1460–1462.
31. Laurent TC, Fraser JR, Laurent UB, Engstrom-Laurent A. Hyaluronan in inflammatory joint disease. Acta Orthop Scand Suppl 1995; 266:116–120.
32. Laurent TC, Laurent UB, Fraser JR. Serum hyaluronan as a disease marker. Ann Med 1996; 28:241–253.
33. Trochon V, Mabilat C, Bertrand P, Legrand Y, Smadja-Joffe F, Soria C, Delpech B, Lu H. Evidence of involvement of CD44 in endothelial cell proliferation, migration and angiogenesis *in vitro*. Int J Cancer 1996; 66:664–668.
34. West DC, Kumar S. Hyaluronan and angiogenesis. Ciba Found Symp 1989; 143: 187–201.
35. West DC, Kumar S. The effect of hyaluronate and its oligosaccharides on endothelial cell proliferation and monolayer integrity. Exp Cell Res 1989; 183: 179–196.
36. Prevo R, Banerji S, Ferguson DJ, Clasper S, Jackson DG. Mouse LYVE-1 is an endocytic receptor for hyaluronan in lymphatic endothelium. J Biol Chem 2001; 276:19420–19430.
37. Moffat FL Jr, Han T, Li ZM, Peck MD, Falk RE, Spalding PB, Jy W, Ahn YS, Chu AJ, Bourguignon LY. Involvement of CD44 and the cytoskeletal linker protein

ankyrin in human neutrophil bacterial phagocytosis. J Cell Physiol 1996; 168: 638–647.
38. Lesley J, Hascall VC, Tammi M, Hyman R. Hyaluronan binding by cell surface CD44. J Biol Chem 2000; 275:26967–26975.
39. Aruffo A. CD44: one ligand, two functions. J Clin Investig 1996; 98:2191–2192.
40. Abattangelo G, Weigel P. New Frontiers in Medical Sciences: Redefining Hyaluronan. Amsterdam: Elsevier, 2000.
41. Knudson C, Knudson W. Hyaluronan binding proteins in development, tissue homeostasis and disease. FASEB J 1993; 7:1233–1241.
42. Toole BP. Hyaluronan and its binding proteins, the hyaladherins. Curr Opin Cell Biol 1990; 2:839–844.
43. Slevin M, Kumar S, Gaffney J. Angiogenic oligosaccharides of hyaluronan induce multiple signaling pathways affecting vascular endothelial cell mitogenic and wound healing responses. J Biol Chem 2002; 277:41046–41059.
44. Vincent T, Jourdan M, Sy MS, Klein B, Mechti N. Hyaluronic acid induces survival and proliferation of human myeloma cells through an interleukin-6-mediated pathway involving the phosphorylation of retinoblastoma protein. J Biol Chem 2001; 276:14728–14736.
45. Jiang H, Peterson RS, Wang W, Bartnik E, Knudson CB, Knudson W. A requirement for the CD44 cytoplasmic domain for hyaluronan binding, pericellular matrix assembly, and receptor-mediated endocytosis in COS-7 cells. J Biol Chem 2002; 277:10531–10538.
46. Sugiyama K, Komada Y, Deguchi T, Zhang XL, Azuma E, Ido M, Yamamoto H, Sakurai M. CD3-mediated T cell activation is inhibited by anti-CD44 monoclonal antibodies directed to the hyaluronan-binding region. Immunol Investig 1999; 28: 185–200.
47. Rafi A, Nagarkatti M, Nagarkatti PS. Hyaluronate-CD44 interactions can induce murine B-cell activation. Blood 1997; 89:2901–2908.
48. Ishida O, Tanaka Y, Morimoto I, Takigawa M, Eto S. Chondrocytes are regulated by cellular adhesion through CD44 and hyaluronic acid pathway. J Bone Miner Res 1997; 12:1657–1663.
49. Bourguignon GJ, Iida N, Welsh CF, Zhu D, Kongrad A, Pasquale D. Involvement of CD44 and its variant isoforms in membrane-cytoskeletal interactions, cell adhesion and tumor metastasis. J Neurooncol 1995; 26:201–208.
50. Bourguignon LY, Zhu H, Shao L, Chen YW. CD44 interaction with tiam1 promotes Rac1 signaling and hyaluronic acid-mediated breast tumor cell migration. J Biol Chem 2000; 275:1829–1838.
51. Teder P, Vandivier RW, Jiang D, Liang J, Cohn L, Pure E, Henson PM, Noble PW. Resolution of lung inflammation by CD44. Science 2002; 296:155–158.
52. van der Voort R, Manten-Horst E, Smit L, Ostermann E, van den Berg F, Pals ST. Binding of cell-surface expressed CD44 to hyaluronate is dependent on splicing and cell type. Biochem Biophys Res Commun 1995; 214:137–144.
53. Orian-Rousseau V, Chen L, Sleeman JP, Herrlich P, Ponta H. CD44 is required for two consecutive steps in HGF/c-Met signaling. Genes Dev 2002; 16:3074–3086.
54. Day AJ, Prestwich GD. Hyaluronan-binding proteins: tying up the giant. J Biol Chem 2002; 277:4585–4588.
55. McKee CM, Penno MB, Cowman M, Burdick MD, Strieter RM, Bao C, Noble PW. Hyaluronan (HA) fragments induce chemokine gene expression in alveolar macrophages. The role of HA size and CD44. J Clin Investig 1996; 98:2403–2413.

56. Miyake K, Underhill CB, Lesley J, Kincade PW. Hyaluronate can function as a cell adhesion molecule and CD44 participates in hyaluronate recognition. J Exp Med 1990; 172:69–75.
57. Turley EA, Noble PW, Bourguignon LY. Signaling properties of hyaluronan receptors. J Biol Chem 2002; 277:4589–4592.
58. Morrison H, Sherman LS, Legg J, Banine F, Isacke C, Haipek CA, Gutmann DH, Ponta H, Herrlich P. The NF2 tumor suppressor gene product, merlin, mediates contact inhibition of growth through interactions with CD44. Genes Dev 2001; 15: 968–980.
59. Lesley J, Kincade PW, Hyman R. Antibody-induced activation of the hyaluronan receptor function of CD44 requires multivalent binding by antibody. Eur J Immunol 1993; 23:1902–1909.
60. Lesley J, Hyman R, Kincade PW. CD44 and its interaction with extracellular matrix. Adv Immunol 1993; 54:271–335.
61. Lesley J, English NM, Gal I, Mikecz K, Day AJ, Hyman R. Hyaluronan binding properties of a CD44 chimera containing the link module of TSG-6. J Biol Chem 2002; 277:26600–26608.
62. Gal I, Lesley J, Ko W, Gonda A, Stoop R, Hyman R, Mikecz K. Role of the extracellular and cytoplasmic domains of CD44 in the rolling interaction of lymphoid cells with hyaluronan under physiologic flow. J Biol Chem 2003; 278: 11150–11158.
63. Perschl A, Lesley J, English N, Trowbridge I, Hyman R. Role of CD44 cytoplasmic domain in hyaluronan binding. Eur J Immunol 1995; 25:495–501.
64. Lokeshwar VB, Fregien N, Bourguignon LY. Ankyrin-binding domain of CD44(GP85) is required for the expression of hyaluronic acid-mediated adhesion function. J Cell Biol 1994; 126:1099–1109.
65. Liu D, Sy MS. A cysteine residue located in the transmembrane domain of CD44 is important in binding of CD44 to hyaluronic acid. J Exp Med 1996; 183:1987–1994.
66. Liu D, Liu T, Sy MS. Identification of two regions in the cytoplasmic domain of CD44 through which PMA, calcium, and foskolin differentially regulate the binding of CD44 to hyaluronic acid. Cell Immunol 1998; 190:132–140.
67. Liu D, Zhang D, Mori H, Sy MS. Binding of CD44 to hyaluronic acid can be induced by multiple signals and requires the CD44 cytoplasmic domain. Cell Immunol 1996; 174:73–83.
68. Liu D, Liu T, Li R, Sy MS. Mechanisms regulating the binding activity of CD44 to hyaluronic acid. Front Biosci 1998; 3:D631–D636.
69. Pure E, Camp RL, Peritt D, Panettieri RA Jr, Lazaar AL, Nayak S. Defective phosphorylation and hyaluronate binding of CD44 with point mutations in the cytoplasmic domain. J Exp Med 1995; 181:55–62.
70. Bates RC, Edwards NS, Burns GF, Fisher DE. A CD44 survival pathway triggers chemoresistance via lyn kinase and phosphoinositide 3-kinase/Akt in colon carcinoma cells. Cancer Res 2001; 61:5275–5283.
71. Fitzgerald KA, Bowie AG, Skeffington BS, O'Neill LA. Ras, protein kinase C zeta, and I kappa B kinases 1 and 2 are downstream effectors of CD44 during the activation of NF-kappa B by hyaluronic acid fragments in T-24 carcinoma cells. J Immunol 2000; 164:2053–2063.
72. Fujita Y, Kitagawa M, Nakamura S, Azuma K, Ishii G, Higashi M, Kishi H, Hiwasa T, Koda K, Nakajima N, Harigaya K. CD44 signaling through focal adhesion kinase and its anti-apoptotic effect. FEBS Lett 2002; 528:101–108.

73. Ranuncolo SM, Ladeda V, Gorostidy S, Morandi A, Varela M, Lastiri J, Loria D, Del Aguila R, Joffe EB, Pallotta G, Puricelli L. Expression of CD44s and CD44 splice variants in human melanoma. Oncol Rep 2002; 9:51–56.
74. Lokeshwar VB, Iida N, Bourguignon LY. The cell adhesion molecule, GP116, is a new CD44 variant (ex14/v10) involved in hyaluronic acid binding and endothelial cell proliferation. J Biol Chem 1996; 271:23853–23864.
75. Hunter T. Oncoprotein networks. Cell 1997; 88:333–346.
76. Neri LM, Borgatti P, Capitani S, Martelli AM. The nuclear phosphoinositide 3-kinase/AKT pathway: a new second messenger system. Biochim Biophys Acta 2002; 1584:73–80.
77. Parsons JT. Focal adhesion kinase: the first ten years. J Cell Sci 2003; 116: 1409–1416.
78. Day RM, Cioce V, Breckenridge D, Castagnino P, Bottaro DP. Differential signaling by alternative HGF isoforms through c-Met: activation of both MAP kinase and PI 3-kinase pathways is insufficient for mitogenesis. Oncogene 1999; 18: 3399–3406.
79. Madrid L, Baldwin A. Regulation of NF-kappaB by oncoproteins and tumor suppressor proteins. Methods Mol Biol 2003; 223:523–532.
80. Bourguignon LY, Singleton PA, Diedrich F, Zhu H. Hyaluronan-mediated CD44 interaction with RhoGEF and Rho-kinase promotes Gab-1 phosphorylation and PI3 kinase signaling leading to cytokine (M-CSF) production breast tumor progression. J Biol Chem 2003; 14:14.
81. Skubitz KM, Campbell KD, Skubitz AP. Tyrosine kinase activity is associated with CD44 in human neutrophils. FEBS Lett 1998; 439:97–100.
82. Rozsnyay Z. Signaling complex formation of CD44 with src-related kinases. Immunol Lett 1999; 68:101–108.
83. Bourguignon LYW, Zhu D, Zhu H. CD44 isoform–cytoskeleton interaction in oncogenic signaling and tumor progression. Front Biosci 1998; 3:D637–D649.
84. Su W, Sin M, Darrow A, Sherman LS. Malignant peripheral nerve sheath tumor cell invasion is facilitated by Src and aberrant CD44 expression. Glia 2003; 42: 350–358.
85. Kauffman EC, Robinson VL, Stadler WM, Sokoloff MH, Rinker-Schaeffer CW. Metastasis suppression: the evolving role of metastasis suppressor genes for regulating cancer cell growth at the secondary site. J Urol 2003; 169:1122–1133.
86. Bretscher A, Reczek D, Berryman M. Ezrin: a protein requiring conformational activation to link microfilaments to the plasma membrane in the assembly of cell surface structures. J Cell Sci 1997; 110:3011–3018.
87. Bourguignon LY, Lokeshwar VB, Chen X, Kerrick WG. Hyaluronic acid-induced lymphocyte signal transduction and HA receptor (GP85/CD44)–cytoskeleton interaction. J Immunol 1993; 151:6634–6644.
88. Welsh CF, Zhu D, Bourguignon LY. Interaction of CD44 variant isoforms with hyaluronic acid and the cytoskeleton in human prostate cancer cells. J Cell Physiol 1995; 164:605–612.
89. Hirao M, Sato N, Kondo T, Yonemura S, Monden M, Sasaki T, Takai Y, Tsukita S. Regulation mechanism of ERM (ezrin/radixin/moesin) protein/plasma membrane association: possible involvement of phosphatidylinositol turnover and Rho-dependent signaling pathway. J Cell Biol 1996; 135:37–51.
90. Bretscher A. Regulation of cortical structure by the ezrin–radixin–moesin protein family. Curr Opin Cell Biol 1999; 11:109–116.

91. Bennett V. Ankyrins. Adaptors between diverse plasma membrane proteins and the cytoplasm. J Biol Chem 1992; 267:8703–8706.
92. Luna EJ, Hitt AL. Cytoskeleton–plasma membrane interactions. Science 1992; 258:955–964.
93. Zhu D, Bourguignon LY. Interaction between CD44 and the repeat domain of ankyrin promotes hyaluronic acid-mediated ovarian tumor cell migration. J Cell Physiol 2000; 183:182–195.
94. Zhu D, Bourguignon LY. The ankyrin-binding domain of CD44s is involved in regulating hyaluronic acid-mediated functions and prostate tumor cell transformation. Cell Motil Cytoskeleton 1998; 39:209–222.
95. Bourguignon LY, Zhu H, Shao L, Zhu D, Chen YW. Rho-kinase (ROK) promotes CD44v(3,8-10)-ankyrin interaction and tumor cell migration in metastatic breast cancer cells. Cell Motil Cytoskeleton 1999; 43:269–287.
96. Singleton PA, Bourguignon LY. CD44v10 interaction with Rho-kinase (ROK) activates inositol 1,4,5-triphosphate (IP3) receptor-mediated Ca^{2+} signaling during hyaluronan (HA)-induced endothelial cell migration. Cell Motil Cytoskeleton 2002; 53:293–316.
97. Bourguignon LY, Zhu H, Shao L, Chen YW. CD44 interaction with c-Src kinase promotes cortactin-mediated cytoskeleton function and hyaluronic acid-dependent ovarian tumor cell migration. J Biol Chem 2001; 276:7327–7336.
98. Sohara Y, Ishiguro N, Machida K, Kurata H, Thant AA, Senga T, Matsuda S, Kimata K, Iwata H, Hamaguchi M. Hyaluronan activates cell motility of v-Src-transformed cells via Ras-mitogen-activated protein kinase and phosphoinositide 3-kinase-Akt in a tumor-specific manner. Mol Biol Cell 2001; 12:1859–1868.
99. Ilangumaran S, Briol A, Hoessli DC. CD44 selectively associates with active Src family protein tyrosine kinases Lck and Fyn in glycosphingolipid-rich plasma membrane domains of human peripheral blood lymphocytes. Blood 1998; 91: 3901–3908.
100. Li R, Wong N, Jabali MD, Johnson P. CD44-initiated cell spreading induces Pyk2 phosphorylation, is mediated by Src family kinases, and is negatively regulated by CD45. J Biol Chem 2001; 276:28767–28773.
101. Ilangumaran S, Borisch B, Hoessli DC. Signal transduction via CD44: role of plasma membrane microdomains. Leuk Lymphoma 1999; 35:455–469.
102. van der Voort R, Taher TE, Wielenga VJ, Spaargaren M, Prevo R, Smit L, David G, Hartmann G, Gherardi E, Pals ST. Heparan sulfate-modified CD44 promotes hepatocyte growth factor/scatter factor-induced signal transduction through the receptor tyrosine kinase c-Met. J Biol Chem 1999; 274:6499–6506.
103. Sherman LS, Rizvi TA, Karyala S, Ratner N. CD44 enhances neuregulin signaling by Schwann cells. J Cell Biol 2000; 150:1071–1084.
104. Bourguignon LY, Zhu H, Chu A, Iida N, Zhang L, Hung MC. Interaction between the adhesion receptor, CD44, and the oncogene product, p185HER2, promotes human ovarian tumor cell activation. J Biol Chem 1997; 272:27913–27918.
105. Kaya G, Rodriguez I, Jorcano JL, Vassalli P, Stamenkovic I. Selective suppression of CD44 in keratinocytes of mice bearing an antisense CD44 transgene driven by a tissue-specific promoter disrupts hyaluronate metabolism in the skin and impairs keratinocyte proliferation. Genes Dev 1997; 11:996–1007.
106. Sherman L, Wainwright D, Ponta H, Herrlich P. A splice variant of CD44 expressed in the apical ectodermal ridge presents fibroblast growth factors to limb mesenchyme and is required for limb outgrowth. Genes Dev 1998; 12:1058–1071.

107. Taher TE, Smit L, Griffioen AW, Schilder-Tol EJ, Borst J, Pals ST. Signaling through CD44 is mediated by tyrosine kinases. Association with p56lck in T lymphocytes. J Biol Chem 1996; 271:2863–2867.
108. Rubin JS, Bottaro DP, Aaronson SA. Hepatocyte growth factor/scatter factor and its receptor, the c-met proto-oncogene product. Biochim Biophys Acta 1993; 1155: 357–371.
109. Jeffers M, Rong S, Vande Woude GF. Enhanced tumorigenicity and invasion-metastasis by hepatocyte growth factor/scatter factor-met signalling in human cells concomitant with induction of the urokinase proteolysis network. Mol Cell Biol 1996; 16:1115–1125.
110. Jeffers M, Rong S, Anver M, Vande Woude GF. Autocrine hepatocyte growth factor/scatter factor-Met signaling induces transformation and the invasive/metastastic phenotype in C127 cells. Oncogene 1996; 13:853–856.
111. Jeffers M, Rao MS, Rulong S, Reddy JK, Subbarao V, Hudson E, Vande Woude GF, Resau JH. Hepatocyte growth factor/scatter factor-Met signaling induces proliferation, migration, and morphogenesis of pancreatic oval cells. Cell Growth Differ 1996; 7:1805–1813.
112. Naldini L, Weidner KM, Vigna E, Gaudino G, Bardelli A, Ponzetto C, Narsimhan RP, Hartmann G, Zarnegar R, Michalopoulos GK. Scatter factor and hepatocyte growth factor are indistinguishable ligands for the MET receptor. EMBO J 1991; 10:2867–2878.
113. Komada M, Miyazawa K, Ishii T, Kitamura N. Characterization of hepatocyte-growth-factor receptors on Meth A cells. Eur J Biochem 1992; 204:857–864.
114. Arakaki N, Hirono S, Ishii T, Kimoto M, Kawakami S, Nakayama H, Tsubouchi H, Hishida T, Daikuhara Y. Identification and partial characterization of two classes of receptors for human hepatocyte growth factor on adult rat hepatocytes in primary culture. J Biol Chem 1992; 267:7101–7107.
115. Day RM, Soon L, Breckenridge D, Bridges B, Patel BK, Wang LM, Corey SJ, Bottaro DP. Mitogenic synergy through multilevel convergence of hepatocyte growth factor and interleukin-4 signaling pathways. Oncogene 2002; 21: 2201–22011.
116. Schwall RH, Chang LY, Godowski PJ, Kahn DW, Hillan KJ, Bauer KD, Zioncheck TF. Heparin induces dimerization and confers proliferative activity onto the hepatocyte growth factor antagonists NK1 and NK2. J Cell Biol 1996; 133:709–718.
117. Sakata H, Stahl SJ, Taylor WG, Rosenberg JM, Sakaguchi K, Wingfield PT, Rubin JS. Heparin binding and oligomerization of hepatocyte growth factor/scatter factor isoforms. Heparan sulfate glycosaminoglycan requirement for Met binding and signaling. J Biol Chem 1997; 272:9457–9463.
118. Gunthert U, Stauder R, Mayer B, Terpe HJ, Finke L, Friedrichs K. Are CD44 variant isoforms involved in human tumour progression? Cancer Surv 1995; 24: 19–42.
119. Rudy W, Hofmann M, Schwartz-Albiez R, Zoller M, Heider KH, Ponta H, Herrlich P. The two major CD44 proteins expressed on a metastatic rat tumor cell line are derived from different splice variants: each one individually suffices to confer metastatic behavior. Cancer Res 1993; 53:1262–1268.
120. Olayioye MA, Neve RM, Lane HA, Hynes NE. The ErbB signaling network: receptor heterodimerization in development and cancer. EMBO J 2000; 19: 3159–3167.

121. Suzuki M, Raab G, Moses MA, Fernandez CA, Klagsbrun M. Matrix metalloproteinase-3 releases active heparin-binding EGF-like growth factor by cleavage at a specific juxtamembrane site. J Biol Chem 1997; 272: 31730–31737.
122. Izumi Y, Hirata M, Hasuwa H, Iwamoto R, Umata T, Miyado K, Tamai Y, Kurisaki T, Sehara-Fujisawa A, Ohno S, Mekada E. A metalloprotease-disintegrin, MDC9/meltrin-gamma/ADAM9 and PKCdelta are involved in TPA-induced ectodomain shedding of membrane-anchored heparin-binding EGF-like growth factor. EMBO J 1998; 17:7260–7272.
123. Yu WH, Woessner JF Jr, McNeish JD, Stamenkovic I. CD44 anchors the assembly of matrilysin/MMP-7 with heparin-binding epidermal growth factor precursor and ErbB4 and regulates female reproductive organ remodeling. Genes Dev 2002; 16: 307–323.
124. Bourguignon LY, Zhu H, Zhou B, Diedrich F, Singleton PA, Hung MC. Hyaluronan promotes CD44v3-Vav2 interaction with Grb2-p185(HER2) and induces Rac1 and Ras signaling during ovarian tumor cell migration and growth. J Biol Chem 2001; 276:48679–48692.
125. Hardwick C, Hoare K, Owens R, Hohn HP, Hook M, Moore D, Cripps V, Austen L, Nance DM, Turley EA. Molecular cloning of a novel hyaluronan receptor that mediates tumor cell motility. J Cell Biol 1992; 117:1343–1350.
126. Lokeshwar VB, Selzer MG. Differences in hyaluronic acid-mediated functions and signaling in arterial, microvessel, and vein-derived human endothelial cells. J Biol Chem 2000; 275:27641–27649.
127. Dowthwaite GP, Edwards JC, Pitsillides AA. An essential role for the interaction between hyaluronan and hyaluronan binding proteins during joint development. J Histochem Cytochem 1998; 46:641–651.
128. Hall CL, Turley EA. Hyaluronan: RHAMM mediated cell locomotion and signaling in tumorigenesis. J Neurooncol 1995; 26:221–229.
129. Mohapatra S, Yang X, Wright JA, Turley EA, Greenberg AH. Soluble hyaluronan receptor RHAMM induces mitotic arrest by suppressing Cdc2 and cyclin B1 expression. J Exp Med 1996; 183:1663–1668.
130. Hall CL, Lange LA, Prober DA, Zhang S, Turley EA. pp60(c-src) is required for cell locomotion regulated by the hyaluronanreceptor RHAMM. Oncogene 1996; 13:2213–2224.
131. Entwistle J, Zhang S, Yang B, Wong C, Li Q, Hall CL, Day AJ, Mowat M, Greenberg AH, Turley EA. Characterization of the murine gene encoding the hyaluronan receptor RHAMM. Gene 1995; 163:233–238.
132. Sherman L, Sleeman J, Herrlich P, Ponta H. Hyaluronate receptors: key players in growth, differentiation, migration and tumor progression. Curr Opin Cell Biol 1994; 6:726–733.
133. Zhang S, Chang MC, Zylka D, Turley S, Harrison R, Turley EA. The hyaluronan receptor RHAMM regulates extracellular-regulated kinase. J Biol Chem 1998; 273: 11342–11348.
134. Lynn BD, Turley EA, Nagy JI. Subcellular distribution, calmodulin interaction, and mitochondrial association of the hyaluronan-binding protein RHAMM in rat brain. J Neurosci Res 2001; 65:6–16.
135. Assmann V, Jenkinson D, Marshall JF, Hart IR. The intracellular hyaluronan receptor RHAMM/IHABP interacts with microtubules and actin filaments. J Cell Sci 1999; 112:3943–3954.

136. Assmann V, Marshall JF, Fieber C, Hofmann M, Hart IR. The human hyaluronan receptor RHAMM is expressed as an intracellular protein in breast cancer cells. J Cell Sci 1998; 111:1685–1694.
137. Maxwell CA, Keats JJ, Crainie M, Sun X, Yen T, Shibuya E, Hendzel M, Chan G, Pilarski LM. RHAMM is a centrosomal protein that interacts with dynein and maintains spindle pole stability. Mol Biol Cell 2003; 14:2262–2276.
138. Yang B, Zhang L, Turley EA. Identification of two hyaluronan-binding domains in the hyaluronan receptor RHAMM. J Biol Chem 1993; 268:8617–8623.
139. Yang B, Yang BL, Savani RC, Turley EA. Identification of a common hyaluronan binding motif in the hyaluronan binding proteins RHAMM, CD44 and link protein. EMBO J 1994; 13:286–296.
140. Turley EA, Belch AJ, Poppema S, Pilarski LM. Expression and function of a receptor for hyaluronan-mediated motility on normal and malignant B lymphocytes. Blood 1993; 81:446–453.
141. Pilarski LM, Pruski E, Wizniak J, Paine D, Seeberger K, Mant MJ, Brown CB, Belch AR. Potential role for hyaluronan and the hyaluronan receptor RHAMM in mobilization and trafficking of hematopoietic progenitor cells. Blood 1999; 93: 2918–2927.
142. Turley EA. Hyaluronan and cell locomotion. Cancer Metastasis Rev 1992; 11: 21–30.
143. Kornovski BS, McCoshen J, Kredentser J, Turley E. The regulation of sperm motility by a novel hyaluronan receptor. Fertil Steril 1994; 61:935–940.
144. Savani RC, Cao G, Pooler PM, Zaman A, Zhou Z, DeLisser HM. Differential involvement of the hyaluronan (HA) receptors CD44 and receptor for HA-mediated motility in endothelial cell function and angiogenesis. J Biol Chem 2001; 276: 36770–36778.
145. Rein DT, Roehrig K, Schondorf T, Lazar A, Fleisch M, Niederacher D, Bender HG, Dall P. Expression of the hyaluronan receptor RHAMM in endometrial carcinomas suggests a role in tumour progression and metastasis. J Cancer Res Clin Oncol 2003; 129:161–164.
146. Greiner J, Ringhoffer M, Taniguchi M, Schmitt A, Kirchner D, Krahn G, Heilmann V, Gschwend J, Bergmann L, Dohner H, Schmitt M. Receptor for hyaluronan acid-mediated motility (RHAMM) is a new immunogenic leukemia-associated antigen in acute and chronic myeloid leukemia. Exp Hematol 2002; 30:1029–1035.
147. Crainie M, Belch AR, Mant MJ, Pilarski LM. Overexpression of the receptor for hyaluronan-mediated motility (RHAMM) characterizes the malignant clone in multiple myeloma: identification of three distinct RHAMM variants. Blood 1999; 93:1684–1696.
148. Kong QY, Liu J, Chen XY, Wang XW, Sun Y, Li H. Differential expression patterns of hyaluronan receptors CD44 and RHAMM in transitional cell carcinomas of urinary bladder. Oncol Rep 2003; 10:51–55.
149. Turley EA, Austen L, Moore D, Hoare K. Ras-transformed cells express both CD44 and RHAMM hyaluronan receptors: only RHAMM is essential for hyaluronan-promoted locomotion. Exp Cell Res 1993; 207:277–282.
150. Masellis-Smith A, Belch AR, Mant MJ, Turley EA, Pilarski LM. Hyaluronan-dependent motility of B cells and leukemic plasma cells in blood, but not of bone marrow plasma cells, in multiple myeloma: alternate use of receptor for hyaluronan-mediated motility (RHAMM) and CD44. Blood 1996; 87:1891–1899.

151. Pilarski LM, Miszta H, Turley EA. Regulated expression of a receptor for hyaluronan-mediated motility on human thymocytes and T cells. J Immunol 1993; 150:4292–4302.
152. Samuel SK, Hurta RA, Spearman MA, Wright JA, Turley EA, Greenberg AH. TGF-beta 1 stimulation of cell locomotion utilizes the hyaluronan receptor RHAMM and hyaluronan. J Cell Biol 1993; 123:749–758.
153. Savani RC, Wang C, Yang B, Zhang S, Kinsella MG, Wight TN, Stern R, Nance DM, Turley EA. Migration of bovine aortic smooth muscle cells after wounding injury. The role of hyaluronan and RHAMM. J Clin Investig 1995; 95: 1158–1168.
154. Hall CL, Wang C, Lange LA, Turley EA. Hyaluronan and the hyaluronan receptor RHAMM promote focal adhesion turnover and transient tyrosine kinase activity. J Cell Biol 1994; 126:575–588.
155. Turner CE. Paxillin and focal adhesion signalling. Rev Nat Cell Biol 2000; 2: E231–E236.
156. Gares SL, Pilarski LM. Balancing thymocyte adhesion and motility: a functional linkage between beta1 integrins and the motility receptor RHAMM. Dev Immunol 2000; 7:209–225.
157. Anneren C, Lindholm CK, Kriz V, Welsh M. The FRK/RAK-SHB signaling cascade: a versatile signal-transduction pathway that regulates cell survival, differentiation and proliferation. Rev Curr Mol Med 2003; 3:313–324.
158. Brecht M, Mayer U, Schlosser E, Prehm P. Increased hyaluronate synthesis is required for fibroblast detachment and mitosis. Biochem J 1986; 239:445–450.
159. Lynn BD, Li X, Cattini PA, Nagy JI. Sequence, protein expression and extracellular-regulated kinase association of the hyaladherin RHAMM (receptor for hyaluronan mediated motility) in PC12 cells. Neurosci Lett 2001; 306: 49–52.
160. Rock FL, Hardiman G, Timans JC, Kastelein RA, Bazan JF. A family of human receptors structurally related to *Drosophila* Toll. Proc Natl Acad Sci 1998; 95: 588–593.
161. Rescigno M, Granucci F, Ricciardi-Castagnoli P. Molecular events of bacterial-induced maturation of dendritic cells. J Clin Immunol 2000; 20:161–166.
162. Poltorak A, He X, Smirnova I, Liu MY, van Huffel C, Du X, Birdwell D, Alejos E, Silva M, Galanos C, Freudenberg M, Ricciardi-Castagnoli P, Layton B, Beutler B. Defective LPS signaling in C3H/HeJ and c57BL/10ScCr mice: mutations in TLR4 gene. Science 1998; 282:2085–2088.
163. Takeuchi O, Hoshino K, Kawai T, Sanjo H, Takada H, Ogawa T, Takeda K, Akira S. Differential roles of TLR2 and TLR4 in recognition of gram-negative and gram positive bacterial cell wall components. Immunity 1999; 11:443–451.
164. Beutler B. TLR4 as the mammalian endotoxin sensor. Curr Top Microbiol Immunol 2002; 270:109–120.
165. Bocker U, Yezerskyy O, Feick P, Manigold T, Panja A, Kalina U, Herweck F, Rossol S, Singer MV. Responsiveness of intestinal epithelial cell lines to lipopolysaccharide is correlated with Toll-like receptor 4 but not Toll-like receptor 2 or CD14 expression. Int J Colorectal Dis 2003; 18:25–32.
166. Chen LY, Zuraw BL, Zhao M, Liu FT, Huang S, Pan ZK. Involvement of protein tyrosine kinase in Toll-like receptor 4-mediated NF-kappa B activation in human peripheral blood monocytes. Am J Physiol Lung Cell Mol Physiol 2003; 284: L607–L613.

167. Miyake K, Nagai Y, Akashi S, Nagafuku M, Ogata M, Kosugi A. Essential role of MD-2 in B-cell responses to lipopolysaccharide and Toll-like receptor 4 distribution. J Endotoxin Res 2002; 8:449–452.
168. Ronni T, Agarwal V, Haykinson M, Haberland ME, Cheng G, Smale ST. Common interaction surfaces of the toll-like receptor 4 cytoplasmic domain stimulate multiple nuclear targets. Mol Cell Biol 2003; 23:2543–2555.
169. Termeer CC, Hennies J, Voith U, Ahrens T, Weiss JM, Prehm P, Simon JC. Oligosaccharides of hyaluronan are potent activators of dendritic cells. J Immunol 2000; 165:1863–1870.
170. Paik YH, Schwabe RF, Bataller R, Russo MP, Jobin C, Brenner DA. Toll-like receptor 4 mediates inflammatory signaling by bacterial lipopolysaccharide in human hepatic stellate cells. Hepatology 2003; 37:1043–1055.
171. Shimazu R, Akashi S, Ogata H, Nagai Y, Fukudome K, Miyake K, Kimoto M. MD-2, a molecule that confers lipopolysaccharide responsiveness on Toll-like receptor 4. J Exp Med 1999; 189:1777–1782.
172. Nagai Y, Akashi S, Nagafuku M, Ogata M, Iwakura Y, Akira S, Kitamura T, Kosugi A, Kimoto M, Miyake K. Essential role of MD-2 in LPS responsiveness and TLR4 distribution. Nat Immunol 2002; 3:667–672.
173. Jefferies CA, Doyle S, Brunner C, Dunne A, Brint E, Wietek C, Walch E, Wirth T, O'Neill LA. Bruton's tyrosine kinase is a Toll/interleukin-1 receptor domain-binding protein that participates in nuclear factor kappaB activation by Toll-like receptor 4. J Biol Chem 2003; 278:26258–26264.
174. Takeuchi O, Takeda K, Hoshino K, Adachi O, Ogawa T, Akira S. Cellular responses to bacterial cell wall components are mediated through MyD88-dependent signaling cascades. Int Immunol 2000; 12:113–117.
175. Medzhitov R, Preston-Hurlburt P, Kopp E, Stalden A, Chen C, Ghosh S, Janeway CAJ. Myd88 is an adaptor protein in the hToll/IL1 receptor family signaling pathways. Mol Cell 1998; 2:253–258.
176. Heine H, El-Samalouti VT, Notzel C, Pfeiffer A, Lentschat A, Kusumoto S, Schmitz G, Hamann L, Ulmer AJ. CD55/decay accelerating factor is part of the lipopolysaccharide-induced receptor complex. Eur J Immunol 2003; 33:1399–1408.
177. Akira S, Hoshino K. Myeloid differentiation factor 88-dependent and -independent pathways in toll-like receptor signaling. J Infect Dis 2003; 187:S356–S363.
178. Zhang H, Tay P, Cao W, Li W, Lu J. Integrin-nucleated Toll-like receptor (TLR) dimerization reveals subcellular targeting of TLRs and distinct mechanisms of TLR4 activation and signaling. FEBS Lett 2002; 532:171–176.
179. Vogel SN, Fenton M. Toll-like receptor 4 signalling: new perspectives on a complex signal-transduction problem. Biochem Soc Trans 2003; 31:664–668.
180. Yamamoto M, Sato S, Mori K, Hoshino K, Takeuchi O, Takeda K, Akira S. Cutting edge: a novel Toll/IL-1 receptor domain-containing adapter that preferentially activates the IFN-beta promoter in the Toll-like receptor signaling. J Immunol 2002; 169:6668–6672.
181. O'Reilly SM, Moynagh PN. Regulation of Toll-like receptor 4 signalling by A20 zinc finger protein. Biochem Biophys Res Commun 2003; 303:586–593.
182. Borowsky ML, Hynes RO. Layilin, a novel talin-binding transmembrane protein homologus with C-type lectins, is localized in membrane ruffles. J Cell Biol 1998; 143:429–442.
183. Bono P, Rubin K, Higgins JM, Hynes RO. Layilin, a novel integral membrane protein, is a hyaluronan receptor. Mol Biol Cell 2001; 12:891–900.

184. Hemmings L, Rees DJ, Ohanian V, Bolton SJ, Gilmore AP, Patel B, Priddle H, Trevithick JE, Hynes RO, Critchley DR. Talin contains three actin-binding sites each of which is adjacent to vinculin-binding site. J Cell Sci 1996; 109:2715–2726.
185. Bolton SJ, Barry ST, Mosley H, Jockusch BM, Wilkinson JM, Critchley DR. Monoclonal antibodies recognizing the N- and C-terminal of talin disrupt actin stress fibers when microinjected into human fibroblasts. Cytoskeleton 1997; 36: 363–376.
186. Lathrop WF, Carmichael EP, Myles DG, Primakoff P. cDNA cloning reveals the molecular structure of a sperm surface protein, PH-20, involved in sperm–egg adhesion and the wide distribution of its gene among mammals. J Cell Biol 1990; 111:2939–2949.
187. Gmachl M, Sagan S, Ketter S, Kreil G. The human sperm protein PH-20 has hyaluronidase activity. FEBS Lett 1993; 336:545–548.
188. Cherr GN, Yudin AI, Overstreet JW. The dual functions of GPI-anchored PH-20: hyaluronidase and intracellular signaling. Matrix Biol 2001; 20:515–525.
189. Ohno S, Ijuin C, Doi T, Yoneno K, Tanne K. Expression and activity of hyaluronidase in human periodontal ligament fibroblasts. J Periodontol 2002; 73: 1331–1337.
190. Beech DJ, Madan AK, Deng N. Expression of PH-20 in normal and neoplastic breast tissue. J Surg Res 2002; 103:203–207.
191. Zhang H, Martin-Deleon PA. Mouse epididymal Spam1 (PH-20) is released in the luminal fluid with its lipid anchor. J Androl 2003; 24:51–58.
192. Godin DA, Fitzpatrick PC, Scandurro AB, Belafsky PC, Woodworth BA, Amedee RG, Beech DJ, Beckman BS. PH20: a novel tumor marker for laryngeal cancer. Arch Otolaryngol Head Neck Surg 2000; 126:402–404.
193. Zheng Y, Deng X, Martin-DeLeon PA. Lack of sharing of Spam1 (Ph-20) among mouse spermatids and transmission ratio distortion. Biol Reprod 2001; 64: 1730–1738.
194. Baba D, Kashiwabara S, Honda A, Yamagata K, Wu Q, Ikawa M, Okabe M, Baba T. Mouse sperm lacking cell surface hyaluronidase PH-20 can pass through the layer of cumulus cells and fertilize the egg. J Biol Chem 2002; 277:30310–30314.
195. Vines CA, Li MW, Deng X, Yudin AI, Cherr GN, Overstreet JW. Identification of a hyaluronic acid (HA) binding domain in the PH-20 protein that may function in cell signaling. Mol Reprod Dev 2001; 60:542–552.
196. Laurent UB, Fraser JR, Laurent TC. An experimental technique to study the turnover of concentrated hyaluronan in the anterior chamber of the rabbit. Exp Eye Res 1988; 46:49–58.
197. Lebel L, Fraser JR, Kimpton WS, Gabrielsson J, Gerdin B, Laurent TC. A pharmacokinetic model of intravenously administered hyaluronan in sheep. Pharm Res 1989; 6:677–682.
198. Frost SJ, McGary CT, Raja RH, Weigel PH. Specific intracellular hyaluronic acid binding to isolated rat hepatocytes is membrane-associated. Biochim Biophys Acta 1988; 946:66–74.
199. Allen SJ, Fraser RE, Laurent UB, Reed RK, Laurent TC. Turnover of hyaluronan in the rabbit pleural space. J Appl Physiol 1992; 73:1457–1460.
200. Philipson LH, Schwartz NB. Subcellular localization of hyaluronate synthetase in oligodendroglioma cells. J Biol Chem 1984; 259:5017–5023.
201. Prehm P. Synthesis of hyaluronate in differentiated teratocarcinoma cells. Mechanism of chain growth. Biochem J 1983; 211:191–198.

202. Scott LJ, Merrilees MJ. Stimulation of smooth muscle cell glycosaminoglycan synthesis by cultured endothelial cells is dependent on endothelial cell density. Atherosclerosis 1987; 63:145–152.
203. Honda A, Noguchi N, Takehara H, Ohashi Y, Asuwa N, Mori Y. Cooperative enhancement of hyaluronic acid synthesis by combined use of IGF-I and EGF, and inhibition by tyrosine kinase inhibitor genistein, in cultured mesothelial cells from rabbit pericardial cavity. J Cell Sci 1991; 98:91–98.
204. Suzuki M, Asplund T, Yamashita H, Heldin CH, Heldin P. Stimulation of hyaluronan biosynthesis by platelet-derived growth factor-BB and transforming growth factor-beta 1 involves activation of protein kinase C. Biochem J 1995; 307:817–821.
205. Fraser JR, Kimpton WG, Laurent TC, Cahill RN, Vakakis N. Uptake and degradation of hyaluronan in lymphatic tissue. Biochem J 1988; 256:153–158.
206. Fraser JR, Laurent TC. Turnover and metabolism of hyaluronan. Ciba Found Symp 1989; 143:41–53.
207. Raja RH, McGary CT, Weigel PH. Affinity and distribution of surface and intracellular hyaluronic acid receptors in isolated rat liver endothelial cells. J Biol Chem 1988; 263:16661–16668.
208. Reed RK, Laurent UB, Fraser JR, Laurent TC. Removal rate of [3H]hyaluronan injected subcutaneously in rabbits. Am J Physiol 1990; 259:H532–H535.
209. Tengblad A, Laurent UB, Lilja K, Cahill RN, Engstrom-Laurent A, Fraser JR, Hansson HE, Laurent TC. Concentration and relative molecular mass of hyaluronate in lymph and blood. Biochem J 1986; 236:521–525.
210. Csoka AB, Frost GI, Stern R. The six hyaluronidase-like genes in the human and mouse genomes. Matrix Biol 2001; 20:499–508.
211. Roden L, Campbell P, Fraser JR, Laurent TC, Pertoft H, Thompson JN. Enzymic pathways of hyaluronan catabolism. Ciba Found Symp 1989; 143:60–76.
212. Banerji S, Ni J, Wang S. LYVE-1 a new homologue of the CD44 glycoprotein is a lymph specific receptor for hyaluronan. J Cell Biol 1999; 144:789–801.
213. Carreira CM, Nasser SM, di Tomaso E. LYVE-1 is not restricted to the lymph vessels; expression for normal liver blood sinusoids and down-regulation in human liver cancer and cirrhosis. Cancer Res 2001; 61:8079–8084.
214. Zhou B, Oka JA, Singh A, Weigel PH. Purification and subunit characterization of the rat liver endocytic hyaluronan receptor. J Biol Chem 1999; 274: 33831–33834.
215. Zhou B, Weigel JA, Fauss L, Weigel PH. Identification of the hyaluronan receptor for endocytosis (HARE). J Biol Chem 2000; 275:37733–37741.
216. Weigel JA, Raymond RC, Weigel PH. The hyaluronan receptor for endocytosis (HARE) is not CD44 or CD54 (ICAM-1). Biochem Biophys Res Commun 2002; 294:918–922.
217. Zhou B, Weigel JA, Saxena A, Weigel PH. Molecular cloning and functional expression of the rat 175-kDa hyaluronan receptor for endocytosis. Mol Biol Cell 2002; 13:2853–2868.
218. Wang WC, Zinn K, Bjorkman PJ. Expression and structural studies of fasciclin I, an insect cell adhesion molecule. J Biol Chem 1993; 268:1448–1455.
219. Yannariello-Brown J, Frost SJ, Weigel PH. Identification of the Ca(2 +)-independent endocytic hyaluronan receptor in rat liver sinusoidal endothelial cells using a photoaffinity cross-linking reagent. J Biol Chem 1992; 267: 20451–20456.

220. Yannariello-Brown J, Zhou B, Ritchie D, Oka JA, Weigel PH. A novel ligand blot assay detects different hyaluronan-binding proteins in rat liver hepatocytes and sinusoidal endothelial cells. Biochem Biophys Res Commun 1996; 218:314–319.
221. Weigel JA, Raymond RC, McGary C, Singh A, Weigel PH. A blocking antibody to the hyaluronan receptor for endocytosis (HARE) inhibits hyaluronan clearance by perfused liver. J Biol Chem 2003; 278:9808–9812.
222. Yannariello-Brown J, McGary CT, Weigel PH. The endocytic hyaluronan receptor in rat liver sinusoidal endothelial cells is Ca(+2)-independent and distinct from a Ca(+2)-dependent hyaluronan binding activity. J Cell Biochem 1992; 48:73–80.
223. Smedsrod B, Malmgren M, Ericsson J, Laurent TC. Morphological studies on endocytosis of chondroitin sulphate proteoglycan by rat liver endothelial cells. Cell Tissue Res 1988; 253:39–45.
224. McGary CT, Raja RH, Weigel PH. Endocytosis of hyaluronic acid by rat liver endothelial cells. Evidence for receptor recycling. Biochem J 1989; 257:875–884.
225. Higgins MK, McMahon HT. Snap-shots of clathrin-mediated endocytosis. Trends Biochem Sci 2002; 27:257–263.
226. Fraser JR, Pertoft H, Alston-Smith J, Laurent TC. Uptake of hyaluronan in hepatic endothelial cells is not directly affected by endotoxin and associated cytokines. Exp Cell Res 1991; 197:8–11.

Chemistry and Biology of Hyaluronan
H.G. Garg and C.A. Hales (editors)

Chapter 8

Structural and Functional Diversity of Hyaluronan-Binding Proteins

CHARLES D. BLUNDELL,
NICHOLAS T. SEYFRIED and
ANTHONY J. DAY

MRC Immunochemistry Unit,
Department of Biochemistry, University of Oxford,
South Parks Road, Oxford OX1 3QU, UK

I. Introduction

Hyaluronan (HA) is an unbranched glycosaminoglycan (GAG) comprised entirely of a repeating disaccharide of D-glucuronic acid and *N*-acetyl-D-glucosamine. Unlike other GAGs, it is not sulphated at any position and is not covalently attached to a core protein. Molecules of HA are generally of very high molecular mass, ranging from about 10^5 to 10^7 Da (depending on the tissue), although they can also exist as smaller fragments and oligosaccharides under certain physiological or pathological conditions (1). HA is found in the extra-cellular matrix of all tissues from adult vertebrates (2), with particularly high concentrations being present in skin, synovial fluid and the eye vitreous (where HA was first described (3)).

Many diverse biological functions have been attributed to HA, such as an involvement in the physiological processes of development, ovulation and wound repair (2,4), and in various diseases such as coronary atherosclerosis, inflammatory bowel disease and cancer (5,6). This diversity of roles may seem surprising for such a simple polysaccharide, and whilst free HA forms solutions that provide space-filling, lubricating and filtering functions (e.g., in synovial fluid), the wider functional diversity of HA is in fact likely to be generated by its interactions with specific HA-binding proteins (7). These 'hyaladherins', of which there are increasing numbers, differ in their tissue expression, cellular localisation, affinity and regulation (8). For example, the multi-molecular complexes formed between HA and extracellular matrix hyaladherins act as vital structural components of many tissues (e.g., aggrecan and HAPLN-1 in cartilage). On the other hand, cell-surface HA-binding receptors (e.g., CD44 and LYVE-1) have been shown to be involved in immune cell adhesion and activation (9) and endocytosis (10).

In solution, HA exists as a highly dynamic ensemble of interchanging semi-ordered states (7). Recent structural studies on HA oligosaccharides in complex with HA-digesting enzymes indicate that even short lengths of HA can be stabilised in a variety of conformations on binding to proteins. These

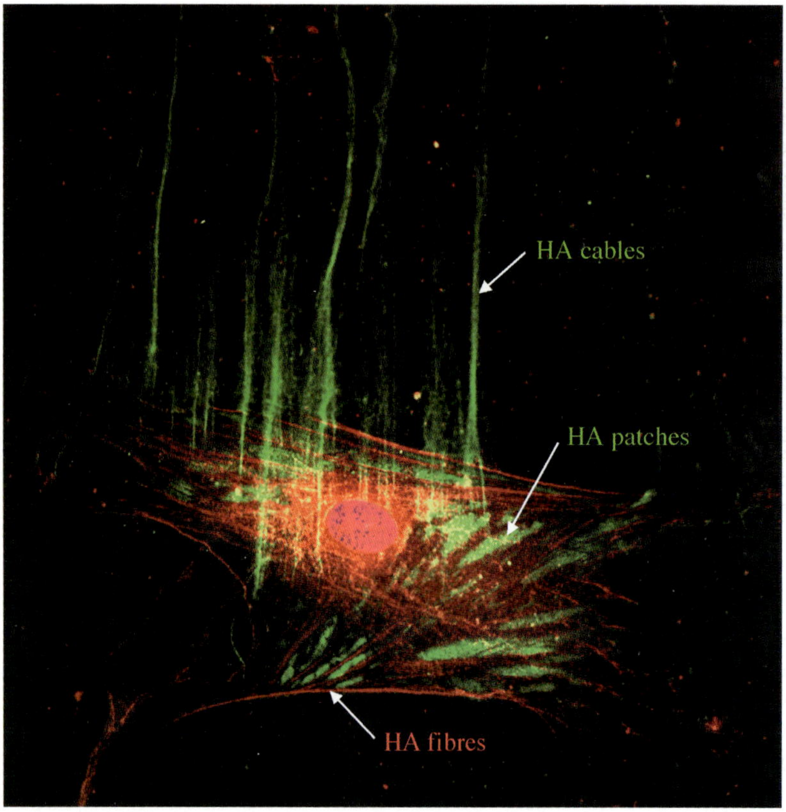

Figure 1 Three different populations of HA one cell! This confocal micrograph shows that a human mucosal smooth muscle cell, after stimulation with a viral mimic, can display at least three different types of HA organisation. HA (green; stained with HABP) is localised in both long cables that project from the cell surface and discrete patches. Polyclonal antisera to the HA-binding protein TSG-6 (red) reveal the presence of surface HA filaments intersecting these patches (de la Motte CA and Day AJ, unpublished).

observations have led to the hypothesis that hyaladherins could act as organisers of HA, capturing and propagating distinct conformers of the polymer, resulting in the formation of differing higher order structures with specific functional activities (7). This is illustrated in Fig. 1 where three distinct populations of HA are visible on a single mucosal smooth muscle cell following stimulation with a viral mimic. These include HA patches on the cell surface, which are probably associated with the HA receptor CD44, and long cable-like structures that protrude from the cell that contain the hyaladherins versican and inter-α-inhibitor (11,12). On the cell surface there are also rod-like HA fibres that are immunoreactive with antibodies against TSG-6 (de la Motte CA and Day AJ, unpublished). Interestingly, mononuclear leukocytes adhere to the HA cables (via CD44 on their surface) but do not associate with the HA patches or fibres, indicating that these structurally different populations of HA also have distinct functional activities.

It seems likely that the structural organisation of a particular HA–protein complex will depend upon the precise molecular details of the interaction between HA and the hyaladherin(s) involved. For this reason, characterisation of the molecular basis of HA recognition by proteins is essential if we are to understand the biology of this important and intriguing polysaccharide. In this regard, more than half of the currently known hyaladherins bind to HA via a protein domain of about 100 amino acids, termed a Link module. This chapter will review recent advances in our knowledge of HA–protein interactions mediated by members of the Link module superfamily.

II. The Link Module Superfamily

The human Link module superfamily currently has 14 members and the majority of these contain a variety of other domain types (Fig. 2). Previously we have suggested (8) that this superfamily can be divided into three subgroups according to the known or expected size of their HA-binding domains (HABDs) (see Fig. 3). Type A domains are comprised of a single, independently folding Link module, as found in TSG-6 (13,14). The proteins KIA0527, Stabilin-1 and Stabilin-2, which all contain a single Link module (Fig. 2), could also belong to this subgroup. Type B domains, however, require N- and C-terminal extensions flanking the Link module for correct folding and functional activity, and these are typified by the ~150 amino acid HABD of CD44 (15); the HA receptor found on lymph vessel endothelium (i.e., LYVE-1) may also have a HABD of this type. The third class of HABD (Type C) is comprised of a contiguous pair of Link modules, which in some cases may also require an N-terminal immunoglobulin module for fold stabilisation. Type C domains are found in the G1-domains of the chondroitin sulphate proteoglycans (CSPGs) aggrecan, versican, neurocan and brevican, and in the HAPLN proteins.

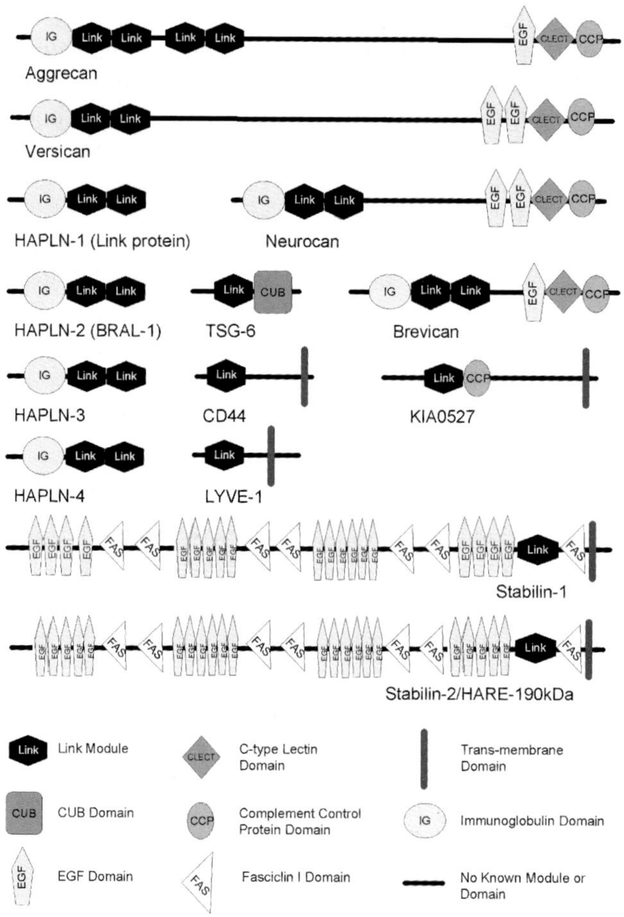

Figure 2 The mosaic nature of the Link module superfamily. The symbols depicting the various modules and the domain organisations are derived from the SMART database (http://smart.embl-heidelberg.de/).

A. HAPLN Proteins and CSPGs

The Link module was first identified in cartilage link protein isolated from rat chondrosarcoma (16). Recently a link protein gene family consisting of four members has been identified (17) and denoted 'HAPLN' (HA and Proteoglycan LiNk protein family). In this new nomenclature, cartilage link protein corresponds to HAPLN-1, and the gene product of *BRAL1* (Brain Link Protein 1 (18)) is renamed HAPLN-2. Interestingly, in the vertebrate genome the four HAPLN genes were all found to be located immediately adjacent to one of the four CSPG genes, i.e., forming HAPLN-1/versican, HAPLN-2/brevican, HAPLN-3/aggrecan and HPLN-4/neurocan gene pairs (17). While brain-specific HAPLN-2 and HAPLN-4 were co-expressed with the two brain-specific CSPGs

Type	Protein(s)	Binding Domain	Structure
A	TSG-6 (KIA0527) (Stabilin 1/2)	[Link] ~100aa	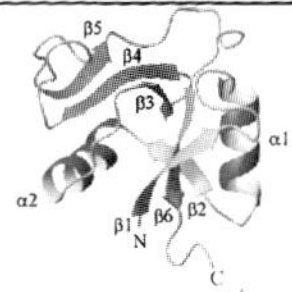
B	CD44 (LYVE-1)	[N Link C] ~150aa	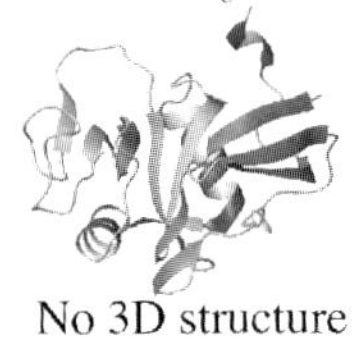
C	HAPLN 1-4 Aggrecan Versican Neurocan Brevican	[Link Link] ~200aa	No 3D structure

Figure 3 The Link module superfamily can be divided into three subgroups on the basis of domain size. The tertiary structures for Type A and B HA-binding domains have been determined from TSG-6 and CD44, respectively (see text). To date, there are no structural data for Type C domains.

(brevican and neurocan), the expression profiles for HAPLN-1 and 3 did not match their corresponding gene partner. For example, HAPLN-1 is mainly restricted to cartilage, where it associates cooperatively with the CSPG aggrecan forming stable complexes with HA (19), while versican is more widely expressed.

The four encoded HAPLN proteins share 45–52% overall amino acid identity, where the highest degree of similarity is in the Link module domains (17). At present the specificity of HAPLN–CSPG interactions is not known. It would seem likely that both HAPLN-2 and 4 form 'link protein'-stabilised complexes with neurocan and/or brevican, given their common tissue expression. However, HAPLN-2 has been reported to co-localise with the V2 splice variant of versican in the brain (20), revealing the possibility that a particular HAPLN protein may associate with more than one CSPG. Clearly, much further work is needed to clarify this question.

The interaction between HAPLN-1 and aggrecan is believed to be mediated via the association of their N-terminal immunoglobulin domains (21–23). However, a recent paper (23) reports that versican also interacts with HAPLN-1, but in this case via its Link module domains. This might indicate that there is considerable complexity in the details of link protein/CSPG interactions leading to aggregates with HA of diverse quaternary structure.

All four HAPLN proteins are predicted to interact with HA based on analyses of their Link module sequences, and these have many features in common with cartilage link protein and the well-characterised hyaladherin TSG-6 (14,17,24). In aggrecan and HAPLN-1, both Link modules in the Type C domains are known to be involved in the interaction with HA (22,25). However, to date there are no structural data for any of these proteins, so that the exact way in which the two Link modules dock together to form a Type C HA-interaction domain is not clear.

B. CD44

The major HA receptor, CD44, contains a Type B HA-binding domain (8). As can be seen from Fig. 3, this consists of a single Link module with N- and C-terminal extensions. Unlike the Type A HABD of TSG-6 (i.e., a single Link module), which folds independently (13,26), these flanking sequences are essential for the structural integrity of the Type B domain (27). This observation has recently been accounted for by the determination of solution and X-ray structures for the HABD from human CD44 (15), which reveals that the extensions come together in space to form an additional 'lobe', comprised of β-strands that extend the Link module β-sheet structure. Residues within both the Link module (28) and the extensions (29) have been implicated in HA binding by site-directed mutagenesis. It is well established that the ability of CD44 to interact with HA is regulated by a variety of factors, including receptor clustering and changes in glycosylation of the extracellular domain (see Ref. 8). The determination of the 3D structure for the CD44 HABD has provided considerable insight into the molecular basis of how glycosylation might directly modulate HA binding and CD44 self-association (15).

C. Stabilin-1 and Stabilin-2: Splitting HAREs!

Stabilin-1 and Stabilin-2 are two new members of the Link module superfamily (30–32). Both genes code for homologous transmembrane proteins (~40% amino acid identity) that contain one Link module, seven fasciclin-like adhesion domains and multiple EGF/EGF-like domains (see Fig. 2). Since they are both expressed in the liver and lymph nodes it has been suggested that they may act as receptors for HA clearance. It should be noted, however, that it is not clear as yet whether Stabilin-1 is a functional HA-binding protein (31).

Stabilin-2 is closely related to the 190 kDa subunit of human HARE, the Hyaluronan Receptor for Endocytosis (33), which is also expressed in liver and lymph nodes (34). HARE is involved in clearing circulating HA but, unlike other HA receptors (e.g., CD44, LYVE-1) it is also specific for chondroitin sulphate (35). It seems likely that HARE (190 kDa) is a proteolytic derivative of Stabilin-2, however, this is somewhat controversial; the reader is referred to recent published correspondence for full discussion of this issue (36).

D. TSG-6

The HA-binding properties of TSG-6, which contains a Type A domain (Fig. 3), are probably the best characterised of any hyaladherin; this is covered in detail in Section III). TSG-6, the gene product of tumor necrosis factor (TNF)-stimulated gene-6, is an ~35 kDa secreted protein comprised almost entirely of a Link module and a CUB module (see Refs. 37 and 38). Its amino acid sequence is very highly conserved between species, with the mouse and human proteins being >94% identical.

TSG-6 is not constitutively produced in healthy adult tissues, but its expression is induced in a wide variety of cell types in response to inflammatory mediators and growth factors (38). Not surprisingly, therefore, it has been found to be associated with inflammatory conditions such as arthritis and bacterial sepsis. However, it is also produced during certain normal physiological processes that can be defined as 'inflammation-like' such as ovulation (39,40) and cervical ripening (41). Recent studies have revealed that TSG-6 protects against cartilage matrix destruction (42,43) and has potent anti-inflammatory effects (44,45) in mouse models of arthritis. These studies suggest that TSG-6 is part of a negative feedback loop capable of down-regulating the inflammatory response and initiating tissue repair. Furthermore, extracellular matrix remodelling appears to be a key feature of most, if not all, sites of TSG-6 expression, indicating that it is likely to participate in this process. One example of this is the crucial role of TSG-6 in stabilising the nascent HA-rich matrix formed during cumulus–oocyte complex (COC) expansion, which is a prerequisite for successful ovulation and fertilisation (46,47).

Female TSG-6 null mice are infertile due to an inability to form and expand a stable cumulus matrix (46). The reasons for this are somewhat complex, but much recent progress has been made in our understanding of this process. Mukhopadhyay et al. (40) have shown that TSG-6 forms covalent complexes with the heavy chains (HCs) of inter-α-inhibitor (IαI) in the context of the cumulus matrix. IαI, a serum protein that enters the ovarian follicle when the blood follicle barrier breaks down, is also necessary for cumulus expansion (48,49); it is comprised of three polypeptides, two HCs (HC1 and HC2) and bikunin, linked via a chondroitin sulphate moiety (50). The HCs become covalently linked to HA via an ester linkage between a carboxylate group of a C-terminal aspartic acid in each of the HCs and a C6 hydroxyl of *N*-acetylglucosamine residues in the HA polysaccharide (51). These HA-linked HCs are believed to provide structural integrity to the growing matrix by cross-linking individual HA chains (48,52). In the TSG-6 knockout mouse there were no HCs linked to HA, revealing that TSG-6 is an essential cofactor in HC attachment to HA (46). Fulop et al. (46) hypothesised that TSG-6-HC complexes act as intermediates in the transfer of HCs to HA. A recent study has revealed that a monoclonal antibody that blocks both HA binding to TSG-6 and TSG-6-HC complex formation severely inhibited cumulus expansion (47), providing further evidence for this. In this regard, we have shown that TSG-6-HC complexes can be formed *in vitro* by mixing TSG-6 and IαI under appropriate conditions (45,47, 53), and that these complexes do indeed act as intermediates in the transfer reaction (Rugg MS, Fries E and Day AJ, unpublished data). From a structural perspective, it would seem likely that HA binding to the TSG-6 Link module, in the context of the HC-TSG-6 complex, serves to orient HA in the correct position relative to the linked HC and may also be responsible for activating the sugar, thus facilitating the transfer reaction (see Ref. 14). All TSG-6's HA-binding activity is likely to reside within the Link module domain (13,53,54); see later.

In addition to being produced in ovulation, TSG-6-IαI and HC–HA complexes have also been detected in the synovial fluids of arthritis patients and may correlate with disease severity (52,55,56). It should be noted that the composition of this particular TSG-6-IαI complex(es) has not yet been established, but it would seem likely that it will be the same as those produced in ovulation and will be involved in the attachment of HCs to HA. Such complexes may be a general feature of inflammation, forming wherever there is TSG-6 expression, free HA and ingress of IαI from serum. At present the precise role of HC–HA complexes in inflammation is not known, but it would be reasonable to propose that they may provide extracellular matrix stabilisation as they do in ovulation.

III. Insights into the Molecular Basis of HA Binding

A. The Structure of the TSG-6 Link Module

The 3D structure of the Link module from human TSG-6 (often referred to as Link_TSG6) has been determined by NMR spectroscopy in solution, thus defining the consensus fold for the Link module superfamily as a whole (13,14). As illustrated in Fig. 4, the structure consists of two α-helices and two anti-parallel triple-stranded β-sheets (joined by contacts between β3 and β6 to form a continuous six-stranded sheet), which are arranged around a large hydrophobic core. The fold of this domain is very similar to that of the C-type lectin module and these structures are likely to have a common evolutionary origin, which is noteworthy since both are involved in carbohydrate recognition (see Refs. 13,14, 57 and 58). The TSG-6 Link module is most similar in structure to the C-type lectin from eosinophil granule major basic protein that binds to the GAG heparin in a calcium ion independent manner (14). As noted previously, the Link module lacks the long Ca^{2+}-binding loop found in classical C-type lectins (13,58).

On the basis of our initial NMR structure for the Link module, we predicted the position of the HA-binding site (13), and subsequent studies (described later) have revealed that this was surprisingly accurate (14,24,54). Recently, we have determined a new solution structure of free Link_TSG6 (14). A much larger number of structural restraints were used than in our previous study, and this has enabled the determination of a more accurate structure for the free protein. It should be noted that while this structure has a very similar fold to that determined previously, there is considerably improved reliability in the precise definition of secondary structure elements, loop geometries and side-chain orientations.

More importantly, we have also determined the structure of the TSG-6 Link module in its HA-bound conformation (14); this was done using identical NMR methodology to that employed for the free protein apart from the presence of an HA octasaccharide (HA_8), shown previously to be close to the minimum size of oligomer that binds optimally (54). Comparison of the free and HA_8-bound

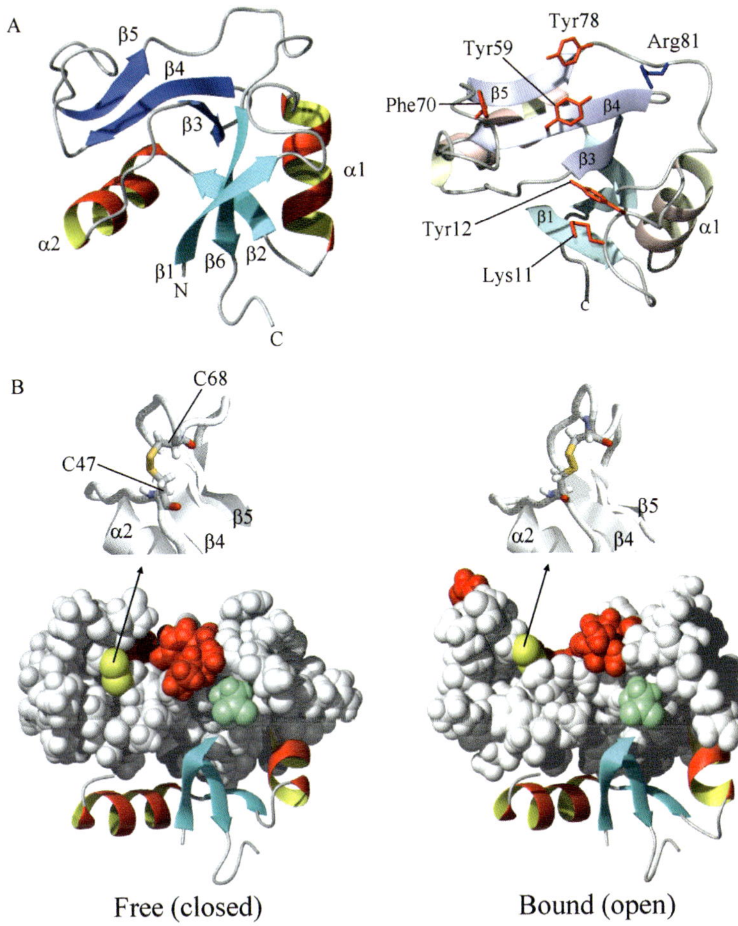

Figure 4 The 3D structure of the Link module from human TSG-6. (A) (left) Solution structure of the Link module in its free state (14) and (right) the bound conformation showing key binding residues determined by site-directed mutagenesis (red) and NMR (blue). (B) Space-filling depiction of the free ('closed') and HA_8-bound ('open') structures, in the same orientation, with the bottom 'half' of each structure shown in a ribbon representation. The conformational change of the β4–β5 loop opens a groove, exposing the key HA-binding residues (red); Glu6 is shown in green (see text). The two states differ principally in the geometry of the disulphide bridge (sulphur atoms in yellow) linking the loop (Cys68) to the rigid connection between α2 and β4 (Cys47), as shown in the insets above the structures.

proteins revealed that there is no gross alteration to the Link module structure on its interaction with HA (14). However, significant structural differences were observed in the region of Link_TSG6 where the five critical HA-binding residues, established previously by site-directed mutagenesis (24,45), are located. These residues (i.e., Lys11, Tyr12, Tyr59, Phe70 and Tyr78; depicted in red in Fig. 4A)

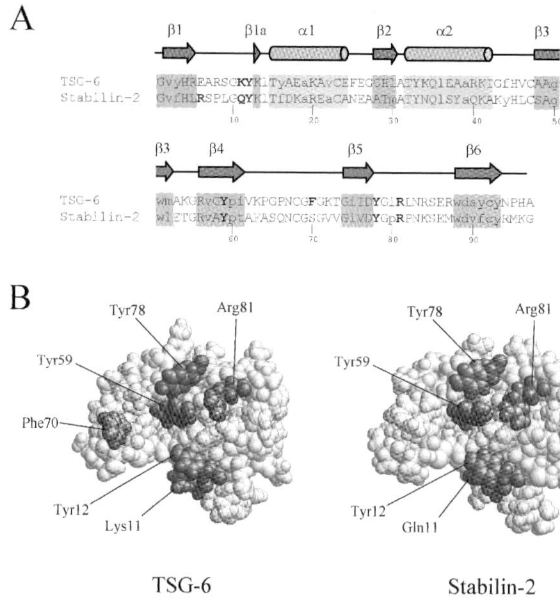

Figure 5 Modelling of the Stabilin-2 Link module. (A) Sequence alignment of the Link modules from TSG-6 and Stabilin-2, numbered according to Link_TSG6. Core residues are shown in lower case and HA-binding residues (known or predicted) are denoted in bold type. Elements of secondary structure are shown above the alignment. (B) Comparison of HA-binding sites in TSG-6 (determined by site-directed mutagenesis and NMR) and the homology model of Stabilin-2. Both modules are shown in the same relative orientation.

along with Arg81 (determined to be involved in binding based on the NMR studies; shown in blue) are brought together from different parts of the primary sequence (Fig. 5) to form a surface patch on one face of the Link module (14). When these amino acids are mapped onto the Link module in its HA-bound conformation they all line a shallow groove on the protein surface. From Fig. 4B it can be seen that the groove is effectively closed in the free protein but opens on interaction with HA; as described later, this open state is generated by a combination of both subtle side-chain rearrangements and a conformational change in the loop between β4 and β5 strands. An HA_8 oligosaccharide can be docked into the binding groove, indicating that it is of a size and shape (~20 Å long, ~10 Å wide and ~10 Å deep) that would allow good intermolecular van der Waals contacts and favourable glycosidic bond angles in a bound HA molecule (14).

B. Getting into the Groove

The side-chains of the key functional residues assume different positions in the free and bound structures (14). For example, Lys11, which is likely to form a salt-bridge to a carboxylic acid moiety in HA (24,45), not only becomes ordered on

HA binding but also changes its orientation. Tyr59, which tends to protrude into solution in the free structure, lies flat against the protein surface in the bound state. The hydroxyl group of this residue almost certainly forms a hydrogen bond with HA, which is also likely to be the case for Tyr12 and Tyr78 (see Ref. 14). Interestingly, the aromatic rings of Tyr59 and Tyr78 appear to have considerably restricted rotation in the HA_8-bound Link module, indicating that they are involved in making stacking interactions with the sugar rings. Such interactions have been observed previously in the crystal structures of hyaluronate lyases in complex with HA oligosaccharides (59,60). Arg81, though not highly resolved in either state, is likely to be involved in forming a salt-bridge with the HA since it experiences significant chemical shift perturbations only towards the end of the side-chain (14). In this regard, isothermal titration calorimetry experiments performed at a range of NaCl concentrations indicate that Link_TSG6 makes between 1 and 2 salt-bridges with HA_8 (61). Consideration of the available mutagenesis data (24) indicates that Lys11 and Arg81 are the best candidates for these interactions, especially given that they lie at either end of the binding groove (Fig. 4A).

The emerging picture, therefore, of the mode of HA binding to Link_TSG6 is one in which interactions with aromatic rings (Tyr12, Tyr59, Phe70, Tyr78) play a major role. Basic residues (Lys11, Arg81) are also crucial, but ionic interactions may only account for about 25% of the binding energy (61). Both these types of interactions are found in other HA–protein complexes; in CD44, for example, basic residues and tyrosines are also crucial for HA binding (28).

C. Turning it on

On HA binding, Tyr59 and Tyr78 become flat against the protein surface and the β4–β5 loop (holding Phe70) withdraws away from them (14); these rearrangements combine to open a previously closed groove on the surface of the protein (Fig. 4B). This loop is effectively hinged at either end by proline and glycine residues, and is opened by a change in the geometry of the disulphide bridge between Cys47 and Cys68 (see Fig. 4). This disulphide bridge is found in all Link modules except KIA0527 (57), and the residues that provide the hinges on which the loop moves (Pro60 and Gly74) are very highly conserved, indicating that the conformational change seen for TSG-6 is likely to occur in most members of the superfamily (14). We have hypothesised, therefore, that these features may comprise a conserved 'switch', providing a mechanism for the allosteric regulation of HA binding. For example, the interaction of Link modules with other ligands could either inhibit or activate HA binding by 'locking shut' or 'forcing open' the groove.

D. Modelling for Beginners

Our recent determination of a high resolution structure for the TSG-6 Link module in its HA-bound form has allowed homology modelling of other Link module containing proteins in their active conformations, thus aiding identification of important functional residues (Almond A, Blundell CD and Day AJ, unpublished data). Previously we have reported how a comparison of mutagenesis data for TSG-6 and CD44 can be used to identify key sequence positions involved in HA binding across the superfamily (24). This analysis can now be revisited in the light of our new models. For example, here we make a prediction for the key HA-binding residues in the Link module of Stabilin-2. A sequence alignment of these two Link modules (Fig. 5A) shows that they are highly homologous (45% identity), and have either identical or conservative replacements for all the critical core residues. It is therefore extremely likely that the structure of the Stabilin-2 Link module is essentially identical to that of TSG-6. In this regard, Fig. 5B shows a homology model based on the coordinates of the HA_8-bound form of Link_TSG6. The three critical tyrosine residues (i.e., Tyr12, Tyr59, Tyr78) involved in HA binding in TSG-6 are found in identical sequence positions in Stabilin-2 (Fig. 5A), as is Arg81 (which has been proposed to be making an important salt-bridge, see earlier); these four residues are therefore probably critical to HA binding in Stabilin-2. Lys11 is replaced by a glutamine, which would be unable to make an ionic interaction. However, this residue could be involved in HA binding by forming hydrogen bonds with HA. Phe70 is replaced by a serine in Stabilin-2 and we, therefore, do not expect it to be involved. In Link_TSG6, mutation of Glu6 (coloured green in Fig. 4B) to lysine causes an ~4-fold increase in the affinity for HA_8 (45), apparently by extending the binding site and forming an additional ionic interaction (14). Interestingly, Stabilin-2 has an arginine residue at this position (Fig. 5A) and thus may utilise such an extended binding surface.

IV. Summary

There has been much recent progress in our knowledge of the molecular basis of HA–protein interactions, in particular with regard to Link module containing proteins. However, to date no high resolution 3D structures for any Link module/HA complexes have been determined. In the next few years, it is hoped that information of this type will become available as this is clearly essential if we are to fully understand the molecular mechanisms underlying the diverse biology of this important GAG, which are mediated in large part by members of the Link module superfamily.

Acknowledgements

We would like to thank Carol de la Motte for kindly providing the image in Fig. 1 and acknowledge the financial support of the Medical Research Council and Arthritis Research Campaign. CDB was the recipient of a Yamanouchi Research Institute Scholarship.

References

1. Tammi MI, Day AJ, Turley EA. Hyaluronan and homeostasis: a balancing act. J Biol Chem 2002; 277:4581–4584.
2. Fraser JRE, Laurent TC, Laurent UBG. Hyaluronan: its nature, distribution, functions and turnover. J Intern Med 1997; 242:27–33.
3. Meyer K, Palmer JW. The polysaccharide of the vitreous humour. J Biol Chem 1934; 107:629–634.
4. Camenisch TD, McDonald JA. Hyaluronan: is bigger better? Am J Respir Cell Mol Biol 2000; 23:431–433.
5. Toole BP, Wight TN, Tammi MI. Hyaluronan–cell interactions in cancer and vascular disease. J Biol Chem 2002; 277:4593–4596.
6. de la Motte CA, Hascall VC, Calabro A, Yen-Lieberman B, Strong SA. Mononuclear leukocytes preferentially bind via CD44 to hyaluronan on human intestinal mucosal smooth muscle cells after virus infection or treatment with poly I:C. J Biol Chem 1999; 274:30747–30755.
7. Day AJ, Sheehan JK. Hyaluronan: polysaccharide chaos to protein organisation. Curr Opin Struct Biol 2001; 11:617–622.
8. Day AJ, Prestwich GD. Hyaluronan-binding proteins: tying up the giant. J Biol Chem 2002; 277:4585–4588.
9. Ponta H, Sherman L, Herrlich PA. CD44: from adhesion molecules to signalling regulators. Nat Rev Mol Cell Biol 2003; 4:33–45.
10. Prevo R, Banerji S, Ferguson DJ, Clasper S, Jackson DG. Mouse LYVE-1 is an endocytic receptor for hyaluronan in lymphatic endothelium. J Biol Chem 2001; 276:19420–19430.
11. de la Motte CA, Hascall VC, Drazba J, Bandyopadhyay SK, Strong SA. Mononuclear leukocytes bind to specific hyaluronan structures on colon mucosal smooth muscle cells treated with polyinosinic acid:polycytidylic acid: inter-alpha-trypsin inhibitor is crucial to structure and function. Am J Pathol 2003; 163:121–133.
12. Majors AK, Austin RC, de la Motte CA, Pyeritz RE, Hascall VC, Kessler SP, Sen G, Strong SA. Endoplasmic reticulum stress induces hyaluronan deposition and leukocyte adhesion. J Biol Chem 2003; 278:47223–47231.
13. Kohda D, Morton CJ, Parkar AA, Hatanaka H, Inagaki FM, Campbell ID, Day AJ. Solution structure of the link module: a hyaluronan-binding domain involved in extracellular matrix stability and cell migration. Cell 1996; 86:767–775.
14. Blundell CD, Mahoney DJ, Almond A, DeAngelis PL, Kahmann JD, Teriete P, Pickford AR, Campbell ID, Day AJ. The Link module from ovulation- and inflammation-associated protein TSG-6 changes conformation on hyaluronan binding. J Biol Chem 2003; 278:49261–49270.
15. Teriete P, Banerji S, Noble M, Blundell CD, Wright AJ, Pickford AR, Lowe E, Mahoney DJ, Tammi MI, Kahmann JD, Campbell ID, Day AJ, Jackson DG. Structure of the regulatory hyaluronan-binding domain in the inflammatory leukocyte homing receptor CD44. Mol Cell 2004; 13:483–496.
16. Neame PJ, Christner JE, Baker JR. The primary structure of link protein from rat chondrosarcoma proteoglycan aggregate. J Biol Chem 1986; 261:3519–3535.
17. Spicer AP, Joo A, Bowling RA Jr. A hyaluronan binding link protein gene family whose members are physically linked adjacent to chondroitin sulfate proteoglycan core protein genes: the missing links. J Biol Chem 2003; 278: 21083–21091.

18. Hirakawa S, Oohashi T, Su WD, Yoshioka H, Murakami T, Arata J, Ninomiya Y. The brain link protein-1 (BRAL1): cDNA cloning, genomic structure, and characterization as a novel link protein expressed in adult brain. Biochem Biophys Res Commun 2000; 276:982–989.
19. Morgelin M, Paulsson M, Heinegard D, Aebi U, Engel J. Evidence of a defined spatial arrangement of hyaluronate in the central filament of cartilage proteoglycan aggregates. Biochem J 1995; 307:595–601.
20. Oohashi T, Hirakawa S, Bekku Y, Rauch U, Zimmermann DR, Su WD, Ohtsuka A, Murakami T, Ninomiya Y. Bral1, a brain-specific link protein, colocalizing with the versican V2 isoform at the nodes of Ranvier in developing and adult mouse central nervous systems. Mol Cell Neurosci 2002; 1:43–57.
21. Perin JP, Bonnet F, Thurieau C, Jolles P. Link protein interactions with hyaluronate and proteoglycans. Characterization of two distinct domains in bovine cartilage link proteins. J Biol Chem 1987; 262:13269–13272.
22. Grover J, Roughley PJ. The expression of functional link protein in a baculovirus system: analysis of mutants lacking the A, B and B′ domains. Biochem J 1994; 300: 317–324.
23. Matsumoto K, Shionyu M, Go M, Shimizu K, Shinomura T, Kimata K, Watanabe H. Distinct interaction of versican/PG-M with hyaluronan and link protein. J Biol Chem 2003; 278:41205–41212.
24. Mahoney DJ, Blundell CD, Day AJ. Mapping the hyaluronan-binding site on the link module from human tumor necrosis factor-stimulated gene-6 by site-directed mutagenesis. J Biol Chem 2001; 276:22764–22771.
25. Watanabe H, Cheung SC, Itano N, Kimata K, Yamada Y. Identification of hyaluronan-binding domains of aggrecan. J Biol Chem 1997; 272: 28057–28065.
26. Day AJ, Aplin RT, Willis AC. Overexpression, purification, and refolding of link module from human TSG-6 in *Escherichia coli*: effect of temperature, media, and mutagenesis on lysine misincorporation at arginine AGA codons. Protein Expr Purif 1996; 8:1–16.
27. Banerji S, Day AJ, Kahmann JD, Jackson DG. Characterization of a functional hyaluronan-binding domain from the human CD44 molecule expressed in *Escherichia coli*. Protein Expr Purif 1998; 14:371–381.
28. Bajorath J, Greenfield B, Munro SB, Day AJ, Aruffo A. Identification of CD44 residues important for hyaluronan binding and delineation of the binding site. J Biol Chem 1998; 273:338–343.
29. Peach RJ, Hollenbaugh D, Stamenkovic I, Aruffo A. Identification of hyaluronic acid binding sites in the extracellular domain of CD44. J Cell Biol 1993; 122: 257–264.
30. Jackson DG, Prevo R, Ni J, Banerji S. Novel endothelial hyaluronan receptors. In: Kennedy JF, Philips GO, Williams PA, Hascall VC, eds. Hyaluronan. Abington, Cambridge, UK: Woodhead Publishing, 2002; vol. 1:355–364.
31. Politz O, Gratchev A, McCourt PA, Schledzewski K, Guillot P, Johansson S, Svineng G, Franke P, Kannicht C, Kzhyshkowska J, Longati P, Velten FW, Johansson S, Goerdt S. Stabilin-1 and -2 constitute a novel family of fasciclin-like hyaluronan receptor homologues. Biochem J 2002; 362:155–164.
32. Zhou B, Weigel JA, Saxena A, Weigel PH. Molecular cloning and functional expression of the rat 175-kDa hyaluronan receptor for endocytosis. Mol Biol Cell 2002; 13:2853–2868.

33. Zhou B, McGary CT, Weigel JA, Saxena A, Weigel PH. Purification and molecular identification of the human hyaluronan receptor for endocytosis. Glycobiology 2003; 13:339–349.
34. Zhou B, Weigel JA, Fauss L, Weigel PH. Identification of the hyaluronan receptor for endocytosis (HARE). J Biol Chem 2000; 275:37733–37741.
35. McGary CT, Raja RH, Weigel PH. Endocytosis of hyaluronic acid by rat liver endothelial cells. Evidence for receptor recycling. Biochem J 1989; 257:875–884.
36. Smedsrod B, Johansson S, Goerdt S. Shooting HARE. Glycobiology 2003; 13:11G–12G (Author reply: Weigel PH. HARE raising. Glycobiology 2003; 13:12G–13G).
37. Wisniewski HG, Vilcek J. TSG-6: an IL-1/TNF-inducible protein with anti-inflammatory activity. Cytokine Growth Factor Rev 1997; 8:143–156.
38. Milner CM, Day AJ. TSG-6: a multifunctional protein associated with inflammation. J Cell Sci 2003; 116:1863–1873.
39. Carrette O, Nemade RV, Day AJ, Brickner A, Larsen WJ. TSG-6 is concentrated in the extracellular matrix of mouse cumulus oocyte complexes through hyaluronan and inter-alpha-inhibitor binding. Biol Reprod 2001; 65:301–308.
40. Mukhopadhyay D, Hascall VC, Day AJ, Salustri A, Fulop C. Two distinct populations of tumor necrosis factor-stimulated gene-6 protein in the extracellular matrix of expanded mouse cumulus cell–oocyte complexes. Arch Biochem Biophys 2001; 394:173–181.
41. Fujimoto T, Savani RC, Watari M, Day AJ, Strauss JF. Induction of the hyaluronic acid-binding protein, tumor necrosis factor-stimulated gene-6, in cervical smooth muscle cells by tumor necrosis factor-alpha and prostaglandin E2. Am J Pathol 2002; 160:1495–1502.
42. Glant TT, Kamath RV, Bardos T, Gal I, Szanto S, Murad YM, Sandy JD, Mort JS, Roughley PJ, Mikecz K. Cartilage-specific constitutive expression of TSG-6 protein (product of tumor necrosis factor alpha-stimulated gene 6) provides a chondroprotective, but not antiinflammatory, effect in antigen-induced arthritis. Arthritis Rheum 2002; 46:2207–2218.
43. Mindrescu C, Dias AA, Olszewski RJ, Klein MJ, Reis LF, Wisniewski HG. Reduced susceptibility to collagen-induced arthritis in DBA/1J mice expressing the TSG-6 transgene. Arthritis Rheum 2002; 46:2453–2464.
44. Wisniewski HG, Hua JC, Poppers DM, Naime D, Vilcek J, Cronstein BN. TNF/IL-1-inducible protein TSG-6 potentiates plasmin inhibition by inter-alpha-inhibitor and exerts a strong anti-inflammatory effect *in vivo*. J Immunol 1996; 156: 1609–1615.
45. Getting SJ, Mahoney DJ, Cao T, Rugg MS, Fries E, Milner CM, Perretti M, Day AJ. The link module from human TSG-6 inhibits neutrophil migration in a hyaluronan- and inter-alpha-inhibitor-independent manner. J Biol Chem 2002; 277: 51068–51076.
46. Fulop C, Szanto S, Mukhopadhyay D, Bardos T, Kamath RV, Rugg MS, Day AJ, Salustri A, Hascall VC, Glant TT, Mikecz K. Impaired cumulus mucification and female sterility in tumor necrosis factor-induced protein-6 deficient mice. Development 2003; 130:2253–2261.
47. Ochsner SA, Day AJ, Rugg MS, Breyer RM, Gomer RH, Richards JS. Disrupted function of tumor necrosis factor-alpha-stimulated gene 6 blocks cumulus cell–oocyte complex expansion. Endocrinology 2003; 144:4376–4384.
48. Zhuo L, Yoneda M, Zhao M, Yingsung W, Yoshida N, Kitagawa Y, Kawamura K, Suzuki T, Kimata K. Defect in SHAP–hyaluronan complex causes severe female

infertility. A study by inactivation of the bikunin gene in mice. J Biol Chem 2001; 276:7693–7696.

49. Sato H, Kajikawa S, Kuroda S, Horisawa Y, Nakamura N, Kaga N, Kakinuma C, Kato K, Morishita H, Niwa H, Miyazaki J. Impaired fertility in female mice lacking urinary trypsin inhibitor. Biochem Biophys Res Commun 2001; 281:1154–1160.
50. Enghild JJ, Thogersen IB, Cheng F, Fransson LA, Roepstorff P, Rahbek-Nielsen H. Organization of the inter-alpha-inhibitor heavy chains on the chondroitin sulfate originating from Ser(10) of bikunin: posttranslational modification of IαI-derived bikunin. Biochemistry 1999; 38:11804–11813.
51. Zhao M, Yoneda M, Ohashi Y, Kurono S, Iwata H, Ohnuki Y, Kimata K. Evidence for the covalent binding of SHAP, heavy chains of inter-alpha-trypsin inhibitor, to hyaluronan. J Biol Chem 1995; 270:26657–26663.
52. Yingsung W, Zhuo L, Morgelin M, Yoneda M, Kida D, Watanabe H, Ishiguro N, Iwata H, Kimata K. Molecular heterogeneity of the SHAP–hyaluronan complex. Isolation and characterization of the complex in synovial fluid from patients with rheumatoid arthritis. J Biol Chem 2003; 278:32710–32718.
53. Nentwich HA, Mustafa Z, Rugg MS, Marsden BD, Cordell MR, Mahoney DJ, Jenkins SC, Dowling B, Fries E, Milner CM, Loughlin J, Day AJ. A novel allelic variant of the human TSG-6 gene encoding an amino acid difference in the CUB module. Chromosomal localization, frequency analysis, modeling, and expression. J Biol Chem 2002; 277:15354–15362.
54. Kahmann JD, O'Brien R, Werner JM, Heinegard D, Ladbury JE, Campbell ID, Day AJ. Localization and characterization of the hyaluronan-binding site on the link module from human TSG-6. Structure 2000; 8:763–774.
55. Wisniewski HG, Maier R, Lotz M, Lee S, Klampfer L, Lee TH, Vilcek J. TSG-6: a TNF-, IL-1-, and LPS-inducible secreted glycoprotein associated with arthritis. J Immunol 1993; 151:6593–6601.
56. Kida D, Yoneda M, Miyaura S, Ishimaru T, Yoshida Y, Ito T, Ishiguro N, Iwata H, Kimata K. The SHAP–HA complex in sera from patients with rheumatoid arthritis and osteoarthritis. J Rheumatol 1999; 26:1230–1238.
57. Day AJ. Understanding hyaluronan–protein interactions. 2001; http://www.glycoforum.gr.jp/science/hyaluronan/HA16/HA16E.html.
58. Kogelberg H, Feizi T. New structural insights into lectin-type proteins of the immune system. Curr Opin Struct Biol 2001; 11:635–643.
59. Jedrzejas MJ, Mello LV, de Groot BL, Li S. Mechanism of hyaluronan degradation by *Streptococcus pneumoniae* hyaluronate lyase. Structures of complexes with the substrate. J Biol Chem 2002; 277:28287–28297.
60. Nukui M, Taylor KB, McPherson DT, Shigenaga MK, Jedrzejas MJ. The function of hydrophobic residues in the catalytic cleft of *Streptococcus pneumoniae* hyaluronate lyase; kinetic characterization of mutant enzyme forms. J Biol Chem 2003; 278:3079–3088.
61. Blundell CD, Kahmann JD, Perczel A, Mahoney DJ, Cordell MR, Teriete P, Campbell ID, Day AJ. Getting to grips with HA–protein interactions. In: Kennedy JF, Philips GO, Williams PA, Hascall VC, eds. Hyaluronan. Abington, Cambridge, UK: Woodhead Publishing, 2002; vol. 1:161–172.

Chemistry and Biology of Hyaluronan
H.G. Garg and C.A. Hales (editors)

Chapter 9

Biological Function of SHAP–Hyaluronan Covalent Complex

LISHENG ZHUO and NAOKI ITANO

Institute for Molecular Science of Medicine, Aichi Medical University, Nagakute, Aichi 480-1195, Japan
CREST, Science and Technology Agency, Kawaguchi, Saitama 332-0012, Japan

TSUNEMASA NONOGAKI

Tokai College of Medical Engineering Nishi-Kamo, Aichi 470-0203, Japan

LI SHEN, JIWEN WU, HIDETO WATANABE and KOJI KIMATA

Institute for Molecular Science of Medicine, Aichi Medical University, Nagakute, Aichi 480-1195, Japan

I. Introduction

Hyaluronan is structurally the simplest glycosaminoglycan consisting of *N*-acetylglucosamine–glucuronic acid disaccharide units repeating in a linear fashion. Hyaluronan is distributed ubiquitously with high concentrations in most soft connective tissues (1), and is involved in a broad spectrum of physiological and pathological events, such as embryonic development, angiogenesis, inflammation, tissue remodeling, and tumor malignancy (2–4). Factors that bridge the great gap between the structural simplicity and functional diversity of hyaluronan include the concentration, chain length, and hyaluronan-binding proteins (HABPs) (5). A large body of HABPs has been identified, and in fact the number is still growing (6–8). A variety of the HABPs identified to date are bound to hyaluronan via non-covalent interactions, which may either be very strong and stable, for example, the aggrecan–hyaluronan interaction strengthened by the link protein in cartilage matrix, or be weak and subtle, for example, the hyaluronan–CD44 interaction that has a complicated regulation via mRNA splicing, sulfation, phosphorylation, glycosylation, and dimerization of CD44 (9,10). Some conserved motifs, for example, the link module and BX7B motif, have been identified in the hyaluronan-binding domains of a number of HABPs, the former having been well characterized. However, for many HABPs, the interaction with hyaluronan still remains unclear. On the other hand, the presence of HABP bound covalently to hyaluronan has also been suggested: the hyaluronateprotein (11), IαI (12,13) and IgG (14,15) in synovial fluid, cell-membrane protein (16,17), and others (18). Such proteins represent an attractive category of HABPs; however, unfortunately, in most cases there is no direct evidence for a covalent linkage, or the protein was not identified. The only exception is serum-derived hyaluronan-associated protein (SHAP) that was first described by us in 1990 (19) (Fig. 1). In the ensuing 10 years, several important breakthroughs have been made in understanding the structure and function of the SHAP–hyaluronan complex. In this chapter, an introduction to the formation of the SHAP–hyaluronan complex and the related physiological significance will be given, followed by a discussion, based on the latest results, of the possible roles in pathological processes. The main point is that the association of SHAP changes the properties of hyaluronan.

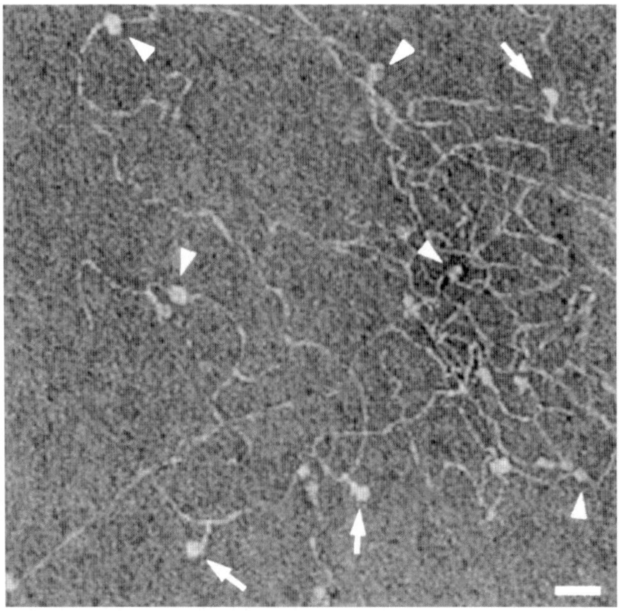

Figure 1 Negative staining electron micrographic images of the SHAP–hyaluronan complex. The SHAP–hyaluronan complex is purified from rheumatoid arthritic synovial fluid. Hyaluronan is visualized as a fibrous network and SHAP as globular structure (diameter at 11 nm in average) (arrows and arrowheads) with a thin tail that attaches to hyaluronan fiber (arrows). The morphology matches that of IαI heavy chains reported previously (20,21).

II. Formation of the SHAP–Hyaluronan Complex

A. Isolation and Identification of SHAP

When cultured in dishes, fibroblasts form a hyaluronan-rich extracellular matrix in the surrounding pericellular space. From the matrix, hyaluronan was isolated together with a protein of ~85 kDa even in the presence of 6 M guanidine hydrochloride and detergents, a condition known to disrupt most non-covalent interactions (19). The protein was derived from the serum added to the culture medium because it was not metabolically labeled, and thus was designated SHAP at that time. Then the partial amino acid sequencing of SHAP revealed that it was identical to the heavy chains of the serum protein inter-α (trypsin) inhibitor (IαI), which was first isolated about 30 years ago (22). Although the term IαI heavy chain may be more familiar to the research community, we suggest using the term SHAP specifically for heavy chains that have formed a complex with hyaluronan so as to clearly discriminate them from those of IαI, which, as mentioned below, is necessary when discussing the function of the proteins.

IαI and related proteins occurring mainly in blood and urine are now collectively known as the IαI family (23,24). The family molecules share a common light chain, bikunin, which is a classic proteoglycan possessing a single low-sulfated (~30%) chondroitin-4-sulfate chain (which is relatively short, consisting of 15 disaccharide units on average) and having a molecular mass of ~40 kDa. The bikunin portion alone occurs mainly in urine, where it is also called urinary trypsin inhibitor. The coupling of one or two heavy chains with bikunin gives other members of the family, such as pre-α inhibitor (PαI, single heavy chain) and IαI (two heavy chains) itself, which occur mainly in blood. Three highly related heavy chains (Mr from 65 to 90 kDa) have been found to assemble with bikunin, the genes of which are obviously derived from a common ancestor (25). The combination of heavy chains exhibits a species-specific pattern; that is, human IαI predominantly consists of heavy chains 1 and 2, bovine IαI of heavy chains 2 and 3 and rodent IαI of all three heavy chains (26–28). The heavy chain of PαI is usually heavy chain 3 in the species examined to date except the bovine PαI, which is associated with heavy chain 2 (26). The reason for such a preference is not clear, but all the three heavy chains are found in a complex with hyaluronan. Although there may be a slight difference in the reaction rate, the IαI family members can be considered as similar donors of SHAP. It is noteworthy that, in addition to the three bikunin-associating heavy chains, there is also a fourth heavy chain (29). Heavy chain 4 shows high sequence similarity with the others in the N-terminal part, but is completely different in the C-terminal part, and is likely to be a consequence of genetic mutation. Heavy chain 4 does not couple to bikunin due to a lack of the important DPHFII motif that is conserved in the three heavy chains in all species examined, and occurs as a free protein in the circulation. It does not appear to form a covalent complex with hyaluronan.

B. The Protein–Glycosaminoglycan Ester Linkage

Mass spectrometry revealed a unique protein–glycosaminoglycan linkage in both IαI family molecules (30–32) and the SHAP–hyaluronan complex (33), where an ester bond is formed between the carboxyl group of the asparate residue at the C-terminus of the heavy chains/SHAPs and the C-6 hydroxyl group of an internal *N*-acetylgalactosamine residue of the bikunin chondroitin sulfate or an *N*-acetylglucosamine residue of hyaluronan (for the SHAP–hyaluronan linkage, see Fig. 2). Based on the similarity of the protein–glycosaminoglycan linkage and the homology of both the N- and C-terminus between the heavy chain and corresponding SHAP, it is concluded that the SHAP–hyaluronan complex is formed via a transesterification reaction in which hyaluronan substitutes for bikunin chondroitin sulfate to form the same type of ester bond with the heavy chains, accompanied by the release of the bikunin proteoglycan (33).

C. Enzyme Catalyzing the Formation of SHAP–Hyaluronan Complex

Incubation of hyaluronan with IαI in a test tube resulted in no complex until a third factor, for example, serum, was added (19,33). The formation of the complex is sensitive to the incubation temperature, and required the presence of divalent cations (19,22). These observations indicate the presence of an enzymatic factor. The factor is as yet unidentified although in addition to serum, the enzymatic activity has been detected in follicle fluid and the media conditioned by follicular granulosa cell, hepatoma, and glioma cells ((34,35), and our unpublished observations). Recently, the product of tumor necrosis factor stimulated gene 6 (TSG6) was suggested to be a candidate because TSG6 protein interacts with both hyaluronan and IαI (36), and, more importantly, TSG6

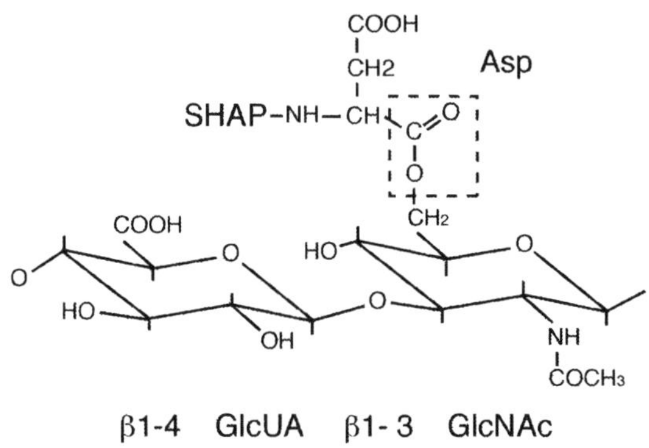

Figure 2 Schematic representation of the structure of the protein–glycosaminoglycan linkage region in the SHAP–hyaluronan complex. A SHAP is linked to hyaluronan via an ester bond (dash frame) between the α-carboxyl group of the C-terminal asparate residue of SHAP and the C-6 hydroxyl group of an internal GlcNAc of hyaluronan (33). The same linkage is also present between the heavy chains and chondroitin sulfate in the IαI family molecules, where a GalNAc replaces the GlcNAc (30–32).

knockout mice exhibited exactly the same phenotype as mice deficient in the formation of the SHAP–hyaluronan complex (37,38). TSG6 protein is a ~35 kDa glycoprotein with a link module in the N-terminal part and a CUB domain in the C-terminal part (39). Recombinant TSG6 protein forms a ~120 kDa complex with IαI, in which TSG6 seems to link to the chondroitin sulfate chain by replacing a heavy chain (36). The TSG6–IαI complex was reported to be present in arthritic synovial fluid and air pouch exudates (40). The pre-ovulatory upregulation of TSG6 in expanding cumulus cells has also been reported (41). In contrast, the cumulus TSG6 seems to link directly to the heavy chain (42). The interaction of TSG6 with IαI remains to be examined. At present, whether TSG6 is the enzyme is still in dispute. It has also been suggested that TSG6 protein itself is not the enzyme, but an enhancing cofactor (43). Consistent with this notion, we have reported that PG-M/versican, a hyaluronan-binding proteoglycan having two tandem link modules, exerts an enhancing effect on the formation of the SHAP–hyaluronan complex in a test tube (44).

III. Physiological Significance of the SHAP–Hyaluronan Complex

A. Protease Inhibition by IαI Family Molecules

IαI family molecules circulate at high concentrations (0.15–0.5 mg/mL). As the name suggests, they were originally isolated and accordingly studied as protease inhibitors (45). The protease inhibitory activity of the family molecules is attributed to the two Kunitz-type domains in bikunin, after which the name bi-kun-in was cast (46). Bikunin shows inhibitory activity against a broad spectrum of proteases. However, the activity is in general so weak that the family of molecules collectively accounts for only 5% of the total protease inhibitory activity in the serum (45). It is still unclear whether the inhibitors target a specific protease, although granulocyte elastase and others are often referred to when the mechanism for the pharmaceutical effects of bikunin in shock and acute pancreatitis is discussed (47). In some environments, the strong inhibition of proteases has been observed: for example, the antiplasmin activity was greatly enhanced by binding with TSG6 during inflammatory responses (40). It has been hypothesized that the inhibitors function as a shuttle that traps proteases and then transfers them to stronger inhibitors like α2-macroglobulin and the α1-inhibitor (48). Recently, it was also reported that bikunin efficiently inhibited granzyme K, the fifth of the lymphocyte granule-stored serine proteases which are implicated in T- and natural killer cell-mediated cytotoxic defense reactions after the recognition of target cells (49).

It is worth noting that the non-inhibitory heavy chain moieties were almost neglected in the above studies. However, the majority (>90%) of circulating bikunin carries the heavy chains. Free bikunin is rapidly excreted in urine with a half-life of about 4 min in mice (50) and 33 min in humans (51). In contrast, the half-life of heavy chain-carrying bikunin is at least several hours (37). Furthermore, urinary bikunin is largely in the heavy chain-free form. These findings imply that the IαI may exert certain functions by involving the heavy chain moieties in extravascular spaces.

B. Stabilization of Hyaluronan-Rich Extracellular Matrix by SHAP

In higher mammalians, upon receiving an ovulatory stimulus, the cumulus–oocyte complex (COC) destined to ovulate undergoes a dramatic morphological change known as cumulus expansion that is characterized by the extensive synthesis and deposition of a hyaluronan-rich extracellular matrix between the cumulus cells. The process can be reproduced *in vitro* by stimulating the COC with follicle-stimulating hormone. It has been noted for some time that fetal bovine serum is required for a successful cumulus expansion. In the 1990s, a serum factor with cumulus matrix-stabilizing activity was isolated and identified as IαI family molecules (52). Direct interaction between IαI and hyaluronan was first suggested as the mechanism underlying the stabilizing effect (53). However, this seems unlikely since the interaction, if it exists, would not be strong enough. An examination of hyaluronan isolated from ovarian follicle fluid indicated that it associated with the heavy chains, whereas bikunin did not, suggesting the formation of a SHAP–hyaluronan complex (54). In 2001, the bikunin gene-knockout mouse was generated, in which the heavy chains were present in the circulation in a bikunin-free form, and were unable to form a complex with hyaluronan (37). The homozygous female mice exhibited a significantly decreased rate of ovulation and fertilization due to the impaired matrix deposition during cumulus expansion. As a consequence, they were infertile. The presence of SHAP, but not bikunin, in the extracellular matrix of the expanded cumulus oophorus was confirmed using western blotting and immunohistochemical staining techniques. The impaired fertilization is cured by administration of purified IαI, but not bikunin (37). Incubation of bikunin with the serum of homozygous mice failed to couple it with the free heavy chains, indicating that, although the blood contains the activity to transfer the coupled heavy chains to hyaluronan, there was no coupling activity (our unpublished observation). This is consistent with the fact that the assembly of IαI family molecules is completed in the Golgi apparatus of hepatocytes (55,56), and explains the failure of bikunin to rescue the impaired fertilization. These results confirm previous *in vitro* observations, and provide conclusive evidence for the physiological significance of the SHAP–hyaluronan complex. It also provides a novel concept as to the function of bikunin, i.e., to provide chondroitin sulfate chains for ester bond formation and present the esterified heavy chains to hyaluronan.

In addition, IαI family molecules were also found to exert a matrix-stabilizing effect on cultured fibroblasts, mesothelial cells (57) and mouse mammary carcinoma cells (58). It is plausible to assume that the formation of the SHAP–hyaluronan complex underlies such an effect because the extracellular matrix of fibroblasts is where the complex was first isolated.

How the SHAPs stabilize the cumulus matrix is now under investigation. Most likely, they, together with other matrix components, contribute to the intercross of hyaluronan chains to form a meshwork structure. The presence of TSG6, link protein, hyaluronan-binding proteoglycans, tenascin-C, laminin, collagen IV, fibronectin, and CD44 in the meshwork has been reported (59). IαI–TSG6 interaction has been well documented (36,42). Indeed, the phenotype of the TSG6 knockout mouse resembles very much that of the bikunin knockout mouse (37,38). The defect is not only due to the absence of TSG6 itself in the cumulus matrix but also due to the impaired formation of the SHAP–hyaluronan complex. The latter finding has led to the assumption that TSG6 is the enzyme catalyzing the transfer of the heavy chains (38). If it were the case, it seems unusual that the enzyme is simultaneously a constitutive component of the matrix.

The formation of the cumulus SHAP–hyaluronan complex is temporally and spatially regulated. The IαI family molecules are excluded from the ovarian follicles until the follicular basal lamina dissolves upon ovulatory stimulation (60). The ovulatory stimulus also evokes a burst of synthesis of hyaluronan, TSG6, and other molecules by cumulus cells (42,61). Then, heavy chains of the infiltrated IαI family molecules are transferred to the newly synthesized hyaluronan by an enzyme that may either be secreted by follicular granulosa cells or have infiltrated from serum together with the IαI family molecules. Such a process is reminiscent of the inflammatory response. Indeed, the ovulatory response parallels the inflammatory response in various respects (62,63).

IV. SHAP–Hyaluronan Complex in Disease

In general, the IαI family of molecules and hyaluronan distribute in different body compartments—the blood circulation and connective tissues, respectively. Because the simple mixing of serum with hyaluronan followed by incubation at 37 °C is sufficient to result in the SHAP–hyaluronan complex, the convergence of serum and hyaluronan seems to be a rate-limiting step for the formation of the complex, which, however, is frequently encountered in inflammatory responses. Therefore, a hypothesis is raised that the SHAP–hyaluronan complex would be formed and play a role in inflammatory diseases (Fig. 3).

A. Serum Level of the SHAP–Hyaluronan Complex

Hyaluronan is produced in peripheral connective tissues. After being partly degraded at local sites, it is carried by lymph to the lymph nodes, where another part is endocytosed and degraded. Finally, a minor part is carried to the general circulation and rapidly cleared by liver sinusoidal endothelial cells (1). The daily turnover of hyaluronan is in the order of 10–100 mg. The level of circulating hyaluronan is very low, at only about 30–40 ng/mL in healthy individuals, but may increase dramatically under pathological conditions (1). The increase can be attributed to upregulated production, impaired uptake and degradation, or both.

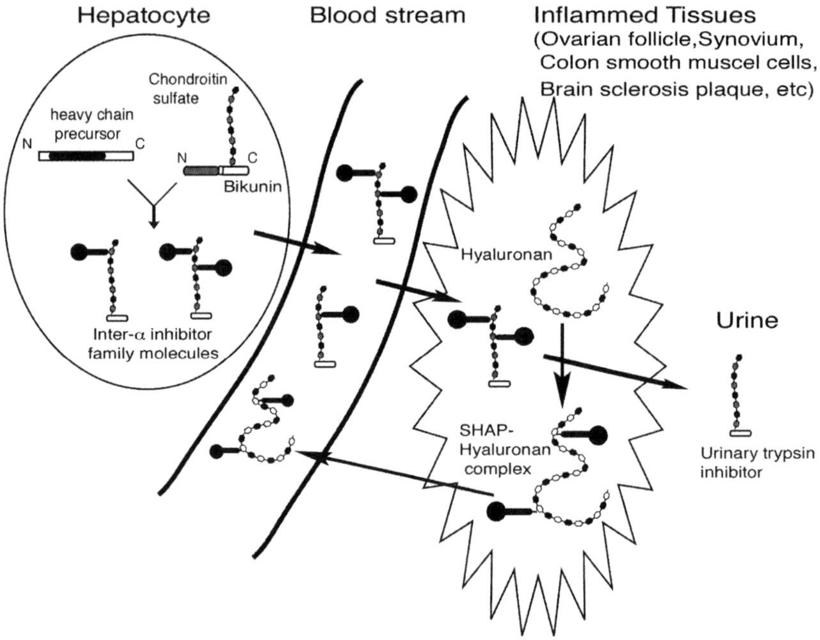

Figure 3 A hypothesis for the function of the SHAP–hyaluronan complex. The inter-α inhibitor family molecules are principally synthesized in hepatocytes, where one or two heavy chains are linked via a unique ester bond to the chondroitin sulfate chain of the light chain, the bikunin proteoglycan. The family molecules are circulating at relatively high concentrations. In response to a proper stimulus, they efflux to extracellular sites to interact with the locally synthesized hyaluronan. As a consequence, the heavy chains are transferred to hyaluronan, while the light chain is released and excreted in urine as urinary trypsin inhibitor. The IαI-recruiting signal may be either physiological, for example, the ovulatory stimulus, or pathological, such as those in chronic inflammatory diseases including rheumatoid arthritis, IBD, and multiple sclerosis. The SHAP–hyaluronan complex participates in the construction of the local extracellular matrix, which may play a role in the regulation of infiltrating inflammatory leukocytes. Some of the SHAP–hyaluronan complex may enter the circulation, providing a marker for the monitoring of pathogenesis.

Clinically, the serum level of hyaluronan is being used to aid diagnosis and monitoring of progress in cases of rheumatoid arthritis and liver cirrhosis (64,65).

An ELISA method has been developed for measuring the level of SHAP–hyaluronan complex in serum as well as in other humoral sources (66). The SHAP–hyaluronan in the samples is first captured by cartilage aggrecan-derived HABP immobilized on a microtiter plate, and then the amount of SHAP is determined with a specific antibody. The method is minimally interfered with by the large amount of IαI family molecules present in the serum. With the assay system, we have measured the levels of the SHAP–hyaluronan complex in serum samples from patients under various disease conditions. A significant correlation was generally observed between the levels of hyaluronan and of SHAP in diseases with elevated serum hyaluronan levels (66–68), providing evidence for the formation of the SHAP–hyaluronan complex under these conditions. In some other diseases, an increase in the level of SHAP–hyaluronan complex was found in other sources, e.g., the ascites of gynecological cancer patients and liposarcoma tissues ((69), and our unpublished observations). The results lead us to conclude that hyaluronan is upregulated in disease conditions, particularly in diseases with a significant inflammatory response, which is frequently, if not always, associated with SHAPs.

B. SHAP–Hyaluronan Complex and Chronic Liver Diseases

Chronic infection by hepatitis C virus and/or hepatitis B virus causes slowly progressive inflammation in the liver, which leads to scarring and architectural changes, and finally to cirrhosis. The infection is also associated with a high risk of development of hepatocellular carcinoma (HCC), which is one of the most common human cancers causing death. The serum level of hyaluronan is elevated significantly when the disease progresses from chronic hepatitis to cirrhosis, and is helpful for the diagnosis and monitoring of the progression of cirrhosis (64). The development of HCC seems to cause no further change to the serum hyaluronan level. In contrast, the serum SHAP level was found to be significantly higher in HCC patients than in liver cirrhosis patients (68). This finding suggests that the formation of the SHAP–hyaluronan complex is enhanced by the development of HCC. Since in the case of chronic liver diseases, the elevated serum hyaluronan level is largely due to decreased clearance, it is possible that the formation of the complex is merely a consequence of the prolonged accumulation of hyaluronan in the circulation since the situation resembles a test tube reaction. In this regard, the finding provides evidence for a direct relationship between the formation of the complex and the pathogenesis of HCC, although the underlying mechanism is unclear at present.

C. SHAP–Hyaluronan Complex and Rheumatoid Arthritis

On the other hand, the formation of the SHAP–hyaluronan complex is clearly associated with the pathogenesis of rheumatoid arthritis because: [1] the complex is formed within the synovial cavity and present at a concentration far higher than that in the circulation (66,67), and [2] in rheumatoid arthritis the half-life of circulating hyaluronan is unchanged (70), but that of the synovial hyaluronan increases significantly (71). The synovium is a highly vascularized tissue lacking basement membrane-like structure, allowing the synoviocyte-secreted hyaluronan to easily enter the joint cavity, where it is one of the major components of the synovial fluid. It is very likely that an inflamed synovium is the principal site for the formation of the synovial SHAP–HA complex.

The effect of protein association on the physicochemical properties of the synovial hyaluronan was first noticed half a century ago. The interaction of serum-derived IαI and hyaluronan in pathological (rheumatoid arthritis, septic arthritis, gout, psoriatic arthritis, and acute rheumatic fever) synovial fluid was found as early as 1965 (12,72). It was then clear that IαI has associated with hyaluronan in the form of SHAP (33). Examination of the SHAP–hyaluronan complex purified from rheumatoid arthritic synovial fluid indicated that the complex is heterogeneous in the chain length of hyaluronan as well as the SHAP-to-hyaluronan molecular ratio: a hyaluronan chain of 2000 kDa carries 3–5 SHAP proteins on average (20). Some altered properties of pathological synovial hyaluronan, such as gelation and rapid sedimentation at pH 4.5, have been observed, which may relate to the association of IαI (SHAPs) because it was not found in normal synovial hyaluronan (12,73). We have recently made a similar observation that part of the purified pathological synovial SHAP–hyaluronan complex forms a macromolecular aggregate (20) (Fig. 3). In addition, a preventive effect of IαI/SHAPs on the degradation of hyaluronan by free radicals has also been suggested (13).

Collagen-induced arthritis in mice resembles rheumatoid arthritis in humans and is often used as an animal model (74). The underlying mechanism is not fully understood, but it is generally believed that the formation of an immune complex in joints initiates the pathological process. When mice deficient in the SHAP–hyaluronan complex (bikunin knockout mice in DBA 1 genetic background) were immunized with bovine type II collagen, an antibody production comparable with that of wild type mice was observed, in particular the production of IgG2a subtype, which is thought to be closely related to the onset of arthritis. However, the arthritic score by macroscopic examination was significantly decreased in the knockout mice (Fig. 4). Histological examination revealed leukocyte infiltration, synovium hyperplasia and cartilage erosion in the knockout mice, but all to a lesser extent than in the control mice. After the onset of arthritis, the serum level of the SHAP–hyaluronan complex was found to be elevated in the control mice, but not in the knockout mice, while the serum levels of hyaluronan were comparable between the two groups (our unpublished observations). The results argue for a positive role of the SHAP–hyaluronan complex in the inflammatory response in arthritis. We hypothesize that the complex participates in the regulation of adhesion and activation of infiltrated leukocytes as discussed in detail below.

D. SHAP–Hyaluronan Complex and Inflammatory Bowel Disease

Crohn's disease and ulcerative colitis are the major chronic inflammatory diseases of the gastrointestinal tract. They are often referred to together as inflammatory bowel disease (IBD). The main pathological changes include an increase in intestinal mucosal mononuclear leukocytes and a dramatic hyperplasia of muscularis mucosae (75). The interaction between recruited leukocytes and mesenchymal smooth muscle cells is thought to be important in

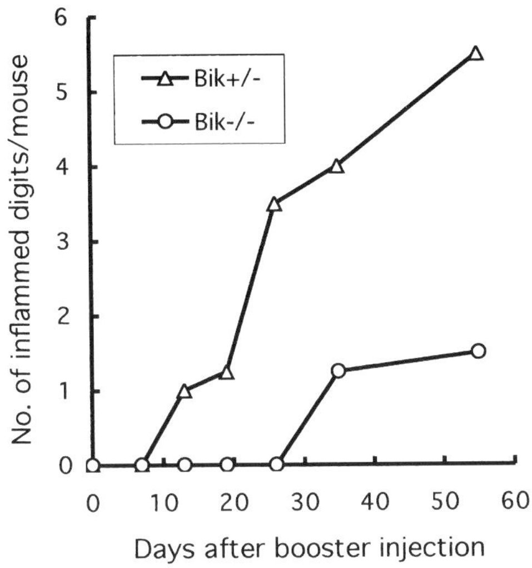

Figure 4 The collagen-induced arthritis is ameliorated in mice deficient in the SHAP–hyaluronan complex. The mice were injected subcutaneously at the tail root with 0.2 mg of bovine type II collagen emulsified with complete Freund's adjuvant, then given a booster injection with 0.1 mg of bovine type II collagen at the same site 3 weeks later. The result shown is representative of two experiments.

the development and propagation of IBD. The etiology of IBD is multifactorial including viral infection. Using a cell-culturing system, De La Motte et al. (76,77) showed that viral infection or treatment with virus mimic polyinosinic acid/polycytidylic acid upregulated the production of hyaluronan in colon smooth muscle cells, and the hyaluronan was deposited in the extracellular space to form the pericellular 'coat' structure and the 'cable' structure spanning several cell lengths in contrast to the small patchy structure of hyaluronan before the stimulation. Again, the SHAP–hyaluronan complex was found to form in these upregulated hyaluronan molecules (77). Cell adhesion assay revealed that peripheral mononuclear cells or histiocytic lymphoma U937 cells bound specifically to the hyaluronan cables via their cell surface hyaluronan receptor CD44. Similar results were obtained with inflamed colon samples. The findings suggested that the SHAP–hyaluronan complex formed in inflammatory tissues plays a role in the activation/regulation of the infiltrating leukocytes. Interestingly, the hyaluronan coat structure exhibited no leukocyte adhesion activity although it was also positive for IαI. Two possibilities may be considered. First, the IαI associated with hyaluronan coat structures was not converted to the SHAP form because the antibody used for staining recognizes the heavy chain of IαI and SHAP and, second, a highly organized structure may be required for the leukocytes to adhere.

Another disease that may involve the formation of the SHAP–hyaluronan complex is multiple sclerosis (MS). MS is an inflammatory/demyelinating disease characterized by discrete acute and chronic lesions, plaques, but the mechanism of both the inflammatory/demyelinating and the neurodegenerative components of its pathogenesis are largely unknown. Because the disease progression is accompanied by a complex alteration of the extracellular matrix as a consequence of the breakdown of the blood–brain barrier, release and activation of extracellular proteases and synthesis of extracellular matrix, the relation between the dynamic alteration of the extracellular matrix and the pathogenesis is attracting more and more attention (78). Normal brain tissue as well as nearly acellular old MS plaques are negative for SHAP immunoreactivity, but relatively fresh plaques at the active phase of inflammation are positive for the SHAP–hyaluronan complex, implying a role for the complex in the pathogenesis of MS (our unpublished observations). As observed in the diseases mentioned above, CD44 is critical to the secondary leukocyte recruitment in experimental encephalomyelitis, an animal model for human MS (79).

E. SHAP–Hyaluronan Complex in Cell Adhesion

To evaluate the effect of SHAP on the CD44–hyaluronan interaction in a simpler system, we compared the cell adhesion activity of the SHAP–hyaluronan complex purified from pathological synovial fluid with that of free hyaluronan. The CD44-positive cutaneous T cell lymphoma cell line Hut78, known to bind strongly to hyaluronan, was used for the adhesion assay. The SHAP–hyaluronan complex or free hyaluronan was immobilized on dishes pre-coated with HABP. The cells were then added, and incubated at room temperature for 30 min. To observe significant cell adhesion, the concentration of immobilizing hyaluronan concentration had to be higher than 0.5 mg/mL; however, significant adhesion was still observed even when the concentration of the SHAP–hyaluronan complex was reduced to 0.02 mg/mL (our unpublished results). The results showed clearly that the presence of SHAP enhanced the cell adhesion. Although control experiments are necessary, the results are in line with the observation made by De La Motte et al. and further highlight the role of SHAP in cell–matrix interactions, which may be important in the pathogenesis of rheumatoid arthritis, IBD, MS, and other diseases.

V. Future Prospects

About 10 years after the isolation of the first protein covalently bound to hyaluronan in 1990, the protein was identified, the chemical structure of the protein–glycosaminoglycan linkage was clarified, knowledge about the biochemical reaction of complex formation was considerably accumulated, and furthermore the physiological significance of the complex was uncovered. These efforts will no doubt encourage the search for proteins that modify hyaluronan by forming a covalent complex.

The relation of hyaluronan to pathogenesis in various diseases has been widely and extensively studied. Unfortunately, in most cases hyaluronan has been considered as a free glycosaminoglycan chain. The latest findings have shown that an associating protein to hyaluronan may exert a critical effect on the properties of hyaluronan, and thus strongly argue for the necessity to include these components, for example SHAP, in the examination of hyaluronan in order to understand hyaluronan and related abnormalities correctly. The exact role of the SHAP–hyaluronan complex in the pathogenesis of diseases is being uncovered. Given the results accumulated to date and the powerful tools available, for example, the SHAP–hyaluronan complex-deficient mouse (bikunin knockout mouse) and the TSG6 knockout mouse, obtaining a clear answer will not be long delayed.

In contrast to bikunin, much less is known about the heavy chain/SHAP. Although some motif/domain structures have been found by comparing sequence homology (80), in most cases a functional relation remains to be confirmed. The latest findings indicate that the heavy chain moiety may be physiologically more important than bikunin (81). The structural and functional study of the heavy chain will greatly help uncover the mechanisms underlying the function of SHAP. On the other hand, the clarification of the mechanism for the formation of the SHAP–hyaluronan complex is also important, because it will be a valuable target for the development of pharmaceutical agents against related diseases.

Acknowledgements

This work was supported in part by a preparatory grant for research at the Matrix Glycoconjugate group, Research Center for Infectious Disease, Aichi Medical University; Grant-in-Aid for Scientific Research on Priority Areas (14082206) from the Ministry of Education, Culture, Sports, Science, and Technology of Japan; a grant-in-Aid for Scientific Research (B) from Japan Society for the Promotion of Science (JSPS); grants from the CREST of JST (Japan Science and Technology Agency); Health Science Research Grants on Comprehensive Research on Aging and Health from Ministry of Health, Labor and Welfare; and a special research fund from Seikagaku Corporation.

We would like to thank Dr Reiji Kannagi, Aichi Cancer Center for providing Hut78 cells, and Dr Zhenxin Li, Shanghai Huashan Hospital, and Dr Jeong-Beom Lee, Soonchunhyang University School of Medicine, for helpful discussion and cooperation.

References

1. Laurent TC. Biochemistry of hyaluronan. Acta Otolaryngol Suppl 1987; 442:7–24.
2. Laurent TC, Fraser JR. Hyaluronan. FASEB J 1992; 6(7):2397–2404.
3. Knudson CB, Knudson W. Hyaluronan-binding proteins in development, tissue homeostasis, and disease. FASEB J 1993; 7(13):1233–1241.
4. McDonald J, Hascall VC. Hyaluronan minireview series. J Biol Chem 2002; 277: 4575–4579.
5. Balazs EA. Why hyaluronan has so many biological activities. In: Abatangelo G, Weigel PH, eds. New Frontiers in Medical Sciences: Redefining Hyaluronan. Amsterdam: Elsevier, 2000:3–10.
6. Toole BP. Hyaluronan and its binding proteins, the hyaladherins. Curr Opin Cell Biol 1990; 2(5):839–844.

7. Heinegard D, Bjornsson S, Morgelin M, Sommarin Y. Hyaluronan-binding matrix proteins. In: Laurent TC, ed. The Chemistry, Biology and Medical Applications of Hyaluronan and Its Derivatives. London: Portland Press, 1998:113–122.
8. Day AJ, Glenn D. Hyaluronan-binding proteins: tying up the giant. J Biol Chem 2002; 277:4585–4588.
9. Lesley J, Hyman R. CD44 structure and function. Front Biosci 1998; 3: D616–D630.
10. Borland G, Ross JA, Guy K. Forms and functions of CD44. Immunology 1998; 93(2):139–148.
11. Sandson J, Hamerman D. Isolation of hyaluronateprotein from normal synovial fluid. J Clin Investig 1962; 41:1817–1830.
12. Sandson J, Hamerman D, Schwick G. Altered properties of pathological hyaluronate due to a bound inter alpha trypsin inhibitor. Trans Assoc Am Physicians 1965; 78:304–313.
13. Hutadilok N, Ghosh P, Brooks PM. Binding of haptoglobulin, inter-alpha-trypsin inhibitor and alpha-1-proteinase inhibitor to synovial fluid hyaluronate and the influence of these proteins on its degradation by oxygen derived free radicals. Ann Rheum Dis 1988; 47:377–385.
14. How MJ, Long VJ, Stanworth DR. The association of hyaluronic acid with protein in human synovial fluid. Biochim Biophys Acta 1969; 194:81–91.
15. Prehm P. Synovial hyaluronate in rheumatoid arthritis binds C1q and is covalently bound to antibodies: a model for chronicity. Ann Rheum Dis 1995; 54(5):408–412.
16. Mikuni-Takagaki Y, Toole BP. Hyaluronate–protein complex of rous sarcoma virus-transformed chick embryo fibroblasts. J Biol Chem 1981; 256(16): 8463–8469.
17. Green SJ, Underhill CB. Hyaluronate appears to be covalently linked to the cell surface. J Cell Physiol 1988; 134(3):376–386.
18. Wang HS, Underhill CB. Hyaluronan can be non-enzymatically linked to protein through an alkali sensitive bond. Connect Tissue Res 1992; 28(1–2):29–48.
19. Yoneda M, Suzuki S, Kimata K. Hyaluronic acid associated with the surfaces of cultured fibroblasts is linked to a serum-derived 85-kDa protein. J Biol Chem 1990; 265:5247–5257.
20. Yingsung W, Zhuo L, Morgelin M, Yoneda M, Kida T, Watanabe H, Ishiguro N, Iwata H, Kimata K. Molecular heterogeneity of the SHAP–hyaluronan complex: isolation and characterization of the complex in synovial fluid from patients with rheumatoid arthritis. J Biol Chem 2003; 278:32710–32718.
21. Blom AM, Morgelin M, Oyen M, Jarvet J, Fries E. Structural characterization of inter-alpha-inhibitor. Evidence for an extended shape. J Biol Chem 1999; 274: 298–304.
22. Huang L, Yoneda M, Kimata K. A serum-derived hyaluronan-associated protein (SHAP) is the heavy chain of the inter alpha-trypsin inhibitor. J Biol Chem 1993; 268:26725–26730.
23. Salier JP. Inter-alpha-trypsin inhibitor: emergence of a family within the Kunitz-type protease inhibitor superfamily. Trends Biochem Sci 1990; 15(11):435–439.
24. Salier JP, Rouet P, Raguenez G, Daveau M. The inter-alpha-inhibitor family: from structure to regulation. Biochem J 1996; 315:1–9.
25. Salier JP, Verga V, Doly J, Diarra-Mehrpour M, Erickson RP. The genes for the inter-alpha-inhibitor family share a homologous organization in human and mouse. Mamm Genome 1992; 2(4):233–239.

26. Castillo GM, Templeton DM. Subunit structure of bovine ESF (extracellular-matrix stabilizing factor(s)). A chondroitin sulfate proteoglycan with homology to human I alpha I (inter-alpha-trypsin inhibitors). FEBS Lett 1993; 318:292–296.
27. Enghild JJ, Thogersen IB, Pizzo SV, Salvesen G. Analysis of inter-alpha-trypsin inhibitor and a novel trypsin inhibitor, pre-alpha-trypsin inhibitor, from human plasma. Polypeptide chain stoichiometry and assembly by glycan. J Biol Chem 1989; 264:15975–15981.
28. Yamamoto T, Yamamoto K, Sinohara H. Inter-alpha-trypsin inhibitor and its related proteins in Syrian hamster urine and plasma. J Biochem (Tokyo) 1996; 120: 145–152.
29. Saguchi K, Tobe T, Hashimoto K, Sano Y, Nakano Y, Miura NH, Tomita M. Cloning and characterization of cDNA for inter-alpha-trypsin inhibitor family heavy chain-related protein (IHRP), a novel human plasma glycoprotein. J Biochem (Tokyo) 1995; 117(1):14–18.
30. Enghild JJ, Salvesen G, Hefta SA, Thogersen IB, Rutherfurd S, Pizzo SV. Chondroitin 4-sulfate covalently cross-links the chains of the human blood protein pre-alpha-inhibitor. J Biol Chem 1991; 266:747–751.
31. Enghild JJ, Salvesen G, Thogersen IB, Valnickova Z, Pizzo SV, Hefta SA. Presence of the protein–glycosaminoglycan–protein covalent cross-link in the inter-alpha-inhibitor-related proteinase inhibitor heavy chain 2/bikunin. J Biol Chem 1993; 268:8711–8716.
32. Morelle W, Capon C, Balduyck M, Sautiere P, Kouach M, Michalski C, Fournet B, Mizon J. Chondroitin sulphate covalently cross-links the three polypeptide chains of inter-alpha-trypsin inhibitor. Eur J Biochem 1994; 221:881–888.
33. Zhao M, Yoneda M, Ohashi Y, Kurono S, Iwata H, Ohnuki Y, Kimata K. Evidence for the covalent binding of SHAP, heavy chains of inter-alpha-trypsin inhibitor, to hyaluronan. J Biol Chem 1995; 270:26657–26663.
34. Chen L, Zhang H, Powers RW, Russell PT, Larsen WJ. Covalent linkage between proteins of the inter-alpha-inhibitor family and hyaluronic acid is mediated by a factor produced by granulosa cells. J Biol Chem 1996; 271:19409–19414.
35. Odum L, Andersen CY, Jessen TE. Characterization of the coupling activity for the binding of inter-alpha-trypsin inhibitor to hyaluronan in human and bovine follicular fluid. Reproduction 2002; 124(2):249–257.
36. Wisniewski HG, Burgess WH, Oppenheim JD, Vilcek J. TSG-6, an arthritis-associated hyaluronan binding protein, forms a stable complex with the serum protein inter-alpha-inhibitor. Biochemistry 1994; 33:7423–7429.
37. Zhuo L, Yoneda M, Zhao M, Yingsung W, Yoshida N, Kitagawa Y, Kawamura K, Suzuki T, Kimata K. Defect in SHAP–hyaluronan complex causes severe female infertility. A study by inactivation of the bikunin gene in mice. J Biol Chem 2001; 276:7693–7696.
38. Fulop C, Szanto S, Mukhopadhyay D, Bardos T, Kamath RV, Rugg MS, Day AJ, Salustri A, Hascall VC, Glant TT, Mikecz K. Impaired cumulus mucification and female sterility in tumor necrosis factor-induced protein-6 deficient mice. Development 2003; 130(10):2253–2261.
39. Lee TH, Wisniewski HG, Vilcek J. A novel secretory tumor necrosis factor-inducible protein (TSG-6) is a member of the family of hyaluronate binding proteins, closely related to the adhesion receptor CD44. J Cell Biol 1992; 116:545–557.
40. Wisniewski HG, Hua JC, Poppers DM, Naime D, Vilcek J, Cronstein BN. TNF/IL-1-inducible protein TSG-6 potentiates plasmin inhibition by inter-alpha-inhibitor

and exerts a strong anti-inflammatory effect *in vivo*. J Immunol 1996; 156: 1609–1615.

41. Fulop C, Kamath RV, Li Y, Otto JM, Salustri A, Olsen BR, Glant TT, Hascall VC. Coding sequence, exon-intron structure and chromosomal localization of murine TNF-stimulated gene 6 that is specifically expressed by expanding cumulus cell-oocyte complexes. Gene 1997; 202:95–102.
42. Mukhopadhyay D, Hascall VC, Day AJ, Salustri A, Fulop C. Two distinct populations of tumor necrosis factor-stimulated gene-6 protein in the extracellular matrix of expanded mouse cumulus cell-oocyte complexes. Arch Biochem Biophys 2001; 394:173–181.
43. Jessen TE, Odum L. Role of tumour necrosis factor stimulated gene 6 (TSG-6) in the coupling of inter-alpha-trypsin inhibitor to hyaluronan in human follicular fluid. Reproduction 2003; 125(1):27–31.
44. Yoneda M, Zhao M, Zhuo L, Watanabe H, Yamada Y, Huang L, Nagasawa S, Nishimura H, Shinomura T, Isogai Z, Kimata K. Roles of inter-α-trypsin inhibitor and hyaluronan-binding proteoglycans in hyaluronan-rich matrix formation. In: Abatangelo G, Weigel PH, eds. New Frontiers in Medical Sciences: Redefining Hyaluronan. Amsterdam: Elsevier, 2000:21–30.
45. Steinbuch M. The inter-alpha-trypsin inhibitor. Methods Enzymol 1976; 45: 760–771.
46. Gebhard W, Hochstrasser K, Fritz H, Enghild JJ, Pizzo SV, Salvesen G. Structure of inter-alpha-inhibitor (inter-alpha-trypsin inhibitor) and pre-alpha-inhibitor: current state and proposition of a new terminology. Biol Chem Hoppe Seyler 1990; 371(Suppl):13–22.
47. Ohnishi H, Kosuzume H, Ashida Y, Kato K, Honjo I. Effects of urinary trypsin inhibitor on pancreatic enzymes and experimental acute pancreatitis. Dig Dis Sci 1984; 29(1):26–32.
48. Pratt CW, Pizzo SV. *In vivo* metabolism of inter-alpha-trypsin inhibitor and its proteinase complexes: evidence for proteinase transfer to alpha-2-macroglobulin and alpha-1-proteinase inihibitor. Arch Biochem Biophys 1986; 248(2): 587–596.
49. Wilharm E, Parry MA, Friebel R, Tschesche H, Matschiner G, Sommerhoff CP, Jenne DE. Generation of catalytically active granzyme K from *Escherichia coli* inclusion bodies and identification of efficient granzyme K inhibitors in human plasma. J Biol Chem 1999; 274(38):27331–27337.
50. Sugiki M, Sumi H, Maruyama M, Yoshida E, Mihara H. Clearance and distribution of acid-stable trypsin inhibitor (ASTI). Enzyme 1989; 42(1):31–38.
51. Jonsson-Berling BM, Ohlsson K. Distribution and elimination of intravenously injected urinary trypsin inhibitor. Scand J Clin Lab Investig 1991; 51(6): 549–557.
52. Chen L, Mao SJT, Larsen WJ. Identification of a factor in fetal bovine serum that stabilizes the cumulus extracellular matrix: a role for a member of the inter-alpha-trypsin inhibitor family. J Biol Chem 1992; 267:12380–12386.
53. Chen L, Mao SJ, McLean LR, Powers RW, Larsen WJ. Proteins of the inter-alpha-trypsin inhibitor family stabilize the cumulus extracellular matrix through their direct binding with hyaluronic acid. J Biol Chem 1994; 269:28282–28287.
54. Jessen TE, Odum L, Johnsen AH. *In vivo* binding of human inter-alpha-trypsin inhibitor free heavy chains to hyaluronic acid. Biol Chem Hoppe Seyler 1994; 375: 521–526.

55. Bratt T, Olsson H, Sjoberg EM, Jergil B, Akerstrom B. Cleavage of the alpha-1-microglobulin-bikunin precursor is localized to the golgi apparatus of rat liver cells. Biochim Biophys Acta 1993; 1157:147–154.
56. Thuveson M, Fries E. The low pH in trans-golgi triggers autocatalytic cleavage of pre-alpha-inhibitor heavy chain precursor. J Biol Chem 2000; 275: 30996–31000.
57. Blom A, Pertoft H, Fries E. Inter-alpha-inhibitor is required for the formation of the hyaluronan-containing coat on fibroblasts and mesothelial cells. J Biol Chem 1995; 270:9698–9701.
58. Zhao M, Yoneda M, Zhuo L, Huang L, Watanabe H, Yamada Y, Nagasawa S, Nishimura H, Kimata K. Proteoglycan enhances the formation of the SHAP–HA complex and its effect in hyaluronan-rich matrix formation. In: Kennedy JF, Phillips GO, Williams PA, Hascall VC, eds. Hyaluronan, Chemical, Biochemical and Biological Aspects. Cambridge: Woodhead Publishing, 2002; vol. 1:497–500.
59. Zhuo L, Kimata K. Cumulus oophorus extracellular matrix: its construction and regulation. Cell Struct Funct 2001; 26:189–196.
60. Powers RW, Chen L, Russell PT, Larsen WJ. Gonadotropin-stimulated regulation of blood-follicle barrier is mediated by nitric oxide. Am J Physiol 1995; 269: E290–E298 (Endocrinol. Metab. 32).
61. Fulop C, Salustri A, Hascall VC. Coding sequence of a hyaluronan synthase homologue expressed during expansion of the mouse cumulus–oocyte complex. Arch Biochem Biophys 1997; 337:261–266.
62. Espey LL. Current status of the hypothesis that mammalian ovulation is comparable to an inflammatory reaction. Biol Reprod 1994; 50(2):233–238.
63. Richards JS, Russell DL, Ochsner S, Espey LL. Ovulation: new dimensions and new regulators of the inflammatory-like response. Annu Rev Physiol 2002; 64: 69–92.
64. Laurent TC. Hyaluronan as a clinical marker of pathological processes. In: Laurent TC, ed. The Chemistry, Biology and Medical Applications of Hyaluronan and Its Derivatives. London: Portland Press, 1998:305–314.
65. Engstrom-Laurent A. Hyaluronan analysis as a tool in evaluating rheumatic diseases. In: Laurent TC, ed. The Chemistry, Biology and Medical Applications of Hyaluronan and Its Derivatives. London: Portland Press, 1998:315–317.
66. Kida D, Yoneda M, Miyaura S, Ishimaru T, Yoshida Y, Ito T, Ishiguro N, Iwata H, Kimata K. The SHAP–HA complex in sera from patients with rheumatoid arthritis and osteoarthritis. J Rheumatol 1999; 26(6):1230–1238.
67. Yingsung W, Zhuo L, Yoneda M, Ishiguro N, Iwata H, Kimata K. The covalent complex formation of hyaluronan with heavy chains of inter-alpha-trypsin inhibitor family is important for its functions. In: Hascall VC, Kuettner KE, eds. The Many Faces of Osteoarthritis. Basel: Birkhauser Verlag, 2002:207–212.
68. Shen L, Zhuo L, Okumura A, Ishikawa T, Miyaji M, Owa Y, Nagata M, Nonami A, Kakumu S, Kimata K. The SHAP–HA complex in serum from patients with chronic liver diseases (manuscript in preparation).
69. Tabata I, Nishida Y, Sugiura H, Zhuo L, Yoneda M, Kimata K, Nakashima H, Ishiguro N, A study of inter-alpha-trypsin inhibitor chains expression in liposarcomas. Eur J Surg Oncol 2003; 29:665–669.
70. Fraser JR, Engstrom-Laurent A, Nyberg A, Laurent TC. Removal of hyaluronic acid from the circulation in rheumatoid disease and primary biliary cirrhosis. J Lab Clin Med 1986; 107(1):79–85.

71. Fraser JR, Kimpton WG, Pierscionek BK, Cahill RN. The kinetics of hyaluronan in normal and acutely inflamed synovial joints: observations with experimental arthritis in sheep. Semin Arthritis Rheum 1993; 22(6 Suppl 1):9–17.
72. Becker A, Sandson J. The source of the inter-alpha trypsin inhibitor in pathologic hyaluronateprotein. Arthritis Rheum 1971; 14:764–766.
73. Hamerman D, Sandson J. Unusual properties of hyaluronateprotein isolated from pathological synovial fluids. J Clin Investig 1963; 42:1882–1889.
74. Holmdahl R, Andersson M, Goldschmidt TJ, Gustafsson K, Jansson L, Mo JA. Type II collagen autoimmunity in animals and provocations leading to arthritis. Immunol Rev 1990; 118:193–232.
75. Knigge KL. Inflammatory bowel disease. Clin Cornerstone 2002; 4(4):49–60.
76. De La Motte CA, Hascall VC, Calabro A, Yen-Lieberman B, Strong SA. Mononuclear leukocytes preferentially bind via CD44 to hyaluronan on human intestinal mucosal smooth muscle cells after virus infection or treatment with poly(I.C). J Biol Chem 1999; 274(43):30747–30755.
77. De La Motte CA, Hascall VC, Drazba J, Bandyopadhyay SK, Strong SA. Mononuclear leukocytes bind to specific hyaluronan structures on colon mucosal smooth muscle cells treated with polyinosinic acid:polycytidylic acid: inter-alpha-trypsin inhibitor is crucial to structure and function. Am J Pathol 2003; 163(1): 121–133.
78. Sobel RA, Ahmed AS. White matter extracellular matrix chondroitin sulfate/ dermatan sulfate proteoglycans in multiple sclerosis. J Neuropathol Exp Neurol 2001; 60(12):1198–1207.
79. Brocke S, Piercy C, Steinman L, Weissman IL, Veromaa T. Antibodies to CD44 and integrin alpha4, but not L-selectin, prevent central nervous system inflammation and experimental encephalomyelitis by blocking secondary leukocyte recruitment. Proc Natl Acad Sci USA 1999; 96(12):6896–6901.
80. Chan P, Risler JL, Raguenez G, Salier JP. The three heavy-chain precursors for the inter-alpha-inhibitor family in mouse: new members of the multicopper oxidase protein group with differential transcription in liver and brain. Biochem J 1995; 306: 505–512.
81. Zhuo L, Salustri A, Kimata K. A physiological function of serum proteoglycan bikunin: the chondroitin sulfate moiety plays a central role. Glycoconj J 2003; 19: 241–247.

Chemistry and Biology of Hyaluronan
H.G. Garg and C.A. Hales (editors)

Chapter 10

Hyaluronan and Associated Proteins in the Visual System

MASAHIRO ZAKO

*Department of Ophthalmology,
Aichi Medical University,
Nagakute, Aichi 480-1195, Japan*

MASAHIKO YONEDA

*Aichi Prefectural College of Nursing
and Health, Nagoya,
Aichi 463-8502, Japan*

I. Introduction

Hyaluronan, a main component of the vitreous body, is a key macromolecule in visual science. Due to its viscoelastic properties it is used in various ophthalmic surgeries. Hyaluronan has been implicated in several biological processes such as cell adhesion, migration and proliferation. Recent studies on synthases (1), binding proteins (2) and receptors in signaling (3,4) provide new insights into understanding the actions of hyaluronan in the visual system. Hyaluronan binding proteins described in this chapter are SPACR, SPACRCAN, versican, aggrecan, link protein, neurocan, brevican, CD44, LYVE-1 and CD38; some of these proteins with hyaluronan form a macromolecular scaffold as an extracellular matrix. This chapter describes recent findings and roles of both hyaluronan and its binding proteins in the ocular physiological system, and also discusses them with regard to the pathogenesis of several ocular diseases.

II. Physiological Role of Hyaluronan and Its Binding Proteins in Ocular Tissues

Fig. 1 shows the eye cut in horizontal section to reveal the ocular structures described in this chapter.

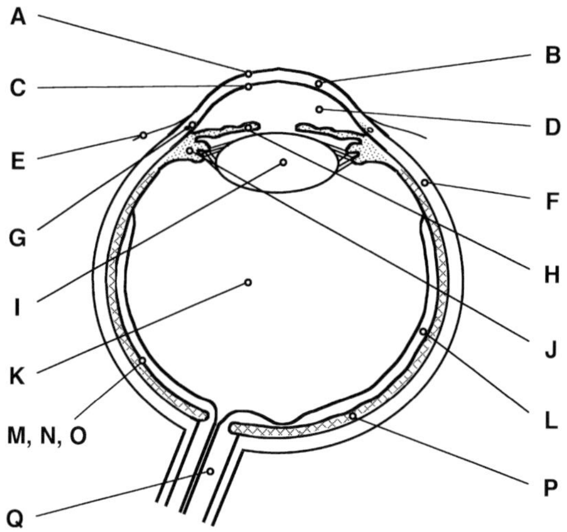

Figure 1 Schematic diagram of horizontal section of the eye indicating each ocular component. A, corneal epithelium; B, keratocyte; C, corneal endothelium; D, aqueous humor; E, conjunctiva; F, sclera; G, trabecular meshwork; H, iris; I, lens; J, ciliary body; K, vitreous; L, retina; M, interphotoreceptor matrix; N, retinal pigment epithelium; O, Bruch's membrane; P, choroid; Q, optic nerve.

A. Corneal Epithelium

Hyaluronan synthase (5,6) and hyaluronan are expressed in corneal epithelial cells (7–9). Hyaluronan stimulates corneal epithelial cell proliferation (10) and migration (11,12), while others, such as chondroitin sulfate, keratan sulfate and heparan sulfate, do not increase epithelial migration (13). Hyaluronan enhances the formation of hemidesmosomes (14), and has a positive influence on the epithelial re-surfacing during the healing phase in corneal wounds (15), with an increase of corneal hyaluronan occurring in the corneal healing region (16,17). Further, the process of corneal epithelial wound healing is modified by fibronectin or matrix metalloprotease (MMP) (18–20). Versican, a large extracellular matrix proteoglycan, is distributed in corneal epithelium (21). In normal corneas, the hyaluronan binding protein CD44 is predominantly expressed on the membranes of basal epithelial cells (9,22,23), and its expression is closely correlated with corneal re-epithelialization (24). Hyaluronan, CD44 and fibronectin collectively play important roles in corneal epithelial wound healing (25,26).

B. Keratocyte

Hyaluronan synthase (6) and hyaluronan (7) have been detected in keratocytes, but under normal conditions the expression associated with keratocyte plasma membranes is missing or insignificant (8,27). Hyaluronan expression is transiently increased during the corneal healing period (28), but completely disappears after the corneal structures become normal (29). Hyaluronan has been shown to have a temporary positive influence on the stromal healing (15,30). Hyaluronan added to corneal keratocytes produces the upregulation of MMP expression and activation, suggesting a crucial role for MMP in the corneal re-modeling process (31). Mice that lack CD44 expression in corneal epithelium display abnormal hyaluronan accumulation in the corneal stroma (32).

Immunohistochemical study has shown that link protein is expressed in the cornea (33). Corneal proteoglycans decrease the extent of fibril formation in corneal stroma, and the widths of fibrils are either unaltered or slightly decreased in the presence of link protein (34). In normal corneas, CD44 is predominantly expressed on the keratocytes (22). Keratan sulfate inhibits MMP-2 activation, while dermatan sulfate and heparan sulfate increase MMP-2 activation in corneal explant cultures. All these effects are inhibited by the addition of CD44 antibodies, suggesting that the effects are mediated by the CD44 receptor (35). Human lymphangiogenesis in the subepithelium and the stroma of corneas, shown by staining with LYVE-1 antibody, appears to be correlated with the degree of corneal hemangiogenesis (36).

C. Corneal Endothelium

Hyaluronan is seen on the apical and lateral membranes of the corneal endothelium (8,37), and three HAS isoforms are expressed in the endothelium (5,6,38,39). The expression of HAS2 is upregulated by TGF-beta 1 through Smad family members (39). Hyaluronan affects endothelial healing during the early repair process after corneal wounds (30). The localization of hyaluronan in the apical surface of the endothelium is associated with that of CD44 (7,40–42). CD44 is related to corneal inflammatory reactions, and the induction of CD44 on the endothelium assists the compensatory processes when the endothelium is injured (22). In addition, corneal endothelial hyaluronidase is involved in hyaluronan metabolism (42).

D. Aqueous Humor

The source of hyaluronan in aqueous humor is unclear, but could be either from the anterior segment production of hyaluronan or from the anterior diffusion of vitreous hyaluronan. Adult aqueous humor has a higher hyaluronan concentration than pediatric aqueous humor (43). There is an increase of corneal hyaluronan after corneal wounds, but aqueous hyaluronan is not significantly increased (16). Corticosteroids suppress trauma-induced hyaluronan production in cornea and aqueous humor (44).

E. Conjunctiva

Hyaluronan synthase is detected in the conjunctival epithelium (5,6). Hyaluronan is found in conjunctival epithelium, stroma, subconjunctival connective tissue, and limbal conjunctiva (9,27,37). By flow cytometric analysis, CD44 has been found in limbal conjunctiva and cultured normal conjunctival fibroblasts (9,45), but not in conjunctival epithelial cells (46).

F. Sclera

Important areas within the choroid, sclera and perimysial connective tissue of extraocular muscle are positive for hyaluronan in mice (47). Human scleral proteoglycans include aggrecan, biglycan and decorin (48). Whereas the relative amounts of newly synthesized and total accumulated aggrecan increase with age, newly synthesized and total accumulated biglycan and decorin decrease (49). Immunohistochemical study has suggested that link protein is expressed in sclera (33,50). Scleral proteoglycans decrease the extent of fibril formation; and the width of fibrils is either unaltered or slightly decreased in the presence of link protein (34).

G. Trabecular Meshwork

In the trabecular meshwork, hyaluronan has potential roles in the regulation of the physiological aqueous outflow resistance, in the maintenance of the outflow channels or both. Intraocular pressure and outflow facility, however, are not simply controlled by the amounts of hyaluronan because they remain unchanged following intracameral injections of hyaluronidase (51). Pronounced hyaluronan staining is observed in the various layers of the trabecular meshwork (27). Hyaluronan is associated with endothelial cells lining the trabecular beams (52), and hyaluronan synthase is expressed at the protein level in trabecular meshwork (6,38). Hyaluronan is the predominant glycosaminoglycan produced at first in trabecular meshwork cell cultures (53) and its expression is regulated by TGF-β and PDGF-BB (54). It has been reported that, in cultured trabecular meshwork cells, exogenous hyaluronan stimulates hyaluronan synthesis (55), and that hyaluronan synthesis is significantly decreased following treatment with dexamethasone (56). Additionally, trabecular meshwork cells express thyroid hormone receptors and modulate hyaluronan production in response to thyroid hormone (57). While aggrecan transcripts are not detectable, trabecular cells contain mRNA coding for versican (58), which is found in the composition of extracellular matrix materials of the juxtacanalicular tissue of normal human eyes (59). Schlemm's canal cell isolates react with antibodies specific for CD44 (60); at least three isoforms of CD44 are expressed in the human trabecular meshwork cells, which suggests that CD44 may play a role there in the binding and turnover of hyaluronan (61).

H. Iris

Hyaluronan is found in iris stroma, but not in the root of the iris (27), and is synthesized by both normal and traumatized iris (62). A transient increase in the expression of hyaluronan is seen in traumatized iris tissues (63); treatment with

COX-2 inhibitors prolongs the trauma-induced elevation of endogenous hyaluronan in the iris (64). Cervical sympathetic denervation results in a moderate increase of the hyaluronan content in the iris and does not appear to influence the hyaluronan response of the iris to trauma (65).

I. Lens

Both hyaluronan and CD44 are detected in the extracellular matrix accumulated on the inner surface of the lens capsule (66). Lens epithelial cells exhibit immunoreactivity to CD44 and not to hyaluronan (67). During eye development, suppression of lens stalk cell apoptosis by hyaluronan leads to faulty separation of the lens vesicle (68). CD44 is involved in lens epithelial cells attachment and growth on collagen and laminin *in vitro*, and may be involved in adhesion of lens epithelial cells to extracellular matrix components of the lens capsule (69). Link protein is also expressed in lens epithelium (33) while immunohistochemical staining of the lens has shown the expression of CD38 in the lenticular epithelium and lens cells (70).

J. Ciliary Body

Hyaluronan is seen in ciliary processes (37), but not in ciliary stroma (27), and hyaluronan synthase is expressed in ciliary epithelium (6,38), which performs the hyaluronan synthesis (71). Hyaluronan and chondroitin sulfate proteoglycans are co-localized in the ciliary zonule (72). The posterior ciliary body plays a role in the biosynthesis of vitreous humor (73).

Prostaglandin analogs down-regulate the expression of versican in the human ciliary muscle (74). Using immunofluorescence microscopy, reactions for both link protein and proteoglycan have been observed in the anterior uveal tract (50). Immunohistochemical staining of the ciliary body shows the expression of CD38 in both pigmented and non-pigmented epithelium (70).

K. Vitreous

Hyaluronan in the vitreous humor has been well described in Refs. 75–77. Hyaluronan appears very early in developing vitreous, and staining is observed first and predominantly in the equatorial portion of the vitreous (78). Hyaluronan staining of the internal layers of the retinal epithelium is detected in the presumptive ciliary body region and in the more posterior retina (78). Vitreous hyaluronan, other glycosaminoglycans and collagens seem to be produced by mesenchymal cells at an early stage and by the retina and hyaloid vessels during the middle and late developmental stages (79). Free radicals cause an increase in the high-molecular weight components and insolubilization of vitreous collagen, and a decrease in the molecular weight of hyaluronan resulting in photo-induced vitreous liquefaction (80,81). Hyaluronidase is present in human vitreous, and is involved in hyaluronan catabolism in the vitreous (82).

Versican and link protein are present in approximately 1:1 molar ratios in mammalian vitreous, but hyaluronan is present in a molar excess of 150 times (83). Versican-like proteoglycan has been demonstrated in vitreous gel (84) having a molecular mass of 380 kDa, and representing a small percentage (about 5%) of the total protein content (85).

L. Retina

Hyaluronan is detected in retinal glial cell plasma membranes (66,86,87) while Müller cells synthesize hyaluronan in the embryonic retina (88,89). Immunohistochemical study shows that link protein is expressed in the membranes of the retina (33).

Six forms of versican/PG-M have been made by alternative splicing in chick retina (Fig. 2) (90,91). Versican/PG-M V0, the most chondroitin sulfate-enriched form, is involved in neurite outgrowth from ganglion cells during retinal development, suggesting that an inhibitory effect against retinal neurite outgrowth of its chondroitin sulfate chains may be important for retinal development (Fig. 3) (91). Neurocan also has an inhibitory effect against retinal neurite outgrowth (92,93). There is spatiotemporal regulation of expression of neurocan and its proteolytic variant during retinal development, indicating its roles in neural network formation (94). The coordinated inhibition of cadherin and integrin

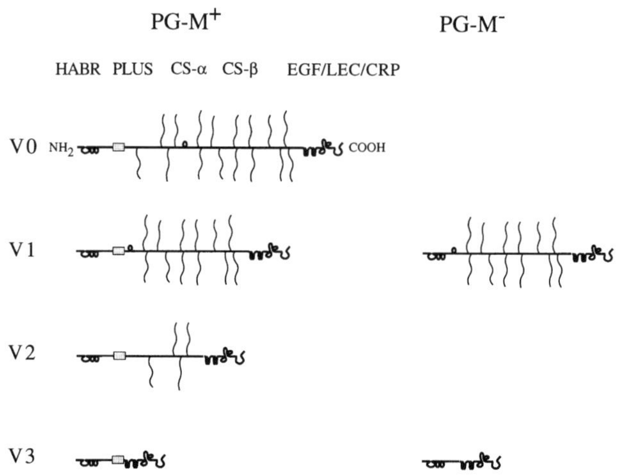

Figure 2 Schematic representation for alternatively spliced multi-forms of chick versican/PG-M, PG-M$^+$ (V0, V1, V2 and V3), and PG-M$^-$ (V1 and V3) in retina. Hyaluronan-binding domain (HABR), epidermal growth factor-like domains (EGF), lectin-like domain (LEC) and complementary regulatory protein-like domain (CRP) are present in all forms. Chondroitin sulfate-attachment domains (CS-α and CS-β) and PLUS domain in the middle of the core proteins are regulated by alternative splicing. Thick lines represent core proteins. Thin lines represent chondroitin sulfate side chains. Fig. 1 in Ref. 91.

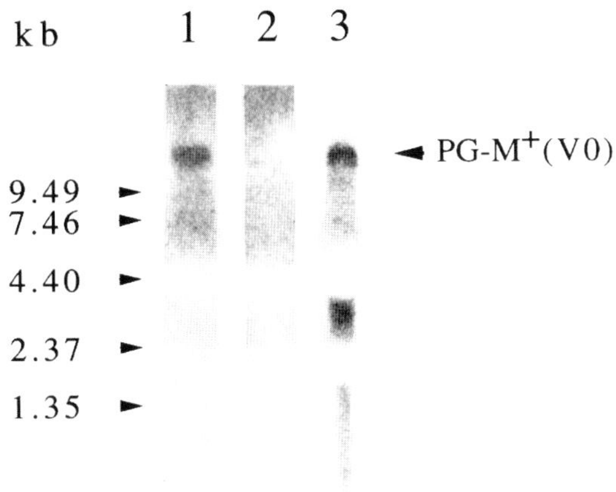

Figure 3 Northern blot analysis of mRNA encoding chick versican/PG-M core proteins on embryonic retinas. Total RNAs from embryonic day 14 retina (lane 1), embryonic day 20 retina (lane 2) and embryonic day 9 whole embryo as a control (lane 3) are analyzed. Versican/PG-M$^+$ (V0) is detected in embryonic day 14 retina and embryonic day 9 whole embryo (lanes 1 and 3). Sizes of molecular markers for calibration are indicated on the left in kilobases. Fig. 3 in Ref. 91.

functions on the interaction of neurocan with its receptor also prevents cell and neurite migration across boundaries in developing chick retina (95).

CD44 has been found on the outer limiting membrane and is specifically localized to the Müller cell; this suggests that CD44 might play a role in mediating the attachment of the neural retina to components of the interphotoreceptor matrix (IPM) (96–98). CD44 is expressed in Müller cells at a late stage of fetal development, and in fetal, infant and adult astrocytes, suggesting its importance in the morphogenesis and homeostasis of the neural retina (99). Müller cells express CD38 (100), which is detected in three distinct layers: the ganglion cell layer, the inner nuclear layer and the pigmented epithelium (70).

M. Interphotoreceptor Matrix

Except in mice, hyaluronan is a prominent constituent of the IPM, where it may serve to organize the matrix by functioning as a basic scaffold to which other macromolecules in the insoluble IPM are attached (47,101). Interactions between SPACR, a hyaluronan binding protein, and hyaluronan serve to form the basic macromolecular scaffold that comprises the insoluble IPM (102). Expression of SPACR increases with developmental age, paralleling the adhesion between neural retina and retinal pigment epithelium (RPE), indicating that SPACR might be involved in a system that mediates adhesion between neural retina and RPE

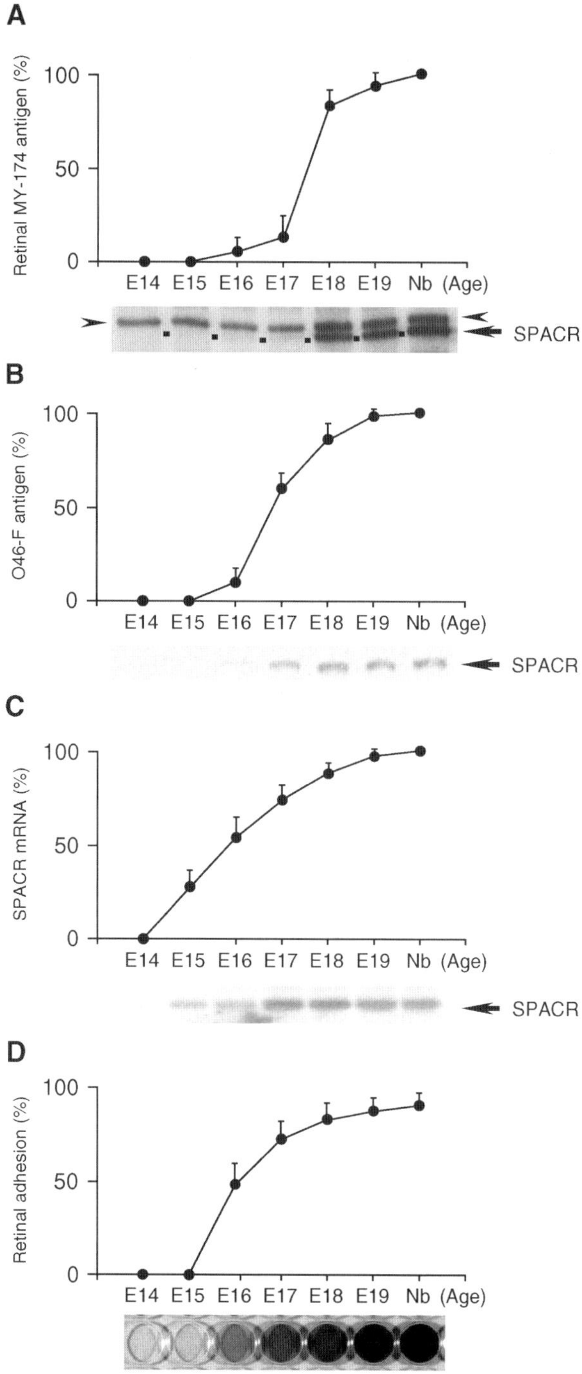
A
Retinal MY-174 antigen (%)
100
50
0
E14 E15 E16 E17 E18 E19 Nb (Age)
SPACR
B
O46-F antigen (%)
100
50
0
E14 E15 E16 E17 E18 E19 Nb (Age)
SPACR
C
SPACR mRNA (%)
100
50
0
E14 E15 E16 E17 E18 E19 Nb (Age)
SPACR
D
Retinal adhesion (%)
100
50
0
E14 E15 E16 E17 E18 E19 Nb (Age)

(Fig. 4) (103). Associations between SPACRCAN, another hyaluronan binding protein, and hyaluronan are also involved in the organization of the insoluble IPM (104).

N. Retinal Pigment Epithelium

Hyaluronan has been identified in the media of cultured RPE (105) where it is secreted preferentially from the apical surface of RPE, suggesting that RPE is an important source of the hyaluronan present in IPM (106). Retinal glia and RPE are the principal sources of glycosaminoglycan components in retina *in vitro*, and endogenous neurotrophic growth factors greatly modify glycosaminoglycan synthesis (107). Human fetal RPE is a direct target for thyroid hormones shown by measuring the accumulation of hyaluronan in RPE culture media (108). Cultured human RPE cells express the standard form of CD44 (termed CD44s) and variant isoforms containing exon v6 or v10, which are preferentially expressed by proliferating human RPE cells (109).

O. Bruch's Membrane

The glycosaminoglycans present in Bruch's membrane have been identified and were found to be heparan sulfate with small amounts of chondroitin and/or dermatan sulfate and hyaluronan (47,110). RPE–stromal interactions modulate hyaluronan depositions in the region of Bruch's membrane (111).

P. Choroid

Hyaluronan is detected in important areas within the choroid, sclera and perimysial connective tissue of extraocular muscle in mice eyes (47). Fetal and adult human eye tissues show the localization of hyaluronan in the chorioretinal complex, which disappears after the fifth decade of life, suggesting it might play a role in aging and age-related retinal disorders (112). Hyaluronan-induced choroid

---

Figure 4 Occurrence of chick SPACR and retinal adhesiveness during development. (A) SPACR expression was measured with MY-174, a monoclonal antibody against specific glycoconjugates of SPACR, on embryonic day 14 (E14) to newborn (Nb) retinal samples. At E16, a 150 kDa band first appears and expression increases with developmental age (arrow). The densities at the positions of 150 kDa bands in E14 and newborn retinas are defined as 0% and 100%, respectively. (B) SPACR expression was measured with O46-F, a polyclonal antibody against SPACR, on E14 to newborn (Nb) retinal samples. (C) The 6.0 kb of mRNA corresponding to chick SPACR was analyzed by Northern blot analysis. At E15, a band is first detected and expression increases with developmental age. (D) Retinal adhesiveness of these samples was measured. Retinal adhesiveness is initially detected at E16 and increases with developmental age. Homogenized samples of peeled retina corresponding to these days are also shown. The amounts of pigmentation derived from retinal pigment epithelium in homogenized samples demonstrate retinal adhesiveness. Fig. 7 in Ref. 103.

fibroblasts invasion into type I collagen gels is markedly concentration-dependent *in vitro*, and is reduced at both high and low concentrations (113). Immunohistochemical study shows link protein in the choroid (33), which is one of the major sites of CD44 expression in mice eyes (114).

Q. Optic Nerve, Chiasm and Tract

Hyaluronan in the retrolaminar optic nerve appears to decrease with age and is further reduced in primary open-angle glaucoma (POAG) (115). Glial hyaluronate-binding protein (GHAP) is a naturally occurring versican degradation product (116), and forms a delicate mesh surrounding myelinated optic nerve axons although no or only faint staining of GHAP is observed in the optic nerve head (117). Hyaluronan and GHAP disappear from hyaluronidase-injected optic nerves, optic chiasm and contralateral optic nerves (118). In hyaluronidase-injected crushed optic nerves, regenerated axons are able to grow for short distances into the distal stump undergoing Wallerian degeneration (118). Versican V2 is identified as a major inhibitor of axonal growth in the extracellular matrix of the mature central nervous system including optic nerves (119). Versican and aggrecan have been shown to be present in the embryonic rat optic tract (120).

Neurocan inhibits retinal axon growth *in vitro*, and is enriched in the hypothalamus and epithalamus, suggesting that neurocan might partly control retinal axon patterning in the embryonic diencephalon (92). Link protein is observed in the endoneurium of the optic nerve (50), which contains high amounts of GPI-linked brevican (121). CD44 serves as an anatomical template for retinal ganglion cell axons to form the optic chiasm (122,123).

R. Visual Cortex

Mature extracellular matrix consisting of chondroitin sulfate proteoglycan and neurocan is inhibitory for experience-dependent plasticity, but degradation of chondroitin sulfate proteoglycans with chondroitinase-ABC reactivates plasticity in the visual cortex (124).

S. Lacrimal System

Hyaluronan is found in tears (125), and an increase of hyaluronan concentration in tears is associated with spontaneous corneal epithelial healing (126). Immunohistochemical study has shown the expressions of hyaluronan and CD44 in the human lacrimal gland (127).

III. Ocular Diseases Involving Hyaluronan and Its Binding Proteins

A. Corneal Disorders

Hyaluronan is not highly expressed in healthy corneas, but is found in the epithelium, stroma and endothelium, and with various intensities across the entire spectrum of corneal disorders, suggesting that hyaluronan might be essential for remodeling of the corneal matrix (128–130). Gene expression using DNA micro arrays shows an upregulation of versican in keratoconus samples (131). In normal corneas, CD44 is predominantly expressed on the membranes of basal epithelial cells and on the keratocytes, but not on corneal endothelial cells. Enhanced expression of CD44 is observed on the epithelium of corneas with inflammation, and on remaining endothelial cells in a number of pathological conditions (22).

B. Primary Open-Angle Glaucoma

In the POAG iris, ciliary body and anterior sclera, hyaluronic content is less and chondroitin sulfate content is more, suggesting that a depletion of hyaluronan and an accumulation of chondroitin sulfate might increase aqueous outflow resistance in the POAG trabecular meshwork (132). Addition of ascorbic acid to the trabecular meshwork cell culture medium results in the dose-dependent stimulation of hyaluronan-synthesis and secretion, which is relatively stronger in cells from glaucomatous human eyes than in cells from normal ones (133). Hyaluronan in the retrolaminar optic nerve appears to decrease with age, and is further reduced in POAG (115). A significant decrease in CD44s content (extracellular domain) is observed in POAG regions compared with normal regions such as ciliary muscle, ciliary stroma, anterior iris, iris root and trabecular meshwork (134). Aqueous humor in POAG contains an increased level of soluble ectodomain of CD44 (135), suggesting that CD44 isoforms are influenced by the POAG process.

C. Pseudoexfoliation Syndrome

Hyaluronan levels of aqueous humor in pseudoexfoliation syndrome are significantly higher than those in healthy individuals (136,137). Hyaluronan is found to coat the fibrillar exfoliation material on the lens, the zonules, the iris epithelium, the ciliary body and the capsular bag (138).

D. Cataracts

Lens epithelial cells exhibit no immunoreactivity to hyaluronan (67), though the reactive production of hyaluronan is found in rabbit lens following an anterior lens wound (139). CD44 is also detected in cultured lens epithelial cells of human cataracts (69). Both hyaluronan and CD44 might be involved in the formation of cataracts.

E. Retinal Disorders

Subretinal fluid from patients with primary rhegmatogenous retinal detachment shows either hyaluronan or hyaluronidase without hyaluronan (140).

Neurocan immunostaining is generally detected over the nerve fiber layer, the plexiform layers, the photoreceptor outer segments region and the ciliary epithelium; labeling throughout the plexiform layers decreases with age (141).

In RCS rats, however, conspicuous labeling is also seen in association with retinal vessels from post-natal day 15 onward (141).

A normal labeling pattern for CD44 is observed on Müller cells at an early age, prior to photoreceptor degeneration. During the time course of the retinal degeneration, Müller cells respond to damage by increased ectopic expression of the CD44 antigen (142–144). The inherited retinal degeneration exhibited by the retinal degeneration slow (rds) mice leads to an upregulation in the expression of CD44s, but no change in the expression of retinal CD44 isoforms (145). Gene array analysis reveals a prominent upregulation of CD44 in wounded RPE cultures by using unwounded RPE cultures as controls (146). CD44 is involved in leukocyte–endothelial interaction *in vivo* and influences the trafficking of primed leukocytes to the retina and their overall survival (147). The failure of memory (high CD44) CD4 T cells to recognize their target antigen in retinas might produce autoimmune uveoretinitis because activated T cells recognize antigen in retinas, an immune-privileged tissue, and may mediate autoimmune diseases (148).

F. Experimental Choroidal Neovascularization

Histopathologically, choroidal neovascularization (CNV) was first observed at 7 days post-photocoagulation in a rat model; CD44 was maximally induced at 3–5 days post-laser photocoagulation, and was localized to RPE, choroidal vascular endothelial and inflammatory cells (149).

G. Sjögren's Syndrome

The CD44 variant with the v6 exon is selectively detected from infiltrating lymphocytes in glands with lymphoproliferative disorders, but not from infiltrating lymphocytes in normal lacrimal glands, suggesting that the CD44 v6 variant exon is closely associated with the development of lymphoproliferative disorders in Sjögren's syndrome (150).

H. Thyroid-Associated Ophthalmopathy

Orbital extracellular matrix exhibits a significant increase in tissue fractions of hyaluronan and chondroitin sulfate in patients with thyroid-associated ophthalmopathy (TAO) (151). Immunoglobulin G of patients with TAO markedly stimulates hyaluronan secretion from retrobulbar fibroblasts (152) and such patients exhibit significantly greater antibody values against hyaluronan (153). Lymphocytes on retrobulbar fibroblasts show a tendency for TAO patients' lymphocytes to enhance the synthesis of hyaluronan (154). Hyaluronan expression is also increased at the extraocular muscle level in patients with TAO (155).

Insulin-like growth factor-1 (IGF-1), platelet-derived growth factor (PDGF) and cAMP increase the synthesis of hyaluronan and chondroitin sulfate proteoglycan by retroocular tissue fibroblasts (156,157). HMC-1, an established human mast cell line, can activate human orbital fibroblasts to produce increased levels of prostaglandin E2 and hyaluronan when co-cultured (158). Interleukin-1beta (IL-1beta) regulates the expression of HAS genes in orbital fibroblasts (159). Cytosolic Ca (2 +) and PKC betaII are involved in IL-1beta-induced hyaluronan synthesis in cultured orbital fibroblasts from patients with Graves' ophthalmopathy (160).

I. Form Deprivation Myopia

Increased synthesis and accumulation of aggrecan, which increases the volume of extracellular matrix in the posterior sclera, are responsible for the ocular enlargement observed in form-deprived chick myopia (161). Synthesis of scleral proteoglycans including aggrecan is higher during the day than at night, but there are no significant differences between rhythms in scleras from normal and form-deprived eyes (162). The turnover rate of scleral proteoglycans is vision-dependent and is accelerated in the posterior sclera during the development of experimental myopia (163). Treatment with beta-xyloside, a specific inhibitor of proteoglycan synthesis, results in a significant reduction in the axial length, vitreous chamber depth and rate of axial elongation of form-deprived eyes (164).

J. Wagner Disease

Wagner syndrome, an autosomal dominant vitreoretinopathy characterized by chorioretinal atrophy, cataract and retinal detachment, is linked to 5q14.3. Within the critical region lie genes encoding two extracellular macromolecules, link protein and versican, and these can represent candidates for Wagner syndrome (165).

IV. Summary and Conclusion

In recent years, significant advances have been made in understanding the roles of hyaluronan and its binding proteins in the field of visual science. Each ocular tissue physiologically expresses these molecules in a tissue-specific manner, and disordered expressions of these molecules are involved in the pathogenesis of ocular diseases. Further investigations into the molecular mechanisms of hyaluronan and its binding proteins, which underlie the diverse homeostatic and morbid reactions in this organ, are essential for better understanding and diagnosis of various diseases, and to plan treatment.

References

1. Itano N, Kimata K. Mammalian hyaluronan synthases. IUBMB Life 2002; 54: 195–199.
2. Day AJ, Prestwich GD. Hyaluronan-binding proteins: tying up the giant. J Biol Chem 2002; 277:4585–4588.
3. Turley EA, Noble PW, Bourguignon LY. Signaling properties of hyaluronan receptors. J Biol Chem 2002; 277:4589–4592.

4. Ponta H, Sherman L, Herrlich PA. CD44: from adhesion molecules to signalling regulators. Nat Rev Mol Cell Biol 2003; 4:33–45.
5. Rittig M, Lutjen-Drecoll E, Prehm P. Immunohistochemical localization of hyaluronan synthase in cornea and conjunctive of cynomolgus monkey. Exp Eye Res 1992; 54:455–460.
6. Usui T, Suzuki K, Kaji Y, Amano S, Miyata K, Heldin P, Yamashita H. Hyaluronan synthase expression in bovine eyes. Investig Ophthalmol Vis Sci 1999; 40:563–567.
7. Asari A, Miyauchi S, Takahashi T, Kohno K, Uchiyama Y. Localization of hyaluronic acid, chondroitin sulfate, and CD44 in rabbit cornea. Arch Histol Cytol 1992; 55:503–511.
8. Eggli PS, Graber W. Ultrastructural distribution of hyaluronan in rat cornea. Exp Eye Res 1993; 56:693–699.
9. Lerner LE, Schwartz DM, Hwang DG, Howes EL, Stern R. Hyaluronan and CD44 in the human cornea and limbal conjunctiva. Exp Eye Res 1998; 67:481–484.
10. Inoue M, Katakami C. The effect of hyaluronic acid on corneal epithelial cell proliferation. Investig Ophthalmol Vis Sci 1993; 34:2313–2315.
11. Nishida T, Nakamura M, Mishima H, Otori T. Hyaluronan stimulates corneal epithelial migration. Exp Eye Res 1991; 53:753–758.
12. Sugiyama T, Miyauchi S, Machida A, Miyazaki K, Tokuyasu K, Nakazawa K. The effect of sodium hyaluronate on the migration of rabbit corneal epithelium. II. The effect of topical administration. J Ocul Pharmacol 1991; 7:53–64.
13. Nakamura M, Hikida M, Nakano T. Concentration and molecular weight dependency of rabbit corneal epithelial wound healing on hyaluronan. Curr Eye Res 1992; 11:981–986.
14. Chung JH, Kim WK, Lee JS, Pae YS, Kim HJ. Effect of topical Na-hyaluronan on hemidesmosome formation in *n*-heptanol-induced corneal injury. Ophthal Res 1998; 30:96–100.
15. Chung JH, Fagerholm P, Lindstrom B. Hyaluronate in healing of corneal alkali wound in the rabbit. Exp Eye Res 1989; 48:569–576.
16. Molander N, Lindquist U, Stenevi U, von Malmborg A, Ehinger B. Influence of radial keratotomy on endogenous hyaluronan in cornea and aqueous humour. Refract Corneal Surg 1993; 9:358–365.
17. Drubaix I, Legeais JM, Robert L, Renard G. Corneal hyaluronan content during post-ablation healing: evidence for a transient depth-dependent contralateral effect. Exp Eye Res 1997; 64:301–304.
18. Nakamura M, Mishima H, Nishida T, Otori T. Binding of hyaluronan to plasma fibronectin increases the attachment of corneal epithelial cells to a fibronectin matrix. J Cell Physiol 1994; 159:415–422.
19. Nakamura M, Nishida T. Synergistic effects of hyaluronan and fibronectin on epithelial migration in rabbit cornea *in vitro*. Cornea 1999; 18:686–692.
20. Sakuma M, Miyachi S, Sakamoto T. The effect of sodium hyaluronate on the expression of gelatinases in rabbit corneal epithelial wound healing. Jpn J Ophthalmol 2000; 44:475–481.
21. Yamagata M, Shinomura T, Kimata K. Tissue variation of two large chondroitin sulfate proteoglycans (PG-M/versican and PG-H/aggrecan) in chick embryos. Anat Embryol (Berl) 1993; 187:433–444.
22. Zhu SN, Nolle B, Duncker G. Expression of adhesion molecule CD44 on human corneas. Br J Ophthalmol 1997; 81:80–84.

23. Huang AJ, Li CH, Chen YW, Bourguignon LY. Co-expression of bcl-2 and CD44s in basal layers of human ocular surface epithelia. Adv Exp Med Biol 1998; 438: 527–531.
24. Yu FX, Guo J, Zhang Q. Expression and distribution of adhesion molecule CD44 in healing corneal epithelia. Investig Ophthalmol Vis Sci 1998; 39:710–717.
25. Asari A, Morita M, Sekiguchi T, Okamura K, Horie K, Miyauchi S. Hyaluronan, CD44 and fibronectin in rabbit corneal epithelial wound healing. Jpn J Ophthalmol 1996; 40:18–25.
26. Miyazaki T, Miyauchi S, Nakamura T, Takeshita S, Horie K. The effect of sodium hyaluronate on the growth of rabbit corneal epithelial cells *in vitro*. J Ocul Pharmacol Ther 1996; 12:409–415.
27. Gong H, Underhill CB, Freddo TF. Hyaluronan in the bovine ocular anterior segment, with emphasis on the outflow pathways. Investig Ophthalmol Vis Sci 1994; 5:4328–4332.
28. Molander N, Ehinger B, Rolfsen W, Harfstrand A, Stenevi U. Histochemical identification of hyaluronan in the rabbit anterior segment during the healing period after extracapsular lens extraction. Acta Ophthalmol (Copenh) 1993; 71: 336–346.
29. Podskochy A, Fagerholm P. Cellular response and reactive hyaluronan production in UV-exposed rabbit corneas. Cornea 1998; 17:640–645.
30. Chung J, Park Y, Paek S, Chong Y, Kim W. Effect of Na-hyaluronan on stromal and endothelial healing in experimental corneal alkali wounds. Ophthal Res 1999; 31:432–439.
31. Isnard N, Legeais JM, Renard G, Robert L. Effect of hyaluronan on MMP expression and activation. Cell Biol Int 2001; 25:735–739.
32. Kaya G, Rodriguez I, Jorcano JL, Vassalli P, Stamenkovic I. Selective suppression of CD44 in keratinocytes of mice bearing an antisense CD44 transgene driven by a tissue-specific promoter disrupts hyaluronate metabolism in the skin and impairs keratinocyte proliferation. Genes Dev 1997; 11: 996–1007.
33. Tsonis PA, Goetinck PF. Expression of cartilage-matrix genes and localization of their translation products in the embryonic chick eye. Exp Eye Res 1998; 46: 753–764.
34. Chandrasekhar S, Kleinman HK, Hassell JR, Martin GR, Termine JD, Trelstad RL. Regulation of type I collagen fibril assembly by link protein and proteoglycans. Coll Relat Res 1984; 4:323–337.
35. Isnard N, Robert L, Renard G. Effect of sulfated GAGs on the expression and activation of MMP-2 and MMP-9 in corneal and dermal explant cultures. Cell Biol Int 2003; 27:779–784.
36. Cursiefen C, Schlotzer-Schrehardt U, Kuchle M, Sorokin L, Breiteneder-Geleff S, Alitalo K, Jackson D. Lymphatic vessels in vascularized human corneas: immunohistochemical investigation using LYVE-1 and podoplanin. Investig Ophthalmol Vis Sci 2002; 43:2127–2135.
37. Lutjen-Drecoll E, Schenholm M, Tamm E, Tengblad A. Visualization of hyaluronic acid in the anterior segment of rabbit and monkey eyes. Exp Eye Res 1990; 51:55–63.
38. Rittig M, Flugel C, Prehm P, Lutjen-Drecoll E. Hyaluronan synthase immunoreactivity in the anterior segment of the primate eye. Graefes Arch Clin Exp Ophthalmol 1993; 231:313–317.

39. Usui T, Amano S, Oshika T, Suzuki K, Miyata K, Araie M, Heldin P, Yamashita H. Expression regulation of hyaluronan synthase in corneal endothelial cells. Investig Ophthalmol Vis Sci 2000; 41:3261–3267.
40. Harfstrand A, Molander N, Stenevi U, Apple D, Schenholm M, Madsen K. Evidence of hyaluronic acid and hyaluronic acid binding sites on human corneal endothelium. J Cataract Refract Surg 1992; 18:265–269.
41. Forsberg N, Von Malmborg A, Madsen K, Rolfsen W, Gustafson S. Receptors for hyaluronan on corneal endothelial cells. Exp Eye Res 1994; 59:689–696.
42. Schwartz DM, Jumper MD, Lui GM, Dang S, Schuster S, Stern R. Corneal endothelial hyaluronidase: a role in anterior chamber hyaluronic acid catabolism. Cornea 1997; 16:188–191.
43. Jumper JM, Chang DF, Hoyt CS, Hall JL, Stern R, Schwartz DM. Aqueous hyaluronic acid concentration: comparison in pediatric and adult patients. Curr Eye Res 1997; 16:1069–1071.
44. Molander N, Ehinger B, Stenevi U, Lindquist U, Lind L. Corticosteroid suppression of trauma-induced hyaluronan in rabbit cornea and aqueous. J Refract Surg 1995; 11:260–266.
45. Lee SB, Li DQ, Tan DT, Meller DC, Tseng SC. Suppression of TGF-beta signaling in both normal conjunctival fibroblasts and pterygial body fibroblasts by amniotic membrane. Curr Eye Res 2000; 20:325–334.
46. Fujihara T, Takeuchi T, Saito K, Tsubota K. Flow cytometric analysis of surface antigens on human conjunctival epithelial cells. Ophthal Res 1997; 29:103–109.
47. Hollyfield JG, Rayborn ME, Tammi R. Hyaluronan localization in tissues of the mouse posterior eye wall: absence in the interphotoreceptor matrix. Exp Eye Res 1997; 65:603–608.
48. Rada JA, Achen VR, Perry CA, Fox PW. Proteoglycans in the human sclera. Evidence for the presence of aggrecan. Investig Ophthalmol Vis Sci 1997; 38: 1740–1751.
49. Rada JA, Achen VR, Penugonda S, Schmidt RW, Mount BA. Proteoglycan composition in the human sclera during growth and aging. Investig Ophthalmol Vis Sci 2000; 41:1639–1648.
50. Poole AR, Pidoux I, Reiner A, Coster L, Hassell JR. Mammalian eyes and associated tissues contain molecules that are immunologically related to cartilage proteoglycan and link protein. J Cell Biol 1982; 93:910–920.
51. Hubbard WC, Johnson M, Gong H, Gabelt BT, Peterson JA, Sawhney R, Freddo T, Kaufman PL. Intraocular pressure and outflow facility are unchanged following acute and chronic intracameral chondroitinase ABC and hyaluronidase in monkeys. Exp Eye Res 1997; 65:177–190.
52. Lerner LE, Polansky JR, Howes EL, Stern R. Hyaluronan in the human trabecular meshwork. Investig Ophthalmol Vis Sci 1997; 38:1222–1228.
53. Crean EV, Tyson SL, Richardson TM. Factors influencing glycosaminoglycan synthesis by calf trabecular meshwork cell cultures. Exp Eye Res 1986; 43: 365–374.
54. Usui T, Nakajima F, Ideta R, Kaji Y, Suzuki Y, Araie M, Miyauchi S, Heldin P, Yamashita H. Hyaluronan synthase in trabecular meshwork cells. Br J Ophthalmol 2003; 87:357–360.
55. Binninger EA, Schachtschabel DO, Rohen JW. Exogenous glycosaminoglycans stimulate hyaluronic acid synthesis by cultured human trabecular-meshwork cells. Exp Eye Res 1987; 45:169–177.

56. Engelbrecht-Schnur S, Siegner A, Prehm P, Lutjen-Drecoll E. Dexamethasone treatment decreases hyaluronan-formation by primate trabecular meshwork cells *in vitro*. Exp Eye Res 1987; 64:539–543.
57. Duncan KG, Jumper MD, Ribeiro RC, Bailey KR, Yen PM, Sugawara A, Patel A, Stern R, Chin WW, Baxter JD, Schwartz DM. Human trabecular meshwork cells as a thyroid hormone target tissue: presence of functional thyroid hormone receptors. Graefes Arch Clin Exp Ophthalmol 1999; 237:231–240.
58. Wirtz MK, Bradley JM, Xu H, Domreis J, Nobis CA, Truesdale AT, Samples JR, Van Buskirk EM, Acott TS. Proteoglycan expression by human trabecular meshworks. Curr Eye Res 1997; 16:412–421.
59. Ueda J, Wentz-Hunter K, Yue BY. Distribution of myocilin and extracellular matrix components in the juxtacanalicular tissue of human eyes. Investig Ophthalmol Vis Sci 2002; 43:1068–1076.
60. Stamer WD, Roberts BC, Howell DN, Epstein DL. Isolation, culture, and characterization of endothelial cells from Schlemm's canal. Investig Ophthalmol Vis Sci 1998; 39:1804–1812.
61. Jumper MD, Stern R, Lui GM, Formby B, Schwartz DM. Expression of CD44 isoforms in cultured human trabecular meshwork cells. Ophthal Res 1998; 30: 314–320.
62. Koralewska-Makar A, Sommarin Y, Bruun A, Molander N, Stenevi U, Ehinger B. Synthesis of hyaluronan by normal and wounded rabbit iris. Graefes Arch Clin Exp Ophthalmol 2000; 238:520–524.
63. Koralewska-Makar A, Molander N, Madsen K, Lind ML, Stenevi U, Ehinger B. Endogenous hyaluronan in the normal and traumatized rabbit iris. Acta Ophthalmol Scand 1998; 76:391–395
64. Koralewska-Makar A, Johnsson C, Bruun A, Stenevi U, Ehinger B. COX-2 inhibitors prolong trauma-induced elevations of iris hyaluronan. J Ocul Pharmacol Ther 2003; 19:385–395.
65. Koralewska-Makar A, Johnsson C, Bruun A, Stenevi U, Ehinger B. Effects of sympathetic denervation on the hyaluronan content of the anterior segment in the normal and traumatized rabbit eye. Acta Ophthalmol Scand 2002; 80: 327–331.
66. Peterson PE, Pow CS, Wilson DB, Hendrickx AG. Localisation of glycoproteins and glycosaminoglycans during early eye development in the macaque. J Anat 1995; 186:31–42.
67. Saika S, Kawashima Y, Miyamoto T, Okada Y, Tanaka S, Yamanaka O, Ohnishi Y, Ooshima A, Yamanaka A. Immunolocalization of hyaluronan and CD44 in quiescent and proliferating human lens epithelial cells. J Cataract Refract Surg 1998; 24:1266–1270.
68. Ozeki H, Ogura Y, Hirabayashi Y, Shimada S. Suppression of lens stalk cell apoptosis by hyaluronic acid leads to faulty separation of the lens vesicle. Exp Eye Res 2001; 72:63–70.
69. Nishi O, Nishi K, Akaishi T, Shirasawa E. Detection of cell adhesion molecules in lens epithelial cells of human cataracts. Investig Ophthalmol Vis Sci 1997; 38: 579–585.
70. Khoo KM, Chang CF. Characterization and localization of CD38 in the vertebrate eye. Brain Res 1999; 821:17–25.
71. Schachtschabel DO, Berghoff K, Rohen JW. Synthesis and composition of glycosaminoglycans by explant cultures of human ciliary body and ciliary

processes in serum-containing and serum-free defined media. Graefes Arch Clin Exp Ophthalmol 1984; 221:207–209.
72. Chan FL, Choi HL, Underhill CB. Hyaluronan and chondroitin sulfate proteoglycans are colocalized to the ciliary zonule of the rat eye: a histochemical and immunocytochemical study. Histochem Cell Biol 1997; 107: 289–301.
73. Bishop PN, Takanosu M, Le Goff M, Mayne R. The role of the posterior ciliary body in the biosynthesis of vitreous humour. Eye 2002; 16:454–460.
74. Zhao X, Pearson KE, Stephan DA, Russell P. Effects of prostaglandin analogues on human ciliary muscle and trabecular meshwork cells. Investig Ophthalmol Vis Sci 2003; 44:1945–1952.
75. Meyer K, Palmer J. The polysaccharide of the vitreous humor. J Biol Chem 1934; 107:629–634.
76. Brewton RG, Mayne R. Mammalian vitreous humor contains networks of hyaluronan molecules: electron microscopic analysis using the hyaluronan-binding region (G1) of aggrecan and link protein. Exp Cell Res 1992; 198:237–249.
77. Bishop PN. Structural macromolecules and supramolecular organisation of the vitreous gel. Prog Retin Eye Res 2000; 19:323–344.
78. Bremer FM, Rasquin F. Histochemical localization of hyaluronic acid in vitreous during embryonic development. Investig Ophthalmol Vis Sci 1998; 39: 2466–2469.
79. Azuma N, Tajima S, Konomi H, Hida T, Akiya S, Uemura Y. Glycosaminoglycan and collagen distribution in the developing human vitreous. Graefes Arch Clin Exp Ophthalmol 1998; 236:679–687.
80. Lapcik L Jr, Omelka L, Kubena K, Galatik A, Kello V. Photodegradation of hyaluronic acid and of the vitreous body. Gen Physiol Biophys 1990; 9:419–429.
81. Akiba J, Ueno N, Chakrabarti B. Mechanisms of photo-induced vitreous liquefaction. Curr Eye Res 1994; 13:505–512.
82. Schwartz DM, Shuster S, Jumper MD, Chang A, Stern R. Human vitreous hyaluronidase: isolation and characterization. Curr Eye Res 1996; 15:1156–1162.
83. Reardon A, Heinegard D, McLeod D, Sheehan JK, Bishop PN. The large chondroitin sulphate proteoglycan versican in mammalian vitreous. Matrix Biol 1998; 17:325–333.
84. Noulas AV, Theocharis AD, Feretis E, Papageorgakopoulou N, Karamanos NK, Theocharis DA. Pig vitreous gel: macromolecular composition with particular reference to hyaluronan-binding proteoglycans. Biochimie 2002; 84: 295–302.
85. Theocharis AD, Papageorgakopoulou N, Feretis E, Theocharis DA. Occurrence and structural characterization of versican-like proteoglycan in human vitreous. Biochimie 2002; 84:1235–1241.
86. Morris JE, Hopwood JJ, Dorfman A. Biosynthesis of glycosaminoglycans in the developing retina. Dev Biol 1977; 58:313–327.
87. Eggli PS, Graber W. Ultrastructural association of hyaluronan with rat unmyelinated nerve fibres. J Neurocytol 1996; 25:79–87.
88. Threlkeld A, Adler R, Hewitt AT. Proteoglycan biosynthesis by chick embryo retina glial-like cells. Dev Biol 1989; 132:559–568.
89. Azuma N, Hida T, Akiya S, Uemura Y, Kohsaka S, Tsukada Y. Histochemical studies on hyaluronic acid in the developing human retina. Graefes Arch Clin Exp Ophthalmol 1990; 228:158–160.

90. Zako M, Shinomura T, Kimata K. Alternative splicing of the unique "PLUS" domain of chicken PG-M/versican is developmentally regulated. J Biol Chem 1997; 272:9325–9331.
91. Zako M, Shinomura T, Miyaishi O, Iwaki M, Kimata K. Transient expression of PG-M/versican, a large chondroitin sulfate proteoglycan in developing chicken retina. J Neurochem 1997; 69:2155–2161.
92. Tuttle R, Braisted JE, Richards LJ, O'Leary DD. Retinal axon guidance by region-specific cues in diencephalon. Development 1998; 125:791–801.
93. Inatani M, Honjo M, Otori Y, Oohira A, Kido N, Tano Y, Honda Y, Tanihara H. Inhibitory effects of neurocan and phosphacan on neurite outgrowth from retinal ganglion cells in culture. Investig Ophthalmol Vis Sci 2001; 42:1930–1938.
94. Inatani M, Tanihara H, Oohira A, Honjo M, Honda Y. Identification of a nervous tissue-specific chondroitin sulfate proteoglycan, neurocan, in developing rat retina. Investig Ophthalmol Vis Sci 1999; 40:2350–2359.
95. Li H, Leung TC, Hoffman S, Balsamo J, Lilien J. Coordinate regulation of cadherin and integrin function by the chondroitin sulfate proteoglycan neurocan. J Cell Biol 2000; 149:1275–1288.
96. Duguid IG, Boyd AW, Mandel TE. The expression of adhesion molecules in the human retina and choroid. Aust N Z J Ophthalmol 1991; 19:309–316.
97. Chaitin MH, Wortham HS, Brun-Zinkernagel AM. Immunocytochemical localization of CD44 in the mouse retina. Exp Eye Res 1994; 58:359–365.
98. Rich KA, Figueroa SL, Zhan Y, Blanks JC. Effects of Müller cell disruption on mouse photoreceptor cell development. Exp Eye Res 1995; 61:235–248.
99. Nishina S, Hirakata A, Hida T, Sawa H, Azuma N. CD44 expression in the developing human retina. Graefes Arch Clin Exp Ophthalmol 1997; 235:92–96.
100. Esguerra M, Miller RF. CD38 expression and NAD^+-induced intracellular Ca^+ mobilization in isolated retinal Müller cells. Glia 2002; 39:314–319.
101. Hollyfield JG. Hyaluronan and the functional organization of the interphotoreceptor matrix. Investig Ophthalmol Vis Sci 1999; 40:2767–2769.
102. Acharya S, Rodriguez IR, Moreira EF, Midura RJ, Misono K, Todres E, Hollyfield JG. SPACR, a novel interphotoreceptor matrix glycoprotein in human retina that interacts with hyaluronan. J Biol Chem 1998; 273:31599–31606.
103. Zako M, Iwaki M, Yoneda M, Miyaishi O, Zhao J, Suzuki Y, Takeuchi M, Miyake G, Ikagawa H, Kimata K. Molecular cloning and characterization of chick sialoprotein associated with cones and rods, a developmentally regulated glycoprotein of interphotoreceptor matrix. J Biol Chem 2002; 277:25592–25600.
104. Acharya S, Foletta VC, Lee JW, Rayborn ME, Rodriguez IR, Young WS 3rd, Hollyfield JG. SPACRCAN, a novel human interphotoreceptor matrix hyaluronan-binding proteoglycan synthesized by photoreceptors and pinealocytes. J Biol Chem 2000; 275:6945–6955.
105. Stramm LE. Synthesis and secretion of glycosaminoglycans in cultured retinal pigment epithelium. Investig Ophthalmol Vis Sci 1987; 28:618–627.
106. deS Senanayake P, Calabro A, Nishiyama K, Hu JG, Bok D, Hollyfield JG. Glycosaminoglycan synthesis and secretion by the retinal pigment epithelium: polarized delivery of hyaluronan from the apical surface. J Cell Sci 2001; 114: 199–205.
107. Normand G, Hicks D, Dreyfus H. Neurotrophic growth factors stimulate glycosaminoglycan synthesis in identified retinal cell populations *in vitro*. Glycobiology 1998; 8:1227–1235.

108. Duncan KG, Bailey KR, Baxter JD, Schwartz DM. The human fetal retinal pigment epithelium: A target tissue for thyroid hormones. Ophthal Res 1999; 31:399–406.
109. Liu NP, Roberts WL, Hale LP, Levesque MC, Patel DD, Lu CL, Jaffe GJ. Expression of CD44 and variant isoforms in cultured human retinal pigment epithelial cells. Investig Ophthalmol Vis Sci 1997; 38:2027–2037.
110. Robey PG, Newsome DA. Biosynthesis of proteoglycans present in primate Bruch's membrane. Investig Ophthalmol Vis Sci 1997; 24:898–905.
111. Gross-Jendroska M, Lui GM, Song MK, Stern R. Retinal pigment epithelium-stromal interactions modulate hyaluronic acid deposition. Investig Ophthalmol Vis Sci 1992; 33:3394–3399.
112. Tate DJ Jr, Oliver PD, Miceli MV, Stern R, Shuster S, Newsome DA. Age-dependent change in the hyaluronic acid content of the human chorioretinal complex. Arch Ophthalmol 1993; 111:963–967.
113. Docherty R, Forrester JV, Lackie JM, Gregory DW. Glycosaminoglycans facilitate the movement of fibroblasts through three-dimensional collagen matrices. J Cell Sci 1989; 92:263–270.
114. Kennel SJ, Lankford TK, Foote LJ, Shinpock SG, Stringer C. CD44 expression on murine tissues. J Cell Sci 1993; 104:373–382.
115. Gong H, Ye W, Freddo TF, Hernandez MR. Hyaluronic acid in the normal and glaucomatous optic nerve. Exp Eye Res 1997; 64:587–595.
116. Perides G, Asher RA, Lark MW, Lane WS, Robinson RA, Bignami A. Glial hyaluronate-binding protein: a product of metalloproteinase digestion of versican? Biochem J 1995; 312:377–384.
117. Perides G, Asher R, Dahl D, Bignami A. Glial hyaluronate-binding protein (GHAP) in optic nerve and retina. Brain Res 1990; 512:309–316.
118. Tona A, Bignami A. Effect of hyaluronidase on brain extracellular matrix *in vivo* and optic nerve regeneration. J Neurosci Res 1993; 36:191–199.
119. Schmalfeldt M, Bandtlow CE, Dours-Zimmermann MT, Winterhalter KH, Zimmermann DR. Brain derived versican V2 is a potent inhibitor of axonal growth. J Cell Sci 2000; 113:807–816.
120. Popp S, Andersen JS, Maurel P, Margolis RU. Localization of aggrecan and versican in the developing rat central nervous system. Dev Dyn 2003; 227:143–149.
121. Seidenbecher CI, Gundelfinger ED, Bockers TM, Trotter J, Kreutz MR. Transcripts for secreted and GPI-anchored brevican are differentially distributed in rat brain. Eur J Neurosci 1998; 10:1621–1630.
122. Sretavan DW, Feng L, Pure E, Reichardt LF. Embryonic neurons of the developing optic chiasm express L1 and CD44, cell surface molecules with opposing effects on retinal axon growth. Neuron 1994; 12:957–975.
123. Lin L, Chan SO. Perturbation of CD44 function affects chiasmatic routing of retinal axons in brain slice preparations of the mouse retinofugal pathway. Eur J Neurosci 2003; 17:2299–2312.
124. Pizzorusso T, Medini P, Berardi N, Chierzi S, Fawcett JW, Maffei L. Reactivation of ocular dominance plasticity in the adult visual cortex. Science 2002; 298: 1248–1251.
125. Frescura M, Berry M, Corfield A, Carrington S, Easty DL. Evidence of hyaluronan in human tears and secretions of conjunctival cultures. Biochem Soc Trans 1994; 22:228S.
126. Miyauchi S, Morita M, Kuramoto K, Horie K. Hyaluronan and chondroitin sulfate in rabbit tears. Curr Eye Res 1996; 15:131–135.

127. Yoshida K, Nitatori Y, Uchiyama Y. Localization of glycosaminoglycans and CD44 in the human lacrimal gland. Arch Histol Cytol 1996; 59: 505–513.
128. Yue BY, Baum JL, Silbert JE. The synthesis of glycosaminoglycans by cultures of corneal stromal cells from patients with keratoconus. J Clin Investig 1979; 63: 545–551.
129. Fitzsimmons TD, Molander N, Stenevi U, Fagerholm P, Schenholm M, von Malmborg A. Endogenous hyaluronan in corneal disease. Investig Ophthalmol Vis Sci 1994; 35:2774–2782.
130. Plaas AH, West LA, Thonar EJ, Karcioglu ZA, Smith CJ, Klintworth GK, Hascall VC. Altered fine structures of corneal and skeletal keratan sulfate and chondroitin/dermatan sulfate in macular corneal dystrophy. J Biol Chem 2001; 276:39788–39796.
131. Bochert A, Berlau J, Koczan D, Seitz B, Thiessen HJ, Guthoff R. Gene expression in keratoconus. Initial results using DNA microarrays. Ophthalmologe 2003; 100: 545–549.
132. Knepper PA, Goossens W, Hvizd M, Palmberg PF. Glycosaminoglycans of the human trabecular meshwork in primary open-angle glaucoma. Investig Ophthalmol Vis Sci 1996; 37:1360–1367.
133. Schachtschabel DO, Binninger E. Stimulatory effects of ascorbic acid on hyaluronic acid synthesis of *in vitro* cultured normal and glaucomatous trabecular meshwork cells of the human eye. Z Gerontol 1993; 26:243–246.
134. Knepper PA, Goossens W, Mayanil CS. CD44H localization in primary open-angle glaucoma. Investig Ophthalmol Vis Sci 1998; 39:673–680.
135. Knepper PA, Mayanil CS, Goossens W, Wertz RD, Holgren C, Ritch R, Allingham RR. Aqueous humor in primary open-angle glaucoma contains an increased level of CD44S. Investig Ophthalmol Vis Sci 2002; 43:133–139.
136. Lamari F, Katsimpris J, Gartaganis S, Karamanos NK. Profiling of the eye aqueous humor in exfoliation syndrome by high-performance liquid chromatographic analysis of hyaluronan and galactosaminoglycans. J Chromatogr B Biomed Sci Appl 1998; 709:173–178.
137. Gartaganis SP, Georgakopoulos CD, Exarchou AM, Mela EK, Lamari F, Karamanos NK. Increased aqueous humor basic fibroblast growth factor and hyaluronan levels in relation to the exfoliation syndrome and exfoliative glaucoma. Acta Ophthalmol Scand 2001; 79:572–575.
138. Fitzsimmons TD, Fagerholm P, Wallin O. Hyaluronan in the exfoliation syndrome. Acta Ophthalmol Scand 1997; 75:257–260.
139. Fagerholm P, Fitzsimmons T, Harfstrand A, Schenholm M. Reactive formation of hyaluronic acid after small and large lens injury. Acta Ophthalmol Suppl 1992; 205:58–64.
140. Hayasaka S, Shiono T, Hara S, Mizuno K. Lysosomal hyaluronidase in the subretinal fluid of patients with rhegmatogenous retinal detachments. Am J Ophthalmol 1982; 94:58–63.
141. Zhang Y, Rauch U, Perez MT. Accumulation of neurocan, a brain chondroitin sulfate proteoglycan, in association with the retinal vasculature in RCS rats. Investig Ophthalmol Vis Sci 2003; 44:1252–1261.
142. Kuhrt H, Hartig W, Grimm D, Faude F, Kasper M, Reichenbach A. Changes in CD44 and ApoE immunoreactivities due to retinal pathology of man and rat. J Hirnforsch 1997; 38:223–229.

143. Chaitin MH, Ankrum MT, Wortham HS. Distribution of CD44 in the retina during development and the rds degeneration. Brain Res Dev Brain Res 1996; 94:92–98.
144. Chaitin MH, Brun-Zinkernagel AM. Immunolocalization of CD44 in the dystrophic rat retina. Exp Eye Res 1998; 67:283–292.
145. Krishnamoorthy R, Agarwal N, Chaitin MH. Upregulation of CD44 expression in the retina during the rds degeneration. Brain Res Mol Brain Res 2000; 77: 125–130.
146. Singh S, Zheng JJ, Peiper SC, Mclaughlin BJ. Gene expression profile of ARPE-19 during repair of the monolayer. Graefes Arch Clin Exp Ophthalmol 2001; 239: 946–951.
147. Xu H, Manivannan A, Liversidge J, Sharp PF, Forrester JV, Crane IJ. Involvement of CD44 in leukocyte trafficking at the blood–retinal barrier. J Leukoc Biol 2002; 72:1133–1141.
148. Gregerson DS, Xiao J. Failure of memory (CD44 high) CD4 T cells to recognize their target antigen in retina. J Neuroimmunol 2001; 120:34–41.
149. Shen WY, Yu MJ, Barry CJ, Constable IJ, Rakoczy PE. Expression of cell adhesion molecules and vascular endothelial growth factor in experimental choroidal neovascularisation in the rat. Br J Ophthalmol 1998; 82:1063–1071.
150. Hiraoka M, Saito I, Tsubota K, Sugai S, Miyasaka N. Augmented expression of CD44 splice variants in lymphoproliferative disorder of the lacrimal gland in Sjögren's syndrome. Jpn J Ophthalmol 1997; 41:312–318.
151. Kahaly G, Forster G, Hansen C. Glycosaminoglycans in thyroid eye disease. Thyroid 1998; 8:429–432.
152. Kahaly G, Stover C, Otto E, Beyer J, Schuler M. Glycosaminoglycans in thyroid-associated ophthalmopathy. Autoimmunity 1992; 13:81–88.
153. Kahaly G, Hansen CE, Stover C, Beyer J, Otto E. Glycosaminoglycan antibodies in endocrine ophthalmopathy. Horm Metab Res 1993; 25:637–639.
154. Kahaly G, Stover C, Beyer J, Otto E. *In vitro* synthesis of glycosaminoglycans in endocrine ophthalmopathy. Acta Endocrinol (Copenh) 1992; 127: 397–402.
155. Pappa A, Jackson P, Stone J, Munro P, Fells P, Pennock C, Lightman S. An ultrastructural and systemic analysis of glycosaminoglycans in thyroid-associated ophthalmopathy. Eye 1998; 12:237–244.
156. Imai Y, Odajima R, Inoue Y, Shishiba Y. Effect of growth factors on hyaluronan and proteoglycan synthesis by retroocular tissue fibroblasts of Graves' ophthalmopathy in culture. Acta Endocrinol (Copenh) 1992; 126:541–552.
157. Imai Y, Ibaraki K, Odajima R, Shishiba Y. Effects of dibutyryl cyclic AMP on hyaluronan and proteoglycan synthesis by retroocular tissue fibroblasts in culture. Endocr J 1994; 41:645–654.
158. Smith TJ, Parikh SJ. HMC-1 mast cells activate human orbital fibroblasts in coculture: evidence for up-regulation of prostaglandin E2 and hyaluronan synthesis. Endocrinology 1999; 140:3518–3525.
159. Kaback LA, Smith TJ. Expression of hyaluronan synthase messenger ribonucleic acids and their induction by interleukin-1beta in human orbital fibroblasts: potential insight into the molecular pathogenesis of thyroid-associated ophthalmopathy. J Clin Endocrinol Metab 1999; 84:4079–4084.
160. Wong YK, Tang KT, Wu JC, Hwang JJ, Wang HS. Stimulation of hyaluronan synthesis by interleukin-1beta involves activation of protein kinase C betaII in

fibroblasts from patients with Graves' ophthalmopathy. J Cell Biochem 2001; 82: 58–67.

161. Rada JA, Thoft RA, Hassell JR. Increased aggrecan (cartilage proteoglycan) production in the sclera of myopic chicks. Dev Biol 1991; 147:303–312.
162. Nickla DL, Rada JA, Wallman J. Isolated chick sclera shows a circadian rhythm in proteoglycan synthesis perhaps associated with the rhythm in ocular elongation. J Comp Physiol A 1999; 185:81–90.
163. Rada JA, Achen VR, Rada KG. Proteoglycan turnover in the sclera of normal and experimentally myopic chick eyes. Investig Ophthalmol Vis Sci 1998; 39: 1990–2002.
164. Rada JA, Johnson JM, Achen VR, Rada KG. Inhibition of scleral proteoglycan synthesis blocks deprivation-induced axial elongation in chicks. Exp Eye Res 2002; 74:205–215.
165. Perveen R, Hart-Holden N, Dixon MJ, Wiszniewski W, Fryer AE, Brunner HG, Pinkners AJ, van Beersum SE, Black GC. Refined genetic and physical localization of the Wagner disease (WGN1) locus and the genes CRTL1 and CSPG2 to a 2- to 2.5-cM region of chromosome 5q14.3. Genomics 1999; 57: 219–226.

Chemistry and Biology of Hyaluronan
H.G. Garg and C.A. Hales (editors)

Chapter 11

Hyaluronan in the Pulmonary Alveolus and Interstitium

BONNIE ANDERSON BRAY

*James P. Mara Center for Lung Disease,
St. Luke's-Roosevelt Hospital Center,
Columbia University College of Physicians
and Surgeons, New York, USA*

I. Introduction

The pulmonary alveolus is the final result of the branching of the air passages in the lung. It is spherical, with a diameter of about 250 μm, in order to present the most surface for air exchange to occur. There are approximately 3×10^8 alveoli in the adult human lung and the total surface area they present is about 80 m^2. In order for the path of oxygen and carbon dioxide to be as short as possible, each alveolus is in apposition to a blood capillary and the area of the alveolar–capillary interface is about 70 m^2. Thus, the alveolar surface provides an area of contact with the environment greater than that of the skin (1) and in addition to its gas exchange functions, this surface presents the first line of defense against air-borne toxins and infectious agents. However, this also means that the elements of the extracellular matrix, e.g., hyaluronan (HA) in the alveolus and its surrounding interstitium may be degraded by air-borne toxins.

An oxygen molecule on its journey from the air interface inside the pulmonary alveolus to the red blood cell in the blood capillary must traverse various barriers, some of which are unique to the lung. First, there is the lining of the alveolus consisting of surfactant on its underlying aqueous subphase, which covers a layer of epithelial cells (Type I and Type II). The epithelial cells rest on an epithelial basement membrane, which is fused to an endothelial basement membrane. On the other side of the endothelial basement membrane is the endothelial cell layer of the capillary wall. The path that oxygen and carbon dioxide must traverse is made short by the fusion of the basement membranes and the fact that the cell layers are stretched thin where gas exchange occurs (2). The total length of the path is about 2 μm (3). The surfactant layer on its aqueous subphase and the fusion of the two basement membranes are peculiar to the lung, and the type II epithelial cells are specialized to produce surfactant. Important to our discussion they also secrete HA and proteoglycans into the aqueous subphase.

Recently (4) a theory was advanced that HA in the aqueous subphase of the alveolus contributes to the structure of the surfactant lining layer by interacting with itself and with proteins to form a hydrophilic gel. At the air interface the components are so dilute that a layer of water is presented upon which surfactant phospholipids spread (Fig. 1). It was also suggested that direct interactions of HA and phospholipids (5) and/or hydrophobic surfactant proteins could contribute

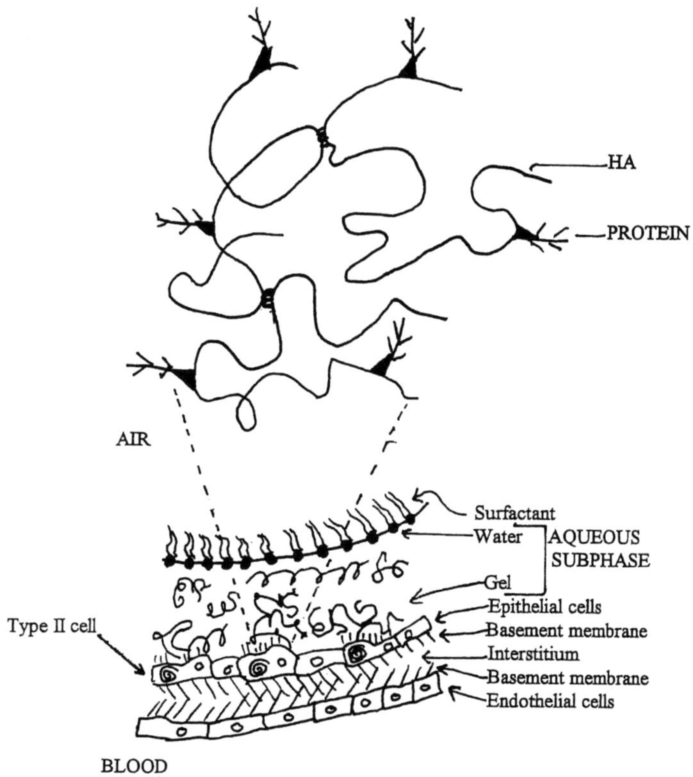

Figure 1 Schematic diagram of the alveolar wall (air–blood barrier). In the subphase of the lining layer of the alveolus while still tethered to the Type II cell, HA binds to proteins and then self-aggregates to form networks. The networks are more concentrated near the cells. Away from the cells, water predominates and the surfactant phospholipids spread on the water. (From Ref. 4 with permission).

to the stability of the surfactant layer. It is now appreciated that HA is an immunomodulator (6) and its location in the alveolus will enable it to quickly activate immune cells when bacteria enter the airway.

The supporting structure of the wall of the alveolus is provided by the fine elastic fibers and small bundles of collagen fibers, which are present in areas where the fused basement membranes separate. The alveoli are suspended in an intricate 3D lacework of connective tissue fibers, which connect the alveoli to the alveolar duct. For the purpose of this chapter this lacework is part of the interstitium. Pathological situations illustrate the importance of these fibers. For example, in emphysema air spaces are enlarged due to the destruction of elastin and collagen in the alveolar wall, whereas in interstitial fibrosis the air spaces become filled with collagen. The cells in this interstitium are largely fibroblasts and alveolar macrophages and these cells have the ability to synthesize and degrade HA. The protection of the elastin and collagen fibers in the interstitium by HA is described in this chapter. Turino and Cantor (7) have reviewed the evidence that HA does exert a protective effect against injury in a number of respiratory diseases and in an animal model of emphysema.

II. Lung Hyaluronan

A. Hyaluronan Content in the Lung

Hyaluronan, a repeating polymer of *N*-acetylglucosamine linked β1,4 to glucuronic acid which is linked β1,3 to the next *N*-acetylglucosamine is ubiquitous in soft tissues. It is increased when more open structures are required for cell migration as in embryogenesis and during repair. In the lung, the HA content correlates with extravascular water (8,9). HA is present in all major structural components of the lung including alveoli, bronchi, pleura and blood vessels and is synthesized by the following lung cells: fibroblasts, alveolar macrophages, mesothelial, endothelial and epithelial cells, including the Type II epithelial cells in the alveolar wall. HA is obvious around larger blood vessels and in alveolar macrophages in experiments using a biotinylated hyaluronan-binding protein (HABP) as a probe (10). There is much less HA in the air exchange tissue—75 ± 19 SEM ng/mg dry weight compared to 435 ± 54 ng/mg in pulmonary arteries and 322 ± 57 ng/mg in pulmonary veins (Bray et al., unpublished). It has been detected in bronchoalveolar lavage (BAL) where its content is 8–9.9 μg/L in rats (9,11) and 9 μg/L in healthy human controls (12). HA in BAL is markedly increased in many human diseases (reviewed in (13)) and in the animal model of fibrosis induced by bleomycin injury (14). It is also increased in the interstitium in bleomycin injury in rats (9,15) and hamsters (13). These experiments of nature and animal models have been instructive in showing that HA is present in the alveolus, where it is near the air/water interface, and in the interstitium. Under normal circumstances the amount in the interstitium is too small to be detected by histological methods (10,15).

The increased synthesis of HA in the lung after bleomycin injury involves lung fibroblasts and alveolar macrophages. The macrophages become activated and release factors which stimulate HA production by lung fibroblasts (16). Impaired removal of HA by alveolar macrophages also contributes to the accumulation of HA after bleomycin injury (16,17). The cell receptor for HA, CD44, is critical for the resolution of the inflammation generated as a result of the bleomycin injury (18).

B. Properties of Hyaluronan Related to Function in the Lung

Although the basic chemistry of HA has been presented in other chapters in this book, it may be useful to emphasize the properties of HA that will affect its function in the alveolus and interstitium and to point out special characteristics of lung HA. The molecular weight of HA varies with the tissue source. This depends not only upon which HA synthases are expressed in the tissue but on other factors as well. The genes encoding HA synthases 2 and 3 are expressed in the lung (19,20), but the product in lung is small, about 220 kDa in BAL (9) and in tissue (Bray, unpublished). HA in normal synovial fluid is 7000 kDa (21) and 2000 kDa in cartilage (22) and varies from 40 to 560 kDa in the vitreous humor (23). The molecular weight of HA will affect the properties that allow it to form hydrophilic gels in the vitreous humor and synovial fluid and, by analogy, in the aqueous subphase of the alveolar lining, and to form protective networks around collagen and elastin fibers. These properties, which will be discussed below are: [1] hydration, [2] viscosity, [3] self-aggregation and [4] binding to many proteins to form networks.

1. In solution, HA is highly hydrated, containing approximately 1000-fold more water than polymer (24). The negatively charged carboxyl groups contribute to the expansion of the coiled molecule and to the volume of water it enfolds and the volume is largely determined by the molecular weight of the HA. Further, Heatley and Scott have suggested that water molecules are hydrogen-bonded to the HA chain with some regularity (25).

2. Solutions of HA are quite viscous and the viscosity is proportional to the molecular weight of HA and its concentration in the solution (23,26). Even small HA segments of 15 and 20 kDa have measurable intrinsic viscosities (27). It is possible to arrive at an estimate of the intrinsic viscosity of lung HA from published data. From the data of Cleland and Wang (23), who obtained fractions of various molecular weights from vitreous humor, their sample of 210 kDa had an intrinsic viscosity of 556 mL/g. Segments of rooster comb HA prepared by sonication and subsequent isolation were studied by Coleman et al. (26). From a plot of their data, lung HA of 220 kDa molecular weight would have an intrinsic viscosity of 600 mL/g, which is in reasonable agreement with the data of Cleland and Wang.

3. Self-aggregation of HA has been demonstrated by rotary shadowing-electron microscopy (28) and by tapping mode atomic force microscopy (29).

Of direct relevance to the aqueous subphase, self-aggregation of HA has been shown by rotary shadowing-electron microscopy to be present in the gel-like matrix of the vitreous humor (30). The theory behind such an aggregation is based on a tape-like, two-fold helix structure demonstrated in NMR studies (31). Earlier it was shown that a molecule of water participates in the secondary structure of HA leading to hydrophilic regions on the molecule and to hydrophobic patches on other areas (25). Self-aggregation is thought to occur at the hydrophobic patches. Antiparallel double-stranded segments are possible with both sides being identical. Thus, each HA molecule could be aligned against another HA molecule, making it an efficient networking molecule. In the study of Cowman et al. (29) in which only high molecular weight HA was used, loops stabilized by antiparallel double-stranded segments were observed. However, the property of self-aggregation is not limited to high molecular weight HA. Scott et al. (28) saw the phenomenon in a preparation of molecular weight as low as 350–400 kDa. Turner et al. (27) found indication of concentration-dependent intermolecular association of short segments of approximately 5–7 kDa and intramolecular association (hairpin formation) by segments of more than 15–20 kDa.

4. By its non-covalent binding to proteins known as hyaladherins, HA becomes a central player in the structure of tissues not only forming networks in the extracellular matrix but connecting these networks to the cell through the cell surface receptor CD44. The hyaladherins and especially CD44 are discussed in other chapters of this book. The following examples of network formation by HA will be informative for the role of HA in the gel of the aqueous subphase and as a protective coating for collagen and elastin fibers.

The best studied example is the formation in cartilage of the large aggregates that give this tissue the required resilience. Molecules of aggrecan, the large cartilage proteoglycan, line up on HA (32) and the binding of each molecule is stabilized by a molecule of link protein which has properties that allow it to bind to both HA (33) and aggrecan (34). It is now known that the A, B and B′ subdomains of aggrecan function together during the binding to HA (35). There is an immediate increase in viscosity upon formation of the aggregate (32). The viscosity is maximum at a ratio of HA to proteoglycan monomer of 1:75 (w/w) (36). The increase in viscosity corresponds to a large increase in the size of the proteoglycan as assessed by gel chromatography. The number of proteoglycans bound and thus the size of the aggregate is proportional to the molecular weight of the HA and the number of aggregates depends upon the concentration of HA in the solution. Thus, the hydrophilic gels that HA forms in the aqueous subphase of the alveolus and in the interstitium are influenced by both the molecular weight of the HA and its concentration. Since the molecular weight of lung HA (220 kDa) is less than that of HA in cartilage (2000 kDa), the size of aggregates formed will be smaller than the ones in cartilage.

Another example of the organization of proteins by HA occurs in the medium of 3T3-L1 preadipocyte cultures (37). During differentiation, the viscosity of the medium increased to twice that of the control medium. *Streptomyces* hyaluronidase, which degrades only HA, abolished the gel-like properties. Since the content of HA in the medium was too low to account for the increased viscosity, the authors suggested there was a highly organized network in the medium.

From the examples cited it is clear that even a small amount of HA can have a large effect in the aqueous subphase of the alveolus and in the extracellular matrix of the interstitium.

C. Interaction of Hyaluronan with Versican

The known interaction of HA with versican, a high molecular weight proteoglycan with chondroitin sulfate side chains, will be relevant in the lung (38,39). Versican is synthesized by human lung fibroblasts, which provided the mRNA for the first cDNA clones of versican (40). The coding sequence was completed using a placental library (41) and it predicted an HA binding region in the amino terminal portion of the molecule. This interaction of versican with HA (38,39) and, additionally, its binding to link protein (39) have now been demonstrated. Versican also binds to fibrillin-1, which is a component of the microfibrillar covering of elastin fibers (42), and this will be discussed in detail further on. Both versican and aggrecan are capable of forming intermolecular disulfide bonds that can provide additional stabilization to the matrix (43).

The importance of both versican and heparan sulfate proteoglycans to the structure of the interstitium was illustrated in an experiment of Passi et al. (44). They induced edema in rabbit lungs by saline infusion or by a bolus of pancreatic elastase and then isolated the proteoglycans from the lungs. Gel filtration experiments showed a large decrease in the versican family of proteoglycans in the saline-infused lungs. In contrast, elastase only partially affected the versican proteoglycans but markedly decreased the heparan sulfate proteoglycans.

III. Hyaluronan Oligosaccharides

A. Examples of Biological Activity

Often when HA is involved in a biological process, rapid changes in its content and in its size occur. Much has been written about the biological effects of low molecular weight fragments of HA when compared to high molecular weight HA and these are a few examples. Forrester and Balazs (45) showed a stimulation of phagocytosis by macrophages with HA of molecular weight 100 kDa or lower, but an inhibition of phagocytosis by HA of molecular weight of 1×10^3–2×10^3 kDa. Fragments of 4–25 disaccharide units (approximately 1.6–10 kDa) induced angiogenesis in a chick chorioallantoic membrane assay whereas intact HA did not (46). These studies on angiogenesis continued and soon it was shown that fragments of 3–10 disaccharides of HA (approximately 1.2–4 kDa) stimulated endothelial cell proliferation (47) and that this occurred by induction of protein tyrosine kinase activity (48). The same size fragments (1.2–3.6 kDa) activate messenger RNAs for collagen synthesis in an X-irradiation model of lung fibrosis (20). Fragments of tetra- and hexasaccharide size (approximately 0.8–1.2 kDa), but not intermediate size nor high molecular weight HA, induced immunophenotypic maturation of human monocyte-derived dendritic cells (49). HA oligomers consisting mainly of 5–6 disaccharide units, around 2 kDa, inhibited B16F10 melanoma growth (50). HA fragments less than 500 kDa stimulated inflammatory genes which direct the production of cytokines (51–53) whereas polymers of 3×10^3–6×10^3 kDa did not. Fragments of HA are produced by the action of hyaluronidases and by free radical reactions and these mechanisms will now be discussed.

B. Production by Lung Hyaluronidases

Hyaluronan is the natural substrate of hyaluronidases, a family of enzymes (54,55) which are discussed in a separate chapter of this book. At least two hyaluronidases are expressed in the lung. They are HYAL1 (56) and HYAL2 (57), which was first described as a product of lung fibroblasts (58) and is expressed in many tissues. Both these enzymes are lysosomal enzymes and they degrade HA differently than testicular hyaluronidase and provide a large piece of approximately 20 kDa. The genes HYAL1, -2 and -3 are located on the chromosomal region 3p21. This region is deleted in many small cell lung cancer lines. In fact, these genes were known as LuCa-1, -2 and -3 before it was realized that hyaluronidases could result from their expression (57). Another peculiarity concerning the lung relates to HYAL2. Rai et al. (59) expressed HYAL2 in NIH 3T3 cells and could not detect hyaluronidase activity, whereas a construct of HYAL1 in the same cell system did produce hyaluronidase. Furthermore, using other cell systems they describe HYAL2 as a glycosylphosphatidylinositol (GPI)-anchored cell-surface receptor for Jaagsiekte sheep retrovirus. This retrovirus causes a contagious lung cancer in sheep. These data suggest a broader role for HYAL2 in biological systems. Perhaps, there are isoforms of HYAL2 which are expressed only in response to differentiation signals or after injury and whose products have activities under physiological conditions which are unrelated to its function in the lysosome.

The content of both HA and hyaluronidase in the chick embryo lung change with development (60), and in humans HA is higher in fetal lung than it is after birth or in adult lung (61). Hyaluronidase activity also changes, showing a rapid increase in rat lung immediately after birth (62). Presumably, the removal of HA is necessary to lower the water content of the lung, with which HA has a direct correlation (8,9), in order to facilitate breathing in air.

Under normal conditions hyaluronidases proceed to depolymerize HA in predictable ways and they are tightly regulated. We have shown this to be true for hamster lungs by measuring both HA and hyaluronidase in individual, normal lungs. A plot of total hyaluronidase against total HA for the individual animals gave points that clustered around a narrow range (Bray and Turino, unpublished). Interestingly, not only HA but hyaluronidase also is increased in the lung after bleomycin injury (13) and after oxygen injury (62) but the concept of a well-regulated process does not apply. One factor that leads to this difference is random degradation of HA to lower molecular weight fragments by reactive oxygen (63) and nitrogen species (64) that have been generated as a result of the injury.

C. Production by Free Radicals

It has been known for many decades that free radicals generated in a variety of chemical systems lowered the viscosity of HA solutions (65–67). Free radicals are also generated by cellular processes such as defense mechanisms of leukocytes (68). The burst of oxidative metabolism which occurs when polymorphonuclear leukocytes (PMNs) are stimulated generates superoxide, a highly reactive free radical (68–70) and secondarily, singlet oxygen and hydroxyl radical are formed. Peroxynitrite can be formed under physiological conditions from nitric oxide, which is synthesized by both neutrophils (71) and macrophages (72), and superoxide (73). Cleavage of HA is also brought about by the OCl^- generated by the myeloperoxidase system (74). McNeil et al. (63) showed that in three systems—autooxidation of ferrous ions, the action of xanthine oxidase on hypoxanthine and stimulated PMNs—the products generated were polydisperse. The major fraction was reduced to a size of 10^4 Da and was not reduced further by exposure to a second oxy radical flux, suggesting that the process of depolymerization was orderly. It is clear that even smaller fragments were created because hexuronic acid was lost when the reaction products from similar experiments were dialyzed (75). These small fragments can initiate the kinds of biological processes mentioned above.

Workers have come to the consensus that in the depolymerization of HA by both oxygen-derived (75) and peroxynitrite-derived (64) free radicals the hydroxyl radical ($OH^{\cdot}$) is the active species. Al-Assaf et al. assert that additional chain breaks are produced by ONOOH and that this may be more effective *in vivo*. They also point out the reaction of peroxynitrite anion with carbon dioxide and this may lead to the production of carbonate radicals. At pH values of 7 or greater and at high carbon dioxide concentrations this reaction would be the dominant one. This is certainly a possibility in the pulmonary alveolus where the carbon dioxide concentration is high and the pH of deaerated, exhaled airway vapor condensate is pH 7.65 (76). McKee et al. (77) showed that fragmentation of HA induces nitric oxide synthase in murine macrophages, which could lead to production of peroxynitrite and therefore, by fragmentation of more HA molecules, to an ongoing inflammatory state. Because of its extracellular location, HA in tissues does not have the protection afforded by the cytoplasmic superoxide reductase, catalase or the glutathione–glutathione reductase systems and HA in the alveolus and interstitium of the lung is especially vulnerable because of direct exposure to air-borne free radicals.

IV. Hyaluronan in the Alveolus

A. Structure of the Alveolus

Inside the alveolus at the air/water interface there are two layers, an aqueous subphase which is overlaid with surfactant (78–80). Irregularities in the epithelial cell layer are smoothed over by the aqueous subphase and the surfactant layer covers the subphase (Fig. 1). By low-temperature scanning electron microscopy, Bastacky et al. (81) established the continuity of the aqueous subphase and its independence from the cell layer. The subphase is very thin (0.14 μm) over flat alveolar walls, but much thicker at alveolar wall junctions (0.89 μm) with an average thickness of 0.2 μm. At the air/water interface surface tension would be high due to the attraction of the water molecules for each other and the alveoli would collapse during expiration were it not for the overlay of surfactant. Surfactant lowers the surface tension, thereby stabilizing the structure of the alveolus. Indeed, if surfactant is deactivated by endotracheal lavage with Tween 20 the alveoli do collapse (82).

B. Surfactant

Surfactant is produced by Type II cells in the wall of the alveolus (83,84), as is HA (85,86). BAL has provided surfactant for biochemical analyses and these reveal it is a complex mixture whose biochemical composition has been reviewed recently (87). The composition is the same for many species and all the components are required for optimal activity. Surfactant is 90% lipid and 10% protein. Phosphatidylcholines, primarily dipalmitoylphosphatidyl choline (DPPC), are the predominant lipids. Lipids spread on water by inserting their hydrophilic heads into the water. Further along the argument will be made that HA, which can bind up to 1000-fold its weight in water, may be helpful in providing this aqueous layer in the alveolus.

There are four known proteins in surfactant. These are surfactant protein A (SP-A), surfactant protein B (SP-B), surfactant protein C (SP-C) and surfactant protein D (SP-D). Of the four, SP-A and SP-D are hydrophilic and SP-B and SP-C are hydrophobic. SP-A and SP-D are lectins of the subgroup called ‘collectins’ and they have both a carbohydrate recognition domain and an amino-terminal collagen-like domain. Both of the domains of SP-D are involved with its interaction with decorin (88), a proteoglycan that decorates collagen fibers. By use of the carbohydrate recognition domains, surfactant proteins A and D are able to bind glycoconjugates on microorganisms and inhibit infection. They enhance phagocytosis by alveolar macrophages and neutrophils and also have direct bactericidal effects (89).

In the type II cell surfactant is packed into lamellar bodies which are secreted into the aqueous subphase. There the lamellar bodies are changed into a structure called tubular myelin, which is thought to be the immediate donor of fresh surfactant to the surfactant layer (90,91). The transit through the aqueous subphase must occur in a matter of seconds and the mechanism of this rapid transfer is not understood, but is a matter of intense research.

C. Hyaluronan and Proteoglycans in the Aqueous Subphase

Hyaluronan is secreted into the aqueous subphase by the type II cell, which synthesizes both HA (85,86) and surfactant (83,84) as well as proteoglycans (86,92). The cell-associated proteoglycans are enriched for heparan sulfate whereas those secreted into the medium contain chondroitin sulfate and are of high molecular weight (>200 kDa) and are probably versican. Thus HA and sulfated proteoglycans are in the aqueous subphase and can easily interact to form an aggregate gel. Near the cell surface the aggregates will be most concentrated and some will still be tethered to the cell. Near the surfactant layer there will be fewer aggregates and the surface of the aqueous subphase will essentially be water and that is what is required for surfactant spreading. Fig. 1 shows these interactions in the wall of the alveolus.

D. Interaction of Hyaluronan with Phospholipids

Hyaluronan and phospholipids interact in an energetically favorable interaction and the type of lipid structures formed is influenced by the molecular weight of HA (5). Even HA of 170 kDa, closer to the molecular weight of lung HA (220 kDa), organized 15–30 nm wide unilamellar vesicles into large aggregates. The aggregates were never observed in the absence of HA. In Fig. 2 is shown a model of how the water molecules bound to HA could interact directly with surfactant phospholipids in the lining of the alveolus.

Thus HA may have two structural functions in the lining of the alveolus. It may hold in place the layer of water upon which surfactant spreads and might even stabilize the surfactant layer by a direct interaction with the phospholipids or with the hydrophobic surfactant proteins B and C. Recent data from our laboratory demonstrate that HA interacts with an unidentified component(s) of surfactant. Half of lung HA solubilized by Pronase digestion in dilute Tris buffer did not inhibit binding of a cartilage HABP to HA-coated wells (93). This could be largely corrected by prior extraction of the lungs with acetone. A lung HA fraction whose HA content was determined by the radioactive cytochrome c method (94) was compared with an equal amount of umbilical cord HA in the standard inhibition assay using cartilage HABP. The lung HA reached a plateau at 66% inhibition compared to 88% inhibition by the umbilical cord sample. These data suggested that a portion of lung HA was already bound to an endogenous HABP and that the HABP was acetone soluble. An acetone extract

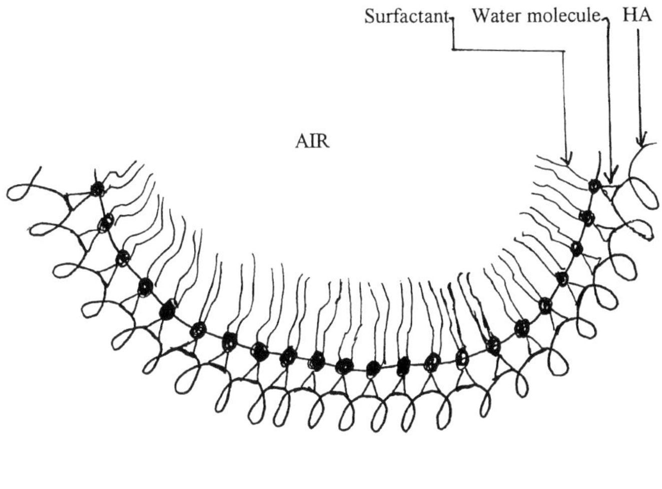

Figure 2 Could the water which is hydrogen-bonded to HA stabilize the surfactant layer? (From Ref. 4 with permission).

of hamster lungs was evaporated to dryness to remove the acetone and was dissolved in water. This preparation stabilized bubbles, a characteristic of surfactant, and it contained a 66 kDa protein. Pre-incubation of umbilical cord HA with this extract markedly affected the inhibition curve with cartilage HABP, suggesting a strong interaction of HA with a surfactant component(s).

E. Hyaluronan as an Immunomodulator

A third role of HA in the alveolus relates to the function of HA as an immunomodulator (6). HA oligosaccharides, but not high molecular weight HA, directly activate dendritic cells (49) through the Toll-like receptor 4 (TLR4) complex (95). Mummert et al. (96) have shown that dendritic cells express the three known HA synthase genes HAS1, -2 and -3 and four hyaluronidase genes HYAL1-4 and T cells constitutively express HAS1 and -3 and the hyaluronidase HYAL3. Termeer et al. (6) suggest this raises the possibility that T cells might be able to regulate their own activation in an autocrine manner. In the example of macrophages, it has already been suggested that HA fragments generated in inflammation may induce peroxynitrite production, which would generate more fragments and create an ongoing inflammatory state (77). Thus the functions of dendritic cells, T cells and macrophages are affected by HA fragments. This role of HA has been brought to a practical use. HA administered subcutaneously to patients with chronic bronchitis reduced the number of infectious exacerbations of disease compared to placebo-treated patients (97).

In summary, HA may have three functions in the alveolus. [1] By forming a gel in the subphase it may hold in place the layer of water upon which surfactant phospholipids can spread. [2] It might stabilize the surfactant layer by a direct interaction with the phospholipids and with hydrophobic surfactant proteins. [3] Fragments of HA will activate immune cells in the alveolus.

V. Hyaluronan in the Pulmonary Interstitium

As in all other tissues HA in the interstitium provides an open, hydrated gel when it is needed for cells to migrate during embryogenesis and repair after injury. Again, pathological conditions suggest there may be yet another role in normal tissue which is to participate in a protective coating for collagen and elastin fibers.

A. Both Collagen and Elastin are Affected in Pulmonary Emphysema

In emphysema, major disruptions of the extracellular matrix occurs with a resulting enlargement of the air spaces. For many years, the focus has been on elastin since the most prominent histological change is a loss of elastin fibers particularly at junction points in the alveolar walls (98). Chemically, there are changes in all components of the connective tissue including elastin (99) and the total elastin content of emphysematous lungs is less than for normal lungs (100). It has been assumed that an imbalance between proteolytic enzymes, especially elastase, and proteinase inhibitors such as α-1-antitrypsin results in the loss of elastin fibers (101,102). This assumption is strongly supported by five facts: [1] Individuals homozygous for α-1-antitrypsin deficiency often go on to develop emphysema (103). [2] Cigarette smoking, which alters α-1-antitrypsin chemically (104) and which leads to an increase in neutrophil elastase-derived fibrinopeptides in the serum of smokers (105), is a risk factor for the disease. [3] α-1-antitrypsin ameliorates cigarette smoke-induced emphysema in mice (106). [4] Metalloelastase (MMP-12) knockout mice do not develop emphysematous changes in the lung as do wild-type animals upon exposure to cigarette smoke (107). [5] Transgenic mice with the IL-13 gene (108) or the Ifn-gamma gene (109) induce activation of matrix metalloproteinases (MMPs), among them elastase (MMP-12), and cause emphysema in the adult transgenic murine lung.

Recently, it is becoming apparent that collagen is also affected. Mice expressing a human collagenase transgene in their lungs developed changes similar to those observed in human emphysema (110). The studies in transgenic mice have progressed to the point that the target of MMP-1 has been shown to be Type III collagen, which is the predominant type of collagen in the alveolar walls of these mice (111,112). Further, interstitial collagenase (MMP-1) RNA, protein and enzymatic activity were present in lung parenchyma of patients with emphysema and not in the lungs of normal, control subjects (113). Metalloelastase expression was absent in the lungs of the patients.

The fact that both elastase and collagenase can produce pathology resembling emphysema in the lung makes it necessary to consider that both enzymes, perhaps at different time points, are important in the etiology of emphysema. Therefore, one must look for an antecedent event that affects both elastin and collagen. One explanation would be that a protective coating around the fibers was removed making them more susceptible to degradation by elastases and collagenases.

B. Hyaluronan in Emphysema

Evidence relating HA to emphysema is accumulating. Konno et al. (114) found HA to be decreased in emphysematous lungs. Data from animal models show that HA is involved in cigarette smoke-induced emphysema and in elastase-induced emphysema. Guinea pigs exposed to tobacco smoke have reduced levels of lung HA (115). Cantor et al. (116) showed that instillment of hyaluronidase and the addition of 60% oxygen, which is a non-toxic concentration of oxygen, produced air-space enlargement. They also showed that prior hyaluronidase treatment increased the effect of elastase instillment (117), an observation that was confirmed by Murakami et al. (118). Further, HA protected elastin fibers *in vitro* and lead to a decrease in air-space enlargement caused by elastase instillment *in vivo* (119).

C. Proteoglycans in Emphysema

There is already strong evidence that proteoglycans, especially heparan sulfate proteoglycans, are affected in emphysema. In the animal model of emphysema, intratracheal instillation of pancreatic elastase, van de Lest et al. (120) showed a decrease in heparan sulfate content in the alveoli of rat lungs. Concurrently, there was an increase of heparan sulfate and dermatan sulfate in the urine. This urinary increase correlated positively with the extent of emphysema that developed after 40 days. Subsequently, van Straaten et al. (121) showed that heparan sulfate proteoglycan staining was diminished in the respiratory air-space walls of patients with emphysema and they suggested that the loss of interstitial proteoglycans might be crucial for elastic recoil loss and subsequent bronchiolar obstruction seen in the disease.

D. Protective Coating of Elastin Fibers

Recent data make it possible to picture HA and proteoglycans as part of the covering of elastic fibers. It has been known for a long time that amorphous elastin is covered with microfibrils. In early embryonic development, which can be as early as 23 h in the chick, elastin colocalized with fibrillin-1 and with fibulin-1 (122). Fibrillin-2 was also expressed at that stage and all three are known to be associated with cross-linked elastin. The elastic fiber assembly process involves association with fibrillin and the hydrophobic sequence in exon 30 is a major element (123). Versican variant V3, when over expressed in smooth muscle cells, increases expression of tropoelastin and the formation of elastic fibers (124) By electron microscopic immunolocalization of antiversican antibodies the C-terminal region of the proteoglycan versican was shown to bind to fibrillin microfibrils (42). The localization appeared to be on or near the beads on the microfibrils. Other data suggested that the bond was covalent. The direct interactions of versican with HA (38,39) and link protein (39) and the fact that versican has been shown to form intermolecular disulfide bonds (43) have already been mentioned. A model for a protective coating of elastin fibers consisting of a proteoglycan and its bound HA, both of which have been detected within normal elastic fibers of human dermis (125), can now be outlined (Fig. 3). A network of fibrillin surrounds the elastin fiber. Versican binds to the fibrillin through its C-terminal region and to HA through its amino terminal region. HA can align itself to another HA molecule which can bind proteoglycans, possibly heparan sulfate proteoglycans (not shown), and this can lead to the formation of larger networks. This presents HA in a strategic structural position.

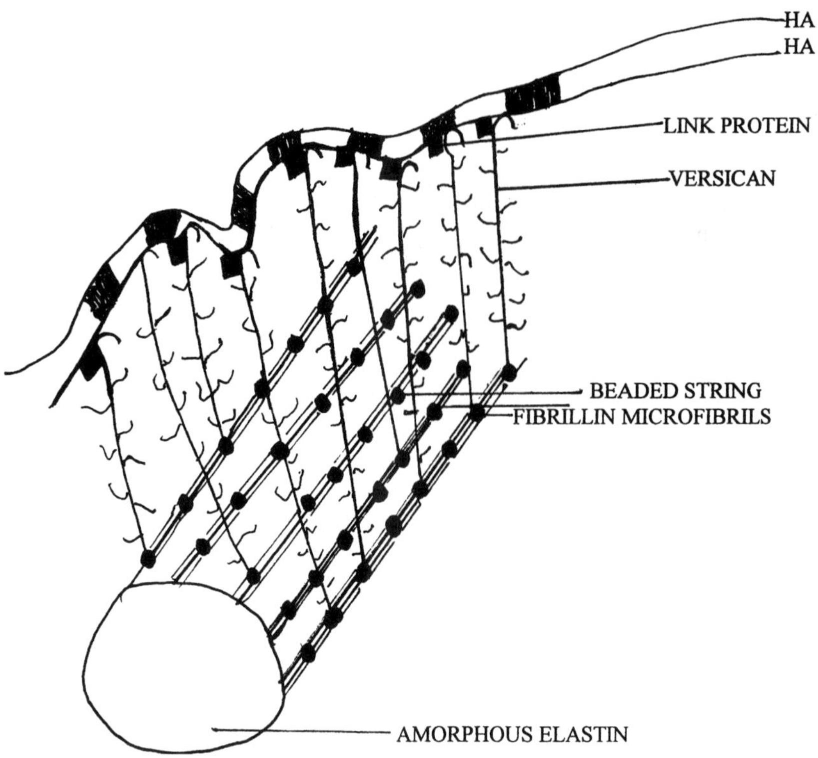

Figure 3 Proposed model for HA association with an elastin fiber. Versican binds to the beaded string fibrillin microfibrils and to HA. A second molecule of HA aligns itself onto the HA. Visualize the network as surrounding the fiber on all sides and forming networks with other components, e.g., heparan sulfate proteoglycans, that can bind to the second molecule of HA.

E. Protective Coating of Collagen Fibers

Collagens in the lung interstitium and in alveolar basement membrane bind proteoglycans (44,126,127) and the proteoglycans also bind HA (44). Type VI collagen, which is widespread in connective tissues, interacts with HA (128) in what appears to be a structural relationship (129). In the rabbit synovium, HA has a major organizational role within the collagen bundle, where it affects the spacing of the fibrils within the bundle (130). HA has also been seen associated with collagen fibers in the cornea and in the vitreous (131) where it was proposed that HA was an ambidexteran and was binding proteoglycan aggregates from both sides of the molecule. Therefore, the same kind of model as described above would apply without the need of a fibrillin-like molecule. The model places HA in a pivotal role in the structure of a protective coating for the fibers and it also shows how added HA could anneal breaks in the chain through the ability of two molecules of HA to align against each other.

F. Hyaluronidase in Emphysema

There are many ways in which HA can be degraded—by free radicals and hyaluronidases (see earlier) and cigarette smoke degrades HA by a free radical mechanism (132). An interesting question to ask is whether or not the presence of too much hyaluronidase can be a cause of emphysema. Pecora et al. (133) found that hyaluronidase in emphysematous lungs without inflammation was more than 1.5 times greater than that of the control group. The method they used for hyaluronidase measurement, production of reducing *N*-acetylglucosamine groups during digestion of HA, is valid and in use today. Study of the expression of HYAL1 and -2 in emphysematous lungs and the use of transgenic mice with increased expression of HYAL1 and -2 could answer the question as to whether hyaluronidases are also involved in the etiology of emphysema. We have shown that extracts of hamster lung degrade high molecular weight HA to approximately 27 kDa and that the activity is partially inhibited by chondroitin sulfate (Bray and Turino, unpublished). Rao et al. (134) showed that Arteparon, a supersulfated derivative of chondroitin sulfate prevented acute injury and emphysema in hamsters when administered before instillment of human leukocyte elastase. They also mentioned that heparan sulfate protected lungs from acute injury caused by human leukocyte elastase. Although the mechanism there was thought to be inhibition of elastase, these sulfated glycosaminoglycans would also inhibit any excess hyaluronidases generated by the injury.

VI. Conclusion

In conclusion, HA has many potential functions in the alveolus and interstitium in addition to the ones already proven. Various studies designed to clarify these roles are already underway in many laboratories. As mentioned previously, the tools of molecular biology will enable the investigators to ask specific questions about HA synthases and hyaluronidases. A method is now available to provide HA oligosaccharides of an exact size with no contamination with other sizes (135) and the use of these oligosaccharides will give definitive answers concerning the biological effects of HA fragments.

Acknowledgements

The author gratefully acknowledges support by funds from the James P. Mara Center for Lung Disease, the Charles A. Mastronardi Foundation, the Ned Doyle Foundation, the Franklyn Bracken Fund and the Alpha One Foundation.

References

1. Murray JF. The Normal Lung: The Basis for Diagnosis and Treatment of Pulmonary Disease. Philadelphia: W.B. Saunders, 1976; 46–48.
2. Weibel ER. Morphological basis of alveolar-capillary gas exchange. Physiol Rev 1973; 53(2):419–495.
3. Purves WK, Orians GH, Heller HC. Life, the Science of Biology, 4th ed. Sunderland: Sinauer Associates, 1995; 944.
4. Bray BA. The role of hyaluronan in the pulmonary alveolus. J Theor Biol 2001; 210:121–130.
5. Pasquali-Ronchetti I, Quaglino D, Mori G, Bacchelli B, Ghosh P. Hyaluronan–phospholipid interactions. J Struct Biol 1997; 120:1–10.
6. Termeer C, Sleeman JP, Simon JC. Hyaluronan—magic glue for the regulation of the immune response? Trends Immunol 2003; 24:112–114.
7. Turino GM, Cantor JO. Hyaluronan in respiratory injury and repair. Am J Respir Crit Care Med 2003; 167:1169–1175.
8. Bhattacharya J, Cruz T, Bhattacharya S, Bray BA. Hyaluronan affects extravascular water in lungs of unanesthetized rabbits. J Appl Physiol 1989; 66: 2595–2599.
9. Nettelbladt O, Tengblad A, Hällgren R. Lung accumulation of hyaluronan parallels pulmonary edma in experimental alveolitis. Am J Physiol 1989; 257: L379–L384.
10. Green SJ, Tarone G, Underhill CB. Distribution of hyaluronate and hyaluronate receptors in the adult lung. J Cell Sci 1988; 89:145–156.
11. Hernnäs J, Nettelbladt O, Bjermer L, Särnstrand B, Malmström A, Hällgren R. Alveolar accumulation of fibronectin and hyaluronan precedes bleomycin-induced pulmonary fibrosis in the rat. Eur Respir J 1992; 5:404–410.
12. Bjermer L, Lundgren R, Hällgren R. Hyaluronan and type III procollagen peptide concentrations in bronchoalveolar lavage fluid in idiopathic pulmonary fibrosis. Thorax 1989; 44:126–131.
13. Bray BA, Sampson PM, Osman M, Giandomenico A, Turino GM. Early changes in lung tissue hyaluronan (hyaluronic acid) and hyaluronidase in bleomycin-induced alveolitis in hamsters. Am Rev Respir Dis 1991; 143:284–288.
14. Nettelbladt O, Hällgren R. Hyaluronan (hyaluronic acid) in bronchoalveolar lavage fluid during the development of bleomycin-induced alveolitis in the rat. Am Rev Respir Dis 1989; 140:1028–1032.

15. Nettelbladt O, Bergh J, Schenholm M, Tengblad A, Hällgren R. Accumulation of hyaluronic acid in the alveolar interstitial tissue in bleomycin-induced alveolitis. Am Rev Respir Dis 1989; 139:759–762.
16. Teder P, Nettelbladt O, Heldin P. Characterization of the mechanism involved in bleomycin-induced increased hyaluronan production in rat lung. Am J Respir Cell Mol Biol 1995; 12:181–189.
17. Teder P, Heldin P. Mechanism of impaired local hyaluronan turnover in bleomycin-induced lung injury in rat. Am J Respir Cell Mol Biol 1997; 17: 376–385.
18. Teder P, Vandivier RW, Jiang D, Liang J, Cohn L, Pure E, Henson PM, Noble PW. Resolution of lung inflammation by CD44. Science 2002; 296:155–158.
19. Jacobson A, Brinck J, Briskin MJ, Spicer AP, Heldin P. Expression of human hyaluronan synthases in response to external stimuli. Biochem J 2000; 348:29–35.
20. Li Y, Rahmanian M, Widstrom C, Lepperdinger G, Frost GI, Heldin P. Irradiation-induced expression of hyaluronan (HA) synthase 2 and hyaluronidase 2 genes in rat lung tissue accompanies active turnover of HA and induction of types I and III collagen gene expression. Am J Respir Cell Mol Biol 2000; 23(3):411–418.
21. Fraser JRE, Laurent TC, Laurent UBG. Hyaluronan: its nature, distribution, functions and turnover. J Intern Med 1997; 242:27–33.
22. Holmes MWA, Bayliss MT, Muir H. Hyaluronic acid in human articular cartilage. Age-related changes in content and size. Biochem J 1988; 250:435–441.
23. Cleland RL, Wang JL. Ionic polysaccharides. III. Dilute solution properties of hyaluronic acid fractions. Biopolymers 1970; 9:799–810.
24. Laurent TC. Structure of hyaluronic acid. In: Balazs EA, ed. Chemistry and Molecular Biology of the Intercellular Matrix. London: Academic Press, 1970: 703–732.
25. Heatley F, Scott JE. A water molecule participates in the secondary structure of hyaluronan. Biochem J 1988; 254:489–493.
26. Coleman PJ, Scott D, Mason RM, Levick JR. Role of hyaluronan chain length in buffering interstitial flow across synovium in rabbits. J Physiol 2000; 526: 425–434.
27. Turner RE, Lin P, Cowman MK. Self-association of hyaluronate segments in aqueous NaCl solution. Arch Biochem Biophys 1988; 265(2):484–495.
28. Scott JE, Cummings C, Brass A, Chen Y. Secondary and tertiary structures of hyaluronan in aqueous solution, investigated by rotary shadowing-electron microscopy and computer simulation. Biochem J 1991; 274:699–705.
29. Cowman MK, Li M, Balazs EA. Tapping mode atomic force microscopy of hyaluronan: extended and intramolecularly interacting chains. Biophys J 1998; 75: 2030–2037.
30. Brewton RG, Mayne R. Mammalian vitreous humor contains networks of hyaluronan molecules: electron microscopic analysis using the hyaluronan-binding region (G1) of aggrecan and link protein. Exp Cell Res 1992; 198:237–249.
31. Scott JE, Heatley F. Hyaluronan forms specific stable tertiary structures in aqueous solution: a 13C NMR study. Proc Natl Acad Sci USA 1999; 96:4850–4855.
32. Hardingham TE, Muir H. The specific interaction of hyaluronic acid with cartilage proteoglycans. Biochim Biophys Acta 1972; 279:401–405.
33. Heinegard D, Hascall VC. Aggregation of cartilage proteoglycans. III. Characteristics of the proteins isolated from trypsin digests of aggregates. J Biol Chem 1974; 249:4250–4256.

34. Caterson B, Baker J. The interaction of link proteins with proteoglycan monomers in the absence of hyaluronic acid. Biochem Biophys Res Commun 1978; 80: 496–503.
35. Watanabe H, Cheung SC, Itano N, Kimata K, Yamada Y. Identification of hyaluronan-binding domains of aggrecan. J Biol Chem 1997; 272:28057–28065.
36. Hardingham TE, Muir H. Binding of oligosaccharides of hyaluronic acid to proteoglycans. Biochem J 1973; 135:905–908.
37. Calvo JC, Gandjbakhche AH, Nossal R, Hascall VC, Yanagishita M. Rheological effects of the presence of hyaluronic acid in the extracellular media of differentiated 3T3-L1 preadipocyte cultures. Arch Biochem Biophys 1993; 302:468–475.
38. LeBaron RG, Zimmermann DR, Ruoslahti E. Hyaluronate binding properties of versican. J Biol Chem 1992; 267:10003–10010.
39. Matsumoto K, Shionyu M, Go M, Shimizu K, Shinomura T, Kimata K, Watanabe H. Distinct interaction of versican/PG-M with hyaluronan and link protein. J Biol Chem 2003; 278:41205–41212.
40. Krusius T, Gehlsen KR, Ruoslahti E. A fibroblast chondroitin sulfate proteoglycan core protein contains lectin-like and growth factor-like sequences. J Biol Chem 1987; 262:13120–13125.
41. Zimmermann DR, Ruoslahti E. Multiple domains of the large fibroblast proteoglycan, versican. EMBO J 1989:2975–2981.
42. Isogai Z, Aspberg A, Keene DR, Ono RN, Reinhardt DP, Sakai LY. Versican interacts with fibrillin-1 and links extracellular microfibrils to other connective tissue networks. J Biol Chem 2002; 277:4565–4572.
43. Chen L, Yang BL, Wu Y, Yee A, Yang BB. G3 domains of aggrecan and PG-M/versican form intermolecular disulfide bonds that stabilize cell–matrix interaction. Biochemistry 2003; 42:8332–8341.
44. Passi A, Negrini D, Albertini R, DeLuca G, Miserocchi G. Involvement of lung interstitial proteoglycans in development of hydraulic- and elastase-induced edema. Am J Physiol 1998; 275:L631–L635 (Lung Cell Mol Physiol 19).
45. Forrester JV, Balazs EA. Inhibition of phagocytosis by high molecular weight hyaluronate. Immunology 1980; 40:435–446.
46. West DC, Hampson IN, Arnold F, Kumar S. Angiogenesis induced by degradation products of hyaluronic acid. Science 1985; 228:1324–1326.
47. West DC, Kumar S. The effect of hyaluronate and its oligosaccharides on endothelial cell proliferation and monolayer integrity. Exp Cell Res 1989; 183: 179–196.
48. Slevin M, Krupinski J, Kumar S, Gaffney J. Angiogenic oligosaccharides of hyaluronan induce protein tyrosine kinase activity in endothelial cells and activate a cytoplasmic signal transduction pathway resulting in proliferation. Lab Investig 1998; 78:987–1003.
49. Termeer CC, Hennies J, Voith U, Ahrens T, Weiss JM, Prehm P, Simon JC. Oligosaccharides of hyaluronan are potent activators of dendritic cells. J Immunol 2000; 165:1863–1870.
50. Zeng C, Toole BP, Kinney SD, Kuo J-W, Stamenkovic I. Inhibition of tumor growth *in vivo* by hyaluronan oligomers. Int J Cancer 1998; 77:396–401.
51. McKee CM, Penno MB, Cowman M, Burdick MD, Strieter RM, Bao C, Noble PW. Hyaluronan (HA) fragments induce chemokine gene expression in alveolar macrophages. The role of HA size and CD44. J Clin Investig 1996; 98: 2403–2413.

52. Noble PW, McKee CM, Cowman M, Shin HS. Hyaluronan fragments activate an NF-kB/I-kBalpha autoregulatory loop in murine macrophages. J Exp Med 1996; 183:2373–2378.
53. Horton MR, Burdick MD, Strieter RM, Bao C, Noble PW. Regulation of hyaluronan-induced chemokine gene expression by IL-10 and IFN-gamma in mouse macrophages. J Immunol 1998; 160:3023–3030.
54. Kreil G. Hyaluronidases—a group of neglected enzymes. Protein Sci 1995; 4: 1666–1669.
55. Frost GI, Csoka T, Stern R. The hyaluronidases: a chemical, biological, and clinical overview. Trends Glycosci Glycotechnol 1996; 8:419–434.
56. Frost GI, Csoka TB, Wong T, Stern R. Purification, cloning, and expression of human plasma hyaluronidase. Biochem Biophys Res Commun 1997; 236: 10–15.
57. Lepperdinger G, Strobl B, Kreil G. HYAL2, a human gene expressed in many cells, encodes a lysosomal hyaluronidase with a novel type of specificity. J Biol Chem 1998; 273:22466–22470.
58. Sampson PM, Rochester CL, Freundlich B, Elias JA. Cytokine regulation of human lung fibroblast hyaluronan (hyaluronic acid) production. Evidence for cytokine-regulated hyaluronan (hyaluronic acid) degradation and human lung fibroblast-derived hyaluronidase. J Clin Investig 1992; 90:1492–1503.
59. Rai SK, Duh F-M, Vigdorovich V, Danilkovitch-Miagkova A, Lerman MI, Miller AD. Candidate tumor suppressor HYAL2 is a glycosylphosphatidylinositol (GPI)-anchored cell-surface receptor for jaagsiekte sheep retrovirus, the envelope protein of which mediates oncogenic transformation. Proc Natl Acad Sci USA 2001; 98: 4443–4448.
60. Stabellini G, Calvitti M, Baroni T, Marinucci L, Calastrini C, Carinci P, Becchetti E. Glycosidases during chick embryo lung development and their colocalization with proteoglycans and growth factors. Eur J Histochem 2002; 46:41–52.
61. Schmid K, Grundboeck-jusco J, Kimura A, Tschopp FA, Zollinger R, Binette JP, Lewis W, Hayashi S. The distribution of the glycosaminoglycans in the anatomic components of the lung and the changes in concentration of these macromolecules during development and aging. Biochim Biophys Acta 1982; 716:178–187.
62. Thet LA, Howell AC, Han G. Changes in lung hyaluronidase activity associated with lung growth, injury and repair. Biochem Biophys Res Commun 1983; 117: 71–77.
63. McNeil JD, Wiebkin OW, Betts WH, Cleland LG. Depolymerisation products of hyaluronic acid after exposure to oxygen-derived free radicals. Ann Rheum Dis 1985; 44:780–789.
64. Al-Assaf S, Navaratnam S, Parsons BJ, Phillips GO. Chain scission of hyaluronan by peroxynitrite. Arch Biochem Biophys 2003; 411:73–82.
65. Matsumura G, Herp A, Pigman W. Depolymerization of hyaluronic acid by autoxidants and radiations. Radiation Res 1966; 28:735–752.
66. Moore JS, Phillips GO, Davies JV. Reactions of connective tissue and related polyanions with hydrated electrons and hydroxyl radicals. Carbohydr Res 1970; 12:253–260.
67. Harris MJ, Herp A, Pigman W. Depolymerization of polysaccharides through the generation of free radicals at a platinum surface: a novel procedure for the controlled production of free-radical oxidations. Arch Biochem Biophys 1971; 142:615–622.

68. Babior BM, Kipnes RS, Curnutte JT. Biological defense mechanisms: the production by leukocytes of superoxide, a potential bactericidal agent. J Clin Investig 1973; 52:741–744.
69. Goldstein IM, Roos D, Kaplan HB, Weissmann G. Complement and immunoglobulins stimulate superoxide production by human leukocytes independently of phagocytosis. J Clin Investig 1975; 56:1155–1163.
70. Johnston RB, Lehmeyer JE, Guthrie LA. Generation of superoxide anion and chemiluminescence by human monocytes during phagocytosis and on contact with surface-bound immunoglobulin G. J Exp Med 1976; 143: 1551–1556.
71. McCall TB, Broughton-Smith NK, Palmer RMJ, Whittle BJR, Moncada S. Synthesis of nitric oxide from L-arginine by neutrophils: release and interaction with superoxide anion. Biochem J 1989; 261:293–296.
72. Marletta MA, Yoon PS, Iyengar R, Leaf CD, Wishnok JS. Macrophage oxidation of L-arginine to nitrite and nitrate: nitric oxide is an intermediate. Biochemistry 1988; 27:8706–8711.
73. Beckman JS, Beckman TW, Chen J, Marshall PA, Freeman BA. Apparent hydroxyl radical production by peroxynitrite: implications for endothelial injury from nitric oxide and superoxide. Proc Natl Acad Sci USA 1990; 87:1620–1624.
74. Green SP, Baker MS, Lowther DA. Depolymerization of synovial fluid hyaluronic acid (HA) by the complete myeloperoxidase (MPO) system may involve the formation of a HA–MPO ionic complex. J Rheum 1990; 17:1670–1675.
75. Greenwald RA, Moy WW. Effect of oxygen-derived free radicals on hyaluronic acid. Arthritis Rheum 1980; 23:455–463.
76. Hunt JF, Fang K, Malik R, Snyder A, Malhotra N, Platts-Mills TAE, Gaston B. Endogenous airway acidification: implications for asthma pathophysiology. Am J Respir Crit Care Med 2000; 161:694–699.
77. McKee CM, Lowenstein CJ, Horton MR, Wu J, Bao C, Chin BY, Choi AMK, Noble PW. Hyaluronan fragments induce nitric-oxide synthase in murine macrophages through a nuclear factor kB-dependent mechanism. J Biol Chem 1997; 272:8013–8018.
78. Weibel ER, Gil J. Electron microscopic demonstration of an extracellular duplex lining layer of alveoli. Respir Physiol 1968; 4:42–57.
79. Manabe T. Freeze-fracture study of alveolar lining layer in adult rat lung. J Ultrastruct Res 1979; 69:86–97.
80. Schurch S, Green FHY, Bachofen H. Formation and structure of surface films: captive bubble surfactometry. Biochim Biophys Acta 1998; 1408:180–202.
81. Bastacky J, Lee CYC, Goerke J, Koushafar H, Yager D, Kenaga L, Speed TP, Chen Y, Clements JA. Alveolar lining layer is thin and continuous: low-temperature scanning electron microscopy of rat lung. J Appl Physiol 1995; 79: 1615–1628.
82. Halter JM, Steinberg JM, Schiller HJ, DaSilva M, Gatto LA, Landas S, Nieman GF. Positive end-expiratory pressure after a recruitment maneuver prevents both alveolar collapse and recruitment/derecruitment. Am J Respir Crit Care Med 2003; 167:1620–1626.
83. Phelps DS, Floros J. Localization of surfactant protein synthesis in human lung by in situ hybridization. Am Rev Respir Dis 1988; 137:939–942.
84. Whitsett JA, Glasser SW. Regulation of surfactant protein gene transcription. Biochim Biophys Acta 1998; 1408:303–311.

85. Sahu SC, Tanswell AK, Lynn WS. Isolation and characterization of glycosaminoglycans secreted by human foetal lung type II pneumocytes in culture. J Cell Sci 1980; 42:183–188.
86. Skinner SJM, Post M, Torday JS, Stiles AD, Smith BT. Characterization of proteoglycans synthesized by fetal rat lung type II pneumonocytes *in vitro* and the effects of cortisol. Exp Lung Res 1987; 12:253–264.
87. Creuwels LAJM, van Golde LMG, Haagsman HP. The pulmonary surfactant system: biochemical and clinical aspects. Lung 1997; 175:1–39.
88. Nadesalingam J, Bernal AL, Dodds AW, Willis AC, Mahoney DJ, Day AJ, Reid KB, Palaniyar N. Identification and characterization of a novel interaction between pulmonary surfactant protein D and decorin. J Biol Chem 2003; 278: 25678–25687.
89. Wu H, Kuzmenko A, Wan S, Schaffer L, Weiss A, Fisher JH, Kim KS, McCormack FX. Surfactant proteins A and D inhibit the growth of gram-negative bacteria by increasing membrane permeability. J Clin Investig 2003; 111: 1589–1602.
90. Goerke J. Pulmonary surfactant: functions and molecular composition. Biochim Biophys Acta 1998; 1408:79–89.
91. Perez-Gil J, Keough KMW. Interfacial properties of surfactant proteins. Biochim Biophys Acta 1998; 1408:203–217.
92. Maniscalco WM, Campbell MH. Alveolar type II cells synthesize hydrophobic cell-associated proteoglycans with multiple core proteins. Am J Physiol 1992; 263:L348–L356.
93. Bray BA, Hsu W, Turino GM. Lung hyaluronan as assayed with a biotinylated hyaluronan-binding protein. Exp Lung Res 1994; 20:317–330.
94. Sampson PM, Heimer R, Fishman A. Detection of glycosaminoglycans at the one-nanogram level by 125 I-cytochrome c. Anal Biochem 1985; 151:304–308.
95. Termeer C, Benedix F, Sleeman J, Fieber C, Voith U, Ahrens T, Miyake K, Freudenberg M, Galanos C, Simons E. Oligosaccharides of hyaluronan activate dendritic cells via Toll-like receptor 4. J Exp Med 2002; 195:99–111.
96. Mummert ME, Mummert D, Edelbaum D, Hui F, Matsue H, Takashima A. Synthesis and surface expression of hyaluronan by dendritic cells and its potential role in antigen presentation. J Immunol 2002; 169:4322–4331.
97. Venge P, Pedersen B, Hakansson L, Hallgren R, Lindblad G, Dahl R. Subcutaneous administration of hyaluronan reduces the number of infectious exacerbations in patients with chronic bronchitis. Am J Respir Crit Care Med 1996; 153:312–316.
98. Wright RR. Elastic tissue of normal and emphysematous lungs. A tridimensional histologic study. Am J Pathol 1961; 39:355–367.
99. Karlinsky J, Fredette J, Davidovits G, Catanese A, Snider R, Faris B, Snider GL, Franzblau C. The balance of lung connective tissue elements in elastase-induced emphysema. J Lab Clin Med 1983; 102:151–162.
100. Chrzanowski P, Keller S, Cerreta J, Mandl I, Turino GM. Elastin content of normal and emphysematous lung parenchyma. Am J Med 1980; 69:351–359.
101. Janoff A. Elastases and emphysema: Current assessment of the protease-antiprotease hypothesis. Am Rev Respir Dis 1985; 132:417–433.
102. Fujita J, Nelson NL, Daughton DM, Dobry CA, Spurzem JR, Irino S, Rennard SI. Evaluation of elastase and antielastase balance in patients with chronic bronchitis and pulmonary emphysema. Am Rev Respir Dis 1990; 142:57–62.

103. Laurell C-B, Eriksson S. The electrophoretic alpha 1-globulin pattern of serum in alpha 1-antitrypsin deficiency. Scand J Clin Lab Investig 1963; 15: 132–140.
104. Beatty K, Matheson N, Travis J. Kinetic and chemical evidence for the inability of oxidized alpha 1-proteinase inhibitor to protect lung elastin from elastolytic degradation. Hoppe-Seyler Z Physiol Chem 1984; 365:731–736.
105. Weitz JI, Crowley KA, Landman SL, Lipman BI, Yu J. Increased neutrophil elastase activity in cigarette smokers. Ann Intern Med 1987; 107:680–682.
106. Churg A, Wang RD, Xie C, Wright JL. Alpha-1-antitrypsin ameliorates cigarette smoke-induced emphysema in the mouse. Am J Respir Crit Care Med 2003; 168: 199–207.
107. Hautamaki RD, Kobayashi DK, Senior RM, Shapiro SD. Requirement for macrophage elastase for cigarette smoke-induced emphysema in mice. Science 1997; 277:2002–2004.
108. Zheng T, Zhu Z, Wang Z, Homer RJ, Ma B, Riese RJ Jr, Chapman HAJ, Shapiro SD, Elias JA. Inducible targeting of IL-13 to the adult lung causes matrix metalloproteinase- and cathepsin-dependent emphysema. J Clin Investig 2000; 106: 1081–1093.
109. Wang Z, Zheng T, Zhu Z, Homer RJ, Riese RJ, Chapman HA Jr, Shapiro SD, Elias JA. Interferon gamma induction of pulmonary emphysema in the adult murine lung. J Exp Med 2000; 192:1587–1599.
110. D'Armiento J, Dalal SS, Okada Y, Berg RA, Chada K. Collagenase expression in the lungs of transgenic mice causes pulmonary emphysema. Cell 1992; 71: 955–961.
111. Shiomi T, Okada Y, Foronjy R, Schiltz J, Jaenish R, Krane S, D'Armiento J. Emphysematous changes are caused by degradation of type III collagen in transgenic mice expressing MMP-1. Exp Lung Res 2003; 29:1–15.
112. Foronjy RF, Okada Y, Cole R, D'Armiento J. Progressive adult-onset emphysema in transgenic mice expressing human MMP-1 in the lung. Am J Physiol Lung Cell Mol Physiol 2003; 284:L727–L737.
113. Imai K, Dalal SS, Chen ES, Downey R, Schulman LL, Ginsburg M, D'Armiento J. Human collagenase (matrix metalloproteinase-1) expression in the lungs of patients with emphysema. Am J Respir Crit Care Med 2001; 163: 786–791.
114. Konno K, Arai H, Motomiya M, Nagai H, Ito M, Sato H, Satoh K. A biochemical study on glycosaminoglycans (mucopolysaccharides) in emphysematous and in aged lungs. Am Rev Respir Dis 1982; 126:797–801.
115. Dabrowski R, Maslinski C. Collagen, glycosaminoglycans and histamine in lungs of guinea pigs exposed chronically to cigarette smoke. Bull Acad Pol Sci Biol 1975; 23:125–128.
116. Cantor JO, Cerreta JM, Armand G, Keller S, Turino GM. Pulmonary air-space enlargement induced by intratracheal instillment of hyaluronidase and concomitant exposure to 60% oxygen. Exp Lung Res 1993; 19:177–192.
117. Cantor JO, Cerreta JM, Keller S, Turino GM. Modulation of airspace enlargement in elastase-induced emphysema by intratracheal instillment of hyaluronidase and hyaluronic acid. Exp Lung Res 1995; 21:423–436.
118. Murakami H, Yoshida M, Aritomi T, Shiraishi M, Ishibashi M, Watanabe K. Effects of hyaluronidase on procine pancreatic elastase-induced lung injury. J Jpn Respir Soc 1998; 36:577–584.

119. Cantor JO, Shteyngart B, Cerreta JM, Liu M, Armand G, Turino GM. The effect of hyaluronan on elastic fiber injury *in vitro* and elastase-induced airspace enlargement *in vivo*. Proc Soc Exp Biol Med 2000; 225:65–71.
120. van de Lest CHA, Versteeg EMM, Veerkamp JH, van Kuppevelt TH. Digestion of proteoglycans in porcine pancreatic elastase-induced emphysema in rats. Eur Respir J 1995; 8:238–245.
121. van Straaten JFM, Coers W, Noordhoek JA, Huitema S, Flipsen JTM, Kauffman HF, Timens W, Postma DS. Proteoglycan changes in the extracellular matrix of lung tissue from patients with pulmonary emphysema. Mod Pathol 1999; 12: 697–705.
122. Visconti RP, Barth JL, Keeley FW, Little CD. Codistribution analysis of elastin and related fibrillar proteins in early vertebrate development. Matrix Biol 2003; 22:109–121.
123. Kozel BA, Wachi H, Davis EC, Mecham RP. Domains in tropoelastin that mediate elastin deposition *in vitro* and *in vivo*. J Biol Chem 2003; 278:18491–18498.
124. Merrilees MJ, Lemire JM, Fischer JW, Kinsella MG, Braun KR, Clowes AW, Wight TN. Retrovirally mediated overexpression of versican v3 by arterial smooth muscle cells induces tropoelastin synthesis and elastic fiber formation *in vitro* and in neointima after vascular injury. Circ Res 2002; 90:481–487.
125. Baccarani-Contri M, Vincenzi D, Cicchetti F, Mori G, Pasquali-Ronchetti I. Immunocytochemical localization of proteoglycans within normal elastin fibers. Eur J Cell Biol 1990; 53:305–312.
126. Vaccaro CA, Brody JS. Ultrastructural localization and characterization of proteoglycans in the pulmonary alveolus. Am Rev Respir Dis 1979; 120:901–910.
127. van Kuppevelt THMSM, Cremers FPM, Domen JGW, van Beuningen HM, van den Brule AJC, Kuyper CMA. Ultrastructural localization and characterization of proteoglycans in human lung alveoli. Eur J Cell Biol 1985; 36:74–80.
128. McDevitt CA, Marcelino J, Tucker L. Interaction of intact type VI collagen with hyaluronan. FEBS Lett 1991; 294:167–170.
129. Kielty CM, Whittaker SP, Grant ME, Shuttleworth CA. Type VI collagen microfibrils: evidence for a structural association with hyaluronan. J Cell Biol 1992; 118:979–990.
130. Coleman PJ. Evidence for a role of hyaluronan in the spacing of fibrils within collagen bundles in rabbit synovium. Biochim Biophys Acta 2002; 1571: 173–182.
131. Scott JE. The chemical morphology of the vitreous. Eye 1992; 6:553–555.
132. McDevitt CA, Beck GJ, Ciunga MJ, O'Brien J. Cigarette smoke degrades hyaluronic acid. Lung 1989; 167:237–245.
133. Pecora LJ, Manne WR, Baum GL, Feldman DP, Recavarren S. Biochemical study of ground substance in normal and emphysematous lungs. Am Rev Respir Dis 1967; 95:623–630.
134. Rao NV, Kennedy TP, Rao G, Ky N, Hoidal JR. Sulfated polysaccharides prevent human leukocyte elastase-induced acute lung injury and emphysema in hamsters. Am Rev Respir Dis 1990; 142:407–412.
135. DeAngelis PL, Oatman LC, Gay DF. Rapid chemoenzymatic synthesis of monodisperse hyaluronan oligosaccharides with immobilized enzyme reactors. J Biol Chem 2003; 278:35199–35203.

Chemistry and Biology of Hyaluronan
H.G. Garg and C.A. Hales (editors)

Chapter 12

Hyaluronan in Ventilator-Induced Lung Injury

HARI G. GARG, DEBORAH A. QUINN, MARCELLA M. MASCARENHAS and CHARLES A. HALES

Massachusetts General Hospital, Harvard Medical School, Boston, Massachusetts, USA

I. Introduction

The lung is a complex organ composed of multiple cells and tissue types containing hyaluronan as well as other proteoglycans. So far, human lung proteoglycans are not fully characterized. Positive pressure mechanical ventilation is a life-saving treatment for patients with acute respiratory distress syndrome (ARDS). Moderate to high tidal volume ventilation strategies can lead to ventilator-induced lung injury (VILI). The properties of hyaluronan after an insult or injury of various connective tissues are discussed in several chapters. This chapter presents the changes in the properties of hyaluronan caused by VILI.

II. Hyaluronan

A. Arrangement of Sugar Residues

Hyaluronic acid containing glucuronic and *N*-acetylglucosamine was extracted and purified from bovine vitreous in 1934 by Meyer and Palmer (1). This macromolecule was precipitated under acidic conditions and it was thought, therefore, to be an acid. Because it contained uronic acid, Meyer named this compound as hyaluronic acid as it was isolated from hyalos (= glassy, vitreous) and contained uronic acid (2). It took almost 20 years before

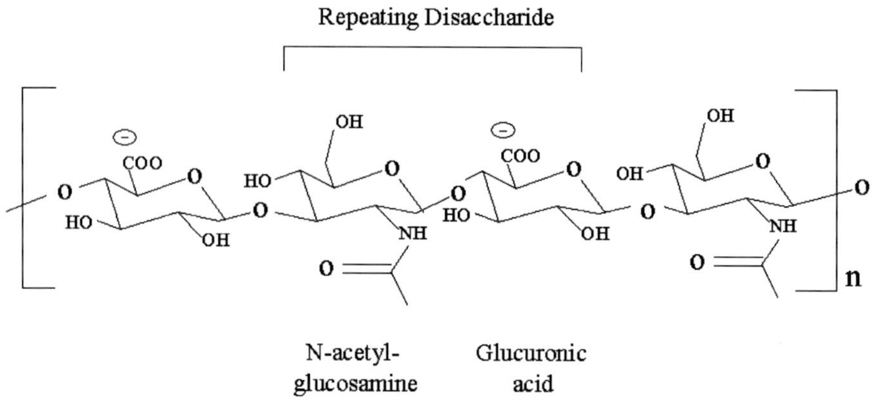

Figure 1 Structure of disaccharide units of hyaluronan.

Meyer could link the two sugar residues together in the disaccharide repeat unit correctly: glucuronic acid-β-1,3–*N*-acetylglucosamine-β-1,4 that forms glycosaminoglycan (Fig. 1) (3).

Repeating disaccharide units of different glycosaminoglycans are summarized in Table 1. The resolution is based on the combination of their migration properties in a barium acetate buffer and their differential sensitivity to ethanol precipitation. The migration of the glycosaminoglycan chains on cellulose acetate plate electrophoresis stained with Alcian Blue occurs (4) in the following order HP < DS < HS < HA < CS (Fig. 2). At physiological pH the COOH groups of uronic acid in HA are dissociated. Therefore, Balazs et al. (5) changed its name from hyaluronic acid to hyaluronan.

B. Role in Different Connective Tissues

Hyaluronan can retain water almost 100 times its weight. It is believed that hyaluronan plays a structural role, i.e., able to regulate water homeostasis and

Table 1 Arrangement of Sugars in Different Glycosaminoglycans

Glycosaminoglycan	Saccharide backbone	Nac	SO_3^-
Hyaluronan	→ 4)-β-D-GlcA(1 → 3)-β-D-GlcN(1 →	1	0
Chondroitin sulfate	→ 4)-β-D-GlcA(1 → 3)-β-D-GalN(1 →	1	1
Dermatan sulfate	→ 4)-α-L-IdoA(1 → 3)-β-D-GalN(1 → [→ 4)-β-D-GlcA)(1 →]	1	1
Heparan sulfate	→ 4)-α-L-IdoA(1 → 4)-α-D-GlcN(1 → [→ 4)-β-D-GlcA)(1 →]	<1	0–2
Heparin	→ 4)-α-L-IdoA(1 → 4)-α-D-GlcN(1 → [→ 4)-β-D-GlcA)(1 →]	≪ 1	2

GlcA, glucuronic acid, residue shown in brackets is minor component; IdoA, iduronic acid; GlcN, glucosamine; GalN, galactosamine. (Source: Ref. 6).

Figure 2 Migration of glycosaminoglycan chains on cellulose acetate plate. CS, chondroitin sulfate; HA, Hyaluronan; HA, Heparan sulfate; DS, dermatan sulfate; HP, heparin.

transport in the interstitium (7,8). Its ability to sequester water, to self-aggregate and to bind to many proteins make hyaluronan an ideal molecule to organize a net work in the aqueous subphase, e.g., in vitreous humor, lung interstitium, cartilage etc. In other cases, it is elastic, i.e., it absorbs energy and bounces back to its original shape, e.g., helping knees in absorbing the impact of jumping. In addition to these functions, hyaluronan works as Silly Putty. The various roles of hyaluronan in connective tissues are summarized in Table 2.

C. Distribution in Mammalian Organs

Hyaluronan analysis of the rat body tissues are summarized in Table 3. The data suggest that about 50% of hyaluronan of the total body is present in the skin, 25% in the skeleton and joints, and the remaining 25% is distributed in other parts of the body (9,10).

D. Physical and Chemical Properties

Almost all hyaluronan preparations isolated from different connective tissues have the same properties, i.e., electrophoretic mobility, infrared spectrum,

Table 2 Physiological Functions of Hyaluronan in Connective Tissue

Physiological functions	Connective tissue
Protection and lubrication	Synovial fluid
Shock absorption	Blood vessels
Maintenance of structural integrity	Vitreous, lung and articular cartilage
Distribution of molecules	Vitreous, intercellular matrix and cartilage

Table 3 Composition of Hyaluronan in Different Rat Tissues

Body part	Weight (g)	Hyaluronan (mg)	Percent
Skin	40.2	33.8	56
Muscles	35.7	4.69	8
Skeleton and supporting tissues	57.6	16.2	27
Intestines and stomach	15.8	0.50	1
Remaining internal organs	43.4	5.25	9

Data taken from Ref. 10.

absence of sulfate, susceptibility to cleavage by hyaluronidase and the presence of eqimolar amounts of glucuronic acid and *N*-acetylglucosamine. In contrast to these unique chemical properties, physical measurements show great variations in molecular weight of hyaluronan isolated from different connective tissues (11). These are summarized in Table 4.

The molecular weight of hyaluronan determines its function. High molecular weight (HMW) has a structural role (7,8) where as low molecular weight (LMW), less than 500 kDa, has been shown to function as a signaling molecule in lung inflammation (12–14). See Chapter 7 for details of the signal transduction properties of hyaluronan.

E. Biosynthesis

Hyaluronan is biosynthesized in plasma membrane by a membrane bound protein whose genetic code has been studied in bacteria, mice and humans (15–18). This adds sugar residues from nucleotide precursors to the chain of the cytoplasmic aspect of the membrane and translocates the growing chain to the pericellular space (10). In contrast with the synthesis of other connective tissue polysaccharides, the growth of hyaluronan chain occurs at the reducing end.

F. Hyaluronan in Lung Disease

Hyaluronan has to been shown to be increased and to play a possible role in many forms of lung disease (24), including cystic fibrosis (25), asthma (26), alveolar

Table 4 Molecular Weight of Hyaluronan Extracted from Various Tissues

Source	Molecular weight ($\times 10^6$)	Reference
Ox synovial fluid	13	19
Human synovial fluid	0.9	20
Human umbilical cord	3.4	21
Rabbit vitreous	2–3	22
Bovine vitreous	1.27	23
Rooster comb	0.6–0.7	20

proteinosis (27), sarcoidosis (28), farmer's lung (29), idiopathic pulmonary fibrosis (30) bleomycin-induced lung injury (31–33), smoke-inhalation injury (34), lung injury from diesel fuel (35) and emphysema (36,37).

Hyaluronan was measured in bronchoalveolar lavage fluid (BAL) and serum from 12 patients with ARDS and 28 controls by Hallgren and associates (28). They found that the median BAL HA concentration was six times higher in the ARDS patients and the median serum HA concentration was 20 times higher than control patients. The three patients who died with ARDS had the highest serum HA concentrations, and two of them also had the highest BAL HA concentrations. The molecular weight of the HA present was not measured. In normal individuals only very low concentrations of HA are found in BAL (38). We therefore explored changes in HA in our rat model of VILI.

III. Ventilator-Induced Lung Injury

Treatment of patients with acute lung injury (ALI) or the more severe form termed ARDS often requires the use of mechanical ventilation with high levels of oxygen, in order to adequately oxygenate the brain and other vital organs. ALI is a general term that refers to damage to the lungs that occurs in a number of different situations, including infection of the blood, lungs or abdomen, aspiration of stomach contents into the lungs, pancreatitis, multiple blood transfusions, trauma, drug overdose or near-drowning. ALI is an inhomogeneous disease (39, 40). Mechanical ventilation with large tidal volumes is used in order to recruit diseased areas of the lung with low compliance. This unfortunately leads to overdistension of normal areas of lung that display normal compliance. In severely damaged lungs, in which air space is reduced by up to 60% (41), the use of even low tidal volumes, calculated on the basis of the patient's size, may lead to the overdistension of the remaining normal lung. In a large clinical trial (800 patients) of large volume ventilation versus small volume ventilation in ALI/ ARDS, there were 22% fewer deaths in the patients ventilated with smaller tidal volumes (42).

The damage to the normal areas of the lung by overdistension in ALI/ARDS has become known as VILI and by some authors as ventilator-associated lung injury. VILI, a form of ALI, is characterized by non-cardiogenic pulmonary edema, production of inflammatory cytokines and subsequent influx of neutrophils (43). Several investigators have attempted to mimic in animal models the large stretch administered to the normally compliant areas of lungs in patients with ALI/ ARDS. These investigators have used mechanical ventilation of the whole lung with large tidal volumes to produce VILI in normal animals (43–45).

A. Hyaluronan Alterations in Rat Model of Ventilator-Induced Lung Injury

In the rats ventilated with large tidal volumes (V_T 20 cc/kg) for 2 h at a rate of 20 cycles/min there was a significant increase in the total amount of HA in

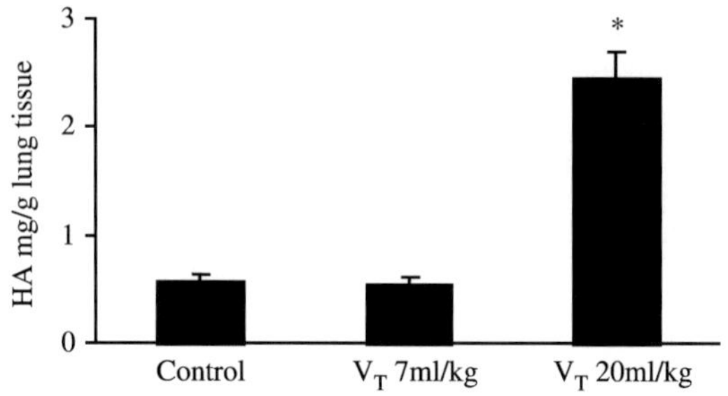

Figure 3 Ventilation of rats with high tidal volumes (V_T 20 mL/kg) significantly increased the amount of total hyaluronan (HA) in the lung tissue as compared to rats ventilated with low tidal volumes (V_T 7 mL/kg) and control non-ventilated rats. $*p < 0.05$ versus control and V_T 7 mL/kg.

the lung tissue as compared to rats ventilated at a smaller tidal volume (V_T 7 cc/kg) and normal rats without ventilation (Fig. 3) (40). HA standards and agarose gel electrophoresis were used to determine the molecular weights of HA that accumulated in the lungs of animals ventilated at V_T 20 cc/kg, of animals ventilated at V_T 7 cc/kg, and of lungs from control, non-ventilated animals (Table 5). A standard graph was plotted between log molecular weight of standards of HA versus their relative electrophoretic mobility. The molecular weight of the lung HA was then interpolated from this standard curve. In V_T 20 cc/kg rat lungs, two LMW (MWs 180 and 370 kDa) forms and one HMW (MW 3100 kDa) form of HA accumulated. This result contrasted with findings in rats ventilated at 7 cc/kg (HA MW = 2730 kDa) and control non-ventilated animals (MW = 3100 kDa) in which only the HMW form was found (46).

Table 5 Molecular Weight of Hyaluronan from Stretched Lungs *In Vivo* and Stretched Cells *In Vitro*

Source	Molecular weight (kDa)	
	High	Low
Non-ventilated rat lung	3100[a]	ND
Rat lung ventilated at 7 mL/kg	2730[a]	ND
Rat lung ventilated at 20 mL/kg	3100[a]	180, 370[a]
Non-stretched fetal lung fibroblast	>1600[b]	ND
Stretched fetal lung fibroblast	1600[b]	178[b]
Non-stretched adult lung fibroblast	1600–600[b]	ND
Stretched adult lung fibroblast	1600–600[b]	219[b]

ND, not detected.
[a]Analyzed by agarose gel electrophoresis.
[b]Analyzed on Sepharose CL-4B column.

B. Hyaluronan Alterations in an *In Vivo* Model of Ventilator-Induced Lung Injury

To examine at the cellular level the effects of ventilator-induced stretch, we have developed an *in vitro* model of VILI (47). We use a cell-stretching device that uniformly applies biaxial strain to flexible cell culture membranes. Primary lung fetal fibroblasts (IMR 90, Coriell Repository, Camden, NJ) and normal adult lung fibroblasts (Clonetics, Walkerville, MD) were grown on fibronectin-coated silicone elastomeric membranes and exposed to 15% strain at 60 cycles/min using our cell stretch model (Fig. 4).

The supernatants of cultured non-stretched fibroblasts contained only HMW HA (MW 1600 kDa, and a very HMW HA that was present only in the void volume) whereas supernatants from stretched fibroblasts contained LMW HA (Sepharose CL-4B gel column fractionation range is between 10^4 and 10^7 kDa). Stretched fetal fibroblasts produced a LMW HA of 178 kDa and a HMW HA of >1600 kDa, whereas normal adult lung fibroblasts produced LMW HA of 219 kDa and HMW between 600 and1600 kDa (Table 5) (46).

Other changes in the proteoglycans in the lung have also been found in VILI. With high tidal volume there was an increased amount of versican, basement membrane heparan sulfate and biglycan. Heparan sulfate and versican were prominent in the alveolar wall and airspace, whereas biglycan was localized in the airway wall (48).

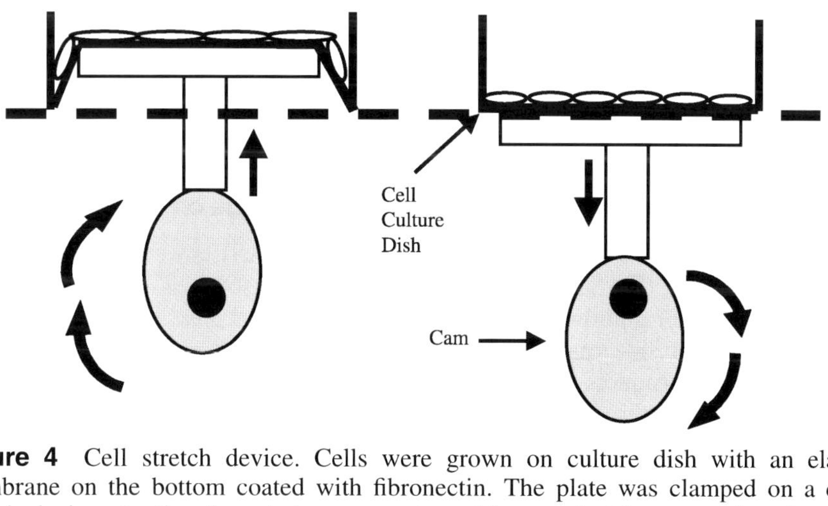

Figure 4 Cell stretch device. Cells were grown on culture dish with an elastic membrane on the bottom coated with fibronectin. The plate was clamped on a cells stretch device. As the elongated cam was turned by an electric motor, the plate was displaced upward in a cyclic manner which produced cyclic stretch on the elastic membrane. Figure designed by Behrouz Jafari.

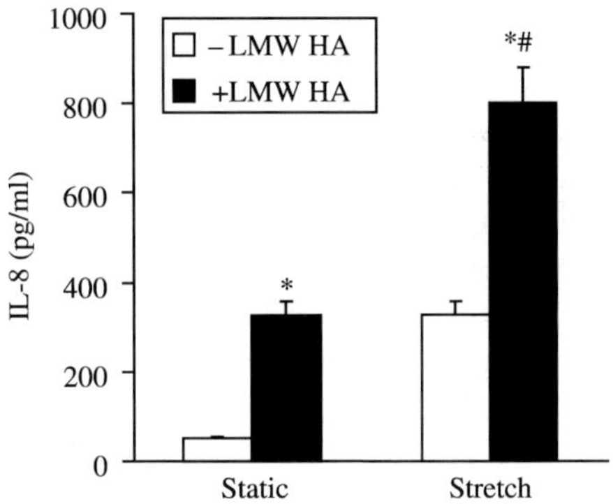

Figure 5 LMW HA increased the production of IL-8 in type II-like alveolar epithelial cells. *$p < 0.05$ versus static cells without low molecular weight hyaluronan (LMW HA); #$p < 0.05$ versus stretch without LMW HA.

C. Low Molecular Weight Forms of Hyaluronan Produced During Lung Cell Stretch are Pro-Inflammatory

Low molecular weight forms of HA have been shown to have inflammatory properties by binding to CD 44 and activating the NFkB pathway (12,14). See Chapter 7 for details of the LMWHA receptors and signaling pathways. We explored the effects of LMW HA isolated from stretched primary lung fibroblasts and HMW HA from static fibroblasts (46). LMW HA from lung fibroblasts caused a significant increase in IL-8 production in a type II-like alveolar epithelial cell line (A549 cells), which are a source of IL-8 in the lung (Fig. 5), whereas HMW HA did not. Cyclic stretch of the A549 cells augmented LMW HA induced IL-8 production. These data were consistent with a pro-inflammatory effect of LMW HA. LWM HA may play an important role in inflammation in ALI such as VILI and ARDS. Under these conditions, overstretched normal or near normal alveoli produce LMW HA, which induces IL-8 secretion with subsequent attraction of neutrophils into the uninjured alveoli.

D. Stretch-Induced Production of LMW HA Depends on HA Synthase 3

Hyaluronan can be synthesized by HA synthase (HAS), an enzyme that exists as three isoforms (HAS1, HAS2 and HAS3). The isoforms are distinct from each other in their stabilities, the rates at which they cause elongation of HA, and the range of size distribution of their HA products. HAS3 forms LMW HA, while the products of HAS1 and HAS2 form HMW HA *in vitro* cell culture (49). The three isoforms of HAS have been cloned and sequenced (17,50–52). All the three isoforms of HAS mRNA are downregulated by dexamethasone and by cyclohexamide, non-specific inhibitors of HA (53). In our *in vivo* and *in vitro* models of lung cell stretch we have found that cyclic stretching increases HAS3 mRNA expression, but not HAS1 and HAS2. Stretch-induced HA3 mRNA expression and HA production was inhibited

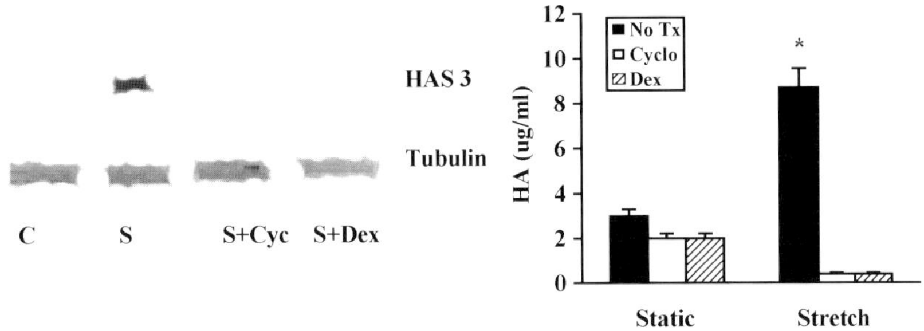

Figure 6 Inhibition of hyaluronan synthase 3 mRNA expression (HAS3) with dexamethasone (Dex) or cyclohexamide (Cyclo) blocked stretch (S)-induced HA production in lung fibroblasts. C, control static fibroblasts. $*p < 0.05$ versus all other groups (5 in each group).

by dexamethasone and cyclohexamide, non-specific inhibitors of HAS (Fig. 6). Therefore, VILI may have involved increased production of LMW HA stemming from upregulated HAS3 expression.

E. Inhibition of LMW HA Production as a Possible Treatment Strategy for Acute Lung Injury

We have shown that lung cell stretch *in vivo* and *in vitro* produced a LMW form of HA that was pro-inflammatory. We have shown one mechanism of stretch-induced LMW HA production was through *de novo* synthesis—by HAS3. LWM HA could also have been produced by breakdown of HMW HA by oxidants to LMW forms (54–56). Lung cell stretch has been shown to cause oxidant injury and blocking oxidant injury inhibited stretch-induced lung cytokine production and neutrophil influx (57,58). Thus breakdown of HMW HA was another possible mechanism of LMW HA production in ALI.

Phosphodiesterase inhibitors such as vesnarinone have been shown to inhibit HA production by HAS (59) and inhibit inflammation in humans (60,61). The inhibition of LMW HA synthesis may be a potential treatment strategy for ALI. Alternatively, treatments that inhibit oxidant injury may also decrease breakdown of HMA to LMW HA.

IV. Conclusion

Hyaluronan has both structural and inflammatory properties in the lung. LMW HA may have an important role in inflammation as found in forms of ALI, including ARDS and VILI. Further study of LMW HA in forms of ALI may lead to new treatment options.

References

1. Meyer K, Palmer J. The polysaccharide of the vitreous humor. J Biol Chem 1934; 107:629–634.
2. Laurent T. The biology of hyaluronan introduction. Ciba Found Symp 1989; 143: 1–5.
3. Weissman B, Meyer K. The structure of hyalobiuronic acid and of hyaluronic acid from umbilical cord. J Am Chem Soc 1954; 76:1753–1757.
4. Cappelletti R, Del Ross M, Chiarugi VP. A new electrophoretic method for the complete separation of all known animal glycosaminoglycans in a monodimensional run. Anal Biochem 1979; 99:311–315.
5. Balazs EA, Laurant TC, Jeanloz RW. Nomenclature of hyaluronic acid. Biochem J 1986; 235:903.
6. Garg HG, Linhardt RJ, Hales CA. Heparin as a potential therapeutic agent to reverse vascular remodeling. In: Garg HG, Roughley PJ, Hales CA, eds. Proteoglycans in Lung Disease. New York: Marcel Dekker, 2002:377–398.
7. Comper WD, Laurent TC. Physiological function of connective tissue polysaccharides. Physiol Rev 1978; 58:255–315.
8. Laurent TC. Structure, function and turnover of the extracellular matrix. In: Staub NC, Hogg JC, Hargens AR, eds. Interstitial Lymphatic Liquid and Solute Movement, Advanced Microcirculation. Basel: Karger, 1987; vol. 13:15–34.
9. Reed RK, Lilja K, Laurent TC. Hyaluronan in the rat with special reference to the skin. Acta Physiol Scand 1988; 134:405–411.
10. Fraser JR, Laurent TC, Laurent UB. Hyaluronan: its nature, distribution, functions and turnover. J Intern Med 1997; 242:27–33.
11. Laurent TC. Structure of hyaluronic acid. In: Balazs EA, ed. Chemistry and Molecular Biology of the Intercellular Matrix. London: Academic Press, 1970; vol. 2:703–732.
12. Nobel PW, McKee CM, Cowman M, Shin HS. Hyaluronan fragments activate an NF-kappaB/Ikappabeta autoregulatory loop in murine macrophages. J Exp Med 1996; 183:2373–2378.
13. Hamilton S, Turley EA, Wang FS. Hyaluronan and hyaladherin signaling in the lung. In: Garg HG, Roughley PJ, Hales CA, eds. Proteoglycans in Lung Disease. New York: Marcel Dekker, 2002:73–106.
14. Itano N, Sawai T, Yoshida M, Lenas P, Yamada Y, Imagawa M, Shinomura T, Hamaguchi M, Yoshida Y, Ohnuki Y, Miyauchi S, Spicer AP, McDonald JA, Kimata K. Three isoforms of mammalian hyaluronan synthase have distinct enzymatic properties. J Biol Chem 1999; 274:25085–25092.
15. DeAngelis PL, Weigel PH. Characterization of the recombinant hyaluronic acid synthase from Streptococcus pyogenes. In: Ferretti JJ, Gilmore MS, eds. Dev Biol Stand No 85. Basel: Karger, 1995:225–229.
16. Spicer AP, Augustine ML, McDonald JA. Molecular cloning and characterization of a putative mouse hyaluronan synthase. J Biol Chem 1996; 271:23400–23406.
17. Shyjan AM, Heldin P, Butcher EC, Yoshino T, Briskin MJ. Functional cloning of the cDNA for a human hyaluronan synthase. J Biol Chem 1996; 271: 23395–23399.
18. Watanabe K, Yamaguchyi Y. Molecular identification of a putative human hyaluronan synthase. J Biol Chem 1996; 271:22945–22948.

19. Preston BN, Davies M, Ogston AG. The composition and physiochemical properties of hyaluronic acid prepared from Ox synovial fluid and from a case of mesothelioma. Biochem J 1965; 96:449–474.
20. Cleland RL. Ionic polysaccharide III. Dilute solution properties of hyaluronic acid fraction. Biopolymer 1970; 9:799–810.
21. Laurent TC, Gergely J. Light scattering studies on hyaluronic acid. J Biol Chem 1955; 212:325–333.
22. Laurent UB, Granath KA. The molecular weight of hyaluronate in aqueous humor and vitreous body of rabbit and cattle eye. Exp Eye Res 1983; 36:481–492.
23. Rowen JW, Brunish R, Bishop FW. Form and dimensions of isolated hyaluronic acid. Biochem Biophys Acta 1956; 19:480–489.
24. Savani RC, DeLisser HM. Hyaluronan and its receptors in lung health and disease. In: Garg HG, Roughley PJ, Hales CA, eds. Proteoglycans in Lung Disease. New York: Marcel Dekker, 2002:23–36.
25. Sahu SC. Hyaluronic acid. An indicator of pathological conditions of human lungs. Inflammation 1980; 4:107–112.
26. Sahu S, Lynn WS. Hyaluronic acid in pulmonary secretion of patients with asthma. Biochem J 1978; 173:565–568.
27. Sahu S, Lynn WS. HA in pulmonary secretion of patients with alveolar proteinosis. Inflammation 1978; 3:149–158.
28. Hallgren R, Eklund A, Engstrom-Laurent A, Aechmekel B. Hyaluronan in bronchoalveolar lavage fluid, a new marker in sarcoidosis reflecting pulmonary disease. Br Med J 1985; 290:1778–1781.
29. Bjermer L, Engstrom-Laurent A, Laundger R, Rosenhall L, Hallgren R. Hyaluronan and type III procollagen peptide concentration in bronchoalveolar lavage fluid as marker of disease activity in farmers lung. Br Med J 1987; 295: 803–806.
30. Bjermer L, Lundgren R, Hallgren R. HA in type III collagen peptide concentration in BAL in idiopathic pulmonary fibrosis. Thorax 1989; 44:126–131.
31. Savani RC, Hou G, Liu P, Wang C, Simons E, Grimm PC, Stern R, Greenberg AH, DeLisser HM, Khalil N. A role for hyaluronan in macrophage accumulation and collagen deposition after bleomycin-induced lung injury. Am J Respir Cell Mol Biol 2000; 23:475–484.
32. Bray BA, Sampson PM, Osman M, Giandomenico A, Turino GM. Early changes in lung tissue hyaluronan (hyaluronic acid) and hyaluronidase in bleomycin-induced alveolitis in hamsters. Am Rev Respir Dis 1991; 143:284–288.
33. Teder P, Vandivier RW, Jiang D, Liang J, Cohn L, Pure E, Henson PM, Nobel PW. Resolution of lung inflammation by CD44. Science 2002; 296:155–158.
34. Bergstrom CE, Eklund A, Skold M, Tornling G. Bronchoalveolar lavage findings in firefighters. Am J Ind Med 1997; 32:332–336.
35. Sato H, Onose J, Toyoda H, Imanari T, Sagai M, Nishimura N, Aoki Y. Quantitative changes in glycosaminoglycans in the lungs of rats exposed to diesel exhaust. Toxicology 2001; 166:119–128.
36. Cantor JO, Cerreta JM, Armand G, Turino GM. Further investigation of the use of intratracheally administered hyaluronic acid to ameliorate elastase-induced emphysema. Exp Lung Res 1997; 23:229–244.
37. Karlinsky JB. Glycosaminoglycans in emphysematous and fibrotic hamster lung. Am Rev Respir Dis 1982; 125:85–88.

38. Hallgren R, Samelson T, Laurent TC, Modig J. Accumulation of HA in lung in adult respiratory distress syndrome. Am Rev Respir Dis 1989; 139:682–687.
39. Maunder RJ, Shuman WP, McHugh JW, Marglin SI, Butler J. Preservation of normal lung regions in the adult respiratory distress syndrome. Analysis by computed tomography. JAMA 1986; 255:2463–2465.
40. Gattinoni L, Presenti A, Bombino M, Baglioni S, Rivolta M, Rossi F, Rossi G, Fumagalli R, Marcolin R, Mascheroni D, Torresin A. Relationships between lung computed tomographic density, gas exchange, and PEEP in acute respiratory failure. Anesthesiology 1988; 69:824–832.
41. Pelosi P, D'Andrea L, Vitale G, Pesenti A, Gattinoni L. Vertical gradient of regional lung inflation in adult respiratory distress. Am J Respir Crit Care Med 1994; 149:8–13.
42. The A.R.D.S Network, Ventilation with lower tidal volumes as compared with traditional tidal volumes for acute lung injury and the acute respiratory distress syndrome. N Engl J Med 2000; 342:1301–1308.
43. Dreyfuss D, Saumon G. Ventilator-induced lung injury: lessons from experimental studies. Am J Respir Crit Care Med 1998; 157:294–323.
44. Tremblay L, Valenza F, Ribeiro S, Li J, Slutsky A. Injurious ventilatory strategies increase cytokines and c-fos mRNA expression in an isolated rat lung model. J Clin Investig 1997; 99:944–952.
45. Quinn DA, Moufarrej RK, Volokhov A, Hales CA. Interactions of lung stretch, hyperoxia, and MIP-2 production in ventilator-induced lung injury. J Appl Physiol 2002; 93:517–525.
46. Mascarenhas MM, Day RA, Garg HG, Hales CA, Quinn DA. Possible mechanism of cytokine release in ventilator induced lung injury (VILI): stretch induced low molecular weight hyaluronan (LMW HA). Am J Respir Cell Mol Biol 2004; 30:51–60.
47. Li LF, Ouyang B, Choukroun G, Matyal R, Mascarenhas M, Jafari B, Bonventre JV, Force T, Quinn DA. Stretch-induced Il-8 depends on c-jun n-terminal and nuclear factor-{kappa}b-inducing kinases. Am J Physiol Lung Cell Mol Physiol 2003; 285: L464–L475.
48. Al-Jamal R, Ludwig MS. Changes in proteoglycans and lung tissue mechanics during excessive mechanical ventilation in rats. Am J Physiol Lung Cell Mol Physiol 2001; 281:L1078–L1087.
49. McKee CM, Penno MG, Cowan M, Burdick MD, Strieter RM, Bao C, Nobel PW. Hyaluronan (HA) fragments induce chemokine gene expression in alveolar macrophages. The role of HA size and CD 44. J Clin Investig 1996; 98: 2403–2413.
50. Spicer AP, Olson JS, McDonald JA. Molecular cloning and characterization of cDNA encoding the third putative mammalian hyaluronan synthase. J Biol Chem 1997; 272:8957–8961.
51. Spicer AP, Augustine MI, McDonald JA. Molecular cloning and characterization of a putative mouse hyaluronan synthase. J Biol Chem 1996; 271:23400–23406.
52. Itano N, Kimata K. Expression cloning and molecular characterization of HAS protein, a eukaryotic hyaluronan synthase. J Biol Chem 1996; 271:9875–9878.
53. Kaback LA, Smith TJ. Expression of hyaluronan synthase messenger ribonucleic acids and their induction by interleukin-beta in human orbital fibroblasts: potential insight into molecular pathogenesis of thyroid-associated opthalmopathy. J Clin Endcrinol Metab 1999; 84:4079–4084.

54. Sampson PM, Rochester CL, Freundlich B, Elias JA. Cytokine regulation of human lung fibroblast hyaluronan (hyaluronic acid) production. Evidence for cytokine-regulated hyaluronan (hyaluronic acid) degradation and human lung fibroblast-derived hyaluronidase. J Clin Investig 1992; 90:1492–1503.
55. Saari H. Oxygen derived free radicals and synovial fluid hyaluronan. Ann Rheum Dis 1991; 50:389–392.
56. Sztrolovics R, Recklies AD, Roughley PJ, Mort JS. Hyaluronate degradation as an alternative mechanism for proteoglycan release from cartilage during interleukin-1β-stimulated catabolism. Biochem J 2002; 362:473–479.
57. Matthay MA, Bhattacharya S, Gaver D, Ware LB, Lim LH, Syrkina O, Eyal F, Hubmayr R. Ventilator-induced lung injury: *in vivo* and *in vitro* mechanisms. Am J Physiol Lung Cell Mol Physiol 2002; 283:L678–L782.
58. Jafari B, Ouyang B, Li LF, Hales CA, Quinn DA. Intracellular glutathione in stretch-induced cytokine release from alveolar type-2 like cells. Respirology 2004; 9:43–53.
59. Ueki N, Taguchi T, Takahashi M, Adachi M, Ohkawa T, Amuro Y, Hada T, Higashino K. Inhibition of hyaluronan synthesis by vesnarinone in culture human myofibroblasts. Biochem Biophys Acta 2000; 1495:160–167.
60. Takeuchi K, del Nido PJ, Ibrahim AE, Cao-Danh H, Friehs I, Glynn P, Poutias D, Cowan DB, McGowan FX Jr. Vesnarinone and amrinone reduce systemic inflammatory response syndrome. J Thorac Cardiovasc Surg 1999; 117:375–382.
61. Boldt J, Brosch C, Suttner S, Piper SN, Lehmann A, Werling C. Prophylactic use of phosphodiesterase III inhibitor, enoximone, in elderly cardiac surgery patients: effect on hemodynamics, inflammation and markers of organ function. Intensive Care Med 2002; 28:1462–1469.

Chemistry and Biology of Hyaluronan
H.G. Garg and C.A. Hales (editors)

Chapter 13

The Role of Hyaluronan in Cancer

SONAL PATEL

Piramed Ltd,
957 Buckingham Avenue,
Slough SL1 4NL, UK

MARTIN J. PAGE

Oncogene Sciences Inc.,
Watlington Road,
Oxford OX4 6LT, UK

I. Introduction

Hyaluronan (HA) or hyaluronic acid is a large glycosaminoglycan comprising repeating disaccharides of glucuronic acid and *N*-acetylglucosamine. HA is a major component of the matrix and tissue fluids and is found in most vertebrate tissues where it has key roles in the maintenance of osmotic balance and tissue hydration (1,2). In tissues such as synovial fluid, vitreous and dermis it offers a structural role based on its unique physiochemical properties. However, with the discovery of HA receptors (e.g., CD44, RHAMM) on the cell surface, a much more complex role for HA in dynamic cellular behaviour has been proposed (3,4). Increased HA deposition in the matrix has been associated with stages of embryonic development (5), wound healing (6), and inflammatory conditions (7), which have been presented in other chapters.

However, in addition to roles within normal cellular physiological processes, HA has been associated with aberrant cell behaviour in cancer development and progression. The enrichment of HA is evident in a number of tumours or in the stroma surrounding the tumours (8). This chapter will focus on the role of HA in cancer progression, the interaction of HA with its receptors and the subsequent downstream effects on signalling cascades in tumours. Finally, studies that have perturbed the effects of HA in cancer and how these could be developed as potential strategies for therapeutic benefit in cancer will be presented.

II. Clinical Indications of Hyaluronan in Cancer

Extensive studies have examined HA levels in clinical tumours and in body fluids such as serum from cancer patients. A summary of the clinical tumours assessed for HA levels is shown in Table 1 with representative references. The number of these types of studies has increased steadily as methods have developed and become available to investigate HA in tumours and in their surrounding stroma. Evaluation of HA has employed techniques such as HPLC (17) or electrophoresis and enzyme digestion (15) to assess the HA content of tumours, as well as direct tumour staining for HA. Using a biotinylated affinity probe specific for HA, Anttila et al. (20) characterised HA staining in 309 epithelial ovarian tumours and 45 matched metastatic tumours. The high stromal staining seen in 98 carcinomas was significantly correlated with poor differentiation, serous histological type, advanced stage and large primary residual tumour. The 5 yr follow-up of the disease showed that overall survival and recurrence-free survival decreased with increased stromal staining of HA.

The increased accumulation of HA may thus provide an independent prognostic marker of ovarian cancer. The molecular basis for increased HA was not precisely defined but the action of growth factors and direct cellular contact with local mesenchymal cells was thought likely to have a role. Interestingly this is supported by previous studies which have shown that murine ovarian cancer cells stimulate HA production on murine mesenteric surfaces and tumour cell clumps (28). The evaluation of epithelial ovarian tumours was extended further by Hiltunen et al. (21), where HA accumulation alone, without hyaluronidase activation (the enzyme family which breaks down HA), was shown to correlate with the aggressiveness of ovarian cancer. In a prostate cancer study Posey et al. (25) evaluated the potential of HA and the

Table 1 Tumours Shown to Have Increased Levels of Hyaluronan

Tumour type	References
Bladder carcinoma	9
Breast adenocarcinoma	10,11
Colon carcinoma	12
Glioma	13
Head and neck cancer	14
Hepatic carcinoma	15
Lung carcinoma	16
Mesothelioma	17
Nephroblastoma	18,19
Ovarian carcinoma	20,21
Pancreatic carcinoma	22
Prostate cancer	23–25
Stomach carcinoma	26
Thyroid carcinoma	27

hyaluronidase family member, hyal-1, as prognostic markers in 70 clinical prostate cancer samples using a biotinylated HA-binding protein and anti-hyal-1 antibody. As seen in ovarian cancers, the tumour stroma in prostate cancer also stained positive for HA and in addition 40% of tumour cells also expressed HA. Hyal-1, however, was exclusively expressed in tumour cells. In a 5 yr follow-up study hyal-1 expression levels together with values for extra-prostatic extension and positive margin were shown to be an independent prognostic indicator of prostate cancer progression. Similarly, tumour cell associated HA was shown to be an unfavourable prognostic factor in a study of tumours from 202 colorectal adenocarcinoma cancer patients (12). In breast cancer, a high risk group of patients was identified based on stromal myxoid changes with high HA content that strongly associated with positive nodes, tumour grade and lymphatic emboli (11).

Even though a large number of studies appear to show correlation of tumour progression with increased HA levels directly in tumour cells or tumour stroma, there are a few reports to the contrary. For example, a conclusive correlation of HA and CD44 expression with the biologic behaviour of different grades of salivary gland tumours was not found (29), suggesting that the tumour cell type may be an important factor and that the associated increase observed in some cancers involves a complex process. In addition, it is important to consider the number of tumour samples examined and size of patient group that is assessed in the conclusion of findings in correlative type studies.

In the main, many clinical cancers appear to be associated with high levels of HA, but to dissect a 'cause and effect' role for the overproduction of HA and increased tumorigenicity presents many challenges. These investigations were largely hampered since HA does not contain any protein components, which could readily be manipulated by molecular biology techniques. However, with the cloning of the mammalian HA synthases (30), HA expression levels could be altered and the consequences examined using *in vitro* cancer models. Kosaki et al. (31) transfected the hyaluronan synthase 2 gene (HAS2) into HT1080 fibrosarcoma cells which promoted anchorage-independent growth *in vitro* and subsequently increased tumorigeniciy in nude mice. Interestingly, this transformation to a malignant phenotype was not associated with an increased proliferative effect in the HT1080 cells grown as monolayers, but HA exerted its growth effects when in a 3D environment, which may support the earlier observation of increased HA in murine tumour cell clumps (28). However, to investigate HA effects further and test its ability to cause transformation, Itano et al. (32) used non-transformed 3Y1-1B6 fibroblast cells transfected with HAS isoforms to examine the effects of overproduction of HA on cells. Although this study showed that the overproduction of HA alone in non-transformed cells did not enhance contact inhibited growth or the formation of subcutaneous tumours, there was a partial reduction of contact inhibited cell growth, and increased cellular motility. This suggested that overproduction of HA resulted in some of the hallmarks of tumour cell behaviour but was insufficient to cause transformation alone.

Studies designed to understand the 'cause or effect' of increased HA in cancers suggests that in clinical tumours HA overproduction alone may not directly be tumorigenic but that HA may potentially provide a favourable environment which supports tumour viability and function.

III. Hyaluronan Receptors

HA can influence cell behaviour by several different mechanisms. This includes the migration properties of many cell types including tumour cells. Due to the unique physical properties of HA, free HA has a direct effect on the biomechanical properties of extracellular and pericellular matrices to which cells are exposed. One of the foremost concepts suggests that the ability of HA to bind large amounts of water causes hydrous channels to be created in the extracellular matrix (ECM), which facilitates cell movement (5). In addition to this mechanism, HA receptor mediated effects on cell movement were proposed and soon thereafter a number of receptors for HA were identified. These include CD44, the first receptor identified for HA (33–35), RHAMM (36), LYVE-1 (37) which was thought to be exclusively expressed on lymphatic endothelium but has since been found in liver sinusoids (38), and further HA receptors, namely HARE, layilin and Toll-4 (39,40). Furthermore, HA is also known to interact with cells by attaching to HAS (40,41). The scope of this section will focus primarily on CD44 and to a lesser extent on RHAMM, and their interactions with HA. Research in these areas has advanced rapidly, clearly marking the importance of HA receptor mediated effects on cell behaviour.

The HA–CD44 interaction is tightly regulated and can mediate cell–cell and cell–ECM interactions. CD44 exists in an active ligand binding form or an inactive non-binding form (42). CD44 has many isoforms generated by alternative splicing and glycosylation (42,43) and these have been correlated with cancers. In particular the over-expression of CD44v5 correlated significantly with metastatic potential of osteosarcoma and survival rates were shown to be markedly lower in osteosarcoma patients who had CD44v5 positive tumours. However, as with many correlative marker type studies, these data are likely to be dependent on many factors such as age, grade and type of tumour. In the case of oro-pharyngeal squamous cell carcinoma, all variant exons of CD44 were expressed and were not fundamentally altered (44) suggesting that in this cancer type CD44 is not a critical indicator of survival.

The receptor for HA-mediated motility (RHAMM) was the second HA receptor that was discovered following an investigation of HA regulation of the locomotion of ras transformed cells (36). The mechanism of HA–RHAMM binding and its effect on motility of tumour cells are complex (45) and not solely due to HA binding. Further to this, RHAMM was also discovered to have an intracellular localisation since antibodies to RHAMM did not give a cell surface staining pattern on human breast cancer cells (46). The history of RHAMM was further complicated with the subsequent isolation of the full-length cDNA for RHAMM which encodes a 95 kDa protein distinct from the original RHAMM protein that had been found. A proposal to rename RHAMM to 'intracellular HA binding protein' (IHABP) was therefore suggested (47). Multiple alternatively spliced forms of CD44 and RHAMM receptors are now thought to be present, localised on the cell surface or intracellular implying different functions at these sites (48). A recent study profiled RHAMM receptor staining based on overall combined cytoplasmic, tumour periphery and intratumoral expression (49). Of 89 clinical endometrial carcinomas, 54 showed positive staining with a high correlation to tumour grade. Additionally, 100% of patients with lymph node positive tumours had a tumour positive status for RHAMM expression compared to 50.7% in patients with negative lymph nodes. The immunohistochemical staining pattern was highly variable; however, all the low-grade tumours showed a focal expression in the periphery of the tumour. These data correlate RHAMM expression with differentiation, invasion and lymph node metastasis. RHAMM has also been identified as a new immunogenic antigen in acute and chronic myeloid leukaemia and in solid tumours following sereologic screening of a cDNA expression library (50). Recently, both CD44 and RHAMM have presented a differential expression pattern in transitional cell carcinoma of bladder (51).

IV. Hyaluronan in Adhesion, Migration, and Invasion of Cancer

CD44 and HA are involved in a variety of biological processes and CD44 is often up-regulated with HA at sites of inflammation (52) and tumour invasion (8). There are now extensive data to indicate that interaction of HA with CD44 and the RHAMM receptors are involved in tumour cell adhesion, migration, invasion, and tumour growth. Adhesion of tumour cells to stromal cells is an important step in tumour establishment and recently expression of CD44v9 on myeloma plasma cells, which is correlated to poor prognosis, was shown to facilitate binding to bone marrow stromal cells (53). This resulted in the induction of IL-6 which is also associated with adverse prognosis in multiple myeloma.

In cancers such as prostate cancer, which specifically metastasise to the bone, the circulating tumour cells undergo an adhesion process to the endothelial cells lining the bone marrow vasculature (54) followed by transmigration through the endothelial cell barrier and subsequent establishment in the stroma. Interestingly, HA has already been shown in murine anterior prostate gland to be a prerequisite for androgen stimulated ductal branching morphogenesis (55). Further to this, the role of HA in prostate cancer cell adhesion to bone marrow derived endothelial cell line (BMEC-1) has been demonstrated (56). In this study, highly metastatic PC3 and PC3M-LN4 showed a rapid adherence to BMEC-1 but not to endothelial cells derived from human vein. Adhesion was inhibited by the addition of excess HA or by pre-treatment of cells with hyaluronidase which digested away pericellular HA. Of note, pericellular HA was also correlated with increased level of HA synthesis and HA synthase expression in these cells. These findings using the above model were investigated further by Simpson et al. (57). PC3M-LN4 cells stably transfected with full length HAS2 or HAS3 failed to form pericellular matrices and showed a significant reduction in adhesion to BMECs. Conversely, when full length HAS2 or HAS3 was transfected into non-adherent LNCaP prostate cells they showed retention of pericellular HA and adhered to BMEC. These data provide direct evidence for tumour cell associated HA and up-regulation of HA synthase on the metastatic potential of prostate cancer cells.

Along with cell adhesion, migration and proliferation are also key factors for tumour progression. HA in its native form is a large MW molecule and in this form has many effects on cell behaviour, some of which have been described here. However, lower MW, or fragmented HA has also been found to have effects on angiogenesis (see below). Further work demonstrates that the low molecular weight forms of HA but not the high molecular weight forms induce proliferation and migration of tumour cells (58). In this study, mesothelioma cells had increased proliferation and migration in response to low MW HA in a CD44-dependent interaction. The proliferation and migration of mesothelioma cells by low MW HA were inhibited by up to 40 and 35%, respectively, with anti-CD44 antibody. This study highlights the role of low MW HA in the localised propagation of tumour growth and the effect of HA on increased mesothelioma cell migration. Similarly, this observation was also corroborated recently in a clonal variant of T-cell murine lymphoma (59).

Cancer metastasis requires genetic and cellular changes to the tumour cells which facilitate their invasion into surrounding tissues, entry into the lymphatic system and/or bloodstream followed by establishment and colonisation of the tumour at the secondary site. Several studies have correlated HA on the surface of tumour cells with metastatic behaviour of cells and have shown that this is dependent on CD44–HA interaction. However, there are many mechanisms for cells to acquire invasive and metastatic properties and generation of this cell behaviour may be cell type and cell environment specific highlighting an underlying complex process. For example, other studies have also demonstrated that CD44 variant isoforms confer a metastatic phenotype in pancreatic carcinoma which is independent of HA binding (60). In melanoma, HA and HA recognition have been closely studied and data have shown positive association of HA–CD44 mediation of melanoma cell line migration and invasion (61,62). Further to this, stable expression of HA synthase in melanoma caused enhanced cell motility, which was inhibited by anti-CD44 antibodies (63). In a study by Ahrens et al. (64), both CD44 and RHAMM/IHABP showed increased expression in melanoma progression, but CD44 was the principal HA surface receptor on melanoma cells which mediated the specific HA dose-dependent increase in melanoma cell line proliferation and release of TGF-B1.

Rodent models have extended these *in vitro* studies and provided important insight into the relevance of HA and HA–CD44 interaction *in vivo*.

Two subsets of B16-F1 mouse melanoma cell lines, which had different rates of HA synthesis and consequently a 32-fold difference in surface HA, were injected into the tail vein of mice (62). The melanoma cell line expressing high levels of HA formed a greater number of nodules in the lung and increased rate of mortality compared to the lower HA expressing melanoma cells. The higher HA expressing cells also showed an enhanced interaction with CD44 expressing endothelial cells. Similarly, Itano et al. (65) compared the metastatic potential of a murine mammary carcinoma FM3A HA1 with HA deficient mutant cells. The mutant clones lacked the ability to form HA-rich pericellular coats and had a decreased ability to form metastases compared to the parental cells. Additionally the number of lung metastases by the HA-deficient cells was increased upon rescue of HA levels by transfecting in HAS1. HA and HAS expression have also been examined in human primary tumours and in metastases, if present, in models of breast, colon, ovarian and small cell lung cancer transplanted into SCID mice (66). These data showed intense staining of HA and HAS in the periphery of tumours derived from highly metastatic cell lines (HT29, MCF-7). In addition, even small lung metastases showed focal staining of HA and HAS at the host–tumour interface closely correlating with the invasiveness and metastatic potential of these tumours. It would be interesting to assess the CD44 and RHAMM receptor expression and interaction with HA in such studies.

The cell invasion process leading to a metastatic phenotype and the role of CD44–HA interaction has been investigated using unique murine models. The TA3 murine mammary carcinoma cell line shows CD44-dependent HA binding, branching morphogenesis and invasion (67). Using this model, E-cadherin was demonstrated to negatively regulate CD44-HA function. The increased levels of E-cadherin displayed a weaker binding affinity between CD44 and HA which were manifested by blocking the spread of TA3 cells on HA substratum and ultimately the CD44-mediated branching morphogenesis and tumour cell invasion. This highlights that a balanced coordination of CD44 and E-cadherin may be required for normal epithelial cell function and that imbalance in the up-regulation of CD44–HA interaction or down-regulation of E-cadherin may facilitate tumour progression (67).

The direct mechanisms for increased invasion and metastasis of some cancers with increased HA is poorly understood but taken together, studies with *in vitro* models, syngeneic and human xenograft models have shown that HA may indeed have a critical role in the process of tumour metastasis. However, it appears likely that CD44-mediated tumour progression and metastasis in particular, is not exclusively due to HA–CD44 interactions but clearly involves several mechanisms. For instance, up-regulation of CD44v2-10, which has previously been shown to be preferentially expressed in colorectal liver metastases, contributes an important site of attachment for heparan sulphate (68), which may also promote the metastatic phenotype by sequestration and presentation of heparin binding growth factors.

V. Interaction of Hyaluronan with the Extracellular Matrix

Parallel to the progress in HA receptor research, the interaction of HA with the remodelling of the ECM offered another important mechanism of HA-mediated tumour cell motility and invasion. With the use of 3D collagen gels Docherty et al. (69) showed that HA could aid the movement of fibroblasts in a collagen fibre network, which was in part mediated by HA effects on the spacing of the collagen fibrils. It was subsequently shown that the 3D structure of fibrin gels was critical for endothelial cell migration (70). Since tumours were known to produce increased amounts of HA and also grow in a fibrin-rich environment, it has been hypothesised by Hayen et al. (71) that tumour-derived HA caused increased cell motility by altering the fibrin fibre structure in the ECM. Indeed, this study showed that the fibrin architecture of fibrin gels was altered by HA to allow increased migration of tumour cells and an increased permeability of fibrin clots. Furthermore, HA induction of cell migration was prevented by antibodies to αv and β1 integrin, and the disintegrin echistatin, but not by anti-CD44 antibodies. Therefore in a 3D fibrin substrate model the primary effect of HA on cell migration appeared to be the modulation of fibrin polymerization.

As mentioned earlier, CD44 and HA, in addition to tumour progression are involved in a variety of normal biological processes such as wound repair (72), inflammatory immune response (73), lymphocyte homing and adhesion (74) and embryonic development (75). Not surprisingly, these events would require a coordinated rearrangement of the actin cytoskeleton as a prerequisite of cell adhesion and migration. Oliferenko et al. (76) showed that the GTP binding protein Rac-1 was activated by HA binding to CD44 in murine mammary epithelial cells and lamellipodial extensions were also promoted by local application of HA directly to a passive cell edge. This morphological change could be prevented by prior injection of cells with dominant-negative N17Rac recombinant protein, or by pre-treatment of cells with anti-CD44 antibodies which interfered with HA binding. These data suggest a direct involvement of CD44 in signalling to Rac-1, actin cytoskeleton re-arrangement and in cellular orientation. Furthermore, the cytoskeletal protein ankyrin is believed to interact with the cytoplasmic domain of CD44 (77), signal via the Rho GTPases and correlate with a tumour phenotype (78). Increasingly the ERM (ezrin, radixin, moesin) family of proteins has received attention since they have been found to act as linker proteins connecting the cytoplasmic domains of transmembrane proteins and actin based cortical cytoskeleton (79). Recently, the control of directional cellular motility by the CD44–ezrin complex was shown to be regulated by activation of protein kinase C which triggered a transition from phosphorylation of CD44 at ser-325 to ser-291 (80). Also, HA has been shown to promote signalling interaction between CD44 and TGFβ receptor thereby activating multiple signalling pathways. These include involvement of ankyrin membrane interaction leading to tumour cell motility and oncogenic events such as smad2/smad3 phosphorylation and parathyroid hormone related protein production in metastatic breast cancer (81). Also recently, exciting studies (described below) have crucially shown HA's effects on key cell signalling pathways which brings our understanding closer to identifying some of the downstream consequences of HA–cell interactions.

VI. Hyaluronan and Angiogenesis

A key aspect of tumour establishment and growth involves angiogenesis. Without neovascularization, essential nutrients cannot be supplied to solid tumours, preventing their ability to proliferate, invade or metastasise. HA has been shown to have an important role in vasculature and the angiogenic process (48,82). West and Kumar (83) first showed that HA oligosaccharides increased angiogenesis and that administration of high MW HA inhibited this process (82). This may appear paradoxical to findings that show the production of high MW HA by HAS (84) is correlated to tumour progression, and that inhibition of HAS reduced prostate tumour vascularity (85). More recently, some of the key intermediates such as the tyrosine phosphorylation and membrane recruitment of PLC-γ1 were shown to be activated in bovine aortic endothelial cells by HA oligos (86).

There is little doubt that the involvement of HA in the angiogenic process is complex and requires further work for clarification. However, these observations may in part be explained by the HA degradative enzymes such as the hyaluronidase family of enzymes. HA is known to have a rapid turnover in the body (87), which is regulated by the action of hyaluronidases. Hyal-1 in particular has been well documented to correlate with tumour progression in a range of clinical cancers. Hyal-1 itself has been shown to confer an advantage to metastasis of prostate tumour cells in an orthotopic model of prostate cancer (88). The hyaluronidases break down HA into low MW fragments, which in turn stimulate angiogenesis and ultimately tumour growth (23,89). Therefore, it may be important to assess the endogenous activity of HA degradative enzymes in cell studies which use native high MW HA or which generate HAS. This may reflect closer the observations which are assigned only to the action of intact HA and not to lower MW HA fragment, which may be generated *in situ* depending on the activity of the HA degradative enzymes. In support of the key role of hyaluronidases in cancer, increased HA production by HAS2 over-expression alone in glioma cells, which lacked hyaluronidase activity, was found not to enhance their tumorigenic potential (90). The role of hyaluronidase in cancer biology is also a parallel area of research and it is important to consider both the role of HA and the family of HA degradative enzymes in a wider context and not in isolation. There is continuing debate on the roles of these two players in cancer progression and as our understanding increases, potential strategies are being developed for therapeutic targeting in cancer. The biology of hyaluronidase in cancer is beyond the scope of this chapter but excellent references are available (91,92). Strategies to target HA for therapeutic benefit in cancer are presented below.

VII. Hyaluronan-Mediated Signalling Mechanisms in Cancer

The mechanisms whereby HA receptor mediated effects are translated into cellular signals which coordinate cell communication, movement, growth, survival and transformation are being studied intensely by many research groups. Sohara et al. (93) previously reported that the enhancement of HA-dependent cell movement was affected by the actions of the pan PI3-kinase inhibitor LY294002 or wortmannin suggesting that activation of PI3-kinase by the HA–CD44 interaction was required for cell motility. This observation was tested further to evaluate the effect of PI3-kinase inhibitors on cellular transformation by HAS2 transfectants (32). The inhibition of the PI3-kinase pathway by LY294002 or wortmannin resulted in the HAS2 transfectants reverting from fibroblast shaped cells, which formed overlapping cell layers, to a normal control phenotype. These data highlighted the role of PI3-kinase in the regulation of diminishing contact inhibition induced by formation of increased HA matrix. It will be important to confirm these observations with selective PI3-kinase inhibitors (94).

Other signalling pathways have also been implicated in the HA–CD44 dependent interaction. Cellular transformation by Rous sarcoma virus is mediated via the V-src gene product (95). One of the cellular changes caused by V-src noted many years ago was an accumulation of HA (96) although its role was not clear. Sohara et al. (93) investigated the production of HA on cell motility in cell lines expressing the V-src mutants. The initial observation was that transformation of 3Y1 fibroblast cells by V-src alone activated HA secretion. Additionally, HA treatment caused significant increase in motility of V-src transformed 3Y1 fibroblast cells, which interestingly was inhibited by expression of a dominant negative Ras or treatment with a Ras farnesyltransferase inhibitor. Similarly, a neutralising anti-CD44 antibody also blocked the activation of cell motility and HA-dependent phosphorylation of mitogen activated protein kinase (MAPK) and Akt. This study implicated the simultaneous activation of the Ras-MAPK pathway and the PI3-kinase pathway in HA–CD44 dependent cellular migration.

Increasingly a number of guanine exchange factors (GEFs) have been identified (97) as downstream components of HA-mediated signalling. Evidence of Rac-1 signalling upon binding of HA with CD44 and its effect on tumour cell activation have been described above. Additionally Tiam 1, which is another GEF, was reported to interact with CD44v3 and to up-regulate Rac-1 signalling and cytoskeletal-mediated metastatic breast cancer progression (98). In a continued search for other CD44 isoform-linked GEFs, which correlated with tumour metastasis, Vav2 was identified (99). This group carefully dissected the interaction of CD44v3 and Vav2 and proposed that CD44v3–Vav2 interacts with Grb2-p185HER2 to form a signalling complex that had a pivotal role in promoting cross-talk between RAC1 and Ras signalling pathways, ultimately causing the migration and growth of ovarian cancer.

With the elucidation of CD44 isoforms, and the identification of an increasing number of GEFs, it is apparent that specific CD44 isoforms mediate different functions, including malignant transformation (e.g., CD44v) in different cancers by interactions with specific GEFs. A unique mechanism involving CD44–HA interaction with RhoGEF and Rho kinase was described by Bourguignon et al. (100), which showed that this complex stimulated Gab-1 phosphorylation and membrane localisation. This in turn caused PI3-kinase and Akt activation, and ultimately macrophage colony stimulating factor production in breast cancer cells.

The importance of CD44–HA interaction and signalling in tumorigenesis is clearly an emerging and important area of research and has primarily been studied in the context of cellular growth and motility. However, the impact of CD44–HA interaction on the destruction of the cellular matrix is also being realised. Matrix degrading enzymes such as the family of matrix metalloproteases have a critical role in invasion and metastasis and matrix components such as fibronectin are known to activate matrix metalloprotease-9 (MMP-9) secretion via MEK1–MAPK and the PI3-kinase/Akt signalling pathways (101). Similarly, HA as a major component of the ECM was shown to activate MMP-2 secretion in a focal adhesion kinase (FAK)–MAPK dependent manner in the QG90 lung carcinoma cell line (102). This lung carcinoma is known to express large amounts of CD44s and interestingly, HA-dependent MMP-2 secretion and subsequent cell invasion were inhibited by several methods including anti-CD44 antibody treatment, expression of antisense CD44 or by the pan PI3-kinase inhibitor, wortmannin (103). It appears from these studies that HA-dependent invasion and MMP-2 secretion requires dual signalling pathways, MEK1–MAPK and PI3-kinase. The regulation of HA induced MMP activity is poorly understood but recently it was proposed that the tumour suppressor gene, PTEN may have a role in reducing HA induced MMP-9 secretion in glioblastoma cells by dephosphorylation of FAK in U87MG glioblastoma cells (104). PTEN is a lipid phosphatase that degrades phosphoinositide 3,4,5-triphosphate, a signalling product of the action of PI3-kinase. There is some debate, however, on the importance of the lipid phosphatase activity of PTEN in the regulation of MMP secretion and potential invasion of glioma cells. Studies have shown that this activity is essential (105); conversely other studies have shown that PTEN lipid phosphatase is not required for invasion of glioma cells (106). An overview of some of the HA-mediated signalling pathways involved in cancer are shown in Fig. 1. Clearly major advances in the understanding of HA-mediated signalling events and tumour cell progression have been made; however, further identification of the downstream signalling molecules, their context in pathways and the extent of cross talk of signals that are involved in HA-mediated tumour invasion have yet to be determined.

This chapter has focused primarily on reports using native HA. It is, however, known that fragmented HA also has specific effects on tumour cells. The association of low MW HA on tumour cell proliferation, migration, and angiogenesis was presented above. Additionally it has been shown that fragmented HA also has effects on signalling pathways. CD44 stimulated by fragmented HA induced up-regulation of tyrosine phosphorylation of the c-Met

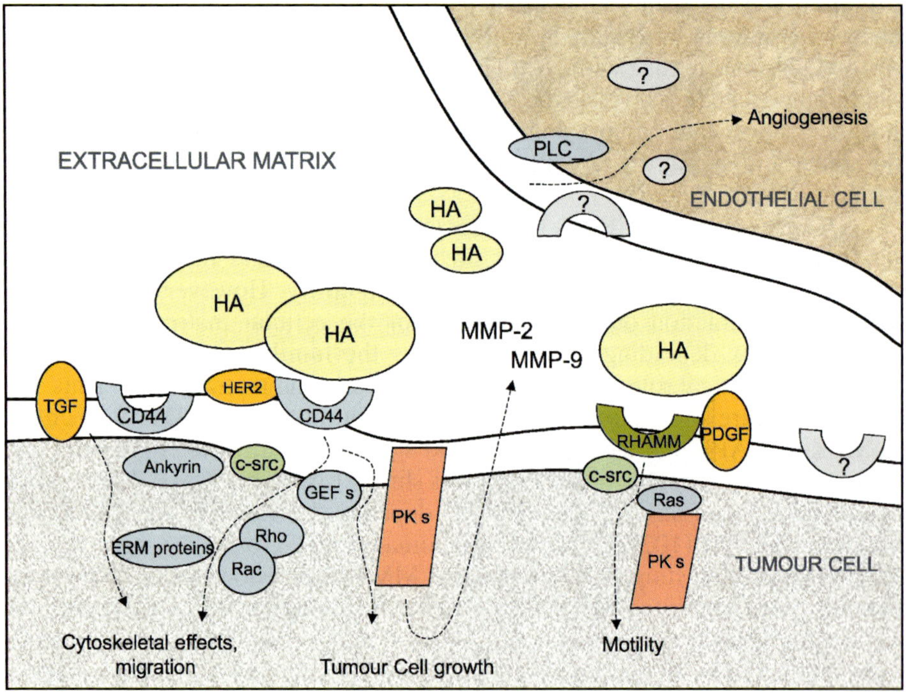

Figure 1 Overview of hyaluronan-mediated signalling mechanisms in cancer. Hyaluronan has been shown to have many effects on signalling cascades. Both native and low MW hyaluronan have been shown to elicit signalling events via hyaluronan receptors such as CD44 or RHAMM or by undefined mechanisms which ultimately affect tumour cell growth, migration, or angiogenesis. HA: hyaluronan; TGFβ: transforming growth factor beta; ERM: ezrin, radixin, moesin family of proteins; PKs: protein kinases, e.g., PI3-kinase pathway; FAK–MAP: kinase pathway; GEFs: guanine exchange factors, e.g., Tiam 1, Vav2; MMP: matrix metalloprotease; PLCγ: phospholipase C gamma; PDGF: platelet-derived growth factor receptor.

receptor (107), activation of MAPK with subsequent enhancement of urokinase-type 1 plasminogen activator and its receptor which ultimately facilitated invasion of human chondrosarcoma cells (108). Clearly fragmented HA, similar to native HA, is also able to initiate signal transduction cascades or promote cross talk originating from CD44–HA interactions.

VIII. Manipulation of Hyaluronan Function as Potential Therapeutic Strategies

The studies described above, and many more, have shown that HA has an intimate role in cancer development, both in cancer cell growth and in metastasis.

Not surprisingly, as research has extended our understanding of the biological role of HA in disease, strategies have also taken shape to manipulate HA function for potential therapeutic benefit.

Even though HA oligos have demonstrated angiogenic effects in tumours, paradoxically, Zeng et al. (109) have shown that HA oligo administration *in vivo*, inhibited melanoma tumour growth and these observations are now being extended to ovarian cancer. The tumour inhibitory effects of HA oligos in this instance are thought to arise from competition for endogenous polymeric HA and replacing high affinity multivalent interactions with weaker low affinity low valency interactions (110). The growth inhibitory effects of HA oligos was further supported by Ghatak et al. (111). In this study HA oligos were shown to inhibit anchorage independent growth in TA3/St murine mammary carcinoma and HCT116 colon adenocarcinoma models. Furthermore, this inhibition was shown to correlate with inhibition of PI3-kinase and phosphorylation of Akt, both of which exert strong anti-apoptotic signals. In addition, HA oligos also stimulated expression of PTEN, which caused downstream decrease in phosphorylation of Akt and consequent activation of pro-apoptotic mediators such as BAD and Forkhead transcription factor (FKHR). However, studies need to be conducted to clarify the critical role of HA oligos in different cancer models.

Another consideration is the potential role of small MW endogenous HA levels generated by the activity of local hyaluronidases, their interaction with HA receptors and possible influence on PI3-kinase signalling. However, such promising data on the use of HA oligos in perturbing HA function is clearly an area under investigation for therapeutic benefit. More recently, HA oligosaccharides have also been reported to suppress the PI3-kinase signalling pathway in multidrug resistant cells (112). Multidrug resistance is a common feature of cancers and arises by several mechanisms such as drug export by ATP-dependent efflux pumps (113). HA oligos were shown to sensitise doxorubicin resistant MCF-7/Adr (MCF-7 breast cancer cells resistant to adriamycin) cells to doxorubicin by 55-fold and to a range of other chemotherapeutics such as taxol, vincristine, BCNU (1,3-bis(2-chloroethyl)-1-nitrosourea). If HA oligos could sensitise drug resistant cells to chemotherapeutics then, increased HA production was hypothesised to induce drug resistance. This was investigated, and indeed transfection of HAS2 into MCF7 cells, which resulted in increased HA production, also induced 10–12-fold increase in resistance to doxorubicin. Interestingly, HA in the ECM of human bone marrow was shown to mediate dexamethasone resistance in multiple myeloma (114). The relation of HA–cell interaction and potential drug resistance may be supported by earlier findings showing that hyaluronidase enhances the effects of chemotherapeutic drugs (115). Effective therapeutic intervention in drug resistant patients is clearly an area of unmet clinical need and strategies using HA oligos, which appear to have a dual therapeutic benefit (i.e., show direct anti-tumour effects *in vivo* and induce apoptosis of drug resistant cancer cells), is highly desirable and should be explored further. The discovery of HA effects on signalling cascades involved in tumour survival have led to other approaches to perturb these effects. Ward et al. (116) have shown that over-expression of soluble HA binding proteins act as a competitive sink for interaction with endogenous HA in glioma which leads to attenuated signalling, inhibition of anchorage independent growth and invasion in matrigel.

Another innovative anticancer strategy employing the physico-chemical properties of HA is under investigation by academic groups (http://www.monash.edu.au/pubs/monmag/issue6-2000/pg12.html) and also industry (e.g., Meditech Research, see http://www.mrl.com.au). These groups are using HA as a sensitising drug delivery vehicle to target drugs to the tumour. The physico-chemical properties of HA allow it to form a meshwork filter at low concentration, which traps drug molecules. Additionally drug targeting is considered to be achieved via interaction of HA receptors which are also up-regulated in tumours. Eliaz et al. (117) have demonstrated that liposomes incorporating HA oligos which encapsulate doxorubicin were targeted to CD44 expressing melanoma cells illustrating the promise of this approach to treat CD44-expressing tumours. Advances exploiting a HA drug delivery strategy will no doubt be closely followed by researchers.

The studies described above, spanning decades of research, have shown that HA is an important component of the ECM which has influence on cell behaviour. In cancers such aberrant cell behaviour is associated with altered HA–cell interactions. Recent work has made outstanding progress in dissecting some of the underlying signalling mechanism involved; however, these studies will clearly need to be extended to fully understand the effects of HA in cancer cells and how to modulate these for potential clinical benefit.

References

1. Tammi MI, Day AJ, Turley EA. Hyaluronan and homeostasis: a balancing act. J Biol Chem 2002; 277:4581–4584.
2. Toole BP. Hyaluronan is not just a goo! J Clin Investig 2000; 106:335–336.
3. Laurent TC, Fraser JR. Hyaluronan. FASEB J 1992; 6:2397–2404.
4. Laurent C, Hellstrom S. Extracellular matrix components reflect the dynamics of a healing tympanic membrane perforation—a histochemical study. Int J Biochem Cell Biol 1997; 29:221–229.
5. Toole BP. Glycosaminoglycans in morphogenesis. In: Hay ED, ed. Cell Biology of Extracellular Matrix. New York: Plenum Press, 1981:259–294.
6. Yung S, Thomas GJ, Davies M. Induction of hyaluronan metabolism after mechanical injury of human peritoneal mesothelial cells *in vitro*. Kidney Int 2000; 58:1953–1962.
7. Gerdin B, Hallgren R. Dynamic role of hyaluronan (HYA) in connective tissue activation and inflammation. J Int Med 1997; 242:49–55.
8. Knudson W. Tumor-associated hyaluronan. Providing an extracellular matrix that facilitates invasion. Am J Pathol 1996; 148:1721–1726.

9. Hautmann SH, Lokeshwar VB, Schroeder GL, Civantos F, Duncan RC, Gnann R, Friedrich MG, Soloway MS. Elevated tissue expression of hyaluronic acid and hyaluronidase validates the HA–HAase urine test for bladder cancer. J Urol 2001; 165:2068–2074.
10. Bertrand P, Girard N, Delpech B, Duval C, d'Anjou J, Dauce JP. Hyaluronan (hyaluronic acid) and hyaluronectin in the extracellular matrix of human breast carcinomas: comparison between invasive and non-invasive areas. Int J Cancer 1992; 52:1–6.
11. Wernicke M, Pineiro LC, Caramutti D, Dorn VG, Raffo MM, Guixa HG, Telenta M, Morandi AA. Breast cancer stromal myxoid changes are associated with tumor invasion and metastasis: a central role for hyaluronan. Mod Pathol 2003; 16: 99–107.
12. Ropponen K, Tammi M, Parkkinen J, Eskelinen M, Tammi R, Lipponen P, Agren U, Alhava E, Kosma VM. Tumor cell-associated hyaluronan as an unfavorable prognostic factor in colorectal cancer. Cancer Res 1998; 58:342–347.
13. Delpech B, Maingonnat C, Girard N, Chauzy C, Maunoury R, Olivier A, Tayot J, Creissard P. Hyaluronan and hyaluronectin in the extracellular matrix of human brain tumour stroma. Eur J Cancer 1993; 29A(7):1012–1017.
14. Franzmann EJ, Schroeder GL, Goodwin WJ, Weed DT, Fisher P, Lokeshwar VB. Expression of tumor markers hyaluronic acid and hyaluronidase (HYAL1) in head and neck tumors. Int J Cancer 2003; 106:438–445.
15. Kojima J, Nakamura N, Kanatani M, Omori K. The glycosaminoglycans in human hepatic cancer. Cancer Res 1975; 35:542–547.
16. Hiltunen EL, Anttila M, Kultti A, Ropponen K, Penttinen J, Yliskoski M, Horai T, Nakamura N, Tateishi R, Hattori S. Glycosaminoglycans in human lung cancer. Cancer 1981; 48:2016–2021.
17. Roboz J, Greaves J, Silides D, Chahinian AP, Holland JF. Hyaluronic acid content of effusions as a diagnostic aid for malignant mesothelioma. Cancer Res 1985; 45: 1850–1854.
18. Hopwood JJ, Dorfman A. Glycosaminoglycan synthesis by Wilms' tumor. Pediatr Res 1978; 12:52–56.
19. Lovvorn HN 3rd, Savani RC, Ruchelli E, Cass DL, Adzick NS. Serum hyaluronan and its association with unfavorable histology and aggressiveness of hetero-transplanted Wilms' tumor. J Pediatr Surg 2000; 35:1070–1078.
20. Anttila MA, Tammi RH, Tammi MI, Syrjanen KJ, Saarikoski SV, Kosma VM. High levels of stromal hyaluronan predict poor disease outcome in epithelial ovarian cancer. Cancer Res 2000; 60:150–155.
21. Hiltunen EL, Anttila M, Kultti A, Ropponen K, Penttinen J, Yliskoski M, Kuronen AT, Juhola M, Tammi R, Tammi M, Kosma VM. Elevated hyaluronan concentration without hyaluronidase activation in malignant epithelial ovarian tumors. Cancer Res 2002; 62:6410–6413.
22. Theocharis AD, Tsara ME, Papageorgacopoulou N, Karavias DD, Theocharis DA. Pancreatic carcinoma is characterized by elevated content of hyaluronan and chondroitin sulfate with altered disaccharide composition. Biochim Biophys Acta 2000; 1502:201–206.
23. Lokeshwar VB, Rubinowicz D, Schroeder GL, Forgacs E, Minna JD, Block NL, Nadji M, Lokeshwar BL. Stromal and epithelial expression of tumor markers hyaluronic acid and HYAL1 hyaluronidase in prostate cancer. J Biol Chem 2001; 276:11922–11932.

24. Aaltomaa S, Lipponen P, Tammi R, Tammi M, Viitanen J, Kankkunen JP, Kosma VM. Strong stromal hyaluronan expression is associated with PSA recurrence in local prostate. Cancer Urol Int 2002; 69:266–272.
25. Posey JT, Soloway MS, Ekici S, Sofer M, Civantos F, Duncan RC, Lokeshwar VB. Evaluation of the prognostic potential of hyaluronic acid and hyaluronidase (HYAL1) for prostate cancer. Cancer Res 2003; 63:2638–2644.
26. Garcia I, Vizoso F, Suarez C, Sanz L, Rodriguez JC, Roiz C, Garcia-Munz JL. Relationship of tumoral hyaluronic acid and cathepsin D contents with histological type of gastric carcinoma. Int J Biol Markers 2000; 15:215–218.
27. Bohm J, Niskanen L, Tammi R, Tammi M, Eskelinen M, Pirinen R, Hollmen S, Alhava E, Kosma VM. Hyaluronan expression in differentiated thyroid carcinoma. J Pathol 2002; 196:180–185.
28. Yeo TK, Nagy JA, Yeo KT, Dvorak HF, Toole BP. Increased hyaluronan at sites of attachment to mesentery by CD44-positive mouse ovarian and breast tumor cells. Am J Pathol 1996; 148:1733–1740.
29. Xing R, Regezi JA, Stern M, Shuster S, Stern R. Hyaluronan and CD44 expression in minor salivary gland tumors. Oral Dis 1998; 4:241–247.
30. Watanabe K, Yamaguchi Y. Molecular identification of a putative human hyaluronan synthase. J Biol Chem 1996; 271:22945–22948.
31. Kosaki R, Watanabe K, Yamaguchi Y. Overproduction of hyaluronan by expression of the hyaluronan synthase Has2 enhances anchorage-independent growth and tumorigenicity. Cancer Res 1999; 59:1141–1145.
32. Itano N, Atsumi F, Sawai T, Yamada Y, Miyaishi O, Senga T, Hamaguchi M, Kimata K. Abnormal accumulation of hyaluronan matrix diminishes contact inhibition of cell growth and promotes cell migration. Proc Natl Acad Sci USA 2002; 99:3609–3614.
33. Underhill CB, Thurn AL, Lacy BE. Characterization and identification of the hyaluronate binding site from membranes of SV-3T3 cells. J Biol Chem 1985; 260:8128–8133.
34. Underhill CB, Green SJ, Comoglio PM, Tarone G. The hyaluronate receptor is identical to a glycoprotein of Mr 85,000 (gp85) as shown by a monoclonal antibody that interferes with binding activity. J Biol Chem 1987; 262: 13142–13146.
35. Aruffo A, Stamenkovic I, Melnick M, Underhill CB, Seed B. CD44 is the principal cell surface receptor for hyaluronate. Cell 1990; 61:1303–1313.
36. Turley EA, Austen L, Vandeligt K, Clary C. Hyaluronan and a cell-associated hyaluronan binding protein regulate the locomotion of ras-transformed cells. J Cell Biol 1991; 112:1041–1047.
37. Banerji S, Ni J, Wang SX, Clasper S, Su J, Tammi R, Jones M, Jackson DG. LYVE-1, a new homologue of the CD44 glycoprotein, is a lymph-specific receptor for hyaluronan. J Cell Biol 1999; 144:789–801.
38. Mouta Carreira C, Nasser SM, di Tomaso E, Padera TP, Boucher Y, Tomarev SI, Jain RK. LYVE-1 is not restricted to the lymph vessels: expression in normal liver blood sinusoids and down-regulation in human liver cancer and cirrhosis. Cancer Res 2001; 61:8079–8084.
39. Day AJ, Prestwich GD. Hyaluronan-binding proteins: tying up the giant. J Biol Chem 2002; 277:4585–4588.
40. Turley EA, Noble PW, Bourguignon LY. Signaling properties of hyaluronan receptors. J Biol Chem 2002; 277:4589–4592.

41. Lee JY, Spicer AP. Hyaluronan: a multifunctional, megaDalton, stealth molecule. Curr Opin Cell Biol 2000; 12:581–586.
42. Isacke CM, Yarwood H. The hyaluronan receptor, CD44. Int J Biochem Cell Biol 2002; 34:718–721.
43. Screaton GR, Bell MV, Jackson DG, Cornelis FB, Gerth U, Bell JI. Genomic structure of DNA encoding the lymphocyte homing receptor CD44 reveals at least 12 alternatively spliced exons. Proc Natl Acad Sci USA 1992; 89: 12160–12164.
44. Bloor BK, Rajarajan A, Jaafary-Haghighat K, Odell EW. Transcription and expression of CD44 variant exons by oro-pharyngeal squamous cell carcinomas. Int J Oncol 2002; 21:907–913.
45. Sherman L, Sleeman J, Herrlich P, Ponta H. Hyaluronate receptors: key players in growth, differentiation, migration and tumor progression. Curr Opin Cell Biol 1994; 6:726–733.
46. Assmann V, Marshall JF, Fieber C, Hofmann M, Hart IR. The human hyaluronan receptor RHAMM is expressed as an intracellular protein in breast cancer cells. J Cell Sci 1998; 111:1685–1694.
47. Hofmann M, Fieber C, Assmann V, Gottlicher M, Sleeman J, Plug R, Howells N, von Stein O, Ponta H, Herrlich P. Identification of IHABP, a 95 kDa intracellular hyaluronate binding protein. J Cell Sci 1998; 111:1673–1684.
48. Toole BP, Wight TN, Tammi MI. Hyaluronan–cell interactions in cancer and vascular disease. J Biol Chem 2002; 277:4593–4596.
49. Rein DT, Roehrig K, Schondorf T, Lazar A, Fleisch M, Niederacher D, Bender HG, Dall P. Expression of the hyaluronan receptor RHAMM in endometrial carcinomas suggests a role in tumour progression and metastasis. J Cancer Res Clin Oncol 2003; 129:161–164.
50. Greiner J, Ringhoffer M, Taniguchi M, Schmitt A, Kirchner D, Krahn G, Heilmann V, Gschwend J, Bergmann L, Dohner H, Schmitt M. Receptor for hyaluronan acid-mediated motility (RHAMM) is a new immunogenic leukemia-associated antigen in acute and chronic myeloid leukemia. Exp Hematol 2002; 30: 1029–1035.
51. Kong QY, Liu J, Chen XY, Wang XW, Sun Y, Li H. Differential expression patterns of hyaluronan receptors CD44 and RHAMM in transitional cell carcinomas of urinary bladder. Oncol Rep 2003; 10:51–55.
52. Yasuda M, Nakano K, Yasumoto K, Tanaka Y. CD44: functional relevance to inflammation and malignancy. Histol Histopathol 2002; 17:945–950.
53. Van Driel M, Gunthert U, van Kessel AC, Joling P, Stauder R, Lokhorst HM, Bloem AC. CD44 variant isoforms are involved in plasma cell adhesion to bone marrow stromal cells. Leukemia 2002; 16:135–143.
54. Haq M, Goltzman D, Tremblay G, Brodt P. Rat prostate adenocarcinoma cells disseminate to bone and adhere preferentially to bone marrow-derived endothelial cells. Cancer Res 1992; 52:4613–4619.
55. Gakunga P, Frost G, Shuster S, Cunha G, Formby B, Stern R. Hyaluronan is a prerequisite for ductal branching morphogenesis. Development 1997; 124: 3987–3997.
56. Simpson MA, Reiland J, Burger SR, Furcht LT, Spicer AP, Oegema TR Jr, McCarthy JB. Hyaluronan synthase elevation in metastatic prostate carcinoma cells correlates with hyaluronan surface retention, a prerequisite for rapid adhesion to bone marrow endothelial cells. J Biol Chem 2001; 276:17949–17957.

57. Simpson MA, Wilson CM, Furcht LT, Spicer AP, Oegema TR Jr, McCarthy JB. Manipulation of hyaluronan synthase expression in prostate adenocarcinoma cells alters pericellular matrix retention and adhesion to bone marrow endothelial cells. J Biol Chem 2002; 277:10050–10057.
58. Nasreen N, Mohammed KA, Hardwick J, Van Horn RD, Sanders K, Kathuria H, Loghmani F, Antony VB. Low molecular weight hyaluronan induces malignant mesothelioma cell (MMC) proliferation and haptotaxis: role of CD44 receptor in MMC proliferation and haptotaxis. Oncol Res 2002; 13:71–78.
59. Alaniz L, Cabrera PV, Blanco G, Ernst G, Rimoldi G, Alvarez E, Hajos SE. Interaction of CD44 with different forms of hyaluronic acid. Its role in adhesion and migration of tumor cells. Cell Commun Adhes 2002; 9:117–130.
60. Sleeman JP, Arming S, Moll JF, Hekele A, Rudy W, Sherman LS, Kreil G, Ponta H, Herrlich P. Hyaluronate-independent metastatic behavior of CD44 variant-expressing pancreatic carcinoma cells. Cancer Res 1996; 56:3134–3141.
61. Thomas L, Byers HR, Vink J, Stamenkovic I. CD44H regulates tumor cell migration on hyaluronate-coated substrate. J Cell Biol 1992; 118:971–977.
62. Zhang L, Underhill CB, Chen L. Hyaluronan on the surface of tumor cells is correlated with metastatic behaviour. Cancer Res 1995; 55:428–433.
63. Ichikawa T, Itano N, Sawai T, Kimata K, Koganehira Y, Saida T, Taniguchi S. Increased synthesis of hyaluronate enhances motility of human melanoma cells. J Investig Dermatol 1999; 113:935–939.
64. Ahrens T, Assmann V, Fieber C, Termeer C, Herrlich P, Hofmann M, Simon JC. CD44 is the principal mediator of hyaluronic-acid-induced melanoma cell proliferation. J Investig Dermatol 2001; 116:93–101.
65. Itano N, Sawai T, Miyaishi O, Kimata K. Relationship between hyaluronan production and metastatic potential of mouse mammary carcinoma cells. Cancer Res 1999; 59:2499–2504.
66. Jojovic M, Delpech B, Prehm P, Schumacher U. Expression of hyaluronate and hyaluronate synthase in human primary tumours and their metastases in scid mice. Cancer Lett 2002; 188:181–189.
67. Xu Y, Yu Q. E-cadherin negatively regulates CD44-hyaluronan interaction and CD44-mediated tumor invasion and branching morphogenesis. J Biol Chem 2003; 278:8661–8668.
68. Barbour AP, Reeder JA, Walsh MD, Fawcett J, Antalis TM, Gotley DC. Expression of the CD44v2-10 isoform confers a metastatic phenotype importance of the heparan sulfate attachment site CD44v3. Cancer Res 2003; 63:887–892.
69. Docherty R, Forrester JV, Lackie JM, Gregory DW. Glycosaminoglycans facilitate the movement of fibroblasts through three-dimensional collagen matrices. J Cell Sci 1989; 92:263–270.
70. Nehls V, Herrmann R. The configuration of fibrin clots determines capillary morphogenesis and endothelial cell migration. Microvasc Res 1996; 51:347–364.
71. Hayen W, Goebeler M, Kumar S, Riessen R, Nehls V. Hyaluronan stimulates tumor cell migration by modulating the fibrin fiber architecture. J Cell Sci 1999; 112:2241–2251.
72. Noble PW. Hyaluronan and its catabolic products in tissue injury and repair. Matrix Biol 2002; 21:25–29.
73. Gunthert U. Importance of CD44 variant isoforms in mouse models for inflammatory bowel disease. Curr Top Microbiol Immunol 1999; 246: 307–312.

74. Siegelman MH, DeGrendele HC, Estess P. Activation and interaction of CD44 and hyaluronan in immunological systems. J Leukoc Biol 1999; 66:315–321.
75. Sherman L, Wainwright D, Ponta H, Herrlich P. A splice variant of CD44 expressed in the apical ectodermal ridge presents fibroblast growth factors to limb mesenchyme and is required for limb outgrowth. Genes Dev 1998; 12: 1058–1071.
76. Oliferenko S, Kaverina I, Small JV, Huber LA. Hyaluronic acid (HA) binding to CD44 activates Rac1 and induces lamellipodia outgrowth. J Cell Biol 2000; 148: 1159–1164.
77. Zhu D, Bourguignon LY. Interaction between CD44 and the repeat domain of ankyrin promotes hyaluronic acid-mediated ovarian tumor cell migration. J Cell Physiol 2000; 183:182–195.
78. Bourguignon LY, Zhu H, Shao L, Zhu D, Chen YW. Rho-kinase (ROK) promotes CD44v(3,8-10)-ankyrin interaction and tumor cell migration in metastatic breast cancer cells. Cell Motil Cytoskeleton 1999; 43:269–287.
79. Tsukita S, Yonemura S. Cortical actin organization: lessons from ERM (ezrin/radixin/moesin) proteins. J Biol Chem 1999; 274:34507–34510.
80. Legg JW, Lewis CA, Parsons M, Ng T, Isacke CM. A novel PKC-regulated mechanism controls CD44 ezrin association and directional cell motility. Nat Cell Biol 2002; 4:399–407.
81. Bourguignon LY, Singleton PA, Zhu H, Zhou B. Hyaluronan promotes signaling interaction between CD44 and the transforming growth factor beta receptor I in metastatic breast tumor cells. J Biol Chem 2002; 277:39703–39712.
82. Rooney P, Kumar S, Ponting J, Wang M. The role of hyaluronan in tumour neovascularization. Int J Cancer 1995; 60:632–636.
83. West DC, Kumar S. Hyaluronan and angiogenesis. Ciba Found Symp 1989; 143: 187–201.
84. Itano N, Sawai T, Yoshida M, Lenas P, Yamada Y, Imagawa M, Shinomura T, Hamaguchi M, Yoshida Y, Ohnuki Y, Miyauchi S, Spicer AP, McDonald JA, Kimata K. Three isoforms of mammalian hyaluronan synthases have distinct enzymatic properties. J Biol Chem 1999; 274:25085–25092.
85. Simpson MA, Wilson CM, McCarthy JB. Inhibition of prostate tumor cell hyaluronan synthesis impairs subcutaneous growth and vascularization in immunocompromised mice. Am J Pathol 2002; 161:849–857.
86. Slevin M, Kumar S, Gaffney J. Angiogenic oligosaccharides of hyaluronan induce multiple signaling pathways affecting vascular endothelial cell mitogenic and wound healing responses. J Biol Chem 2002; 277:41046–41059.
87. Fraser JR, Kimpton WG, Laurent TC, Cahill RN, Vakakis N. Uptake and degradation of hyaluronan in lymphatic tissue. Biochem J 1988; 256:153–158.
88. Patel S, Turner PR, Stubberfield C, Barry E, Rohlff CR, Stamps A, McKenzie E, Young K, Tyson K, Terrett J, Box G, Eccles S, Page MJ. Hyaluronidase gene profiling and role of hyal-1 overexpression in an orthotopic model of prostate cancer. Int J Cancer 2002; 97:416–424.
89. Liu D, Pearlman E, Diaconu E, Guo K, Mori H, Haqqi T, Markowitz S, Willson J, Sy MS. Expression of hyaluronidase by tumor cells induces angiogenesis *in vivo*. Proc Natl Acad Sci USA 1996; 93:7832–7837.
90. Enegd B, King JA, Stylli S, Paradiso L, Kaye AH, Novak U. Overexpression of hyaluronan synthase-2 reduces the tumorigenic potential of glioma cells lacking hyaluronidase activity. Neurosurgery 2002; 50:1311–1318.

91. Csoka TB, Frost GI, Stern R. Hyaluronidases in tissue invasion. Invasion Metastasis 1997; 17:297–311.
92. Junker N, Latini S, Petersen LN, Kristjansen PE. Expression and regulation patterns of hyaluronidases in small cell lung cancer and glioma lines. Oncol Rep 2003; 10:609–616.
93. Sohara Y, Ishiguro N, Machida K, Kurata H, Thant AA, Senga T, Matsuda S, Kimata K, Iwata H, Hamaguchi M. Hyaluronan activates cell motility of v-Src-transformed cells via Ras-mitogen-activated protein kinase and phosphoinositide 3-kinase-Akt in a tumor-specific manner. Mol Biol Cell 2001; 12:1859–1868.
94. Wymann MP, Zvelebil M, Laffargue M. Phosphoinositide 3-kinase signalling—which way to target? Trends Pharmacol Sci 2003; 24:366–376.
95. Jove R, Hanafusa H. Cell transformation by the viral src oncogene. Annu Rev Cell Biol 1987; 3:31–56.
96. Kabat EA. A polysaccharide in tumours due to a virus of leucosis and sarcoma of fowls. J Biol Chem 1939; 130:143–147.
97. Whitehead IP, Campbell S, Rossman KL, Der CJ. Dbl family proteins. Biochim Biophys Acta 1997; 1332:1–23.
98. Bourguignon LY, Zhu H, Shao L, Chen YW. Ankyrin-Tiam1 interaction promotes Rac1 signaling and metastatic breast tumor cell invasion and migration. J Cell Biol 2000; 150:177–191.
99. Bourguignon LY, Zhu H, Zhou B, Diedrich F, Singleton PA, Hung MC. Hyaluronan promotes CD44v3-Vav2 interaction with Grb2-p185(HER2) and induces Rac1 and Ras signaling during ovarian tumor cell migration and growth. J Biol Chem 2001; 276:48679–48692.
100. Bourguignon LY, Singleton PA, Zhu H, Diedrich F. Hyaluronan-mediated CD44 interaction with RhoGEF and Rho kinase promotes Grb2-associated binder-1 phosphorylation and phosphatidylinositol 3-kinase signaling leading to cytokine (macrophage-colony stimulating factor) production and breast tumor progression. J Biol Chem 2003; 278:29420–29434.
101. Thant AA, Nawa A, Kikkawa F, Ichigotani Y, Zhang Y, Sein TT, Amin AR, Hamaguchi M. Fibronectin activates matrix metalloproteinase-9 secretion via the MEK1-MAPK and the PI3K-Akt pathways in ovarian cancer cells. Clin Exp Metastasis 2000; 18:423–428.
102. Zhang Y, Thant AA, Hiraiwa Y, Naito Y, Sein TT, Sohara Y, Matsuda S, Hamaguchi M. A role for focal adhesion kinase in hyluronan-dependent MMP-2 secretion in a human small-cell lung carcinoma cell line, QG90. Biochem Biophys Res Commun 2002; 290:1123–1127.
103. Zhang Y, Thant AA, Machida K, Ichigotani Y, Naito Y, Hiraiwa Y, Senga T, Sohara Y, Matsuda S, Hamaguchi M. Hyaluronan-CD44s signaling regulates matrix metalloproteinase-2 secretion in a human lung carcinoma cell line QG90. Cancer Res 2002; 62:3962–3965.
104. Park MJ, Kim MS, Park IC, Kang HS, Yoo H, Park SH, Rhee CH, Hong SI, Lee SH. PTEN suppresses hyaluronic acid-induced matrix metalloproteinase-9 expression in U87MG glioblastoma cells through focal adhesion kinase dephosphorylation. Cancer Res 2002; 62:6318–6322.
105. Koul D, Parthasarathy R, Shen R, Davies MA, Jasser SA, Chintala SK, Rao JS, Sun Y, Benvenisite EN, Liu TJ, Yung WK. Suppression of matrix metalloproteinase-2 gene expression and invasion in human glioma cells by MMAC/PTEN. Oncogene 2001; 20:6669–6678.

106. Maier D, Jones G, Li X, Schonthal AH, Gratzl O, Van Meir EG, Merlo A. The PTEN lipid phosphatase domain is not required to inhibit invasion of glioma cells. Cancer Res 1999; 59:5479–5482.
107. Suzuki M, Kobayashi H, Kanayama N, Nishida T, Takigawa M, Terao T. CD44 stimulation by fragmented hyaluronic acid induces upregulation and tyrosine phosphorylation of c-Met receptor protein in human chondrosarcoma cells. Biochim Biophys Acta 2002; 1591:37–44.
108. Kobayashi H, Suzuki M, Kanayama N, Nishida T, Takigawa M, Terao T. CD44 stimulation by fragmented hyaluronic acid induces upregulation of urokinase-type plasminogen activator and its receptor and subsequently facilitates invasion of human chondrosarcoma cells. Int J Cancer 2002; 102:379–389.
109. Zeng C, Toole BP, Kinney SD, Kuo JW, Stamenkovic I. Inhibition of tumor growth *in vivo* by hyaluronan oligomers. Int J Cancer 1998; 77:396–401.
110. Lesley J, Hascall VC, Tammi M, Hyman R. Hyaluronan binding by cell surface CD44. J Biol Chem 2000; 275:26967–26975.
111. Ghatak S, Misra S, Toole BP. Hyaluronan oligosaccharides inhibit anchorage-independent growth of tumor cells by suppressing the phosphoinositide 3-kinase/Akt cell survival pathway. J Biol Chem 2002; 277:38013–38020.
112. Misra S, Ghatak S, Zoltan-Jones A, Toole BP. Regulation of multidrug resistance in cancer cells by hyaluronan. J Biol Chem 2003; 278:25285–25288.
113. Gottesman MM, Fojo T, Bates SE. Multidrug resistance in cancer: role of ATP-dependent transporters. Nat Rev Cancer 2002; 2:48–58.
114. Vincent T, Molina L, Espert L, Mechti N. Hyaluronan, a major non-protein glycosaminoglycan component of the extracellular matrix in human bone marrow, mediates dexamethasone resistance in multiple myeloma. Br J Haematol 2003; 121:259–269.
115. Baumgartner G, Gomar-Hoss C, Sakr L, Ulsperger E, Wogritsch C. The impact of extracellular matrix on the chemoresistance of solid tumors—experimental and clinical results of hyaluronidase as additive to cytostatic chemotherapy. Cancer Lett 1998; 131:85–99.
116. Ward JA, Huang L, Guo H, Ghatak S, Toole BP. Perturbation of hyaluronan interactions inhibits malignant properties of glioma cells. Am J Pathol 2003; 162: 1403–1409.
117. Eliaz RE, Szoka FC Jr. Liposome-encapsulated doxorubicin targeted to CD44: a strategy to kill CD44-overexpressing tumor cells. Cancer Res 2001; 61: 2592–2601.

Chemistry and Biology of Hyaluronan
H.G. Garg and C.A. Hales (editors)

Chapter 14

Hyaluronan in Atherosclerosis and Restenosis

THOMAS N. WIGHT and STEPHEN EVANKO

The Hope Heart Institute Seattle, Washington, U.S.A.

FRANK KOLODGIE, ANDREW FARB and RENU VIRMANI

Armed Forces Institute of Pathology Washington, D.C., U.S.A.

I. Introduction

Hyaluronan is present in low amounts in arteries and veins and associated with the surface of vascular cells as well as dispersed throughout the vascular extracellular space (1,2). Veins usually contain more hyaluronan than arteries (3) possibly contributing to the tortuous and dilated nature of these vessels. However, in some arteries, such as the umbilical artery, the hyaluronan content can be as high as 40% of the total glycosaminoglycans (4). Since these vessels are embedded in tissues experiencing constant deforming forces, elevated hyaluronan may allow these vessels to be flexible under changing pressures.

The amount of hyaluronan can change as the normal functional demand of the artery changes. For example, the ductus arteriosus is a fetal blood vessel that closes off at birth in order to shunt blood directly to the lungs. To facilitate this closure, endothelial cells synthesize and secrete large amounts of hyaluronan into the subendothelial space. This serves to create a swelling pressure and expand this tissue space to allow smooth muscle cells to migrate into this space from the underlying medial layer. This causes the blood vessel to thicken and eventually close (5). Such changes are not unlike those that occur in small and medium-sized arteries that occlude as a result of atherosclerosis and restenosis (see later).

While the amount of hyaluronan in arteries is usually low, hyaluronan content in non-diseased portions of human aorta from patients with type 2 diabetes is elevated above normal (6). In fact, there is a significant correlation between aortic hyaluronan content and duration of diabetes, but not between aortic hyaluronan content and age. Interestingly, sera from type 1 and type 2 diabetic patients increase the synthesis of hyaluronan by cultured human arterial smooth muscle cells (ASMCs) (7,8) but the factor(s) responsible for this simulation are not known. Diabetic factors such as high plasma glucose leads to increased hyaluronan production in the glomeruli of rats (9) and recent studies have shown that glucose stimulates the production of hyaluronan by mesangial cells and promotes macrophage retention *in vitro* (10) but similar responses have not been observed in ASMCs. Hyaluronan can effect the synthesis of plasminogen activator-1 by ASMCs (11), which could eventually contribute to reduced fibrinolytic activity seen in the blood vessels of diabetic patients.

II. Hyaluronan in Atherosclerosis

Atherosclerosis is a disease of medium and large-sized arteries that claims more lives in the western world than any other disease (12). The disease involves the progressive thickening of blood vessels over several decades of life due to accumulations of cells, components of the extracellular matrix (ECM) and deposits of lipids in the form of lipoproteins with the eventual formation of the atherosclerotic plaque. With time, the plaques weaken and rupture leading to thrombosis and occlusion (13,14).

Atherosclerosis is characterized by changes in the content and distribution of hyaluronan. For example, the hyaluronan content of human atherosclerotic plaques generally decreases with increasing severity of atherosclerosis (15–17). However, morphological studies indicate that hyaluronan is present throughout both early and late human atherosclerotic lesions in defined locations (18,19).

Hyaluronan is also present in regions of atherosclerotic lesions that contain inflammatory cells such as macrophages and lymphocytes (18,20) (Fig. 1).

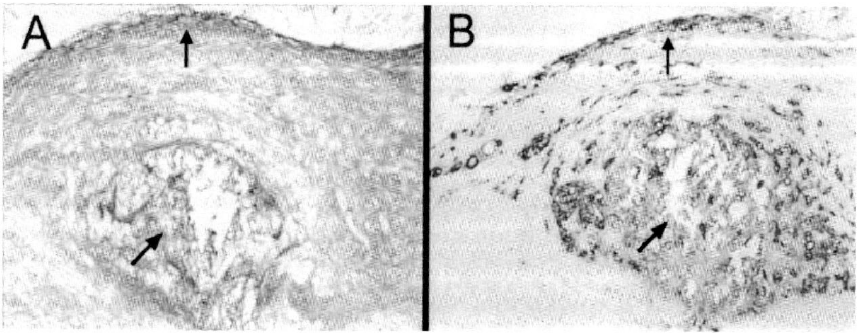

Figure 1 Sections from an atherosclerotic lesion from hypercholesterolemic non-human primates stained for hyaluronan using a biotinylated probe for hyaluronan (A) and an antibody to macrophages (HAM 56). Arrows denote areas of positive staining and indicate staining overlap for hyaluronan and macrophages, × 160. From Ref. 20.

Consistent with this, the extravasation of leukocytes from the blood into the vascular wall involves hyaluronan anchored to the surface of the endothelial cells by CD44 (21) or RHAMM (22) and is mediated by CD44 on the surface of the leukocytes (23–28). These findings place hyaluronan at the beginning of the inflammatory response, thought to be a critical step in the formation of the atherosclerotic lesion (29). Not only is hyaluronan important in the initial stages of leukocyte extravasation but also its accumulation in the early lesions may promote inflammatory cell retention by serving as a substrate for these cells! The presence of hyaluronan in macrophage-rich regions of the plaque (20) supports this possibility. This association of macrophages with hyaluronan appears to occur early since more advanced human atherosclerotic plaques are often devoid of hyaluronan in macrophage-rich regions (Kolodgie, Wight and Virmani, unpublished observations). Macrophages are present in hyaluronan-rich regions in other inflammatory tissues such as in ulcerative colitis through associations with CD44 (30,31). These studies also highlight the importance of hyaluronan binding proteins such as inter α trypsin inhibitor as critical factors in the attachment of leukocytes to hyaluronan. In fact, early studies identified hyaluronan as an agglutinating factor for macrophages (32,33). The importance of the hyaluronan–CD44 connection in developing atherosclerotic lesions is further highlighted by studies that show blocking CD44 receptors on monocytes and lymphocytes by the exogenous administration of hyaluronan prevents their accumulation in developing lesions and markedly reduces the severity of experimental atherosclerosis (34). Furthermore, CD44-null mice crossed with atherosclerosis-prone apo E-deficient mice have a 50–70% reduction in aortic lesions compared to wild-type littermates (35). These lesions are characterized by significant decreases in macrophage content. Furthermore, we have recently found that overexpressing one of the enzymes responsible for hyaluronan synthesis, HAS 1, by retrovial transduction in ASMCs promotes macrophage retention within the hyaluronan-enriched ECM produced by these cells (Wilkinson T, Bressler S and Wight TN, unpublished observations). The mechanism(s) promoting this interaction is under study.

Another aspect for the importance of hyaluronan on macrophage function is the induction of pro-inflammatory cytokines and chemokine expression by hyaluronan degradation products (36–42). Thus, hyaluronan may drive the inflammatory response by not only retaining inflammatory cells but also partly regulating inflammatory cell activation!

Hyaluronan is also present in areas of atherosclerotic lesions that contain extracellular lipid deposits (18,20). In fact, lipoprotein–hyaluronan complexes have been isolated from human atherosclerotic lesions (43) and *in vitro* studies have shown that hyaluronan does interact with phospholipids through hydrophobic interactions (44). Furthermore, lipoproteins influence hyaluronan production by cultured smooth muscle cell-like mesangial cells (45) but it remains to be shown whether lipoproteins affect the biosynthesis of hyaluronan by vascular cells of the atherosclerotic plaque. However, there does appear to be a lipid connection with regards to hyaluronan metabolism in the vascular system.

Experimental animal models of atherosclerosis induced by lipid feeding frequently have elevated levels of hyaluronan associated with developing vascular disease (20,46). Furthermore, atherosclerotic lesions present in apo E-deficient or LDL receptor negative mice and/or rabbits are enriched in hyaluronan and macrophages. In addition, skin fibroblasts taken from patients with familial hypercholesterolemia that lack the LDL receptor exhibit elevated levels of hyaluronan synthesis *in vitro* (Gouëffic, Sakr, Potter-Perigo and Wight, unpublished observations). Such results indicate that lipids may modify hyaluronan production in such a way as to promote a pro-inflammatory pro-atherosclerotic condition. Thus, it is clear that early lesions that contain excess lipid are usually enriched in hyaluronan. Such a concentration of molecules that soften and swell the tissue could very well weaken the plaque and predispose the plaque to rupture.

Hyaluronan is also present in advanced atherosclerotic plaques with acute thrombi (Fig. 2) (47). In fact, hyaluronan together with versican, a chondroitin sulfate proteoglycan that interacts with hyaluronan, is present at the plaque thrombus interface together with CD44-positive immunostaining. It may be that hyaluronan is mediating the adhesion of platelets through a CD44-dependent mechanism promoting thrombosis (48). It is also of interest that hyaluronan accelerates fibrin polymerization (49) suggesting additional roles in the thrombotic process.

Another event that contributes to the growth of atherosclerotic and restenotic plaques and, in part, dictates their severity is the formation of neovessels within the developing lesions (50–53). Hyaluronan may influence multiple events that contribute to the growth of these new blood vessels. For example, fragments of hyaluronan stimulate CD44-mediated endothelial migration and proliferation (54–57). Furthermore, hyaluronan fragments stimulate endothelial cell synthesis of ECM molecules such as type I and VIII collagens, which are macromolecules associated with the angiogenic phenotype (58). Furthermore, hyaluronan fragments promote the formation of new blood vessels *in vivo* (59,60).

III. Hyaluronan in Restenosis

Vessels that become blocked can be treated by removing the plaque by percutaneous transluminal angioplasty, which surgically splits or dissects the plaque to reopen the blood vessel (61). However, in a large percentage of these patients, the treated vessels reocclude in a remarkably short period of time (3–6 months), necessitating further surgery. This process of reocclusion is called restenosis and is thought to involve a combination of blood vessel thickening and tissue shrinkage (62).

Data from experimental animal studies show that hyaluronan is dramatically increased as lesions begin to develop in response to vascular injury (63,64). In the early lesions, hyaluronan is especially enriched around proliferating and

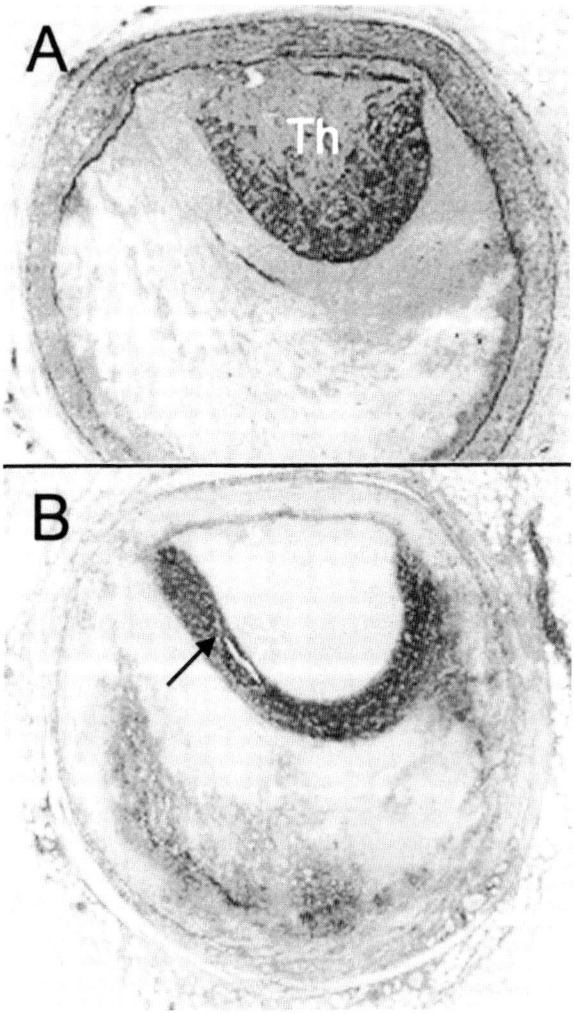

Figure 2 Sections from a human coronary artery that has undergone thrombosis (Th) stained with a Movats stain to reveal extracellular matrix layers (A) and a hyaluronan probe (B). Note a discrete hyaluronan-rich layer (arrow) at the plaque–thrombus interface, × 20. From Ref. 47.

migrating ASMCs (65–69) (Fig. 3). Factors such as insulin also promote hyaluronan accumulation following vascular injury but it is not clear whether these changes influence ASMC phenotype (67,70). The accumulation of hyaluronan in early stages following vascular injury is often accompanied by increases in molecules that associate with hyaluronan such as versican (71–74), TSG-6 (75) and CD44 (76). These findings suggest that hyaluronan plays a role in the early ASMC proliferative and migratory phases of vascular disease.

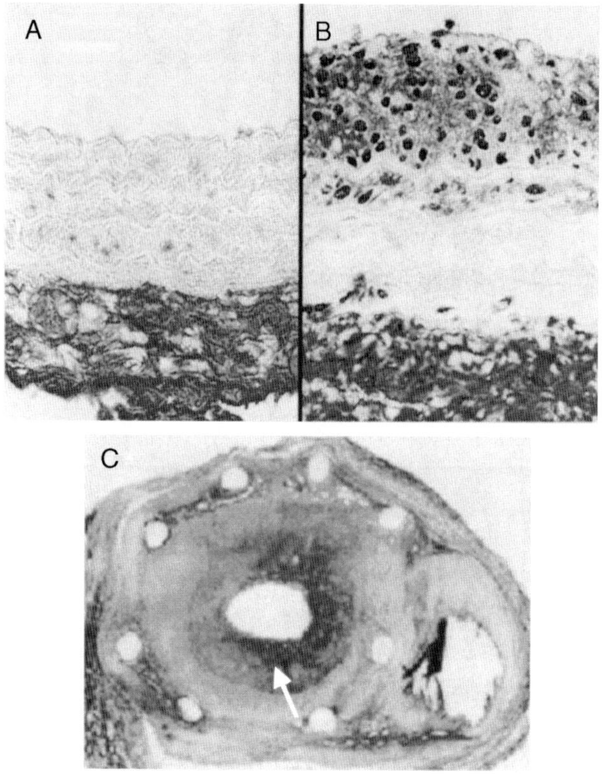

Figure 3 (A) Sections from a normal rat carotid artery (A) and an artery 7 days following balloon injury (B) probed for hyaluronan and immunostained with an antibody to proliferating cell nuclear antigen (PCNA). Note in the normal artery that the hyaluronan was confined to the adventitia with no PCNA positive cells (A). However, in the 7-day injured-vessel (B) intense hyaluronan staining is seen in the thickened intima as well as in the first layer of the media surrounding a number of PCNA positive cells, × 200. From Ref. 68. (C) A section probed for hyaluronan from a stented human coronary artery exhibiting a significant restenotic lesion. The most lumenal aspects of the lesion stained intensely for hyaluronan (arrow), × 20.

The mitogen, PDGF, stimulates hyaluronan synthesis by ASMCs (77,78) (Fig. 4) and promotes the formation of pericellular coats as these cells divide and migrate (79) (Fig. 5). Interference with the binding of hyaluronan to the surface of ASMCs by using either competitive oligosaccharides (79) or blocking antibodies to hyaluronan receptors such as RHAMM (80) blocks ASMC proliferation and migration (Fig. 6). Hyaluronan is also enriched inside proliferating ASMCs (81) suggesting an intracellular role for hyaluronan in this process. The fact that there are multiple intracellular proteins that exhibit hyaluronan binding characteristics supports an intracellular role for this molecule.

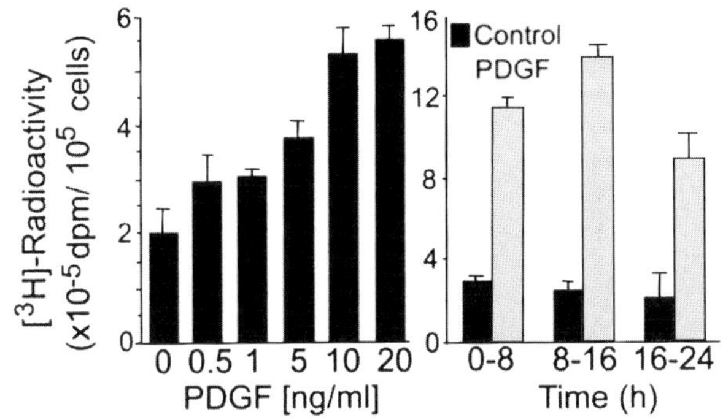

Figure 4 A dose response experiment showing that PDGF increases hyaluronan synthesis by arterial smooth muscle cells (left panel). A time course experiment of hyaluronan synthesis over 8 h intervals following PDGF treatment (right panel). Note that hyaluronan synthesis is elevated early after PDGF stimulation. From Ref. 78.

Hyaluronan also increases when human vessels are subjected to balloon angioplasty during surgical procedures to open blocked arteries. Hyaluronan is a prominent component of ASMC rich in both stented and non-stented human restenotic arteries (68,82,83) (Farb, Kolodgie, Virmani and Wight, unpublished observations). Like the experimental lesions, hyaluronan-binding molecules such as versican accumulate in these lesions as well (84,85). Tissues enriched in hyaluronan have the tendency to trap water and swell. The rapid expansion of

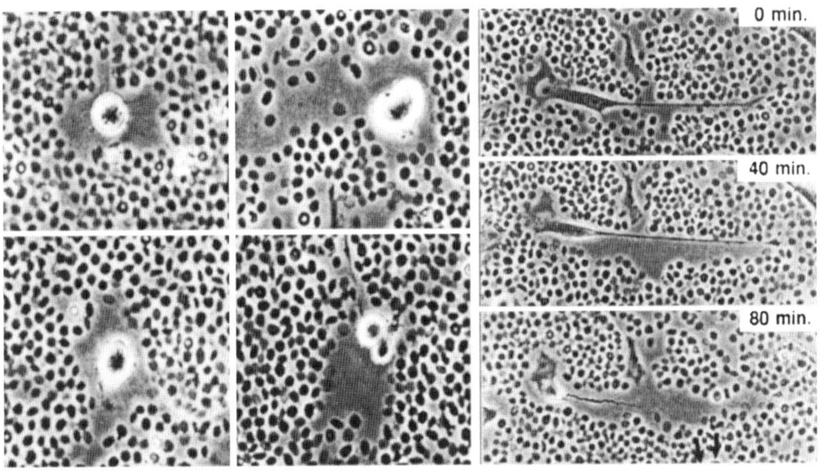

Figure 5 Human arterial smooth muscle cells were cultured in the presence of fixed red blood cells and imaged using video microscopy. The red blood cells are excluded from the pericellular matrix enriched in hyaluronan of a dividing cell (shown in the left four panels) and from a migrating cell (right three panels) using time-lapse video microscopy. Note that the hyaluronan-rich pericellular matrix expands from the sides of the cell as the cell migrates, × 10. From Ref. 79.

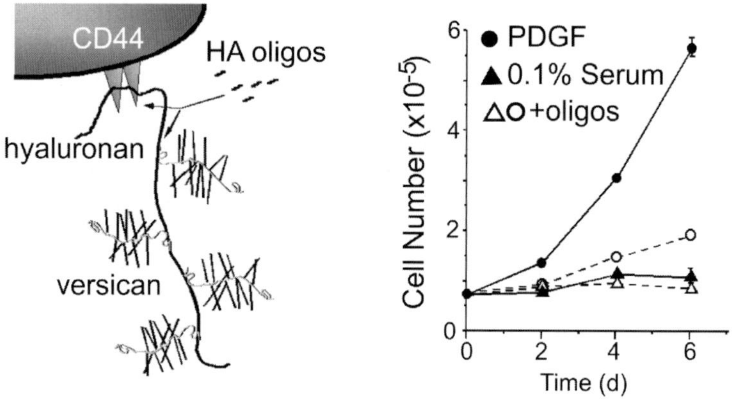

Figure 6 The effect of hyaluronan oligosaccharides or arterial smooth muscle cell proliferation. Left panel is a model of the interaction of hyaluronan with the arterial smooth muscle cell surface through CD44. Hyaluronan oligosaccharides (shown as short lines) can compete for hyaluronan binding sites on CD44 and eliminate hyaluronan and versican binding to the cell surface. The right panel shows a typical growth pattern of human arterial smooth muscle cells stimulated with PDGF in the presence or absence of oligosaccharides. Note that hyaluronan oligosaccharides block smooth muscle cell proliferation in response to PDGF. From Ref. 79.

restenotic lesions could, in large part, be due to edematous changes created by hyaluronan and associated molecules. On the other hand, loss or breakdown of hyaluronan as restenotic lesions' remodel could lead to expulsion of water and tissue shrinkage with reduction in arterial circumference, a condition seen in restenotic lesions. Thus, this conversion may involve a waterlogged ECM becoming a cicatrix that shrinks and contracts the artery, causing loss of lumen diameter. Furthermore, hyaluronan may promote vessel shrinkage following angioplasty by influencing the contraction of the ECM by ASMCs. For example, collagen gels impregnated with hyaluronan show CD44-dependent enhanced contraction when populated by ASMCs (86). Thus, hyaluronan may play significant roles in both the hyperplastic and remodeling phases of human restenosis. It is clear that this molecule could be a useful target in attempts to therapeutically modify the events associated with restenotic lesion progression.

IV. Hyaluronan in Other Vascular Diseases

Hyaluronan also increases in varicose veins (87–89) and may predispose veins to varicosity and thrombosis. Additionally, veins are more susceptible to tumor invasion than arteries and the elevated content of hyaluronan may influence several events in tumorgenesis (90).

V. Conclusions

Hyaluronan is a critical ECM component in atherosclerosis and restenosis. Not only does it contribute to blood vessel wall thickening but also effects the biology of the vascular cells involved in vascular lesion progression. Attention needs to be given to understand the mechanism(s) responsible for hyaluronan accumulation in blood vessel disease as well as the molecular and signaling events that regulate the impact of hyaluronan on vascular cell phenotype. Targeting hyaluronan and/or associated molecules would seem to be a reasonable strategy to limit or prevent atherosclerosis and restenosis.

Acknowledgements

The authors thank past and present members of the laboratory as well as collaborators whose works are cited in this review. This work was supported by NIH grants HL-18645 (DK02456) and HL-71148. The authors also thank Ms Ellen Briggs for her skillful editing and preparation of this manuscript.

References

1. Antonelli A, D'Amore PA. Density-dependent expression of hyaluronic acid binding to vascular cells *in vitro*. Microvasc Res 1991; 41:239–251.
2. Eggli PS, Graber W. Association of hyaluronan with rat vascular endothelial and smooth muscle cells. J Histochem Cytochem 1995; 43:689–697.
3. Varghese ET, Sudhakaran PR, Kurup PA. Studies on the macromolecular components of the bovine aortic and venous tissue. Artery 1986; 14:52–65.
4. Cardoso LE, Erlich RB, Rudge MC, Peracoli JC, Mourao PA. A comparative analysis of glycosaminoglycans from human umbilical arteries in normal subjects and in pathological conditions affecting pregnancy. Lab Investig 1992; 67: 588–595.
5. de Reeder EG, Girard N, Poelmann RE, Van Munsteren JC, Patterson DF, Gittenberger-de Groot AC. Hyaluronic acid accumulation and endothelial cell detachment in intimal thickening of the vessel wall. The normal and genetically defective ductus arteriosus. Am J Pathol 1988; 132:574–585.
6. Heickendorff L, Ledet T, Rasmussen LM. Glycosaminoglycans in the human aorta in diabetes mellitus: a study of tunica media from areas with and without atherosclerotic plaque. Diabetologia 1994; 37:286–292.
7. Jäirveläinen H, Ronnemaa T, Lehtonen A. Effect of sera from male type I (insulin-dependent) diabetics on human aortic smooth muscle cells in culture. Acta Endocrinol 1987; 114:362–370.
8. Järveläinen H, Ronnemaa T, Viikari J. Effect of sera from male type 2 (non-insulin-dependent) diabetics on human aortic smooth muscle cells in culture. Med Biol 1986; 64:361–366.
9. Mahadevan P, Larkins RG, Fraser JR, Fosang AJ, Dunlop ME. Increased hyaluronan production in the glomeruli from diabetic rats: a link between glucose-induced prostaglandin production and reduced sulphated proteoglycan. Diabetologia 1995; 38:298–305.

10. Wang A, Hascall VC. Hyaluronan structures synthesized by rat mesangial cells in response to hyperglycemia induce monocyte adhesion. J Biol Chem 2004; 279: 10279–10285.
11. Marutsuka K, Woodcock-Mitchell J, Sakamoto T, Sobel BE, Fujii S. Pathogenetic implications of hyaluronan-induced modification of vascular smooth muscle cell fibrinolysis in diabetes. Coron Artery Dis 1998; 9:177–184.
12. Kannel WB. Incidence, prevalence, and mortality and coronary artery disease. In: Fuster V, Topol EJ, eds. Atherosclerosis and Coronary Artery Disease. Philadelphia: Lippincott-Raven, 1996; vol. 1:13–24.
13. Benditt EP, Schwartz SM. Blood vessels In: Rubin E, Farber J, eds. Pathology. Philadelphia: J.P. Lippincott, 1988; vol. 1:452–495.
14. Ross R, Fuster V. The pathogenesis of atherosclerosis. In: Fuster V, Topol EJ, eds. Atherosclerosis and Coronary Artery Disease. Philadelphia: Lippincott-Raven, 1996; vol. 1:441–462.
15. Klynstra FB, Bottcher CJ, van Melsen JA, van der Laan EJ. Distribution and composition of acid mucopolysaccharides in normal and atherosclerotic human aortas. J Atheroscler Res 1967; 7:301–309.
16. Tammi M, Seppala PO, Lehtonen A, Mottonen M. Connective tissue components in normal and atherosclerotic human coronary arteries. Atherosclerosis 1978; 29: 191–194.
17. Yla-Herttuala S, Sumuvuori H, Karkola K, Mottonen M, Nikkari T. Glycosaminoglycans in normal and atherosclerotic human coronary arteries. Lab Investig 1986; 54:402–407.
18. Levesque H, Girard N, Maingonnat C, Delpech A, Chauzy C, Tayot J, Courtois H, Delpech B. Localization and solubilization of hyaluronan and of the hyaluronan-binding protein hyaluronectin in human normal and arteriosclerotic arterial walls. Atherosclerosis 1994; 105:51–62.
19. Papakonstantinou E, Roth M, Block LH, Mirtsou-Fidani V, Argiriadis P, Karakiulakis G. The differential distribution of hyaluronic acid in the layers of human atheromatic aortas is associated with vascular smooth muscle cell proliferation and migration. Atherosclerosis 1998; 138:79–89.
20. Evanko S, Raines EW, Ross R, Gold LI, Wight TN. Proteoglycan distribution in lesions of atherosclerosis depends on lesion severity, structural characteristics and the proximity of PDGF and TGF-β1. Am J Pathol 1998; 152:533–546.
21. Nandi A, Estess P, Siegelman MH. Hyaluronan anchoring and regulation on the surface of vascular endothelial cells is mediated through the functionally active form of CD44. J Biol Chem 2000; 275:14939–14948.
22. Lokeshwar VB, Selzer MG. Differences in hyaluronic acid-mediated functions and signaling in arterial, microvessel, and vein-derived human endothelial cells. J Biol Chem 2000; 275:27641–27649.
23. DeGrendele HC, Estess P, Siegelman MH. Requirement for CD44 in activated T cell extravasation into an inflammatory site. Science 1997; 278:672–675.
24. DeGrendele HC, Kosfiszer M, Estess P, Siegelman MH. CD44 activation and associated primary adhesion is inducible via T cell receptor stimulation. J Immunol 1997; 159:2549–2553.
25. Estess P, DeGrendele HC, Pascual V, Siegelman MH. Functional activation of lymphocyte CD44 in peripheral blood is a marker of autoimmune disease activity (see comments). J Clin Investig 1998; 102:1173–1182.

26. Levesque MC, Haynes BF. Cytokine induction of the ability of human monocyte CD44 to bind hyaluronan is mediated primarily by TNF-alpha and is inhibited by IL-4 and IL-13. J Immunol 1997; 159:6184–6194.
27. Mohamadzadeh M, DeGrendele H, Arizpe H, Estess P, Siegelman M. Proinflammatory stimuli regulate endothelial hyaluronan expression and CD44/HA-dependent primary adhesion. J Clin Investig 1998; 101:97–108.
28. Siegelman MH, DeGrendele HC, Estess P. Activation and interaction of CD44 and hyaluronan in immunological systems. J Leukoc Biol 1999; 66:315–321.
29. Ross R. Atherosclerosis—an inflammatory disease. N Engl J Med 1999; 340: 115–126.
30. de La Motte CA, Hascall VC, Calabro A, Yen-Lieberman B, Strong SA. Mononuclear leukocytes preferentially bind via CD44 to hyaluronan on human intestinal mucosal smooth muscle cells after virus infection or treatment with poly(I.C). J Biol Chem 1999; 274:30747–30755.
31. de La Motte CA, Hascall VC, Drazba J, Bandyopadhyay SK, Strong SA. Mononuclear leukocytes bind to specific hyaluronan structures on colon mucosal smooth muscle cells treated with polyinosinic acid:polycytidylic acid:inter-alpha-trypsin inhibitor is crucial to structure and function. Am J Pathol 2003; 163:121–133.
32. Love SH, Shannon BT, Myrvik QN, Lynn WS. Characterization of macrophage agglutinating factor as a hyaluronic acid–protein complex. J Reticuloendothelial Soc 1979; 25:269–282.
33. Lynn WS, Somayajulu RS, Sahu S, Selph J. JI, Quie PG, eds. Characterization of chemotactic agents produced in experimental pleural inflammation. In: Gallin JI, Quie PG, eds. Leukocyte Chemotaxis. New York: Raven Press, 1978; vol. 93: 299–306.
34. Ferns GAA, Konneh M, Rutherford C, Woolaghan E, Anggard EE. Hyaluronan (HYAL-BV 5200) inhibits neo-intimal macrophage influx after balloon-catheter induced injury in the cholesterol-fed rabbit. Atherosclerosis 1995; 114: 157–164.
35. Cuff CA, Kothapalli D, Azonobi I, Chun S, Zhang Y, Belkin R, Yeh C, Secreto A, Assoian RK, Rader DJ, Pure E. The adhesion receptor CD44 promotes atherosclerosis by mediating inflammatory cell recruitment and vascular cell activation. J Clin Investig 2001; 108:1031–1040.
36. Horton MR, Burdick MD, Strieter RM, Bao C, Noble PW. Regulation of hyaluronan-induced chemokine gene expression by IL-10 and IFN-gamma in mouse macrophages. J Immunol 1998; 160:3023–3030.
37. Horton MR, McKee CM, Bao C, Liao F, Farber JM, Hodge-DuFour J, Pure E, Oliver BL, Wright TM, Noble PW. Hyaluronan fragments synergize with interferon-gamma to induce the C-X-C chemokines mig and interferon-inducible protein-10 in mouse macrophages. J Biol Chem 1998; 273:35088–35094.
38. McKee CM, Lowenstein CJ, Horton MR, Wu J, Bao C, Chin BY, Choi AM, Noble PW. Hyaluronan fragments induce nitric-oxide synthase in murine macrophages through a nuclear factor kappaB-dependent mechanism. J Biol Chem 1997; 272: 8013–8018.
39. McKee CM, Penno MB, Cowman M, Burdick MD, Strieter RM, Bao C, Noble PW. Hyaluronan (HA) fragments induce chemokine gene expression in alveolar macrophages. The role of HA size and CD44. J Clin Investig 1996; 98: 2403–2413.

40. Noble PW, Lake FR, Henson PM, Riches DW. Hyaluronate activation of CD44 induces insulin-like growth factor-1 expression by a tumor necrosis factor-alpha-dependent mechanism in murine macrophages. J Clin Invest 1993; 91:2368–2377.
41. Noble PW, McKee CM, Cowman M, Shin HS. Hyaluronan fragments activate an NF-kappa B/I-kappa B alpha autoregulatory loop in murine macrophages. J Exp Med 1996; 183:2373–2378.
42. Noble PW. Hyaluronan and its catabolic products in tissue injury and repair. Matrix Biol 2002; 21:25–29.
43. Srinivasan SR, Yost K, Radhakrishnamurthy B, Dalferes ER Jr, Berenson GS. Lipoprotein–hyaluronate associations in human aorta fibrous plaque lesions. Atherosclerosis 1980; 36:25–37.
44. Pasquali-Ronchetti I, Quaglino D, Mori G, Bacchelli B, Ghosh P. Hyaluronan–phospholipid interactions. J Struct Biol 1997; 120:1–10.
45. Chana RS, Wheeler DC, Thomas GJ, Williams JD, Davies M. Low-density lipoprotein stimulates mesangial cell proteoglycan and hyaluronan synthesis. Nephrol Dial Transplant 2000; 15:167–172.
46. Dattilo JB, Dattilo MP, Yager DR, Makhoul RG. Hypercholesterolemia alters the gene expression of novel components of the extracellular matrix in experimental vein grafts. Ann Vasc Surg 1998; 12:168–173.
47. Kolodgie FD, Burke AP, Farb A, Weber DK, Kutys R, Wight TN, Virmani R. Differential accumulation of proteoglycans and hyaluronan in culprit lesions: insights into plaque erosion. Arterioscler Thromb Vasc Biol 2002; 22:1642–1648.
48. Koshiishi I, Shizari M, Underhill CB. CD44 can mediate the adhesion of platelets to hyaluronan. Blood 1994; 84:390–396.
49. Weigel PH, Frost SJ, LeBoeuf RD, McGary CT. The specific interaction between fibrin(ogen) and hyaluronan: possible consequences in haemostasis, inflammation and wound healing. Ciba Found Symp 1989; 143:248–261 (discussion 261–264).
50. Ignatescu MC, Gharehbaghi-Schnell E, Hassan A, Rezaie-Majd S, Korschineck I, Schleef RR, Glogar HD, Lang IM. Expression of the angiogenic protein, platelet-derived endothelial cell growth factor, in coronary atherosclerotic plaques: *in vivo* correlation of lesional microvessel density and constrictive vascular remodeling. Arterioscler Thromb Vasc Biol 1999; 19:2340–2347.
51. Komatsu R, Ueda M, Naruko T, Kojima A, Becker AE. Neointimal tissue response at sites of coronary stenting in humans: macroscopic, histological, and immunohistochemical analyses. Circulation 1998; 98:224–233.
52. O'Brien ER, Garvin MR, Dev R, Stewart DK, Hinohara T, Simpson JB, Schwartz SM. Angiogenesis in human coronary atherosclerotic plaques. Am J Pathol 1994; 145:883–894.
53. Zhang Y, Cliff WJ, Schoefl GI, Higgins G. Immunohistochemical study of intimal microvessels in coronary atherosclerosis. Am J Pathol 1993; 143:164–172.
54. Montesano R, Kumar S, Orci L, Pepper MS. Synergistic effect of hyaluronan oligosaccharides and vascular endothelial growth factor on angiogenesis *in vitro*. Lab Investig 1996; 75:249–262.
55. Slevin M, Krupinski J, Kumar S, Gaffney J. Angiogenic oligosaccharides of hyaluronan induce protein tyrosine kinase activity in endothelial cells and activate a cytoplasmic signal transduction pathway resulting in proliferation. Lab Invest 1998; 78:987–1003.

56. Trochon V, Mabilat C, Bertrand P, Legrand Y, Smadja-Joffe F, Soria C, Delpech B, Lu H. Evidence of involvement of CD44 in endothelial cell proliferation, migration and angiogenesis *in vitro*. Int J Cancer 1996; 66:664–668.
57. West DC, Kumar S. The effect of hyaluronate and its oligosaccharides on endothelial cell proliferation and monolayer integrity. Exp Cell Res 1989; 183: 179–196.
58. Rooney P, Wang M, Kumar P, Kumar S. Angiogenic oligosaccharides of hyaluronan enhance the production of collagens by endothelial cells. J Cell Sci 1993; 105:213–218.
59. Feinberg RN, Beebe DC. Hyaluronate in vasculogenesis. Science 1983; 220: 1177–1179.
60. West DC, Hampson IN, Arnold F, Kumar S. Angiogenesis induced by degradation products of hyaluronic acid. Science 1985; 228:1324–1326.
61. Nikkari ST, Clowes AW. Restenosis after vascular reconstruction. Ann Med 1994; 26:95–100.
62. Isner JM. Vascular remodeling. Circulation 1994; 89:2937–2941.
63. Helin P, Lorenzen I, Garbarsch C, Mattiessen ME. Repair in arterial tissue. Morphological and biochemical changes in rabbit aorta after a single dilation injury. Circ Res 1971; 29:542–554.
64. Wight TN, Curwen KD, Litrenta MM, Alonso DR, Minick CR. Effect of endothelium on glycosaminoglycan accumulation in injured rabbit aorta. Am J Pathol 1983; 113:156–164.
65. Chajara A, Delpech B, Courel MN, Leroy M, Basuyau JP, Levesque H. Effect of aging on neointima formation and hyaluronan, hyaluronidase and hyaluronectin production in injured rat aorta. Atherosclerosis 1998; 138:53–64.
66. Chajara A, Levesque H, Courel MN, Chauzy C, Maingonnat C, Bertrand P, Delpech B. Hyaluronan and hyaluronectin production in injured rat thoracic aorta. Atherosclerosis 1996; 125:193–207.
67. Chajara A, Raoudi M, Delpech B, Leroy M, Basuyau JP, Levesque H. Increased hyaluronan and hyaluronidase production and hyaluronan degradation in injured aorta of insulin-resistant rats. Arterioscler Thromb Vasc Biol 2000; 20: 1480–1487.
68. Riessen R, Wight TN, Pastore C, Henley C, Isner JM. Distribution of hyaluronan during extracellular matrix remodeling in human restenotic arteries and balloon-injured rat carotid arteries. Circulation 1996; 93:1141–1147.
69. Savani RC, Turley EA. The role of hyaluronan and its receptors in restenosis after balloon angioplasty: development of a potential therapy. Int J Tissue React 1995; 17:141–151.
70. Chajara A, Raoudi M, Delpech B, Courel M, Leroy M, Basuyau JP, Levesque H. Effects of diabetes and insulin treatment of diabetic rats on hyaluronan and hyaluronectin production in injured aorta. J Vasc Res 1999; 36:209–221.
71. Geary RL, Nikkari ST, Wagner WD, Williams JK, Adams MR, Dean RH. Wound healing: a paradigm for lumen narrowing after arterial reconstruction. J Vasc Surg 1998; 27:96–106.
72. Geary RL, Williams JK, Golden D, Brown DG, Benjamin ME, Adams MR. Time course of cellular proliferation, intimal hyperplasia, and remodeling following angioplasty in monkeys with established atherosclerosis. A non-human primate model of restenosis. Arterioscler Thromb Vasc Biol 1996; 16:34–43.

73. Nikkari ST, Järveläinen HT, Wight TN, Ferguson M, Clowes AW. Smooth muscle cell expression of extracellular matrix genes after arterial injury. Am J Pathol 1994; 144:1348–1356.
74. Wolf YG, Rasmussen LM, Ruoslahti E. Antibodies against transforming growth factor-β1 suppress intimal hyperplasia in a rat model. J Clin Investig 1994; 93: 1172–1178.
75. Ye L, Mora R, Akhayani N, Haudenschild CC, Liau G. Growth factor and cytokine-regulated hyaluronan-binding protein TSG-6 is localized to the injury-induced rat neointima and confers enhanced growth in vascular smooth muscle cells. Circ Res 1997; 81:289–296.
76. Jain M, He Q, Lee WS, Kashiki S, Foster LC, Tsai JC, Lee ME, Haber E. Role of CD44 in the reaction of vascular smooth muscle cells to arterial wall injury. J Clin Investig 1996; 98:877.
77. Papakonstantinou E, Karakiulakis G, Roth M, Block LH. Platelet-derived growth factor stimulates the secretion of hyaluronic acid by proliferating human vascular smooth muscle cells. Proc Natl Acad Sci USA 1995; 92:9881–9885.
78. Evanko SP, Johnson PY, Braun KR, Underhill CB, Dudhia J, Wight TN. Platelet-derived growth factor stimulates the formation of versican-hyaluronan aggregates and pericellular matrix expansion in arterial smooth muscle cells. Arch Biochem Biophys 2001; 394:29–38.
79. Evanko SP, Angello JC, Wight TN. Formation of hyaluronan and versican-rich pericellular matrix is required for proliferation and migration of vascular smooth muscle cells. Arterioscler Thromb Vasc Biol 1999; 19:1004–1013.
80. Savani RC, Wang C, Yang B, Zhang S, Kinsella MG, Wight TN, Stern R, Nance DM, Turley EA. Migration of bovine aortic smooth muscle cells after wounding injury: the role of hyaluronan and RHAMM. J Clin Investig 1995; 95:1158–1168.
81. Evanko SP, Wight TN. Intracellular localization of hyaluronan in proliferating cells. J Histochem Cytochem 1999; 47:1331–1342.
82. Riessen R, Henley C, Brogi E, Karsch KR, Wight TN, Isner JM. Hyaluronic acid is a characteristic constituent of the extracellular matrix in human restenotic coronary and peripheral arteries. Eur Heart J 1994; 15(Suppl):247.
83. Chung IM, Gold HK, Schwartz SM, Ikari Y, Reidy MA, Wight TN. Enhanced extracellular matrix accumulation in restenosis of coronary arteries after stent deployment. J Am Coll Cardiol 2002; 40:2072–2081.
84. Matsuura R, Isaka N, Imanaka-Yoshida K, Yoshida T, Sakakura T, Nakano T. Deposition of PG-M/versican is a major cause of human coronary restenosis after percutaneous transluminal coronary angioplasty. J Pathol 1996; 180:311–316.
85. Wight TN, Lara S, Reissen R, LeBaron R, Isner J. Selective deposits of versican in the extracellular matrix of restenotic lesions from human peripheral arteries. Am J Pathol 1997; 151:963–973.
86. Travis JA, Hughes MG, Wong JM, Wagner WD, Geary RL. Hyaluronan enhances contraction of collagen by smooth muscle cells and adventitial fibroblasts: role of CD44 and implications for constrictive remodeling. Circ Res 2001; 88:77–83.
87. Drubaix I, Maraval M, Robert L, Robert AM. Hyaluronic acid (hyaluronan) levels in pathological human saphenous veins. Effects of procyanidol oligomers. Pathol Biol (Paris) 1997; 45:86–91.
88. Drubaix I, Viljanen-Tarifa E, Robert AM, Robert L. Role of glycosoaminoglycans in venous disease. Mode of action of some flavonoid drugs. Pathol Biol (Paris) 1995; 43:461–470.

89. Niebes P, Engels E, Jegerlehner ML. Studies on normal and varicose human saphenous veins. II. Differences in the composition of collagen and glycosaminoglycans. Bibl Anat 1977; 16:301–303.
90. Zeng C, Toole BP, Kinney SD, Kuo JW, Stamenkovic I. Inhibition of tumor growth *in vivo* by hyaluronan oligomers. Int J Cancer 1998; 77:396–401.

Chemistry and Biology of Hyaluronan
H.G. Garg and C.A. Hales (editors)

Chapter 15

Hyaluronan in the Airways

ROSANNA FORTEZA and MATTHIAS SALATHE

Division of Pulmonary and Critical Care Medicine, University of Miami School of Medicine, 1600 NW 10th Ave, RMSB 7058, Miami, FL 33136, USA

GREGORY E. CONNER

Division of Pulmonary and Critical Care Medicine, University of Miami School of Medicine, 1600 NW 10th Ave, RMSB 7058, Miami, FL 33136, USA
Department of Cell Biology, University of Miami School of Medicine, Miami, Florida, USA

I. Introduction

Although hyaluronan has been known to be a component of normal airway secretions for about three decades, the importance of luminal hyaluronan at the airway surface has only been recently recognized. Thus, the discussion here covers a topic that is incompletely understood, but currently under increasing scrutiny.

Hyaluronan is a non-sulfated linear polysaccharide of repeating disaccharide subunits composed of glucuronic acid and *N*-acetylglucosamine, linked by β1,4 and β1,3 glycosidic bonds. Hyaluronan is widely distributed in tissues of all vertebrate species. It is synthesized by hyaluronan synthases (HAS) at the inner face of the plasma membrane. There are three known HAS isoforms. All isoforms make hyaluronan by adding sugar units from nucleotide precursors to a growing disaccharide chain, translocating it directly into the extracellular space (1). Depending on its location, hyaluronan removal occurs through several mechanisms. In the interstitial space, the following two are mainly responsible for hyaluronan removal: [a] cellular uptake and subsequent degradation by lysosomal hyaluronidases and [b] uptake into the lymphatic system and subsequent clearance by the reticuloendothelial system, mainly in the liver. In the epidermis on the other hand, removal of hyaluronan seems to rely mainly on reactive oxygen species (ROS)-induced hyaluronan breakdown (2).

Although most of the knowledge gained about hyaluronan's structure, synthesis, and function has been gathered from studies involving connective tissues and extracellular matrix materials, hyaluronan is secreted by glands in several organs including the airway. This chapter will therefore focus on the current knowledge about the presence and function of hyaluronan in the airway lumen.

II. Airway Hyaluronan: General Aspects

In the airway, hyaluronan is found in the submucosal, connective tissue compartment but also on the luminal, ciliated surface of the airway epithelium as shown by histochemistry of human and ovine tracheal sections using a biotinylated hyaluronan-binding protein (3). Not all tissue sections examined revealed apical staining for hyaluronan, perhaps due to differences in tissue sources and processing, possibly explaining a dissenting report (4). Digestion of tissue sections with hyaluronidase eliminated apical staining for hyaluronan but did not remove all glycoconjugates as seen by retention of Alcian-blue-PAS positive material (3). In addition, chondroitinase ABC at pH 7.5 did not eliminate staining for hyaluronan from the apical surface. Although chondroitinase ABC has hyaluronidase activity, the chondroitinase activity is 10-fold higher than the hyaluronidase activity at the used pH of 7.5 (5). However, when chondroitinase ABC was used at pH 5.6, where it has high hyaluronidase activity, apical staining for hyaluronan was eliminated. Thus, hyaluronan is present at the ciliated border of the airway epithelium (3), but it can also be found in airway secretions (6) and in supernatants of airway epithelial cells grown at the air–liquid interface (ALI) in culture (7).

While hyaluronan is certainly made by airway submucosal gland cells (8,9), it is possible that cells of the superficial epithelium, including ciliated cells, contribute to its production since confluent monolayers of human epithelial cells grown at the ALI culture express HAS 2 and 3 (7). The finding of apically and luminally 'secreted' hyaluronan is not unique to the airway. Polarized release of hyaluronan by epithelial cells at the apical surface has been described in retinal (10), endometrial (11), and mammary epithelial cells (12).

Removal of hyaluronan from the airway lumen is complex, as at least part of luminal hyaluronan seems protected from mucociliary clearance (3). Initial breakdown of hyaluronan is required for its removal and this is likely accomplished by ROS. In fact, ROS may regulate many functions of hyaluronan in the airway lumen and these aspects will be discussed in detail below.

Numerous publications in the early 1970s examined the content of hyaluronan in human bronchoalveolar lavage (BAL) from normal and diseased lungs. Increased hyaluronan levels were found in asthmatic and chronic bronchitic patients, as well as in patients suffering from alveolar proteinosis and adult respiratory distress syndromes (13–15). Since hyaluronan is found in the extracellular matrix of the lung parenchyma, many of these studies interpreted an increase of soluble hyaluronan in BAL as a marker of interstitial lung damage

(16,17). Increases in hyaluronan contents of BAL fluid found in chronic airway diseases were interpreted as a sign of tissue 'remodeling', i.e., release of hyaluronan into the airway lumen from the submucosal connective tissue after epithelial damage. In the 1980s, however, hyaluronan was discovered to be actively secreted from airway gland cells (18–20) and was used as a marker of gland differentiation under various culture conditions (18,21–23). As discussed above, hyaluronan has now also been found on the apical surface of the airway epithelium. Thus, the early interpretations of linking increased BAL hyaluronan content to tissue damage missed the fact that hyaluronan is a component of normal airway secretions (8,19) and that it is present on the airway surface (3). The presence of luminal hyaluronan in secretions is not unique to the lower airways: it is a known component of saliva (24), seminal fluid (25), secretions from the small intestine (26), and secretions from the nasal mucosa (19,27).

The role of hyaluronan in the airway is likely multifold. Originally, it was assumed that it has the only function of conferring optimal physical properties to airway secretions. Hyaluronan forms a continuous three-dimensional chain network with the ability to absorb and release water, thereby contributing to the hydration and thus rheological properties of mucus, critical for proper mucociliary function. However, hyaluronan plays other roles in the connective tissue, providing a concentration-dependent osmotic pressure and regulating macromolecular movement by acting as a size-exclusion 'gel'. Such hyaluronan functions can also be expected in the airways. Finally, hyaluronan has been shown to inhibit and immobilize certain enzymes at the airway epithelial surface as well as regulate airway ciliary beat frequency (CBF) as discussed below (3).

III. Airway Host Defense and Hyaluronan

The lungs fulfill the important task of gas exchange and are therefore uniquely exposed to the outside environment. As human beings ventilate up to 20,000 L of air every day during quiet breathing, they may inhale up to 200,000 bacteria and a big load of dust, even breathing air that is considered clean to EPA standards. Thus, the airways have developed a sophisticated defense mechanism against airborne material of a variety of sizes and composition. The major defense mechanism is the mucociliary escalator, consisting of a mucous blanket on top of the ciliated epithelium. The mucus blanket traps and transports inhaled and deposited particles and chemicals out of the airways by means of coordinated ciliary action. The cilia beat below the mucus blanket, surrounded by periciliary fluid. Ciliary action moves not only the mucus blanket but also the whole periciliary fluid layer, surprisingly at the same speed as the overlying mucus (28).

Both the mucus blanket and the periciliary fluid provide additional defenses to assist the mechanical removal of the surface fluid/mucus that contains the foreign particles. For instance, a variety of antibacterial products, including lactoperoxidase (LPO), lysozyme, lactoferrin, or antiproteases, such as secretory leukoprotease inhibitor (SLPI) and tissue inhibitors of metalloproteases (TIMPs), are present in this layer and work in concert with mucociliary clearance to maintain airway sterility and homeostasis.

Because of the constant removal of the periciliary fluid and mucus blanket from the airway lumen, it would be expected that the defense molecules and enzymes mentioned above will be continuously cleared from the airway. Thus, it was assumed that the secretion of these products into the airway lumen mainly regulates their availability and activity on the airway epithelial surface. Such a paradigm would also apply to other mucosal surfaces where secretions are rapidly and continuously removed, such as tears removed by blinking or intestinal secretions removed by peristalsis. However, the presence of hyaluronan and possibly other glycosaminoglycans on the airway surface changes this situation dramatically. Before explaining why hyaluronan changes this assumption, we will have to review the properties of certain enzymes secreted into the airway lumen. Here we will focus mainly on two, namely tissue kallikrein and LPO.

IV. Tissue Kallikrein and Hyaluronan: Interaction and Enzyme Inhibition

Tissue kallikrein (TK) is a serine protease that generates lysyl-bradykinin in the airways by cleaving kininogens. Bradykinin is an important mediator of airway inflammation and has been implicated in the pathophysiology of asthma (29,30). Tissue kallikrein is made in submucosal gland cells and secreted into the airway, albeit with inhibited enzymatic activity. The inhibition of TK activity in the airway was a puzzle as it is relatively insensitive, at least *in vitro*, to known serine protease inhibitors found in the bronchial lumen (31). As TK activity should be suppressed in the airway lumen under normal conditions, it was concluded that enzyme activity was likely regulated by substrate availability. We have now shown, however, that hyaluronan inhibits bronchial TK activity by binding to it (32). Therefore, hyaluronan functions as a natural inhibitor for bronchial TK in the airway lumen. This finding was novel for bronchial TK in the airways, but hyaluronan and other glycosaminoglycans such as heparin have been shown to modulate the activity of several proteases and protease inhibitors, directly or indirectly. For example, hyaluronan binds to elastic fibers thereby protecting them from proteolytic degradation by elastase (33). In addition, heparin restores the activity of oxidized secreted leukocyte protease inhibitor (SLPI), thereby enhancing its antiprotease activity (34). Heparin also increases the activity of protein C inhibitor to inhibit many serine proteases (35).

The fact that hyaluronan binding to TK inhibits its enzymatic activity suggest a specific interaction between these two molecules. Although there are no specific amino acid sequences shown conclusively to be binding 'motifs' for hyaluronan in general, many hyaluronan-binding proteins contain 'link modules' that are domains of approximately 100 amino acids with four cysteine disulfide-bonds and other highly conserved residues (36). TK does not contain such a link module, however. On the other hand, there are a growing number of hyaluronan-binding proteins that are unrelated to each other and do not have a link module. In fact, the crystal structure of TK (37) reveals surface clusters of basic amino acids close to the catalytic site. These amino acids are good candidates for hyaluronan-binding. We are currently investigating whether these amino acids are in fact responsible for the specific interaction between TK and hyaluronan.

V. Hyaluronan Serves as an Anchor for Secreted Proteins, Preventing Their Removal by Mucociliary Clearance

Many of the discussed molecules that assist mucociliary clearance in host defense are produced and secreted from submucosal gland cells. Tissue kallikrein is also produced there as shown by positive immunostaining in secretory granules of gland cells in tracheal sections (32,38,39). To our surprise, however, we also found staining for TK along the ciliated border of the airway epithelium both in airway sections (3) as well as along cilia in primary cultures of ovine airway epithelial cells, which contain submucosal gland cells. The immunostaining along the ciliary border was not unique to TK: LPO, another host defense enzyme secreted into the airways (40), can also be visualized along the ciliary border (3). In addition, super oxide dismutase has been reported to be expressed around airway epithelial cells as well (41) and can be found by immunohistochemistry at the ciliary surface (42). Since there are no known receptors for these enzymes on the epithelium and since we have shown that hyaluronan is present at the same location, we treated airway sections with hyaluronidase and probed them again for the presence of TK and LPO after treatment. Hyaluronidase, but not chondroitinase ABC (at pH 7.5) or heparinase, eliminated apical staining for LPO and TK, together with the previously mentioned staining for hyaluronan (3). These data suggested that hyaluronan was responsible for retention of TK and LPO at the ciliary border of the airway epithelium, either directly or indirectly. While we know that hyaluronan interacts with TK specifically, LPO does not seem to have a specific hyaluronan-binding site. LPO has an alkaline p*I* of 9.0, however, and it is therefore possible that it binds to hyaluronan by ionic interactions. This interaction can be shown *in vitro*. In contrast to the specific interaction of TK with hyaluronan, the non-specific interaction of LPO with hyaluronan does not affect LPO's activity. Whether LPO truly interacts with hyaluronan *in vivo*, is currently an unanswered question. Since hyaluronan does not have the highest charge density of glycosaminoglycans present in the airways (it is non-sulfated in contrast to other glycosaminoglycans such as chondroitin sulfate), it is also possible that LPO binds through ionic interaction to another glycosaminoglycan that is associated with hyaluronan. In any case, these data suggest that nascent or cell membrane-bound hyaluronan is responsible for the retention of TK, LPO, and possibly other molecules at the airway epithelial surface.

These findings still do not rule out that the material is continuously cleared from the surface by mucociliary action. We therefore examined the transport of

TK and LPO applied to the surface of ovine tracheas under ex vivo conditions. Fluorescently labeled TK (with fluorescein) was immobilized at the site of application whereas rodhamine labeled bovine serum albumin (BSA) moved by mucociliary action (3). Fluorescently labeled LPO was also immobilized at the location of application. Both LPO and TK were immobilized by the presence of apical hyaluronan as hyaluronidase treatment of the tracheal surface allowed mucociliary TK and LPO movement at the same speed as BSA.

These data show that hyaluronan actually immobilizes molecules important for host defense at the airway epithelial surface, thereby protecting them from removal by mucociliary clearance. Enzymatic activity of certain enzymes (such as TK) is inhibited while the activity of others (such as LPO) is unchanged. Thus, secretion of these molecules is not the only determining factor for their availability and activity on the apical side of the airway epithelium.

VI. Receptor for Hyaluronan-Mediated Motility is Expressed at the Apical Border of Epithelial Cells

The mechanisms that retain hyaluronan at the airway epithelial surface are not known, but possibilities include interactions with HAS (43) at the apical membrane (44–46) or interactions of hyaluronan with one of its receptors, possibly CD168 (receptor for hyaluronan-mediated motility, RHAMM). Previous reports indicated that CD44 is found on the basolateral surface of the normal airway epithelium, but not on the apex of normal, ciliated airway epithelial cells; thus, CD44 is an unlikely candidate (47). Since RHAMM is expressed on sperm flagella, which share major ultrastructural features with cilia, and since RHAMM regulates sperm flagellar motility, we examined the expression of RHAMM in ovine tracheal epithelia. Immunohistochemistry with anti-RHAMM antibodies (48) revealed specific staining at the apex of ciliated cells, but no staining in goblet cells (3). To further support that the epithelial epitope recognized by the antibody was in fact RHAMM, we used our ovine tracheal mucosal cDNA library and specific primers. PCR reactions yielded bands of expected sizes (249 bp), co-migrating with an amplified control from mouse RHAMM cDNA. The fragment was sequenced (Genbank #AF310973) and the deduced ovine amino-acid sequence was 91% identical to the human and 81% to the mouse sequence (3). Human airway epithelial cells also express RHAMM: using RNA isolated from cells re-differentiated at the ALI (using both passage 1 and passage 2 of expanded cells) for RT-PCR, RHAMM mRNA expression could be shown in these cells as well.

VII. Ciliary Beating and Hyaluronan

Since hyaluronan enhances sperm motility, we studied the effect of hyaluronan on CBF *in vitro* using digital video-microscopy. Exposure of cultured ciliated epithelial cells to hyaluronan had no effect on CBF. However, when the endogenous hyaluronan on the apical surface of these cells was removed using hyaluronidase (in the presence of protease inhibitors), exposure to 50–100 μg/mL hyaluronan with an average molecular size of ~200 kDa increased CBF by about 15% above baseline (49). This increase was independent of the commercial hyaluronan source. Hyaluronan digested for an extensive period of time with hyaluronidase (18 h) to yield disaccharides, however, had no effect on CBF (3).

The next question was whether RHAMM is involved in signaling cilia to beat faster upon exposure to medium-sized hyaluronan. In fact, functionally blocking anti-RHAMM antibodies completely abolished the ciliary response to hyaluronan. Cells exposed to non-specific, control rabbit anti-chicken IgG responded to hyaluronan with an increase in CBF indistinguishable from untreated controls. These experiments reveal a clear functional role of RHAMM in increasing CBF. RHAMM's exact function in this process is unclear, however, since it lacks a transmembrane signaling component. Future studies are necessary to unravel this interesting problem, especially since the few signaling pathways known to be activated by RHAMM (e.g., ERK) have not been reported to regulate CBF and are usually slower than the observed responses of CBF to mid-sized hyaluronan.

VIII. Hyaluronan Size and Airway Pathophysiology

From the above given data, it is clear that hyaluronan plays an important role in the airway lumen: it immobilizes certain molecules at the apical border of airway epithelial cells, thereby protecting them from mucociliary clearance. It also inhibits the activity of certain enzymes, e.g., TK. While being present there, it also does not seem to influence ciliary beating, a fact supported by a previous study showing that large molecular weight hyaluronan (>1000 kDa), the size likely secreted into the airway lumen as well (see below), had no influence on nasal cell CBF (50).

As has been shown in other systems, the size of hyaluronan is critical for its biological function. Medium to low-sized hyaluronan (200–300 kD) has been demonstrated to stimulate cell proliferation and to initiate signaling cascades involving inflammation (51,52). The same molecular size of hyaluronan has been shown to stimulate sperm motility (53,54) and, as discussed here, CBF (3,49). On the other hand, high molecular weight hyaluronan (>1000 kD) inhibits cell proliferation (55) and does not stimulate CBF (50). Thus, two questions remain to be answered: [1] what size of hyaluronan is present in normal airways and [2] how can large molecular weight hyaluronan be broken down into smaller sized hyaluronan? In other systems, ROS and reactive nitrogen species (RNS) have been shown to be potent inducers of hyaluronan depolymerization to yield smaller sized molecules (56,57). Degradation of hyaluronan by ROS and RNS have therefore been implicated in the development and maintenance of inflammatory events in the lung interstitium (58,59), arthritic joints (56) and the eye (60,61).

In the many airway pathologies, ROS and RNS production is increased and could therefore lead to hyaluronan degradation. Allergen challenge, for instance, causes bronchoconstriction, at least in part, via oxidative stress (62). We, therefore, examined the effects of segmental allergen challenge on the average airway hyaluronan size recovered in BAL from six human volunteers with allergic asthma in comparison with six healthy subjects (samples were kindly provided by Drs Hastie A and Peters S, Thomas Jefferson University, Philadelphia, PA). If our hypothesis that hyaluronan is degraded by oxidative stress applies as in other tissues, the size of recovered hyaluronan should decrease but the amount of soluble hyaluronan should increase due to ROS-mediated cleavage and at least partial release from the cell surface.

Hyaluronan sizes before and 24 h after allergen challenge were determined by agarose gel electrophoresis followed by transfer to a nylon membrane and labeling with a biotinylated hyaluronan-binding protein (63), hyaluronan content was estimated using a biotinylated hyaluronan-binding protein (49,64). The average size of soluble hyaluronan in baseline BAL was ~800 kD and decreased to ~125 kD after allergen (Fig. 1) while hyaluronan concentrations increased from a baseline of 24.3±4.9 to 59.2 ± 18.9 ng/mL ($p < 0.05$). In normal subjects, on the other hand, neither the average hyaluronan size nor its concentration in BAL changed (9.0 ± 1.0 vs. 7.1 ± 4.4 ng/mL, $p > 0.05$). These

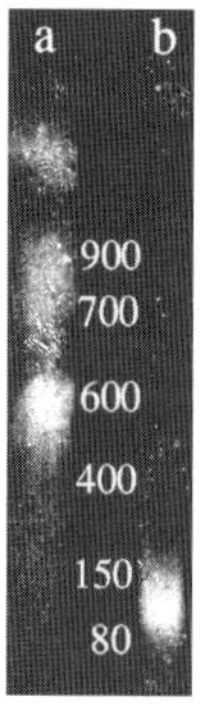

Figure 1 Changes in hyaluronan size after allergen challenge: BAL from asthmatic patients was collected before (a) and 24 h after (b) segmental allergen challenge. The lavages were run on agarose gels, transferred to a nylon membrane, probed with hyaluronan-binding protein coupled to biotin, and visualized with an alkaline phosphatase substrate. Molecular size was estimated using defined DNA molecules (65); their calculated molecular weight was marked on the membrane after transfer using UV illumination and ethidium bromide. Before challenge, there was a wide distribution of hyaluronan sizes (a) with an estimated average size of ~800 kDa. After challenge (b), the average size decreased dramatically and some smaller fractions may have run off the gel. A negative image is shown to provide better visual contrast.

results confirmed data by other investigators showing that asthmatics have a higher hyaluronan content in their BAL (17) and provide novel evidence that hyaluronan is degraded in the airway after allergen challenge, perhaps due to oxidative stress generated by epithelial cells or by phagocytes recruited into the airways. Other possibilities would include hyaluronan degradation by hyaluronidases from resident cellular sources, or, during infection, from bacterial sources.

What are the consequences of hyaluronan degradation in the airways? One response will be the stimulation of CBF, likely increasing mucociliary clearance at the same time, at least initially until ROS may overwhelm cellular defenses and become cytotoxic leading to decreased clearance. The initial response, however, would fit well into an innate host defense response, i.e., an attempt to remove noxious stimuli from the airways. Since TK is bound to and inhibited by hyaluronan in the airways (32), hyaluronan depolymerization by hyaluronidase or ROS would also immediately increase the availability of active TK in the airway lumen. During allergic bronchoconstriction, for instance, hyaluronan degradation may be, at least in part, responsible for TK activation and kinin generation. In favor of this hypothesis we showed that in the BAL of asthmatics after allergen challenge, active TK increased from 15.4 ± 3.6 to 99.3 ± 57.0 ng/mL ($p < 0.001$). In normal subjects, however, TK activity remained stable after allergen challenge compared to baseline (12.4 ± 3.47 vs. 18.5 ± 3.0 ng/mL, $p > 0.05$). We also showed that aerosolized hyaluronan *in vivo* prevented TK-mediated bronchoconstriction in an animal model of asthma (66).

These results confirm and extend previously published reports that TK activity increases in asthmatics after antigen challenge (67) and that soluble hyaluronan is elevated in BAL from chronic asthmatics when compared with normal subjects (17). More importantly, they provide evidence that hyaluronan is degraded after allergen challenge, as shown by a change in its average size. The results also suggest that the increase of TK activity is likely due to TK release from smaller sized hyaluronan. While inhibition constants are difficult to assess for large polymers of hyaluronan, preliminary experiments using recombinant TK suggest that fragments of at least 10 disaccharides can inhibit TK activity.

IX. Concluding Remarks and Outlook

The commonly held notion that enzymes secreted onto epithelial surfaces are rapidly cleared by mechanical action has been challenged by the results summarized in this chapter: some luminal hyaluronan is immobilized at the apex of the airway epithelium where it can bind and retain secreted proteins. Since other mucosal secretions contain hyaluronan and possibly other glycosaminoglycans, these observations in the airway may also be relevant to other epithelia bathed in secretions and cleared by mechanical processes.

How hyaluronan is immobilized at the airway surface remains unknown. The two possibilities include retention by its synthase or binding to RHAMM (although not signaling when bound to high molecular size HA). Large polymers of hyaluronan also immobilize enzymes at the airway surface, including TK and LPO. The three-dimensional structure of TK contains basic amino acids close to the active site of TK that could serve as the specific binding site for hyaluronan. On the other hand, LPO's amino acid sequence does not contain a link module or a potential hyaluronan-binding site. Since LPO has an alkaline p*I* it could therefore interact by non-specific ionic interactions with glycosaminoglycans. In fact, hyaluronan (or other glycosaminoglycans/proteins interacting with hyaluronan) may act in general as cation exchangers that could bind several other cationic proteins present in the airway, many of which are antibacterial proteins, and peptides such as defensins (68).

Under conditions of increased ROS and RNS production (commonly seen in many airway diseases), hyaluronan will break down (Fig. 2). This degradation will have several consequences. First, inactive TK will be released as active TK, generating a pro-inflammatory peptide (lysyl-bradykinin). Since TK has been shown to cleave pro-EGF in other tissues, it is also possible that active TK could initiate a cascade of EGF receptor activation under oxidative stress, potentially contributing to airway metaplasia. Second, smaller hyaluronan fragments will interact with RHAMM to increase CBF. This response can be seen as a fight and flight response; an attempt to remove potentially noxious stimuli out of the

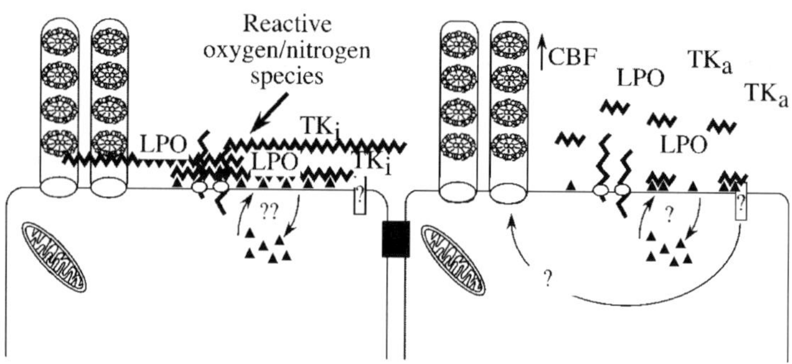

Figure 2 Model: the apical aspect of a ciliated cell depicted on the left reveals the hypothetical situation at rest, where hyaluronan (⩘) is made by hyaluronan synthase (○) in the apical membrane at a high molecular size. This high molecular form does not allow RHAMM (▲) to associate with its unknown transmembrane signaling molecule ([?]). When reactive oxygen/nitrogen species depolymerize hyaluronan, the smaller sized hyaluronan will allow interaction of RHAMM with its signaling molecule; this association, requiring the presence of low molecular weight hyaluronan, initiates signaling to stimulate CBF. Also, high molecular hyaluronan retains lactoperoxidase (LPO) and tissue kallikrein (TK) at the apical surface, protecting these enzymes from mucociliary clearance. In addition, the enzymatic activity of TK is inhibited (TK_i). Upon hyaluronan breakdown, TK is released from hyaluronan and thereby activated (TK_a). TK_a can produce bradykinins in the airways and potentially release growth factors by processing their pro-forms.

airways by increasing the rate of mucociliary clearance. Multiple other functions are possible.

In summary, we propose that hyaluronan and potentially other glycosaminoglycans serve a previously unrecognized pivotal role in mucosal host defense. The data discussed here suggest a new paradigm of airway mucosal defense that involves epithelial-bound hyaluronan retaining a pool of molecules important in host defense, 'ready for use' and protected from ciliary clearance.

References

1. Weigel PH, Hascall VC, Tammi M. Hyaluronan synthases. J Biol Chem 1997; 272: 13997–14000.
2. Agren UM, Tammi RH, Tammi MI. Reactive oxygen species contribute to epidermal hyaluronan catabolism in human skin organ culture. Free Radic Biol Med 1997; 23:996–1001.
3. Forteza R, Lieb T, Aoki T, Savani RC, Conner GE, Salathe M. Hyaluronan serves a novel role in airway mucosal host defense. FASEB J 2001; 15:2179–2186.
4. Pirinen RT, Tammi RH, Tammi MI, Paakko PK, Parkkinen JJ, Agren UM, Johansson RT, Viren MM, Tormanen U, Soini YM, Kosma VM. Expression of hyaluronan in normal and dysplastic bronchial epithelium and in squamous cell carcinoma of the lung. Int J Cancer 1998; 79:251–255.
5. Yamagata T, Saito H, Habuchi O, Suzuki S. Purification and properties of bacterial chondroitinases and chondrosulfatases. J Biol Chem 1968; 243: 1523–1535.
6. Forteza R, Abraham J, Hastie A, Peters S, Salathe M, Conner G. Segmental allergen challenge increases hyaluronic acid and tissue kallikrein in bronchoalveolar lavage of asthmatics. Am J Respir Crit Care Med 2000; 161:A599.
7. Casalino-Matsuda M, Aoki T, Delgado A, Nlend M, Salathe M, Conner G, Forteza R. Reactive oxygen species (ROS) activate epidermal growth factor in the airway surface epithelium through a hyaluronan/tissue kallikrein (TK)-mediated effect. Am J Respir Crit Care Med 2003; 167:A126.
8. Basbaum CB, Finkbeiner WE. Airway secretion: a cell-specific analysis. Horm Metab Res 1988; 20:661–667.
9. Paul A, Picard J, Mergey M, Veissiere D, Finkbeiner WE, Basbaum CB. Glycoconjugates secreted by bovine tracheal serous cells in culture. Arch Biochem Biophys 1988; 260:75–84.
10. De SSP, Calabro A, Nishiyama K, Hu JG, Bok D, Hollyfield JG. Glycosaminoglycan synthesis and secretion by the retinal pigment epithelium: polarized delivery of hyaluronan from the apical surface. J Cell Sci 2001; 114:199–205.
11. Mani SK, Carson DD, Glasser SR. Steroid hormones differentially modulate glycoconjugate synthesis and vectorial secretion by polarized uterine epithelial cells *in vitro*. Endocrinology 1992; 130:240–248.
12. Parry G, Lee EY, Farson D, Koval M, Bissell MJ. Collagenous substrata regulate the nature and distribution of glycosaminoglycans produced by differentiated cultures of mouse mammary epithelial cells. Exp Cell Res 1985; 156: 487–499.
13. Sahu S, Lynn WS. Hyaluronic acid in the pulmonary secretions of patients with asthma. Biochem J 1978; 173:565–568.

14. Sahu S, Lynn WS. Hyaluronic acid in the pulmonary secretions of patients with alveolar proteinosis. Inflammation 1978; 3:149–158.
15. Hallgren R, Samuelsson T, Laurent TC, Modig J. Accumulation of hyaluronan (hyaluronic acid) in the lung in adult respiratory distress syndrome. Am Rev Respir Dis 1989; 139:682–687.
16. Song W, Zhang A, Pang Y, Liu L, Zhao J, Deng S, Zhang S. Fibronectin and hyaluronan in bronchoalveolar lavage fluid from young patients with chronic obstructive pulmonary diseases. Respiration 1995; 62:125–129.
17. Vignola AM, Chavez P, Campbell AM, Souques F, Lebel B, Enander I, Bousquet J. Airway inflammation in mild intermitent and persistent asthmatics. Am J Respir Crit Care Med 1998; 157:403–409.
18. Wu R, Yankaskas J, Cheng E, Knowles MR, Boucher R. Growth and differentiation of human nasal epithelial cells in culture. Serum-free, hormone-supplemented medium and proteoglycan synthesis. Am Rev Respir Dis 1985; 132: 311–320.
19. Baraniuk JN, Shizari T, Sabol M, Ali M, Underhill CB. Hyaluronan is exocytosed from serous, but not mucous cells, of human nasal and tracheobronchial submucosal glands. J Investig Med 1996; 44:47–52.
20. Basbaum CB, Jany B, Finkbeiner WE. The serous cell. Annu Rev Physiol 1990; 52: 97–113.
21. Martin WR, Brown C, Zhang YJ, Wu R. Growth and differentiation of primary tracheal epithelial cells in culture: regulation by extracellular calcium. J Cell Physiol 1991; 147:138–148.
22. Wu R, Wu MM. Effects of retinoids on human bronchial epithelial cells: differential regulation of hyaluronate synthesis and keratin protein synthesis. J Cell Physiol 1986; 127:73–82.
23. Lin H, Carlson DM, St George JA, Plopper CG, Wu R. An ELISA method for the quantitation of tracheal mucins from human and nonhuman primates. Am J Respir Cell Mol Biol 1989; 1:41–48.
24. Pogrel MA, Lowe MA, Stern R. Hyaluronan (hyaluronic acid) in human saliva. Arch Oral Biol 1996; 41:667–671.
25. Geipel U, Kropf J, Krause W, Gressner AM. The concentration pattern of laminin, hyaluronan, and aminoterminal propeptide of type III procollagen in seminal fluid. Andrologia 1992; 24:205–211.
26. Gerdin B, Hallgren R. Localisation of hyaluronan in the human intestinal wall. Gut 1991; 32:760–762.
27. Laurent C, Yoon YJ, Hvidsten I, Hellstrom S. Hyaluronan and alpha-atrial natriuretic polypeptide in human nasal polyps: contributing factors to oedema formation and polyp growth? Acta Otolaryngol 2003; 123:406–412.
28. Matsui T, Sakurai Y, Fujimura Y, Hayashi I, Oh-ishi S, Suzuki M, Hamako J, Yamamoto Y, Yamazaki J, Kinoshita M, Titani K. Purification and amino acid sequence of halystase from snake venom of Agkistrodon halys blomhoffii, a serine protease that cleaves specifically fibrinogen and kininogen. Eur J Biochem 1998; 252:569–575.
29. Proud D. Kinins in the pathogenesis of human airway diseases. Braz J Med Biol Res 1994; 27:2021–2031.
30. Abraham WM, Ahmed A, Cortes A, Soler M, Farmer SG, Baugh LE, Harbeson SL. Airway effects of inhaled bradykinin, substance P, and neurokinin A in sheep. J Allergy Clin Immunol 1991; 87:557–564.

31. Christiansen SC, Zuraw BL, Proud D, Cochrane CG. Inhibition of human bronchial kallikrein in asthma. Am Rev Respir Dis 1989; 139:1125–1131.
32. Forteza R, Lauredo I, Abraham W, Conner G. Bronchial tissue kallikrein activity is regulated by hyaluronic acid binding. Am J Respir Cell Mol Biol 1999; 21: 666–674.
33. Cantor JO, Cerreta JM, Armand G, Turino GM. Further investigation of the use of intratracheally administered hyaluronic acid to ameliorate elastase-induced enphysema. Exp Lung Res 1997; 23:229–244.
34. Fath MA, Wu X, Hileman RE, Linhardt RJ, Kashem MA, Nelson RM, Wright CD, Abraham WM. Interaction of secretory leukocyte protease inhibitor with heparin inhibits proteases involved in asthma. J Biol Chem 1998; 273: 13563–13569.
35. Ecke S, Geiger M, Binder BR. Heparin binding of protein-C inhibitor—analysis of the effect of heparin on the interaction of protein-C inhibitor with tissue kallikrein. Eur J Biochem 1997; 248:475–480.
36. Day A. The structure and regulation of hyaluronan-binding proteins. Biochem Soc Trans 1999; 27:115–121.
37. Katz B, Liu B, Barnes M, Spingman E. Crystal structure of recombinant human tissue kallikrein at 2.0 A resolution. Protein Sci 1998; 7:875–885.
38. Poblete MT, Garces G, Figueroa CD, Bhoola KD. Localization of immunoreactive tissue kallikrein in the seromucous glands of the human and guinea-pig respiratiry tree. Histochem J 1993; 25:834–839.
39. Proud D, Vio CP. Localization of immunoreactive tissue kallikrein in human trachea. Am J Respir Cell Mol Biol 1993; 8:16–19.
40. Gerson C, Sabater J, Scuri M, Torbati A, Coffey R, Abraham JW, Lauredo I, Forteza R, Wanner A, Salathe M, Abraham WM, Conner GE. The lactoperoxidase system functions in bacterial clearance of airways. Am J Respir Cell Mol Biol 2000; 22: 665–671.
41. Oury TD, Chang LY, Marklund SL, Day BJ, Crapo JD. Immunocytochemical localization of extracellular superoxide dismutase in human lung. Lab Investig 1994; 70:889–898.
42. Bowler RP, Crapo JD. Oxidative stress in airways: is there a role for extracellular superoxide dismutase? Am J Respir Crit Care Med 2002; 166:S38–S43.
43. Ahmed T, D'Brot J, Abraham WM, Lucio J, Mendelssohn R, Robinson MJ, Shakir S, Sanpedro B. Heterogeneity of allergic airway responses in sheep: differences in signal tranduction? Am J Respir Crit Care Med 1996; 154:843–849.
44. Itano N, Sawai T, Yoshida M, Lenas P, Yamada Y, Imagawa M, Shinomura T, Hamaguchi Y, Yoshida Y, Ohnuri Y, Miyauchi S, Spicer A, Mc Donald J, Kimata K. Three isoforms of mammalian hyaluronan synthases have distinct enzymatic properties. J Biol Chem 1999; 274:25085–25092.
45. Simpson MA, Wilson CM, Furcht LT, Spicer AP, Oegema TR Jr, McCarthy JB. Manipulation of hyaluronan synthase expression in prostate adenocarcinoma cells alters pericellular matrix retention and adhesion to bone marrow endothelial cells. J Biol Chem 2002; 277:10050–10057.
46. Calabro A, Oken MM, Hascall VC, Masellis AM. Characterization of hyaluronan synthase expression and hyaluronan synthesis in bone marrow mesenchymal progenitor cells: predominant expression of HAS1 mRNA and up-regulated hyaluronan synthesis in bone marrow cells derived from multiple myeloma patients. Blood 2002; 100:2578–2585.

47. Lackie P, Baker J, Gunthert U, Holgate S. Expression of CD44 isoforms is icreased in the airway epithelium of asthmatic subjects. Am J Respir Cell Mol Biol 2001; 16: 14–22.
48. Lovvorn HN III, Cass DL, Sylvester KG, Yang EY, Crombleholme TM, Adzick NS, Savani RC. Hyaluronan receptor expression increases in fetal excisional skin wounds and correlates with fibroplasia. J Pediatr Surg 1998; 33:1062–1069.
49. Lieb T, Forteza R, Salathe M. Hyaluronic acid in cultured ovine tracheal cells and its effect on ciliary beat frequency *in vitro*. J Aerosol Med 2000; 13:231–238.
50. Morimoto K, Yamaguchi H, Iwakura Y, Morisaka K, Ohashi Y, Nakai Y. Effects of viscous hyaluronate-sodium solutions on the nasal absorption of vasopressin and an analogue. Pharm Res 1991; 8:471–474.
51. Horton MR, Olman MA, Bao C, White KE, Choi AM, Chin BY, Noble PW, Lowenstein CJ. Regulation of plasminogen activator inhibitor-1 and urokinase by hyaluronan fragments in mouse macrophages [in process citation]. Am J Physiol Lung Cell Mol Physiol 2000; 279:L707–L715.
52. Horton MR, Shapiro S, Bao C, Lowenstein CJ, Noble PW. Induction and regulation of macrophage metalloelastase by hyaluronan fragments in mouse macrophages. J Immunol 1999; 162:4171–4176.
53. Sbracia M, Grasso J, Sayme N, Stronk J, Huszar G. Hyaluronic acid substantially increases the retention of motility in cryopreserved/thawed human spermatozoa. Hum Reprod 1997; 12:1949–1954.
54. Ranganathan S, Bharadwaj A, Datta K. Hyaluronan mediates sperm motility by enhancing phosphorylation of proteins including hyaluronan binding protein. Cell Mol Biol Res 1995; 41:467–476.
55. Delpech B, Girard N, Bertrand P, Courel MN, Chauzy C, Delpech A. Hyaluronan: fundamental principles and applications in cancer. J Intern Med 1997; 242:41–48.
56. Mapp PI, Grootveld MC, Blake DR. Hypoxia, oxidative stress and rheumatoid arthritis. Br Med Bull 1995; 51:419–436.
57. Frati E, Khatib AM, Front P, Panasyuk A, Aprile F, Mitrovic DR. Degradation of hyaluronic acid by photosensitized riboflavin *in vitro*. Modulation of the effect by transition metals, radical quenchers, and metal chelators. Free Radic Biol Med 1997; 22:1139–1144.
58. McKee CM, Lowenstein CJ, Horton MR, Wu J, Bao C, Chin BY, Choi AM, Noble PW. Hyaluronan fragments induce nitric-oxide synthase in murine macrophages through a nuclear factor kappaB-dependent mechanism. J Biol Chem 1997; 272: 8013–8018.
59. McKee CM, Penno MB, Cowman M, Burdick MD, Strieter RM, Bao C, Noble PW. Hyaluronan (HA) fragments induce chemokine gene expression in alveolar macrophages. The role of HA size and CD44. J Clin Investig 1996; 98:2403–2413.
60. Schmut O, Ansari AN, Faulborn J. Degradation of hyaluronate by the concerted action of ozone and sunlight. Ophthal Res 1994; 26:340–343.
61. Kakehashi A, Ueno N, Chakrabarti B. Molecular mechanisms of photochemically induced posterior vitreous detachment. Ophthal Res 1994; 26:51–59.
62. Lansing MW, Ahmed A, Cortes A, Sielczak MW, Wanner A, Abraham WM. Oxygen radicals contribute to antigen-induced airway hyperresponsiveness in conscious sheep. Am Rev Respir Dis 1993; 147:321–326.
63. Yang B, Yang BL, Savani RC, Turley EA. Identification of a common hyaluronan binding motif in the hyaluronan binding proteins RHAMM, CD44 and link protein. EMBO J 1994; 13:286–296.

64. Bray A, Hsu W, Turino GM. Lung hyaluronan as assayed with a biotinilated hyaluronan-binding protein. Exp Lung Res 1994; 20:317–330.
65. Lee HG, Cowman MK. An agarose gel electrophoretic method for analysis of hyaluronan molecular weight distribution. Anal Biochem 1994; 219:278–287.
66. Scuri M, Abraham WM, Botvinnikova Y, Forteza R. Hyaluronic acid blocks porcine pancreatic elastase (PPE)-induced bronchoconstriction in sheep. Am J Respir Crit Care Med 2001; 164:1855–1859.
67. Christiansen SC, Proud D, Sarnoff RB, Juergens U, Cochrane CG, Zuraw BL. Elevation of tissue kallikrein and kinin in the airways of asthmatic subjects after endobronchial allergen challenge. Am Rev Respir Dis 1992; 145:900–905.
68. Cole AM, Ganz T. Human antimicrobial peptides: analysis and application. Biotechniques 2000; 29:822–831.

Chemistry and Biology of Hyaluronan
H.G. Garg and C.A. Hales (editors)

Chapter 16

Hyaluronan Biology in Vocal Fold Morphology and Biomechanics

SUSAN L. THIBEAULT

Division of Otolaryngology—Head and Neck Surgery, 3C120 School of Medicine, 30 North 1900 East, University of Utah, Salt Lake City, UT 84132, USA

I. Introduction

Vocal folds are complex, multilayered structures that consist of a variety of tissue types including epithelium, connective tissue, striated muscle, nerves, and blood vessels (1). As is shown in Fig. 1, these layers of mucosa and muscle include a nonkeratinized, stratified squamous epithelium, a lamina propria deep to the epithelium, and an underlying muscular layer consisting of thyroarytenoid muscle. The lamina propria plays a critical role in the production of voice as the lamina propria's shape and tension determine the vibratory characteristics of the vocal folds. The vibratory characteristics in turn establish the vocal quality or sound produced. The shape and tension are modified by vocal fold pathologies, which arise from the cover and change the biomechanics and subsequent voice source, causing considerable vocal dysphonias. Hyaluronan (HA) found in the extracellular matrix (ECM) of the lamina propria has significant biological and mechanical impact on the properties of the vocal folds and its subsequent vibration. In the vocal folds, HA influences several different functions including tissue viscosity, tissue flow, tissue osmosis, shock absorption, wound healing, and it is space filling (2–4). These functions are especially important in vocal folds due to the constant trauma caused by the vibratory actions of phonation. In particular, the large, loosely coiled molecular structure of HA allows it to

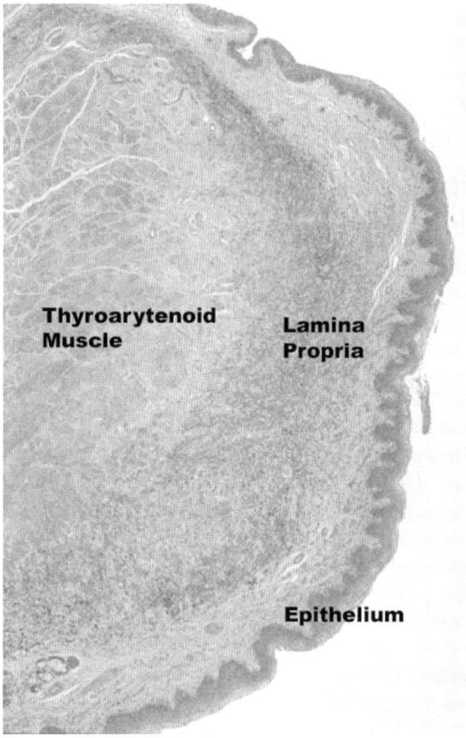

Figure 1 Coronal histological section of a vocal fold, H&E stained at 10× magnification. Thyroarytenoid muscle is the deepest layer with the amorphous lamina propria, which is comprised of fibroblasts and extracellular matrix. The epithelium is the outermost layer.

function as a shock absorber that allows the tissue to resist compression. In this capacity, HA acts as a tissue damper that may protect the vocal fold edges from the oscillatory trauma experienced during phonation. Moreover, the osmotic, viscoelastic and space-filling properties of HA are important in voice because they directly affect the thickness and viscosity of the vocal fold (5,6). This wide range of properties HA modulates suggest that there are pharmacological and/or structural manipulations of HA that may produce valuable treatment options for a number of vocal fold diseases and pathologies.

II. Hyaluronan Localization in the Vocal Fold

In the vocal folds, HA is produced by both fibroblasts and macrophages and likely has a half-life of 3–5 days (7). Gray and co-workers (4) showed that the concentration and distribution of HA in vocal folds is gender-specific with men having a higher average concentration of HA than women. Furthermore, they showed that male vocal folds have a relatively consistent distribution pattern throughout the vocal fold, whereas female vocal folds have a more variable distribution, with HA being less concentrated in the superficial regions and more concentrated in the deeper regions of the lamina propria (Fig. 2). No age differences exist for either sex. Hammond et al. (8) and Gray et al. (3) further noted that HA levels in some cadavers seem more intense in the immediate infrafold area as opposed to the concentration at the exact leading edge of the vocal fold, in both sexes. The localization differences discussed earlier in HA distribution may have clinical significance. It has been suggested that female vocal folds may have a reduced capacity to withstand vibratory trauma and repair damaged tissue because of less HA in the superficial region, a finding that agrees with the clinical observation that female patients have more vocal fold injuries due to voice-use than male patients (4). Furthermore, the higher concentration of HA in the infrafold region corresponds to the region where the mucosal wave begins its vertical travel upward. Biomechanically, the presence of HA in the region makes it ideal to endure the vibratory collision to which the vocal folds are subjected to.

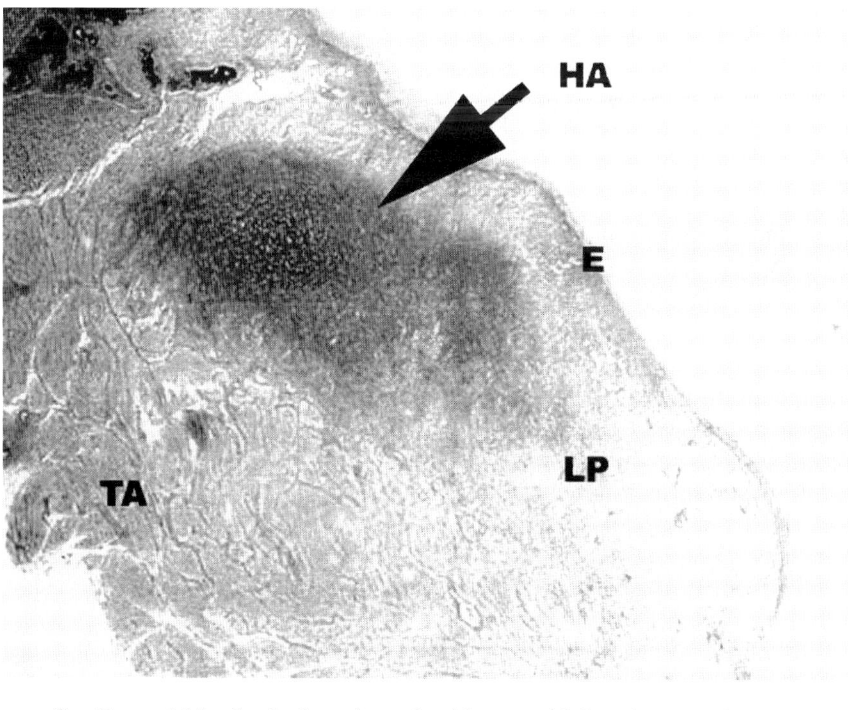

Figure 2 Coronal histological section of a 43-year-old female vocal fold, stained for hyaluronic acid (HA), at 10× magnification. HA is concentrated mainly in the middle lamina propria with an infrafold bulge noted. TA, thyroarytenoid muscle; LP, lamina propria; E, epithelium.

III. Hyaluronan Localization in Vocal Fold Pathologies

Although most vocal fold pathologies result in an altered ECM (9,10) of the lamina propria, neither the presence nor absence of HA appears to be associated with any particular vocal fold pathology. There may indeed be such an association, however, there is a paucity of research in which HA levels have been measured in vocal fold pathologies. Dikkers and Nikkels (11) have reported an increase in HA around the blood vessels of polyps in one-third of male patients. Edlin (12) has reported high levels of HA in vocal fold nodules. Using reverse transcriptase polymerase chain reaction (RT-PCR), Thibeault et al. (13) found similar levels of hyaluronic acid synthase 2 and hyaluronidase in polyps and Reinke's edema.

There has been keen interest in the measurement of HA levels in the lamina propria of vocal folds that have been scarred given the possibility of using exogenous HA to treat the resultant dysphonia. Vocal fold scarring causes devastating vocal dysphonia and there are suboptimal treatment options. HA is the most prominent glycosaminoglycan in the fetal ECM (14), which heals without scar. The role played by HA in influencing the scarless nature of fetal wound healing has been well documented (15,16). Whether or not the use of exogenous HA will decrease the incidence or decrease scar formation in the vocal folds remains to be documented, but offers a stimulating area of wound-healing research.

To date, HA levels have only been measured in two animal models, the rabbit and canine (17–21). At time points representing chronic scar (2 and 6 months after scar induction), utilizing histological measures, HA levels were not significantly different between scar and control. HA levels immediately post-induction have been shown to decrease immediately, with a transient increase at day 5 such that the level of HA was not significantly different than that found in normal vocal folds (19). After this increase, a fall in HA concentration has been reported. This transient increase in HA content in early wounds has been suggested to be a vital event necessary for successive remodeling of the ECM and has also been found in tissue morphogenesis, and limb regeneration, both being biological processes in which the ECM is sequentially remodeled. Combining the reported findings in the literature, at some point between 15 and 60 days HA concentration returns to a normative quantity.

IV. Biomechanical Properties of Hyaluronan

The ability of HA to form highly polarized chains in the ECM allows it to attract and regulate water content, which affects several biomechanical properties of the vocal fold. Vocal fold tissues have been described as viscoelastic, demonstrating both viscous and elastic properties. Viscosity and elasticity are essential to voice because they directly affect the initiation and maintenance of phonation (6,22) as well as the vocal fold fundamental frequency. It has been demonstrated theoretically (23,24) and empirically that phonation threshold pressure, which is an objective indication of perceived vocal effort and ease of phonation is linearly related to the viscous shear properties of the vocal fold cover. Rheological methodology is one way to investigate the viscoelastic shear properties of vocal fold tissues *in vitro* (25,26). For a viscoelastic material, elastic shear modulus (μ or G') is a quantification of the energy storage component of the material in shear deformation (the material's stiffness in shear), while dynamic viscosity (η or η') is a quantification of the energy loss component of the material (the material's resistance to shear flow). Chan and Titze (22,25,27), related shear stiffness and viscosity to the ease of relative displacement and slippage between molecules of the material, which are determined by different kinds of intramolecular and intermolecular interactions.

To determine the rheological characterization of vocal folds, a parallel plate on plate rheometer has been used, as is shown in Fig. 3. This rheometer has a stationary lower plate and a rotating upper plate with a variable gap size (distance between the plates). The vocal fold sample is placed in the gap between the two plates and is subjected to precisely controlled sinusoidal torque. A sensitive transducer monitors the resulting angular displacement and angular velocity of the upper plate as functions of time. Shear stress, shear strain and strain rate associated with the oscillatory shear deformation are computed from the prescribed torque and the measured angular velocity by a computer and viscoelastic data are obtained based on these functions. The viscoelastic data can then be plotted for visualization of the resultant data (Fig. 4).

The viscoelastic shear properties of HA were first measured by Chan and Titze comparing it to that of normal vocal folds (27). It should be noted that increasing the molecular weight of HA or increasing its concentration leads to greater tissue viscosity because either change results in greater entanglement of HA and other matrix molecules (28). Chan and Titze have been interested in finding the HA concentration which has the most similar viscous shear properties

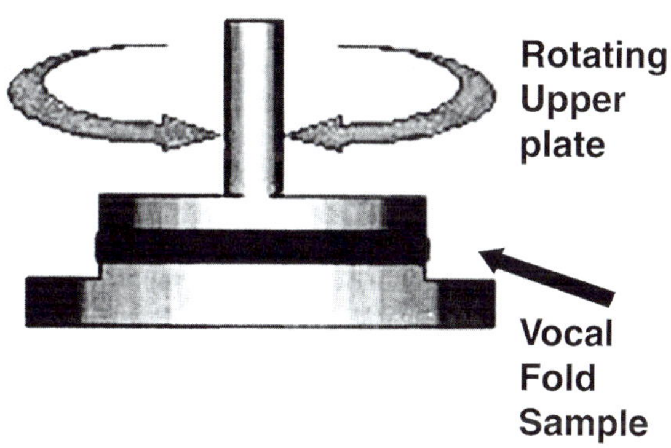

Figure 3 Schematic of a parallel plate on plate rheometer employed to determine viscoelastic measurements. The upper plate rotates, the lower plate is stable. Vocal fold tissue is placed between the two adjustable plates that accommodate various thicknesses.

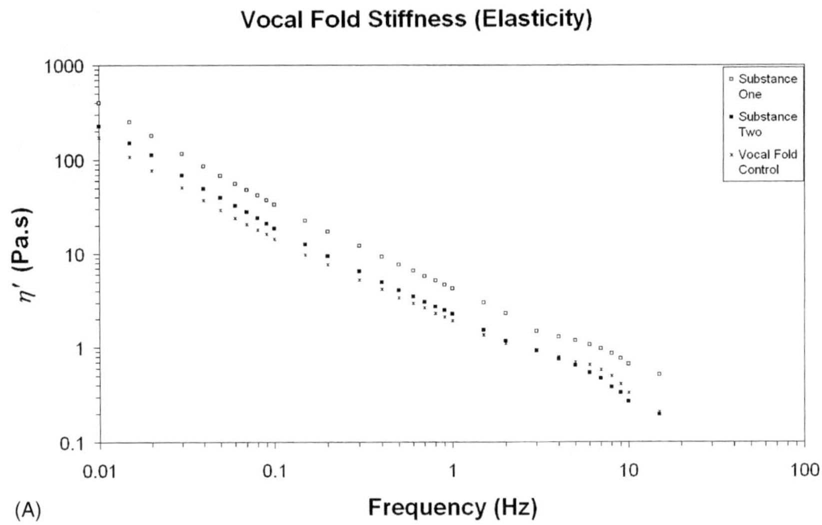

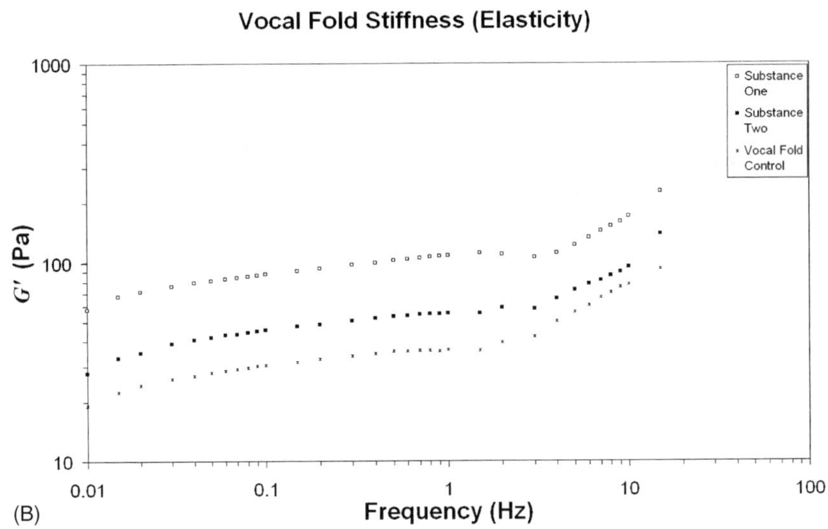

Figure 4 (A) Example diagram of the mean dynamic viscosity as a function of frequency for vocal fold and two hypothetical bioimplantable substances. (B) Example diagram of the mean elasticity as a function of frequency for vocal fold and two hypothetical bioimplantable substances.

of human vocal fold mucosa. *In vitro* rheological measures found that at concentrations of 0.5–1% HA was found to have similar viscous shear properties to normal vocal fold mucosa (22). To further quantify the effects of HA on the viscoelastic properties of HA in the vocal folds, Chan and co-workers compared the biomechanical properties of vocal folds with and without the presence of HA (6,22).

They found that removal of HA from the vocal folds resulted in a 35% average reduction in the stiffness of the vocal fold cover and a 70% average increase in the viscosity of the vocal fold cover at high frequencies (>1 Hz). There was considerable variability in the biomechanical effect of hyaluronidase treatment on individual larynges. These results verify the essential nature of HA in creating optimal conditions for initiation of phonation and vocal fundamental frequency. Vocal fundamental frequency is largely determined by the effective stiffness of the vocal fold, which is regulated both actively by contractile tissue stress (muscular contraction) and passively by noncontractile tissue stress (tissue elastic properties). Furthermore, these results illustrate one of the most interesting properties of HA, the so-called shear thinning effect.

Shear thinning, which refers to a decrease in a material's viscosity as the flow rate or frequency of oscillation of the material increases, is important for both physiological and practical reasons. Physiologically, shear thinning has been postulated to be responsible for the wide range of frequencies produced by the larynx because the viscosity or stiffness of the vocal folds change with frequency (6). From a practical standpoint for an injectable agent, shear thinning is important because the decrease in viscosity that occurs as the flow rate increases allows HA solutions to be passed through a smaller bore needle without a corresponding increase in resistance.

An attempt to alter the *in vivo* viscoelastic properties of mucosal tissue with dilute (0.05%) and concentrated (0.5%) HA, Gray and co-workers (unpublished data) found that HA injections could possibly affect tissue viscosity. However, it appears that while increasing tissue viscosity and stiffness were possible, decreasing tissue viscosity and stiffness through HA injections were difficult. The viscoelastic effect of the dilute HA was very short, less than 24 h, possibly because of rapid diffusion away from the sight of injection. Concentrated HA did lead to a greater viscoelastic effect. This research demonstrates that optimization of HA (purity, rheological properties, molecular weight, concentration, etc.) is required and that HA is unlikely the sole factor in regulating tissue viscoelasticity. The unlikelihood of HA being the sole factor in regulating vocal fold tissue viscoelasticity is further supported by the findings of unaltered HA levels in vocal fold scar measured concomitantly with decreased viscoelastic properties (17–21). It is rather likely that HA is a strategic regulator of a multitude of other proteins that are together responsible for vocal fold tissue viscoelasticity.

V. Therapeutic Uses of Hyaluronan in the Vocal Folds

A substance that could be used to augment the vocal folds would ideally be nonimmunogenic, nontoxic, noninflammatory, and easily injectable. HA meets many of these qualifications and is currently used in a variety of settings including eye and ear surgery, treatment of anthropathies, adhesion management, and wound healing (29). Preliminary studies discussed previously in this chapter indicate that HA may be a favorable implant for the restoration of the biomechanical function of vocal folds. There are two potential vocal fold insufficiency applications for exogenous HA injection into the muscle and injection into the ECM of the lamina propria. It very well may be that completely different HA preparations (rheological properties, concentration, molecular weight, etc.) will be determined to be efficacious for these two applications in the vocal folds—muscle versus ECM of the lamina propria. Premature use of HA in the vocal folds without regard to these factors will more than likely produce conflicting results.

The use of HA to treat laryngeal incompetence, a condition in which the vocal folds do not come together completely, raises several interesting issues. The medial part of the vocal folds, known as the lamina propria and to a lesser extent the medial part of the thyroarytenoid muscle oscillates at voice frequencies to produce sound. Traditionally, surgeons have avoided injecting or placing biomaterials or biological grafts into this tissue because if the biomechanical properties of the tissue were not a near perfect match, the oscillatory pattern of the vocal folds was perturbed, and this would result in a poor voice production. Consequently, materials or grafts such as Teflon, alloderm, fascia and cartilage have been injected or implanted laterally to the oscillating portion of the vocal folds. The introduction of collagen and autogenous fat for injection gave surgeons additional tools, but these still lacked ideal properties for placement into the oscillatory part of the vocal folds. Currently, there are no biomaterials or biological grafts that match the biomechanical properties of vocal folds needed for easy tissue oscillation. HA or modified HA may be a possible injectable material whose biomechanical properties may be potentially matched to vocal folds.

Therapeutic studies of HA in the vocal folds have been limited; to date, all human clinical trials have taken place in Europe. The first study that used HA therapeutically was completed by Hallen et al. (7), which injected a mixture of dextranomere microspheres in sodium hyaluronan solution (1.0%) (DiHA) into the thyroarytenoid muscle of rabbits. A weak inflammatory reaction was observed with HA lasting 7 days. At 6 months, the dextranomeres appeared to recruit fibroblasts with newly generated collagen, resulting in endogenous soft tissue augmentation. In a subsequent human study, Hallen et al. (30) completed unilateral thyroarytenoid injections for patients with unilateral vocal fold paralysis and bowed vocal folds. They measured improved vocal fold closure in five of eight patients and improved vocal quality in three of eight patients. These were encouraging yet there is concern over the slight inflammatory reaction and use of a foreign material (dextranomeres).

The same research group have investigated with use of a cross-linked hyaluronan, Hylan B gel (Hylaform, Genezyme Biosurgery Inc.) in both animal and human trials. Hylan B gel is a viscoelastic, cross-linked (with divinyl sulfone) and insoluble HA derivative, which behaves as a soft gel and has been used for intradermal implantation in plastic surgery. Because Hylan B is cross-linked, it is highly resistant to degradation and migration (31). In an animal study, rabbit thyroarytenoid muscle and the ECM of the lamina propria were injected with

Hylan B gel. Within the first 12 months there were no signs of inflammatory reaction. From 1 to 12 months, there was a gradual ingrowth of connective tissue mainly of collagen around and into the Hylan B injection. No rheological measurements were made of the Hylan B gel *in vitro* or *in vivo*. In humans, Hertegard et al. (32) performed a randomized trial between the cross-linked hyaluronan derived Hylan B gel and bovine collagen for patients with unilateral vocal fold paresis or atrophy. Additionally, a nonrandomized portion of the study included treatment of patients with unilateral vocal fold paresis or vocal fold scar with Hylan B gel only. All injections were made into the thyroarytenoid muscle. Three patients in the HA arm of the study had a short (less than 30 days) episode of reddening or inflammation of the injected vocal fold. Subjective voice ratings indicated more favorable effects on vocal fold function for the patients treated with Hylan B gel. Overall, there was less resorption of the Hylan B gel compared to collagen. Male patients with atrophy had higher levels of resorption yet better glottal closure. In the nonrandomized arm, only one patient demonstrated improvement in glottal parameters. These results indicate a potential promising treatment for medialization into the thyroarytenoid muscle with guarded potential for treatment of the ECM of the lamina propria with Hylan B gel.

To assess the viscoelastic properties of Hylan B gel, Hertegard et al. (33) injected the thyroarytenoid muscles of euthanized rabbits. Comparing the injected vocal folds to nontreated vocal folds, the dynamic viscosity of the vocal folds injected with Hylan B gel was very similar to those of the normal vocal folds. An *in vitro* study assessed the viscoelastic properties of Hylan B gel and DiHA in rabbits 6 months after injection augmentation (34). The dynamic viscosity of Hylan B gel and DiHA-injected vocal folds was similar to that of normal vocal folds with no difference between the two types of injections, 6 months post-injection augmentation.

Though not in laryngology, but in many medical applications that use HA, the viscoelastic properties of HA are important but precise engineering of those properties is not a factor in its medical use. The clinical trials reported to date for treatment of vocal fold diseases place additional considerations on the ideal properties of HA substances. Primarily, the HA used could and should be used to either match the properties of the host tissue, or modified in beneficial ways, such as making the injected soft tissue feel the same or softer than surrounding tissue with functional properties that match the host tissue.

As previously mentioned, there has been considerable interest in the therapeutic use of HA for the treatment of vocal fold scarring. To date, there have been no published animal or human trials using HA to treat vocal fold scarring or use of HA as a prophylaxis to vocal fold scarring.

VI. *In Vivo* Alteration of Hyaluronan Production

Preliminary research has been undertaken with the goal of stimulating fibroblast production of HA with various growth factors. There is a multitude of clinical applications in voice that would benefit from such a treatment including vocal fold scar, paralysis, paresis, and atrophy. All of the work to date has been accomplished *in vitro*. Hirano et al. (35) have demonstrated that treatment of normal canine laryngeal fibroblasts with hepatocyte growth factor, a potent mitogen of hepatocyte and modulates hepatic stellate cell proliferation, collagen formation and the expression of transforming growth factor beta 1, stimulated HA production up to 48 h. In a supplementary study, Hirano et al. (36), found an increase in HA levels from normal canine laryngeal fibroblast cultures up to 7 days with hepatocyte growth factor, epidermal growth factor, basic fibroblast growth factor, and transforming growth factor beta1.

This line of voice research is intriguing, unique and offer therapeutic potential for various vocal fold pathologies. One of the most significant advantages of this line of research is the production of *in vivo* autologous HA that would mimic the tissue's inherent viscoelastic properties. There remains a considerable number of unanswered questions in regard to the line of research that is acknowledged by Hirano et al. (35). Further investigation into this area of study requires incorporation of phenotypically altered fibroblasts (i.e., vocal fold scar fibroblasts), further utilization of other growth factors, and *in vivo* experimentation which incorporates the effects of mechanical forces.

VII. Future Directions

One major drawback in using HA as a lamina propria bioimplant for the treatment of vocal fold disorders is that its residence time within vocal folds is short, its half-life in rabbit vocal folds is only 3–5 days (7). The half-life is most likely shorter in human vocal folds because of the effects of vibration and mechanical forces causing the glycoaminoglycan to breakdown faster. To overcome this obstacle, the HA molecular structure must be modified in order to have any meaningful residence time (6). Various strategies, including chemical, enzymatic, and mechanical cross-linking and cleavage, need to be employed to modify HA into forms that have increased residence times within vocal folds. Unfortunately, modified HA molecules have a longer half-life, but often have physico-chemical properties that are significantly different from unmodified HA. The main problem with HA that has been modified to have longer tissue residence times is that its viscosity is greatly increased beyond that which is acceptable in the thyroarytenoid muscle and ECM of the lamina propria. Furthermore, the shear thinning properties that are necessary for changes in vocal fold vibration and which is inherent of HA may be lost. Further research is being performed to determine how to preserve the viscoelastic properties of unmodified HA while maintaining a longer half-life. The ideal HA modification is one that properly balances the benefits of the longer half-life with the disadvantages of different biomechanical properties. Multiple types, concentrations, and molecular sizes of HA may be necessary for treatment of the numerous types of vocal fold disorders. With continued research, HA, or modified forms of it, is likely to be a commonly used implant, combined with other proteins for the augmentation of vocal folds based on its nonantigenicity, biocompatibility, and biomechanical properties.

References

1. Gray SD. Cellular physiology of the vocal folds. Otolaryngol Clin North Am 2000; 33:679–698.
2. Laurent TC, Laurent UB, Fraser JR. Functions of hyaluronan. Ann Rheum Dis 1995; 54:429–432.
3. Gray SD, Titze IR, Chan RW, Hammond TH. Vocal fold proteoglycans and their influence on biomechanics. Laryngoscope 1999; 109:845–854.
4. Butler JE, Hammond TH, Gray SD. Gender-related differences of hyaluronic acid distribution in the human vocal fold. Laryngoscope 2001; 111:907–911.
5. Evered D, Whelan J. Hyaluronic acid and its functions. In: CIBA Foundation Symposium, ed. Functions of the Proteoglycans. New York: Wiley, 1996; 267–282.
6. Chan RW, Gray SD, Titze IR. The importance of hyaluronic acid in vocal fold biomechanics. Otolaryngol Head Neck Surg 2001; 124:607–614.
7. Hallen L, Dahlqvist A, Laurent C. Dextranomeres in hyaluronan (DiHA): a promising substance in treating vocal cord insufficiency. Laryngoscope 1998; 108: 393–397.
8. Hammond TH, Zhou R, Hammond EH, Pawlak A, Gray SD. The intermediate layer: a morphologic study of the elastin and hyaluronic acid constituents of normal human vocal folds. J Voice 1997; 11:59–66.
9. Murphy G, Docherty AJ. The matrix metalloproteinases and their inhibitors. Am J Respir Cell Mol Biol 1992; 7:120–125.
10. Dikkers FG, Nikkels PG. Benign lesions of the vocal folds: histopathology and phonotrauma. Ann Otol Rhinol Laryngol 1995; 104:698–703.
11. Dikkers FG, Nikkels PG. Lamina propria of the mucosa of benign lesions of the vocal folds. Laryngoscope 1999; 109:1684–1689.
12. Edlin GP. A histochemical study of simple laryngeal polyps. J Clin Pathol 1980; 33: 526–527.
13. Thibeault SL, Gray SD, Li W, Ford CN, Smith ME, Davis RK. Genotypic and phenotypic expression of vocal fold polyps and Reinke's edema: a preliminary study. Ann Otol Rhinol Laryngol 2002; 111:302–309.
14. Mackool RJ, Gittes GK, Longaker MT. Scarless healing: the fetal wound. Clin Plast Surg 1998; 25:357–365.
15. Adzick N, Longaker MT. Scarless wound healing in the fetus: the role of the extracellular matrix. Prog Clin Biol Res 1991; 365:177–192.
16. Moriarty KP, Cromblehome TM, Kerry GE, O'Donnell C. Hyaluronic acid-dependent pericellular matrices in fetal fibroblasts: implications for scar-free wound repair. Wound Repair Regen 1996; 4:346–352.
17. Rousseau B, Hirano S, Scheidt T, Welham NV, Thibeault SL, Chan RW, Bless DM. Characterization of vocal fold scarring in a canine model. Laryngoscope 2003; 113: 620–627.
18. Rousseau B, Hirano S, Chan RW, Welham NV, Thibeault SL, Ford CN, Bless DM. Characterization of chronic vocal fold scarring in a rabbit model. J Voice 2004; 18(1):116–124.

19. Thibeault SL, Rousseau B, Welham NV, Hirano S, Bless DM. Hyaluronan levels in acute vocal fold scar. Laryngoscope; in press.
20. Thibeault SL, Bless DM, Gray SD. Interstitial protein alterations in rabbit vocal fold with scar. J Voice 2003; 17:376–382.
21. Rousseau B, Hirano S, Welham NV, Thibeault SL, Bless DM, Ford CN, Chan RW. Histological characterization of chronic vocal fold scarring in a rabbit model. In 6th International Voice Symposium of Australia; 2002.
22. Chan RW, Titze IR. Hyaluronic acid (with fibronectin) as a bioimplant for the vocal fold mucosa. Laryngoscope 1999; 109:1142–1149.
23. Titze IR. Phonation threshold pressure: a missing link in glottal aerodynamics. J Acoust Soc Am 1992; 91:2926–2935.
24. Chan RW, Titze IR, Titze MR. Further studies of phonation threshold pressure in physical model of the vocal fold mucosa. J Acoust Soc Am 1997; 101:3722–3727.
25. Chan RW, Titze IR. Viscoelastic shear properties of human vocal fold mucosa: measurement methodology and empirical results. J Acoust Soc Am 1999; 106: 2008–2021.
26. Chan RW, Titze IR. Viscoelastic shear properties of human vocal fold mucosa: theoretical characterization based on constitutive modeling. J Acoust Soc Am 2000; 107:565–580.
27. Chan RW, Titze IR. Viscosities of implantable biomaterials in vocal fold augmentation surgery. Laryngoscope 1998; 108:725–731.
28. Bothner H, Wik O. Rheology of hyaluronate. Acta Otolaryngol (Stockh) 1987; 442: 25–30.
29. Bulpitt P, Aeschlimann D. New strategy for chemical modification of hyaluronic acid: preparation of functionalized derivatives and their use in the formation of novel biocompatible hydrogels. J Biomed Mat Res 1999; 47:152–169.
30. Hallen L, Testad P, Sederholm E, Dahlquist A, Laurent C. DiHA (dextranomers in hyaluronan) injections for treatment of insufficient closure of the vocal folds: early clinical experience. Laryngoscope 2001; 111:1063–1067.
31. Larsen NE, Pollack CT, Reiner K, Leshchiner E, Balazs EA. Hylan gel biomaterial: dermal and immunologic compatibility. J Biomed Mat Res 1993; 27:1129–1134.
32. Hertegard S, Hallen L, Laurent C, Lindstrom E, Olofosson K, Testad P, Dahlquist A. Cross linked hyaluronan used as augmentation substance for treatment of glottal insufficiency: safety aspects and vocal fold function. Laryngoscope 2002; 112: 2211–2219.
33. Hertegard S, Dahlqvist A, Laurent C, Borzacchiello A, Ambrosio L. Viscoelastic properties of rabbit vocal folds after augmentation. Otolaryngol Head Neck Surg 2003; 128:401–406.
34. Dahlqvist A, Garskog O, Laurent C, Hertegard S, Ambrosio L, Borzacchiello A. Viscoelasticity of rabbit vocal folds after injection augmentation. Laryngoscope 2004; 114(1):138–142.
35. Hirano S, Bless DM, Heisey D, Ford CN. Roles of hepatocyte growth factor and transforming growth factor beta 1 in production of extracellular matrix by canine vocal fold fibroblasts. Laryngoscope 2003; 113:144–148.
36. Hirano S, Bless DM, Heisey D, Ford CN. Effect of growth factors on hyaluronan production by canine vocal fold fibroblasts. Ann Otol Rhinol Laryngol 2003; 112: 617–624.

Chemistry and Biology of Hyaluronan
H.G. Garg and C.A. Hales (editors)

Chapter 17

Hyaluronan in Aging

MARÍA O. LONGAS

Department of Chemistry and Physics, Purdue University Calumet, Hammond, Indiana, USA

I. Introduction

Hyaluronan (HA) is an acidic glycosaminoglycan (GAG) found in most mammalian connective tissue. It is an unbranched polymer of the disaccharide [4)-β-D-GlcA-(1 → 3)-β-D-GlcNAc-(1 →]*n* (1–5). It forms intramolecular hydrogen bonds (Fig. 1) (4,5), and occurs as a free chain (6–8) and covalently bonded to protein (9–11). HA serves as the core for the proteoglycan aggregate, which is a macromolecular arrangement that involves non-covalently bonded HA, individual proteoglycans and link proteins (12–15) (Fig. 2). HA is found in the extracellular space (1–5) as well as inside the cells (16–20). The difference in concentration appears to depend on the cell state. During mitosis and cell migration, when the production of HA is higher, its concentration inside and outside the cell is about the same (21,22). Other work has indicated that mitotic and migrating cells have much more intracellular HA (17). Several studies indicate that HA is synthesized, released in the extracellular space and then internalized. However, it is not known if the biosynthesis of the extracellular and the intracellular GAG is catalyzed by the same enzyme(s).

The chemical structure of HA we know is mainly that of a polymer isolated from a whole tissue that has been subjected to exhaustive degradation by proteolytic enzymes and/or denaturing reagents (1,23,24). Knowledge of intracellular vs. extracellular HA, exact sugar-sequence would shed light on its destination. Like proteins, HA may have its final destination written on its

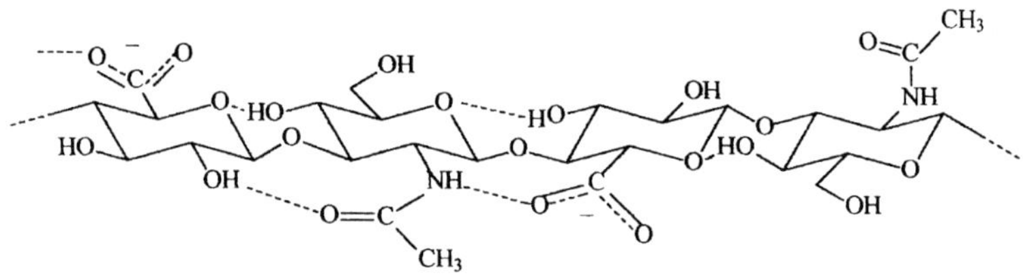

Figure 1 Tetrasaccharide of hyaluronan showing the intramolecular hydrogen bonds proposed by Scott and co-workers (1,3,4).

primary structure (sugar-sequence) (25). Data indicating that most, but not all intracellular HA has lower molecular weight than the extracellular one (26) support the hypothesis that extracellular and intracellular HAs are different (26).

HA is a polyanion whose functional groups make it so hydrophilic that it binds 1000 times more water than is predicted from its molar mass (27). The heterogeneity and hydrophilicity of HA facilitate its interaction with a variety of tissue constituents inside and outside the cells. In the extracellular space, HA controls the retention of water, ionic and molecular diffusion, and provides 3D-structural meshwork (27–32). Its interaction with the cell surface receptors: CD44, RHAMM (receptor for HA-mediated motility), TSG-6 (tumor necrosis factor-stimulated gene-6) and HABP1 (HA-binding protein) is known to affect cellular functions (33–35).

The functions of extracellular, free-chain HA have been assigned based on experimental evidence, but little is known about the functions of the

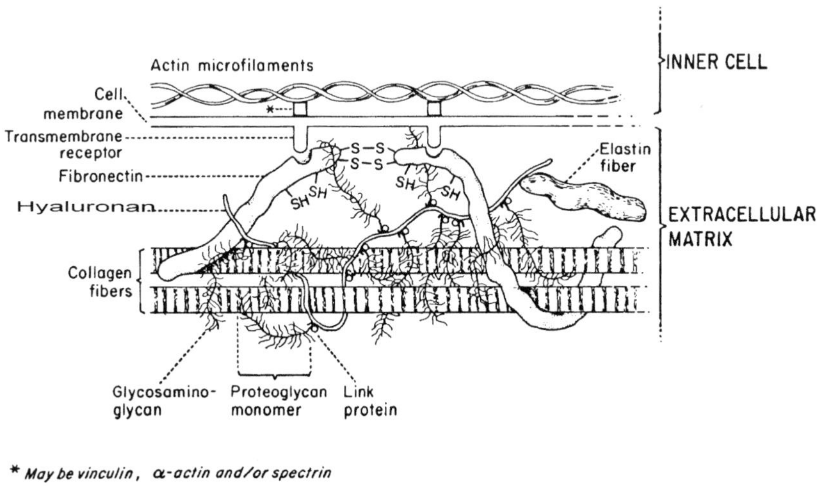

Figure 2 Classical proteoglycan aggregate (12–14) in a simplified, hypothetical arrangement where it intertwines with the collagen fibers and intermingles with other extracellular macromolecules.

serum-derived HA-associated protein (SHAP)–HA complex or the intracellular GAG. The SHAP–HA complex has been implicated in various biologic processes (36). Although the secondary and tertiary structures of HA covalently bonded to protein should differ from those of the free chain, little has been done to explain the effect or effects of this covalent bonding, and of the protein on these structures, and on the biologic roles they may have inside and outside the cells.

II. Hyaluronan Structures in Solution: Relevance to Tissue Biologic Functions and Aging

A. Extracellular HA

Extracellular HA is a polydisperse molecule whose molecular weight ranges from 300 to $>10^3$ kDa (37,38). It displays both flexible and stiff domains in solution, suggesting deviation of its primary structure from that elucidated by Meyer and co-workers, which is an unbranched chain of alternating D-GlcA and D-GlcNAc in a 1:1 molar ratio (1,2,39). The secondary and tertiary structures of HA may be determined by its primary structure (sugar-sequence) like in proteins (40). If this is the case, the flexible and stiff domains that HA forms in solution are functions of its primary structure. Therefore, the stiff segments, which comprise 55–70% of the molecule and are not affected by changes in temperature, ionic strength, denaturing reagents like urea or pH, may originate from a primary structure different from the one indicated above (41). It is worth noting that some reports suggest that basic reagents relieve stiffness (42).

Stiffness can be relieved by base, whether it originates from hydrogen bonding or hydrophobic interactions. The hydrogen bond network would be disrupted as the base (e.g., OH^-) replaces the hydrogen acceptor of the existing hydrogen bonds. Thus, the size needed for stiffness is altered and relaxation occurs (42). The intramolecular hydrogen bond pattern of HA proposed by Scott and co-workers (3,4) appears in Fig. 1. If stiffness originates from hydrophobic interactions of the acetamido methyl of D-GlcNAc, a basic reagent may introduce relaxation, as it abstracts methyl protons of the acetamido moiety. Such process would break the van der Waals interactions responsible for the stiffness (43). Methyl protons of the acetamido moieties are acidic, because they are α to the carbonyl (C=O) group (44).

Hydrophobic (stiff) regions of about eight sequential CH groups have been reported in HA solutions of dimethylsulfoxide (DMSO) (3). They would form if the acetamido moieties of D-GlcNAc positioned themselves in such a way as to accommodate DMSO in between as shown in Fig. 3. Such an arrangement would be facilitated if HA coiled to bring acetamido moieties closer, with at least one DMSO holding their methyl groups in a van der Waals association. The helical structures proposed for HA support this hypothesis (45). Hydrophobic patches may also result from sequences of D-GlcNAc only as shown in Fig. 4. These patches might hinder the spectroscopic detection of carboxylate moieties (3), but

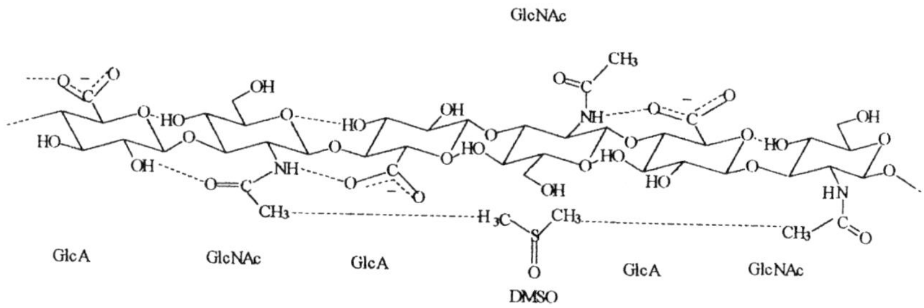

Figure 3 Hyaluronan hexasaccharide (1) in DMSO showing the hypothetical location of DMSO between D-GlcNAc moieties.

should facilitate the association of HA with the hydrophobic regions of lipids (46), proteins (32,47) and other HA molecules (48).

HA-stiff regions of at least 60 disaccharides, each distributed along the polymer and separated by flexible regions, have been proposed based on their susceptibility to hyaluronidase (41). It has also been reported that HA-stiff segments with fewer than 60 disaccharides form in solution, suggesting that the fine structures responsible for this 2D-arrangement occur along the entire HA chain (41).

Scott and co-workers (3,4) have proposed a model for high molecular weight HA in aqueous solutions, based on NMR spectroscopy, X-ray crystallography, rotary shadowing and electron microscopy data. In this model they suggest that HA forms "2-fold helices with gentle curves in the polymer backbone in two planes at right angles, with hydrophobic patches on alternate sides of the polymer" (4). This model is facilitated by an anti-parallel arrangement of the HA chains that provides the greatest proximity for groups on adjacent chains to interact. Thus, hydrogen bonds of the polar and ionic groups, and van der Waals interactions of the methyl groups of the acetamido moieties could hold many HA chains together in extensive networks of sizes

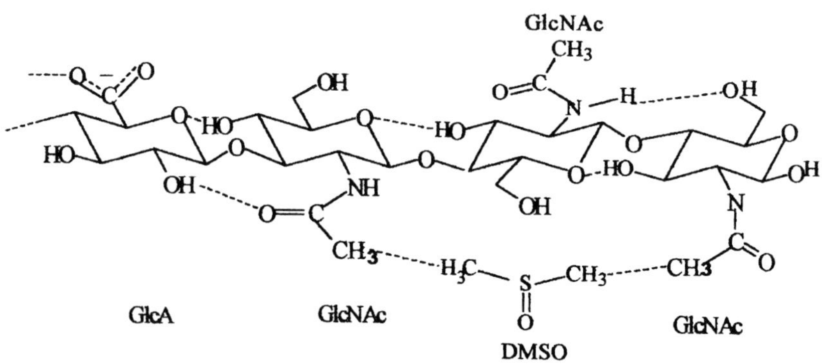

Figure 4 A hypothetical tetrasaccharide of hyaluronan showing three adjacent D-GlcNAcs where two of them are connected through DMSO by van der Waals bonds.

limited mainly by HA concentration (3,4). Such macromolecular aggregates should affect the biologic functions of the tissues containing HA.

1. Highly Hydrophilic Hyaluronan Sequences

In aqueous, HA solutions, basic (D-GlcN) (Fig. 5) and acidic (D-GlcA) (Fig. 6) sugar-sequences should form stronger hydrogen bond networks with the solvent than the alternating sequence of D-GlcNAc and D-GlcA, where the methyl moieties of the acetamido groups would interrupt the hydrogen bond meshwork (43). Water would form cages around the basic and acidic regions making them appear stiff. This dense, hydrogen bond web may not be disrupted by low concentrations of urea, since this reagent may hydrogen bond with the water present in the medium without disturbing the HA–water bonds. The presence of basic or acidic regions on the HA chain would support the hypothesis that HA stiffness in aqueous solution originates from highly dense, hydrogen bond webs that resist cleavage by low concentrations of urea (41).

2. Biological Importance of Highly Dense, Hyaluronan–Water, Hydrogen Bond Networks

Densely hydrogen bonded HA in water should be relevant in numerous physiologic functions, as the hydrogen bonding contributes to its viscosity (42). In synovial fluid, a highly viscous HA–water medium offers the efficient support required for healthy joints (49). In the corneal endothelium, viscous solutions of Na-HA in water are known to provide protection from mechanical damage (50). The viscosity of the vitreous body of the eye should, at least partially, be a function of the degree of hydrogen bonding provided by HA in water (38). Fetal skin contains highly hydrated HA in gel-like structures that may result mainly from the HA–water hydrogen bonds and may be needed for cell differentiation (51). The 3D-structural support that HA provides in the extracellular space of connective tissue should, at least partly, be a function of its degree of hydration and hydrogen bonding (27–32,42). The SHAP–HA complex has been found in rheumatoid arthritis, not in normal joints, suggesting its involvement in this

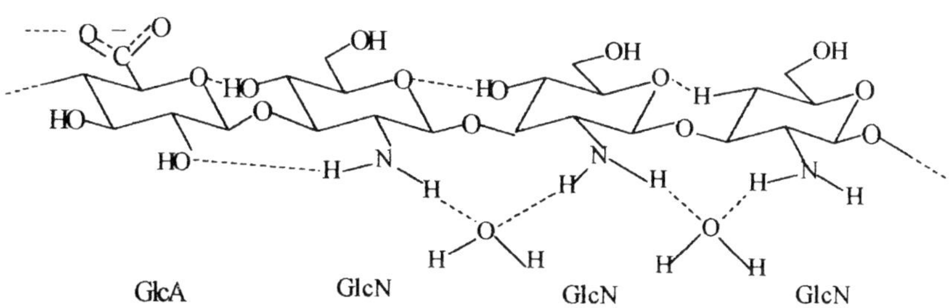

Figure 5 Hypothetic, basic tetrasaccharide of hyaluronan showing one D-GlcA followed by a sequence of three D-GlcN moieties.

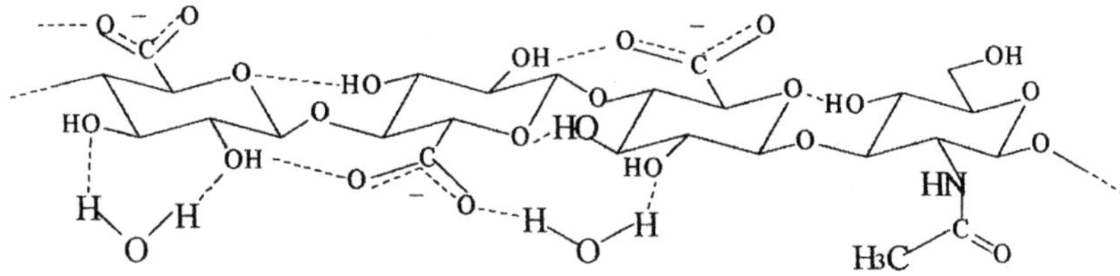

Figure 6 Hypothetic, acidic tetrasaccharide of hyaluronan showing three adjacent D-GlcA moieties with a terminal D-GlcNAc.

disorder (52). In addition, the number of therapeutic and medical uses of HA is rapidly growing (42).

The interaction of extracellular HA with cells *in vivo* should be affected by its degree of hydration and the density of the hydrogen bond meshwork it forms. A highly dense, hydrogen bond network would hinder HA interactions with the polar groups on the cell surface. The degree of hydrogen bonding HA undergoes *in vivo* should also affect the nature of the classical proteoglycan aggregates it is known to form (Fig. 2) (12–15).

Proton-NMR spectroscopy data indicate that not all hexosamine in HA is N-acetylated (Longas et al., unpublished work). Therefore, the universally accepted, alternating sequence of equimolar amounts of D-GlcA and D-GlcNAc, shown in Fig. 1 (1–5), may incorporate some D-GlcN. Other regions may have sequences of D-GlcNAc only followed by at least one D-GlcA (Fig. 4); basic (Fig. 5), acidic (Fig. 6) and D-GlcNAc segments alternating with D-GlcN (Fig. 7) are also possible.

In HA covalently bonded to protein (SHAP–HA), the specific sugar-sequences under consideration may or may not occur *in vivo*, since the covalent binding and the protein should give different senses to the secondary and tertiary structures of the GAG. Deviations from the alternate sugar-sequence (1) are possible in HA, since dermatan sulfate, which is related to HA, has been reported to display sequential D-GlcA (53).

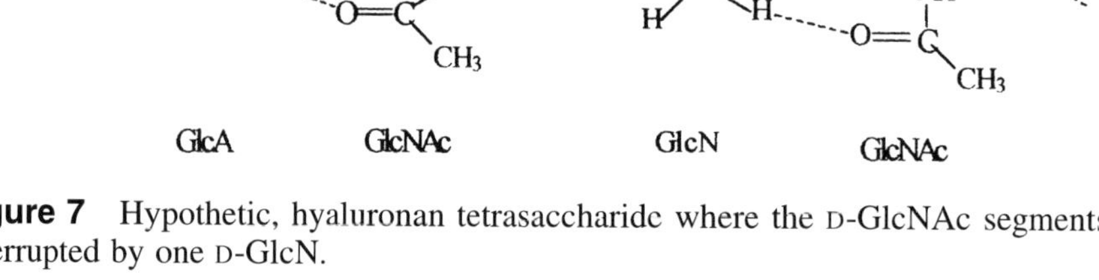

Figure 7 Hypothetic, hyaluronan tetrasaccharide where the D-GlcNAc segments are interrupted by one D-GlcN.

3. Relevance in Aging

The highly hydrophilic HA segments (Figs. 5 and 6) should render HA more hydrophilic than the alternating sequence of D-GlcA and D-GlcNAc (Fig. 1). This should be important in aging, because human skin, for example, loses 77% (w/w) of its HA content at 75 years (Table 1) (54). Besides, HA becomes N-deacetylated at this age (Fig. 8) (54,55). Perlish and his collaborators (56) found a significant loss of GAGs in the salt-soluble extracts of human dermis, but the quantitation of HA in human skin extracts carried out by other workers showed no significant age-related changes (57). In the dermis of ISh rats, HA does not appear to change with age either (58). Fasted rats lose the GAGs from their skin, suggesting that fasting fatigues the system and makes it function as an aged one (59). The discrepancy in the data may originate from the different species analyzed and from the purity of the GAGs used. The work of Longas et al. (54,55) was performed with highly purified molecules.

The reason (or reasons) for the loss of HA with aging has not been elucidated. Some studies suggest that it is due to depolymerization caused by natural free radicals produced during metabolism (62). The finding that free radical scavengers inhibit HA fragmentation *in vitro* suggests that its depolymerization *in vivo* follows a free radical pathway (63). HA has also been described as a free radical scavenger and antioxidant (64,65). Its cleavage *in vitro* upon exposure to irradiation (66–68) also supports the free radical pathway for its depolymerization.

In vivo, the UV irradiation of hairless mice resulted in an increase of their skin GAGs including HA (69). Also, the UVA irradiation of albino rats yielded abnormally elevated GAG composition in their skin (70). Because the latter effects were reversed by adding vitamin E, a free radical scavenger, to the rat's diet, the data suggest a free radical involvement (70). Overall, UV-light irradiation stimulates new GAG biosynthesis, while destroying the old ones.

The age-mediated depolymerization of HA has been demonstrated in rat skin (71), while other work has indicated no change in HA size as a function of age (57). Regardless of the discrepancies, there is enough evidence in support of the hypothesis that HA functions *in vivo* as a free radical scavenger that protects

Table 1 Effect of Age on the Concentration of Human Skin Hyaluronan

Age group (years)	Hyaluronan[a] (%)	SD[b] (%)	Decrease
19 ± 2.5	0.030	0.005	
35 ± 3.5	0.030	0.005	
47 ± 1.7	0.030	0.006	
60 ± 0.8	0.015	0.003	50
75 ± 5.0	0.007	0.001	77

[a]Mean percent (w/w) based on whole, surgically defatted, wet mastectomy skin from four different people of every age group.
[b]SD, standard deviation of the mean (54).

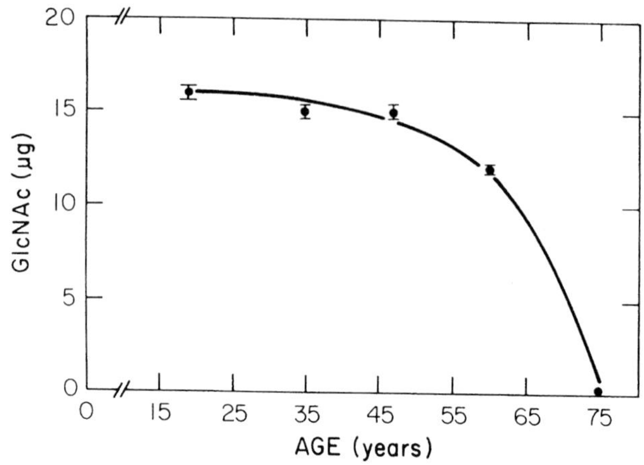

Figure 8 Effect of age on the composition of reducing 2-acetamido-2-deoxyglucose generated from hyaluronan upon digestion by Streptomyces hyaluronidase. HA (100 μg) and 2 turbidity units of enzyme were used under the conditions described previously (54). D-GlcNAc was quantified by utilizing the Morgan–Elson method (60) as modified by Rissing et al. (61).

the skin from endogenous and exogenous, free radicals. Hydrolytic enzymes, whose catalytic activity is known to increase with aging, may cleave HA into small fragments that are then removed from the tissues (72).

4. *In Vitro Aging and Hyaluronan*

Fibroblast cells in culture have been studied with regard to the fate of HA concentration during cell aging (cell passage) with and without UVA irradiation. The results show that the biosynthesis of HA increases with the number of cell passages (73) in an UVA dose-dependent way (74), but the newly synthesized HA is rapidly degraded (73). These findings suggest that the biosynthesis of the respective hydrolytic enzymes also increases with cell passage (73,74). In the absence of UV irradiation, hyaluronidase stimulates cell proliferation during *in vitro* aging, and the biosynthesis of HA decreases with the number of cell passages (75,76).

Although fibroblast cultures from human skin of 75-year-old subjects synthesize HA with D-GlcNAc as its hexosamine component (Longas MO, unpublished observations), an enzyme that appears to be induced with aging and becomes highly active in the seventh decade of life is present in the skin of these individuals and cleaves the acetyl ($-COCH_3$) group from N at position 2 of the hexosamine (77). These findings are relevant in aging, because N-deacetylated HA is more hydrophilic than the one found in young skin. The formation of wrinkles is believed to originate, at least partially, from the loss of GAG in the extracellular space and the water they retain (27,54). Highly hydrophilic HA may be needed in aged skin to retain water, if no other molecules are made to replace it, because about half the water content of the skin is apparently bonded to this GAG (54,55,78,79). The age-related loss of water has also been demonstrated in the skin of Sprague–Dawley rats (80).

5. *Hyaluronan, Body Fluids and Aging*

Data available on the effect of age on the HA concentration in body fluids indicate that it increases in human serum (81). Based on its concentration in human urine, HA has been utilized as a marker for aging (82).

6. *Hyaluronan in Aging Disorders*

Conclusive data on HA alterations in disorders that resemble premature aging like progeria and Down's syndrome are scarce. In progeria patients, urinary HA is abnormally elevated (83,84). In this disorder, HA has been postulated as the culprit for the lack of vasculogenesis, characteristic of these patients (84). In the serum of patients with Down's syndrome, HA is only slightly higher than in the normal serum (85). It would be of utmost importance to elucidate the exact chain sequence of HA from these patients. Chances are that the HA of senescent human skin, which lacks its *N*-acetyl moieties (54,55,76), plays a role in these pathologic states (83,84).

B. Intracellular Hyaluronan

In the intracellular space, HA finds nucleic acids, nuclear proteins and cellular organelles, among others, with which to associate. As in the extracellular space, the nature of the HA sugar-sequence(s), the degree of hydrogen bonding, and its secondary and tertiary structures should affect its interactions with other intracellular constituents and thus the biologic functions of the cells. High molecular weight HA appears to form highly dense, hydrogen bond networks *in vitro* (3,4) that may be needed by some cells, while the smaller fragments identified intracellularly may be needed by others (20–26). Cell functions that originate intracellularly like mitosis appear to be affected by the interactions of extracellular HA with the cell surface receptors: CD44, TSG-6, RHAMM and HABP1 (33–35). Obviously, intracellular and extracellular functions mediated by HA are interconnected.

III. Summary and Conclusion

A. Summary

HA is an acidic GAG of D-GlcA and D-GlcNA that is found in mammalian connected tissue. It is a polydisperse, unbranched chain whose molecular weight varies from 300 to $>10^3$ kDa. HA occurs as a free chain and covalently bonded to protein as the SHAP–HA complex. It can be found inside and outside the cells. Its secondary and tertiary structures in solution display both stiff and flexible domains that suggest deviations of its primary structure (sugar-sequence) from the alternating sequence of the sugars indicated earlier. In this review, the secondary and tertiary structures of HA in solution are analyzed, in an attempt to explain their relevance to normal biologic functions and aging. Hydrophilic sequences of D-GlcN and D-GlcA are used to explain the stiff and flexible domains detected in HA solutions. The hydrophilic segments should be more densely hydrogen bonded with water than the regions of alternating D-GlcA and D-GlaNAc, as water would form cages around them.

Stiff segments can originate from hydrogen bond networks or from van der Waals associations of the methyl moieties of D-GlcNAc in a sequence of this sugar only or in the classical alternating sequence of D-GlcA and D-GlcNAc. Stiff segments originating from acidic (D-GlcA) and basic (D-GlcN) sugars are disrupted by base (e.g., OH^-), as it becomes the hydrogen acceptor in the hydrogen bonds HA makes with water. If stiffness results from hydrophobic interactions, a basic reagent relaxes it, as it removes methyl protons from the acetamido moiety of D-GlcNAc, since these protons are acidic.

Viscous solutions of HA in water are the result of highly dense, HA–water, hydrogen bond networks. These solutions appear to be involved in a variety of biologic functions such as the efficient support of healthy joints by synovial fluid; the protection of corneal endothelium from mechanical damage; cell differentiation in fetal tissue; fluidity in the vitreous body of the eye and many others. The therapeutic and medical uses of highly viscous HA are presently diverse and are growing rapidly.

In aging, the hypothetic D-GlcN sequence was used in an attempt to explain the N-deacetylation of human skin HA at 75 years of age. Because the skin composition of HA decreases by 77% (w/w) also at 75 years, and the basic sugar-sequence makes HA more hydrophilic than the alternating sequence of D-GlcA and D-GlcNAc, it was concluded that nature compensates for the loss of HA concentration in aging by having an enzyme that removes the *N*-acetyl groups, making HA more hydrophilic. Highly hydrophilic HA is needed in aged human skin, if no other molecules are made to retain the water, since about 50% (w/w) of this reagent is bonded to this GAG.

The reasons for the loss of HA from the skin were attributed to depolymerization induced by both endogenous and exogenous free radicals. This hypothesis is supported by findings that indicate that free radical scavengers inhibit UV-light-mediated cleavage of HA *in vitro*. It was found that the UV-light irradiation of skin *in vivo* resulted in increased GAG biosynthesis in hairless mice and albino rats; but providing vitamin E in the rat's diet reversed the latter results. Because vitamin E is a natural free radical scavenger, the data support a free radical involvement. In the absence of UV-light irradiation, depolymerization may be caused by endogenous free radicals, which are natural products of metabolism, or by enzymatic degradation. Results on this area are conflicting; some workers reported HA depolymerization in rat skin, while others found no significant age-related changes in HA size.

Analysis of the literature on the age-related effect of HA in cell aging (cell passage) *in vitro* revealed that the biosynthesis of HA increased with the number of cell passages in an UVA-dose-dependent way, and that there was an increment in hydrolytic enzymes. In the absence of UV-irradiation, the concentration of HA decreased as the number of cell passages increased.

Literature pertaining to the effect of age on the concentration of HA in body fluids has been investigated. It revealed that the HA content increased with aging in human serum.

Data on the alterations of HA in accelerated aging diseases like progeria and Down's syndrome are inconclusive. Nevertheless, it was found that urinary HA was increased in progeria patients. In the serum of patients with Down's syndrome, HA content was only slightly higher than in the normal serum.

B. Conclusion

HA is an essential heteropolysaccharide of mammalian connective tissue. It is a free radical scavenger that protects the system from endogenous and exogenous free radicals. Its broad involvement in biologic functions, aging, medicine and therapeutic processes requires that its exact sugar-sequence be elucidated. Such knowledge would shed light on its mechanism of action and facilitate: the development of ways to retard human aging; therapies to prevent or cure premature aging-like disorders as progeria and Down's syndrome; and drugs for the treatment or prevention of other diseases of the aged in which HA appears to play a role.

Acknowledgements

Scholarly Research Release Award from Purdue University Calumet supported this work.

References

1. Meyer K. Chemical structure of hyaluronic acid. Fed Proc 1958; 17:1075–1077.
2. Longas MO, Meyer K. Sequential hydrolysis of hyaluronate by β-*N*-acetylhexosaminidase and β-glucuronidase. Biochem J 1981; 197:275–282.
3. Scott JE. Chemical morphology of hyaluronan. In: Laurent TC, ed. The Chemistry, Biology and Medical Applications of Hyaluronan and Its Derivatives. London: Portland Press, 1998:7–15.
4. Scott JE, Heatley F. Hyaluronan forms specific stable tertiary structures in aqueous solution: a ^{13}C NMR study. Proc Natl Acad Sci USA 1999; 96:4850–4855.
5. Laurent TC. Structure of hyaluronic acid. In: Balazs EA, ed. Chemistry and Molecular Biology of the Intercellular Matrix. New York: Academic Press, 1970; vol. 2:703–732.
6. Piepkorn M, Hovingh P, Linker A. Glycosaminoglycan free chains. External plasma membrane components distinct from the membrane proteoglycans. J Biol Chem 1989:8662–8669.
7. Fessler JH, Fessler LJ. Electron microscopic visualization of the polysaccharide hyaluronic acid. Proc Natl Acad Sci USA 1966; 56:141–147.

8. Varma R, Varma RS, Allen WS, Wardi AH. On the carbohydrate–protein linkage groups in vitreous humor hyaluronate. Biochim Biophys Acta 1974; 362: 584–588.
9. Yoneda M, Suzuki S, Kimata K. Hyaluronic acid associated with the surfaces of cultured fibroblasts is linked to a serum-derived 85-kDa protein. J Biol Chem 1990; 265:5247–5257.
10. Zhao M, Yoneda M, Ohashi Y, Kurono S, Iwata H, Ohnuki Y, Kimata K. Evidence for the covalent binding of SHAP, heavy chains of inter-alpha-trypsin inhibitor, to hyaluronan. J Biol Chem 1995; 270:26657–26663.
11. Chem L, Zhang H, Powers RW, Russell PT, Larsen WJ. Covalent linkage between proteins of the inter-alpha-inhibitor family and hyaluronic acid is mediated by a factor produced by granulosa cells. J Biol Chem 1996; 271:19409–19414.
12. Rosenberg L, Hellmann W, Kleinschmidt AK. Electron microscopic studies of proteoglycan aggregates from bovine articular cartilage. J Biol Chem 1975; 250: 1877–1883.
13. Hascall VC, Heingard D. Aggregation of cartilage proteoglycans I. The role of hyaluronic acid. J Biol Chem 1974; 249:4232–4241.
14. Hascall VC, Heinegard D. Aggraegation of cartilage proteoglycans II. Oligosaccharide competitors of the proteoglycan–hyaluronic acid interaction. J Biol Chem 1974; 249:4242–4249.
15. Hascall VC, Heinegard D. The structure of cartilage proteoglycans. In: Slavkin HC, Greulich R, eds. Extracellular Matrix Influences on Gene Expression. New York: Academic Press, 1975:423–434.
16. Ripellino JA, Bailo M, Margolis RU, Margolis RK. Light and electron microscopic studies on the localization of hyaluronic acid in developing rat cerebellum. J Cell Biol 1988; 106:845–855.
17. Bertolami CN, Berg S, Messadi DV. Binding and internalization of hyaluronate by human cutaneous fibroblasts. Matrix 1992; 11:11–21.
18. Eggli PS, Graber W. Association of hyaluronan with rat vascular endothelial cells and smooth muscle cells. J Histochem Cytochem 1995; 43:689–697.
19. Evanko SP, Wight TN. Intracellular localization of hyaluronan in proliferating cells. J Histochem Cytochem 1999; 47:1331–1341.
20. Tammi R, Rilla K, Pienimaki J-P, MacCallum DK, Luukkonen M, Hascall VC, Tammi M. Hyaluronan enters keratinocytes by a novel endocytic route for catabolism. J Biol Chem 2001; 276:35111–35122.
21. Brecht M, Mayer U, Schlosser E, Prehm P. Increased hyaluronate synthesis required for fibroblast detachment and mitosis. Biochem J 1986; 239:445–450.
22. Tammi R, Tammi M. Correlations between hyaluronan and epidermal proliferation as studied by [^{3}H]glucosamine and [^{3}H]thymidine incorporations and staining of hyaluronan on mitotic keratinocytes. Exp Cell Res 1991; 195: 524–527.
23. Ueno N, Chakrabarti B, Garg HG. Hyaluronic acid of human skin and post-burn scar: heterogeneity in primary structure and molecular weight. Biochem Int 1992; 26:787–796.
24. Rodén L. Structure and metabolism of connective tissue proteoglycans. In: Lennarz WJ, ed. The Biochemistry of Glycoproteins and Proteoglycans. New York: Plenum Press, 1980:267–373.
25. Garrett RH, Grisham CM. Protein synthesis and degradation, signal peptide sequence. In: Garrett RH, Grishman CM, eds. Biochemistry, 2nd ed. New York:

Saunders College Publishing/Harcourt Brace College Publishers, 1999: 1089–1127.
26. McGuire PG, Castellot JJ, Orkin RW. Size-dependent hyaluronate degradation by cultured cells. J Cell Physiol 1987; 133:267–276.
27. Hascall VC, Hascall GK. Proteoglycans. In: Hay ED, ed. Cell Biology of Extracellular Matrix. New York: Plenum Press, 1981:39–63.
28. Fessler JH. A structural function of mucopolysaccharides in connective tissue. Biochem J 1960; 76:124–132.
29. Laurent TC, Bjork I, Pietruszkiewicz A, Persson H. On the interaction between polysaccharides and macromolecules. II. The transport of globular proteins through hyaluronic acid solutions. Biochim Biophys Acta 1963; 78:351–359.
30. Toole BP. Morphogenic role of glycosaminoglycans (acid mucopolysaccharides) in brain and other tissues. In: Barondes SH, ed. Neuronal Recognition. New York: Plenum Press, 1976:275–329.
31. Chakrabarti B, Park JW. Glycosaminoglycas: structure and interaction. CRC Rev Biochem 1980; 8:225–313.
32. Burd DAR, Siebert JW, Ehrlich HP, Garg HG. Human skin and post-burn scar hyaluronan: demonstration of the association with collagen and other proteins. Matrix 1989; 9:322–327.
33. Hall CL, Wang C, Lange LA, Turley EA. Hyaluronan and the hyaluronan receptor RHAMM promote focal adhesion turnover and transient tyrosine kinase activity. J Cell Biol 1994; 126:575–588.
34. Knudson W, Knudson CB. The hyaluronan receptor, CD44. In: Laurent TC, ed. The Chemistry, Biology and Medical Applications of Hyaluronan and Its Derivatives. London: Portland Press, 1998:169–179.
35. Kumar R, Choudhury NR, Salunke DM, Datta K. Evidence for clustered mannose as a new ligand for hyaluronan binding protein (HABP1) from human fibroblasts. J Biosci 2001; 26:325–332.
36. Zhuo L, Yoneda M, Zhao M, Yingsung W, Yoshida N, Kitagawa Y, Kawamura K, Suzuki T, Kimata K. Defect in SHAP–hyaluronan complex causes severe female infertility. A study by inactivation of the bikunin gene in mice. J Biol Chem 2001; 276:7693–7696.
37. Hascall VC, Sajdera SW. Physical properties and polydispersity of proteoglycan from bovine nasal cartilage. J Biol Chem 1970; 245:4920–4930.
38. Laurent TC, Ryan M, Pietruszkiewicz A. Fractionation of hyaluronic acid. The polydispersity of hyaluronic acid from bovine vitreous body. Biochim Biophys Acta 1960; 42:476–485.
39. Laurent TC, Fraser RE. Hyaluronan. FASEB J 1992; 6:2397–2404.
40. Garrett RH, Grisham CM. Proteins: their biologic functions and primary structure. In: Garrett RH, Grishman CM, eds. Biochemistry, 2nd ed. New York: Saunders College Publishing/Harcourt Brace College Publishers, 1999:158–208.
41. Darke A, Finer EG, Moorhouse R, Rees DA. Studies of hyaluronate solutions by nuclear magnetic relaxation measurements. Detection of covalently defined, stiff segments within the flexible chains. J Mol Biol 1975; 99:477–486.
42. Lubomír L Jr, Lubomír L, Desmedt S, Demeester J, Chabrecek P. Hyaluronan: preparation, structure, properties, and applications. Chem Rev 1998; 98:2663–2684.
43. Solomons TW, Fryhle CB. van der Waals forces. In: Solomons TW, Fryhle CB, eds. Organic Chemistry. 7th ed. upgrade New York: Wiley, 2002:73–74 see also pp. 152–153.

44. Solomons TW, Fryhle CB. The acidity of the α hydrogens of carbonyl compounds: enolate anions. In: Solomons TW, Fryhle CB, eds. 7th ed. upgrade, Organic Chemistry. New York: Wiley, 2002:767–776.
45. Atkins EDT, Sheehan JK. Hyaluronates: relation between molecular conformations. Science 1973; 179:562–564.
46. Pasqualli-Ronchetti I, Quaglino D, Mori G, Bacchelli B. Hyaluronan–phospholipid interactions. J Struct Biol 1997; 120:1–10.
47. Lee GM, Johnstone B, Jacobson K, Caterson B. The dynamic structure of the pericellular matrix of living cells. J Cell Biol 1993; 123:1899–1907.
48. Sheehan J, Brass A, Almond A. The conformation of hyaluronan in aqueous solution: comparison of theory and experiment. Biochem Soc Trans 1999; 2:121–124.
49. Balazs EA, Watson D, Duff IF, Roseman S. Hyaluronic acid in synovial fluid. I. Molecular parameters of hyaluronic acid in normal and arthritic human fluids. Arthritis Rheum 1967; 10:357–375.
50. Graue EL, Polack FM, Balazs EA. The protective effect of Na-hyaluronate to corneal endothelium. Exp Eye Res 1980; 31:119–127.
51. Tammi R, Ripellino JA, Margolis RU, Tammi M. Localization of epidermal hyaluronic acid using the hyaluronate binding region of cartilage proteoglycan as a specific probe. J Investig Dermatol 1988; 90:412–414.
52. Becker A, Sandson J. The source of the inter-alpha trypsin inhibitor in pathologic hyaluronate protein. Arthritis Rheum 1971; 14:764–766.
53. Fransson L-A, Malmström A. Structure of pigskin dermatan sulfate. I. Distribution of D-glucuronic acid residues. Eur J Biochem 1971; 18:422–430.
54. Longas MO, Russell CS, He X-Y. Evidence for structural changes in dermatan sulfate and hyaluronic acid with aging. Carbohydr Res 1987; 159:127–136.
55. Longas MO, Russell CS, He X-Y. Chemical alterations of hyaluronic acid and dermatan sulfate detected in aging human skin by infrared spectroscopy. Biochim Biophys Acta 1986; 884:265–269.
56. Perlish JS, Longas MO, Fleischmajer R. The role of glycosaminoglycans in aging of the skin. In: Balin AK, Kligman AM, eds. Aging and the Skin. New York: Raven Press, 1989:153–165.
57. Meyer LJ, Stern R. Age-dependent changes of hyaluronan in human skin. J Investig Dermatol 1994; 102:385–389.
58. Koichi S. Biochemical analysis of collagen, glycosaminoglycans and elastin in ISh rat dermis. Jpn J Dermatol 1992; 102:425–431.
59. Cechowskapasko M, Palka J. Age-dependent changes in glycosaminoglycan content in the skin of fasted rats. A possible mechanism. Exp Toxicol Path 2000; 52:127–131.
60. Morgan WTJ, Elson LA. A colorimetric method for the determination of *N*-acetylglucosamine and *N*-acetylchondrosamine. Biochem J 1934; 28:988–995.
61. Rissing JL, Strominger JL, Leloir LF. A modified colorimetric method for the estimation of *N*-acetylaminosugars. J Biol Chem 1955; 217:959–966.
62. Gilchrest BA. Aging of the skin. In: Gilchrest BA, ed. Skin and Aging Processes. Boca Raton: CRC Press, 1984:5–36.
63. Deguine V, Menasche M, Ferrari P, Fraisse L, Pouliquen Y, Robert L. Free radical depolymerization of hyaluronan by maillard reaction products. Role in liquefaction of aging vitreous. Biol Macromol 1998; 22:17–22.
64. Sato H, Takahashi T, Ide H, Fukushima T, Tabata M, Sekine F, Kobayashi K, Negishi M, Niwa Y. Antioxidant activity of synovial fluid, hyaluronic acid, and two subcomponents of hyaluronic acid. Arthritis Rheum 1988; 31:63–71.

65. Foschi D. Hyaluronan, a radical scavenger. Int J Tissue React XII 1990; 6:333–339.
66. Deeble DJ, Bothe E, Schuchmann H-P, Parsons BJ, Phillips GO, von Sonntag C. The kinetics of hydoxyl-radical-indiced strand breakage of hyaluronic acid. A pulse radiolysis study using conductometry and laser-light-scattering. Z Naturforsch 1990; 45c:1031–1043.
67. Lapcik L, Schurz J. Photochemical degradation of hyaluronic acid by singlet oxygen. Colloid Polym Sci 1991; 269:633–635.
68. Rehakova M, Bakos D, Soldan M, Vizarova K. Depolymerization reactions of hyaluronic acid in solution. Int J Biol Macromol 1994; 16:121–124.
69. Chatterjee R, Benzinger MJ, Ritter JL, Bissett DL. Chronic ultraviolet B radiation-induced biochemical changes in the skin of hairless mice. Photochem Photobiol 1990; 51:991–997.
70. Longas MO, Bhuyan DK, Bhuyan KC, Gutsch CM, Breitweiser KO. Dietary vitamin E reverses the effect of ultraviolet light on rat skin glycosaminoglycans. Biochim Biophys Acta 1993; 1156:239–244.
71. Miyamoto I, Nagase S. Changes in the molecular weight of hyaluronic acid from rat skin. Exp Anim 1984; 33:481–485.
72. Verdugo ME, Ray J. Age-related changes in activity of specific lysosomal enzymes in the human retinal pigment epithelium. Exp Eye Res 1997; 65:231–240.
73. Fodil-Bourahla I, Drubaix I, Robert L. Effect of *in vitro* aging on the biosynthesis of glycosaminoglycans by human skin fibroblasts. Modulation of the elastin-laminin receptor. Mech Ageing Dev 1999; 106:241–260.
74. Schachtschabel DO, Beerbaum MK. UVA-irradiaton stimulates hyaluronan synthesis of human skin fibroblast *in vitro*. Med Welt 1996; 47:191–195.
75. Sluke G, Schachtschabel DO, Weres J. Age-related changes in distribution pattern of glycosaminoglycans synthesized by cultured human diploid fibroblasts (WI-38). Mech Ageing Dev 1981; 16:19–27.
76. Moczar M, Robert L. Stimulation of cell proliferation by hyaluronidase during *in vitro* aging of human skin fibroblasts. Exp Gerontol 1993; 28:59–68.
77. Longas MO, Burden JD, Lesniak J, Booth RM, McPencow JA, Park JI. Hyaluronic acid *N*-deacetylase assay in whole skin. Biomacromolecules 2003; 4:189–192.
78. Manery JF, Haege LF. The extent to which radioactive chloride penetrates tissues and its significance. Am J Physiol 1941; 134:83–93.
79. Ogston AG. On water binding. Fed Proc 1966; 25:986–989.
80. Jung JW, Cha SH, Lee SC, Chun IK, Kim YP. Age-related changes of water content in the rat skin. J Dermatol Sci 1997; 14:12–19.
81. Lindqvist U, Laurent TC. Serum hyaluronan and amino terminal propeptide of type III procollagen: variation with age. Scand J Clin Lab Investig 1992; 52: 613–621.
82. Goto M. Urinary hyaluronic acid as marker of aging. Kotsu Yobo Igaku Kenkyu Zaidan Kenkyu Hokokusho 1995; 1993:47–53.
83. Sweeney KJ, Weiss AS. Hyaluronic acid in progeria and the aged phenotype. Gerontology 1992; 38:139–152.
84. Brown WT. Progeria: a human-disease model of accelerated aging. Am J Clin Nutr 1992; 55:1222S–1224S.
85. Hutchin T, Martin L, Prasher V, Sinclair AJ. Serum hyaluronic acid in Down's syndrome. Mech Ageing Dev 1998; 106:155–160.

Chemistry and Biology of Hyaluronan
H.G. Garg and C.A. Hales (editors)

Chapter 18

Hyaluronan and Scarring

ANDREW BURD

Division of Plastic and Reconstructive Surgery, Department of Surgery, The Chinese University of Hong Kong, Prince of Wales Hospital, Shatin, Hong Kong SAR, China

I. Introduction

Hyaluronan (HA) was first isolated by Meyer and Palmer in 1934 (1), and has been shown to be a ubiquitous constituent of connective tissues. Because of its composition of hexosamine and uronic acid, and also of its biosynthesis, it has been classified among the glycosaminoglycans such as the dermatan sulfate, chondroitin sulfate, heparin sulfate, heparin, etc., although it is the only polymer of this class of compounds in which a linkage to protein backbone (to form a proteoglycan) has not been demonstrated (2). The history of the structure determination of glycosaminoglycans, where new sugar components or new anomeric configurations have been found repeatedly, suggests that variations in the structure of HA are possible. N-deacetylated HA has been reported in old human skin (3). Although a linear polysaccharide, its macromolecular structure is that of an expanded coil. This is susceptible to enzymatic degradation, resulting in the release of low-molecular weight oligosaccharides. HA and its degradation products have been associated with a number of cell regulatory actions. Considerable attention has been directed towards the role of HA in growth and development (4), and it would appear that the structural configuration of hyaluronic acid facilitates the movement of cells within the extracellular matrix (5). In a review of the biology of HA published in 1989 one chapter was dedicated to the clinical applications of HA (6). Ten years later 10 chapters were devoted to the medical applications and endogenous use of HA in clinical processes (7). Now increasing attention is being directed towards the role of HA in wound healing and scarring.

II. Wound Healing

A. Regeneration

Following injury, the skin has a tremendous capacity to heal. When the injury involves disruption of the dermis, post-natal wounds heal with a macrophage-effected (8,9) fibro-proliferative repair process which results in the laying down of a new collagen-rich dermal matrix over which epidermal continuity can be re-established. This process of repair and the subsequent reorganization of the dermal matrix results in scar formation.

A scar is produced as the result of a reorganization of mesenchymal tissue produced in response to the healing of a wound. A wound is the discontinuity in tissue integrity. Such a discontinuity may be physical, as in a surgical incision, or functional, as in a dermal burn. A complex pathophysiological response results in a cascade of cell–matrix interactions that comprise the biological concept of wound healing. Such healing, as occurs in the post-natal human occurs by two major processes regeneration and repair (Fig. 1). Regeneration is illustrated by the rapid healing of a superficial abrasion of the skin. Complete replacement of the lost tissue occurs rapidly and following a period of hyperaemia and/or hypopigmentation the skin will return to its pre-injured appearance and function. Regeneration has occurred with a replacement of the lost tissue by an identical new tissue.

B. Repair

Repair is very different. Repair occurs when the wound involves mesenchymal connective tissue. The predominant matrix component of connective tissue is collagen and the major cell type is the fibroblast. Skin, the largest organ in the body, is composed of both ectodermally and mesodermally derived layers. The dermis is derived from the mesoderm and when the dermis is damaged it will attempt to heal by the process of repair. The outcome of repair is scar tissue. This is obvious in the skin and produces both deformity and disability. Scarring does, however, have widespread effects throughout the body resulting in an extensive range of morbidity. Connective tissues, which produce the scaffolding for specialized organ tissues, can be deformed and distorted by the process of scarring and can result in respiratory dysfunction due to lung parenchymal fibrosis, liver dysfunction due to cirrhosis, renal dysfunction due to post-glomerulonephritic scarring, adhesions in the bowel after surgery, strictures resulting in infertility after pelvic inflammatory disease affecting the fallopian tubes, urinary dysfunction following urinary tract strictures, post-traumatic

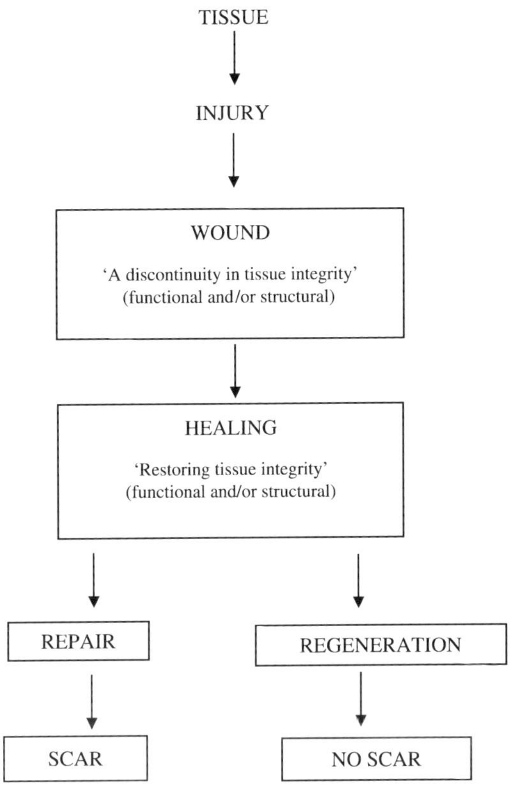

Figure 1 The key difference between regeneration and repair: scar or no scar.

epilepsy, vascular compromise due to vessel narrowing secondary to athero-sclerotic disease. The extent of human pathology and disease associated with scarring is vast. As such, the study of wound healing and the research into the fundamental biological processes of scarring is of major interest throughout the globe. It is evident that research in wound healing and scarring is not carried out in isolation as the biological phenomenon of repair and regeneration have to be closely related to tissue development as well as degeneration and ageing.

C. Fetal

Fetal wounds heal without scarring (10–17). The mechanism for the scarless healing in the fetus is unclear; immuno-histochemical staining (18) and bio-chemical studies (19) indicate that there is a deposition of a collagen-rich repair matrix, which is rapidly organized. In an early-third trimester fetal sheep, there is no observable scar tissue by 2 weeks after wounding. What is notable is that while macrophages are present in these fetal wounds, polymorphonuclear leucocytes (PML) are not. Healing after the mid-third trimester in fetal sheep, is associated with scarring and this time point coincides with the appearance of PMLs in the circulation (10). It appears that earlier fetal wounds have an ability to undertake an autonomous, site-specific repair while classically activated macrophage-effected repair is not site-specific and ends in scarring.

D. The Role of Hyaluronan

There is now evidence to suggest that post-natal wound healing can be modulated, resulting in a decrease in scarring. Hellstrom (20) reported enhanced wound healing in tympanic membrane perforations in a rat model following the topical application of tissue-extracted HA. Membranes treated with HA appeared otomicroscopically normal (translucent) 3 months later in contrast to untreated membranes, which showed extensive opacification. In a follow up study it was reported that this effect was independent of molecular size or rheological properties of the HA but related to the concentration of the preparation (21). But what is the mechanism for the proposed effect of HA in wound healing and scarring? The possibilities are:

1. macrophage-effected post-natal repair process is suppressed,
2. an intrinsic tissue repair process is augmented, or,
3. a combination of the two effects occurs.

III. Hyaluronan and Wound Healing-Hypotheses

Whatever the case, a mechanism must be proposed where HA has a definite role in harnessing and manipulating the natural reparative capacity of tissue fibroblasts. This raises the question whether HA alone has the capacity to control events of such complexity as wound healing and scar maturation contain hydroxyproline. In the early stages of repair, when HA levels are high, cell migration plays a prominent role in restoring cellular continuity (22). Later, when HA concentrations fall and sulfated GAG concentrations are high, cell differentiation, collagen production, and tissue organization occur (23,24). The high molecular weight HA seems to inhibit the formation of capillaries (25) while angiogenesis is induced by degradation products of HA (26). Weigel et al. (27), have proposed a model for the role of HA in wound healing in which a HA–fibrin matrix forms which attracts inflammatory cells into the wound. This matrix is in turn modified by the cells entering the wound as they secrete hyaluronidase and plasminogen activator into the extracellular space to degrade hyaluronate and fibrin. The degradation products from both HA and fibrin are then seen as important regulatory molecules for controlling cellular functions involved in the inflammatory response and inducing new blood vessel formation in the healing wounds. While this is an attractive model it does not suggest the degree of control that would be expected in such complex cell and matrix interactions as occur in wound healing and scar maturation. It also concerns the early events of the inflammatory response and wound healing.

Further evidence of the role of HA in wound healing and scarring has come from more observations of scarless healing in the foetus. HA is a major component of the fetal extracellular matrix and it was thought that HA could play an important role in the process of 'scarless' healing. Studies of HA and HA-stimulating activity in a sheep model at 75, 100 and 120-day gestation fetuses showed an ontogenic transition in wound HA matabolism from fetal to the adult-like phenotype in the 120 day gestation fetuses. This corresponded to an increase in scar formation in these late gestation fetuses (28). These observational studies have been corroborated with other studies of biological processes such as morphogenesis and oncology (29,30), which provide valuable insights into the potential role of HA in tissue remodelling and repair. Another study looking further at the role of HA in the repair matrix used a sponge implant model in mice to which was added exogenous HA. A comparison was made between sponges to which were added HA and those to which were added hyaluronidase. The repeated addition of the HA appeared to mimic the fetal dermal connective tissue with a matrix containing finer bundles of collagen fibres in an organized fashion lying within granulation tissue. The sponges containing hyaluronidase showed elevated levels of granulation tissue and poorly organized collagen fibre depostion (31). Yet another study looking at a mouse limb organ culture indicated that exogenous HA application can promote scarless repair (32). A further study has looked at the role of HA in increased scar tissue response. Keloids are examples of aberrant scar tissue response and immunohistochemical study of keloids, hypertrophic and normal scar showed increased HA staining within the epidermis of keloids compared to normal scars and skin. However, fibroblasts cultured from keloid scar tissue demonstrated significantly lower levels of HA release into the culture media and also diminished accumulation of HA in the keloid derived cells. It is postulated that in the post-natal situation where keloids develop that this is an indication of aberrant signaling between the keloid stroma and the keloid epidermis (33). One further experimental study looked at the effects of increasing HA on collagen fibrillar organization in an *in vitro* model. When high doses (>1 mg/mL) of HA were added to the culture medium the contraction of floating collagen fibrillar matrices (CFM) was significantly reduced. No direct correlation could be found however between viscosity of glycosaminoglycans and their ability to reduce CFM contraction (34). There is now clear evidence to support the role of peptide factors in the control of acute phase healing (35–39) and it is very likely that such protein factors are involved in the longer term processes of remodelling the repair matrix.

IV. Perfect Skin

By 1997, Martin, reviewing the goal of perfect skin regeneration in wound healing, detailed a list of 18 cytokines identified as having a major role in wound healing (40). Within 5 years this number had increased considerably particularly with the realization that previously identified single factors actually represent families of isoforms, thus, e.g., the fibroblast growth factor family currently consists of 22 members, at least five of which have been implicated in wound healing (41). Table 1 shows the current list of endogenous factors associated with wound healing.

This, then is an introduction to a chapter which is going to explore the current views on the role of hyaluronic acid in wound healing and scarring (the intrinsic role) and then look at medical developments of hyaluronic acid that have been used or are proposed to either reduce or eliminate scar development in healing wounds or to reduce the degree of matrix disorganization, a key attribute of scar tissue, in the established scar (the extrinsic role).

V. Properties of Hyaluronan

HA has unique hygroscopic, rheological and viscoelastic properties as well as biological properties. The biological role of HA ranges from a purely structural function in the extracellular matrix to developmental regulation through effects of cellular behavior via control of tissue macro and microenvironments. In addition HA can directly affect gene expression through receptor-mediated interaction. The following is a brief overview of these physicochemical and biological properties.

A. Physicochemical

HA has a considerable affinity for water. In the tissues, this property is of considerable significance during periods of change, e.g., in embryonic

Table 1 A List of the Families of Growth Factors and Cytokines Involved in the Regulation of Wound Healing

Platelet-derived growth factor family
Fibroblast growth factor family
Epidermal growth factor family
Vascular endothelial growth factor family
Angiopoietins
Insulin-like growth factors
Scatter factors
Nerve growth factor
Transforming growth factor-β
Activins
Bone morphogenetic proteins
Connective tissue growth factor/cysteine-rich 61/nephroblastoma overexpressed (CNN) family
Chemokines
Proinflammatory cytokines
Granulocyte-macrophage colony stimulating factor
Leptin
Interleukin-10

development and during the inflammatory response to tissue injury when HA levels are elevated. It is thought that focal HA release can weaken cell anchorage to the extracellular matrix and thereby aid proliferation and migration. The HA molecule is very large and has viscoelastic properties that are exploited in ophthalmic surgery. The rheologic properties of the molecule allow a cushioning and lubricating effect when used as a substitute for aqueous humor in the eye but also as an expander in synovial joints. In the tissues, the HA molecule traps and restricts the movement of water and small molecules. Larger molecules of such proteins are excluded from the HA matrix. The highly viscous nature of HA is also thought to restrict the movement of viruses and bacteria through the tissues.

In physiological conditions, HA is fully ionized and may have a role in cellular signaling through membrane ion channels. HA also acts on a scavenger for free radicals and an antioxident. This may have a protective action against solar radiation and a modulating action in inflammation.

B. Biological

HA has, in addition to these physicochemical properties, the ability to bind to cells and extracellular matrix components in a specific manner which allows the modulation of the biological properties and functions of cells and tissues.

1. Cell Surface Receptors

HA binds to cells via three main classes of cell surface receptor:

1. CD44 (42)
2. RHAMM (receptor for HA-mediated motility) (43,44)
3. ICAM-1 (intracellular adhesion molecule-1) (45)

CD44 is widely distributed in the body. Its association with HA is not fully determined but it appears to be related to cell–cell and cell–substrate adhesion as well as cell migration, proliferation and activation as well as HA uptake and degradation. In the skin, CD44 appears to be involved in the regulation of keratinocyte proliferation in response to extracellular stimuli and the maintenance of local HA synthesis.

2. Binding to Matrix Components

HA is an integral part of the extracellular matrix and forms the backbone for the organization of proteoglycans. It forms associations with collagen, fibrin and other matrix molecules and allows cell migration to occur (46).

3. Hyaluronan-Binding Proteins

HA binding proteins (hyaladherins) are widely distributed in the body and have diverse functions. The binding of HA to cells and other matrix components is effected by hyaladherins. These form complexes with proteinase inhibitors. Through HA–hyaladherin–proteinase inhibitor complexes, HA can help to limit inflammation and stabilize the extracellular matrix in inflammatory processes (47–49).

4. Hyaluronan and Hyaluronan-Oligosaccharide Action on Cell Behavior

Hyaluronan-degradation products, oligosaccharides, have been shown to have an effect on angiogenesis. The mechanism is unknown. Soluble HA fragments have been shown to upregulate the expression of several cytokines such as interleukin 1β, TNF-α and insulin-like growth factor-1. They also induce the expression of several inflammatory genes in macrophages through a CD44 receptor mediated mechanism and promote collagen production by endothelial cells (50).

5. Hyaluronan Synthesis

HA is unique amongst matrix molecules in that following synthesis it is directly secreted into the extracellular space. The production occurs on the cell membrane and the newly extruded HA provides a highly hydrated microenvironment which may further facilitate cell movement (51,52).

VI. Role of Hyaluronan in Scarring

A. Sequence of Events

As HA is a key element of the extracellular matrix it is bound to have a key role in the processes of wound healing which involves new matrix deposition. The sequence of events of wound repair, the form of healing that results in scar formation involves inflammation, cell proliferation, matrix deposition angiogenesis and matrix remodeling (Fig. 2) (53–55).

B. Inflammatory Phase Modulation

HA is implicated in both activation and suppression of inflammation. During the early phases of inflammation, HA is involved in the enhancement of cell infiltration (56). It promotes an increase in pro-inflammatory cytokines TNF-α, IL-1β and IL-8 via a CD44-mediated mechanism (57). HA also facilitates primary adhesion of cytokine activated lymphocytes to the endothelium (58). Suppression of the inflammatory response is effected through the free radical scavenging and antioxident properties of HA. This is a physicochemical characteristic of large polyionic polymers (59–63). Further, HA may also function as a negative feedback loop in inflammatory activation. TSG-6 is an HA-binding protein whose release is stimulated by TNF-α, a peptide factor important in promoting the inflammatory response. TSG-6 forms a stable complex with the serum proteinase inhibitor IαI which has a synergistic action on plasmin-inhibitory activity. Plasmin is involved in actuation of the proteolytic cascade

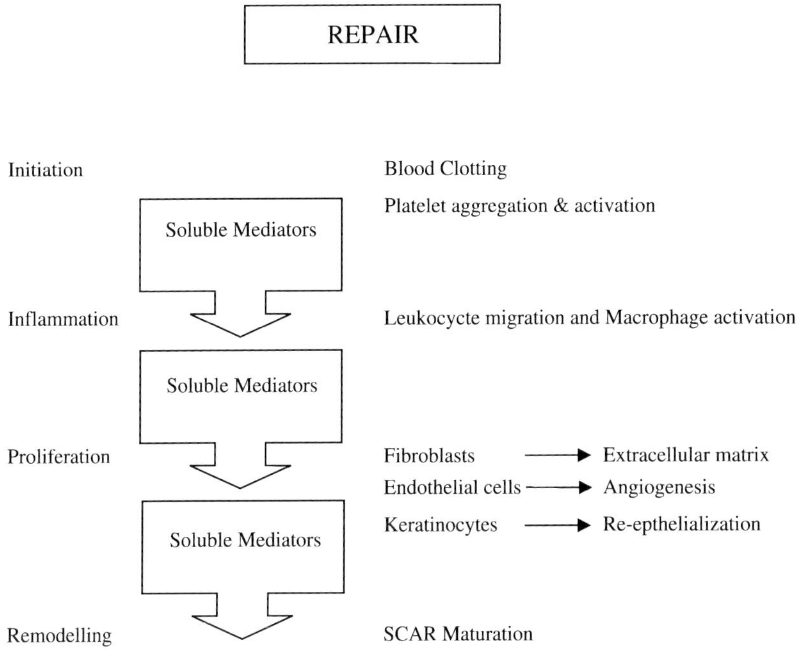

Figure 2 The stages of repair.

of matrix metalloproteinases that lead to further inflammatory tissue damage. As such the TSG-6/IαI complex, further organized by binding the HA may serve as a negative feedback loop to moderate the inflammatory response (64,65). Cell proliferation is an essential part of the repair process. HA is essential for fibroblast detachment from the extracellular matrix and mitosis (66,67). HA does have a direct mitogenic effect on cells but the facilitation of detachment (68) allows mitosis to occur and thus HA indirectly has a role in cell proliferation. Cell migration is promoted in the early phases of wound repair when increased HA synthesis occurs (69–73). The newly deposited extracellular matrix rich in HA provides an open, hydrated environment that facilitates cell migration (74,75). This physicochemical component is also enhanced by the CD44 and RHAMM receptor mediated cell migration (76–83). Angiogenesis is a key feature of the proliferative phase of wound repair. The angiogenic properties of low molecular weight HA oligosaccharides have been well established (84–94).

As the extracellular matrix forms, the peripheral keratinocytes are undergoing increased mitosis and proliferation prior to migrating across the new matrix. The HA-rich matrix supports and promotes the proliferating basal keratinocytes (95–99) and facilitates their migration again through a CD44-mediated mechanism (100). After re-epithelialization has occurred the new matrix undergoes remodeling. This remodeling process can be prolonged and the tissue is referred to as scar. In the HA-rich matrix seen in the fetal wounds,

HA may reduce collagen deposition, enhance remodeling and contribute to reduced scarring (101,102).

VII. Medical Applications

A. Ophthalmology

The original development of HA as a product to be used in clinical medicine is entirely due to Endre Balazs. He derived the main concepts and was the first to prepare HA samples that were of sufficient purity to be tolerated. During the 1950s Balazs concentrated his research on the composition of the vitreous body and started to experiment with vitreous substitutes to be used in surgery of retinal detachment. He introduced the term "viscosurgery" for these medical applications. One of the crucial obstacles for using HA in implants was to prepare HA free of impurities that could cause inflammatory reactions. Balazs solved this problem, and his final preparation was called NIF-NaHA (non-inflammatory fraction of sodium hyaluronate). In 1970 Rydell and Balazs injected HA in the arthritic joints of racehorses with a dramatic positive effect on the clinical symptoms. Two years later, Balazs convinced Pharmacia AB in Uppsala to start production of HA for veterinary and human use. Ten years later, Miller and Stegman, following the advice of Balazs, started to use HA as a device in the implantation of plastic intraocular lenses, and HA became a major product for ophthalmic surgery under the trade name of Healon® (103–106).

B. Tendon Healing

A major problem associated with tendon repair is the formation of adhesions, essentially scar tissue, which restrict the range of movement of the repaired tendon and result in significant dysfunction. The visco-elastic properties of HA were considered to be of benefit to enhancing the gliding of the repaired tendon. Experimental studies in chickens (107–109), rabbits (110–113) and horses (114) have supported the proposal that HA membranes acting as a physicochemical barrier can prevent restrictive adhesions in primary tendon repair.

C. Cosmetic Filler

In skin, there is a need for a tissue 'filler' for cosmetic reasons. Currently, a popular treatment for wrinkles is the use of botulinum toxin. The effect of this agent is to cause the paralysis of muscles, which influences the development of wrinkles. Comparative studies of using botulinum toxin alone or in conjunction with HA in form of Hyalan B indicate that moderate to severe rhytides were treated better by botulinum toxin combined with HA than with botulinum toxin alone (115). This study was repeated with an HA-botulinum toxin combination where the HA was produced by a non-animal source. Restylane is an HA produced by bacterial fermentation incorporating recombinant gene technology.

Again the results of combined therapy were superior to the use of botulinum toxin alone (116). The use of cosmetic fillers has produced quite a significant debate in the literature concerning their safety. The essential debate concerns the source of the materials and whether they produce adverse reactions. It does appear that some individuals can be allergic to commercial preparations of HA (117,118) although in general they do seem to be well tolerated (119). When the safety of injectable, non-animal stabilized HA gel used for soft tissue augmentation, was assessed using a retrospective review of all adverse events data in Europe, Canada, Australia, South America and Asia from 1999 to 2000 it appeared that major adverse events are most likely to be due to impurities from bacterial fermentation and that the incidence of hypersensitivity appears to be declining after the introduction of a more purified HA raw material (120,121). Of course the safety of any new product has to be compared with an established product. Thus a comparison of Restylane (Q-Med Uppsula, Sweden) a non-animal stabilized HA gel compared to bovine collagen (Zyplast) for treatment for sunken nasolabial folds was of particular interest. The conclusion of this prospective, randomized trial was that non-animal stabilized HA provided a more durable aesthetic improvement than bovine collagen and was well tolerated (122).

D. Soft Tissue Augmentation

Another clinical situation where soft tissue augmentation has dramatic functional consequence is the use in the vocal cords. Hyalon B has been assessed in both animal (123,124) and clinical studies (125) and found to be effective and safe when used in patients with glottal insufficiency. Further animal studies have confirmed the potential benefit of HA as a filler in posterior pharyngeal wall insufficiency (126).

E. Post-Operative Adhesions

Another area of clinical concern are the surgical scars that result in post-operative adhesions. Clinical studies have indicated that HA and cross-linked derivatives can be effective in preventing adhesions after endoscopic sinus surgery (127, 128). Experimental studies have assessed the efficacy of a hyaluronate/carboxymethylcellulose gel using a rabbit uterine horn model. The positive benefits of applying the gel were demonstrated (129). Further studies confirm the positive benefits of HA based applications in rat and rabbit models of surgical adhesions (130,131).

VIII. Hyaluronan Derivatives

A. Medical Applications

HA is attractive as a building block for new biocompatible and biodegradable polymers that have applications in drug delivery, and tissue engineering. However, HA has poor biomechanical properties in its native form and a variety of chemical modifications have been devised to provide mechanically and chemically robust materials. The resulting HA derivatives have physicochemical properties that may significantly differ from the native polymer, but most derivatives retain the biocompatibility and biodegradability, and in some cases, the pharmacological properties of native HA. The most commonly used modification of native HA is esterification of the carboxyl groups (132).

B. Esterification

Esterified HA biomaterials have been prepared by alkylation of the tetra (*n*-butyl) ammonium salt of HA with an alkyl halide in dimethylformamide (DMF) solution. At higher percentages of esterification, the resulting HYAFF® materials (Fidia Advanced Biopolymers) became insoluble in water. These HA esters can be extruded to produce membranes and fibers, lyophilized to obtain sponges, or processed by spray-drying, extraction, and evaporation to produce microspheres. These polymers show good mechanical strength when dry, but the hydrated materials are less robust. The degree of esterification influences the size of hydrophobic patches, which produces a polymer chain network that is more rigid and stable, and less susceptible to enzymatic degradation.

C. HYAFF—Structure and Properties

The HYAFF series of semi-synthetic HA derivatives consist of alcohol esters of HA which vary in respect to the pendant alcohol group (benzyl and ethyl) and the level of esterification achieved. X-ray diffraction studies of the partial and total benzyl esters of HA (HYAFF 11p75 and HYAFF 11) indicate that the HYAFF derivatives exhibit a diffraction pattern similar to the unsubstituted HA. Molecular modeling based on the X-ray data demonstrates that the benzyl esters of HA have the same conformity and flexibility as unmodified HA. The ethyl ester of HA is known as HYAFF 7. Extruding HYAFF into organic solvents has allowed the production of fibers and membranes. The fibers can be woven and the membranes perforated to produce a wide range of biomaterials. These HYAFF-based biomaterials are biodegradable. The principle factor that determines the length of time it takes for the materials to degrade is the polymer used. The three most common polymers with increasing resistance to biodegradation are HYAFF 11p75 < HYAFF7 < HYAFF11. The length of time implanted products may remain can range from several weeks to several months and is also influenced by the physical conformation of the product and the site of the implantation (133). Degradation of the HYAFF occurs through an initial hydrolysis of the ester bond, which releases free alcohol. Solubilization if the material then follows and the free alcohol and HA are broken down by their normal metabolic pathway. This means that a wide range of safe and biodegradable medical devices can be developed for specific clinical applications.

D. Animal Studies

1. Otology

Histochemical studies show that HA is a normal constituent of middle ear tissues although the functional significance of the occurrence of HA in the middle ear is still a matter of debate. Most probably it is the physico-chemical property of the molecule acting as a stress-absorber that results in its presence in the normal ear. The application in the abnormal ear has been proposed and experimentally evaluated in chronic tympanic membrane perforations. Topically applied HA applied to experimental tympanic membrane perforations not only accelerated the healing process but also reduced scar tissue in the healed tympanic membrane. Further studies have been performed with pieces of sponge manufactured from HYAFF7, the 100% ethyl ester of HA. Pieces of HYAFF7 sponges were placed into middle ear cavities and compared to cavities similarly treated with Gelfoam—a gelatin sponge. When followed sequentially over 6 months the HYAFF7 sponges slowly degraded leaving cavities that were clean and healthy and contained a foamy material intermingled with a few scattered fibroblasts. In contrast, the Gelfoam filled ears showed a dense collection of connective tissue.

When HYAFF7 sponges were applied to the chronic tympanic membrane perforations they were loaded with unmodified HA allowing a sustained release of HA in the healing environment. Healing was reportedly better with less scarring in ears thus treated compared to membranes that healed without HA application (134).

2. Cutaneous Wounds

Further studies have been performed on cutaneous wounds on pigs. In these studies two biomaterials formulated from the partial benzyl ester of HA (HYAFF11p75) were used. These were a fleece-like material HYALOFILL-F and a rope-like material HYALOFILL-R. Both forms of the material have shown good biocompatibility and were compared with Sorbsan, a calcium alginate preparation, in the healing of deep excisional punch wounds made in Yucatan micro pigs.

The histological assessment of the wound healing process in this comparative study underlined some of the biological advantages of using HA and HA derivatives in wound healing. A principle advantage of HA over other biological materials is that it is characteristically inert or even inhibitory as far as inflammatory and immunologic reactions are concerned. Some macrophages were observed transiently during the phase of degradation but these were presumed to be consuming the soluble biomaterial. These macrophages are therefore acting in a relatively unactivated form as far as expressing cytokines in the classical wound repair context. Thus, in this animal model the speed of healing but also the quality, in terms of reduced scarring, both appeared to be enhanced by using HYAFF formulations (135).

The use of biodegradable products derived from HA derivates has extended into other areas of experimental research. HYAFF11 seeded with chondrocytes was transplanted into cartilage defects in rabbits. The comparison was between the HYAFF alone and the HYAFF plus chondrocytes. Statistically significant differences in the quality of the regenerated tissue were found between the grafts carried out with the HYAFF plus chondrocyte combination compared to the HYAFF alone. This study demonstrates the efficacy of this HA-based scaffold for autologous chondrocyte transplantation (136). Further studies have looked at the incorporation of calcium phosphate and HA sponges in the repair of osteochondral defects indicating that a two phase composite graft may hold potential for the repair of such defects by providing mechanical support that mimics subchondral bone whilst providing a chondrogenic scaffold for the overlying cartilage repair (137). Further studies support the use of HA derivatives in bone and cartilage repair (138–150).

Other studies have looked at the possibility of using HYAFF11 as a carrier for adipocytes. Human preadipocytes were implanted into nude mice using HYAFF11 scaffolds, or collagen sponges. Scaffolds without cells were used as controls. Pre-adipocytes matured earlier *in vitro* when attached to HYAFF11 scaffolds when compared to other carriers. *In vivo* the HYAFF sponges supported the expansion and differentiation of the adipose precursor cells. The carrier was superior to the collagen sponges with regard to cellularity. Modifications of the scaffold using variations in pore size and pre-coating with adipogenic factors were proposed for further evaluation (151).

3. Ocular Surgery

Experimental studies have been performed on rabbits to look at the biocompatibility and biodegredation of intravitreal HA implants. HYAFF7, HYAFF11 and HYAFF11p75 were used. All show good biocompatibility and the implants of the total esters HYAFF7 and HYAFF11 showed biodegradability, 60 and 150 days, respectively. The partial benzyl ester HYAFF11p75 was completely resorbed in 15 days. This study indicates the potential for using intravitreal implants of HA esters for potential drug delivery systems in the treatment of posterior segment disease (152).

4. Urology

Another potential application for HYAFF derived biomaterials has been explored in urethral reconstruction. A 1.5 cm segment of rabbit urethra was resected and replaced with a HYAFF11 tubular graft. By post-operative week three the HYAFF11 guide had disappeared. At 4 weeks the retrograde urethrogram showed good distensibility of the neourethra. Histology showed the regenerated stroma consisted of fibroblastic cells and collagenous and elastic fibers. The neo-epithelium was stratified and showed cuboidal type cells (153).

E. Clinical Studies

Clinical applications of HYAFF formulation have proved to demonstrate considerable promise. Laserskin™ is a membrane made entirely of 100% benzyl ester of low molecular weight HA (HYAFF11). Transparent sheets were manufactured with rows of 40 μm diameter laser drilled holes and 0.5 mm diameter mechanically drilled holes. The concept is that the membrane can act as a carrier for keratinocytes cultured in the laboratory. Traditionally, keratinocytes are cultured on a 3T3 fibroblasts feeder layer. The cells reach confluence and stratify and the delicate film of cultured skin is transferred from the laboratory to the clinical situation for application to the patient. There are several drawbacks in this process not the least being the delicate nature of the cultured skin, but also the time taken for the preparation of the material and the fact that the majority of the cells transferred were non-proliferating. The concept in Laserskin is that a 3T3 fibroblast feeder layer would still be used to 'prime' the Laserskin after which keratinocytes would be added. The feeder layer enhances the attachment of the keratinocytes through deposition of matrix proteins and stimulates cell proliferation through cytokine release. Keratinocyte colonies displace the fibroblasts and proliferating cells begin to migrate through the laser drilled micro holes onto the under surface of the laser skin (Fig. 3). At this stage the membrane can be lifted from the culture dish and transferred to the patient. The macro holes allow for drainage of wound exudates and enhance attachment. The advantages of this technique are easier handling, more rapid preparation and also the application of proliferating keratinocytes to the wound bed (154). This early clinical study has been supported by a series of follow up studies looking at a number of variables within the model. A comparison of 'clinical' take rates in a porcine kerato-dermal model demonstrated a significant reduction in take as a result of halving the keratinocyte seeding density onto the membrane. The take rates of grafts grown on the membrane at a conventional seeding density and then

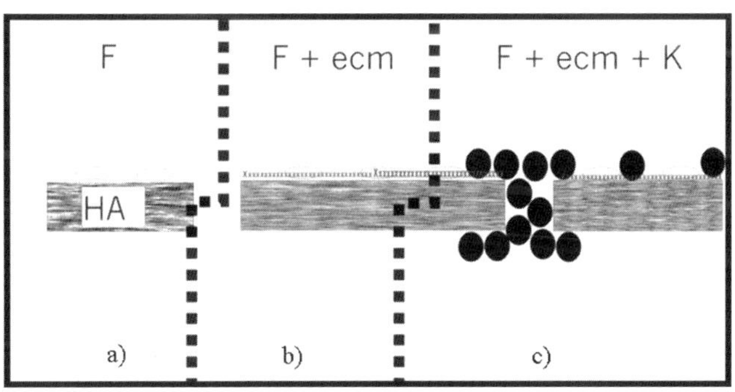

Figure 3 A disgramatic cross-section of Laserskin. (a) HA is the laser skin preseeded with fibroblasts (F). (b) The fibroblasts produce the extracellular matrix (ecm). (c) Keratinocytes (K) attach and proliferate and migrate through the laser dotted holes to the deep surface of the Laserskin, which is applied to the wound surface.

transplanted to the dermal wound bed were comparable, and in some cases superior, than those of keratinocyte sheet grafts (155).

A further modification of this technique is to use a HYAFF scaffold in which fibroblasts are seeded. This is then applied to the wound and allowed to incorporate for several weeks before being overgrafted with the Laserskin carrying keratinocytes. In this manner a bilaminar skin has been formed with both epidermal and dermal components (156–158). The use of Laserskin as a carrier for autologus keratinocytes has also been advocated in chronic wounds. In this situation the keratinocytes appear to stimulate the underlying cells leading to granulation tissue formation and progressive wound healing (159). HYAFF has also been used in wound healing studies. Hyallofill has been used in chronic wound situations and initial clinical studies indicate that it has potential for converting the chronic wound to an acute wound. That is to say that it is effective in changing the cytokine profile and cellular dynamics in the wound and 'kick-starting' the wound healing process towards repair (160,161). One possible explanation of this effect comes from a study of the comparative antioxidant effects of HYAFF11p75, Aquacel (a carboxymethylcellulase material) and HA. Experimental data indicates that HYAFF11p75 has a greater antioxidant capacity towards the superoxide radical due to the esterified benzyl groups providing alternative sites for superoxide radical attack other than the HA backbone of HYAFF11p75 itself (162).

IX. Hyaluronan, Wound Healing and Scarring: A New Perspective

In 1991 a paper was published suggesting that a new perspective should be taken on HA and wound healing (163). The background to this paper was the serendipitous discovery that HA extracted from human scar tissue and highly purified still had an identifiable collagen component using cyanogen bromide digests (164). This led to further studies of other HA preparations which indicated that all HA preparations had some protein 'association' (Table 2). The sources of these preparations were all animals with the exception of the preparation from Genzyme which was produced by bacterial fermentation.

The effect of these HA preparations was then assessed on models of cell proliferation and matrix organization. HA preparations were used with and without prior digestion with hyaluronidase. The conclusion of these studies was that even after hyaluronidase digestion the extracts with a high protein content had significant effects on cell proliferation. It is now possible to review these findings in the context of the accumulating information from experimental and clinical studies. It is remarkable to look at a single clinical publication that aims to provide the current knowledge base within a clinical discipline. Over the past 12 years three issues of Clinics in Plastic Surgery have been devoted to wound healing and they reflect an exponential increase in information regarding the biological phenomenon of wound healing and scarring (165–168). The underlying message of this increase in understanding of the biological processes

Table 2 Amino Acid Analysis (Residues per 1000) of HA Preparations. Cystine and Methionine Have Not Been Included Because of Interference in the Chromatogram Due to a Glucosamine Peak

	Sigma	Genzyme	Pharmacia	Medchem	Epidermis	Dermis	Normal SC
HPRO	74	0	0	0	10	5	22
ASP	101	0	59	21	85	94	83
THR	59	0	11	12	40	44	40
SER	57	0	14	20	85	82	77
GLU	81	25	76	20	116	130	108
PRO	33	0	0	0	34	37	59
GLY	86	38	90	34	153	137	141
ALA	81	0	36	0	66	75	80
VAL	31	0	0	0	166	94	176
ILE	36	0	0	0	31	39	26
LEU	71	0	0	0	70	87	62
TYR	0	0	0	0	16	23	15
PHE	15	41	0	2	27	31	25
HIS	102	361	34	516	17	17	20
HLYS	63	17	130	56	0	1	2
LYS	33	175	549	319	44	57	32
ARG	78	343	0	0	39	47	31
% Protein in HA	0.17	0.06	0.15	0.15	1.83	4.63	1.58

of wound healing and scarring is that it is a very complex process. Indeed the reductionist philosophy of traditional scientific research is no longer applicable as we move into a far more complex and multifactorial world of co-inter dependencies. One analogy that helps to comprehend the enormity of the task of the biological scientist is the field of linguistics. An elemental part of the intonation of all languages is the phoneme. This is a basic sound element. There are a finite number of sound elements, which may be found in all languages. Each language drawing upon a select group of phonemes to produce, morphemes, syllables, words and with the addition of complex grammar, a form of communication is derived. The complexity of language is an indication of the complexity of biological processes. Thus with HA we are defining an elemental part of the communication processes of considerable complexity. The scientist who imbues a single molecule with a relevance of outstanding significance in any biological process, is in fact, a 'dinosaur' in science. The contemporary perspective has to be integrative, not reductionist, and the integration has be complex and multilayered. But what does this mean in real terms? HA is a unique molecule that has a profound impact on the biological behavior of cells and tissues which are under the influence of multiple peptide factors. HA creates a permissive environment for cell proliferation and is able to suppress, to a degree, the inflammatory response of injured tissues. Nevertheless it is essential to realize and establish that HA, alone, is not going to determine the outcome of the wound healing response. The reality is that HA can suppress the inflammatory response to wound healing through receptor mediated interaction; it can promote cell movement by its physicochemical properties and it can deliver the cytokine factors which are the major elements of intercellular communication via its structural conformation. HA facilitates wound healing and plays a role in scarless healing but it is not the primary agent. It may well be that as our understanding of the complexity of biological interactions develops that we will begin to unravel a number of different 'languages'. Thus we may find a language of growth, another of development, another of repair, another of regeneration. HA has a role in all these processes but the defining differences most probably relate to the peptide factors which are active within the differing HA environments. To take this concept into the clinical situation it is likely that there will be a considerable increase in the number of 'HA-plus' preparations, where the 'plus' refers to cytokine combinations that can promote certain types of tissue behavior. One of the favored single peptide factors popularized in wound healing is transforming growth factor-β (TGF-β) (169,170). It has been proposed that application of a neutralizing antibody to TGF-β can control adult scarring (171). Whilst this is too simplistic a concept it is possible that combining HA and such an antibody may have more effect.

The proposal previously made (163) was that HA acted like a 'taxi' in the tissues taking the 'passengers' cytokines from cell of origin to their effector site. This concept may still well be valid in terms of considering a role of HA to be the facilitated distribution of cytokines in the remodeling matrix. It is evident, however, that HA has other potent contributions to play in the complex biological process of wound healing and scarring. Further developments in unraveling of this contribution are likely to focus on the electro-chemical influences on molecular interactions and conformational change (172). The determination of the mechanisms concerning the control, direction and potential extrinsic modulation of this energy is going to be a major challenge in biological research.

References

1. Meyer K, Palmer JW. The polysaccharide of vitreous humor. J Biol Chem 1934; 107:629–634.
2. Comper WD, Laurent TC. Physiological function of connective tissue polysaccharides. Physiol Rev 1978; 58:255–315.
3. Longas MO, Russell CS, He XY. Evidence for structural changes in dermatan sulfate and hyaluronic acid with aging. Carbohydr Res 1987; 159:127–136.
4. Toole BP, Goldberg RL, Chi-Rosso G, Underhill CB, Orkin RW. Hyaluronate cell reactions. In: Trelstad RL, ed. The Role of the Extracellular Matrix in Development. New York: Alan R Liss, 1984:43–66.
5. Docherty R, Forrester JV, Lackie JM, Gregory DW. Glycosaminoglycans facilitate the movement of fibroblasts through three-dimensional collagen matrices. J Cell Sci 1989; 92:263–270.
6. Balazs EA, Denlinger JL. Clinical uses of hyaluronan. In: Laurent TC, ed. The Biology of Hyaluronan. Sussex: Wiley, 1989:265–280.
7. Laurent TC, ed. The Chemistry, Biology and Medical Applications of Hyaluronan and Its Derivatives. London: Portland Press, 1998.

8. Knighton DR, Fiegel UD. The macrophage: effector cell in wound repair. Prog Clin Biol Res 1989; 299:217–226.
9. Chvapil M, Koopman CF Jr. Scar formation: physiology and pathological states. Otolaryngol Clin North Am 1984; 17:265–272.
10. Goss AN. Intrauterine healing of fetal rat oral mucosal, skin and cartilage wounds. J Oral Pathol 1977; 6:35–43.
11. Burrington JD. Wound healing in the fetal lamb. J Pediatr Surg 1971; 6:523–528.
12. Robinson BW, Goss AN. Intrauterine healing of fetal rat cheek wounds. Cleft Palate J 1981; 18:251–255.
13. Rowsell AR. The intra-uterine healing of foetal muscle wounds: experimental study in the rat. Br J Plast Surg 1984; 37:635–642.
14. Hallock GG. In utero cleft lip repair in A/J mice. Plast Reconstr Surg 1985; 75:785–790.
15. Hallock GG, Rice DC, McClure HM. In utero lip repair in the rhesus monkey: an update. Plast Reconstr Surg 1987; 80:855–858.
16. Krummel TM, Nelson JM, Dieglemann RF, Lindblad WJ, Salzberg AM, Greenfield LJ, Cohen IK. Fetal response to injury in the rabbit. J Pediatr Surg 1987; 22:640–644.
17. Siebert JW, Burd DAR, McCarthy J, Ehrlich HP. Fetal wound healing: a biochemical study of scarless healing. Plast Reconstr Surg 1990; 85:495–502.
18. Longaker MT, Whitby DJ, Adzick NS, Crombleholme TM, Langer JC, Duncan BW, Bradley SM, Stern R, Ferguson MWJ, Harrison MR. Studies in fetal wound healing: VI. Second and early third trimester fetal wounds demonstrate rapid collagen deposition without scar formation. J Pediatr Surg 1990; 25:63–68.
19. Burd DAR, Longaker MT, Adzick NS, Harrison MR, Ehrlich HP. Fetal wound healing in a large animal model: the deposition of collagen is confirmed. Br J Plast Surg 1990; 43:571–577.
20. Hellstrom S, Laurent C. Hyaluronan and healing of tympanic membrane perforations. An experimental study. Acta Otolaryngol 1987; 442(Suppl):54–61.
21. Laurent C, Hellstrom S, Fellenius E. Hyaluronan improves the healing of experimental tympanic membrane perforations. Arch Otolaryngol Head Neck Surg 1988; 114:1435–1441.
22. Abatangelo G, Martelli M, Vecchia P. Healing of hyaluronic acid-enriched wounds: histological observations. J Surg Res 1983; 35:410–416.
23. Toole BP, Trelstad RL. Hyaluronate production and removal during corneal development in the chick. Dev Biol 1971; 26:28–35.
24. Toole BP. Hyaluronate turnover during chondrogenesis in the developing chick limb and axial skeleton. Dev Biol 1972; 29:321–329.
25. Feinberg RN, Beebe DC. Hyaluronate in vasculogenesis. Science 1983; 220: 1177–1179.
26. West DC, Hampson IN, Arnold F, Kumar S. Angiogenesis induced by degradation products of hyaluronic acid. Science 1985; 228:1324–1326.
27. Weigel PH, Fuller GM, LeBoeuf RD. A model for the role of hyaluronic acid and fibrin in the early events during the inflammatory response and wound healing. J Theor Biol 1986; 119:219–234.
28. Estes JM, Adzick NS, Harrison MR, Longaker MT, Stern R. Hyaluronate metabolism undergoes an ontogenic transition during fetal development: implications for scar-free wound healing. J Pediatr Surg 1993; 28:1227–1231.
29. Toole BP. Hyaluronan in morphogenesis. J Intern Med 1997; 242:35–40.

30. Delpech B, Girard N, Bertrand P, Courel MN, Chauzy C, Delpech A. Hyaluronan: fundamental principles and applications in cancer. J Intern Med 1997; 242:41–48.
31. Iocono JA, Krummel TM, Keefer KA, Allison GM, Paul H. Repeated additions of hyaluronan alters granulation tissue deposition in sponge implants in mice. Wound Repair Regen 1998; 6:442–448.
32. Iocono JA, Ehrlich HP, Keefer KA, Krummel TM. Hyaluronan induces scarless repair in mouse limb organ culture. J Pediatr Surg 1998; 33:564–567.
33. Meyer LJM, Russell SB, Russell J, Trupin JS, Egbert BM, Shuster S, Stern R. Reduced hyaluronan in keloid tissue and cultured keloid fibroblasts. J Investig Dermatol 2000; 114:953–959.
34. Huang-Lee LL, Wu JH, Nimni ME. Effects of hyaluronan on collagen fibrillar matrix contraction by fibroblasts. J Biomed Mater Res 1994; 28:123–132.
35. Nemeth GG, Bolander ME, Martin GR. Growth factors and their role in wound and fracture healing. In: Barbul A, Pines E, Caldwell M, Hunt TK, eds. Growth Factors and Other Aspects of Wound Healing. New York: Alan R Liss, 1988:1–17.
36. Davidson J, Buckley A, Woodward S, Nichols W, McGee G, Demetriou A. Mechanisms of accelerated wound repair using epidermal growth factor and basic fibroblast growth factor. In: Barbul A, Pines E, Caldwell M, Hunt TK, eds. Growth Factors and Other Aspects of Wound Healing. New York: Alan R Liss, 1988: 63–75.
37. Sprugel KH, McPherson JM, Clowes AW, Ross R. The effects of different growth factors in subcutaneous wound chambers. In: Barbul A, Pines E, Caldwell M, Hunt TK, eds. Growth Factors and Other Aspects of Wound Healing. New York: Alan R Liss, 1988:77–91.
38. Pierce GF, Mustoe TA, Deuel TF. Transforming growth factor β induces increased directed cellular migration and tissue repair in rats. In: Barbul A, Pines E, Caldwell M, Hunt TK, eds. Growth Factors and Other Aspects of Wound Healing. New York: Alan R Liss, 1988:93–102.
39. Spencer EM, Skover G, Hunt TK. Somatomedins: do they play a pivotal role in wound healing? In: Barbul A, Pines E, Caldwell M, Hunt TK, eds. Growth Factors and Other Aspects of Wound Healing. New York: Alan R Liss, 1988:103–106.
40. Martin P. Wound healing-aiming for perfect skin regeneration. Science 1997; 276:75–81.
41. Werner S, Grose R. Regulation of wound healing by growth factors and cytokines. Physiol Rev 2003; 83:835–870.
42. Knudson W, Chow G, Knudson CB. CD44-mediated uptake and degradation of hyaluronan. Matrix Biol 2002; 21:15–23.
43. Hall CL, Wang C, Lange LA, Turley EA. Hyaluronan and the hyaluronan receptor RHAMM promote focal adhesion turnover and transient tyrosine kinase activity. J Cell Biol 1994; 126:575–588.
44. Cheung WF, Gruz TF, Turley EA. Receptor for hyaluronan-mediated motility (RHAMM), a hyaladherin that regulates cell responses to growth factors. Biochem Soc Trans 1999; 27:135–142.
45. van de Stolpe A, van der Saag PT. Intercellular adhesion molecule-1. J Mol Med 1996; 74:13–33.
46. Entwistle J, Hall CL, Turley EA. HA receptors: regulators of signaling to the cytoskeleton. J Cell Biochem 1996; 61:569–577.
47. Toole BP. Hyaluronan and its binding proteins, the hyaladherins. Curr Opin Cell Biol 1990; 2:839–844.

48. Knudson CB, Knudson W. Hyaluronan-binding proteins in development, tissue homeostasis, and disease. FASEB J 1993; 7:1233–1241.
49. Day AJ. The structure and regulation of hyaluronan-binding proteins. Biochem Soc Trans 1999; 27:115–121.
50. Lee JY, Spicer AP. Hyaluronan: a multifunctional, megaDalton, stealth molecule. Curr Opin Cell Biol 2000; 12:581–586.
51. Weigel PH. Functional characteristics and catalytic mechanisms of the bacterial hyaluronan synthases. IUBMB Life 2002; 54:201–211.
52. O'Regan M, Martini I, Crescenzi F, De Luca C, Lansing M. Molecular mechanisms and genetics of hyaluronan biosynthesis. Int J Biol Macromol 1994; 16:283–286.
53. Manuskiatti W, Maibach HI. Hyaluronic acid and skin: wound healing and aging. Int J Dermatol 1996; 35:539–544.
54. Juhlin L. Hyaluronan in skin. J Intern Med 1997; 242:61–66.
55. Oksala O, Salo T, Tammi R, Hakkinen H, Jalkanen M, Inki P, Larjava H. Expression of proteoglycans and hyaluronan during wound healing. J Histochem Cytochem 1995; 43:125–135.
56. Wisniewski HG, Hua JC, Poppers DM, Naime D, Vilcek J, Cronstein BN. TNF/IL-1-inducible protein TSG-6 potentiates plasmin inhibition by inter-α-inhibitor and exerts a strong anti-inflammatory effect *in vivo*. J Immunol 1996; 156: 1609–1615.
57. Kobayashi H, Terao T. Hyaluronic acid-specific regulation of cytokines by human uterine fibroblasts. Am J Physiol 1997; 276:1151–1159.
58. Mohamadzadeh M, DeGrendele H, Arizpe H, Estess P, Siegelman M. Proinflammatory stimuli regulate endothelial hyaluronan expression and CD44/HA-dependent primary adhesion. J Clin Investig 1998; 101:97–108.
59. Presti D, Scott JE. Hyaluronan-mediated protective effect against cell damage caused by enzymatically produced hydroxyl (OH) radicals is dependent on hyaluronan molecular mass. Cell Biochem Funct 1994; 12:281–288.
60. Kvam BJ, Fragonas E, Degrassi A, Kvam C, Matulova M, Pollesello P, Zanetti F, Vittur F. Oxygen-derived free radical (ODFR) action on hyaluronan (HA), on two HA ester derivatives, and on the metabolism of articular chrondrocytes. Exp Cell Res 1995; 218:79–86.
61. Fukuda K, Tanaka S, Kumano F, Asada S, Oh M, Ueno M, Takayama M. Hyaluronic acid inhibits interleukin-1-induced superoxide anion in bovine chondrocytes. Inflamm Res 1997; 46:114–117.
62. Cortivo R, Brun P, Cardarelli L, O'Regan M, Conconi MT, Radice M, Abatangelo G. Antioxidant effects of hyaluronan and its alpha-methyl-prednisolone derivative in chrondrocyte and cartilage cultures. Semin Arthritis Rheum 1996; 26:492–501.
63. Foschi D, Castoldi L, Radaelli E, Abelli P, Calderini G, Rastrelli A, Mariscotti C, Marazzi M, Trabucchi E. Hyaluronic acid prevents oxygen free-radical damage to granulation tissue: a study in rats. Int J Tissue React 1990; 12:333–339.
64. Fries E, Blom AM. The structure and function of inter-α-inhibitor and related proteins. In: Laurent TC, ed. The Chemistry, Biology and Medical Applications of Hyaluronan and Its Derivatives. London: Portland Press, 1998:149–154.
65. Barnhart BJ, Cox SH, Kraemer PM. Detachment variants of Chinese Hamster cells. Hyaluronic acid as a modulator of cell detachment. Exp Cell Res 1979; 119:327–332.

66. Brecht M, Mayer U, Schlosser E, Prehm P. Increased hyaluronate synthesis is required for fibroblast detachment and mitosis. Biochem J 1986; 239:445–450.
67. Mian N. Analysis of cell-growth-phase-related variations in hyaluronate synthase activity of isolated plasma-membrane fractions of cultured human skin fibroblasts. Biochem J 1986; 237:333–342.
68. Abatangelo G, Cortivo R, Martelli M, Vecchia P. Cell detachment mediated by hyaluronic acid. Exp Cell Res 1982; 137:73–78.
69. Morriss-Kay GM, Tuckett F, Solursh F. The effect of Streptomyces hyaluronidase on tissue organization and cell cycle times in rat embryos. J Embryol Exp Morphol 1986; 98:59–70.
70. Turley EA, Austen L, Moore D, Hoare K. ras-Transformed cells express both CD44 and RHAMM hyaluronan receptors: only RHAMM is essential for hyaluronan promoted locomotion. Exp Cell Res 1993; 207:277–282.
71. Chen WYJ, Grant ME, Schor AM, Schor SL. Differences between adult and foetal fibroblasts in the regulation of hyaluronate synthesis: correlation with migratory activity. J Cell Sci 1989; 94:577–584.
72. Ellis I, Grey AM, Schor AM, Schor SL. Antagonistic effects of TGF-β1 and MSF on fibroblast migration and hyaluronic acid synthesis—possible implications for dermal wound healing. J Cell Sci 1992; 102:447–456.
73. Ellis I, Banyard J, Schor SL. Differential response of fetal and adult fibroblasts to cytokines: cell migration and hyaluronan synthesis. Development 1997; 124:1593–1600.
74. Toole BP. Proteoglycans and hyaluronan in morphogenesis and differentiation. In: Hay ED, ed. Cell Biology of Extra-Cellular Matrix, 2nd ed. New York: Plenum Press, 1991:305–341.
75. Toole BP. Hyaluronan in morphogenesis. J Intern Med 1997; 242:35–40.
76. Wang C, Thor AD, Moore DH, Zhao Y, Kerschmann R, Stern R, Watson PH, Turley EA. The overexpression of RHAMM, a hyaluronan-binding protein that regulates ras signaling, correlates with overexpression of mitogen-activated protein kinase and is a significant parameter in breast cancer progression. Clin Cancer Res 1998; 4:567–576.
77. Hall CL, Lange LA, Prober DA, Zhang S, Turley EA. pp 60 (c-src) is required for cell locomotion regulated by the hyaluronan receptor RHAMM. Oncogene 1996; 13:2213–2224.
78. Pratt RM, Larsen MA, Johnston MC. Migration of cranial neural crest cells in a cell-free hyaluronate-rich matrix. Dev Biol 1975; 44:298–305.
79. Turley EA. Hyaluronan-binding proteins and receptors. Adv Drug Deliv Rev 1991; 7:257–264.
80. Thomas L, Byers HR, Vink J, Stamemkovic I. CD44 regulates tumor cell migration on a hyaluronate-coated substrate. J Cell Biol 1992; 118:971–977.
81. Schor SL, Schor AM, Grey AM, Chen WYJ, Rushton G, Grant ME, Ellis I. Mechanism of action of the migration stimulating factor produced by fetal and cancer patient fibroblasts: effect on hyaluronic acid synthesis. *In Vitro* Cell Dev Biol 1989; 25:737–746.
82. Banerjee SD, Toole BP. Hyaluronan-binding protein in endothelial cell morphogenesis. J Cell Biol 1992; 119:643–652.
83. Samuel SK, Hurta RA, Spearmanf MA, Wright JA, Turley EA, Greenberg AH. TGF-β1 stimulation of cell locomotion utilizes the hyaluronan receptor RHAMM and hyaluronan. J Cell Biol 1993; 123:749–758.

84. Rooney P, Wang M, Kumar P, Kumar S. Angiogenic oligosaccharides of hyaluronan enhance the production of collagens by endothelial cells. J Cell Sci 1993; 105:213–218.
85. Noble PW, Lake FR, Henson PM, Riches DWH. Hyaluronate activation of CD44 induces insulin-like growth factor-1 expression by a tumor necrosis factor-α-dependent mechanism in murine macrophages. J Clin Investig 1993; 91: 2368–2377.
86. McKee CM, Penno MB, Cowman M, Burdick MD, Ttrieter RM, Bao C, Noble PW. Hyaluronan (HA) fragments induce chemokine gene expression in alveolar macrophages. The role of HA size and CD44. J Clin Investig 1996; 15:2403–2413.
87. Noble PW, McKee CM, Cowman M, Shin HS. Hyaluronan fragments activate an NF-kB/I-kBa autoregulatory loop in murine macrophages. J Exp Med 1996; 183: 2373–2378.
88. Noble PW, McKee CM, Horton MR. Induction of inflammatory gene expression by low-molecular-weight hyaluronan fragments in macrophages. In: Laurent TC, ed. The Chemistry, Biology and Medical Applications of Hyaluronan and Its Derivatives. London: Portland Press, 1998:219–225.
89. Dvorak HF, Harvey VS, Estrella P, Brown LF, McDonagh J, Dvorak AM. Fibrin containing gels induce angiogenesis. Implications for tumor stroma generation and wound healing. Lab Investig 1987; 57:673–686.
90. West DC, Kumar S. The effect of hyaluronate and its oligosaccharides on endothelial cell proliferation and monolayer integrity. Exp Cell Res 1989; 183: 179–196.
91. Lees VC, Fan TP, West DC. Angiogenesis in a delayed revascularization model is accelerated by angiogenic oligosaccharides of hyaluronan. Lab Investig 1995; 73:259–266.
92. Arnold F, Jia C, Cherry GW, Carbow B, Meyer-Ingold W, Bader D, West DC. Hyaluronan, heterogeneity, and healing: the effects of ultrapure hyaluronan of defined molecular size on the repair of full-thickness pig skin wounds. Wound Repair Regen 1995; 3:299–310.
93. Sattar A, Rooney P, Kumar S, Pye D, West DC, Scott I, Ledger P. Application of angiogenic oligosaccharides of hyaluronan increases blood vessel numbers in rat skin. J Investig Dermatol 1994; 103:576–579.
94. West DC, Shaw DM. Tumour hyaluronan in relation to angiogenesis and metastasis. In: Laurent TC, ed. The Chemistry, Biology and Medical Applications of Hyaluronan and Its Derivatives. London: Portland Press, 1998:227–233.
95. Mast BA, Haynes JH, Krummel TM, Diegelmann RF, Cohen IK. *In vivo* degradation of fetal wound hyaluronic acid results in increased fibroplasias, collagen deposition, and neovascularization. Plast Reconstr Surg 1992; 89: 503–509.
96. Tammi R, Ripellino JA, Margolis RU, Tammi M. Localization of epidermal hyaluronic acid using the hyaluronate-binding region of cartilage proteoglycan as a specific probe. J Investig Dermatol 1988; 90:412–414.
97. Tammi R, Ripellino JA, Margolis RU, Maibach HI, Tammi M. Hyaluronate accumulation in human epidermis treated with retinoic acid in skin organ cultures. J Investig Dermatol 1989; 92:326–332.
98. Wang C, Tammi M, Tammi R. Distribution of hyaluronan and its CD44 receptor in the epithelia of human skin appendages. Histochemistry 1992; 98:105–112.

99. Tuhkanen AL, Tammi M, Pelttari A, Agren UM, Tammi R. Ultrastructural analysis of human epidermal CD44 reveals preferential distribution on plasma membrance domains facing the hyaluronan-rich matrix pouches. J Histochem Cytochem 1998; 46:241–248.
100. Kaya G, Stamenkovic I, Vassalli P, Jorcano JL, Rodriguez I. Selective suppression of CD44 in keratinocytes of mice bearing an antisense CD44 transgene driven by a tissue-specific promoter disrupts hyaluronate metabolism in the skin and impairs keratinocyte proliferation. Genes Dev 1997; 15:996–1007.
101. Longaker MT, Chiu ES, Adzick NS, Stern M, Harrison MR, Stern R. Studies in fetal wound healing. A prolonged presence of hyaluronic acid characterizes fetal wound fluid. Ann Surg 1991; 213:292–296.
102. West DC, Shaw DM, Lorenz P, Adzick NS, Longaker MT. Fibrotic healing of adult and late gestation fetal wounds correlates with increased hyaluronidase activity and removal of hyaluronan. Int J Biochem Cell Biol 1997; 29:201–210.
103. Grav EL, Polack FM, Balazs EA. The protective effect of Na-hyaluronate to corneal endothelium. Exp Eye Res 1979; 31:119–127.
104. Balazs EA, Denlinger JL. Clinical uses of hyaluronan. Ciba Found Symp 1989; 143:265–285.
105. Miller D, Stegmann R. Healon (Sodium Hyaluronate). A Guide to Its Use in Ophthalmic Surgery. New York: Wiley, 1983.
106. Glasser DB, Osborn DC, Joanne FN, Yuan IM. Endothelial protection and viscoelastic retention during phaco-emulsification and intraocular lens implantation. Arch Opthalmol 1991; 109:1438–1440.
107. Yuzawa K. Experimental studies on the healing and restoration of gliding function of the injured digital flexor tendon. Part 9: The use of drugs to prevent adhesion formation of the injured tendon. Nippon Seikeigeka Gakkai Zasshi 1985; 59: 1107–1118.
108. Isik S, Ozturk S, Gurses S, Yetmez M, Guler MM, Selmanpakoglu N, Gunhan O. Prevention of restrictive adhesions in primary tendon repair by HA-membrane: experimental research in chickens. Br J Plast Surg 1999; 52:373–379.
109. Miller JA, Ferguson RL, Powers DL, Burns JW, Shalaby SW. Efficacy of hyaluronic acid/non-steroidal anti-inflammatory drug systems in preventing postsurgical tendon adhesions. J Biomed Mater Res 1997; 38:25–33.
110. Weiss C, Suros JM, Michalow A, Denlinger J, Moore M, Tejeiro W. The role of Na-hylan in reducing postsurgical tendon adhesions. Bull Hosp Jt Dis Orthop Inst 1987; 47:31–39.
111. Meyers SA, Seaber AV, Glisson RR, Nunley JA. Effect of hyaluronic acid/ chondroitin sulfate on healing of full-thickness tendon lacerations in rabbits. J Orthoptera Res 1989; 7:683–689.
112. Salti NI, Tuel RJ, Mass DP. Effect of hyaluronic acid on rabbit profundus flexor tendon healing *in vitro*. J Surg Res 1993; 55:411–415.
113. Xu J, Gu Y, Wang H. Experimental study of sodium hyaluronate products on prevention of tendon adhesion. Zhonghua Wai ke Za Zhi 1995; 33:529–531.
114. Gaughan EM, Nixon AJ, Krook LP, Yeager AE, Mann KA, Mohammed H, Bartel DL. Effects of sodium hyaluronate on tendon healing and adhesion formation in horses. Am J Vet Res 1991; 52:764–773.
115. Carruthers J, Carruthers A, Maberley D. Deep resting glabellar rhytides respond to BTX-A and Hylan B. Dermatol Surg 2003; 29:539–544.

116. Carruthers J, Carruthers A. A prospective, randomized, parallel group study analyzing the effect of BTX-A (Botox) and non-animal sourced hyaluronic acid (NASHA, restylane) in combination compared with NASHA (Restylane) alone in severe glabellar rhytides in adult female subjects: treatment of severe glabellar rhytides with a hyaluronic acid derivative compared with the derivative and BTX-A. Dermatol Surg 2003; 29:802–809.
117. Micheels P. Human anti-hyaluronic acid antibodies: is it possible? Dermatol Surg 2001; 27:185–191.
118. Bergeret-Galley C, Latouche X, Illouz YG. The value of a new filler material in corrective and cosmetic surgery: DermaLive and DermaDeep. Aesthetic Plast Surg 2001; 25:249–255.
119. Lowe NJ, Maxwell CA, Lowe P, Duick MG, Shah K. Hyaluronic acid skin fillers: adverse reactions and skin testing. J Am Acad Dermatol 2001; 45:930–933.
120. Friedman PM, Mafong EA, Kauvar AN, Geronemus RG. Safety data of injectable nonanimal stabilized hyaluronic acid gel for soft tissue augmentation. Dermatol Surg 2002; 28:491–494.
121. Brown LH, Frank PJ. What's new in fillers? J Drugs Dermatol 2003; 2:250–253.
122. Narins RS, Brandt F, Leyden J, Lorenc ZP, Rubin M, Smith S. A randomized, double-blind, multicenter comparison of the efficacy and tolerability of Restylane versus Zyplast for the correction of nasolabial folds. Dermatol Surg 2003; 29: 588–595.
123. Hertegard S, Dahlqvist A, Laurent C, Borzacchiello A, Ambrosio L. Viscoelastic properties of rabbit vocal folds after augmentation. Otolaryngol Head Neck Surg 2003; 128:401–406.
124. Hallen L, Johansson C, Laurent C. Cross-linked hyaluronan (Hylan B gel): a new injectable remedy for treatment of vocal fold insufficiency—an animal study. Acta Otolaryngol 1999; 119:107–111.
125. Hertegard S, Hallen L, Laurent C, Lindstrom E, Olofsson K, Testad P, Dahlqvist A. Cross-linked hyaluronan used as augmentation substance for treatment of glottal insufficiency: safety aspects and vocal fold function. Laryngoscope 2002; 112:2211–2219.
126. Hallen L, Dahlqvist A. Cross-linked hyaluronan for augmentation of the posterior pharyngeal wall: an experimental study in rats. Scand J Plast Reconstr Surg Hand Surg 2002; 36:197–201.
127. Kimmelman CP, Edelstein DR, Cheng HJ. Sepragel sinus (hylan B) as a postsurgical dressing for endoscopic sinus surgery. Otolaryngol Head Neck Surg 2002; 126:331–332.
128. Frenkiel S, Desrosiers MY, Nachtigal D. Use of hylan B gel as a wound dressing after endoscopic sinus surgery. J Otolaryngol 2002; 31(Suppl 1):S41–S44.
129. Leach RE, Burns JW, Dawe EJ, SmithBarbour MD, Diamond MP. Reduction of postsurgical adhesion formation in the rabbit uterine horn model with use of hyaluronate/carboxymethylcellulose gel. Fertil Steril 1998; 69:415–418.
130. Burns JW, Skinner K, Colt MJ, Burgess L, Rose R, Diamond MP. A hyaluronate based gel for the prevention of postsurgical adhesions: evaluation in two animal species. Fertil Steril 1996; 66:814–821.
131. Goldberg EP, Burns JW, Yaacobi Y. Prevention of postoperative adhesions by precoating tissues with dilute sodium hyaluronate solutions. Prog Clin Biol Res 1993; 381:191–204.

132. Campoccia D, Doherty P, Radice M, Brun P, Giovanni A, Williams DF. Semisynthetic resorbable materials from hyaluronan esterification. Biomaterials 1998; 19:2101–2127.
133. Benedetti L, Cortivo R, Berti T, Berti A, Pea F, Mazzo M, Moras M, Abatangelo G. Biocompatibility and biodegradation of different hyaluronan derivatives (Hyaff) implanted in rats. Biomaterials 1993; 14:1154–1160.
134. Hellstrom S, Laurent C, Soderberg O, Spandow O. Endogenous and exogenous hyaluronan in otology. In: Novel Biomaterials Based on Hyaluronic Acid and Its Derivatives. Proceedings of a Workshop Held at the Annual Meeting of the European Society for Biomaterials, Pisa, Italy, September, 1994:38–42.
135. Davidson JM, Beccaro M, Pressato D, Dona M, Pavesio A. Biological response of experimental cutaneous wounds in the pig to hyaluronan ester biomaterials. In Novel Biomaterials Based on Hyaluronic Acid and Its Derivatives. Proceedings of a Workshop Held at the Annual Meeting of the European Society for Biomaterials, Pisa, Italy, September, 1994:44–51.
136. Grigolo B, Roseti L, Fiorini M, Fini M, Giavaresi G, Aldini NN, Giardino R, Facchini A. Transplantation of chondrocytes seeded on a hyaluronan derivative (hyaff-11) into cartilage defects in rabbits. Biomaterials 2001; 22:2417–2424.
137. Gao JD, Solchaga LA, Goldberg VM, Caplan AI. Repair of osteochondral defect with tissue-engineered two-phase composite material of injectable calcium phosphate and hyaluronan sponge. Tissue Eng 2002; 8:827–837.
138. Kim HD, Valentini RF. Retention and activity of BMP-2 in hyaluronic acid-based scaffolds *in vitro*. J Biomed Mater Res 2002; 59:573–584.
139. Lilli C, Marinucci L, Stabellini G, Belcastro S, Becchetti E, Balducci C, Staffolani N, Locci P. Biomembranes enriched with TGFbetal favor bone matrix protein expression by human osteoblasts *in vitro*. J Biomed Mater Res 2002; 63: 577–582.
140. Solchaga LA, Yoo JU, Lundberg M, Dennis JE, Huibregtse BA, Goldberg VM, Caplan AI. Hyaluronan-based polymers in the treatment of osteochondral defects. J Orthoptera Res 2000; 18:773–780.
141. Brun P, Abatangelo G, Radice M, Zacchi V, Guidolin D, Daga GD, Cortivo R. Chondrocyte aggregation and reorganization into three-dimensional scaffolds. J Biomed Mater Res 1999; 46:337–346.
142. Solchaga LA, Dennis JE, Goldberg VM, Caplan AI. Hyaluronic acid-based polymers as cell carriers for tissue-engineered repair of bone and cartilage. J Orthoptera Res 1999; 17:205–213.
143. Aigner J, Tegeler J, Hutzler P, Campoccia D, Pavesio A, Hammer C, Kastenbauer E, Naumann A. Cartilage tissue engineering with novel nonwoven structured biomaterial based on hyaluronic acid benzyl ester. J Biomed Mater Res 1998; 42:172–181.
144. Grigolo B, Lisignoli G, Piacentini A, Fiorini M, Gobbi P, Mazzotti G, Duca M, Pavesio A. Evidence for redifferentiation of human chondrocytes grown on a hyaluronan-based biomaterial (HYAff11): molecular, immunohistochemical and ultrastructural analysis. Biomaterials 2002; 23:1187–1195.
145. Milella E, Brescia E, Massaro C, Ramires PA, Miglietta MR, Fiori V, Aversa P. Physico-chemical properties and degradability of non-woven hyaluronan benzylic esters as tissue engineering scaffolds. Biomaterials 2002; 23:1053–1063.
146. Lisignoli G, Fini M, Giavaresi G, Nicoli AN, Toneguzzi S, Facchini A. Osteogenesis of large segmental radius defects enhanced by basic fibroblast

growth factor activated bone marrow stromal cells grown on non-woven hyaluronic acid-based polymer scaffold. Biomaterials 2002; 23:1043–1051.
147. Grigolo B, Roseti L, Fiorini M, Fini M, Giavaresi G, Aldini NN, Giardino R, Facchini A. Transplantation of chondrocytes seeded on a hyaluronan derivative (hyaff-11) into cartilage defects in rabbits. Biomaterials 2001; 22: 2417–2424.
148. Gao J, Dennis JE, Solchaga LA, Awadallah AS, Goldberg VM, Caplan AI. Tissue-engineered fabrication of an osteochondral composite graft using rat bone marrow-derived mesenchymal stem cells. Tissue Eng 2001; 7:363–371.
149. Lisignoli G, Zini N, Remiddi G, Piacentini A, Puggioli A, Trimarchi C, Fini M, Maraldi NM, Facchini A. Basic fibroblast growth factor enhances in vitr mineralization of rat bone marrow stromal cells grown on non-woven hyaluronic acid based polymer scaffold. Biomaterials 2001; 22:2095–2105.
150. Elvassore N, Baggio M, Pallado P, Bertucco A. Production of different morphologies of biocompatible polymeric materials by supercritical CO(2) antisolvent techniques. Biotechnol Bioeng 2001; 73:449–457.
151. von Heimburg D, Zachariah S, Low A, Pallua N. Influence of different biodegradable carriers on the *in vivo* behavior of human adipose precursor cells. Plast Reconstr Surg 2001; 108:411–420.
152. Avitabile T, Marano F, Castiglione F, Bucolo C, Cro M, Ambrosio L, Ferrauto C, Reibaldi A. Biocompatibility and biodegradation of intravitreal hyaluronan implants in rabbits. Biomaterials 2001; 22:195–200.
153. Campoccia D, Hunt JA, Doherty PJ, Zhong SP, O'Regan M, Benedetti L, Williams DF. Quantitative assessment of the tissue response to films of hyaluronan derivatives. Biomaterials 1996; 17:963–975.
154. Donati L, Veronesi AM, Garbin S, Ordanini MN, Ferrone MP, Falcone L, Marazzi M. *In vitro* development and clinical application in burn patients of keratinocytes cultured on a hyaluronic acid ester membrane. In Novel Biomaterials Based on Hyaluronic Acid and Its Derivatives. Proceedings of a Workshop Held at the Annual Meeting of the European Society for Biomaterials, Pisa, Italy, September 1994:52–61.
155. Myers SR, Grady J, Soranzo C, Sanders R, Green C, Leigh IM, Navsaria HA. A hyaluronic acid membrane delivery system for cultured keratinocytes: clinical "take" rates in the porcine kerato-dermal model. J Burn Care Rehabil 1997; 18:214–222.
156. Harris PA, di Francesco F, Barisoni D, Leigh IM, Navsaria HA. Use of hyaluronic acid and cultured autologous keratinocytes and fibroblasts in extensive burns. Lancet North Am Ed 1999; 353:35–36.
157. Hollander DA, Soranzo C, Falk S, Windolf J. Extensive traumatic soft tissue loss: reconstruction in severely injured patients using cultured hyaluronan-based three-dimensional dermal and epidermal autografts. J Trauma 2001; 50:1125–1136.
158. Galassi G, Brun P, Radice M, Cortivo R, Zanon GF, Genovese P, Abatangelo G. *In vitro* reconstructed dermis implanted in human wounds: degradation studies of the HA-based supporting scaffold. Biomaterials 2000; 21:2183–2191.
159. Bernd A, Hollander D, Pannike A, Kippenberger S, Muller J, Stein M, Kaufmann R. Benzylester hyaluronic acid membranes as substrate for cultivation and transplantation of autologous keratinocytes for the treatment of non-healing wounds. J Investig Dermatol 1996; 107:450.

160. Edmonds M, Bates M, Doxford M, Gough A, Foster A. New treatments in ulcer healing and wound infection. Diabetes Metab Res Rev 2000; 16(Suppl 1): S51–S54.
161. Georgina C. Wound repair: advanced dressing materials. Nurs Stand 2002; 17: 49–53.
162. Moseley R, Leaver M, Walker M, Waddington RJ, Parsons D, Chen WY, Embery G. Comparison of the antioxidant properties of HYAFF-11p75, AQUACEL and hyaluronan towards reactive oxygen species *in vitro*. Biomaterials 2002; 23:2255–2264.
163. Burd DAR, Greco RM, Regauer S, Longaker MT, Siebert JW, Garg HG. Hyaluronan and wound healing: a new perspective. Br J Plast Surg 1991; 44: 579–584.
164. Burd DAR, Siebert JW, Ehrlich HP, Garg HG. Human skin and post-scar hyaluronan; demonstration of association with collagen and other proteins. Matrix 1989; 9:322–327.
165. McGrath MH. Peptide growth factors and wound healing. Clin Plast Surg 1990; 17:421–432.
166. Barbul A. Immune aspects of wound repair. Clin Plast Surg 1990; 17:433–442.
167. Lawrence WT. Physiology of the acute wound. Clin Plast Surg 1998; 25:321–340.
168. Monaco JL, Lawrence WT. Acute wound healing—an overview. Clin Plast Surg 2003; 30:1–12.
169. Sporn MB, Roberts AB. Transforming growth factor-B: recent progress and new challenges. J Cell Biol 1992; 119:1017–1021.
170. Ellis IR, Schor SL. Differential effects of TGF-β1 on hyaluronan synthesis by fetal and adult skin fibroblasts: implications for cell migrationa and wound healing. Exp Cell Res 1996; 228:326–333.
171. Shah M, Foreman DM, Ferguson MWJ. Control of scarring in adult wounds by neutralizing antibody to transforming growth factor B. Lancet North Am Ed 1992; 339:213–214.
172. Toole BP. Hyaluronan. In: Iozzo RV, ed. Proteoglycans. New York: Marcel Dekker, 2000:61–92.

Chemistry and Biology of Hyaluronan
H.G. Garg and C.A. Hales (editors)

Chapter 19

Hyaluronan in the Epidermis and Other Epithelial Tissues

RAIJA H. TAMMI and
MARKKU I. TAMMI

Department of Anatomy, University of Kuopio, P.O. Box 1627, Savilahdentie 9, Kuopio 70211, Finland

I. Introduction

Hyaluronan was until recently considered an extracellular matrix space filler, a gel-forming substance important just for the resilience of soft connective tissues. It was not until the radioisotopic labeling of epithelial tissues (1) and the introduction of specific microscopic staining techniques (2) that the presence of hyaluronan in the usually thin epithelial tissues was realized. Charting the expression of hyaluronan in the different specific epithelia of animals has been conducted. New information on factors that control the expression level of hyaluronan in the epithelial cells, such as injury, inflammation, and malignant transformation, have prompted this first review on hyaluronan in epithelial tissues. How the unique properties of hyaluronan contribute to the distinct functions of different epithelial tissues and cells is discussed in this chapter.

II. Epidermal Hyaluronan

Chemical assays indicate that skin is the largest single source of hyaluronan in the bodies of humans and other mammals (3). The dermal connective tissue constitutes the bulk of the skin, while epidermis, the epithelial compartment, is thinner than 50 μm in most locations of the human body surface. Therefore, most early investigators did not take the trouble of trying to separate the two tissues for quantitative assays of hyaluronan (3). However, as early as 1969, Mier and Wood showed that hyaluronan was not only present in dermis but in the thin sheets of dissected epidermis (4) as well. Specific histological localization of epidermal hyaluronan was achieved with the biotinylated probe prepared from the hyaluronan binding region of cartilage aggrecan, and link protein (bHABR) (2), confirming the results of biochemical assays that had the potential risk of contamination from the abundant hyaluronan pool on the dermal side.

A. Localization of Hyaluronan in the Different Strata of Epidermis

Hyaluronan in the human epidermis is mostly extracellular, localized between the basal and spinous cells, i.e., the live cell layers (Fig. 1a and b). In contrast, when keratinocytes enter the terminal differentiation program and go to the granular cell stage and later develop the cornified cells, no hyaluronan signal remains detectable for the specific histological probe that recognizes hyaluronan (2) (Fig. 1a and b). Interestingly, hyaluronan signal is also missing from the basal side of the basal cells in normal human epidermis (2) (Fig. 1a and b). The epidermal localization of the hyaluronan receptor CD44 completely overlaps with that of hyaluronan; both reside in the pouches between desmosomes connecting adjacent keratinocytes (5).

The hyaluronan deposits in the epidermis are excluded from structures with tight adherence to adjacent cells (corneocytes, desmosomes) or matrix (basal lamina), suggesting that its synthesis is specifically directed to places on the plasma membrane where open extracellular space is needed. On the other hand, it can be speculated that the presence of hyaluronan inhibits the formation of such tight adhesions as desmosomes (6) and those between the upper spinous cells about to enter terminal differentiation (7). *In vitro*, hyaluronan is excluded from the vicinity of the adhesion plaques between dish bottom and keratinocytes, and increased number of these adhesions is associated with a low content of hyaluronan under the cell (8,9).

B. Synthesis and Metabolism of Epidermal Hyaluronan

The currently available data on hyaluronan metabolism in the epidermis is derived from metabolic labeling of pig (10) and human (1) skin explants in organ culture and organotypic, differentiating rat keratinocyte cultures (11). Analysis of ^{3}H-glucosamine incorporation revealed that hyaluronan was the major glycosaminoglycan synthesized in human epidermis (1,11). Furthermore, the heparan sulfate containing isoform of the hyaluronan receptor CD44 is one of the major proteoglycans in this tissue (12), emphasizing the importance of the hyaluronan in the intercellular space between keratinocytes.

Epidermal keratinocytes express all three enzymes responsible for hyaluronan synthesis (Has1, Has2, and Has3), each subject to different regulations (7,13,14). The newly synthesized hyaluronan is of high molecular mass (mostly $>2 \times 10^6$ Da), both in keratinocyte monolayer cultures (13,15) and in structurally normal epidermis (11,16).

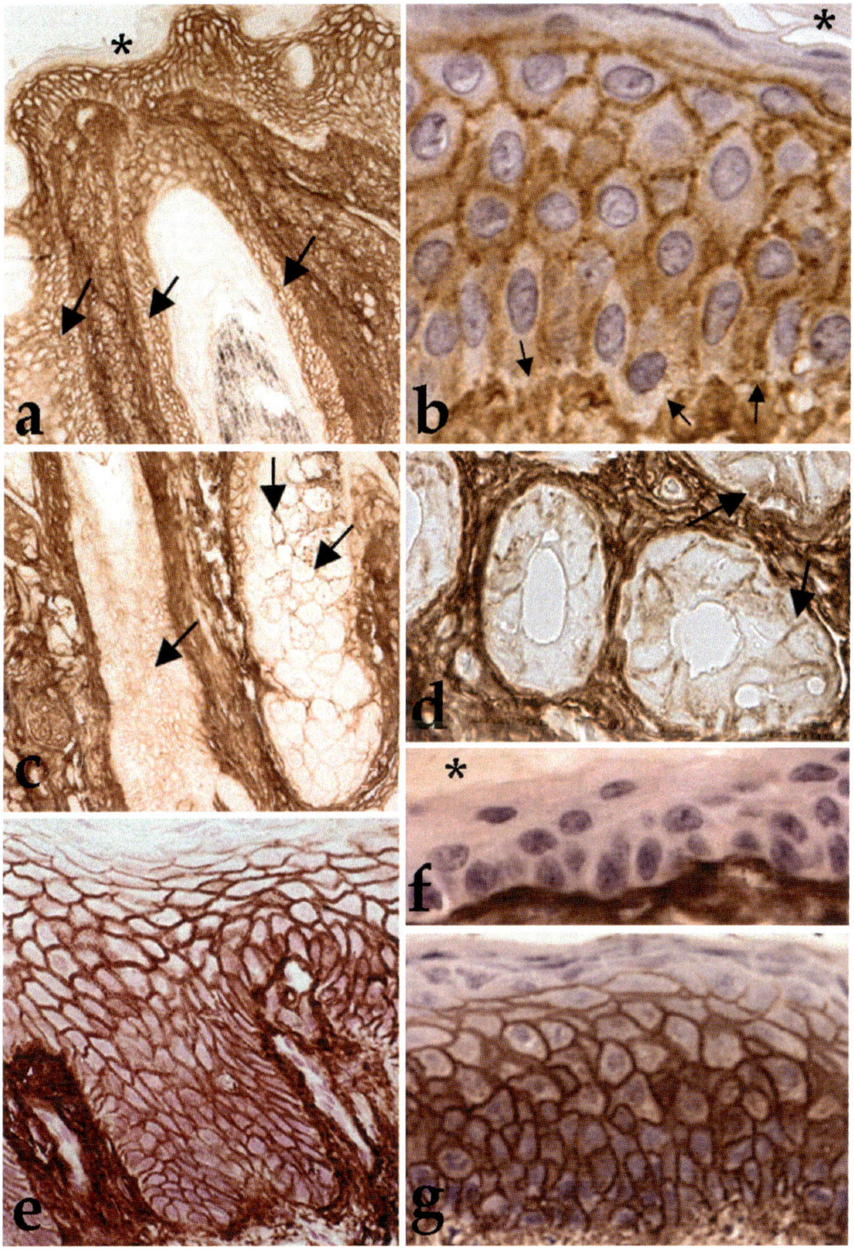

Figure 1 Hyaluronan in skin: hyaluronan was visualized using biotinylated hyaluronan binding complex (bHABC) containing cartilage link protein and aggrecan G1 domain, and avidin biotin-peroxidase technique using DAB as a chromogen. (a–d) Normal adult human skin, (e) psoriatic skin, (f) normal adult mouse skin, (g) hyperplastic mouse skin, induced by tape stripping. Dermal connective tissue shows strong positive staining (a–g).

The vital segment of the epidermis is limited from above by stratum corneum, impermeable even to water, and below by basal lamina, a proteinaceous sheet restricting the movement of cells and large molecules like newly synthesized hyaluronan. The closed tissue structure suggests that hyaluronan is also degraded in the epidermis, at least into fragments small enough to penetrate the limiting barriers. Indeed, hyaluronan disappears from the epidermis of human skin explants and from rat organotypic keratinocyte cultures with a half-life of 1 day, suggesting active local catabolism (11,16).

Scavengers of reactive oxygen species like superoxide dismutase, catalase, and desferroxamine slow down hyaluronan disappearance from the epidermis in human skin organ cultures, suggesting that oxygen-free radicals contribute to the degradation (17). However, there was little support for a direct attack of the glycosidic bonds of the polymer, since the molecular mass distribution of hyaluronan was not changed during the treatment (17). The reactive oxygen species may have a more regulatory role, enhancing the cellular uptake of hyaluronan for lysosomal degradation, perhaps by limited cleavages that release the chain free for endocytosis or by inducing cellular signals that stimulate endocytosis.

The uptake of hyaluronan has been studied in detail in keratinocyte monolayer cultures (15). A large proportion of the keratinocyte cell surface hyaluronan is retracted into cytoplasmic vesicles. An excess of hyaluronan oligosaccharides reduces the hyaluronan vesicles, suggesting that specific receptors mediate a part of the uptake, while another portion apparently enters by pinocytosis without receptors (15). The uptake operates with an unknown machinery, since hyaluronan is not wrapped into vesicles through clathrin coated pits or caveolae, the common vehicles for receptor-mediated endocytosis. The endocytosed hyaluronan had a half-life of 2–3 h, and inhibition of lysosomal functions induced intracellular accumulation of hyaluronan, suggesting that the endosomal hyaluronan was en route to degradation in lysosomes (15). It appears that growth factors like EGF (14) and KGF (18) increase the content of intracellular hyaluronan, probably by increasing endocytosis.

While hyaluronan degradation in the epidermis may be regulated, there are currently no examples of cases where epidermal hyaluronan content changes without parallel alterations in the expression of hyaluronan synthases. Some kind

---

Epidermal cells in normal human skin are surrounded by hyaluronan from the basal up to the granular cell layer (a,b), whereas normal mouse epidermis shows very little hyaluronan (f). Hyperplasia both in human and mouse epidermis is associated with increased hyaluronan expression (e,g). Small arrows in (b), (e), (f), and (g) indicate the location of basal lamina, the asterisk in (a), (b), and (f) the location of stratum corneum. Large arrows in (a) indicate the upper part of the outer epidermal root sheath of hair follicles, which is strongly hyaluronan positive. The arrow on the left in (c) indicate the lower part of the outer epidermal root sheath of hair follicles, which shows weaker hyaluronan staining. The arrows on the right in (c) indicate mature sebocytes, which are intensely stained with bHABC. Large arrows in (d) indicate positive staining in sweat gland acini.

of coupling between the rates of synthesis and degradation is quite likely, since uncontrolled accumulation of hyaluronan in the closed epidermal tissue is expected to ruin the tissue structure and epithelial barrier function.

C. Influence of Hyaluronan on Keratinocyte Growth and Differentiation

Hyaluronan synthesis is generally active in tissues and cells under a phase of active growth. This rule is also valid in the epidermis where hypertrophy is associated with increased hyaluronan concentration and thinning of the epidermis with decreased hyaluronan synthesis. Accordingly, all-*trans* retinoic acid (1), EGF (7) and KGF (18) increase epidermal hyaluronan and stimulate cell proliferation, while hydrocortisone (19), TGFβ (7) and 4-methylumbelliferone (9) inhibit hyaluronan synthesis and reduce keratinocyte proliferation. The exact role of hyaluronan synthesis in the increased numbers of vital cell layers is not known, but it can act as an autocrine signal for cell proliferation and facilitate the diffusion of nutrients to the upper cell layers by keeping open the intercellular spaces. In the human skin disorder psoriasis, epidermis shows hyperplasia, which is associated with a marked increase in the thickness of the hyaluronan-positive epidermal layer (20) (Fig. 1e).

The increased thickness of the vital epidermis may also be due to a delay of terminal differentiation, which involves extreme flattening of the keratinocytes, disappearance of the cellular organelles and deposition of special lipids between the tightly appositional cells. Hyaluronan as a highly hydrophilic and bulky molecule may interfere with the formation of these intercellular lipid seals and inhibit proper differentiation by impeding the construction of this water barrier (7). It is also possible that receptor signaling by hyaluronan retards the entrance of the upper spinous cells into terminal differentiation. Anyway, agents that increase epidermal hyaluronan tend to hold back normal differentiation by morphological (structure of stratum corneum), biochemical (expression of filaggrin and keratin 10), or functional criteria (permeability) (7). In general, hyaluronan is associated with the pre-differentiation state of the epidermal keratinocytes.

III. Hyaluronan in Epidermal Appendices and Other Derivatives of Ectodermal Epithelium

Stratified layers of keratinocytes constitute the epidermis, but a number of other epithelial cell types differentiate from the same primitive surface ectoderm during the early embryonic period. These cell types include the secretory and conducting channels of sweat glands, the several cells involved in hair formation, the sebaceous glands and sebocytes and the suggested stem cell population located in the hair bulge (21). The epithelial, secretory cells of tear glands, mammary glands and salivary glands are also derived from the surface ectodermal cells, and the enamel organ of teeth is induced from the mouth epithelium (22). Microscopic examination of sections from human skin and the above epithelial cell derivatives show a distinct expression pattern of hyaluronan in each cell type. The expression sites of hyaluronan closely correspond to those of CD44 in most of these cell types (23).

The stratified epithelia in the hair follicle, including outer and inner root sheaths, show hyaluronan in the same pattern as epidermis, abundant around the non-differentiated cells but absent in the differentiated cell layers (23). The fat-laden sebocytes in the sebaceous glands show an interesting pattern, with a cell surface hyaluronan staining shifting into the cells upon maturation (Fig. 1c). This feature may be associated with the fact that maturating sebocytes were one of the few cases in which CD44 was missing from hyaluronan-positive cells (23).

A curious dichotomy was present in the sweat gland acini: the more basal, so-called clear cells were negative for both hyaluronan and CD44, while the serous (dark) cells mostly occupying the luminal surface were strongly positive (Fig. 1d). The basolateral surfaces of the ductal cells of the sweat glands were also hyaluronan-positive (23). The presence of hyaluronan on the luminal side of the serous cells and the fact that sweat collected from human skin has a significant concentration of hyaluronan suggest that hyaluronan is actively secreted by the sweat gland (Ågren et al., unpublished data). Hyaluronan is also present in the serous, but not mucous, cells of the salivary and tear glands (Fig. 2a and b), and in the secretions of these glands (24). Interestingly, hyaluronan concentration increases in the saliva of patients with Sjögren's syndrome, a disease affecting salivary glands. Inflammation in the periodontium also increases hyaluronan concentration in the crevicular fluid and saliva (25). The increased concentrations may result from a stimulation in active epithelial secretion, or leakage from the hyaluronan-rich stroma in glands injured by inflammation.

The ducts of resting human mammary glands are lined with non-stratified epithelial cells normally negative for hyaluronan staining (26), while the acini and ducts of active bovine mammary glands are positive (Ågren et al., unpublished data). Both human and bovine milk contains detectable hyaluronan (Ågren et al., unpublished data). Human breast cancer cells, those from the more aggressive cases in particular, often turn positive for hyaluronan (27).

IV. Gastrointestinal and Respiratory Epithelium

The upper gastrointestinal tract, including mouth, pharynx, and esophagus, is covered by a stratified epithelium, which is not keratinized but otherwise resembling that of epidermis. Hyaluronan fills the interstitium of the lower cell layers of these epithelia and gradually disappears when cells differentiate towards the surface (Fig. 3a and b) (28,29). The decline of hyaluronan associated with the differentiation in the superficial esophageal epithelium is reversed in carcinoma in situ lesions, where a strong hyaluronan expression

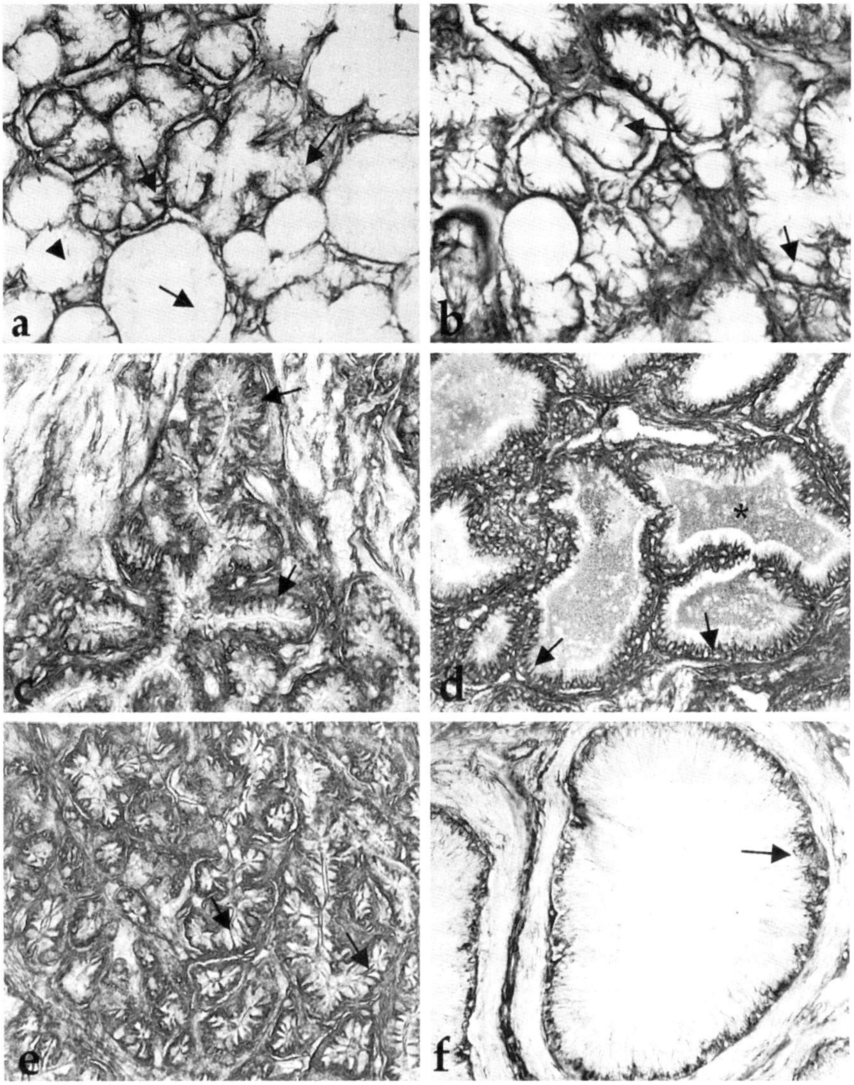

Figure 2 Hyaluronan in glandular epithelia. Hyaluronan in bovine tissues was visualized using bHABC and avidin biotin-peroxidase technique using DAB as a chromogen. (a) Sublingual gland, (b) lacrimal gland, (c) prostate, (d) seminal vesicle, (e) Cowper's gland, (f) epididymis. Arrows indicate the location of hyaluronan in the glandular epithelial tissues (a–e), while in the epididymis the staining intensity is lower and localized around the basal cells. The arrowhead in (a) indicates mucous glands, which are devoid of hyaluronan. The asterisk in (d) indicates secretion inside the gland, showing intense staining with bHABC.

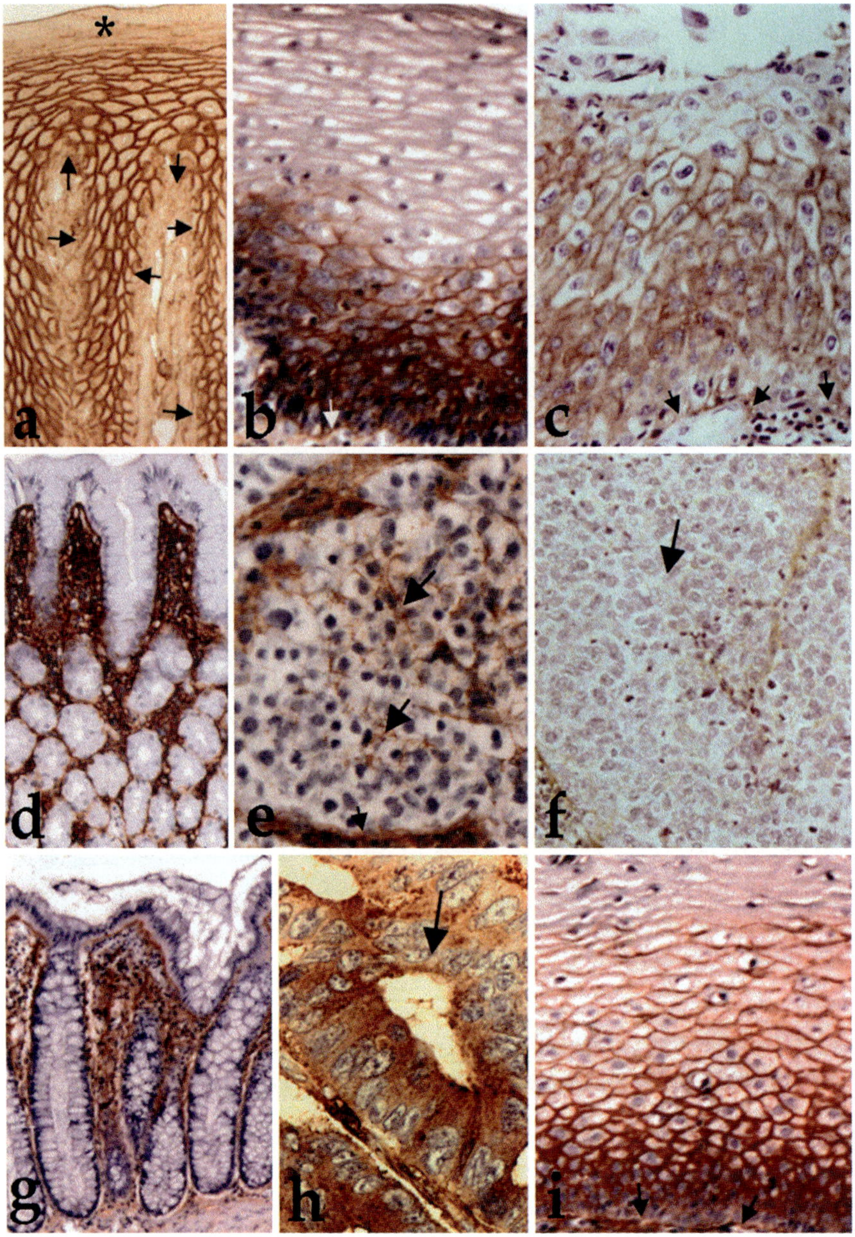

Figure 3 Hyaluronan in normal and malignant human epithelia. Hyaluronan was visualized using bHABC and avidin biotin-peroxidase technique using DAB as a chromogen. (a) Normal gingival epithelium, from an area which cornifies, (b) normal esophageal epithelium, (c) carcinoma in situ of esophagus, (d) normal gastric epithelium, (e) adenocarcinoma of stomach, (f) poorly differentiated squamous carcinoma originating from the esophagus (g) normal colon epithelium, (h) adenocarcinoma of colon, (i) normal

extends up to the surface (Fig. 3c) (29). The continued and even enhanced hyaluronan content is typical also for other early (low-grade) malignancies of stratified (squamous) epithelial cells such as mouth and pharynx cancers (30,31). In contrast, the higher grade squamous cell carcinomas with an unfavorable prognosis show reduced and irregular hyaluronan levels on their cell surface (Fig. 3f) (30,31).

The pseudostratified epithelium in bronchi contains little or no hyaluronan. When detected, it is located around the basal cells while the superficial ciliated cells are HA-negative (32). Squamous metaplasia or dysplasia of bronchial epithelium is associated with upregulation of hyaluronan expression, and squamous cell carcinomas are generally strongly hyaluronan-positive (32). However, like other cancers of stratified origin, further dedifferentiation is associated with reduced hyaluronan content (32,33).

Stomach and the intestines have a non-stratified epithelium rapidly renewed from the bottom of the intestinal crypts or glands. Like many other simple epithelia, these epithelial cells do not show hyaluronan except for a little in the basolateral surfaces of some cells at the bottom of the crypts (Fig. 3d) (34). However, the enterocytes become positive for hyaluronan and CD44 in the immunological injury caused by Crohn's disease and celiac disease (Kemppainen et al., unpublished data). Introduction of allergens increases hyaluronan in the gut lumen (35). Likewise, transformation of the colon epithelial cells is frequently accompanied by the expression of hyaluronan, while normal colon epithelium is virtually hyaluronan free (Fig. 3g and h) (23). In colon cancer cells, the level of hyaluronan is a strong, unfavorable prognostic indicator of the patient survival (36). The emergence of hyaluronan expression on gastric cancer cells (Fig. 3e) shows a similar negative correlation with the patient survival (37). The gastrointestinal epithelium thus offers another example where hyaluronan induction associates with epithelial injury, attempt of regeneration, and malignant transformation.

V. Genital and Urinary Tracts and Mesothelial Cells as Hyaluronan Producers

The expression of hyaluronan in the simple epithelia of the male reproductive tract and its accessory glands is more active than that in the gut and intestine. Distal epididymis, vas deferens, seminal vesicle, prostate, and Cowper's gland

---

epithelium from the cervix of uterus. Stratified epithelia (a,b,i) show strong hyaluronan expression, while normal intestinal epithelia contain little hyaluronan. Poorly differentiated squamous carcinoma is practically hyaluronan-negative, while adenocarcinomas have acquired hyaluronan expression. Small arrows indicate the location of the basal lamina, large arrows in (e) and (h) indicate positive staining in cancer cells and lack of staining in (f). The asterisk in (a) indicates cornified cell layer.

show hyaluronan located mainly on the lower basolateral surface, but with little on the apical surface (Fig. 2c–f) (38). In these epithelia, the location of CD44 correlates with that of hyaluronan on the basolateral surfaces. This distribution suggests that hyaluronan is mostly synthesized and remains bound on the basal side of the adjacent epithelial cells. However, the lumen of the seminal vesicle, prostate and Cowper's gland (Fig. 2c–e) contain hyaluronan, suggesting that it is either secreted from the apical surface, or passes the cell junctions in the lateral sides of the cells. There is an extensive pool of hyaluronan in the underlying stroma of these glands, a potential source of the epithelial and lumenal hyaluronan. Interestingly, testis seems almost devoid of hyaluronan (38), perhaps a finding related to the high level of hyaluronidase in the sperm cells. At ejaculation, the hyaluronan in the product of the accessory sex glands comes in contact with sperm cells.

In the female genital tract, the stratified epithelium of the cervix shows strong hyaluronan expression similar to other stratified epithelia (Fig. 3i), while the simple columnar epithelium covering the lumen of the uterus and the fallopian tubes show little hyaluronan, except the tubal cells close to the uterotubal junction, showing an apical hyaluronan signal (39). The pig tubal fluid contains soluble hyaluronan (39) and some cells in this part of the tuba are positive for CD44, with an elevated expression before ovulation, suggesting that the enrichment of CD44 and hyaluronan may enhance the viability and functionality of sperm stored at this site (40). The sperm cells are thus both in the seminal fluid and in the upper female genital tract in a milieu containing hyaluronan. Interestingly, epithelial cells scraped from the luminal surface of pig oviduct express Has3 mRNA (but not that of Has2 or Has1), suggesting that hyaluronan in the oviduct may be produced by the epithelial cells rather than by the underlying stroma (41).

Most of the normal epithelial cells in the adult kidney, including the loop of Henle, proximal and distal tubules, collecting ducts, and calices of different sizes are completely negative in staining with a probe specific for hyaluronan. The mesenchymal compartment in the kidney cortex and outer medulla is also very weak in hyaluronan staining, in striking contrast to the inner medulla where hyaluronan resides in high concentrations, perhaps as part of the urine concentration process. There is also little, if any, expression of the hyaluronan receptor CD44 in normal kidney. However, both CD44 and hyaluronan are highly upregulated in the kidney cortex after immunological (42) and ischemic (43) injury, CD44 particularly in the tubular epithelial cells, and hyaluronan in the interstitium. Cultured cortical tubular epithelial cells are also capable of hyaluronan synthesis (44) induced by cytokines, high glucose (45), and cell dispersal, while downregulated by cell density-dependent epithelial integrity (46). The inducible hyaluronan synthesis appears a result of increased Has2 expression (45). Hyaluronan is one of the molecules on tubular epithelium supposed to bind developing crystals that may eventually lead to kidney stone formation (47).

VI. Epithelial Barrier Function and Hyaluronan

The bulky hydrophilic hyaluronan, when present between epithelial cells, maintains high permeability for water and low molecular weight solutes. Therefore, hyaluronan is usually scarce or absent between epithelial cells with tight junctions. Good examples are epithelial cells lining the normal gut and urinary tract. However, some epithelia may contain hyaluronan on one side of the tight junctions, like the epidermis below its water barrier around the granular cells and bronchial epithelium on the basal part. This situation may be completely changed in pathological situations when epithelial barriers are inflamed or injured and experience a leak. Then hyaluronan is present, for example, on all surfaces of the gut epithelial cells, often together with induction of CD44, and there may also be increased leakage of hyaluronan from the lamina propria into the lumen of the gut and bronchi. Whether the induction of hyaluronan expression on the normally tightly apposed surfaces between epithelial cells causes the leakage, or whether cell dyshesion allows the appearance of hyaluronan on the normally closed area, remains unknown at present.

VII. Hyaluronan in Wound Healing

It was established long ago that the concentration of hyaluronan increases very rapidly in experimental skin wounds (3). Biochemical assays on experimental wounds indicated that hyaluronan reached its peak concentration 3 days after the wound and started to decline before the peaks of proteoglycans and collagen were reached. It was generally accepted that hyaluronan provides a temporary matrix for the migration of inflammatory cells and proliferation of fibroblasts in the connective tissue, but the contribution of epithelial cells was not considered (3).

More recently, hyaluronan was found to be abundant in the migrating cells of wounded human oral epithelium and in cells flanking the wound site (48), suggesting that the epithelial cells undergo a similar upregulation of hyaluronan synthesis. Experiments in our laboratory have shown that this is even more obvious in the mouse epidermis injured by tape stripping of the epidermis. The epidermis of the adult mouse, containing a relatively weak signal for hyaluronan, undergoes a major activation of hyaluronan synthesis in and around the wounded epidermis (Fig. 1f,g), and this is also detectable in an ELISA assay. This is due to the elevated levels of mRNA for the hyaluronan synthases Has2 and Has3 in keratinocytes isolated from the wound area. Interestingly, the RT-PCRs also showed a parallel induction of CD44 mRNA and a corresponding augmentation of CD44 protein, as detected by immunohistochemistry. The importance of the hyaluronan response in the epidermal keratinocytes was stressed by the finding that mice with inhibited epidermal CD44 expression showed an attenuated response to mitogens (49). *In vitro*, monolayer cultures of CD44−/− keratinocytes maintain less hyaluronan on their surface and migrate slower into a scratch wound introduced into a confluent culture (Tammi et al., unpublished data). Furthermore, overexpression of Has2 increase the migration of keratinocytes in an *in vitro* wounding assay, while suppression of Has2 by transfection of antisense Has2 gene reduces their mobility (14). There are also *in vivo* experiments in which application of hyaluronan enhances skin wound healing (50). Upregulation of hyaluronan synthesis and Has2 expression is also essential for wound healing of the mesothelial lining cells of the peritoneal cavity (51). Hyaluronan is thus an important factor in the normal epithelial wound healing process.

VIII. The Role of Hyaluronan in Transformation and Spreading of Malignant Epithelial Cells

Biochemical assays have indicated high concentrations of hyaluronan in several malignant tumors, as compared with the corresponding normal tissue (52). Specific detection and localization of hyaluronan in tissue sections with the biotinylated hyaluronan binding probe has recently provided large amounts of new information about the role of hyaluronan in a number of different tumors of epithelial origin.

A. Hyaluronan in the Stroma of Epithelial Malignancies

The connective tissue stroma that surrounds most epithelial cancers is enriched in hyaluronan. Examples of tumors with a hyaluronan ‘halo’ include breast (27), prostate (53), ovarian (54), lung (33), gastric (29), colon (36), thyroid (55), and skin squamous cell (56) carcinomas. These hyaluronan deposits can serve as strong indicators of unfavorable prognosis for the patients (27,33,53–55,57) and probably contribute to the spreading of the malignant epithelial cells. Several mechanisms are likely to contribute to the hyaluronan accumulation. The malignant epithelial cells secrete (growth) factors, like TGFβ (58), that stimulate the synthesis of hyaluronan by the stromal cells in a paracrine manner (59,60).

Multiple mechanisms have been implicated in the promotion of malignant growth by hyaluronan (61). By its physicochemical properties, creating hydrated malleable extracellular space, hyaluronan, often associated with matrix proteoglycans like versican, can create an open tissue environment favorable for migration and invasion of the malignant cells. It has become clear that these mechanistic systems are accompanied by finely balanced, specific interactions between hyaluronan and cellular receptors, leading to changes in cell adhesion and signals that govern the phenotype and behavior of the cells (62,63).

B. Expression of Hyaluronan and Hyaluronidase in Cancer Cells of Epithelial Origin

The cancer cells are not always dependent on hyaluronan produced by adjacent stromal cells, but may have transformed into active hyaluronan synthesizers themselves. This is especially obvious in those cancers derived from epithelia normally negative for hyaluronan. Pronounced hyaluronan staining within malignant cells in human breast (27), gastric (Fig. 3e) (37), and colon (Fig. 3h) (36) cancers occurs frequently *in vivo* and makes the prognosis of these patients unfavorable.

Recent experimental research has brought extensive evidence supporting the idea that 'Has' expression and hyaluronan synthesis in tumor cells promote malignant growth (61). Transfected Has3 gene enhances the malignancy of prostate cancer cells (64). Overrepresentation of the chromosomal region 8q23–24 is common in prostate cancer and forms an unfavorable prognostic factor. This region contains Has2, which is overexpressed in an aggressive prostate cancer cell line (65). Accordingly, prostate cancer cells with high Has2 expression have a higher affinity to bone endothelial cells, and form more metastases (66), while antisense inhibition of Has2 or Has3 reduces metastasis and subcutaneous growth of implanted prostate cancer cells (67). Mouse mammary carcinoma cells with a transfected overexpressed Has1 gene showed higher metastatic activity than their parental cells, while an inactivating mutation in HAS1 reduces metastasis (68). Metastasis is also increased by cell surface hyaluronan expression of melanoma cells (69). Furthermore, Has2 transfection and overexpression increases the growth rate of transplantable colon carcinoma cells (70).

It is thus obvious that hyaluronan increases malignant growth of epithelial cells. The signals induced in the cells are different when short oligosaccharides and intact hyaluronan chains are compared, the former being inhibitory for tumor growth (71). Therefore, it is natural that hyaluronidases, enzymes that chop the chain into smaller fragments, are important in all aspects of hyaluronan functions. There are still some ambiguities in the roles of hyaluronidase and hyaluronan oligosaccharides concerning their influence on cancer growth (72,73). For instance, enhanced coexpression of hyaluronan and hyaluronidases greatly increases the spreading risk of prostate (74) and bladder cancer (75), while epithelial ovarian cancers showed no marked change in hyaluronidase activity (76).

Altogether, it is thus likely that increased hyaluronan synthesis somehow releases the epithelial cells from their normal restraints of growth and migration and helps in their survival and metastasis. The molecular mechanisms involve changes in the extracellular milieu and signaling through receptors, both enhancing the invasive capacity.

IX. Summary and Conclusions

It has turned out that hyaluronan is an indispensable product of many epithelial tissues. Stratified epithelia, like that in the epidermis, mouth, oropharynx, esophagus, and vagina contain hyaluronan in all cell strata until terminal differentiation closes the extracellular space and hyaluronan disappears. The interstitial hyaluronan is supposed to aid in the nutrition of the multilayered cells, provide cell mobility and plasticity during the stratification process when cell shape changes, and provide a suitable environment and signals that facilitate re-epithelialization during wound healing. In contrast, many simple non-stratified epithelia are completely negative in hyaluronan staining or express a limited signal on the basolateral surface. However, injury or dispersal of these cells, for example those in the intestinal and renal tubular epithelia, upregulates the expression of CD44 and Has2, and hyaluronan staining appears on the injured cells. Malignant transformation can also turn on hyaluronan synthesis in epithelia normally devoid of detectable hyaluronan and promote cancer progression by enhancing cell proliferation, migration, survival from apoptosis, and possibly through enhanced tumor angiogenesis. Hyaluronan from or through the epithelium of prostate and other male accessory sex glands accompanies ejaculated spermatozoa, which are also greeted by hyaluronan in the fallopian tubes. Hyaluronan provided by these epithelia may have a specific function important for fertilization. Obviously, a number of molecules interacting with hyaluronan (e.g., CD44, IαI, TSG-6, link protein, and hyaluronan binding proteoglycans and hyaluronidases) may be required for completing the epithelial functions of hyaluronan, many of which are still unknown.

Acknowledgements

The work of the authors was financially supported by the Academy of Finland (MT), Finnish Cancer Organizations (RT), The EVO funds of the Kuopio University Hospital (MT, RT), The Technology Centre of Finland (Tekes), and Sigrid Juselius Foundation (MT).

References

1. Tammi R, Tammi M. Influence of retinoic acid on the ultrastructure and hyaluronic acid synthesis of adult human epidermis in whole skin organ culture. J Cell Physiol 1986; 126:389–398.
2. Tammi R, Ripellino JA, Margolis RU, Tammi M. Localization of epidermal hyaluronic acid using the hyaluronate binding region of cartilage proteoglycan as a specific probe. J Invest Dermatol 1988; 90:412–414.
3. Tammi R, Ågren UM, Tuhkanen AL, Tammi M. Hyaluronan metabolism in skin. Prog Histochem Cytochem 1994; 29:1–81.
4. Mier PD, Wood M. Acid mucopolysaccharides of human skin. Brit J Dermatol 1969; 81:528–533.
5. Tuhkanen AL, Tammi M, Pelttari A, Ågren UM, Tammi R. Ultrastructural analysis of human epidermal CD44 reveals preferential distribution on plasma membrane domains facing the hyaluronan-rich matrix pouches. J Histochem Cytochem 1998; 46:241–248.
6. Tammi R, Ripellino JA, Margolis RU, Maibach HI, Tammi M. Hyaluronate accumulation in human epidermis treated with retinoic acid in skin organ culture. J Invest Dermatol 1989; 92:326–332.
7. Pasonen-Seppänen S, Karvinen S, Törrönen K, Hyttinen JM, Jokela T, Lammi MJ, Tammi MI, Tammi R. EGF upregulates, whereas TGF-beta downregulates,

the hyaluronan synthases Has2 and Has3 in organotypic keratinocyte cultures: correlations with epidermal proliferation and differentiation. J Investig Dermatol 2003; 120:1038–1044.
8. Rilla K, Lammi MJ, Sironen R, Törrönen K, Luukkonen M, Hascall VC, Midura RJ, Hyttinen M, Pelkonen J, Tammi M, Tammi R. Changed lamellipodial extension, adhesion plaques and migration in epidermal keratinocytes containing constitutively expressed sense and antisense hyaluronan synthase 2 (Has2) genes. J Cell Sci 2002; 115:3633–3643.
9. Rilla K, Pasonen-Seppänen S, Kultt, A, Rieppo J, Tammi M, Tammi R. The hyaluronan synthesis inhibitor 4-methylumbelliferone prevents keratinocyte activation and epidermal hyperproliferation induced by epidermal growth factor. Submitted for publication.
10. King IA, Tabiowo A, Williams RH. Incorporation of l-[3H]fucose and D-[3H]glucosamine into cell-surface-associated glycoconjugates in epidermis of cultured pig skin slices. Biochem J 1980; 190:65–77.
11. Tammi RH, Tammi MI, Hascall VC, Hogg M, Pasonen S, MacCallum DK. A preformed basal lamina alters the metabolism and distribution of hyaluronan in epidermal keratinocyte "organotypic" cultures grown on collagen matrices. Histochem Cell Biol 2000; 113:265–277.
12. Tuhkanen AL, Tammi M, Tammi R. CD44 substituted with heparan sulfate and endo-beta-galactosidase-sensitive oligosaccharides: a major proteoglycan in adult human epidermis. J Investig Dermatol 1997; 109:213–218.
13. Sayo T, Sugiyama Y, Takahashi Y, Ozawa N, Sakai S, Ishikawa O, Tamura M, Inoue S. Hyaluronan synthase 3 regulates hyaluronan synthesis in cultured human keratinocytes. J Investig Dermatol 2002; 118:43–48.
14. Pienimäki JP, Rilla K, Fülöp C, Sironen RK, Karvinen S, Pasonen S, Lammi MJ, Tammi R, Hascall VC, Tammi MI. Epidermal growth factor activates hyaluronan synthase 2 in epidermal keratinocytes and increases pericellular and intracellular hyaluronan. J Biol Chem 2001; 276:20428–20435.
15. Tammi R, Rilla K, Pienimäki JP, MacCallum DK, Hogg M, Luukkonen M, Hascall VC, Tammi M. Hyaluronan enters keratinocytes by a novel endocytic route for catabolism. J Biol Chem 2001; 276:35111–35122.
16. Tammi R, Säämänen AM, Maibach HI, Tammi M. Degradation of newly synthesized high molecular mass hyaluronan in the epidermal and dermal compartments of human skin in organ culture. J Investig Dermatol 1991; 97: 126–130.
17. Ågren UM, Tammi RH, Tammi MI. Reactive oxygen species contribute to epidermal hyaluronan catabolism in human skin organ culture. Free Radic Biol Med 1997; 23:996–1001.
18. Karvinen S, Pasonen-Seppänen S, Hyttinen JMT, Törrönen K, Jokela TA, Tammi MI, Tammi R. Keratinocyte growth factor stimulates migration and hyaluronan synthesis in the epidermis by activation of hyaluronan synthases 2 and 3. J Biol Chem, 2003; 278: 49495–49504.
19. Ågren UM, Tammi M, Tammi R. Hydrocortisone regulation of hyaluronan metabolism in human skin organ culture. J Cell Physiol 1995; 164:240–248.
20. Tammi R, Paukkonen K, Wang C, Horsmanheimo M, Tammi M. Hyaluronan and CD44 in psoriatic skin. Intense staining for hyaluronan on dermal capillary loops and reduced expression of CD44 and hyaluronan in keratinocyte–leukocyte interfaces. Arch Dermatol Res 1994; 286:21–29.

21. Ågren UM, Tammi M, Ryynänen M, Tammi R. Developmentally programmed expression of hyaluronan in human skin and its appendages. J Investig Dermatol 1997; 109:219–224.
22. Felszeghy S, Hyttinen M, Tammi R, Tammi M, Modis L. Quantitative image analysis of hyaluronan expression in human tooth germs. Eur J Oral Sci 2000; 108: 320–326.
23. Wang C, Tammi M, Tammi R. Distribution of hyaluronan and its CD44 receptor in the epithelia of human skin appendages. Histochemistry 1992; 98:105–112.
24. Tishler M, Yaron I, Shirazi I, Yaron M. Salivary and serum hyaluronic acid concentrations in patients with Sjogren's syndrome. Ann Rheum Dis 1998; 57: 506–508.
25. Utoh E, Okazaki J, Gonda Y. Analysis of hyaluronic acid in human gingival crevicular fluid using high-performance liquid chromatography. J Osaka Dent Univ 1998; 32:1–7.
26. Auvinen PK, Parkkinen JJ, Johansson RT, Ågren UM, Tammi RH, Eskelinen MJ, Kosma VM. Expression of hyaluronan in benign and malignant breast lesions. Int J Cancer 1997; 74:477–481.
27. Auvinen P, Tammi R, Parkkinen J, Tammi M, Ågren U, Johansson R, Hirvikoski P, Eskelinen M, Kosma VM. Hyaluronan in peritumoral stroma and malignant cells associates with breast cancer spreading and predicts survival. Am J Pathol 2000; 156:529–536.
28. Tammi R, Tammi M, Häkkinen L, Larjava H. Histochemical localization of hyaluronate in human oral epithelium using a specific hyaluronate-binding probe. Arch Oral Biol 1990; 35:219–224.
29. Wang C, Tammi M, Guo H, Tammi R. Hyaluronan distribution in the normal epithelium of esophagus, stomach, and colon and their cancers. Am J Pathol 1996; 148:1861–1869.
30. Hirvikoski P, Tammi R, Kumpulainen E, Virtaniemi J, Parkkinen JJ, Tammi M, Johansson R, Ågren U, Karhunen J, Kosma VM. Irregular expression of hyaluronan and its CD44 receptor is associated with metastatic phenotype in laryngeal squamous cell carcinoma. Virchows Arch 1999; 434:37–44.
31. Kosunen A, Ropponen K, Kellokoski J, Pukkila M, Virtaniemi J, Valtonen H, Kumpulainen E, Johansson R, Tammi R, Tammi M, Nuutinen J, Kosma VM. Reduced expression of hyaluronan is a strong indicator of poor survival in oral squamous cell carcinoma. Oral Oncol 2003; 40: 257–263.
32. Pirinen RT, Tammi RH, Tammi MI, Pääkkö PK, Parkkinen JJ, Ågren UM, Johansson RT, Viren MM, Törmanen U, Soini YM, Kosma VM. Expression of hyaluronan in normal and dysplastic bronchial epithelium and in squamous cell carcinoma of the lung. Int J Cancer 1998; 79:251–255.
33. Pirinen R, Tammi R, Tammi M, Hirvikoski P, Parkkinen JJ, Johansson R, Böhm J, Hollmen S, Kosma VM. Prognostic value of hyaluronan expression in non-small-cell lung cancer: Increased stromal expression indicates unfavorable outcome in patients with adenocarcinoma. Int J Cancer 2001; 95:12–17.
34. Alho AM, Underhill CB. The hyaluronate receptor is preferentially expressed on proliferating epithelial cells. J Cell Biol 1989; 108:1557–1565.
35. Knutson TW, Bengtsson U, Dannaeus A, Ahlstedt S, Stalenheim G, Hallgren R, Knutson L. Intestinal reactivity in allergic and nonallergic patients: an approach to determine the complexity of the mucosal reaction. J Allergy Clin Immunol 1993; 91:553–559.

36. Ropponen K, Tammi M, Parkkinen J, Eskelinen M, Tammi R, Lipponen P, Ågren U, Alhava E, Kosma VM. Tumor cell-associated hyaluronan as an unfavorable prognostic factor in colorectal cancer. Cancer Res 1998; 58:342–347.
37. Setälä LP, Tammi MI, Tammi RH, Eskelinen MJ, Lipponen PK, Ågren UM, Parkkinen J, Alhava EM, Kosma VM. Hyaluronan expression in gastric cancer cells is associated with local and nodal spread and reduced survival rate. Br J Cancer 1999; 79:1133–1138.
38. Tammi R, Rönkkö S, Ågren UM, Tammi M. Distribution of hyaluronan in bull reproductive organs. J Histochem Cytochem 1994; 42:1479–1486.
39. Tienthai P, Kjellen L, Pertoft H, Suzuki K, Rodriguez-Martinez H. Localization and quantitation of hyaluronan and sulfated glycosaminoglycans in the tissues and intraluminal fluid of the pig oviduct. Reprod Fertil Dev 2000; 12:173–182.
40. Tienthai P, Yokoo M, Kimura N, Heldin P, Sato E, Rodriguez-Martinez H. Immunohistochemical localization and expression of the hyaluronan receptor CD44 in the epithelium of the pig oviduct during oestrus. Reproduction 2003; 125:119–132.
41. Tienthai P, Kimura N, Heldin P, Sato E, Rodriguez-Martinez H. Expression of hyaluronan synthase-3 in porcine oviducal epithelium during oestrus. Reprod Fertil Dev 2003; 15:99–105.
42. Benz PS, Fan X, Wuthrich RP. Enhanced tubular epithelial CD44 expression in MRL-lpr lupus nephritis. Kidney Int 1996; 50:156–163.
43. Lewington AJ, Padanilam BJ, Martin DR, Hammerman MR. Expression of CD44 in kidney after acute ischemic injury in rats. Am J Physiol Regul Integr Comp Physiol 2000; 278:R247–R254.
44. Feusi E, Sun L, Sibalic A, Beck-Schimmer B, Oertli B, Wuthrich RP. Enhanced hyaluronan synthesis in the MRL-Fas(lpr) kidney: role of cytokines. Nephron 1999; 83:66–73.
45. Jones S, Phillips AO. Regulation of renal proximal tubular epithelial cell hyaluronan generation: implications for diabetic nephropathy. Kidney Int 2001; 59:1739–1749.
46. Verhulst A, Asselman M, Persy VP, Schepers MS, Helbert MF, Verkoelen CF, De Broe ME. Crystal retention capacity of cells in the human nephron: involvement of CD44 and its ligands hyaluronic acid and osteopontin in the transition of a crystal binding- into a nonadherent epithelium. J Am Soc Nephrol 2003; 14:107–115.
47. Asselman M, Verkoelen CF. Crystal–cell interaction in the pathogenesis of kidney stone disease. Curr Opin Urol 2002; 12:271–276.
48. Oksala O, Salo T, Tammi R, Häkkinen L, Jalkanen M, Inki P, Larjava H. Expression of proteoglycans and hyaluronan during wound healing. J Histochem Cytochem 1995; 43:125–135.
49. Kaya G, Rodriguez I, Jorcano JL, Vassalli P, Stamenkovic I. Selective suppression of CD44 in keratinocytes of mice bearing an antisense CD44 transgene driven by a tissue-specific promoter disrupts hyaluronate metabolism in the skin and impairs keratinocyte proliferation. Genes Dev 1997; 11:996–1007.
50. Abatangelo G, Martelli M, Vecchia P. Healing of hyaluronic acid-enriched wounds: histological observations. J Surg Res 1983; 35:410–416.
51. Yung S, Thomas GJ, Davies M. Induction of hyaluronan metabolism after mechanical injury of human peritoneal mesothelial cells *in vitro*. Kidney Int 2000; 58:1953–1962.
52. Knudson W. Tumor-associated hyaluronan. Providing an extracellular matrix that facilitates invasion. Am J Pathol 1996; 148:1721–1726.

53. Lipponen P, Aaltomaa S, Tammi R, Tammi M, Ågren U, Kosma VM. High stromal hyaluronan level is associated with poor differentiation and metastasis in prostate cancer. Eur J Cancer 2001; 37:849–856.
54. Anttila MA, Tammi RH, Tammi MI, Syrjänen KJ, Saarikoski SV, Kosma VM. High levels of stromal hyaluronan predict poor disease outcome in epithelial ovarian cancer. Cancer Res 2000; 60:150–155.
55. Böhm J, Niskanen L, Tammi R, Tammi M, Eskelinen M, Pirinen R, Hollmen S, Alhava E, Kosma VM. Hyaluronan expression in differentiated thyroid carcinoma. J Pathol 2002; 196:180–185.
56. Karvinen S, Kosma VM, Tammi MI, Tammi R. Hyaluronan, CD44 and versican in epidermal keratinocyte tumours. Br J Dermatol 2003; 148:86–94.
57. Aaltomaa S, Lipponen P, Tammi R, Tammi M, Viitanen J, Kankkunen JP, Kosma VM. Strong stromal hyaluronan expression is associated with PSA recurrence in local prostate cancer. Urol Int 2002; 69:266–272.
58. Hirashima Y, Kobayashi H, Suzuki M, Tanaka Y, Kanayama N, Terao T. Transforming growth factor-beta1 produced by ovarian cancer cell line HRA stimulates attachment and invasion through an up-regulation of plasminogen activator inhibitor type-1 in human peritoneal mesothelial cells. J Biol Chem 2003; 278:26793–26802.
59. Hirano S, Bless DM, Heisey D, Ford CN. Effect of growth factors on hyaluronan production by canine vocal fold fibroblasts. Ann Otol Rhinol Laryngol 2003; 112: 617–624.
60. Hauptmann S, Siegert A, Berger S, Denkert C, Kobel M, Ott S, Siri A, Borsi L. Regulation of cell growth and the expression of extracellular matrix proteins in colorectal adenocarcinoma: a fibroblast-tumor cell coculture model to study tumor-host interactions *in vitro*. Eur J Cell Biol 2003; 82:1–8.
61. Toole BP, Wight TN, Tammi MI. Hyaluronan–cell interactions in cancer and vascular disease. J Biol Chem 2002; 277:4593–4596.
62. Turley EA, Noble PW, Bourguignon LY. Signaling properties of hyaluronan receptors. J Biol Chem 2002; 277:4589–4592.
63. Isacke CM, Yarwood H. The hyaluronan receptor, CD44. Int J Biochem Cell Biol 2002; 34:718–721.
64. Liu N, Gao F, Han Z, Xu X, Underhill CB, Zhang L. Hyaluronan synthase 3 overexpression promotes the growth of TSU prostate cancer cells. Cancer Res 2001; 61:5207–5214.
65. Tsuchiya N, Kondo Y, Takahashi A, Pawar H, Qian J, Sato K, Lieber MM, Jenkins RB. Mapping and gene expression profile of the minimally overrepresented 8q24 region in prostate cancer. Am J Pathol 2002; 160: 1799–1806.
66. Simpson MA, Wilson CM, Furcht LT, Spicer AP, Oegema TR Jr, McCarthy JB. Manipulation of hyaluronan synthase expression in prostate adenocarcinoma cells alters pericellular matrix retention and adhesion to bone marrow endothelial cells. J Biol Chem 2002; 277:10050–10057.
67. Simpson MA, Wilson CM, McCarthy JB. Inhibition of prostate tumor cell hyaluronan synthesis impairs subcutaneous growth and vascularization in immunocompromised mice. Am J Pathol 2002; 161:849–857.
68. Itano N, Sawai T, Miyaishi O, Kimata K. Relationship between hyaluronan production and metastatic potential of mouse mammary carcinoma cells. Cancer Res 1999; 59:2499–2504.

69. Zhang L, Underhill CB, Chen L. Hyaluronan on the surface of tumor cells is correlated with metastatic behavior. Cancer Res 1995; 55:428–433.
70. Jacobson A, Rahmanian M, Rubin K, Heldin P. Expression of hyaluronan synthase 2 or hyaluronidase 1 differentially affect the growth rate of transplantable colon carcinoma cell tumors. Int J Cancer 2002; 102:212–219.
71. Zeng C, Toole BP, Kinney SD, Kuo JW, Stamenkovic I. Inhibition of tumor growth *in vivo* by hyaluronan oligomers. Int J Cancer 1998; 77:396–401.
72. Slevin M, Kumar S, Gaffney J. Angiogenic oligosaccharides of hyaluronan induce multiple signaling pathways affecting vascular endothelial cell mitogenic and wound healing responses. J Biol Chem 2002; 277:41046–41059.
73. Ghatak S, Misra S, Toole BP. Hyaluronan oligosaccharides inhibit anchorage-independent growth of tumor cells by suppressing the phosphoinositide 3-kinase/Akt cell survival pathway. J Biol Chem 2002; 277:38013–38020.
74. Posey JT, Soloway MS, Ekici S, Sofer M, Civantos F, Duncan RC, Lokeshwar VB. Evaluation of the prognostic potential of hyaluronic acid and hyaluronidase (HYAL1) for prostate cancer. Cancer Res 2003; 63:2638–2644.
75. Lokeshwar VB, Obek C, Pham HT, Wei D, Young MJ, Duncan RC, Soloway MS, Block NL. Urinary hyaluronic acid and hyaluronidase: markers for bladder cancer detection and evaluation of grade. J Urol 2000; 163:348–356.
76. Hiltunen EL, Anttila M, Kultti A, Ropponen K, Penttinen J, Yliskoski M, Kuronen AT, Juhola M, Tammi R, Tammi M, Kosma VM. Elevated hyaluronan concentration without hyaluronidase activation in malignant epithelial ovarian tumors. Cancer Res 2002; 62:6410–6413.

Chemistry and Biology of Hyaluronan
H.G. Garg and C.A. Hales (editors)

Chapter 20

Viscoelastic Properties of Hyaluronan and Its Therapeutic Use*

ENDRE A. BALAZS

Matrix Biology Institute, 725 River Road, Ste. 205, Edgewater, NJ 07020, USA

I. Elastoviscosity of Hyaluronan Solutions

Hyaluronan is an ubiquitous polysaccharide component of the intercellular matrix of all vertebrate tissues. It is not present in plants and insects, but the capsules of some bacteria contain it. In fluid tissue spaces like the synovial fluid of the joints and the liquid vitreus of the eye hyaluronan is present at a concentration and with an average molecular weight that provides these fluids with elastoviscous properties. This means that the tissue compartments containing such fluids can absorb mechanical impact (energy) by elasticity or dissipate it by viscous flow. Depending on the concentration and the average molecular weight of the hyaluronan as well as the frequency of the mechanical impact, the fluid will behave as an elastic body or a viscous fluid.

The hyaluronan molecules in these fluids are present in random coil conformation and, depending on their molecular weight, they occupy a large hydrated volume. As the concentration increases, the molecular domains (volumes) decrease and the movement of the molecular segments becomes more restricted. Consequently, this crowded molecular system has significant

* In this chapter, *vitreus* will be used as a noun, designating the connective tissue surrounded by the lens, ciliary body and retina. Thus vitreus will replace vitreous humor, vitreous body and vitreous. The two rheological states of the vitreus will be designated as *gel vitreus* and *liquid vitreus*. *Vitreous* will be used, as an adjective, in the following manner: vitreous anomalies, vitreous strands, vitreous implants, etc. From Ref. 189.

viscous and elastic properties that depend on the primary structure and length of the molecular chain, its conformation and concentration (crowding). Because the hyaluronan molecular chain is not branched and is polyanionic, it tends to be extended and somewhat rigid, both of which characteristics greatly contribute to its elastic properties. Because of all these properties, this molecule belongs to the class of polymers called 'super-elastic liquids' which means they are more rubber-like than rubber. Thus, relatively small forces, depending on their impact frequency, can cause a much larger elastic deformation in this liquid than in liquid rubber. The biological role of the native hyaluronan in the vitreous and the joint was interpreted according to these rheological properties (1,2).

Hyaluronan present in the intercellular matrix of connective tissues where the intercellular space is constricted can also be elastoviscous, but not much is known of its real rheological properties. When hyaluronan-rich connective tissues, such as the mucoid layer of the rooster comb or the perivascular connective tissue (Wharton's jelly) of the umbilical cord, are extracted with water, the extraction fluid has very high elastoviscosity and the extracted hyaluronan has a random coil conformation. Hyaluronan extracted from bacterial capsules or cell cultures *in vitro* also has similar properties. It has, however, been suspected for a long time that the extraction process of hyaluronan from these tight tissue spaces drastically changes the conformation of the molecule. It is assumed that hyaluronan may be present in the tight intercellular spaces in a condensed, more structured form than the random coiled form, and therefore can be even more elastic than the extracted fluid itself. One can also speculate that the condensed structure of hyaluronan has some biological functions other than those suggested by its elastoviscous properties.

The reader must be reminded that this chapter is written to review the therapeutic use of hyaluronan, which is primarily based on the molecular structure of a crowded (saturated) molecular chain network that has non-Newtonian flow properties. This is the reason why the rheological properties of these solutions and gels are emphasized in this chapter. It is known that hyaluronan molecular chains are highly negatively charged (polyanionic) and have structural specificity that has been shown to produce specific interactions with certain proteins (soluble, non-soluble) and cell membrane structures. The long hyaluronan chains, especially in crowded conditions, interact with the water structures (hydrophilic). It was also shown that in certain conditions the molecular chain can have a conformation that will exhibit a hydrophobic character (lipid binding).

The broadly accepted therapeutic application of hyaluronan (practical and experimental) today is primarily based on the role of crowded molecular networks that exhibit elastoviscous properties. Therefore this review will concentrate on the properties and uses of such hyaluronan preparations. For other recent review on the strategies used for the therapeutic application of hyaluronan see Ref. 3.

The viscous properties of hyaluronan solutions of various average molecular weights and concentrations are shown in Figs. 1 and 2 (4). The viscous properties depend on the concentration of the hyaluronan in solution and on the average molecular weight of the molecules, as well as on the solvent and the shear rate

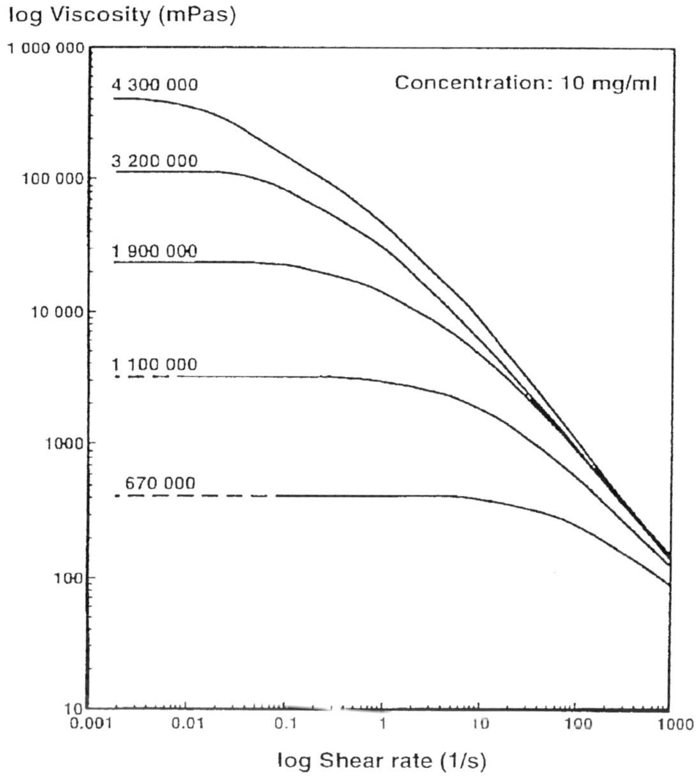

Figure 1 Shear dependence of the viscosity of hyaluronan solution (10 mg/mL, dissolved in 0.15 N NaCl solution). Five hyaluronan preparations with average molecular weights of 0.67, 1.1, 1.9, 3.2 and 4.3 million are shown (from Ref. 4).

at which the measurement is made. When the concentration decreases and the molecules are not crowded or when the average molecular weight is very low, the frequency dependence of the viscosity drastically decreases. This is the reason why hyaluronan in blood, lymph and the aqueous of the eye is not viscous.

The elastic properties of hyaluronan and hyaluronan-containing fluids were studied in dynamic conditions, i.e., in oscillating rheometers in which the viscous and elastic properties could be separately measured over a broad range of frequencies (0.002–10 Hz). The frequency expresses the rate of energy transferred from the environment to the elastoviscous fluid or gel. Fig. 3 (4) demonstrates the proportion between elasticity and viscosity (both 0–100%) of five hyaluronan preparations of various average molecular weights (0.7–4.3 million), one hylan A preparation (6 million) and the gel slurry of hylan B. For further details on the calculation of this proportionality see the legend of Fig. 3. The figure demonstrates that the proportion (percentage) of the elasticity increases with the average molecular weight. Hylan A fluid, because of its high molecular weight, reaches the highest percentage of elasticity at 10 Hz. On the

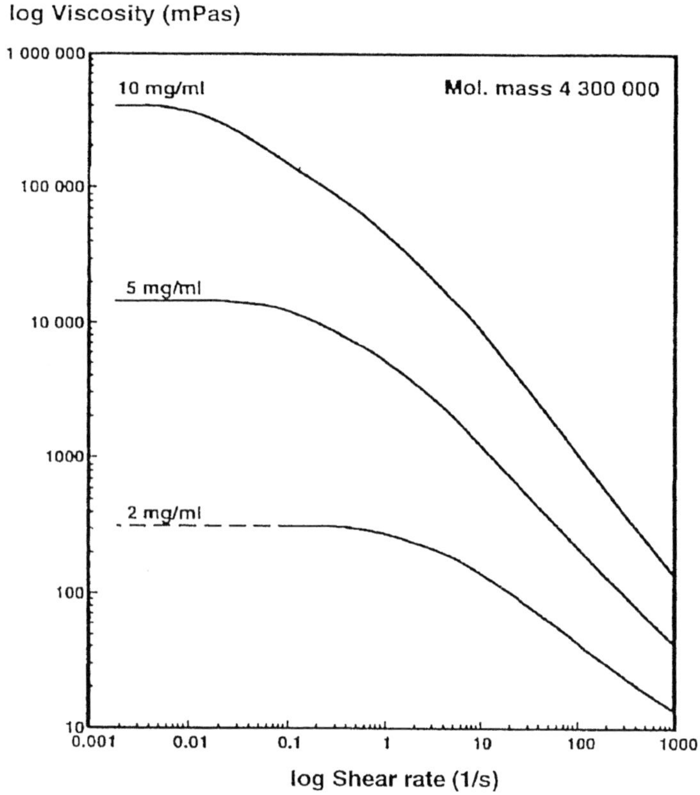

Figure 2 Shear dependence of the viscosity of hyaluronan solutions (average molecular weight 4.3 million) of 2, 5 and 10 mg/mL concentrations (solvent: 0.15 N NaCl) (from Ref. 4).

other hand, hylan B gel slurry behaves practically as an elastic, rubber-like body throughout the entire frequency range.

Fig. 4 demonstrates the elastoviscous behavior of two different hyaluronan crowded molecular networks (4). Both solutions have the same concentration of polymer (10 mg/mL). The difference is in the average molecular weights (hylan A: 6 million; hyaluronan: 0.75 million). The difference between the two solutions is first, in the rigidity (the sum of elastic and viscous properties). The rigidity at 0.1 Hz is approximately 60 times greater for hylan A than for the hyaluronan solution. Second, the hylan A behaves as an elastic body at 0.01 Hz, while the hyaluronan solution does not become an elastic body, even at the highest frequencies (10 Hz). The region before the intersection of the two upper curves representing the dynamic elastic modulus (G') and the dynamic viscous modulus (G'') of hylan A is called the viscous region. Under these experimental conditions, the time scale of the rate of input of mechanical energy that deforms the crowded hyaluronan molecular network is slow and the molecular chains can regain their original (equilibrium) configuration due to their Brownian motion. Under these

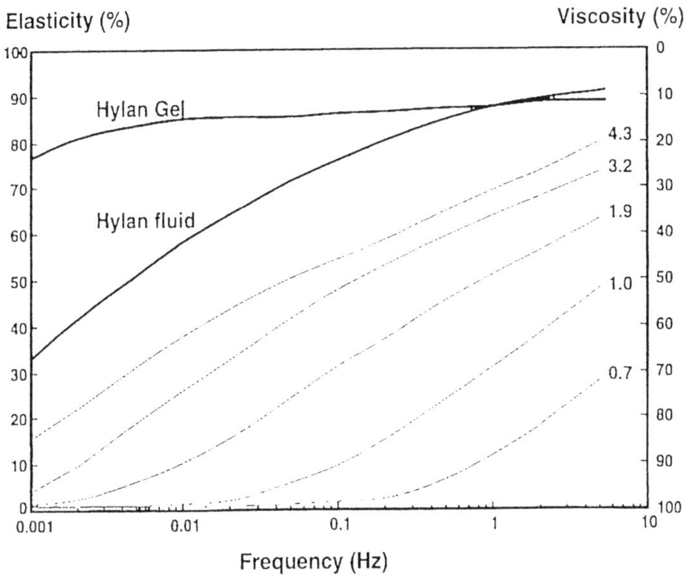

Figure 3 The elastoviscous properties of five hyaluronan preparations and hylan A with average molecular weight of 0.7, 1.0, 1.9, 3.2, 4.3 and 6.0 million and hylan B gel slurry. The viscous and elastic properties are expressed as percentages of the total rigidity of the system. The frequency (in Hz) represents the rate of energy input by oscillating movement during measurements. All solutions represented here are measured at a polymer concentration of 10 mg/mL (dissolved in 0.15 N NaCl). Hylan B gel is hydrated with the same salt solution and the polymer concentration in the gel particles is 5 mg/mL.

The percent elasticity and viscosity can be calculated from the dynamic viscous (G'') and elastic (G') moduli

$$\text{elasticity}(\%) = \frac{G'}{G' + G''} \times 100$$

and

$$\text{viscosity}(\%) = \frac{G''}{G' + G''} \times 100 \quad \text{(from Ref. 4).}$$

conditions complete stress relaxation is attained and the sample appears physically as a viscous solution. Above the intersection, the cross-over point of the two curves of the dynamic moduli, the time scale of the rate of input of mechanical energy that intends to deform the crowded network is too fast, and therefore the network elastically absorbs the energy. Under these conditions, stress relaxation is not attained, the network is not relaxed, and the sample appears physically as an elastic body. Because of the different ways the mechanical energy is lost in viscous flow or temporarily stored (to be dissipated later as heat), the elastic dynamic modulus (G') is also called the storage modulus and the viscous modulus (G''), the loss modulus. The importance of this rapid transition of the same hyaluronan solution from a viscous to an elastic state is that

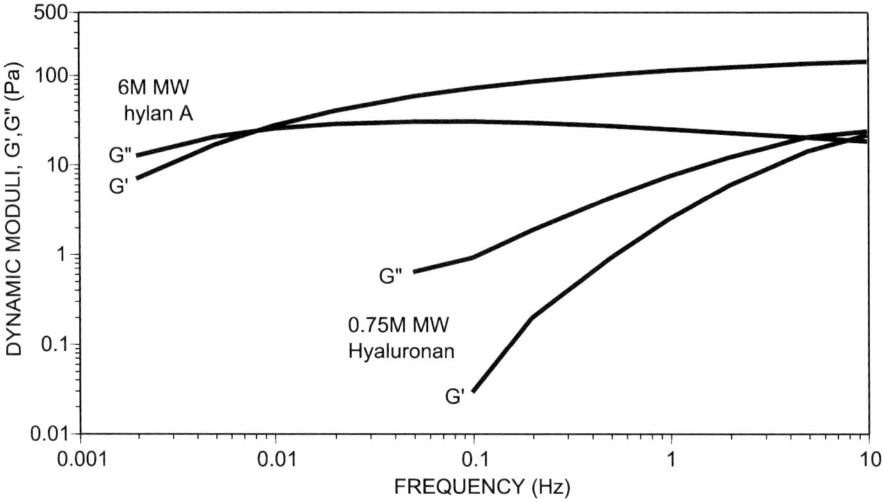

Figure 4 The dependence of dynamic moduli (G' and G'' in Pa) of one hyaluronan and one hylan A preparation on frequency (in Hz). The upper two curves represent the dynamic moduli of hylan A (concentration 10 mg/mL, average molecular weight 6 million). The lower curves represent the dynamic moduli of a hyaluronan preparation (concentration 10 mg/mL) with average molecular weight of 0.75 million. Where the values of G' and G'' are the same it is called the cross-over point.

it occurs at a frequency rate (rate of input of mechanical energy) that is present in joints under various types of movements. For example, when a knee joint is flexed under no load (lying down position) the frequency of the energy input is slow. The synovial fluid behaves as a viscous solution. When the person loads the knee by slowly standing up, the input frequency is still low and the fluid is viscous. However, when the person is walking, running or jumping, the input frequency becomes faster and faster and the fluid behaves as an elastic body. It absorbs the mechanical energy and thereby protects the tissue next to it (cartilage) or embedded in it (synovial cells) from potential mechanical damage. When the elastoviscous hyaluronan surrounds nociceptive fiber terminals, its elastic properties damp the mechanical dislocation of the cell membranes and thereby prevent the mechanical stimulation that leads to perception of movement-related pain. For more details, see Refs. 5 and 6.

These are briefly the experimental findings that constitute the foundation of the therapeutic use of hyaluronan solutions and gels as elastoviscous molecular networks.

II. Therapeutic Applications

Elastoviscous hyaluronan solutions (crowded molecular networks) and elastoviscous hyaluronan gels (cross-linked molecular networks) have been used in several therapeutic areas. In order to emphasize that in all these therapeutic uses the primary mode of action is based on the viscous and elastic properties of a crowded molecular network, we designated the diverse uses as follows: *viscosurgery*, used in ophthalmic surgery to protect delicate eye tissues and provide space during surgical manipulation; *viscoaugmentation*, to fill and augment the intercellular spaces in connective tissues of the skin, sphincter muscles, vocal cords and pharyngeal tissues; *viscoseparation*, to separate connective tissue surfaces traumatized by surgical procedures in order to prevent adhesion and excessive scar formation; *viscosupplementation*, to supplement or replace tissue fluids and restore homeostasis in certain pathological conditions such as replacing synovial fluid in painful arthritis to alleviate pain and promote the healing of intra-articular wounds; and *viscoprotection*, to protect healthy, injured or wounded tissue surfaces from dryness and noxious environmental agents (conjunctiva or cornea) or promote the wound healing of such surfaces as well as of mucous membranes and skin. Finally, we will review briefly the use of elastoviscous hyaluronan solutions and solids used for delivery of drugs and in tissue engineering.

Very dilute solutions made of hyaluronan of any molecular weight or concentrated solutions of hyaluronan of very low molecular weight are not used in the above-described therapeutic applications because they do not have significant elastoviscous properties. We will not review here the emerging therapeutic fields for the possible uses of condensed hyaluronan molecular systems, hyaluronan copolymers with proteins, glycosaminoglycans or lipids, or the potential use of oligosaccharides and small hyaluronan fragments.

III. Viscosurgery

A. Viscosurgical Use in the Eye

The concept and the early experiments using elastoviscous hyaluronan, and the development of the first purified preparation to replace the vitreus after surgery and to be used as a 'soft tool' to manipulate the retina in retinal detachment surgery as well as to be used as a viscoelastic protector of the corneal endothelium in corneal transplantation, were the work of the author and his co-investigators (7,8) and a group of eye surgeons, who carried out the first clinical trials in the late 1960s and early 1970s (9–12). By the end of the 1970s the use of the first highly purified, non-inflammatory fraction of Na-hyaluronan, NIF-NaHA (manufactured under the trade name Healon®, Biotrics, Inc., Arlington, MA) in retinal surgery and to protect the corneal endothelium during corneal transplantation was well established (13). At the same time the increasing use of intraocular plastic lenses to replace cataractous lenses was slowed down by the problem presented by accidental contact of the corneal endothelium with the plastic lenses. The contact with the plastic stripped off the corneal endothelial cells and, since these cells do not regenerate, the endothelium could not properly continue to function as a water pump for the corneal stroma, resulting in swelling and cloudiness of this tissue and, eventually, blindness. Since it was well established that the elastoviscous hyaluronan solutions used to cover the corneal endothelium during corneal transplantation act as a mechanical buffer over this sensitive cell layer, it was logical to use these solutions in cataract surgery to protect the endothelium (14). The elastoviscous hyaluronan solution was expected to be useful for maintaining a deep anterior chamber during the extraction of the cataractous lens and to facilitate insertion of the plastic lens. This made the surgery safer and faster. The first clinical results were reported in 1979 to ophthalmic surgeons specializing in surgery of the anterior segment of the eye (15). In rapid succession several papers were published using animal studies to explain how the corneal endothelium is protected (16) and how the use of hyaluronan decreases the damage to the endothelium in clinical practice (17,18). In the early 1980s, when the first books were published in German (19) and in English (20,21) on viscosurgery in the eye, the use of the first therapeutic hyaluronan preparation (Healon®, developed and first manufactured by Biotrics, Inc., Arlington, MA and later by Pharmacia AB, Uppsala, Sweden) was extended to vitrectomy; it was used during excision of preretinal membranes (22) or removal of foreign bodies (23) from the vitreus. Use of Healon® was reported in difficult cases of retinal detachment surgery (20) and corneal transplantation (24), in trabeculectomy to prevent flat anterior chamber (17) and in extra or intracapsular lens extraction and for the implantation of plastic lenses using a variety of surgical techniques (for reviews, see Refs. 25–28).

During the next two decades the use of ophthalmic viscosurgical devices (OVDs) and the surgical techniques developed to take advantage of their availability exponentially increased. At the same time clinical scientists increasingly utilized the rheological properties of various formulations of hyaluronan products, taking advantage of these properties for the best therapeutic use of OVDs. While Pharmacia (now Pfizer, Inc.) developed new Healon® formulations with greater molecular weight and elastoviscosity (Healon®GV and Healon®5), other companies marketed lower viscosity products as well as low viscosity hyaluronan mixed with chondroitin sulfate (Viscoat®, Alcon Laboratories, Inc.). Today, more than a dozen hyaluronan preparations with a variation of average molecular weight from 0.1 to 7 million and with a concentration of 1–3% hyaluronan compete for the attention of eye surgeons. Naturally the usefulness of this great variety of hyaluronan products depends on their rheological properties and the inventiveness of the ophthalmic surgeons who demand broad applicability and various performances from these OVDs.

It was observed in clinical use that the cohesiveness and solidity of the elastoviscous solutions affect their dispersibility and solubility under the influence of the mixing process provided by the mechanical manipulation of the instruments and the flow of the irrigation fluids (29,30). The OVDs have been recently classified according to their cohesiveness and dispersive properties under the mechanical stresses provided by surgical procedures. The classification was based on their therapeutic performance that was assumed to parallel their extrapolated zero shear dynamic viscosity. This parameter varied between 25,000 and 7 million milliPascal second (mPa s) in 14 hyaluronan products (1 mPa s is the viscosity of water at all shear values) (25). The problem with using zero shear viscosity for characterization purposes is that it neglects completely the elastic properties of the solution. We suggest using the coil overlap parameter of the solution (concentration times average molecular weight), which expresses the crowding of the molecules as the concentration and/or the molecular weight increases. Molecular crowding means the restriction of the domains of neighboring molecules and the resulting decrease of the molecular volume, i.e., the size and possibly the conformation of the individual molecules. These cooperative molecular properties (concentration and molecular weight) determine all the qualities that are important for the optimal viscosurgical performance of OVDs. Applying the coil overlap parameter for the 13 hyaluronan-based OVDs listed by Arshinoff (25) two major categories could be differentiated: five OVDs with a very high overlap parameter (70,000–172,000 g/cc) and eight with much lower values (12,000–20,000 g/cc). The extrapolated zero shear viscosity for the first five OVDs is between 0.5 and 7.0 million mPa s. The OVDs with significantly lower coil overlap parameters have 0.025–0.28 million mPa s viscosities (for details see Ref. 25). As expected, the greater the elastoviscosity (coil overlap parameter) the more cohesive and less dispersive the OVDs.

B. Other Viscosurgical Uses

It was suggested in the early 1980s that elastoviscous hyaluronan solutions should be used as viscosurgical agents in transplantation surgery to protect surfaces of implants and transplants from mechanical damage and loss of water (drying out) (31). It was also proposed that hyaluronan solutions with very high elastic and viscous properties should be used to protect tissues from mechanical damage by surgical tools or to create space for surgical manipulation during arthroscopic surgery (32). The first arthroscopic viscosurgical application of hylan A (0.5% solution) was reported in 1989. These investigators used 3 mL of hylan solution for distention of the capsule in the temporomandibular joints to protect the surfaces of the menisci, cartilage and soft tissues from mechanical damage during diagnostic arthroscopy. The elastoviscous solution provided control of the tissues in the joint spaces and reduced iatrogenic damage especially to the cartilage surfaces. Post-operative pain, mouth opening and range of motion were not adversely effected by the procedure (33). These findings were confirmed by others (for review, see Ref. 34). *In vitro* studies confirmed the mechanical protective effect of elastoviscous hyaluronan. Experiments demonstrated that confluent condrocyte monolayer cultures were protected from mechanical damage (falling weights) by appropriate concentration (10 mg/mL/2 cm^2) of highly elastoviscous hylan A solutions (35).

IV. Viscoaugmentation

A. Hyaluronan Gels

The first hyaluronan gel was prepared by Pharmacia AB (Uppsala, Sweden) in the early 1960s (36) by using 1,2,3,4-diepoxybutane as a cross-linking agent. This chemical has been widely used for cross-linking of polysaccharides such as cellulose and dextran. The epoxy cross-linked hyaluronan was prepared to be used for column chromatography and not for therapeutic use because of its poor biocompatibility. Water-insoluble hyaluronan gels for various therapeutic uses were first synthesized in the 1980s by creating sulfonyl-bis-ethyl cross-links between the polysaccharide chains (37–40). The gel was found to be highly biocompatible, which means it did not elicit allergic, inflammatory, granulation or foreign body reactions (41–43). Because the gel can be made at a very low polymer concentration ($\geq$0.1%) with sufficient solidity and very high elasticity, it is ideal as tissue filler or artificial intercellular matrix. Its high water content (>99%) makes it very permeable to small metabolites, and therefore it does not create a metabolic barrier between cells (44,45). Furthermore it is very stable when injected as gel slurry through a very narrow needle (30 gauge) into connective tissues or implanted as a slab of hydrated gel or as a dry membrane that rehydrates in the tissue (46). When the hyaluronan is injected as gel slurry, cells can grow in between the particles. When it is implanted as a larger piece of gel as a slab or membrane, cells will not degrade it, but they can attach to these surfaces. The half-life tissue of C^{14}-labeled hylan B gel in guinea pig dermis was found to be approximately 12 months (45,47). The very elastic gel particles (0.1–1 mm diameter) deform during the injection and then regain their original form in the tissue. Because the soft particles are not round but polygonal, forces produced by muscle movements and massage of the tissues will cause their fragmentation into very small pieces, which eventually migrate from the site of injection (45,47).

Another important biological property of hylan B gel is its remarkable hemocompatibility (48,49). Thrombocytes and proteins that participate in blood coagulation do not interact with this gel. Hyaluronan in the native form is also extremely compatible with blood, and therefore it is important that the chemical and structural changes introduced by the cross-linking process do not diminish this compatibility (48,49).

Tissue engineering with viscoelastic hyaluronan focused first on its use as a tissue filler or tissue augmentator (viscoaugmentation). Unlike collagen and non-biological tissue fillers, hyaluronan is an extremely elastic molecule and as such provides elasticity to the intercellular spaces into which it is injected. Hylan B gel was first used for viscoaugmentation of the vitreus after retinal detachment surgery, and later for correcting facial wrinkles and depressed scars for vocal cord augmentation in glottal insufficiency and augmentation of the connective tissue in sphincter muscles to treat urinary incontinence.

The second hyaluronan gel to be used for viscoaugmentation was made by using various epoxy compounds (50–53). Because the cross-linking was light, the products—liquid or gels—became biocompatible. Most of the gels used contained more polymer ($\sim \geq 2\%$) than hylan B gel ($\leq 0.5\%$) and consequently they became more solid and less elastic (44). To differentiate this hyaluronan from hylan B gel the developers called it 'stabilized hyaluronan' (54). The rheological properties of these gels have been described recently (55). Despite the considerable differences in the chemical composition and rheological properties of these two gels, their utility in medical therapeutics and medical cosmetics is comparable.

Other cross-linked gels of hyaluronan have been developed in the recent years. The most important requirements of these gels are their biodegradability and biocompatibility. Depending on the intended use, gels that slowly dissolve in the biological environment or those that do not dissolve at all are preferred. Very few of these gels have been tested in clinical experiments.

Hyaluronan cross-linked with ferric ions forms a viscous solution or gel depending on the concentration of the polymer and the cation (Gynecare Intergel®, Lifecore Biomedical Inc., Chaska, MN). Animal studies and clinical trials indicated that this absorbable preparation can be used intraperitoneally after open conservative gynecological surgery to reduce post-surgical adhesions (56).

Slowly soluble (biodegradable) and insoluble hyaluronan gels were produced by ester or ether linkages and tested *in vitro* and in animal studies (57–61).

B. Viscoaugmentation to Fill Facial Wrinkles and Depressed Scars

The first clinical trials using hylan B gel began in the late 1980s in Sweden and continued later in the USA (38,41,62). These trials clearly showed that hylan B gel is a safe and effective material for facial tissue augmentation depending on the location used and, most importantly, on the precision of the application. When the gel is placed too deeply under the dermal fold into the loose connective tissue, the movements of the skin caused by massage and muscle movements for facial expressions will cause the gel to disperse and migrate. The location of the correction is also important. Forehead and cheek lesions have a longer lasting response than areas such as nasolabial folds (63). While subjective evaluation of the success of viscoaugmentation is very important, profilometry was also used in some clinical trials as an objective measure of evaluation (63).

The first clinical trials for epoxy cross-linked hyaluronan were reported in 1998 (54,64) with results comparable to those of hylan B gel. Comparative clinical studies with bovine collagen product (Zyplast®) showed that hyaluronan gel was 'more effective in maintaining cosmetic correction' (65). The safety profile of the epoxy cross-linked hyaluronan gel (stabilized hyaluronan) was reviewed recently, based on data collected worldwide from 1993 to 2000 (more than 400,000 patients). Local adverse reactions were very low (0.06–0.15%), which compared favorably with similar events after collagen skin augmentation (66).

The remaining question is the possible hypersensitivity reaction of the hyaluronan gels. One of the great advantages of the use of hyaluronan instead of collagen is not only its elastic properties (collagen gels are not elastic), but also that hyaluronan has the same chemical and physical properties in all species and tissues. Unlike collagen, it has no structural species or tissue specificity. Therefore if protein impurities are completely removed, hyaluronan cannot cause a hypersensitivity reaction. The two hyaluronan gels discussed above differ not only in the chemical nature of their cross-links, but also in the source of hyaluronan. Hylan B can be prepared from any source of hyaluronan, but the one used for viscoaugmentation was purified from chicken combs, while stabilized hyaluronan was derived from bacterial cell capsules. The developers of the latter product emphasize the lack of animal protein in their preparation but, in fact, the hyaluronan from bacterial fermentation contains proteins originating from the fermentation medium used or from the bacteria themselves. The total protein content in the bacterial hyaluronan in the stabilized hyaluronan products is several times greater than that of the chicken comb hyaluronan (66). Consequently, there is no reason to assume any preferential advantage in a potential allergic reaction for the gel originating from a bacterial source. Alleged allergic reactions and delayed inflammatory skin reactions were reported for both gels (67–69). Analysis of data reported in clinical practice worldwide for both gels indicates that the adverse events, potentially due to the protein impurities present in both preparations, do not differentiate them (70).

C. Viscoaugmentation of the Vocal Cord and the Treatment of Glottal Insufficiency

Hylan B gel slurry, when injected into the healthy vocal fold of rabbits in the course of a 12-month follow-up, did not produce an inflammatory or foreign body reaction. An ingrowth of fibrocytes and deposition of new collagen on hyaluronan was observed between the hylan B gel pieces 1 month after injection until the end of the follow-up (71). Hylan B gel was still present 12 months after injection (72). Hylan B gel injected into the posterior pharyngeal wall of rats showed no chronic inflammatory reaction during a 6 month follow-up (73). These results suggested that this gel could be used for viscoaugmentation of glottal insufficiencies.

An interesting experiment was performed with rabbit vocal fold injected with collagen gel, Teflon® and hylan B gel. The dynamic viscous modulus was measured in oscillation mode between the frequencies of 0.01 and 10 Hz. Both Teflon® and collagen injections decreased the high frequency sensitive elasticity of the injected tissue, but hylan B gel did not change it. This indicates that only the hylan B gel showed similar rheological (elastic) properties to those of the intercellular matrix in the subepithelial space of the vocal fold. In the author's opinion "the hylan B gel rendered the most natural viscoelastic properties of the vocal fold of the substances tested. This should be of importance to restore or preserve the vibratory spacing of the vocal fold with injection treatment in glottal insufficiency" (74).

Another product made of a mixture of equal volumes of hyaluronan solution (1%) and dextran spheres (80–120 nm diameter) was injected into the vocal fold of rabbits. As expected, the hyaluronan disappeared from the injection sites within a week, but the space between the dextran spheres was invaded by fibrocytes and a new collagen was generated (75).

Vocal fold augmentation with collagen and hylan B gel was compared in a 12 month evaluation clinical trial on 70 patients. To evaluate the efficacy of the augmentation, in addition to the patients' subjective ratings, objective methods were used (videostroboscopy, phonetograms, maximum phonation time and phonation quotients). Both products could be used safely for the treatment of glottal insufficiency. The hylan B treated patients showed better vocal fold status and longer maximum phonation time than collagen-treated patients at 12 months after treatment (76).

D. Viscoaugmentation of the Urinary Sphincter

The first animal studies to demonstrate the usefulness of hylan B gel in viscoaugmentation as a tissue bulking agent for urinary sphincter muscle were completed in 1993 (Biomatrix, Inc., Ridgefield, NJ). Later clinical trials demonstrated the beneficial effect of this viscoelastic gel slurry injected intramuscularly to augment the connective tissue between the sphincter muscles of the urethra in patients with certain types of urinary incontinence. Hyaluronan was also used in pilot clinical studies combined with dextran particles for endoscopic treatment of vesicoureteral reflux in children (77) and in women with a history of stress incontinence (78). In this case the dextran particles were suspended in 1% elastoviscous hyaluronan solution to decrease the foreign body reaction caused by the dextran.

V. Viscoseparation

The first *in vivo* (rabbit, guinea pig) use of hyaluronan (NIF-NaHA, average MW 2–3 million) to prevent adhesions between two surgically traumatized connective tissue surfaces was reported in the early 1970s. Fasciae, tendon and tendon sheaths, conjunctival and scleral surfaces were tested in owl monkeys, guinea pigs and rats. Hyaluronan pastes or dried membranes that became hydrated and dissolved were placed between the two damaged surfaces. Importantly, bleeding was well controlled during and after surgery in order to prevent 'washing out' of the hyaluronan. The use of these hyaluronan preparations, if properly applied, decreased post-traumatic granulation and fibrous tissue formation and subsequent adhesions. Hyaluronan pastes were also applied around polyethylene implants, which significantly decreased foreign body reaction, granulation and capsule formation (79,80).

Topical applications of hyaluronan paste under skin allographs (rats) extended their survival time (81). The authors of these pioneering works stipulated that the effect could be explained by several biological effects of the crowded molecular network of high average molecular weight hyaluronan. First, the migration of cells of the lymphomyeloid system is inhibited by such hyaluronan solutions (81). Second, the diffusion of fibrinogen, present in the post-injury exudate, is inhibited by the barrier of the hyaluronan molecular network (80). Third, as reported by several authors (81–83), the fibrin clot formation is delayed or inhibited by hyaluronan. Thus, the objective of the use of hyaluronan pastes or gels between wounded connective tissue surfaces is to prevent the formation of fibrin coagulum that would form a solid intercellular matrix connecting the healing tissue surfaces and provide scaffolding for invasion of cells, leading to the formation of collagen fibers that would permanently connect the two surfaces and form adhesions and excessive scars.

The experimental studies extended to more complicated adhesion models like the flexor tendon wounds in monkey hands in the area called 'no-man's land' (84). Achilles tendon, pelvic and abdominal models were also used, and hyaluronan solutions with varying concentrations and average molecular weight and consequently with very different elastoviscosities were applied (for more detailed review, see Ref. 85). The efficacy of viscoseparation depends significantly on the technique of application and the elastoviscosity and density (concentration) of the molecular network barrier and its residence time at the site of application. We believed that solid hyaluronan gels (non-water soluble barriers) would be more effective than pastes and fluids because their residence time is substantially longer or they are permanently present at the application site. When hylan B gels were developed they were extensively and successfully used for adhesion prevention. Membranes or gel slurries made of hylan B were considerably more efficacious than even the most concentrated hyaluronan or hylan A preparations used previously. Hylan B gels were successfully used in animal models for prevention of tendon-sheath adhesion (86,87), post-laminectomy scar formation (88) and cecal adhesions (89).

These studies not only showed that two tissue surfaces can be successfully separated by hylan B gel, but also that the post-surgical adhesion can be prevented. They also demonstrated the exceptional biocompatibility of this gel (43,90), and its inhibitory effect on migration of lymphocytes, macrophages and granulocytes, on fibrin formation and on the inhibition of superoxide release from granulocytes (for review see Ref. 85). Solid hyaluronan (hylan B gel) not only provides a highly biocompatible molecular barrier between two healing tissue surfaces, but can also down-regulate various inflammatory processes as demonstrated in *in vitro* experiments (43,85).

The discovery of hyaluronan and hylan B anti-adhesion barriers encouraged the development of other anti-adhesion devices. The first such bioabsorbable barrier was made of a carboxymethylcellulose–hyaluronan mixture in membrane form and was used in animal and human studies in the 1990s (91–93). This device was made available in 1996 for use in abdominal and pelvic surgery (Seprafilm® and Sepracoat®, Genzyme Corporation, Cambridge, MA).

VI. Viscosupplementation

A. Hyaluronan in the Joint

The rheological properties of synovial fluid and purified hyaluronan were first systematically studied by the author and his co-workers in the 1960s using both healthy and arthritic human and equine synovial fluids and purified hyaluronan (94–96). Fig. 5 demonstrates the rheological properties of two pooled human synovial fluid samples obtained from healthy knee joints and one from osteoarthritic joints (95). The synovial fluids of healthy young (27–34 years) and old (52–78 years) subjects, when exposed in dynamic experiments to shear stress at low frequencies (<0.1 Hz), behaved as viscous fluids, but when the frequency increased (>0.1 Hz), the fluids became more and more elastic. At a given point of frequency ('cross-over point'), values of the elastic modulus (G')

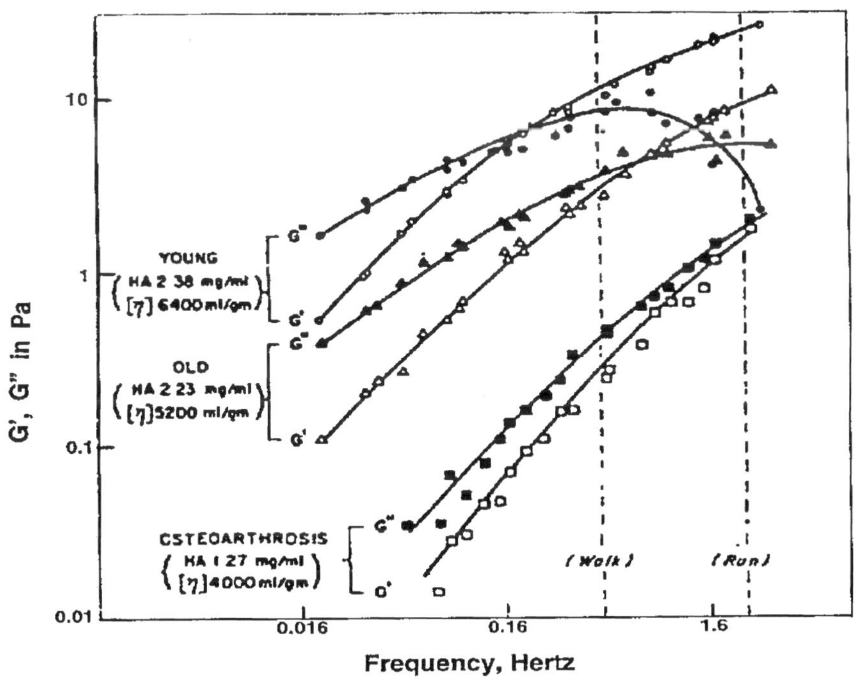

Figure 5 The dependence of dynamic moduli (G' and G'' in Pa) of three human synovial fluid samples on frequency (in Hz). The concentration and limiting viscosity number, $[\eta]$, are given in parentheses. The two vertical dotted lines indicate the transfer of energy (at various frequencies) that the cartilage surfaces and the synovial tissue experience in running, walking or jumping (from Ref. 95).

and the viscous modulus (G'') became equal. This is where the transition from the viscous fluid to elastic body occurs. It is especially obvious at higher frequencies when the viscous modulus was found to be much lower in fluids from young than from old persons. Note also that the fluid from the young donor at all frequencies showed greater elasticity than that of the old (2,94). It was also discovered in these studies (see Fig. 5) that arthritic human synovial fluids lose some or all of their elastic properties at all frequencies because the cross-over point shifts to higher frequencies and it never occurs in the measured frequency range. This is due to the lower concentration, lower average molecular weight or changes in the conformation of the hyaluronan molecules in arthritic joint fluids. Thus the ratio of energy stored elastically to the energy dissipated as viscous flow decreases with aging and, even more significantly, with osteoarthritis. This is particularly true at high frequencies, corresponding to vigorous movement like running or jumping.

These studies also found that the rigidity of the fluids, which is a complex of the two moduli, is extremely dependent on the concentration and molecular weight of the hyaluronan molecule. The logarithm of the rigidity, thus also the elasticity at higher frequencies, is linearly dependent on the product of molecular weight (or intrinsic viscosity) and concentration. A 10-fold increase in the coil-overlap parameter (the product of molecular weight and concentration) results in approximately a 100-fold increase in the rigidity or elasticity of the fluid (2). Most importantly, the protective effect of hyaluronan in healthy joints dramatically increases as the frequency reaches the range that the joint experiences during running and jumping. The gap between the two cartilage surfaces closes faster and faster, resulting in a rapid transfer of mechanical energy from one cartilage surface to the other. The importance of the synovial fluid is to absorb this mechanical energy by its rigidity and elasticity. This loss of rigidity and elastic properties with aging and arthritis was interpreted as a loss in the protective properties of the fluid in between cartilage surfaces and in the soft tissues of the joint. This protection of the joint tissues by the rigidity and elasticity of the synovial fluid is the first of two cardinal principles in explaining the role of hyaluronan in the joint. The second cardinal principle is the elastic protection of the mechanosensitive apparatus of pain receptors (nociceptors) in the synovial tissue.

The discovery of the analgesic effect of intra-articularly injected elastoviscous hyaluronan solutions (NIF-NaHA) in the 1960s in arthritic dogs and racehorses was one of the most important events that led to viscosupplementation as a new therapeutic paradigm. It was also noted that the analgesic effect lasted longer than the residence time of the injected hyaluronan in the joint. It was also an early observation in horses that, parallel with the long-lasting analgesia, the abnormally low rheological properties of the hyaluronan in the affected joints returned to normal (97). This meant that the broad polydispersity and the low concentration of hyaluronan reverted to normal levels parallel with the long-lasting analgesia. The first comprehensive hypothesis on the mode of action of viscosupplementation was based on this observation: the restoration of the homeostasis of the overall metabolism of the joint. This chain reaction is triggered by pain relief, which permits the regular movement of the joint and is essential to restore the normal flow of the synovial fluid. This is the 'sine qua non' for the maintenance of the normal metabolism of joint tissues as well as that of the hyaluronan. The turnkey in this hypothesis is the restoration of joint motion due to the analgesia caused by viscosupplementation (98).

The first behavioral studies on racehorses and dogs were confirmed later by Japanese researchers' work using peritoneal and joint pain models in rats. These studies revealed that the analgesia is related to the concentration and average molecular weight of hyaluronan used (5). The pioneering electrophysiological studies of Belmonte and Schmidt and their co-workers in the 1990s confirmed the analgesic effect discovered in behavioral animal models. Using healthy joints and an acute intra-articular inflammation model in cats and rats, their studies established that the analgesic effect of hyaluronan solution depends on the elastoviscosity of the hyaluronan solution used. They found that only the most elastoviscous hyaluronan (hylan G-F 20) reduced the nerve impulse activity in sensory fibers evoked by movements within or outside the working range of both normal and inflamed joints. This analgesic effect is observable in the animal model only for a few hours after the intra-articular injection of the hyaluronan preparation because, after that time, the model is not maintainable. Therefore, this experiment explains the analgesic effect of viscosupplementation in the short term only. But it establishes that the stretch-activated channels of the nociceptive nerve terminals, the mechanosensory transduction apparatus that detects deflections on the cell membranes and is coupled to the cytoskeleton, are highly sensitive to the elastoviscous environment, in this case, created by hyaluronan solution injected into the joint (6). This interpretation of the mechanosensitive apparatus of the cell membranes to elastoviscous fluids to which it is exposed was confirmed by these authors using the frog oocyte model. In this *in vitro* cellular model the stretch-activated ion channels both on the inside and the outside surface of the cell membranes were exposed to various elastoviscous hyaluronan and DNA solutions. The mechanical stretch of the cell membrane caused deformation and openings of the ion channels, which were recorded by electrical activity that is proportional to the number of channels opened. Only elastoviscous solutions, both hyaluronan and DNA, adjacent to the channels could desensitize them (5,99). For further details on the electrophysiological work using cat, rat and frog oocyte models in evaluating the sensitivity of the mechanosensitive apparatus of nociceptive nerve endings and ion channels of cell membranes to elastoviscous hyaluronan and hylan preparations, see quoted references (5,99,100).

Parallel to the development of the use of hyaluronan in the vitreus the same preparations were used for the first time in the knee joint to eliminate pain (105). Preclinical studies on dogs, monkeys and horses showed that articular pain after traumatic arthritis could be alleviated by intra-articular injection of elastoviscous solutions of hyaluronan (80). Non-elastoviscous solutions of the same concentration of hyaluronan were not effective. The pain-relieving (analgesic) effect of viscoelastic hyaluronan solutions was later studied in rats using behavioral pain models (101,102). It was found that the analgesic effect depends on concentration and average molecular weight, confirming that the elastoviscosity of the solution, which is the product of concentration times average molecular weight, is the active analgesic principle (for review see Ref. 5). The hypothesis, that in arthritis in horses and humans the elastoviscosity of the synovial fluid was always lower than in the healthy joint, was supported by the earlier findings. This is due to the decrease of the concentration of hyaluronan and its average molecular weight in this pathological condition (97).

B. Hyaluronan Preparations Used for Viscosupplementation in the Joints

The first hyaluronan preparation used for intra-articular injection was the so-called non-inflammatory fraction of the natrium salt of hyaluronan (NIF-NaHA) and it had an average molecular weight of 2–3 million. This name was given to this highly purified preparation because it was distinctly different from the other fraction of the molecule that caused inflammatory reactions in various tissues used for testing such as the vitreus of the owl monkey, the peritoneal cavity of mouse and the horse joint. The test carried out in the monkey eye was the most sensitive and reproducible. The acute inflammatory reaction is similar to an endotoxin-induced inflammation but is not a result of the endotoxin content of the hyaluronan preparations because the endotoxin content of these hyaluronan preparations was very low even before they were fractionated. Nor was the very small amount of protein and nucleic acid impurities the cause of inflammation. The monkey vitreous test became the standard between 1970 and 1990 for all hyaluronan preparations used in ophthalmic surgery and for intra-articular injection in human and veterinary medicine (103,104). Because this test was expensive it was replaced in the 1990s with a much less sensitive and less expensive test using injections into the vitreus or anterior chamber of rabbit eyes.

At the time NIF-NaHA, a high molecular weight (average MW 2–3 million) hyaluronan, was developed the biological role of this molecule was perceived in a much broader spectrum than simply in mechanical terms. In 1971 the author of this chapter wrote: "The proposed biological role of hyaluronic acid will be discussed in four areas of matrix function: (a) mechanical function; (b) regulation of transport and distribution of molecules; (c) development and maintenance of matrix assembly; and (d) control of cell assembly" (79). The importance of the mechanical properties meant that 'hyaluronic acid provides the matrix with mechanical shock and vibration absorbing system of high permeability, transparency and deformability'. The regulatory role of hyaluronan in transport and distribution of molecules meant that 'the extended (polyanionic) coil structure of hyaluronic acid can serve as a filter and prevent or slow the passage of water and solutes through its domain. The effect is called the sieve or dynamic filtration effect'. The regulation of matrix assembly by hyaluronan means that it can 'under certain conditions influence the formations and mechanical stability of collagen fibrils' as well as 'the development and maintenance of the ordered structures of matrix' (79). The goal of developing hyaluronan for therapeutic purposes was to engineer the intercellular matrix and thereby influence tissue healing and regeneration. Matrix engineering was coined to describe the implantation or injection of crowded molecular networks of hyaluronan in solutions ($\geq 1\%$) or in the form of dry sheaths and hydrated membranes into the liquid or solid intercellular spaces of tissue compartments where disease or aging has altered or destroyed the normal matrix structure. In the case of the therapeutic application in painful arthritis this meant two steps. The first was to remove the pathological synovial fluid or exudate that contained hyaluronan of lower-than-normal concentration and with molecules of much lower molecular weight and broader polydispersity than present in healthy fluid. The second was to replace the fluid removed with NIF-NaHA made of hyaluronan molecules that, as closely as possible, resembled those present in the healthy fluid. Since the technology in the early 1970s was not available to produce sterile NIF-NaHA with an average molecular weight of 6 million or greater and with a narrow polydispersity comparable to that present in healthy synovial fluid, an adjustment of the concentration had to be made. By increasing the concentration to 10 mg/mL, the rheological properties of the preparation became closer to those of the healthy synovial fluid. In contrast, the concentration in the healthy joint is only 3–4 mg/mL. This was the first hyaluronan preparation marketed for veterinary medicine worldwide in the late 1970s and extensively tested in more than two dozen clinical trials in human osteoarthritis between 1968 and 1980 (Biotrics, Inc., Arlington, MA and Pharmacia AB, Uppsala, Sweden, under the trade names Healon® and Hylartil®). The very early exploratory human and veterinary clinical trials were published (80,105–107), but the later and more extensive human clinical trials were not, and this product was never marketed for treatment of painful osteoarthritis in humans.

More than a decade later, Italian and Japanese researchers started clinical trials with NIF-NaHA that had a very low average molecular weight (0.5–0.8 million) with very broad polydispersity (108,109). Unlike the pathological human synovial fluid this preparation did not have the high molecular weight component but only the very low one. Fig. 6 shows the polydispersity of hyaluronan in the human synovial fluid of a healthy knee joint compared with various hyaluronan preparations used as therapeutics for arthritic pain (110). Note than hylan A (avg MW 6 million) most closely resembles the polydispersity of the healthy synovial fluid; next is Hylartil® (avg MW 2.5 million), then Orthovisc® (avg MW 1.2 million), Artzal® (avg MW 0.7 million) and Hyalgan® (avg MW 0.5 million). Hylartil® was the first hyaluronan in clinical tests for human and equine arthritis and was marketed for veterinary use in the late 1970s. Artz® (or Artzal® or Supartz®) and Hyalgan® (avg MW 0.5 million) were first marketed for human osteoarthritis in 1986 in Japan and Italy (Seikagaku SPH,

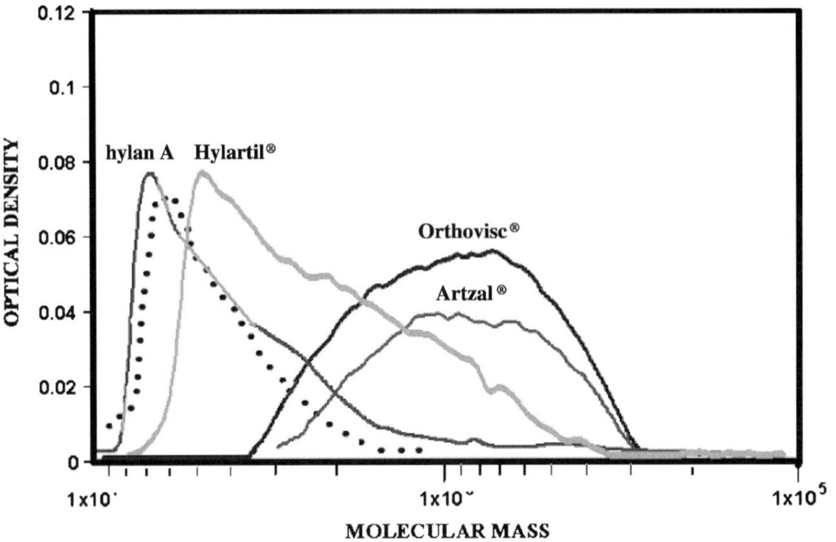

Figure 6 Polydispersity of some hyaluronan products used for viscosupplementation. Hylartil® was the first hyaluronan product clinically tested for the treatment of osteoarthritis in humans and marketed for veterinary medicine. Hylan A is the fluid component of hylan G-F 20, representing 80% per volume of the product (average molecular weight 6 million). The average molecular weight of Orthovisc® is 1.2 million and that of Artzal® is 0.75 million. The dotted line represents the polydispersity of hyaluronan in a healthy human synovial fluid, obtained from the knee joint. The polydispersity was analyzed by agarose gel electrophoresis according to the method described in Ref. 110.

Japan and Fidia SpA, Italy). It is important to note the fundamental differences in the distribution of molecules of various sizes in these preparations and compare them with the hyaluronan in the healthy fluid. Fig. 7 demonstrates the difference in polydispersity of the healthy and osteoarthritic fluids (5). The important fact is that all osteoarthritic synovial fluids ever studied have a considerable amount of hyaluronan with high molecular weight ($\geq$2 million). Therefore, the pathological change, as far as the polydispersity of hyaluronan is concerned, is the appearance of various amounts of hyaluronan molecules with molecular weights between 0.3 and 2 million (110–113).

Fig. 8 (5) demonstrates the proportion of elasticity and viscosity as functions of frequency of the same hyaluronan preparations (Hylartil®, Orthovisc®, Artzal®) as shown in Fig. 6. As expected from the polydispersity data, preparations that have higher and narrower distribution of molecular weight will show greater elastic properties at the higher frequency range. This is only if the concentration of hyaluronan is same or similar.

Neither of the above hyaluronan preparations fulfilled the original criteria for viscosupplementation. Therefore in the early 1980s the author decided to

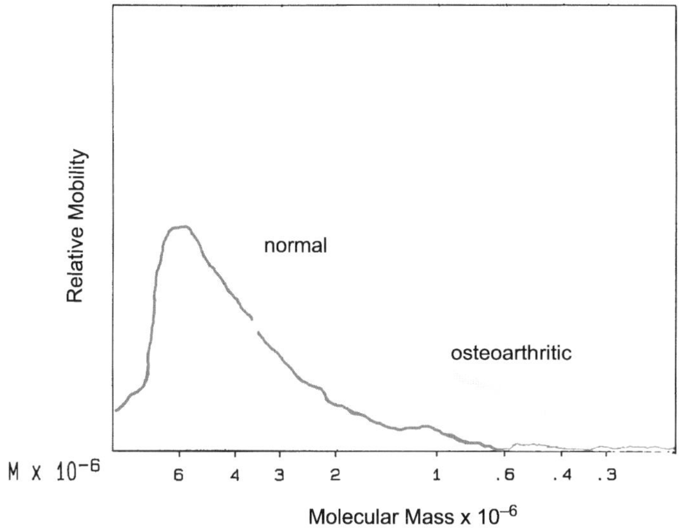

Figure 7 The polydispersity of hyaluronan in the synovial fluid of the knee of a healthy and an osteoarthritic patient demonstrated in agarose gel electrophoresis according to Lee and Cowman (110). The relative mobility, obtained from densitometry measurements and MW distribution, is based on electrophoretic analysis of hyaluronan standards with known average molecular weight (from Ref. 5).

develop hyaluronan with an average molecular weight (with narrow polydispersity) similar to or even greater than that of the native hyaluronan. This new hyaluronan derivative that could be produced with an average molecular weight of 6–25 million was given the generic name of hylan A. A cross-linked gel of hyaluronan called hylan B was also developed (37–39). The mixture of these two hyaluronan derivatives (8 volumes hylan A of 10 mg/mL concentration and 2 volumes of hylan B of 0.5 mg/mL concentration) became the formulation of hylan G-F 20 (Synvisc®, Biomatrix, Inc., Ridgefield, NJ and since 2000, Genzyme Corp., Cambridge, MA). This is today the only viscosupplementation product available worldwide to patients that has rheological properties similar to the healthy human synovial fluid since it was first marketed in Canada in 1993 (98). Shear, extensional and squeeze flow experiments on hyaluronan (as present in the joint) suggest that the hyaluronan in healthy synovial fluid has a combination of rheological properties which make it remarkably effective for tissue and cell protection in the joint. The hylan A and hylan B system (hylan G-F 20) was studied with the same rheological methods (114). The authors of these studies interpreted this special mixture of hylan A and hylan B as "successfully mimicking over long periods, the behavior in joints of healthy synovial fluid".

Many other hyaluronan preparations have recently been made available to patients worldwide, but most of them have elastoviscous properties comparable

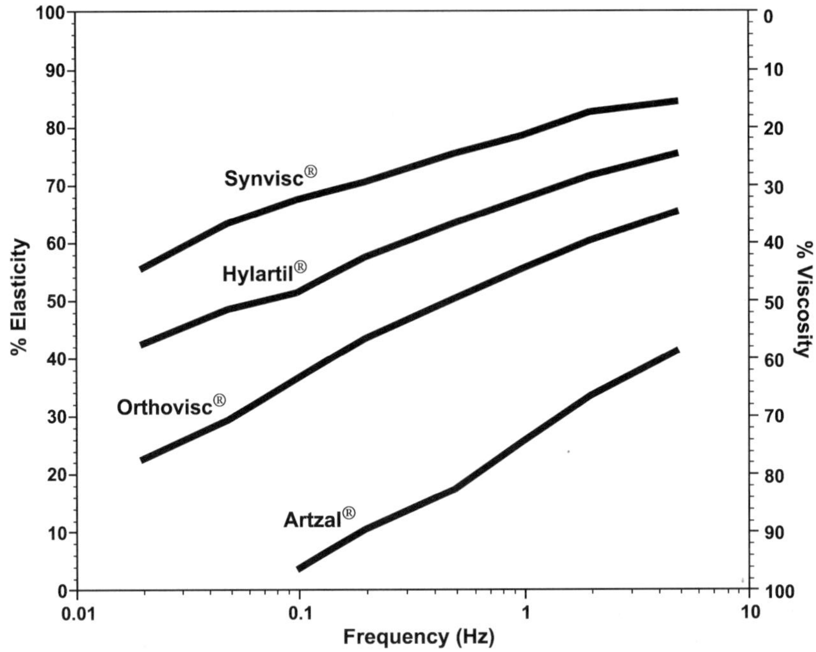

Figure 8 Proportion of elasticity and viscosity in hyaluronan preparations as a function of the frequency at which the measurement of the elastic (G') and viscous (G'') dynamic moduli is made. For more details see legend of Fig. 3 (from Ref. 5).

to those of the lower molecular weight component of the pathological synovial fluid (for review see Refs. 115 and 116).

C. Clinical Use of Hyaluronan for Alleviating Joint Pain

The principal symptom that brings the patient to the doctor is local pain in the joint that may be initiated by movements (with load or not) or present in the resting joint. All treatments by drugs or surgery are aimed at alleviating this pain. Viscosupplementation is the latest, if by now not a very new, therapeutic paradigm targeted as a local analgesic treatment. Since its introduction to veterinary and human medicine by clinical trials and marketing in the late 1970s, viscosupplementation has met with enthusiastic support and also doubtful criticism. While millions of patients have been treated worldwide and dozens of drug companies have conducted and financed laboratory and clinical research to define its efficacy, safety and mode of action, the question of which patients benefit the most from this local treatment remains unanswered. There are several well-defined reasons for this. First is the lack of general recognition that there are fundamental differences between the hyaluronan preparations used today. It is surprising that this fact is seldom noted when the efficacy of various preparations is considered.

It is generally accepted that there is a lower limit of elastoviscosity of a hyaluronan preparation below which it is not effective in viscosupplementation. There have been several clinical trials where hyaluronan (10 mg/mL concentration) with average molecular weight of less than 100,000 was used as control. The developer of one of the lowest average molecular weight hyaluronan (Artz® 750,000) always emphasized that their preparation was efficacious because it was made of 'high molecular weight' hyaluronan (108). This means that the hyaluronan preparation that has efficacy over physiological salt solution must have elastoviscous properties at least at the level of the lowest observed in pathological synovial fluids. In some clinical studies, hyaluronan (1%) with average molecular weight of 100,000 or even lower was used as a control. The difficulty lies in the differentiation of the efficacy levels (effect size) between the various hyaluronan preparations and between a control treatment using physiological salt solutions or low average molecular weight hyaluronan of the same volume and with the same injection frequency. It is also generally accepted that a control treatment is not equivalent to a classic placebo ('sugar pill') because if it is properly used, it combines the removal of the exudate from the joint and its replacement with the physiological salt solution, thereby washing out and diluting pain-causing agents in the joint. Indeed with all hyaluronan preparations used for viscosupplementation a sizable number of patients respond to physiological salt solution treatment with analgesia, which in some cases is long lasting (weeks). This is not surprising in view of the long-lasting analgesic effect of diagnostic arthroscopy, a procedure during which the joint is washed out with a large volume of physiological saline solution.

Another problem with clinical trials using salt solutions as a comparative treatment is the standardization of patients. The well over 100 clinical trials carried out with various hyaluronan products during the past 35 years taught us that in some clinical trials with randomly selected patients, only half respond well to the treatment and the other half respond to the control treatment or do not respond at all. Such a situation does not produce satisfactory statistical results. On the other hand, in other trials with a different group of patients, viscosupplementation shows a statistically and clinically significant benefit over the control treatment. The complexity of the evaluation of efficacy of the various hyaluronan products is shown by the fact that from the very outset, the low elastoviscosity and low average molecular weight products were tested and introduced to the market with 5–10 consecutive weekly injections required for efficacy, while the very high elastoviscous product showed efficacy after only 2–3 weekly injections. This recommended treatment schedule was often changed under commercial pressure (cost and reimbursability) during the past decades, but the bulk of clinical trials published still demonstrates the difference.

Direct comparisons of the various hyaluronan products in one clinical trial are few and not conclusive (117). A very recent meta-analysis (118) compared the effect size (at 95% confidence intervals) of 14 low (0.6–0.9 million), two medium (2–3 million) and three high (6 million) average molecular weight preparations. The authors' conclusion was that only two trials using the high average molecular weight (6 million) products "had an effect size of 1.55 and 1.76, suggesting efficacy equivalent to that of a total knee replacement". The authors (118) were skeptical that the effect size could be so large; however, others have in fact reported that hylan G-F 20 treatment can provide pain relief to postpone replacement therapy (119). All the remaining 17 trials, including one with the highest molecular weight product, had a median effect size varying between 0.04 and 0.58 which, according to the authors, is "equivalent to the effect of non-steroidal anti-inflammatory drugs over that of acetaminophen" (118).

In the course of the past two decades it has become more and more obvious to the author and his co-workers that another evaluation method must be found in order to define the clinical value of viscosupplementation and determine which hyaluronan preparation is the most efficacious. Viscosupplementation by definition cannot be regarded as a classical drug or hormonal therapy, even if the beneficial effect is based on a combination of physicochemical and chemical properties of the hyaluronan molecule. This is because it is a replacement therapy. It supplements a natural polymer in the joint space with the same polymer in highly purified form, only at a different concentration and/or average molecular weight than present in the pathological joint. Furthermore, as we know it today, the long-lasting analgesic effect works only if the hyaluronan supplement is applied intra-articularly and only if the exudate, if present, is first removed, and it certainly is not effective in all patients with joint pain. Therefore it should be classified as a tissue fluid replacement therapy, not very different from blood transfusion (more correctly, infusion).

In recent years, the usefulness of effectiveness studies over efficacy studies was emphasized. Effectiveness studies address broader aspects of the usefulness of therapeutic paradigms and are especially useful from a pharmaco-economical and health policy point of view (120–122). The very high average molecular weight (6 million) hylan G-F 20 preparation was studied in two independent prospective effectiveness clinical trials in Canada (123,124) and France (125). The conclusion of these effectiveness trials was that the incremental improvement in the group of patients receiving hylan G-F 20 injection intra-articularly was statistically and clinically significant for all primary and secondary outcome parameters, including the measuring of pain relief by patient and physician. A statistically significant decrease of the use of non-steroidal anti-inflammatory drugs and steroid injections was also noted. Importantly, significant decreases in gastrointestinal side effects and the use of medication to treat them were also observed.

The question of what the clinically most effective formulation of a hyaluronan product is for viscosupplementation (the combination of concentration, average molecular weight, rheological properties and volume injected) and how to select the most responsive patients will not be answered until similar effectiveness studies are also carried out on very low (<1 million) and medium (2–3 million) average molecular weight products using comparable concentrations of the polymer and comparable doses (frequency of injection and volume injected).

VII. Viscoprotection

A. Surface of the Eye

The first highly elastoviscous hyaluronan available for therapeutic use (Healon®, Biotrics, Inc., Arlington, MA) was found very effective in alleviating discomfort in 'dry eye syndrome'. In the late 1970s it was used in several clinical trials (for review see Refs. 126–129). Hyaluronan is not present in tears, and therefore hyaluronan solutions cannot be classified as true artificial tears. Their usefulness is their water-containing capacity (hydration) and function as elastoviscous barriers between the corneal and conjunctival epithelium and noxious environmental factors (dust, smoke, etc.).

Elastoviscous hyaluronan and hylan solutions were shown to be efficacious in treating patients with dry eye conditions and also increased the comfort of patients wearing hard contact lenses. In the latter case, they provided an elastoviscous buffer between the cornea and hard contact lens (129–131). The first hyaluronan-containing eye drop (0.15%) available for patients with keratoconjunctivitis sicca was Hylashield® (Biomatrix, Inc., Ridgefield, NJ), a 0.15% solution of 6 million average molecular weight hylan A (132,133). The utility of this preparation was also shown as a wetting agent for the cornea during surgery (134).

The importance of the elasticity of hyaluronan or hylan eye drops is demonstrated by their residence time on the surface of the eye (cornea, sclera). The rapid and frequent eyelid movement during blinking will remove any solution that is simply viscous but not elastic, from the surface of the eye. However, solutions like high molecular weight hyaluronan or hylan (~6 million average MW) in the appropriate concentration (≥1%) are highly elastic at the frequency of the eyelid movement. They will elastically deform but not be removed from the surface of the eye with blinking movements. Thus their residence time on the surface of the eye is considerably increased (133).

Another application of elastoviscous hyaluronan and hylan solutions as hydrating and therefore comfort agents is to reduce discomfort and eye fatigue during viewing of video display terminals. The normal blinking frequency (~12 blinks per minute) decreases during strong visual attention to display screens, and this causes discomfort and fatigue which is attenuated by environmental conditions such as low humidity, high temperatures and dust or smoke in the air. The beneficial effects of highly elastic hyaluronan solutions have been well demonstrated under these conditions (135).

B. Wound Healing

The first report of the effect of highly purified (NIF-NaHA), high molecular weight (average MW 1–2 million) hyaluronan on the healing of skin wounds of owl monkeys, rabbits and guinea pigs (32 animals) was reported in 1971.

Two cuts penetrating to the dorsal fascia were made on each side of the dorsal skin. The fascia was traumatized by strokes of gauze pads: one side served as untreated control, and 0.5 mL of 8 mg/mL hyaluronan solution was installed into the other side before the wound was closed aseptically. Eighty percent of the animals showed significantly 'less connective tissue reaction with smoother subcutaneous scar when the wound was treated with hyaluronan' (80).

The role of hyaluronan in the granulation phase of the healing of sterile skin wounds was first studied by the author in the late 1940s (136,137). An accumulation of the polyanionic glycosaminoglycans was observed during the early phase of healing, reaching a maximum on the fourth and fifth day after wounding. Extract made from healing tissue had a fibroblast-proliferating effect in tissue culture (embryonic fibroblasts growing in fibrin coagulum). In this inflammatory phase of wound healing an increased accumulation (increased synthesis) of hyaluronan was observed (138,139). The putative role of native hyaluronan during all four phases of wound healing (inflammation, granulation, re-epithelialization and re-endothelialization) has been studied in the healing of skin and corneal wounds and also in nerve regeneration (140). In this later case, hylan B slurry placed between cut ends of axons prevented invasion of connective tissue cells into the area of nerve regeneration, thereby promoting the healing of axons (140). It was hypothesized that the role of hyaluronan in the granulation phase of wound healing is to regulate (inhibit) the deposition of collagen fibers and thereby inhibit excessive scar formation (141).

Full thickness skin wounds in large pigs were used to evaluate the effect of low concentration (1 mg/mL) hyaluronan of 10,000, 100,000 and 1 million average molecular weight. These molecules at 1 mg/mL concentration have very low viscosity and no elasticity at low frequencies. There were no significant differences in the appearance of the healing wounds, but there was a slight but significant increase in the rate of wound closure and in the tensile strength compared to untreated controls after treatment with medium and high average molecular weight hyaluronan. The treatment with the two high molecular weight hyaluronan solutions resulted in a slight but significant increase in the rate of wound closure and a similar decrease of the tensile strength of the closed wounds (142). The combination of the variability of the wounds investigated, and diversity of the hyaluronan solutions used made it very difficult to evaluate the efficacy of the hyaluronan preparations. It is even more difficult to define an encompassing theory for the mode of action. Since tissue culture conditions are not closely related to *in vivo* wound healing models, to draw conclusions from one and apply to the other is difficult. What the exact role of hyaluronan in the healing process is at this state is not clear, but many hypotheses have been presented based on speculations on the observations in cell culture studies extrapolated to *in vivo* events. Most of these hypotheses are related to the effect of hyaluronan on phagocytosis, cell migration, cell proliferation, detachment and mitosis. Synthesis or inhibited catabolism of hyaluronan and consequent accumulation in the matrix were also studied. It was speculated that this accumulation created space between the newly formed collagen fibers and facilitated the traffic of cells and their commitment to rebuild the tissues.

In the final phase (re-epithelialization and re-endothelialization) hyaluronan was implicated in the regulation of keratinocyte migration, proliferation and therefore in the timely closure of the wound (for review, see Refs. 143 and 144). From the point of view of the therapeutic role of hyaluronan in wound healing, the protective and filter effect of the crowded hyaluronan molecular networks seems to be the cardinal issue. It is well established that high concentration ($\geq 1\%$), high molecular weight (≥ 2 million) hyaluronan solutions form an elastoviscous barrier when applied between wound surfaces before closure or on the surface of open wounds. This barrier remains a barrier in closed wounds as long as the molecules are not washed away by bleeding or excessive exudation. The residence time of hyaluronan gels or pastes is much long lasting, and therefore, they are more effective barriers between tissue surfaces than fluids. On the surface of open wounds, hyaluronan solutions and gels lose water by evaporation, but even when they seem completely dry at body temperature, they will still contain as much as 15–30% water. This means they can control water flow, diffusion and evaporation on the surface of the wound. More importantly, this protective hydrated molecular network forms a barrier against the penetration of microorganisms and, in the opposite direction, prevents the loss of macromolecules or even smaller signal molecules by its specific molecular filter effects. The molecular barrier qualities of concentrated hyaluronan solutions ($\geq 1\%$) or hyaluronan gels ($\geq 0.5\%$) are regarded as the primary functions of this molecule. Naturally specific interactions with signal molecules or cell surface proteins can greatly alter, for better or worse, their therapeutic usefulness.

Hyaluronan preparations have been tested in clinical studies to promote the healing of venous leg ulcers (145) and in the management of chronic wounds (146).

Angiogenesis is important in various wound healing processes. Hyaluronan solutions that contain molecules greater than 100,000 average molecular weight were found to inhibit angiogenesis in various *in vitro* and *in vivo* models. On the other hand, enzymatic degradation products of hyaluronan (so-called hyaluronan fragments) showed promotion of angiogenesis. This observation triggered the concept of the so-called 'angiogenic metastatic switch', which could play a role in the metastatic spread of certain tumors (for review, see Ref. 147).

These concepts may eventually lead to therapeutic exploitation of certain hyaluronan products, and even more likely, its derivatives. What seems to be important in the therapeutic applications of hyaluronan in the broadest sense for the regulation of the healing of internal wounds, including those created by the spread of tumors, is that the therapeutic modality must involve the widest distribution of molecular sizes (from oligosaccharides to 6 million average molecular weight). Consequently its putative regulatory function will be based on its physicochemical function, polyanionic character, specific binding characteristics to proteins, as well as the signaling and signal neutralizing (detoxicating) effects.

For more details on the various hypothetical relationships between observations of hyaluronan effects on wound healing and the effect of this molecule on cell function in cultured cells as well as the current clinical use of hyaluronan for wound healing, we recommend two recent reviews on this subject (144,148).

C. Other Applications

The effect of highly elastoviscous hyaluronan solution (Healon®) on corneal wound healing was studied in the 1980s (127,129,149,150). Later studies extended the scope of the use of hyaluronan of various elastoviscosities for the treatment of alkali burns, keratitis sicca, and for promotion of re-epithelialization (151–155). Hyaluronan applied topically on the wounded surface of the cornea may act as a barrier protector agent against 'drying in' and infection, and may also reduce inflammatory reactions by suppressing the infiltration of granulocytes. This protective pharmacological effect of hyaluronan can influence stromal healing and re-epithelialization.

Preclinical studies as well as clinical investigations demonstrated that hyaluronan solutions of various elastoviscosities promote the healing of perforated tympanic membranes (156–162). Films of hyaluronan esters (HYAFF®, Fidia, Inc., Italy) were also successfully used in chronic ear surgery as well as in endoscopic sinus surgery for chronic sinusitis, re-epithelialization of the mastoid cavity and for prevention of adhesion in these tissues. The rationale of using these hyaluronan derivatives for promotion of wound healing in otorhinolaryngolic procedures is based on a combination of the function of the elastoviscous hyaluronan or its gels: one to provide rapid re-epithelialization of the tympanic membrane and second to prevent the adhesions by its barrier function between two mucous tissues which have not yet re-epithelialized.

Hyaluronan was used in the peritoneal cavity to promote healing and decrease inflammation (163). Hyaluronan was found in the peritoneal fluid in somewhat higher concentration than in the blood (164). Later it was found that peritoneal cells (mesothelial, fibrocytes and macrophages) were responsible for producing hyaluronan in the peritoneal space and that during peritoneal inflammation, the hyaluronan content increases (165–169). It was hypothesized (163) that chronic peritoneal dialysis washes out the natural hyaluronan content of the peritoneal cavity which adversely affects the healing of peritoneal wounds and the natural anti-inflammatory process.

VIII. Drug Delivery with Hyaluronan

Hyaluronan gel for delivery of drugs was first used in the early 1980s after hylan B was developed. This extremely elastic gel with very high water content (≥99.5%) was well suited for delivering various pharmaceutical agents, contained in the water phase of the gel, through a very narrow lumen catheter. Hylan B gel was used for percutaneous embolization of aneurysms, vascular tumors, hemorrhages and arteriovenous malformations. Tantalum was incorporated into the gel to render it radiopaque. Microcrystalline cellulose and hexamethonium chloride were mixed with the gel, and thrombin was added to cause coagulation of the blood at the site of delivery. The gel, delivered intravascularly through small lumen catheters (monitored by X-ray), caused local, rapid arterial coagulation of the blood, and the embolous formed a permanent blockage of blood flow by intravascular fibrous tissue formation (in rabbits and rats) (170–172). Later clinical trials showed the therapeutic utility of this device. Hylan gel loaded with anti-bacterial drugs like gentamicin or others like pilocarpine, betaxolol or seratonin were developed and proven to be effective for controlled and enhanced delivery of these agents. In all cases the drugs, imbibed into the water phase of the gel, dramatically extended the release rate (173–175). Most importantly, since the hylan B gel is blood and tissue compatible, adverse reactions (inflammation, granulation) were not observed at the site of injection (for review of the early studies, see Ref. 173).

Sustained release of human growth factor with hyaluronan was reported (176). Various average molecular weight hyaluronans were tested as delivery vehicles *in vivo* for recombinant human granulocytes-stimulating factor (rHG-CSF). The mixture of the drug with hyaluronan solutions of various viscosities sustained its activity in subcutaneous application. The longest sustained release (5 days) was achieved at the highest concentration (4%) of the largest average molecular weight hyaluronan solutions (4×10^6) (175).

These studies before the early 1980s opened the field for the use of hyaluronan and its derivatives (fluids and solids) for drug delivery. We will not review here the further rapid development of the application of hyaluronan, but direct the reader to some selected recent references and reviews (177–180).

IX. Hyaluronan in Matrix and Tissue Engineering

Tissue engineering with hyaluronan addresses the fundamental issue of how the various intercellular macromolecules, primarily collagens, proteoglycans and hyaluronan affect the assembly of cells by providing intercellular scaffolding. The question is how hyaluronan or its derivatives can allow, or in combination with other structural matrix molecules, can help organize cells *in vitro* to form tissues or help to integrate such cell assemblies into living organs. The role of hyaluronan can be structural or organizational (informational) (181). In the 1970s we called this ‘matrix engineering’ and focused on the role of hyaluronan in the healing process of internal wounds such as chronic intra-articular inflammatory (arthritis) or degenerative (arthrosis) processes (79). During the past decade the signaling and information-transmitting role of hyaluronan brought a new dimension to this field. *In vitro* and animal experiments explored the use of various hyaluronan derivatives and combinations with other substances as scaffolding to reconstruct or graft tissues (cartilage, bone, bone marrow, dermal, epidermal, and adipose tissue) (182–186). Some of these studies have been extended to clinical trials in patients with skin ulcers and severe burns (185,186).

Hyaluronan gels (hylan B) have also been used to develop bioengineered artificial heart valves. In this development the goal is to create elastin-rich tissue *in vitro*. It was found that neonatal aortic smooth muscle cells seeded on hylan B gel formed elastin-rich sheets (187,188).

Tissue engineering with hyaluronan is still in the early development stage. The full utilization of this molecule for matrix and tissue engineering will be achieved when one can integrate the mechanical and chemical (pharmacological) role of this molecule in the maintenance of the healthy tissues and then exploit this knowledge to provide and direct regenerative processes.

Acknowledgements

The author wishes to express his gratitude for the excellent work of Joanne Caha in preparing the manuscript and the help of Janet Denlinger for the critical review and editing.

References

1. Balazs EA. The physical properties of synovial fluid and the special role of hyaluronic acid. In: Helfet A, ed. Disorders of the Knee, 2nd ed. Philadelphia: J.B. Lippincott, 1982:61–74.
2. Balazs EA, Gibbs DA. The rheological properties and biological function of hyaluronic acid. In: Balazs EA, ed. Chemistry and Molecular Biology of the Intercellular Matrix. London: Academic Press, 1970; vol. 3:1241–1253.
3. Band PA. Molecular strategies for the therapeutic utilization of hyaluronan. In: Kennedy JF, Phillips GO, Williams PA, Hascall VC, eds. Hyaluronan. Cambridge, UK: Woodhead Publishing, 2002; vol. 2:427–440.
4. Wik HB, Wik O. Rheology of hyaluronan. In: Laurent TC, ed. The Chemistry, Biology and Medical Applications of Hyaluronan and Its Derivatives, Proceedings of the Wenner-Gren Foundation International Symposium Held in Honor of Endre A Balazs, Stockholm, Sweden, September 18–21, 1996. London: Portland Press, 1998:25–32.
5. Balazs EA. Analgesic effect of elastoviscous hyaluronan solutions and the treatment of arthritic pain. Cells Tissues Organs 2003; 174:49–62.
6. Gomis A, Pawlak M, Balazs EA, Schmidt RF, Belmonte C. Effects of different molecular weight elastoviscous hyaluronan solutions on articular nociceptive afferents. Arthritis Rheum 2004; 50:314–326.
7. Balazs EA, Sweeney DB. The use of hyaluronic acid and collagen preparations in eye surgery. In: Schepens CL, Regan CDJ, eds. Controversial Aspects of the Management of Retinal Detachment. Boston: Little, Brown and Company, 1965; vol. 3:200–202.
8. Balazs EA, Sweeney DB. Replacement of the vitreous body of monkeys by reconstituted vitreous and hyaluronic acid. In: Streiff EB, ed. Modern Problems in Ophthalmology (Surgery of Retinal Vascular Diseases, Amersfoort, 1963). Basel: S. Karger, 1966; vol. 4:230–232.

9. Meyer-Schwickerath G. Further experience with Healon® (hyaluronic acid) in retinal detachment surgery. Mod Probl Ophthal 1974; 12:384.
10. Klöti R. Hyaluronsäure als glaskörpersubstituent. Schweiz Ophthal Ges 1972:165.
11. Regnault F. Acide hyaluronique intravitéen et cryocoagulatin dans le traitement des formes graves de décollement de la rétine. Bull Soc Ophthalmol Fr 1971; 84: 106–112.
12. Algvere P. Intravitreal implantation of a high-molecular hyaluronic acid in surgery for retinal detachment. Acta Ophthalmol 1971; 49:975–976.
13. Balazs EA, Freeman MI, Klöti R, Meyer-Schwickerath G, Regnault F, Sweeney DB. Hyaluronic acid and the replacement of vitreous and aqueous humor. In: Streiff EB, ed. Modern Problems in Ophthalmology (Secondary Detachment of the Retina, Lausanne, 1970). Basel: S. Karger, 1972; vol. 10:3–21.
14. Edmund J. Comments on the clinical use of Healon and a short survey of the use of intraocular injection of hyaluronic acid. In: Irvine AR, O'Malley C, eds. Advances in Vitreous Surgery. Springfield, IL: C. Thomas, 1976:624–625.
15. Balazs EA, Miller D, Stegmann R. Viscosurgery and the Use of Na-Hyaluronate in Intraocular Lens Implantation, International Congress and First Film Festival on Intraocular Implantation, Cannes, France. 1979.
16. Graue EL, Polack FM, Balazs EA. The protective effect of Na-hyaluronate to corneal endothelium. Exp Eye Res 1980; 31:119–127.
17. Pape L, Balazs EA. The use of sodium hyaluronate (Healon®) in human anterior segment surgery. Ophthalmology 1980; 87:699–705.
18. Hoopes PC. Sodium hyaluronate (Healon®) in anterior segment surgery: a review and a new use in extracapsular surgery. Am Intraocular Implant Soc J 1982; 8: 148–154.
19. Meyer-Schwickerath G, ed. Viskochirurgie des Auges: Beiträge des Ersten Nationalen Healon®—Symposiums, October 15 and 16, 1982. 1982.
20. Miller D, Stegmann R, eds. Healon (Sodium Hyaluronate). A Guide to Its Use in Ophthalmic Surgery. New York, NY: Wiley, 1983:234.
21. Eisner G, ed. Ophthalmic Viscosurgery. A Review of Standards, Techniques and Applications. Montreal, Canada: Medicopea, 1986:103.
22. Stenkula S, Ivert L, Gislason I, Törnquist R, Weijdegard L. The use of sodium hyaluronate (Healon®) in treatment of retinal detachment. Ophthal Surg 1981:12.
23. Stenkula S, Törnquist R. Use of Healon in vitrectomy and difficult retinal detachments. In: Miller D, Stegmann R, eds. Healon (Sodium Hyaluronate): A Guide to Its Use in Ophthalmic Surgery. New York: Wiley, 1983:207–221.
24. Miller D, Stegmann R. Use of Na-hyaluronate in corneal transplantation. J Ocular Ther Surg 1981; 1:28.
25. Arshinoff SA. The use of ophthalmic viscosurgical devices in cataract surgery. In: Kennedy JF, Phillips GO, Williams PA, Hascall VC, eds. Hyaluronan. Cambridge, UK: Woodhead Publishing, 2002; vol. 2:119–128.
26. Denlinger JL. Hyaluronan and its derivatives as viscoelastics in medicine. In: Laurent TC, ed. The Chemistry, Biology and Medical Applications of Hyaluronan and Its Derivatives, Proceedings of the Wenner-Gren Foundation International Symposium Held in Honor of Endre A Balazs, Stockholm, Sweden, September 18–21, 1996. London: Portland Press, 1998:235–242.
27. Balazs EA. Sodium hyaluronate and viscosurgery. In: Miller D, Stegmann R, eds. Healon: A Guide to Its Use in Ophthalmic Surgery. New York: Wiley, 1983: 5–28.

28. Balazs EA. The development of sodium hyaluronate (Healon®) as a viscosurgical material in ophthalmic surgery. In: Eisner G, ed. Ophthalmic Viscosurgery—A Review of Standards, Techniques and Applications. Bern: Medicopea, 1986: 3–19.
29. Arshinoff SA. Dispersive and cohesive viscoelastic materials in phacoemulsification. Ophthal Prac 1995; 13:98–104.
30. Arshinoff SA. Dispersive–cohesive viscoelastic soft shell technique. J Cataract Refract Surg 1999; 25:167–173.
31. Balazs EA. Viscosurgery. Transpl/Implant Today 1985; 2:62–64.
32. Weiss C, Balazs EA. Arthroscopic viscosurgery. Arthroscopy 1987; 3:138–139.
33. McCain JP, Balazs EA, de la Rua H. Preliminary studies on the use of a viscoelastic solution in arthroscopic surgery of the temporomandibular joint. J Oral Maxillofac Surg 1989; 47:1161–1168.
34. Weiss C. Role of viscosurgery in arthroscopy. In: Mohan T, Bronstein S, eds. Arthroscopy of the Temporomandibular Joint. Philadelphia, PA: W.B. Saunders, 1991:335–337.
35. Larsen NE, Lombard KM, Balazs EA. The Effect of Hyaluronan on Cartilage and Chondrocyte Response to Mechanical and Biochemical Perturbation, Orthopaedic Research Society, 35th Annual Meeting, Las Vegas, NV, February 6–9, 1989. 1989; 151.
36. Laurent TC, Hellsing K, Gelotte B. Cross-linked gels of hyaluronic acid. Acta Chem Scand 1964; 18:274–275.
37. Balazs EA, Leshchiner EA, Leshchiner A, Band P. Chemically modified hyaluronic acid preparation and method of recovery thereof from animal tissues. US patent 4,713,448. Biomatrix, Inc., Ridgefield, NJ, 1985:1–18.
38. Balazs EA, Leshchiner EA, Leshchiner A, Larsen NE, Band PA. Hylan preparation and method of recovery thereof from animal tissues. US patent 5, 099,013. Biomatrix, Inc., Ridgefield, NJ, 1990.
39. Balazs EA, Leshchiner A. Cross-linked gels of hyaluronic acid and products containing such gels. US patent 4,582,865. Biomatrix, Inc., Ridgefield, NJ, 1984.
40. Balazs EA, Leshchiner EA. Hyaluronan, its crosslinked derivative—hylan—and their medical applications. In: Inagaki H, Phillips GO, eds. Cellulosics Utilization: Research and Rewards in Cellulosics, Proceedings of Nisshinbo International Conference on Cellulosics Utilization in the Near Future. New York: Elsevier, 1989:233–241.
41. Balazs EA, Denlinger JL, Leshchiner E, Band P, Larsen N, Leshchiner A, Morales B. Hylan: Hyaluronan Derivatives for Soft Tissue Repair and Augmentation, Biotech USA Proceedings of the Fifth International Congress on Biotechnology, San Francisco, CA, November 14–16, 1988. 1988; 442–451.
42. Larsen NE, Kling MB, Balazs EA, Leshchiner EA. Hylan Gel for Soft Tissue Augmentation, Society for Biomaterials 16th Annual Meeting, Transactions of Society of Biomaterials XIII, Charleston, SC, May 20–23, 1990. 1990; 302.
43. Larsen N, Pollak C, Reiner K, Leshchiner E, Balazs EA. Hylan gel biomaterial: dermal and immunologic compatibility. J Biomed Mater Res 1993; 27:1129–1134.
44. Biomatrix, Viscoaugmentation: a historical perspective. In: Kennedy JF, Phillips GO, Williams PA, Hascall VC, eds. Hyaluronan. Cambridge, UK: Woodhead Publishing, 2002; vol. 2:41–44.
45. Larsen NE, Leshchiner E, Pollak CT, Balazs EA, Piacquadio D. Evaluation of hylan B (hylan gel) as soft tissue dermal implants. In: Mikos AG, Leong KW,

Radomsky ML, Tamada JA, Yaszemski MJ, eds. Polymers in Medicine and Pharmacy, Materials Research Society, Spring Meeting, San Francisco, CA, April 17–21, 1995. Pittsburgh, PA: Materials Research Society, 1995:193–197.
46. Leshchiner E, Balazs EA, Larsen NE, Leshchiner A. Biocompatible viscoelastic gel slurries, their preparation and use. US patent 5,399,351, 1993.
47. Piacquadio D, Jarcho M, Golz R. Evaluation of hylan b gel as a soft-tissue augmentation implant material. J Am Acad Dermatol 1997; 36:544–549.
48. Balazs EA, Band PA, Denlinger JL, Goldman AI, Larsen NE, Leshchiner EA, Leshchiner A, Morales B. Matrix engineering. Blood Coagul Fibrinolysis 1991; 2:173–178.
49. Larsen NE, Leshchiner EA, Parent EG, Hendrikson-Aho J, Balazs EA, Hilal SK. Hylan gel composition for percutaneous embolization. J Biomed Mater Res 1991; 25:699–710.
50. Ågerup B. Polysaccharide gel composition. US patent 5,827,937. Q Med AB, Uppsala, Sweden, 1995.
51. Malson T, Lindqvist BL. Gel of cross-linked hyaluronic acid for use as a vitreous humor substitute. US patent 4,716,154, 1987.
52. Sakurai K, Ueno Y, Okuyama T. Crosslinked hyaluronic acid and its use. US patent 4,716,224. Seikagaku Kogyo Co. Ltd., Tokyo, Japan, 1987.
53. Malson T, Lindqvist B. Cross-linked hyaluronate gels, their use and method for producing them. WO #90/09401, 1990.
54. Olenius M. The first clinical study using a new biodegradable implant for the treatment of lips, wrinkles, and folds. Aesthetic Plast Surg 1998; 22: 97–101.
55. Wik O, Agerup B, Wik HB. Rheology of hyaluronan products. In: Kennedy JF, Phillips GO, Williams PA, Hascall VC, eds. Hyaluronan. Cambridge, UK: Woodhead Publishing, 2002; vol. 1:201–204.
56. Intergel. FDA premarket approval letter P990015 for gynecare intergel, November 16, 2001. Lifecore Biomedical I, ed., Chaska, MN, 2001.
57. Crescenzi V, Francescangeli A, Renier D, Bellini D. Hyaluronan linear and crosslinked derivatives as potential/actual biomaterials. In: Kennedy JF, Phillips GO, Williams PA, Hascall VC, eds. Hyaluronan. Cambridge, UK: Woodhead Publishing, 2002; vol. 1:261–268.
58. Michielin L, Bevilacqua C, Paoletti S, Gamini A, Toffanin R, Micali F. Novel biomaterials based on cross-linked hyaluronan structural investigations. In: Kennedy JF, Phillips GO, Williams PA, Hascall VC, eds. Hyaluronan. Cambridge, UK: Woodhead Publishing, 2002; vol. 1:269–276.
59. Zhao X, Fraser J, Alexander C. A novel crosslinking process for hyaluronan. In: Kennedy JF, Phillips GO, Williams PA, Hascall VC, eds. Hyaluronan. Cambridge, UK: Woodhead Publishing, 2002; vol. 1:277–284.
60. Barbucci R, Pasqui D, Leone G. Derivatized hyaluronan for gels and nanochemically patterned surfaces. In: Kennedy JF, Phillips GO, Williams PA, Hascall VC, eds. Hyaluronan. Cambridge, UK: Woodhead Publishing, 2002; vol. 1:293–304.
61. Okamoto A, Miyoshi T. A biocompatible gel of hyaluronan. In: Kennedy JF, Phillips GO, Williams PA, Hascall VC, eds. Hyaluronan. Cambridge, UK: Woodhead Publishing, 2002; vol. 1:285–292.
62. Piacquadio DJ. Crosslinked hyaluronic acid (hylan gel) as a soft tissue augmentation material: a preliminary assessment. In: Elson ML, ed. Evaluation and Treatment of the Aging Face. New York: Springer, 1994:304–308.

63. Frank P, Gendler E. Hyaluronic acid for soft-tissue augmentation. Clin Plast Surg 2001; 28:121–126.
64. Duranti F, Salti G, Bovani B, Calandra M, Rosati ML. Injectable hyaluronic acid gel for soft tissue augmentation: a clinical and histological study. Dermatol Surg 1998; 24:1317–1325.
65. Narins R, Brandt F, Leyden J, Lorenc ZP, Rubin M, Smith S. A randomized, double-blind, multicenter comparison of the efficacy and tolerability of Restylane versus Zyplast for the corectin of nasolabial folds. Dermatol Surg 2003; 29:588–595.
66. Manna F, Dentini M, Desideri P, De Pità O, Mortilla E, Maras B. Comparative chemical evaluation of two commercially available derivatives of hyaluronic acid (Hylaform® from rooster combs and Restylane® from Streptococcus) used for soft tissue augmentation. J Eur Acad Dermatol Venereol 1999; 13:183–192.
67. Micheels P. Human anti-hyaluronic acid antibodies: is it possible? Dermatol Surg 2001; 27:185–191.
68. Lowe NJ, Maxwell CA, Lowe PD, Duick MG, Shaw K. Hyaluronic acid skin fillers: adverse reactions and skin testing. J Am Acad Dermatol 2001; 45:930–933.
69. Lupton JR, Alster TS. Cutaneous hypersensitivity reaction to injectable hyaluronic acid gel. Dermatol Surg 2000; 26:135–137.
70. Friedman PM, Mafong EA, Kauvar AN, Geronemus RG. Safety data of injectable nonanimal stabilized hyaluronic acid gel for soft tissue augmentation. Dermatol Surg 2002; 28:491–494.
71. Hallén L, Johannson C, Laurent C. Cross-linked hyaluronan (hylan B gel): a new injectable remedy for treatment of vocal fold insufficiency—an animal study. Acta Otolaryngol (Stockholm) 1999; 119:107–111.
72. Hallén L, Johansson C, Dahlqvist Å, Laurent C. The potential use of hyaluronan-based compounds in laryngeal augmentative surgery. In: Abatangelo G, Weigel P, eds. New Frontiers in Medical Sciences: Redefining Hyaluronan, Proceedings of the Symposium Held in Padua, Italy, June 17–19, 1999. Amsterdam: Elsevier, 2000:353–359.
73. Hallén L, Dahlqvist A. Cross-linked hyaluronan for augmentation of the posterior pharyngeal wall: an experimental study in rats. Scand J Plast Reconstr Surg Hand Surg 2002; 36:197–201.
74. Hertegård S, Dahlqvist Å, Laurent C, Borzacchiello A, Ambrosio L. Viscoelastic properties of rabbit vocal folds after augmentation. Otolaryngol Head Neck Surg 2003; 128:401–406.
75. Hallén L, Dahlqvist A, Laurent C. Dextraneomeres in hyaluronan (DiHA): a promising substance in treating vocal chord insufficiency. Laryngoscope 1998; 108:393–397.
76. Hertegård S, Hallén L, Laurent C, Lindström E, Olofsson K, Testad P, Dahlqvist Å. Cross-linked hyaluronan used as augmentation substance for treatment of glottal insufficiency: safety aspects and vocal fold function. Laryngoscope 2002; 112:2211–2219.
77. Stenberg A, Lackgren G. A new bioimplant for the endoscopic treatment of vesicoureteral reflux: experimental and short-term clinical results. J Urol 1995; 154:800–803.
78. Stenberg A, Larsson G, Heimer G, Johnson P, Ulmsten U. Transurethral endoscopic treatment of urinary stress incontinence in women: materials and results in former and present agents. Acta Obstet Gynecol Scand Suppl 1998; 168: 44–46.

79. Balazs EA. Hyaluronic Acid and Matrix Implantation, 2nd ed. Arlington, MA: Biotrics Inc., 1971.
80. Rydell N, Balazs EA. Effect of intra-articular injection of hyaluronic acid on the clinical symptoms of osteoarthritis and on granulation tissue formation. Clin Orthop 1971; 80:25–32.
81. Balazs EA, Darzynkiewicz Z. The effect of hyaluronic acid on fibroblasts, mononuclear phagocytes and lymphocytes. In: Kulonen E, Pikkarainen J, eds. Biology of the Fibroblast. London: Academic Press, 1973:237–252.
82. LeBoeuf RD, Gregg RR, Weigel PH, Fuller GM. Effects of hyaluronic acid and other glycosaminoglycans on fibrin polymer formation. Biochemistry 1987; 26: 6052–6057.
83. Weigel PH, Frost SJ, LeBoeuf RD, McGary CT. The specific interaction between fibrin(ogen) and hyaluronan: possible consequences in haemostasis, inflammation and wound healing. In: Evered D, Whelan J, eds. The Biology of Hyaluronan, Ciba Foundation Symposium #143. New York: Wiley, 1989:248–264.
84. St Onge R, Weiss C, Denlinger JL, Balazs EA. A preliminary assessment of Na-hyaluronate injection into "no man's land" for primary flexor tendon repair. Clin Orthop 1980; 146:269–275.
85. Larsen NE. Management of adhesion formation and soft tissue augmentation with viscoelastics: hyaluronan derivatives. In: Laurent TC, ed. The Chemistry, Biology and Medical Applications of Hyaluronan and Its Derivatives, Proceedings of the Wenner-Gren Foundation International Symposium Held in Honor of Endre A Balazs, Stockholm, Sweden, September 18–21, 1996. London: Portland Press, 1998:267–282.
86. Weiss C, Suros JM, Michalow A, Denlinger J, Moore M, Tejeiro W. The role of Na-hylan in reducing postsurgical tendon adhesions: part 2. Bull Hosp Joint Dis Orthop Inst 1987; 47:31–39.
87. Weiss C, Levy HJ, Denlinger J, Suros J, Weiss H. The role of Na-hylan in reducing postsurgical tendon adhesions. Bull Hosp Joint Dis Orthop Inst 1986; 46:9–15.
88. Weiss C, Dennis J, Suros JM, Denlinger J, Badia A, Montane I. Sodium Hylan for the Prevention of Postlaminectomy Scar Formation, Orthopaedic Research Society, 35th Annual Meeting, Las Vegas, NV. 1989; 44.
89. Harris ES, Foresman PA, Rodeheaver GT, Larsen NE, Balazs EA. Efficacy of a Resorbable Hylan Barrier Membrane in the Prevention of Adhesions in a Rat Cecal Abrasion Model, Fifth World Biomaterials Congress, Toronto, Canada, May 29–June 2, 1996. 1996; 371.
90. Larsen NE, Leshchiner E, Balazs EA, Belmonte C. Biocompatibility of hylan polymers in various tissue compartments. In: Mikos AG, Leong KW, Radomsky ML, Tamada JA, Yaszemski MJ, eds. Polymers in Medicine and Pharmacy, Materials Research Society Spring Meeting, San Francisco, CA, April 17–21, 1995. Pittsburgh, PA: Materials Research Society, 1995:149–153.
91. Burns JW, Colt MJ, Burgess LS, Skinner KC. Preclinical evaluation of Seprafilm™ bioresorbable membrane. Eur J Surg 1997; 163:40–48.
92. Burns JW, Skinner K, Colt J, Sheidlin A, Bronson R, Yaacobi Y, Goldberg EP. Prevention of tissue injury and postsurgical adhesions by precoating tissues with hyaluronic acid solutions. J Surg Res 1995; 59:644–652.
93. Diamond M. Reduction of postsurgical adhesions by intraoperative precoating with sepracoat (HAL-C) solution: a prospective, randomized, blinded, placebo-controlled multicenter study. Fertil Steril 1998; 69:1067–1073.

94. Gibbs D, Merrill E, Smith K, Balazs EA. Rheology of hyaluronic acid. Biopolymers 1968; 6:777–791.
95. Balazs EA. Viscoelastic properties of hyaluronic acid and biological lubrication. (Symposium: prognosis for arthritis: rheumatology research today and prospects for tomorrow, Ann Arbor, Michigan, 1967). Univ Mich Med Ctr J 1968; 9(Suppl): 255–259.
96. Balazs EA. Some Aspects of the Aging and Radiation Sensitivity of the Intercellular Matrix with Special Regard to Hyaluronic Acid in Synovial Fluid and Vitreous, Thule International Symposium: Aging of Connective and Skeletal Tissue Symposium, October 1–3, 1968. Thule, Sweden: Nordiska Bokhandelns Förlag, Stockholm, 1969.
97. Balazs EA, Denlinger JL. Sodium hyaluronate and joint function. Equine Vet Sci 1985; 5:217–228.
98. Balazs EA, Denlinger JL. Viscosupplementation: a new concept in the treatment of osteoarthritis. J Rheumatol 1993; 20(Suppl 39):3–9.
99. de la Peña E, Sala S, Schmidt RF, Belmonte C. Effects of elastoviscous solutions of hyaluronan derivatives on mechanotransduction. In: Kennedy JF, Phillips GO, Williams PA, Hascall VC, eds. Hyaluronan. Cambridge, UK: Woodhead Publishing, 2002; vol. 2:407–418.
100. Balazs EA, Weiss C. Elastoviscous hyaluronan in the synovium in health and disease. In: Hascall VC, Kuettner KE, eds. The Many Faces of Osteoarthritis. Basel, Switzerland: Birkhäuser, 2002:189–206.
101. Miyazaki K, Gotoh S, Ohkawara H, Yamaguchi T. Studies on analgesic and anti-inflammatory effects of sodium hyaluronate (SPH). Pharmacometrics 1984; 28: 1123–1135.
102. Gotoh S, Miyazaki K, Onaya J, Sakamoto T, Tokuyasu K, Namiki O. Experimental knee pain model in rats and analgesic effect of sodium hyaluronate (SPH). Folia Pharmacol Jpn 1988; 92:17–27.
103. Denlinger JL, El-Mofty AA, Balazs EA. Replacement of the liquid vitreous with sodium hyaluronate in monkeys. II. Long-term evaluation. Exp Eye Res 1980; 31: 101–117.
104. Denlinger JL, Balazs EA. Replacement of the liquid vitreous with sodium hyaluronate in monkeys. I. Short-term evaluation. Exp Eye Res 1980; 31:81–99.
105. Rydell NW, Butler J, Balazs EA. Hyaluronic acid in synovial fluid. VI. Effect of intra-articular injection of hyaluronic acid on the clinical symptoms of arthritis in track horses. Acta Vet Scand 1970; 11:139–155.
106. Weiss C, Balazs EA, St Onge R, Denlinger JL. Clinical studies of the intraarticular injection of Healon (sodium hyaluronate) in the treatment of osteoarthritis of human knees. In: Talbott JH, ed. Seminars in Arthritis and Rheumatism. New York: Grune and Stratton, 1981; vol. 11:143–144.
107. Balazs EA. The physical properties of synovial fluid and the special role of hyaluronic acid. In: Helfet AJ, ed. Disorders of the Knee, 1st ed. Philadelphia: J.B. Lippincott, 1974:61–74.
108. Namiki O, Toyoshima H, Morisaki N. Therapeutic effect of intra-articular injection of high molecular weight hyaluronic acid on osteoarthritis of the knee. Int J Clin Pharmacol Ther Toxicol 1982; 20:501–507.
109. Grecomoro G, Martorana U, Di Marco C. Intra-articular treatment with sodium hyaluronate in gonarthrosis: a controlled clinical trial versus placebo. Pharmatherapeutica 1987; 5:137–141.

110. Lee HG, Cowman MK. An agarose gel electrophoretic method for analysis of hyaluronan molecular weight distribution. Anal Biochem 1994; 219:278–287.
111. Adam N, Ghosh P. Hyaluronan molecular weight and polydispersity in some commercial intra-articular injectable preparations and in synovial fluid. Inflamm Res 2001; 50:294–299.
112. Soltés L, Mendichi R, Lath D, Mach M, Bakos D. Molecular characteristics of some commercial high-molecular-weight hyaluronans. Biomed Chromatogr 2002; 16:459–462.
113. Weiss C, Band P. Basic principles underlying the development of viscosupplementation for the treatment of osteoarthritis. J Clin Rheum 1999; 5:S2–S11.
114. Backus C, Carrington SP, Fisher LR, Odell JA, Rodrigues DA. The roles of extensional and shear flows of synovial fluid and replacement systems in joint protection. In: Kennedy JF, Phillips GO, Williams PA, Hascall VC, eds. Hyaluronan. Cambridge, UK: Woodhead Publishing, 2002; vol. 1:209–218.
115. Peyron JG. Intra-articular hyaluronan injections in the treatment of osteoarthritis: state-of-art review. J Rheumatol 1993; 20:10–15.
116. Biomatrix, Viscosupplementation: a historical perspective. In: Kennedy JF, Phillips GO, Williams PA, Hascall VC, eds. Hyaluronan. Cambridge, UK: Woodhead Publishing, 2002; vol. 2:385–390.
117. Wobig M, Bach G, Beks P, Dickhut A, Runzheimer J, Schwieger G, Vetter G, Balazs E. The role of elastoviscosity in the efficacy of viscosupplementation for osteoarthritis of the knee: a comparison of hylan G-F 20 and a lower-molecular-weight hyaluronan. Clin Ther 1999; 21:1549–1562.
118. Lo GH, LaValley M, McAlindon T, Felson DT. Intra-articular hyaluronic acid in treatment of knee osteoarthritis. JAMA 2003; 290:3115–3121.
119. Weiss C, Waddell D, Miller E. The availability of highly elastoviscous hylan for viscosupplementation can delay knee replacement in patients with advanced osteoarthritis. In: Kennedy JF, Phillips GO, Williams PA, Hascall VC, eds. Hyaluronan. Cambridge, UK: Woodhead Publishing, 2002; vol. 2:391–396.
120. Tunis SR, Stryer DB, Clancy CM. Practical clinical trials: increasing the value of clinical research for decision making in clinical and health policy. JAMA 2003; 290:1624–1632.
121. Fritz JM, Cleland J. Effectiveness versus efficacy: more than a debate over language. J Orthop Sports Phys Ther 2003; 33:163–165.
122. Revicki DA, Frank L. Pharmacoeconomic evaluation in the real world. Pharmacoeconomics 1999; 15:423–434.
123. Raynauld JP, Torrance GW, Band PA, Goldsmith CH, Tugwell P, Walker V. A prospective, randomized, pragmatic, health outcomes trial evaluating the incorporation of hylan G-F 20 into the treatment paradigm for patients with knee osteoarthritis (part 1 of 2): clinical results. Osteoarthritis Cartilage 2002; 10: 506–517.
124. Torrance GW, Raynauld JP, Walker V, Goldsmith CH, Bellamy N, Band PA. A prospective, randomized, pragmatic, health outcomes trial evaluating the incorporation of hylan G-F 20 into the treatment paradigm for patients with knee osteoarthritis (part 2 of 2): economic results. Osteoarthritis Cartilage 2002; 10: 518–527.
125. Kahan A, Lieu PL, Salin L. Prospective randomized study comparing the medicoeconomic benefits of Hylan GF-20 vs. conventional treatment in knee osteoarthritis. Joint Bone Spine 2003; 70:276–281.

126. Biomatrix, Viscoprotection: a historical perspective. In: Kennedy JF, Phillips GO, Williams PA, Hascall VC, eds. Hyaluronan. Cambridge, UK: Woodhead Publishing, 2002; vol. 2:115–118.
127. Polack FM, McNiece MT. The treatment of dry eyes with Na hyaluronate (Healon®). Cornea 1982; 1:133–136.
128. DeLuise VP, Peterson WS. The use of topical Healon® tears in the management of refractory dry-eye syndrome. Ann Ophthalmol 1984; 16:823–824.
129. Stuart JC, Linn JG. Dilute sodium hyaluronate (Healon) in the treatment of ocular surface disorders. Ann Ophthalmol 1985; 17:190–192.
130. Hamano T, Horimoto K, Lee M, Komemushi S. Evaluation of the effect of sodium hyaluronate ophthalmic solution on corneal desiccation due to hard contact lens wear. J Eye 1993; 10:627–630.
131. Itoi M, Kim O, Kimura TKA, Momose T, Kanki K, Yamaguchi T, Ueno Y, Kurokawa M, Komemushi S. The effect of sodium hyaluronate ophthalmic solution on corneal epithelial disorders in contact lens wearers. J Eye 1993; 10:617–626.
132. Arshinoff SA, Fichman S, Laflamme M, Rosen J, Holland S, Hofmann I. Comparative, Double-Masked, Crossover Study Evaluating the Effectiveness of Hylashield® Versus Placebo Eye Drops in Keratoconjunctivitis Sicca, ARVO, Fort Lauderdale, FL, May 1999.
133. Larsen NE, Balazs EA. Hylashield (2.0 Pa elastoviscous hylan fluid 0.15%) protective corneal shield: evaluation of biological and physical properties. Ophthal Pract 1994; 12:137–140.
134. Arshinoff SA, Khoury E. HsS versus a balanced salt solution as a corneal wetting agent during routine cataract extraction and lens implantation. J Cataract Refract Surg 1997; 23:1221–1225.
135. Acosta MC, Gallar J, Belmonte C. The influence of eye solutions on blinking and ocular comfort at rest and during work at video display terminals. Exp Eye Res 1999; 68:663–669.
136. Balazs EA, Holmgren HJ. The basic dye-uptake and the presence of a growth-inhibiting substance in the healing tissue of skin wounds. Exp Cell Res 1950; 1:206–216.
137. Holmgren HJ, Balazs EA. Experimental studies on wound healing. Nord Med 1950; 43:471–474.
138. Oksala O, Salo T, Tammi R, Hakkinen L, Jalkanen M, Inki P, Larjava H. Expression of proteoglycans and hyaluronan during wound healing. J Histol Cytochem 1995; 43:125–135.
139. Weigel PH, Frost SJ, McGary CT, LeBoeuf RD. The role of hyaluronoic acid in inflammation and wound healing. Int J Tissue React 1988; 10:355–365.
140. Gutierrez-Miguelez N, Cuenca N, de Juan J, Belmonte C. El ácido hialuronico no constituye un spoorte adecuado para la regeneración de fibras nerviosas periféricas, Trabajos del Instituto Cajal, Proceedings of Second Congress of Sociedad Española de Neurociencia, Barcelona, Spain. 1987; vol. 86:141.
141. Balazs EA, Denlinger JL. Clinical uses of hyaluronan. In: Evered D, Whelan J, eds. The Biology of Hyaluronan, CIBA Foundation Symposium #143. Chichester, Sussex, UK: Wiley, 1989:265–280.
142. Arnold F, Jia C, He C, Cherry GW, Carbow B, Meyer-Ingold W, Bader D, West DC. Hyaluronan, heterogeneity, and healing: the effects of ultrapure hyaluronan of defined molecular size on the repair of full-thickness pig skin wounds. Wound Repair Regen 1995; 3:299–310.

143. Moore K. Chronic wound healing and inflammation. In: Kennedy JF, Phillips GO, Williams PA, Hascall VC, eds. Hyaluronan. Cambridge, UK: Woodhead Publishing, 2002; vol. 2:137–146.
144. Chen WYJ, Abatangelo G. Functions of hyaluronan in wound repair. Wound Repair Regen 1999; 7:79–89.
145. Ortonne JP. A controlled study of the activity of hyaluronic acid in the treatment of venous leg ulcers. J Dermatolog Treat 1996; 7:75–81.
146. Edmonds M, Foster A. Hyalofill: a new product for chronic wound management. Diabet Foot 2000; 3:29–30.
147. West DC, Chen H. Is hyaluronan degradation an angiogenic/metastatic switch? In: Kennedy JF, Phillips GO, Williams PA, Hascall VC, eds. Hyaluronan. Cambridge, UK: Woodhead Publishing, 2002; vol. 2:165–172.
148. Chen WYJ. Functions of hyaluronan in wound repair. In: Kennedy JF, Phillips GO, Williams PA, Hascall VC, eds. Hyaluronan. Cambridge, UK: Woodhead Publishing, 2002; vol. 2:147–156.
149. Saric D, Reim M. Behandlung von veratzungen des vorderen augenabschnitts mit hochpolymerem Na-Hyaluronat (Healon). Fortschr Ophthalmol 1984; 81: 588–591.
150. Reim M, Lenz V. Behandliung von schweren veratzungen mit hochpolymerer hyaluronsaure (Healon). Fortschr Ophthalmol 1984; 81:323–325.
151. Chung J-H. Therapeutic effect of 1% Na hyaluronan on corneal wound healing. In: Kennedy JF, Phillips GO, Williams PA, Hascall VC, eds. Hyaluronan. Cambridge, UK: Woodhead Publishing, 2002; vol. 2:129–136.
152. Chung J-H, Park Y-K, Paek S-M, Chong Y-H, Kim W-K. Effect of Na-hyaluronan on stromal and endothelial healing in experimental corneal alkali wounds. Ophthal Res 1999; 31:432–439.
153. Chung J-H, Kim W-K, Lee J-S, Pae Y-S, Kim H-J. Effect of topical Na-hyaluronan on hemidesmosome formation in *n*-heptanol-induced corneal injury. Ophthal Res 1998; 30:96–100.
154. Chung J-H. Experimental corneal alkali wound healing. Acta Ophhthalmol 1988; 66(Suppl 187):1–35.
155. Inoue M, Katakami C. The effect of hyaluronic acid on corneal epithelial cell proliferation. Investig Ophthalmol Vis Sci 1993; 34:2313–2315.
156. Laurent C, Söderberg O, Anniko M, Hartwig S. Repair of chronic tympanic membrane perforations using applications of hyaluronan or rice paper prostheses. ORL J Otorhinolaryngol Relat Spec 1991; 53:37–40.
157. Hellström S, Bloom GD, Berghem L, Stenfors L-E, Söderberg M. A comparison of hyaluronan and fibronectin in the healing of tympanic membrane perforations. Eur Arch Otorhinolaryngol 1991; 248:230–235.
158. Stenfors LE. Repair of tympanic membrane perforations with HA. An alternative to myringoplasty. J Laryngol Otol 1989; 103:39–40.
159. Bagger-Sjöbäck D, Holmquist J, Mendel L, Mercke U. Hyaluronic acid in middle ear surgery. Am J Otol 1993; 14:501–506.
160. Stenfors LE. Repair of traumatically ruptured tympanic membrane using HA. Acta Otolaryngol 1987; 442(Suppl):88–91.
161. Laurent C, Hellström S, Fellenius E. Hyaluronan improves the healing of experimental tympanic membrane perforations. Arch Otolaryngol Head Neck Surg 1988; 114:1435–1441.

162. Lacarte MPR, Casasin T, Pumarola F, Alonso A. An alternative treatment for the reductin of tympanic membrane performations: sodium hyaluronate. Acta Otolaryngol 1990; 110:110–114.
163. Breborowicz A, Moberly JB, Pawlaczyk K, Polublinska A, Kuzlan-Pawlaczyk M, Wiecorowska-Tobis K, Ogle K, Martis L, Oreopoulos DG. Effects of hyaluronan used as a supplement in peritoneal dialysis solutions. In: Kennedy JF, Phillips GO, Williams PA, Hascall VC, eds. Hyaluronan. Cambridge, UK: Woodhead Publishing, 2002; vol. 2:453–460.
164. Lipkin GW, Forbes MA, Cooper EH, Turney JH. Hyaluronic acid metabolism and its clinical significance in patients treated by continuous ambulatory peritoneal dialysis. Nephrol Dial Transplant 1993; 8:357–360.
165. Yung S, Coles GA, Williams JD, Davies M. The source and possible significance of hyaluronan in the peritoneal cavity. Kidney Int 1994; 46:527–533.
166. Lai KN, Szeto CC, Lai KB, Lam CWK, Chan DTM, Leung JCK. Increased production of hyaluronan by peritoneal cells and its significance in patients on CAPD. Am J Kidney Dis 1999; 33:318–324.
167. Edelstam GAB, Laurent UBG, Lundkvist OE, Fraser JRE, Laurent TC. Concentration and turnover of intraperitoneal hyaluronan during inflammation. Inflammation 1992; 16:459–469.
168. Yamagata K, Tomida C, Koyama A. Intraperitoneal hyaluronan production in stable continuous ambulatory peritoneal dialysis patients. Perit Dial Int 1999; 19:131–137.
169. Breborowicz A, Wisniewska J, Polubinska AA, Wieczorowska-Tobis K, Martis L, Oreopoulos DG. Role of peritoneal mesothelial cells and fibroblasts in synthesis of hyaluronan during peritoneal dialysis. Perit Dial Int 1998; 18: 382–386.
170. Hilal SK, Leshchiner EA, Larsen NE, Khandji AG, Moser FG, Balazs EA. A Novel Viscoelastic Composition for Percutaneous Embolization, Proceedings of 25th Annual Meeting of American Society of Neuroradiology, New York, NY, May 10–15, 1987. 1987; 80.
171. Larsen NE, Leshchiner EA, Parent EG, Hendrickson-Aho J, Balazs EA. Local Intravascular Drug Delivery Via Percutaneous Embolization with Hylan Gel, Transactions of Society of Biomaterials XII, Society for Biomaterials 15th Annual Meeting, Lake Buena Vista, FL, April 28–May 2, 1989. 1989; 141.
172. Gunson DE, Hilal SK, Leshchiner EA, Larsen NE, Balazs EA. Percutaneous Embolization of Rabbit Auricular Arteries Using Hylan Gel, 23rd Annual Meeting of American Society for Clinical Pathology, Kansas City, MO, October 31–November 4, 1988. 1988.
173. Larsen NE, Balazs EA. Drug delivery systems using hyaluronan and its derivatives. Adv Drug Deliv Rev 1991; 7:279–293.
174. Larsen NE, Parent EG, Balazs EA. Release of the Antibiotic Gentamicin from Hylan Vehicles, Society for Biomaterials 17th Annual Meeting, Transactions of Society of Biomaterials XIV, Scottsdale, AZ, May 1–5, 1991. 1991; 54.
175. Meyer J, Whitcomb L, Treuheit M, Collins D. Sustained *in vivo* activity of recombinant human granulocyte colony stimulating factor (rHG-CSF) incorporated into hyaluronan [and hylan]. J Control Release 1995; 35:67–72.
176. Prisell PT, Camber O, Hiselius J, Norstedt G. Evaluation of hyaluronan as a vehicle for peptide growth factors. Int J Pharm 1992; 85:51–56.

177. Brown MB, Forbes B, Hanpanitcharoen M, Martin GP. The use of hyaluronan in topical drug delivery. In: Kennedy JF, Phillips GO, Williams PA, Hascall VC, eds. Hyaluronan. Cambridge, UK: Woodhead Publishing, 2002; vol. 2:249–256.
178. Kaetsu I, Sutani K, Uchida K, Phillips GO, Matsumura H, Matsubara Y. Preparation of hyaluronan gel and the application to drug delivery. In: Kennedy JF, Phillips GO, Williams PA, Hascall VC, eds. Hyaluronan. Cambridge, UK: Woodhead Publishing, 2002; vol. 2:261–270.
179. Luo Y, Kirker KR, Prestwich GD. Hyaluronic acid hydrogel film: a new biomaterial for drug delivery and wound healing. In: Kennedy JF, Phillips GO, Williams PA, Hascall VC, eds. Hyaluronan. Cambridge, UK: Woodhead Publishing, 2002; vol. 2:271–276.
180. Prestwich GD, Marecak DM, Marecek JF, Vercruysse KP, Ziebell MR. Chemical modification of hyaluronic acid for drug delivery, biomaterials and biochemical probes. In: Laurent TC, ed. The Chemistry, Biology and Medical Applications of Hyaluronan and Its Derivatives, Proceedings of the Wenner-Gren Foundation International Symposium Held in Honor of Endre A Balazs, Stockholm, Sweden, September 18–21, 1996. London, UK: Portland Press, 1998:43–65.
181. Solchaga LA, Goldberg VM, Caplan AI. Hyaluronan and tissue engineering. In: Kennedy JF, Phillips GO, Williams PA, Hascall VC, eds. Hyaluronan. Cambridge, UK: Woodhead Publishing, 2002; vol. 2:45–54.
182. Solchaga LA, Goldberg VM, Caplan AI. Cartilage repair with bone marrow in a hyaluronan-based scaffold. In: Kennedy JF, Phillips GO, Williams PA, Hascall VC, eds. Hyaluronan. Cambridge, UK: Woodhead Publishing, 2002; vol. 2: 63–66.
183. Solchaga LA, Caplan AI. Potential use of novel hyaluronan-based delivery vehicle in bone regeneration. In: Kennedy JF, Phillips GO, Williams PA, Hascall VC, eds. Hyaluronan. Cambridge, UK: Woodhead Publishing, 2002; vol. 2:67–70.
184. Facchini A, Piacentini A, Fiorni M, Scorzoni L, Grigolo B, Lisignoli G, Gobbi P, Mazzotti G, Duca M, Pavesio A. A novel hyaluronan biomaterial (Hyaff®-11) as scaffold for chondrocytes and bone marrow stromal cells. In: Kennedy JF, Phillips GO, Williams PA, Hascall VC, eds. Hyaluronan. Cambridge, UK: Woodhead Publishing, 2002; vol. 2:71–78.
185. Caravaggi C, Faglia E, Paola LD, Clerici G, Giglio RD, Sommariva E, Pritelli C, Mantero M, Caminiti M, Curci V, Fratino P. Hyaluronan based dermal and epidermal grafts in the treatment of diabetic foot ulcers. In: Kennedy JF, Phillips GO, Williams PA, Hascall VC, eds. Hyaluronan. Cambridge, UK: Woodhead Publishing, 2002; vol. 2:79–86.
186. Hollander DA, Krämer S, Hakimi MY, Windolf J. Soft tissue reconstruction in severely traumatized patients using hyaluronan based dermal and epidermal grafts. In: Kennedy JF, Phillips GO, Williams PA, Hascall VC, eds. Hyaluronan. Cambridge, UK: Woodhead Publishing, 2002; vol. 2:87–98.
187. Ramuamurthi A, Vesely I. Ultraviolet light-induced modification of crosslinked hyaluronan gels. J Biomed Mater Res 2003; 66A:317–329.
188. Shi Y, Garg A, Vesely I. Growth of Elastin Sheets on Hylan Gels for Tissue-Engineered Heart Valves, Hyaluronan 2003: An International Conference Celebrating the Research Contributions of Torvard Laurent, Cleveland, OH, October 11–16, 2003. 2003; 114.
189. Balazs EA, Denlinger JL. The vitreus. In: Davson H, ed. The Eye, Vol Ia Vegetative Physiology and Biochemistry. New York: Academic Press, 1984; 533–589.

Chemistry and Biology of Hyaluronan
H.G. Garg and C.A. Hales (editors)

Chapter 21

Medical Application of Hyaluronan

AKIRA ASARI

*Glyco-Research Project,
Central Research Laboratories,
Seikagaku Corporation,
1253, Tateno 3-chome,
Higashiyamato-shi,
Tokyo 207-0021, Japan*

I. Introduction

Hyaluronan (HA) is a ubiquitous constituent of extracellular matrices and has two main characteristics: [1] physicochemical properties and [2] cell biological functions. The physicochemical properties of HA are viscosity, elasticity, lubrication and a high capacity for holding water. HA is implicated in a number of cell biological phenomena via HA receptors, CD44 and RHAMM (including cell motility, cell proliferation, cell differentiation, cell–cell interaction), and in the production of cell physiological substances, such as cytokines, PGE_2 and matrix metalloproteinases (MMPs). Accordingly, medical applications of HA can be classified into three types: application(s) of [1] the physicochemical properties of HA, [2] the cell biological functions of HA and [3] both the physicochemical properties and cell biological functions of HA (Fig. 1). The molecular weight of HA varies from approximately 0.4 kDa (disaccharides) to several thousand kDa. Both the physicochemical properties and cell biological functions of HA depend on the molecular size of HA (Fig. 2) (1–5). HA can be modified to regulate its properties according to its usage: isolation of a certain range of molecular weight of HA, depolymerization of HA to obtain oligosaccharides and linkage of HA molecules to make a sponge, sheet or gel of HA. HA is highly biocompatible and biodegradable, suggesting that it is a suitable and safe biomaterial for many medical applications.

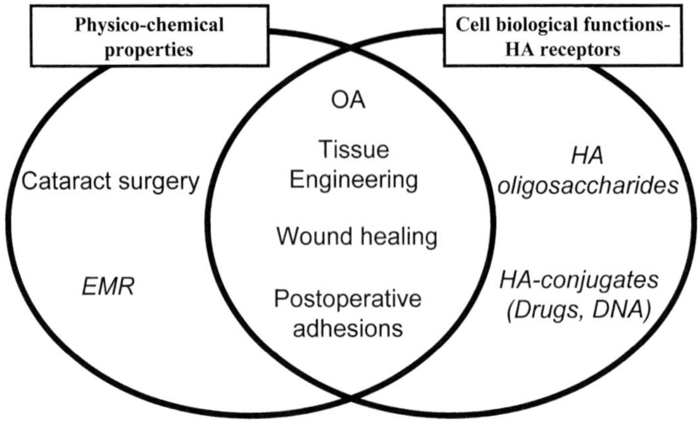

Figure 1 Three types of medical applications of hyaluronan. HA, hyaluronan; OA, osteoarthritis; EMR, endoscopic mucosal resection. The italic indicates future medical applications.

There are many reports regarding the medical applications of HA. However, only a small number of them have been put into practice. This chapter will review the medical applications of HA that have already been used in practice and some of those that will likely be developed in the near future.

II. Applications of Physicochemical Properties of Hyaluronan

A. Cataract Surgery

The use of viscoelastic materials in surgery is based on a general concept of solving problems that occur during cataract surgery (6). In this case, viscoelastic

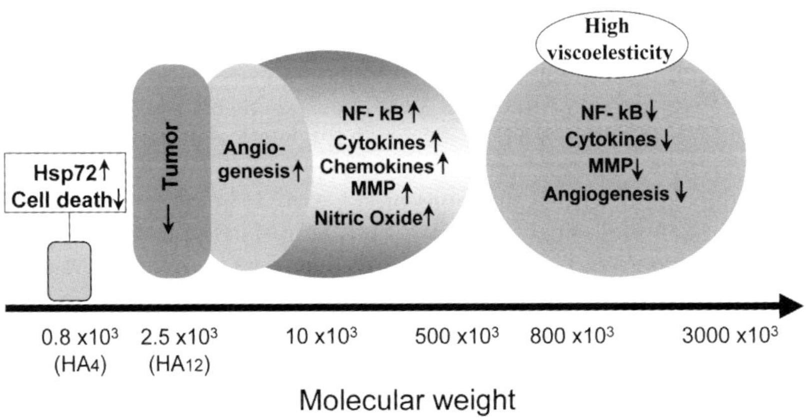

Figure 2 Functions of hyaluronan depending on its molecular weight. Hsp 72, heat shock protein 72; MMP, matrix metalloproteinase; NF-kB, nuclear factor kappa B; HA_4, tetrasaccharides of hyaluronan; HA_{12}, 12-mers of hyaluronan.

materials protect the endothelial layer of the cornea from damage during phacoemulsification and aspiration and maintain operative space. Two types of HA are used in cataract surgery: higher molecular weight HA (1900–3900 kDa) and lower molecular weight HA (600–1200 kDa). The former has high viscoelasticity and is mainly used to maintain operative space. The latter has higher dispersity and is mainly used to protect the endothelial layer. HA has now become indispensable in cataract surgery (7).

B. Endoscopic Mucosal Resection

Endoscopic mucosal resection (EMR) is a useful method for excision of non-polyploid and flat early neoplastic lesions in mucous membranes, and is an effective and minimally invasive method of treatment. Although an early tumor that has not reached the muscle properly can be excised endoscopically, it is difficult to excise flat lesions with a snare. To make excision safer and easier, physiological or hypertonic saline is commonly injected submucosally underneath the target region to create and maintain tissue elevation.

Yamamoto (8,9) reported EMR using submucosal injection of HA in the expectation of an advantage to reduce the risk of perforation and to accomplish complete en bloc resection. Mucosal elevations created by submucosal injection of HA remained for a longer time with a clearer margin compared with those made by saline injection (9). Use of HA instead of saline for EMR could make the procedure easier and more reliable. Onaya et al. (personal communication) evaluated various concentrations of HA as a novel submucosal injectant for EMR. In their study, the pressure of the fluid through the catheter, which was measured by a rheometer, increased with increasing concentrations of injected HA in the submucosal layer of resected porcine stomach at concentrations of 0.1–0.8%. However, it was difficult to measure HA solutions of 0.5% or higher since their resistance force was very high. Therefore, three concentrations of HA (0.2, 0.3 and 0.4%) were selected for injection into the submucosal layer of live rabbit stomachs. During 1 h of observation, the mucosal elevation produced by 0.4% HA solution was more prominent than that with the other two concentrations. In addition, 0.4% HA injection created a significantly higher protrusion at 30 min. Then they compared the height of the protrusion after injection of 0.4% HA, physiological saline, 50% dextrose, hypertonic saline and glycerol, which are used in Japanese clinical practice. Among them, 0.4% HA showed the highest value. Moreover, histological analyses showed no tissue injury after injection of a 0.4% HA solution or physiological saline, while other materials induced tissue injury, apparently caused by their hypertonicity.

These results suggest that an appropriate concentration of HA may be more useful in EMR than saline and other kinds of viscous materials such as dextrose and glycerol.

III. Application of Cell Biological Functions of Hyaluronan

A. Effects of Hyaluronan Oligosaccharides on Tumors

Many studies have demonstrated that HA plays an important role in tumor malignancy. Toole et al. (10) have shown that perturbation of endogenous HA–tumor cell interactions by overexpression of soluble CD44, an HA-binding protein, inhibits tumor growth *in vivo*. Moreover, they have reported that perturbation of endogenous HA–tumor cell interactions by HA oligosaccharides (~2.5 kDa) also inhibits tumor growth *in vivo* (11) as well as anchorage-independent growth of tumors *in vitro* (4). In addition, Ghatak et al. (4) have shown that treatment of malignant tumor cells with HA oligosaccharides induces apoptosis and stimulates caspase-3 activity under anchorage-independent conditions. In addition to that, they found that treatment with HA oligosaccharides stimulates expression of the tensin homolog (PTEN), a phosphatase that degrades the major signaling product of phosphoinositide 3-kinase (PI3-kinase). The ability of tumor cells to grow in an anchorage-independent manner is linked to elevation of the PI3-kinase/Akt cell survival pathway in tumor cells. Suppression of the PI3-kinase/Akt pathway in tumor cells leads to decreased phosphorylation and thus activation of pro-apoptotic mediators, BAD and FKHR, and increased activity of caspase-3.

The interaction of HA with cell surface receptors for HA, CD44 and RHAMM transduces intracellular signals (12). HA oligosaccharides and certain HA-blocking anti-CD44 antisera replace the multivalent, high-affinity interaction between endogenous HA polymers and HA receptors with a low valency, low-affinity interaction, and this would be expected to lead to changes in HA receptor signaling (4). Endogenous HA–CD44 interaction is necessary for maintenance of elevated levels of this pathway in tumor cells and thus their survival. In conclusion, treatment with HA oligosaccharides induces apoptosis in tumor cells through disrupting the endogenous HA–CD44 interaction, followed by enhancement of PTEN expression and inhibition of PI3-kinase signaling (4).

Misra et al. (13) have shown that an increase in HA polymer production induces resistance in drug-sensitive tumor cells, whereas treatment with HA oligosaccharides sensitizes multidrug-resistant tumor cells to a variety of chemotherapeutic drugs, such as doxorubicin and taxol. This is also due to disruption of endogenous HA–CD44 interaction by HA oligosaccharides.

B. Hyaluronan Conjugates

HA binds cell surface receptors for HA such as RHAMM and CD44. HA was shown to be useful as a carrier for a drug delivery system (DSS) to deliver the drug to cells expressing HA receptors. CD44 is expressed in many kinds of cells. However, CD44-positive cells do not necessarily bind HA (14). CD44 can gain HA-binding ability after its activation. Because HA receptors are overexpressed and activated in these cells (15,16), some kinds of tumor cells and lymphocytes in inflammation can strongly bind HA (17,18). Therefore, the application of HA as a DDS is also feasible in these cells.

A *N*-(2-hydroxypropyl) methacrylamide (HPMA)-HA polymeric DSS was used for targeted delivery of doxorubicin to cancer cells (19). In addition, superoxide dismutase (SOD) from bovine erythrocytes was conjugated with sodium hyaluronate with a mean molecular weight of 10^6 to have greater anti-inflammatory activity *in vivo* (20). This conjugate exhibited much higher anti-inflammatory activity than HA or SOD alone in models of inflammatory disease such as ischemic edema of the foot-pad in mice, carrageenin-induced pleurisy and adjuvant arthritis in rats (20). Polyampholyte comb-type copolymers, consisting of a poly main chain, a DNA binding site and HA side chains, were prepared as DNA carriers targeting sinusoidal endothelial cells of the liver that express the HA receptor, HARE (21). This DNA–HA conjugate may be useful for targeted delivery of genes in gene therapy.

IV. Applications of Both Physicochemical Properties and Cell Biological Functions of Hyaluronan

A. Osteoarthritis

A large number of osteoarthritis (OA) patients have already been treated with HA preparations after it was put into clinical practice around the world. The use of HA for the treatment of OA is based on the concept of Balazs and Denlinger (22). Several studies have shown that HA suppresses cartilage degeneration and reduces the release of proteoglycans from the extracellular matrix in cartilage tissue. It also protects the surface of articular cartilage (23,24), normalizes the properties of synovial fluid (25) and reduces pain perception (26,27).

HA influences physical properties such as viscoelasticity and/or lubrication of the stroma. Therefore, HA of a higher molecular weight is considered to be more beneficial for OA treatment because the efficacy of HA as a shock absorber to protect joint tissues is dependent on its average molecular weight. Kikuchi et al. (28) have shown that higher average molecular weight HA (2000×10^3) suppressed cartilage damage in a rabbit anterior cruciate ligament transection (ACLT) model for OA better than a lower average molecular weight preparation (900×10^3). In contrast, Ghosh et al. (29) have shown that the lower average molecular weight HA better protected articular cartilage from damage in a sheep medial meniscectomy model than did HA with the higher average weight. Taken together, the molecular weight-dependent effects of HA vary among experimental models, and the effects of HA on articular cartilage cannot be explained only by molecular weight, i.e., viscoelasticity.

The effects of an HA preparation HA84 (average MW 840×10^3) were compared to HA230 (average MW 2300×10^3) using a canine ACLT model for OA (25). In this model, proliferation and degeneration of synovial cells and increased synovial fluid volume were observed. However, cartilage damage was not detected. These synovial changes were more significantly suppressed in the group treated with HA84 than in those treated with HA230. The exposure of HA to synovial cells was determined by examining the distribution of fluorescein-labeled HA84 or HA230. Many fluorescein particles were scattered in the synovial lining layers, in most cases in the HA84-treated group. In contrast, there were only a few granules of fluorescein in the same location for the HA230-treated group. Thus, HA84 was exposed to synovial cells more effectively than HA230. In addition, we examined the distribution of fluorescein-labeled HA230 or HA90 (average MW 900×10^3) in synovial tissues of 12-month-old guinea pigs 3, 12, 24 h, 3 and 7 days after intra-articular injections of them. It is well known that OA naturally occurs in old Hartley strain guinea pigs (30). Laser-scan microscopy showed that the fluorescein-labeled HA90 was more widely distributed than the fluorescein-labeled HA230 in the synovial tissues at all time points (Fig. 3). HA230 cannot immediately permeate the synovial lining layers, probably because of strong entanglement of the HA230 molecules as well as its larger molecular size. This may be due to the network formation of HA depending on the molecular size (31). These events described above suggest that HA access to synovial cells is inversely dependent on molecular size *in vivo* (Fig. 4).

Prostaglandin E_2 (PGE_2) production is up-regulated, and proliferation of synovial cells occurs in the synovium of patients with OA and rheumatoid arthritis. HA suppresses PGE_2 production (32) and proliferation of synovial cells (33) in a molecular size-dependent manner, the larger sizes being more effective *in vitro*. In contrast, we have shown that HA84 suppresses both the increase in PGE_2 concentration in synovial fluid and the proliferation of synovial cells more significantly than HA230 using the canine ACLT model described earlier (Table 1) (25). It has also been shown that HA84, but not HA of 3600×10^3 average molecular weight, suppressed proliferation of synovial cells in a rabbit ACLT model (34). These also indicate that the molecular size dependency of the effects of HA found *in vitro* could be reversed *in vivo* (Fig. 4). Moreover, Yamashita et al. (35) showed that HA90 suppressed bradykinin-induced knee joint pain in rats better than HA230. Therefore, the molecular size dependency of the effects of HA found *in vitro* does not necessarily coincide with the *in vivo* data (Fig. 4).

PGE_2 is implicated in the progression of arthritis (36). Therefore, PGE_2 suppression by HA treatment leads to improvement of arthritis. Because vascular permeability is enhanced by PGE_2, increased vascular leak may be down-regulated by the suppressive effect of HA on PGE_2 production. Further, the pain relief after intra-articular administration of HA (26,27) may be due to reduced PGE_2 production, since PGE_2 enhances nociceptive activity.

Homandberg et al. (37) showed that HA, of 900×10^3 average molecular weight, suppressed both the alteration in proteoglycan synthesis induced by fibronectin fragments and subsequent cartilage degradation. In addition, HA suppressed the production of free radicals and reversed proteoglycan synthesis induced by interleukin-1β (IL-1β) in cultured bovine articular chondrocytes (23). Further, HA84 suppressed IL-1β production in cultured synovial cells from rheumatoid arthritis patients (38). Takahashi et al. (39) have found that the

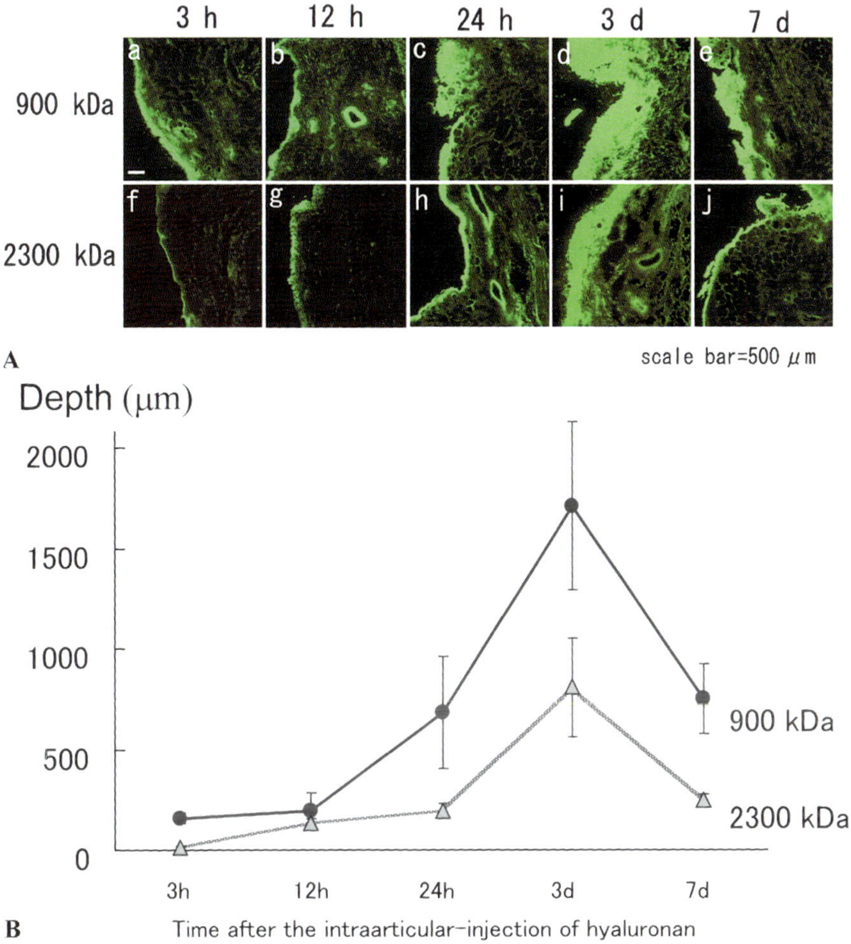

Figure 3 (A) Time course of distribution of fluorescein-labeled HA90 or HA230 in synovial tissues of guinea pigs that were intra-articularly injected. a–e, HA90; f–j, HA230. (B) Depth of fluorecein penetration from synovial surface into synovial tissues at several time points after the injection of fluorescein-labeled HA90 or HA230. 3d, 3 days; 7d, 7 days.

expressions of IL-1β and MMP-3 were suppressed in synovium but not in cartilage tissue of the rabbit ACLT model by treatment with HA84. The roles of PGE_2, MMP-3 and IL-1β in OA, whose production was suppressed by HA treatment, are summarized in Table 2.

In the canine ACLT model described above, treatment with HA84 suppressed the degeneration of synovial cells (25). In this study, we found that heat shock protein 72, which can suppress cell death and degeneration of cells, is up-regulated in synovial cells of the group treated with HA84. However, HA84 did not rescue the rat pheochromocytoma cell line (PC12 cells) from apoptosis

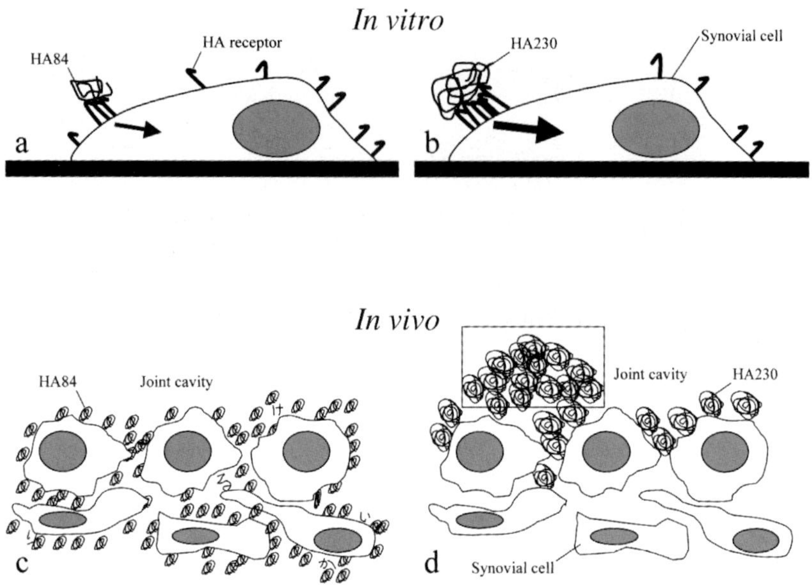

Figure 4 Proposed mechanism of molecular size dependency in the effects of hyaluronate preparations on synovial cells *in vitro* and *in vivo*. *In vitro*, higher molecular weight of HA, HA230 (2300 kDa), forms capping of HA receptors more densely than lower molecular weight of HA, HA84 (840 kDa) (a, b). *In vivo*, HA230 does not easily access synovial cells because of its large molecular sizes and entanglement among these molecules (square) (d), whereas HA84 can penetrate within the synovial tissue and adhere to the synovial cells (c). The sizes of arrows indicate the intensity of effects induced by the HA preparations (a, b). The effects by HA230 are more intense than those by HA84 *in vitro* (a, b). However, *in vivo*, the effects of HA230 are less than those of HA84, because accessibility of HA230 to the synovial cells is lower than that of HA84 *in vivo* (c, d). Molecular size dependency of the effects of HA *in vitro* is reversed *in vivo*.

induced by serum deprivation (3). We then examined many sizes of HA to PC12 cells and, as a result, found that HA 4-mers, and not the other sizes of HA, suppress apoptosis (3). Moreover, only HA4 up-regulated mRNA and protein expression of heat shock protein 72 (Hsp72) in human leukemic cell line K562 cells exposed to

Table 1 Effects of HA on Prostaglandin E_2 Concentration in Synovial Fluids and Proliferation of Synovial Cells of the Canine ACLT Model

Group	Non-operated	PBS	HA84	HA230
PGE_2 (pg/ml)	39.8 ± 4.97	89.8 ± 19.5*	75.0 ± 18.7	95.2 ± 10.6**
Thickness	13.2 ± 0.84	31.0 ± 1.22**	24.3 ± 1.57***	26.4 ± 1.54

Mean ± standard error; PGE_2, prostaglandin E_2; Non-operated, non-operated non-injected group; *$P < 0.05$ versus non-operated group; **$P < 0.01$ versus non-operated group; ***$P < 0.05$ versus PBS group.

Table 2 Roles of PGE2, MMP-3 and IL-1beta in Osteoarthritis

PGE2	Increase in vascular permeability, synovial fluid and pain perception
MMP-3	Degradation of collagen and proteoglycans
IL-1beta	Inflammation, reduction of proteoglycans synthesis

hyperthermia (3). Collectively, these data are consistent with the hypothesis that, in the synovial tissue of the canine ACLT model, HA84 is degraded into HA oligosaccharides that up-regulate Hsp72 expression of synovial cells (25).

In hydroarthrosis, vascular permeability is increased in synovial tissue (40) due to neuropeptides released from fine nerve terminals in neurogenic inflammation, and the concentration of neuropeptides, e.g., of substance P, in joint fluid is increased in arthritis. These events suggest that neurogenic inflammation is involved in arthritis. Kato et al. (personal communication) have shown the suppressive effect of HA90 on Evans blue extravasation and substance P release in the skin of a neurogenic inflammation model induced by electrical stimulation of sensory nerves of the skin. After intravenous injection of HA or PBS in rats, Evans blue was intravenously administered to evaluate extravasation of plasma. The saphenous nerve was electrically stimulated using bipolar electrodes 5 min after Evans blue injection. The amount of dye released into the skin extract was measured, as was the substance P content in the skin extract with an enzyme immunoassay method. Treatment with HA significantly suppressed Evans blue extravasation and substance P release. When HA (10 mg/kg) was injected, the amount of Evans blue extravasation and substance P release was only 37 and 31% of that in the PBS group, respectively. However, further studies are required to elucidate the precise mechanism. It seems that the suppressive effect of HA on neurogenic inflammation is involved in the therapeutic effects of HA on arthritis, in particular, in alteration of pain and hydroarthrosis.

Viscoelasticity of HA preparations is an important factor in the efficacy of HA for OA. In addition, many cell biological functions of HA, e.g., suppression of PGE_2, cytokine release, MMP production, and of substance P release are implicated in the efficacy of HA for OA.

B. Tissue Engineering

HA is abundant in fetal or young tissue extracellular matrix (41), and HA provides a fetal-like environment to cultured cells, stimulating regeneration. Moreover, during embryonic development, tissue regeneration and wound healing, the extracellular matrix surrounding migrating and proliferating cells is rich in HA (42). HA is implicated in morphogenesis (41,42) and plays a role in tissue organization (43). These characteristics of HA provide applications of HA for either tissue engineering or wound healing. In other words, the application of HA to wound healing is closely associated with that for tissue engineering.

Several studies have used HA in tissue engineering. An HA-based scaffold, consisting of chemically modified HA, has been used to support the growth of human chondrocytes (44). Human chondrocytes can proliferate and produce aggrecan and type II collagen but not type I collagen, indicating that the HA-based scaffold maintains the phenotype of chondrocytes. Therefore, the HA-based scaffold may also be used for the repair of articular cartilage defects.

Keratinocytes cultured on the HA-based scaffold form sheets, mimicking epidermis (45). On the other hand, fibroblasts can be cultured in an HA-based scaffold to form a three-dimensional dermis (46). Taken together, an HA-based scaffold can be used to form a skin equivalent.

Chemically modified HA has also been formed into a gel, film, sheet, tube or sponge. These designed forms of HA can be applied for growth or differentiation of many kinds of cells or tissues according to their form. In addition to its physicochemical property as a scaffold, interaction of HA with HA receptors, e.g., CD44 or RHAMM, influences cell behavior (47). Therefore, the application of HA to tissue engineering or wound healing is based on both its physicochemical properties and cell biological functions.

C. Wound Healing

1. Cystitis

Interstitial cystitis is characterized by prolonged pollakiuria and pain in the suprapubic region when the bladder is full of urine (48). For the treatment of interstitial cystitis, sulfated exogenous polysaccharides, sodium pentosanpolysulfate (49), DMSO (50), heparin (51) and other agents are used (52–54). Interstitial cystitis may be related to a primary defect in glycosaminoglycans (GAGs), i.e., HA, heparan sulfate, dermatan sulfate and keratan sulfate in the vesical mucosa (55). A GAG coating of the bladder urothelium may protect the bladder tissue from injury by stimulants, e.g., by microorganisms, crystals and other toxic substances (56). It has been reported that intravesical injection of HA is useful for interstitial cystitis (57,58).

Intravesical HA has been used to treat interstitial cystitis, since this may possibly replenish bladder GAGs. Takahashi et al. (59) have evaluated the effect of 0.1–0.4% HA (average MW 890×10^3) on epithelial healing of the vesical mucosa and vesical fibrosis. They injected three concentrations of HA intravesically into rabbits with cystitis induced by instillation of 5% acetic acid solution. The effect of HA on cystitis was evaluated 7 days after injection. An increase in the capacity of the bladder was observed in HA-treated groups, and the area of the epithelial defect was reduced. Moreover, treatment with HA accelerated epithelial healing of the vesical mucosa and suppressed vesical fibrosis. Treatment with HA alleviated the model of cystitis in a concentration-dependent manner. Further studies are required to elucidate the mechanism of the effects of HA on this model of cystitis. It seems that the injected HA coats the injured area of the epithelium to protect the bladder tissue from injury, in a manner similar to that reported by Erickson et al. (60) for endogenous HA. In addition, the acceleration of epithelial wound healing by treatment with HA may be due to the interaction between HA and epithelial cell CD44, as shown in corneal epithelial wound healing (61). Further, Boucher et al. (62) have shown that immobilization stress induces bladder mast cell activation and the secretion of proinflammatory mediators, histamine and IL-6, which are inhibited by HA. These data suggest that the effect of HA on cystitis is in part due to its cell biological functions.

2. *Corneal Epithelial Wound Healing*

HA has already been developed as a preparation for corneal wound healing in dry eyes. In this case, it seems that HA protects the cornea from dryness through its high capacity for holding water. In addition, interaction between HA and epithelial cell CD44 is observed during wound healing of the corneal epithelium. In a rabbit corneal epithelial wound-healing model induced by *n*-heptanol, we examined the expression of HA and CD44 (61). HA concentration in the cornea gradually increased until day 14 after injury, and then decreased, returning to a normal level by 56 days. Staining of CD44 and HA was examined in corneal tissues by immunohistochemical and histochemical techniques, respectively. HA staining in corneal tissues correlated synchronously with the level of HA determined by biochemical analysis in corneal tissues. Both HA and CD44 were detected in the corneal epithelium and stroma. During the process of epithelial wound healing, a single layer of epithelium first extended to the damaged area 3 days after injury. When the epithelial layer completely covered the damaged area, HA staining was enhanced in both the epithelial layer and the underlying stroma. The enhanced staining of HA in the epithelium gradually decreased until it reached normal thickness. Immunoreactivity for CD44 changed almost concomitantly with that for HA after injury. HA expression returned to a normal level by day 28. This investigation indicates that HA and CD44 synergistically play important roles in corneal epithelial wound healing, and that epithelial cells produce HA as a scaffold for their proliferation and migration. In support of these studies, Miyazaki et al. (63) have reported that treatment with HA enhanced the growth of cultured corneal epithelial cells. Moreover, Miyauchi et al. (64) have shown that the topical administration of HA accelerates corneal wound healing in the rabbit by enhancing the growth of corneal epithelial cells.

D. Postoperative Adhesions

Adhesions remain a significant postoperative complication of abdominal surgery. A sodium hyaluronate and carboxymethylcellulose bioresorbable membrane (65) as well as a cross-linked hyaluronate hydrogel (66) have been shown to reduce postoperative adhesions. These HA derivatives were used as physical barriers in these studies.

Theoretically, it seems possible to reduce adhesion formation by taking the following steps: [1] reducing the initial inflammatory reaction and subsequent exudate release; [2] inhibiting coagulation of this exudate; [3] promoting the removal of fibrin deposition; [4] mechanically separating fibrin-covered surfaces; and [5] inhibiting fibroblast proliferation (67). Weigel et al. (68) have shown that HA binds fibrin and fibrinogen. This suggests that the effects of HA derivatives on the adhesion protection may be due to the binding of HA to fibrin and fibrinogen preventing fibrin from acting as a glue in adhesion formation.

It has been reported that halofuginone, an inhibitor of collagen type I synthesis, prevents postoperative adhesion formation in the rat uterine horn model (69). This indicates that collagen is implicated in postoperative adhesion. Coleman (70) has shown that HA chains have a major organizational role within the collagen bundle. Exogenous HA may prevent the formation of collagen bundles in postoperative adhesions.

Collagen is produced by fibroblasts. It is well known that HA suppresses the proliferation of fibroblasts (33). Therefore, the prevention of adhesions may involve the suppressive effect of HA on the proliferation of fibroblasts. Moreover, HA suppresses inflammation by preventing cytokine production (5). Thus, HA may prevent several steps in postoperative adhesions.

These events described above suggest that the effects of HA on postoperative adhesions are due to the role of HA as a physical barrier based on its physicochemical properties and also due to cell biological or physiological function of HA.

V. Conclusions

Viscoelasticity is the main characteristic of HA. However, some of the effects of HA in medical applications are also based on the cell biological functions. Recently, signal transductions via HA receptors have been elucidated. HA has important roles through its cell biological functions as well as its physicochemical properties. HA preparations can be used as medicines, medical devices or both.

Recently, new HA-binding proteins have been found (71–73); however, most of their functions are still unknown. Moreover, low molecular weight HA and HA oligosaccharides have cell biological functions that high molecular weight HA does not show (1–5). Therefore, HA obtains novel functions after its depolymerization (Fig. 2). The newly found cell biological functions of HA may lead to new medical applications of HA in the next generation. Basic research in the field of HA has important keys to future medical applications of HA.

Acknowledgements

I would like to thank Mr Jun-ichi Onaya, Mr Katsuya Takahashi, Mr Tadahiko Katoh and Ms Hiroko Yamanokuchi at Seikagaku Corporation for their cooperation. I am indebted to Dr Hascall at Cleveland Clinic Foundation and Dr Savani at Children's Hospital of Philadelphia for the review of this manuscript.

References

1. West DC, Shaw DM. Tumour hyaluronan in relation to angiogenesis and metastasis. In: Laurent TC, ed. The Chemistry, Biology and Medical Applications of Hyaluronan and Its Derivatives. London: Portland Press, 1998: 227–233.
2. Noble PW, Lake FR, Henson PM, Riches DW. Hyaluronate activation of CD44 induces insulin-like growth factor-1 expression by a tumor necrosis factor-alpha-dependent mechanism in murine macrophages. J Clin Investig 1993; 91: 2368–2377.
3. Xu H, Ito T, Tawada A, Maeda H, Yamanokuchi H, Isahara K, Yoshida K, Uchiyama Y, Asari A. Effect of hyaluronan oligosaccharides on the expression of heat shock protein 72. J Biol Chem 2002; 277:17308–17314.
4. Ghatak S, Misra S, Toole BP. Hyaluronan oligosaccharides inhibit anchorage-independent growth of tumor cells by suppressing the phosphoinositide 3-kinase/Akt cell survival pathway. J Biol Chem 2002; 277:38013–38020.
5. Neumann A, Schinzel R, Palm D, Riederer P, Munch G. High molecular weight hyaluronic acid inhibits advanced glycation endproduct-induced NF-kappaB activation and cytokine expression. FEBS Lett 1999; 453:283–287.
6. Balazs EA. Sodium hyaluronate and viscosurgery. In: Miller D, Stegmann R, eds. Healon (Sodium Hyaluronate). A Guide to Its Use in Ophthalmic Surgery. New York: Wiley, 1983:5–28.
7. Liesegang TJ. Viscoelastic substances in ophthalmology. Surv Ophthalmol 1990; 34:268–293.
8. Yamamoto H, Kawata H, Sunada K, Satoh K, Kaneko Y, Ido K, Sugano K. Success rate of curative endoscopic mucosal resection with circumferential mucosal incision assisted by submucosal injection of sodium hyaluronate. Gastrointest Endosc 2002; 56:507–512.
9. Yamamoto H, Sekine Y, Higashizawa T, Kihira K, Kaneko Y, Hosoya Y, Ido K, Saito K, Sugano K. Successful en bloc resection of a large superficial gastric cancer by using sodium hyaluronate and electrocautery incision forceps. Gastrointest Endosc 2001; 54:629–632.
10. Yu Q, Toole BP, Stamenkovic I. Induction of apoptosis of metastatic mammary carcinoma cells *in vivo* by disruption of tumor cell surface CD44 function. J Exp Med 1997; 186:1985–1996.
11. Zeng C, Toole BP, Kinney SD, Kuo JW, Stamenkovic I. Inhibition of tumor growth *in vivo* by hyaluronan oligomers. Int J Cancer 1998; 77:396–401.
12. Turley EA, Noble PW, Bourguignon LY. Signaling properties of hyaluronan receptors. J Biol Chem 2002; 277:4589–4592.
13. Misra S, Ghatak S, Zoltan-Jones A, Toole BP. Regulation of multidrug resistance in cancer cells by hyaluronan. J Biol Chem 2003; 278:25285–25288.
14. Lesley J, Hyman R. CD44 can be activated to function as an hyaluronic acid receptor in normal murine T cells. Eur J Immunol 1992; 22:2719–2723.
15. Levesque MC, Haynes BF. Cytokine induction of the ability of human monocyte CD44 to bind hyaluronan is mediated primarily by TNF-alpha and is inhibited by IL-4 and IL-13. J Immunol 1997; 159:6184–6194.
16. Weiss JM, Renkl AC, Ahrens T, Moll J, Mai BH, Denfeld RW, Schopf E, Ponta H, Herrlich P, Simon JC. Activation-dependent modulation of hyaluronate-receptor

expression and of hyaluronate-avidity by human monocytes. J Investig Dermatol 1998; 111:227–232.
17. Yanagisawa N, Mikami T, Mitomi H, Saegusa M, Koike M, Okayasu I. CD44 variant overexpression in gallbladder carcinoma associated with tumor dedifferentiation. Cancer 2001; 91:408–416.
18. Katoh S, Taniguchi H, Matsubara Y, Matsumoto N, Fukushima K, Kadota J, Matsukura S, Kohno S. Overexpression of CD44 on alveolar eosinophils with high concentrations of soluble CD44 in bronchoalveolar lavage fluid in patients with eosinophilic pneumonia. Allergy 1999; 54:1286–1292.
19. Luo Y, Bernshaw NJ, Lu ZR, Kopecek J, Prestwich GD. Targeted delivery of doxorubicin by HPMA copolymer-hyaluronan bioconjugates. Pharm Res 2002; 19:396–402.
20. Sakurai K, Miyazaki K, Kodera Y, Nishimura H, Shingu M, Inada Y. Anti-inflammatory activity of superoxide dismutase conjugated with sodium hyaluronate. Glycoconj J 1997; 14:723–728.
21. Asayama S, Nogawa M, Takei Y, Akaike T, Maruyama A. Synthesis of novel polyampholyte comb-type copolymers consisting of a poly(L-lysine) backbone and hyaluronic acid side chains for a DNA carrier. Bioconjug Chem 1998; 9: 476–481.
22. Balazs EA, Denlinger JL. Clinical uses of hyaluronan. Ciba Found Symp 1989; 143: 265–275.
23. Fukuda K, Dan H, Takayama M, Kumano F, Saitoh M, Tanaka S. Hyaluronic acid increases proteoglycan synthesis in bovine articular cartilage in the presence of interleukin-1. J Pharmacol Exp Ther 1996; 277:1672–1675.
24. Sakamoto T, Mizuno S, Maki T, Suzuki T, Yamaguchi T, Iwata H. Studies on the affinity of hyaluronic acid to the surface of articular cartilage and the suppression of proteoglycan release from matrix. Orthop Res Sci 1984; 11:264.
25. Asari A, Miyauchi S, Matsuzaka S, Ito T, Kominami E, Uchiyama Y. Molecular weight-dependent effects of hyaluronate on the arthritic synovium. Arch Histol Cytol 1998; 61:125–135.
26. Gotoh S, Onaya J, Abe M, Miyazaki K, Hamai A, Horie K, Tokuyasu K. Effects of the molecular weight of hyaluronic acid and its action mechanisms on experimental joint pain in rats. Ann Rheum Dis 1993; 52:817–822.
27. Iwata H. Pharmacologic and clinical aspects of intraarticular injection of hyaluronate. Clin Orthop 1993; 289:285–291.
28. Kikuchi T, Yamada H, Shinmei M. Effect of high molecular weight hyaluronan on cartilage degeneration in a rabbit model of osteoarthritis. Osteoarthritis Cartilage 1996; 4:99–110.
29. Ghosh P, Read R, Armstrong S, Wilson D, Marshall R, McNair P. The effects of intraarticular administration of hyaluronan in a model of early osteoarthritis in sheep. I. Gait analysis and radiological and morphological studies. Semin Arthritis Rheum 1993; 122(6 Suppl 1):18–30.
30. Huebner JL, Hanes MA, Beekman B, TeKoppele JM, Kraus VB. A comparative analysis of bone and cartilage metabolism in two strains of guinea-pig with varying degrees of naturally occurring osteoarthritis. Osteoarthritis Cartilage 2002; 10: 758–767.
31. Yanaki T, Yamaguchi T. Temporary network formation of hyaluronate under a physiological condition. 1. Molecular-weight dependence. Biopolymers 1990; 30: 415–425.

32. Yasui T, Akatsuka M, Tobetto K, Hayaishi M, Ando T. The effect of hyaluronan on interleukin-1 alpha-induced prostaglandin E2 production in human osteoarthritic synovial cells. Agents Actions 1992; 37:155–156.
33. Goldberg RL, Toole BP. Hyaluronate inhibition of cell proliferation. Arthritis Rheum 1987; 30:769–778.
34. Shimizu C, Kubo T, Hirasawa Y, Coutts RD, Amiel D. Histomorphometric and biochemical effect of various hyaluronans on early osteoarthritis. J Rheumatol 1998; 25:1813–1819.
35. Yamashita I, Atsuta Y, Shimazaki S, Miyatsu M. Effects of prostaglandin E2 and sodium hyaluronate on bradykinin induced knee joint pain in rat. Nippon Seikeigeka Gakkai Zasshi—Jpn Orthop Assoc 1995; 69:735–743.
36. Asari A, Mizuno S, Tanaka I, Sunose A, Kuriyama S, Miyazaki K, Namiki O. Suppression of hyaluronan and prostaglandin E2 production in traumatic arthritic synovial cells by NaHA. Connect Tissue 1997; 29:1–5.
37. Homandberg GA, Hui F, Wen C, Kuettner KE, Williams JM. Hyaluronic acid suppresses fibronectin fragment mediated cartilage chondrolysis: I. *In vitro*. Osteoarthritis Cartilage 1997; 5:309–319.
38. Takeshita S, Mizuno S, Kukuchi T, Yamada H, Namiki O, Horie K. The *in vitro* effect of hyaluronic acid on IL-1 production in cultured rheumatoid synovial cells. Biomed Res 1997; 18:187–194.
39. Takahashi K, Goomer RS, Harwood F, Kubo T, Hirasawa Y, Amiel D. The effects of hyaluronan on matrix metalloproteinase-3 (MMP-3), interleukin-1beta (IL-1beta), and tissue inhibitor of metalloproteinase-1 (TIMP-1) gene expression during the development of osteoarthritis. Osteoarthritis Cartilage 1999; 7:182–190.
40. Asari A, Miyauchi S, Sekiguchi T, Machida A, Kuriyama S, Miyazaki K, Namiki O. Hyaluronan, cartilage destruction and hydrarthrosis in traumatic arthritis. Osteoarthritis Cartilage 1994; 2:79–89.
41. Toole BP, Jackson G, Gross J. Hyaluronate morphogenesis: inhibition of chondrogenesis *in vitro*. Proc Natl Acad Sci USA 1972; 69:1384–1386.
42. Toole BP. Hyaluronan–cell interactions in morphogenesis. In: Laurent TC, ed. The Chemistry, Biology and Medical Applications of Hyaluronan and Its Derivatives. London: Portland Press, 1998:155–160.
43. Scott JE. The chemical morphology of the vitreous. Eye 1992; 6:553–555.
44. Grigolo B, Lisignoli G, Piacentini A, \Fiorini M, Roseti L, Major EO, Duca M, Pavesio A, Facchini A. Evidence for re-differentiation of human chondrocytes seeded on an HA derivative scaffold innovative rheumatology: gene and cell therapies of arthritis and related autoimmune disorders. Second International Meeting. Montpellier, France 2001. Arthritis Res 2001; 3:7.
45. Hollander D, Stein M, Bernd A, Windolf J, Pannike A. Autologous keratinocytes cultured on benzylester hyaluronic acid membranes in the treatment of chronic full-thickness ulcers. J Wound Care 1999; 8:351–355.
46. Galassi G, Brun P, Radice M, Cortivo R, Zanon GF, Genovese P, Abatangelo G. *In vitro* reconstructed dermis implanted in human wounds: degradation studies of the HA-based supporting scaffold. Biomaterials 2000; 21:2183–2191.
47. Hall CL, Wang C, Lange LA, Turley EA. Hyaluronan and the hyaluronan receptor RHAMM promote focal adhesion turnover and transient tyrosine kinase activity. J Cell Biol 1994; 126:575–588.
48. Ho N, Koziol JA, Parsons CL. Epidemiology of interstitial cystitis. In: Sant GR, ed. Interstitial Cystitis. Philadelphia: Lipincott-Raven, 1997:9–16 Chap. 2.

49. Parsons CL. The therapeutic role of sulfated polysaccharides in urinary bladder. Urol Clin N Am 1994; 21:93–100.
50. Sant GR, LaRock DR. Standard intravesical therapies for interstitial cystitis. Urol Clin N Am 1994; 21:73–83.
51. Parsons CL, Housley T, Schmidt JD, Lebow D. Treatment of interstitial cystitis with intravesical heparin. Br J Urol 1994; 73:504–507.
52. Edwards L, Bucknall TE, Makin C. Interstitial cystitis: possible cause and clinical study of sodium cromoglycate. Br J Urol 1986; 58:95–96.
53. Khanna OP, Loose JH. Interstitial cystitis treated with intravesical doxorubicin. Urology 1990; 36(Suppl):139–142.
54. Zeidman EJ, Helfrick B, Pollard C, Thompson IM. Bacillus Calmette-Guerin immunotherapy for refractory interstitial cystitis. Urology 1994; 43:121–124.
55. Parsons CL, Lilly JD, Stein P. Epithelial dysfunction in nonbacterial cystitis (interstitial cystitis). J Urol 1991; 145:732–735.
56. Ruggieri MR, Chelsky MJ, Rosen SI, Schikley TJ, Hanno PM. Current findings and future research avenues in the study of interstitial cystitis. Urol Clin N Am 1994; 21:163–176.
57. Morales A, Emerson L, Nickel JC, Lundie M. Intravesical hyaluronic acid in the treatment of refractory interstitial cystitis. J Urol 1996; 156:45–48.
58. Porru D, Campus G, Tudino D, Valdes E, Vespa A, Scarpa RM, Usai E. Results of treatment of refractory interstitial cystitis with intravesical hyaluronic acid. Urol Int 1997; 59:26–29.
59. Takahashi K, Takeuchi J, Takahashi T, Miyauchi S, Horie K, Uchiyama Y. Effects of sodium hyaluronate on epithelial healing of the vesical mucosa and vesical fibrosis in rabbits with acetic acid induced cystitis. J Urol 2001; 166: 710–713.
60. Erickson DR, Sheykhnazari M, Ordille S, Bhavanendan VP. Increased urinary hyaluronic acid and interstitial cystitis. J Urol 1998; 160:1282–1284.
61. Asari A, Morita M, Sekiguchi T, Okamura K, Horie K, Miyauchi S. Hyaluronan, CD44 and fibronectin in rabbit corneal epithelial wound healing. Jpn J Ophthalmol 1996; 40:18–25.
62. Boucher WS, Letourneau R, Huang M, Kempuraj D, Green M, Sant GR, Theoharides TC. Intravesical sodium hyaluronate inhibits the rat urinary mast cell mediator increase triggered by acute immobilization stress. J Urol 2002; 167: 380–384.
63. Miyazaki T, Miyauchi S, Nakamura T, Takeshita S, Horie K. The effect of sodium hyaluronate on the growth of rabbit corneal epithelial cells *in vitro*. J Ocul Pharmacol 1996; 12:409–415.
64. Miyauchi S, Sugiyama T, Machida A, Sekiguchi T, Miyazaki K, Tokuyasu K, Nakazawa K. The effect of sodium hyaluronate on the migration of rabbit corneal epithelium. I. An *in vitro* study. J Ocul Pharmacol 1990; 6:91–99.
65. Ghellai AM, Stucchi AF, Lynch DJ, Skinner KC, Colt MJ, Becker JM. Role of a hyaluronate-based membrane in the prevention of peritonitis-induced adhesions. J Gastrointest Surg 2000; 4:310–315.
66. Osada H, Takahashi K, Fujii TK, Tsunoda I, Satoh K. The effect of cross-linked hyaluronate hydrogel on the reduction of post-surgical adhesion reformation in rabbits. J Int Med Res 1999; 27:233–241.
67. Holtz G. Prevention of postoperative adhesions. J Reprod Med 1980; 24: 141–146.

68. Weigel PH, Frost SJ, LeBoeuf RD, McGary CT. The specific interaction between fibrin(ogen) and HA: possible consequences in haemostasis, inflammation and wound healing. Ciba Found Symp 1989; 143:248–261.
69. Nagler A, Genina O, Lavelin I, Ohana M, Pines M. Halofuginone, an inhibitor of collagen type I synthesis, prevents postoperative adhesion formation in the rat uterine horn model. Am J Obstet Gynecol 1999; 180:558–563.
70. Coleman PJ. Evidence for a role of hyaluronan in the spacing of fibrils within collagen bundles in rabbit synovium. Biochim Biophys Acta 2002; 1571:173–182.
71. Meenakshi J, Anupama, Goswami SK, Datta K. Constitutive expression of hyaluronan binding protein 1 (HABP1/p32/gC1qR) in normal fibroblast cells perturbs its growth characteristics and induces apoptosis. Biochem Biophys Res Commun 2003; 300:686–693.
72. Huang L, Grammatikakis N, Yoneda M, Banerjee SD, Toole BP. Molecular characterization of a novel intracellular hyaluronan-binding protein. J Biol Chem 2000; 275:29829–29839.
73. Vostrov AA, Quitschke WW. Plasma hyaluronan-binding protein is a serine protease. J Biol Chem 2000; 275:22978–22985.

Chemistry and Biology of Hyaluronan
H.G. Garg and C.A. Hales (editors)

Chapter 22

Therapeutic Biomaterials from Chemically Modified Hyaluronan

XIAO ZHENG SHU and GLENN D. PRESTWICH

Department of Medicinal Chemistry, The University of Utah, 419 Wakara Way, Suite 205, Salt Lake City, UT 84108-1257, USA

I. Introduction

Hyaluronan is a non-sulfated glycosaminoglycan present in all connective tissues as a major constituent of the extracellular matrix (ECM) (1–3). Due to its unique physicochemical properties and distinctive biological functions, this polyanionic polymer is an attractive building block for new biocompatible and biodegradable materials in drug delivery, tissue engineering, and viscosupplementation. However, the poor biomechanical properties of this soluble, natural polymer and its rapid degradation *in vivo* preclude many direct clinical applications. Therefore, to obtain materials that are more mechanically and chemically robust, a variety of chemical modification strategies have been explored to produce insoluble or gel-like hyaluronan materials. These hyaluronan derivatives have physicochemical properties that may significantly differ from the native polymer. Most derivatives are biocompatible and biodegradable and retain the pharmacological and therapeutic properties of native hyaluronan. This chapter explores the chemical modification of hyaluronan and the potential uses of these biomaterials in tissue engineering and drug delivery.

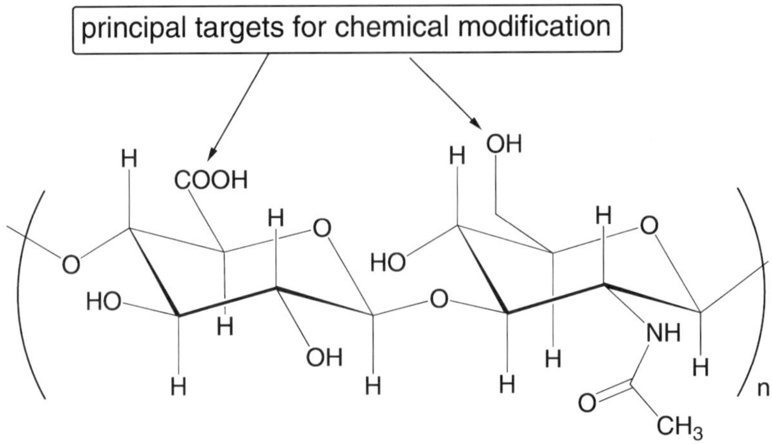

Figure 1 Disaccharide repeat units of hyaluronan.

II. Chemical Modifications

Most chemical modification strategies target the carboxyl and hydroxyl groups of the sugar moieties (Fig. 1). First, the basic chemistry of these modifications will be reviewed. Second, the uses of these and other reactions to produce biomedically useful hyaluronan-derived materials will be explored. Fig. 2 summarizes the chemical details of several of the reactions of hyaluronan carboxyl groups. Selected references to the primary literature detailing the chemical modifications are provided, and the remainder can best be accessed by referring to three recent reviews (4–6).

A. Modification of the Hyaluronan Carboxyl Group

1. Modification of Hyaluronan by Esterification

Esterification of hyaluronan usually involves a two-step procedure: [i] preparation of a quaternary salt of hyaluronan and [ii] subsequent reaction with an esterifying agent in aprotic solvent at a controlled temperature (Fig. 2A). A variety of alcohols can be used for the esterification (aliphatic, benzylic, cycloaliphatic and others). At higher percentages of esterification, the resulting hyaluronan derivatives became insoluble in water (7). These HYAFF® materials (Fidia Advanced Biopolymers) can be extruded to produce membranes and fibers, lyophilized to obtain sponges, or treated by spray-drying, extraction, and evaporation to produce microspheres. From the physicochemical and biological aspects, the ethyl and benzyl esters of hyaluronan, respectively termed HYAFF® 7 and HYAFF® 11 are the two best characterized HYAFF® polymers (8).

In addition to alcohols, the glucuronic acid moiety of hyaluronan can be converted to active ester intermediates for coupling to fluorescent probes,

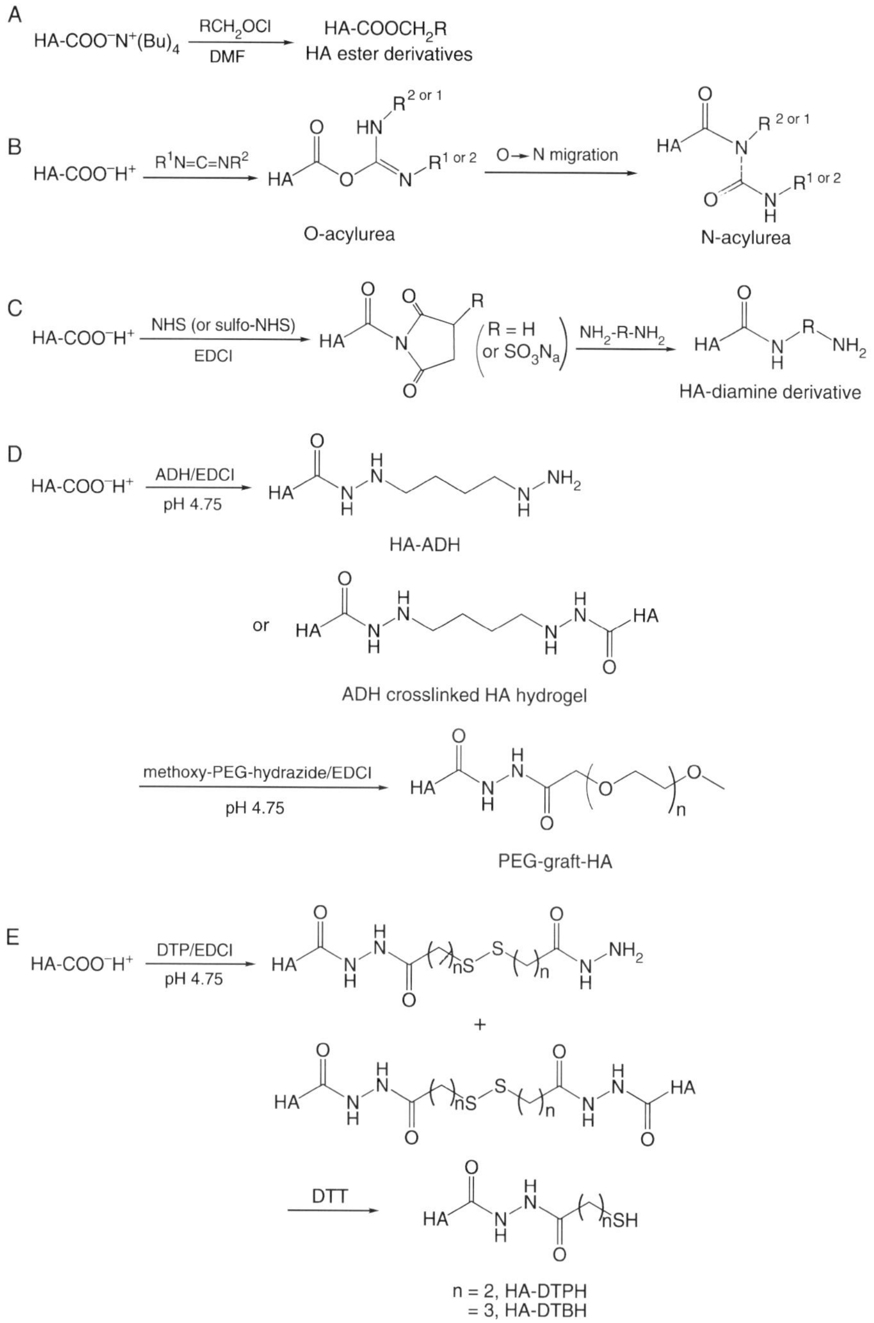

Figure 2 Chemical modifications of carboxylic acid groups of hyaluronan: (A) esterification; (B) carbodiimide modification; (C) EDCI/(sulfo-)NHS modification; (D) EDCI/hydrazide modification; (E) thiol modification.

drug–polymer conjugates, and crosslinked hydrogels. Hyaluronan-*N*-hydroxysuccinimide (HA-NHS) was synthesized by the reaction of *N*-hydroxysuccinimido diphenyl phosphate with the tetrabutylammonium salt of hyaluronan in DMSO. A hydrazide-containing fluorescent probe was condensed with HA-NHS to make a hyaluronan–BODIPY conjugate. This conjugate was used to demonstrate intracellular uptake into human ovarian cancer cells that overexpress cell-surface hyaluronan receptors (9). The activating agent 2-chloro-1-methylpyridinium iodide was used to make a hyaluronan active ester intermediate by reaction with tetrabutylammonium hyaluronate in DMF. The reactive intermediate was crosslinked by a variety of diamine agents (1,3-diaminepropane, 1,6-diaminohexane, poly(ethylene glycol)(PEG)500 diamine, and PEG800 diamine) to produce hydrogels (10). Hydrogels crosslinked by 1,3-diaminepropane improved chondrocyte density and matrix appearance for the treatment of osteoarthritis in a rabbit model (11).

2. *Carbodiimide-Mediated Reactions*

Carbodiimides have been widely used for activation of carboxyl groups of glycosaminoglycans under the acidic conditions necessary for protonation of the carbodiimide nitrogens. After activation, nucleophilic attack of the carboxylate anion at the central carbon gives an initial *O*-acylisourea intermediate that can be captured by nucleophiles or may rearrange to the unreactive *N*-acylurea (12). The modification of hyaluronan by carbodiimides is usually performed at pH 4–6, where carbodiimide nitrogens appear to be sufficiently protonated, while hyaluronan mainly exists in approximately equal amounts of carboxylic acid and carboxylate, depending on the pH (Fig. 2B) (13,14).

If a nucleophile is present, a new conjugate may be formed by a nucleophilic addition–elimination reaction with activated *O*-acylisourea. Perhaps the most common nucleophilic reagents are primary amines, although this chemistry is problematic since the basic amines are protonated at pH 4–6, with less than 0.001% present in the nucleophilic amine form at equilibrium. Nonetheless, a number of hyaluronan–drug bioconjugates have been prepared by using carbodiimide-mediated amine conjugation. For instance, hyaluronan–mitomycin C and hyaluronan–epirubicin conjugates were synthesized through an amide bond using 1-ethyl-3-[*N*,*N*-dimethylaminopropyl]-carbodiimide (EDCI), a water-soluble carbodiimide as the coupling agent. These conjugates were selectively targeted to the parent anti-tumor drugs into regional lymph nodes and cancerous tissues through hyaluronan receptor CD44 (15). Similarly, adriamycin, daunomycin, 5-fluorouracil, cytosine arabinoside, and superoxide dismutase were coupled to hyaluronan (16,17). More recently, hyaluronan was derivatized by *N*-(3-aminopropyl)-methacrylamide hydrochloride with EDCI for hydrogel preparation via photopolymerization (18).

However, the coupling efficiency of most amine groups to hyaluronan through the one-step EDCI chemistry is poor as a result of the minscule fraction of free amine present in the acidic environment, with the possible exception of less basic amines such as 4-vinylaniline (19). For viscous macromolecules, the unstable intermediate *O*-acylisourea can rearrange intramolecularly to a stable *N*-acylurea in seconds, thus outcompeting intermolecular capture by exogenously added amines (12,13,20). One method to improve the conjugation of primary amine to hyaluronan was to increase the stability of the reactive intermediate. Therefore, a two-step EDCI chemistry was developed. The unstable *O*-acylisourea was first converted to a longer-lived second intermediate that survived several hours in aqueous solution. This conversion involved the highly efficient reaction of the hydroxyl group on either sulfo-*N*-hydroxysuccinimide (sulfoNHS) or NHS itself with the *O*-acylisourea. The sulfo group enhances the water solubility of the NHS species and potentially increases subsequent reaction rates. (Sulfo)NHS ester intermediates are active groups that react rapidly with amines on target molecules (Fig. 2C). Therefore, EDCI/(sulfo-)NHS-coupled reactions are efficient and increase the yield of conjugation dramatically over that obtainable solely with EDCI. Similar to (sulfo) NHS, 1-hydroxybenzotriazole (HOBt) has also been used to improve the half-life of the activated carboxylates of hyaluronan (20). Using the HOBt chemistry, hyaluronan amides with diamine reagents, such as ethylenediamine, lysine methyl ester, 1,4-diaminobutane, 1,5-diaminopentane, 1,6-diaminohexane, and GlyLys (20–22).

Although the two-step EDCI chemistry improved the coupling efficiency of primary amine to the glucuronic acid moiety of hyaluronan, this chemistry still has some disadvantages. The degree of substitution is generally low (less than 10%), even using a large excess of EDCI. Moreover, it becomes extremely difficult to control the degree of substitution, and batch-to-batch variability is high.

An alternative approach to improve the conjugation of primary amine-like reagents to hyaluronan, is to increase the fraction of nucleophile present at pH 4–6. An important milestone in hyaluronan modification chemistry was the recognition that hydrazides would retain their nucleophilicity at pH 4–6 and would couple efficiently to carbodiimide-activated glucuronic acid residues of hyaluronan. Most aliphatic hydrazides have much lower pK_a values (usually 2–4) than primary amines (usually $pK_a = 12$) (23,24). Many mono- and polyfunctional hydrazides have been synthesized and used for the modification of hyaluronan (5,25). Hyaluronan grafted poly(ethylene glycol) (PEG-*graft*-HA) was synthesized by coupling methoxy-PEG-hydrazide with EDCI (Fig. 2D). This heterogeneous polymeric mixture was suggested as an injectable therapeutical formulation for ophthalmic or arthritis treatment (26). The use of dihydrazide compounds, such as adipic dihydrazide (ADH), provided multiple pendant hydrazide groups for further derivatization with drugs, biochemical probes, and for crosslinking agents (Fig. 2D) (27,28). Thus, the pendant hydrazide groups of ADH-modified hyaluronan can be loaded with drug molecules and then crosslinked into a hydrogel (23,29).

Alternatively, a limited number of the hydrazide groups could be crosslinked and the remaining hydrazide groups on the resulting hydrogel could be loaded with a therapeutic agent. For example, HA–Taxol® bioconjugates were synthesized by conjugation of HA–ADH and ester activated Taxol® (30).

Labeling of the HA–ADH–Taxol® conjugate with a fluorescent dye allowed monitoring of the uptake of the conjugate by human breast, ovarian, and colon cancer cells, and measurement of the corresponding selective toxicity of the polymeric prodrug (31). Hyaluronan was also coupled to poly(*N*-(2-hydroxypropyl)methacrylamide) (HPMA)–doxorubin (DOX) conjugate as a targeting moiety by the conjugation of HA–ADH to the activated ester of HPMA–DOX. The HPMA–HA–DOX targeted bioconjugate showed higher intake by cancer cells when compared to the non-targeted HPMA–DOX, and thus has potential as a tumor-targeted cancer treatment (32).

Another important advance in hyaluronan modification was the recent synthesis of thiolated hyaluronan derivatives (33). Dithiobis(propanoic dihydrazide) (DTP) and dithiobis(butyric dihydrazide) (DTB) were coupled to hyaluronan through EDCI-mediated chemistry and then the disulfide bonds of the first-formed gels were reduced by dithiothreitol (DTT). The resulting thiolated hyaluronan derivatives (HA–DTPH and HA–DTBH, where the H indicates the reduced form) were obtained (Fig. 2E) (33) as non-viscous solutions that were readily characterized spectroscopically. The side-chain structure of thiolated hyaluronan influenced the reactivity of thiols, and the thiol groups of HA–DTPH were more reactive than those of HA–DTBH. As a result, HA–DTPH was selected for further development of a new class of hyaluronan biomaterials (34, 35). The multiple pendant thiol groups of thiolated hyaluronan could be further derivatized with drugs, biochemical probes, and crosslinking agents. For instance, fluorescently labeled hyaluronan derivatives were readily prepared by the reaction of HA–DTPH with fluorescein-5-maleimide or Texas Red® C_2 maleimide (Shu XZ and Prestwich GD, unpublished results). More recently, thiolated hyaluronan have shown promise in biomedical applications, particularly for *in vivo* tissue regeneration using an in situ crosslinkable hydrogel (36,37).

B. Modification of the HA Hydroxyl Group

1. Sulfation

The coating of hyaluronan on medical devices, including catheters, guide wires and sensors, results in hydrophilic and lubricious surfaces that are by nature biocompatible, abrasion-resistant, and non-thrombogenic (HYDAK®, Biocoat Inc.). In order to obtain a blood-compatible surface on these devices, approaches to sulfation of the hyaluronan hydroxyl groups were devised (38). For example, sulfation of hyaluronan with a sulfur trioxide–pyridine complex in DMF produced $HyalS_x$ with different degrees of sulfation, where $x = 1$–4 sulfates per disaccharide (39). The introduction of sulfate groups made the macromolecule resistant to hyaluronidase and chondroitinase digestion. In addition, $HyalS_x$ showed low cytotoxicity (L929 fibroblast cells) and good cytocompatibility (human umbilical vein endothelial cells, HUVECs) (40). $HyalS_{3.5}$ was then immobilized onto plasma-processed polyethylene (PE) using diamine PEG derivative and water-soluble carbodiimide. The thrombin time test and platelet adhesion behavior yielded promising blood-compatible, anti-thrombotic PE surfaces (41). In addition, $HyalS_x$ was converted to a photolabile azidophenylamino derivative and was photoimmobilized onto a poly(ethylene terephthalate) film (42) and a glass substrate (43). Surfaces coated with $HyalS_x$ exhibited a marked reduction of cellular attachment, fouling, and bacterial growth compared with uncoated surfaces. Moreover, the coating was stable to degradation by chondroitinase and hyaluronidase (44,45).

2. *Esterification*

Butyric acid, which is known to induce cell differentiation and inhibit the growth of a variety of human tumors, was coupled to HA by the dimethyl-aminopyridine-catalyzed reaction between butyric anhydride and the sym-collidinium salt of low molecular weight hyaluronan in DMF (46). Hyaluronan butyrate offered a novel drug-delivery system targeted specifically to tumor cells. In a second example, hydroxyl groups of hyaluronan were esterified with methacrylate anhydride to give a methacrylate derivative. This derivative was used to fabricate a photopolymerized hydrogel (47).

3. *Etherification*

Epoxide and vinyl sulfone functionalities can react with hyaluronan hydroxyl groups under basic condition to create ether bonds. In a classical process, 1,4-butanediol diglycidyl ether and divinyl sulfone were used to crosslink hyaluronan and prepare hydrogels (48–50). The reaction of hyaluronan with glycidyl methacrylate also resulted in a new ether bond and a novel methacrylate derivative of hyaluronan. This glyceryl hyaluronan methacrylate was photo-polymerized to form a hydrogel (51–53).

4. *Isourea Coupling*

The anthracycline antibiotics and daunomycin were coupled to hyaluronan by cyanogen bromide (CNBr) activation (16). This reaction scheme is commonly used to activate oligosaccharides to produce affinity matrices by first generating a highly reactive isourea intermediate (54). The therapeutic agents appear to become attached by a urethane bond to one of the hydroxyl groups of the oligosaccharide or the glycosaminoglycan, but no spectroscopic verification was provided. Moreover, the harshness of the reaction conditions may compromise the integrity and biocompatibility of hyaluronan. Similarly, hyaluronan-collagen matrices were prepared by using CNBr activated hyaluronan to crosslink the amino functionalities of collagen (55).

5. *Periodate Oxidations*

Reactive bisaldehyde functionalities can be generated from the vicinal diol groups of HA by oxidation with sodium periodate. This chemistry is a standard method for chemical activation of glycoproteins for affinity immobilization or conversion to a fluorescent probe. With periodate-activated hyaluronan, reductive coupling with primary amines can result in crosslinking, attachment of peptides containing cell attachment domain, or immobilization to surfaces (56,57). As with the strong basic conditions, however, the harsh oxidative treatment introduces chain breaks and potentially immunogenic linkages into the hyaluronan biomaterials.

C. Other Chemistries

1. *Reducing End Modification*

End-labeling has not otherwise been extensively used for hyaluronan biomaterials or pro-drug applications, since there is only one attachment point per glycosaminoglycan. This severely limits loading and crosslinking possibilities for high molecular weight hyaluronan.

The reducing end of low molecular weight of hyaluronan and ε-amino groups of poly(L-lysine) (PLL) were covalently coupled by reductive amination to obtain the resulting comb-type copolymers (PLL-*graft*-hyaluronan). The polyampholyte comb-type copolymers consisted of a PLL main chain, a DNA binding site, and hyaluronan side chains. By adding cell specific ligands, these polyampholytes have potential as DNA carriers that can target hyaluronan receptor-rich sinusoidal endothelial liver cells (58).

2. *Amide Modifications*

Hydrazinolysis of the *N*-acetyl groups of hyaluronan creates free glucosamine moieties, which can be used for further coupling reactions. However, the hydrazinolysis usually performed in a basic solution at high temperature (100 °C), which also causes considerable degradation of molecular weight by base-induced backbone cleavage and reducing end modification (59,60).

3. *Hyaluronan-Graft-Poly(N-Isopropylacrylamide)*

Thermoresponsive hyaluronan hydrogels were prepared by graft polymerization of *N*-isopropylacrylamide (NIPAM) onto hyaluronan using a dithiocarbamate that acts as an initiator, transfer agent and terminator. First, hyaluronan was transferred into tri-*n*-butylamine salt in *N,N*-dimethylformamide, and then 4-(*N,N*-diethyldithiocarbamoylmethyl)benzoic acid was coupled to GlcNAc C-6 hydroxyls of hyaluronan using dicyclohexylcarbodiimide as the coupling reagent. Then, the thermoresponsive hyaluronan-graft-poly(*N*-Isopropylacrylamide) was synthesized by the irradiation of the solution containing the dithiocarbamate-derivatized hyaluronan and NIPAM (61). From low temperature to 37 °C, the hyaluronan-graft-PNIPAM solution exhibits unique solution–gel change, and the gel supported cell growth. However, the poor degradability of PNIPAM may impede its use for tissue regeneration applications.

III. Hyaluronan-Based Biomaterials

To obtain materials that are mechanically and chemically robust, a variety of crosslinking strategies based on the chemistry described above have been explored to produce insoluble or gel-like hyaluronan materials. We separate hyaluronan-based biomaterials into two categories based on the crosslinking mechanism: physical and chemical. In physical crosslinking, hydrophobic and ionic interactions convert hyaluronan or its derivatives into hydrogels. Chemically crosslinked materials have new covalent inter- and intramolecular bonds that create an infinite molecular network.

A. Esterified Hyaluronan (HYAFF®)

The HYAFF® esterfied hyaluronans are versatile and well-studied biomaterials. Esterfication results in the loss of hydrophilicity and hydrogen bonding by masking the free carboxylic group, and the simultaneous increase of hydrophobilicity by the introduction of alkyl or benzyl groups. As a result, most HYAFF® materials are insoluble in water. Instead, HYAFF® materials can be dissolved in organic solvents and extruded to produce membranes and fibers, lyophilized to obtain sponges, or spray-dried to produce microspheres.

The degree of esterification and the type of alcohol used influence the size of hydrophobic clusters. These clusters create a polymer chain network that is more rigid and stable, and less susceptible to enzymatic degradation. Hyaluronan esters undergo spontaneous de-esterification in an aqueous environment, and the hydrolytic degradation of ester bonds of HYAFF® 7 and HYAFF® 11 was mostly complete within 2–3 months in artificial plasma (62,63). The materials degrade *in vivo* within 110 days, synchronous with neotissue formation (64). The biocompatibility of HYAFF® materials has been extensively documented *in vitro* and *in vivo* and recently reviewed (8).

Drug release from HYAFF®-based devices was examined for entrapped or covalently attached molecules. For example, release of the steroids hydrocortisone and α-methylpredisolone from microspheres fabricated from different HYAFF® materials was examined with the drug either dispersed within or bound to the polymer. However, the release of physically entrapped small molecular weight drug with considerable aqueous solubility from HYAFF®-based devices was too rapid; hydrocortisone diffused out of microspheres in 10 min. The covalent attachment of steroid drugs to HYAFF® dramatically retarded drug release. The release showed zero-order kinetics with a half-time of 100 h, consistent with ester bond hydrolysis (65,66). Preclinical *in vivo* evaluations in a rabbit model demonstrated that an α-methylpredisolone carried by a HYAFF® 11 formulation could maintain its anti-inflammatory and chondroprotective properties (65–67) and increase the residence time of the drug in the tear fluid (68,69).

HYAFF® 11 matrix has more acceptable controlled-release properties for macromolecules. Polypeptides and proteins such as recombinant human granulocyte-macrophage colony-stimulating factor (rhGM-CSF) diffuse out slowly from these matrices (70). Furthermore, the mucoadhesive properties of HYAFF® materials permit drug delivery via nasal, oral, and vaginal mucosal routes. For example, HYAFF® materials have been used for intranasal delivery of insulin in sheep (71), vaginal delivery of calcitonin in rats (72,73), and vaginal delivery of flu vaccine in rodents (74).

Hyaluronan has many roles in biology, particularly during embryonic development (75–77). Therefore, hyaluronan derivatives may create a suitable environment not only for the growth of cells derived from organ biopsy samples, and may stimulate stem or precursor cells to divide and differentiate into specific cell types in this embryo-like environment (78,79).

Hyaluronan and its various chemical derivatives have been used already in wound healing and related tissue repair application. These unique properties of hyaluronan are attributed to its biophysical properties and also its biological properties as an ECM molecule as well as a biological signaling molecule (80,81). HYAFF® scaffolds have been employed to produce hyaluronan-based tissue engineered skin grafts. Epithelialization is achieved by employing autologous keratinocyte grafts delivered on Laserskin®, a HYAFF® micro-perforated transparent membrane designed to facilitate graft handling procedures and to enable grafting in preconfluence. The dermal component of the skin has been addressed by using a 3D HYAFF® fiber mesh scaffold, named Hyalograft 3D®, designed to support fibroblast cultures. Under these conditions, fibroblasts proliferate within the interstices of the scaffolds, depositing extracellular components such as type I, III, and IV collagen, and adhesion molecules such as laminin and fibronectin (82–89). To expedite vascularization and thereby facilitate transport of oxygen and nutrients and the removal of toxins, reconstruction of a dermal-like tissue provided with a capillary-like structure would have important advantages. To this end, HUVECs, either alone or with fibroblasts at 1:1 ratio, were seeded into HYAFF® 11 non-woven mesh. HYAFF® 11 non-woven mesh was found to be a suitable substrate for HUVEC adhesion, proliferation, and reorganization into microcapillary network (90).

Primary hepatocytes were seeded onto HYAFF® 7 and HYAFF® 11 films and non-woven fabrics in order to develop a potential bioartificial liver support device. After *in vitro* culture for 1 week, the cells in HYAFF® 7 non-woven fabrics exhibited an ammonia elimination approximately equal to that of cells on collagen films. The urea synthesis was more than threefold greater than that of adherent cells on collagen films, a widely used substrata for monolayer hepatocyte culture. Thus, HYAFF® 7 non-woven fabrics are promising substrata for hepatocyte culture in bioartificial liver devices (91).

Tissue reconstruction after surgical intervention or trauma tissue reconstruction represents an unresolved clinical problem. A biohybrid construct, consisting of hyaluronic acid-based scaffolds (HYAFF® 11) and human adipocyte precursor cells, was developed and the implantation of such bioconstructs into nude mice resulted in local differentiation of pre-adipocytes, suggesting the potential of the bioconstructs as autologous soft tissue-filler for clinical application in humans (92,93).

The use of HYAFF® as a scaffold for articular cartilage and bone tissue engineering has also been explored (64,94–96). For example, avian chondrocytes were shown to adhere and proliferate onto non-woven meshes of the hyaluronan-based scaffolds (HYAFF® 7 and HYAFF® 11). The adherent cells also synthesized a glycosaminoglycan- and collagen type II-rich ECM (64). HYAFF® 11 also supported the growth of human chondrocytes, and the cells maintained their original phenotype. Human chondrocytes seeded on HYAFF® 11 expressed and produced collagen type II and aggrecan and down-regulated the production of collagen type I, indicating the therapeutic potential of HYAFF® 11 as a delivery vehicle in a tissue-engineering approaches to the repair of articular cartilage defects (97). HYAFF® 11 was an effective human chondrocyte delivery matrix for the treatment of articular cartilage defects (98). Hyaluronan also showed a molecular weight-specific and dose-specific stimulation of osteoblasts, suggesting a mechanism to enhance the osteogenic and osteoinductive properties of bone graft materials (95).

Solchaga and co-workers (99) hypothesized that synthetic matrices based on hyaluronan derivatives could mimic the embryonic hyaluronan-rich environment supportive of progenitor cell development. When used as a delivery vehicle for cell-based therapies in adult organisms, the hyaluronan matrix should promote a recapitulation of these embryonic events and thereby facilitate tissue repair. Scaffolds made from HYAFF® 11 and an autocrosslinked derivative of hyaluronan (ACP™) were loaded with rabbit mesenchymal progenitor cells and compared with a control of porous, fibronectin-coated calcium phosphate ceramic cubes. HYAFF® 11 scaffolds retained 90% more cells than did the ACP™ scaffolds and the fibronectin-coated ceramics. When implanted subcutaneously in nude mice, HYAFF® 11 scaffolds retained their integrity after 3–6 weeks, whereas the ACP® scaffolds rapidly degraded within 7–10 days. Bone, cartilage, and fibrous tissue formation was evident in the pores of the implanted HYAFF® 11 and ceramic scaffolds, with the former showing a 30% increase in the relative amount of bone and cartilage per unit area (96,99). The HYAFF® 11 scaffold was also a good substrata for adhesion and proliferation of bone marrow-derived mesenchymal progenitor cells (100). When a bioconstruct of HYAFF® 11 scaffold seeded with autologous bone marrow-derived mesenchymal progenitor cells was implanted in a full-thickness osteochondral lesion in rabbit, the histological evidence revealed that lesions filled with the biomaterial, either seeded or unseeded with cells, healed faster and more completely than controls.

In bone repair, HYAFF® 11 scaffolds promoted mineralization of rat bone marrow stromal cells and basic fibroblast growth factor (bFGF) further enhanced the mineralization *in vitro* (101). Implantation of HYAFF® 11 scaffold seeded with bone marrow stromal cells, which had been previously grown *in vitro* in medium supplemented with bFGF, was used to treat large segmental radius defects in rat. The presence of bFGF significantly accelerated bone mineralization (102).

In addition to the HYAFF® materials, long alkyl chains (dodecyl and octadecyl) have also been covalently linked to hyaluronan via ester functions with low degree of substitution. In dilute aqueous solutions, these amphiphilic derivatives exhibited the rheological properties typical of hydrophobically associated polymers. These properties included the tremendous enhancement of zero shear rate Newtonian viscosity, steep shear-thinning behavior, and formation of physically crosslinked gel-like networks. Such physically associated polymeric materials could be used for synovial fluid viscosupplementation as well as in cartilage replacement (103).

B. Ionic Crosslinking HA-Based Biomaterials

GYNECARE INTERGEL® (LifeCore) is a viscous formulation of hyaluronan formed by chelation with ferric hydroxide. This reddish gel can be used during gynecological surgery to separate and protect tissues as they heal, and was approved by the FDA in November 2001 (104,105). Adhesions—bands of scar tissue that form during healing—are a common complication of surgery on the female reproductive organs (ovaries, uterus, or fallopian tubes). Adhesions inappropriately connect organs or tissues that should normally be separate, leading to pelvic pain, bowel obstruction, and infertility. In clinical use, the inner tissues were coated with GYNECARE INTERGEL® after surgical procedures were finished, but before the abdominal cavity was closed. The gel was meant to lubricate the tissue surfaces and prevent contact and adhesion formation. After about a week, the tissues absorbed the gel and the metabolites were excreted. However, in 2003 this product caused inflammatory responses during clinical use and was withdrawn from the market.

A thermoreversible hyaluronan gel was fabricated for drug release by simply mixing hyaluronan with doxycycline and a divalent metal cation in an aqueous solution (106). Two kinds of electrostatic interactions were invoked for the crosslinking and drug retention. The desired drug release profiles were achieved by controlling the absolute concentration of hyaluronan and doxycycline and the molecular weight of hyaluronan.

Polycations form robust polyelectrolyte complexes with polyanions. Such complexes have been suggested as scaffolds for tissue regeneration. Chitosan, the only natural positively charged polysaccharide with good biocompatibility and biodegradability, has been widely studied for biomedical applications (107). A chitosan–hyaluronan complex was obtained by simply adding chitosan hydrochloride solution into hyaluronan solution The resulting gel was well tolerated by chondrocytes and keratinocytes *in vitro* and *in vivo* in rats (108,109). However, the results showed that chitosan alone gave better results than the complex. This was attributed to the dissociation of the complex at physiological pH. In another study, chitosan–hyaluronan microspheres were prepared containing gentamycin as model drug, and the combination of hyaluronan with chitosan appeared to combine the mucoadhesive potential of hyaluronan with the penetration enhancing effect of chitosan (110).

Under physiological conditions, PLL has a higher charge density than chitosan, and thus would form stronger electrostatic interactions with hyaluronan. A biocompatible film based on PLL and hyaluronan was developed by deposition of alternating layers of PLL and hyaluronan. The molecular basis for this buildup was investigated by Picart and co-workers (111–113) using fluorescently labeled hyaluronan and PLL. These multilayered polyelectrolyte films may provide a novel approach to surface modification for cell adhesion and growth.

Oxidized polypyrrole (PP) is a polycationic electronic conductor, and recent results have shown that PP is able to support the growth and differentiation of multiple cell types *in vitro*, including neurons and endothelial cells. PP can also facilitate the *in vivo* regeneration of damaged peripheral nerves in rats (114,115). A PP–hyaluronan film was fabricated to combine the biological activity of hyaluronan and the electrical conductivity of PP. These films were promising candidates for tissue engineering and wound-healing applications that could benefit from both electrical stimulation and enhanced vascularization (116).

Alginate sponges loaded with hyaluronan provided an adapted environment for proteoglycan and collagen synthesis by chondrocytes (117). However, the supplemented hyaluronan rapidly diffused out of the beads, thus limiting the *in vitro* development of a cartilage repair material (118). Altenatively, the deacetylation of hyaluronan as described above creates new amino functionalities, thereby increasing the electrostatic interaction between alginate and hyaluronan. Additional biomaterials can be fabricated based on alginate and deacetylated hyaluronan (60).

C. Covalent Crosslinked Hyaluronan-Based Biomaterials

1. Hylans

Originally developed in the early 1980s by Balazs and commercialized by Biomatrix, Inc., Hylan® is produced by chemically crosslinking the hyaluronan containing residual proteins with formaldehyde (soluble gel) or divinyl sulfone (119). These materials are now marketed by Genzyme Corporation. Soluble Hylan A® is a very high molecular weight form (8–23 MDa) of hyaluronan (5–6 MDa) that exhibits enhanced rheological properties compared to native hyaluronan, and is used in ophthalmic viscosurgery (Hylashield®). Hylan B® was the first crosslinked, water-insoluble hyaluronan derivative with appropriate physical and biological properties for clinical use. In an alkaline environment, divinyl sulfone reacts with the hydroxyl groups of hyaluronan, creating bis-ethyl sulfone crosslinks. Hylan B® gel has greater elasticity and viscosity than soluble Hylan A® and retains the biocompatibility of the native hyaluronan (non-immunogenic, non-inflammatory and non-toxic).

Hylan B® gel for viscoaugmentation was developed as a "bulking agent" that is physically and biologically compatible with various soft tissues, including subdermal and sphincter muscle tissue. Hylan B® gel can be injected through a 30 gauge needle as small, elastically deformable gel particles. This gel is available in some countries for soft tissue augmentation to correct dermal wrinkles and depressed scars (Hylaform®) (120,121), and the approval by the FDA for use in the US is pending. Preclinical studies also showed that Hylan B® gel could be used to augment the sphincter muscle for the treatment of urinary stress incontinence. Additionally, the injection of viscoelastic hylan B® gel gave a durable augmentation of the soft tissue in the vocal fold in a rabbit model for the treatment of glottic insufficiency in laryngeal disorders (122).

Hylans have also been used for the treatment of degenerative joint disease and rheumatoid arthritis. The gel protects cartilage and prevents further chondrocyte injury. However, the effect was found to be reversible and viscosity dependent (123). Synvisc® (Hylan G-F 20®) is an FDA-approved local therapy that provides lubrication for the knee joint and acts as a "shock absorber." Successful with treatment Synvisc® can help reduce osteoarthritis knee pain, which can lead to increased mobility (124,125).

Hylan gel (Sepragel®) has been offered as an anti-adhesion in surgery (126), but has not been used in tissue regeneration due to its poor capacity to support cell growth. Cells from eight established cell lines originating from fibroblasts, epithelial or endothelial cells, chondrocytes, tumor cells, and stem cells were seeded on the hylan B® gel surface. Though all but the endothelial-origin cells attached to the gel, only the L929 fibroblasts and stem cells multiplied (127). To enhance cell attachment, coating with PLL (128), or the addition of matrix factors such as collagen type I, laminin, or fibronectin (129), was required.

2. Diepoxide-Crosslinked Hyaluronan

The methodology of using diepoxides to prepare hyaluronan gel was reported by Laurent (48) in the 1960s. Similar to the Hylans, this gel is mainly used in dermal augmentation in Europe under the tradename Restlane® (Q-Med Inc., Sweden). FDA approval for use in the United States is pending. Additionally, several reports have appeared for the use of this gel for drug delivery (130–132) and wound healing therapies after modification by the cell attachment domain sequence Arg–Gly–Asp (RGD) (133,134).

3. Auto-Crosslinked Hyaluronan

The internally crosslinked ACP™ was developed by Fidia Advanced Polymers, Inc. based on the intra-molecular esterfication of the carboxyl groups and hydroxyl groups of hyaluronan. This biomaterial, due to its highly viscous gel formulation, allows application even on vertical areas between adjacent organs. It has been shown to have a longer residence time than native hyaluronan, being slowly reabsorbed from peritioneal surfaces over 6–7 days. Thus, the potential of ACP™ gel as an anti-adhesive barrier after abdominal and gynecological surgery has been investigated (135,136).

In addition, porous ACP™ could behave as a scaffold for tissue regeneration. ACP was investigated as a cell delivery vehicle for the repair of bone and cartilage, but relatively poor cell growth was observed (99,137). The addition of BMP-2, as a bone enhancing osteogenic substrate, to ACP™ clearly increased both the volume and density in critically sized bony defects in a dog model (138). The cell affinity of ACP™ will need to be improved for the further use of ACP™ for tissue regeneration. Recently, studies have indicated that fibronectin coating can enhance the tissue regeneration properties of ACP™ (139).

4. *Photo-Crosslinking Hyaluronan*

Methacrylate and methacryamide derivatized hyaluronans were synthesized, and then used for hydrogel fabrication under UV or visible light with an initiator (18, 47,51–53). The potential uses for these in situ photopolymerized hydrogels include prevention of adhesions and tissue regeneration. The poor cell attachment of these in situ photopolymerized hydrogels can be significantly enhanced by RGD modification (18).

5. *Carbodiimide-Mediated Crosslinking*

Hyaluronan was claimed to be directly crosslinked by a water-soluble carbodiimide through intra-esterfication of the carboxyl groups and hydroxyl groups, but supporting physicochemical data was lacking (140). Recently, flexible, mucoadhesive, biocompatible controlled release films of hyaluronan "crosslinked" with 2 mM EDCI and containing 10% glycerol and 5% paclitaxel were synthesized (141) and examined for prevention of post-surgical adhesion formation. However, the materials degraded quickly, most likely due to the very low incidence of crosslinking relative to *N*-acylurea formation as described above.

The *O*-acyl to *N*-acylurea rearrangement could be used to advantage in a novel crosslinking approach. Thus, using biscarbodiimide coupling agents in the absence of added nucleophiles, hyaluronan was crosslinked by the joining two carboxyl groups in an inter- or intramolecular fashion. In most cases, a water-miscible organic cosolvent was used due to the poor aqueous solubility of biscarbodiimide agents (4). The water-insoluble gels, films and sponges of hyaluronic acid with biscarbodiimide were produced, and may be used as surgical aids to prevent adhesions (INCERT®, Anika Therapeutics Inc.).

As described above, the use of bishydrazide, trishydrazide, or polyvalent hydrazide compounds as nucleophiles in combination with EDCI has become the most versatile methodology for crosslinking hyaluronan. Hydrogels with physico-chemical properties ranging from soft-pourable gels to more mechanically rigid and brittle gels were created by varying reaction conditions and the molar ratios of the reagents involved (5,25). Applications of this chemistry were licensed in the United States by Clear Solutions Biotech, Inc. (Stony Brook, NY).

An alternative strategy to produce hyaluronan gels involves the synthesis of hyaluronan derivatives with unique properties followed by crosslinking with commercially available homobifunctional reagents. Through carbodiimide-mediated chemistry as described above, hyaluronan derivatives with primary amino or hydrazide functionalities were synthesized. The derivatives were then further crosslinked by glutaraldehyde, polyethylene glycol bis(succinimidyl propionate), and polyethylene glycol dialdehyde to produce hydrogels for drug delivery or tissue engineering (5,20–22,142,143). While biocompatible materials could be made, the crosslinking conditions per se were cytotoxic, and precluded the development of cell-seeded materials for in situ crosslinking approaches.

To work towards a cytocompatible crosslinking strategy, new chemistry was required. Thus, at the University of Utah, thiolated hyaluronan derivatives were synthesized using EDCI-hydrazide chemistry (33). Murine L-929 fibroblasts in cell culture medium were added to a solution of thiolated hyaluronan, and air oxidation of the thiols produced disulfide-crosslinked hyaluronan hydrogels. The encapsulated cells remained viable and proliferated slowly *in vitro*. However, the poor cell attachment and slow crosslinking kinetics limited the further use of this hydrogel for tissue regeneration (33). We next investigated the addition of a covalently linked ECM protein to create a covalently crosslinked biomimic of the ECM. Using thiol-modified gelatin, a solubilized collagen derivative, the cell attachment of this hyaluronan hydrogel was significantly improved. Thus, co-crosslinking of a 50:50 mixture of thiol-modified HA and thiol-modified gelatin produced a hydrogel that could be lyophilized to give a macroporous sponge that constituted a synthetic, covalent ECM mimetic. Both hydrogel films and sponges were successfully employed for cell culture *in vitro* and for the growth of new fibrous tissue *in vivo* (34,35). However, while these materials could be pre-seeded with cells prior to air-induced crosslinking, surgical implantation of the seeded scaffold was still required.

To proceed towards an injectable, cytocompatible gel for *in vivo* cell delivery, we sought to accelerate the gelation step. Conjugate addition of a thiol to a thiol acceptor was selected as the desired chemistry, and several different homobifunctional thiol acceptors based on the PEG chain were evaluated. PEG diacrylate (PEGDA) emerged as the optimal choice for crosslinking thiol-modified hyaluronan, both for cytocompatibility and speed of crosslinking—less than 10 min was required for gelation. This approach satisfied the key requirements that we had envisioned for injectable *in vivo* tissue engineering applications (36). Preliminary results showed that T31 human tracheal scar fibroblasts maintained the same phenotype and proliferated in this hydrogel 10-fold during 4 weeks culture *in vitro*. In addition, these human fibroblasts also proliferated and secreted ECM *in vivo* when implanted subcutaneously into nude mice. However, crosslinked hyaluronan alone was too hydrophilic and polyanionic and inhibited cell attachment. This limitation was overcome by covalent attachment of RGD-containing peptides to the hydrogel (Fig. 3). The attachment, spreading, and proliferation of human and murine fibroblasts were significantly enhanced on the surface of PEGDA-crosslinked thiol-modified hydrogels. Moreover, essentially perfect fibrous tissues were formed *in vivo* by injecting the viscous, gelling mixture of NIH 3T3 fibroblasts, thiol-modified HA, and PEGDA subcutaneously into the flanks of nude mice (37).

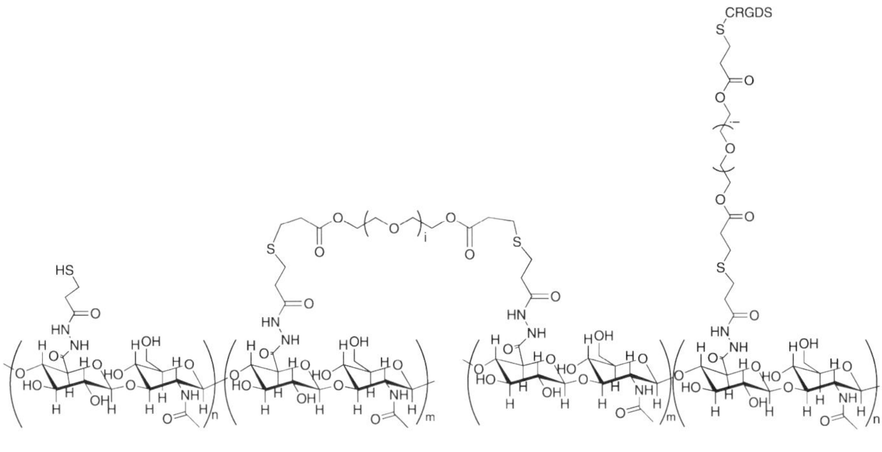

Figure 3 CRGDS peptide modified HA-DTPH hydrogel crosslinked by PEGDA.

6. *Glutaraldehyde-Crosslinked Hyaluronan*

Hyaluronan was crosslinked in glutaraldehyde aqueous solution and then the glutaraldehyde–hyaluronan adduct was resurfaced using a PLL coating to improve the biocompatibility and fibroblast attachment (144). The chemistry of the glutaraldhyde–hyaluronan adduct is that of a hemiacetal and is inherently hydrolytically unstable. Partial stabilization can occur with a polycationic polyamine such as PLL, but the linkages are still hydrolytically labile. In any case, residual aldehyde functionality from monovalent coupling rather than bivalent linkages can be quite cytotoxic and cause an inflammatory response. Recently, the use of glutaraldehyde in vapor phase as a crosslinking agent was reported to improve gel biocompatibility (145).

7. *Multiple-Component Crosslinked Hyaluronan*

Hyaluronan derivatives can be prepared via three- or four-component reactions known as the Passerini reaction and Ugi reaction. While this chemistry is extremely powerful and very versatile, only very limited information concerning the modification of hyaluronan by this chemistry is available (146). The removal of by-products, characterization of the linkages, and immobilization of partial reaction products remain concerns to be addressed before the potential of this chemistry for the fabrication of hyaluronan biomaterials can be realized.

8. *Hyaluronan-Grafted Copolymers*

Three new kinds of hyaluronan-grafted copolymer were described above, and each has been investigated for a specific application. Hyaluronan-grafted PEG was explored as a drug delivery vehicle (26). The temperature-sensitive hyaluronan-*graft*-poly(*N*-isopropylacrylamide) (61) was developed for injectable use, and the PLL-*graft*-hyaluronan has been studied for gene delivery (58).

9. Composites

Hyaluronan has been blended with other materials in order to produce interpenetrating molecular networks with good physicochemical, mechanical, and biocompatible properties. Several of these composites are described below.

a. Hyaluronan–Liposome Composites (147–149).

Unmodified hyaluronan has been incorporated into liposomes, and the solution of hyaluronan and liposomes was developed as a topically administered pharmaceutical agent. This composite was found to effectively treat skin disorders while minimizing systemic circulation. Bioadhesive liposomes, in which hyaluronan is the surface-anchored bioadhesive ligand on a liposome surface, were prepared by the pre-activation of hyaluronan with a carbodiimide. This activated form was then added to a suspension of multi-lamellar liposomes consisting of phosphatidylcholine, phosphatidylethanolamine, and cholesterol. In principle, a hyaluronan-coated liposome functionally resembles the PEG-coated "stealth" liposomes. Liposomes that avoid detection have been investigated for their ability to act as site-adherent and sustained-release carriers of epidermal growth factor for the topical therapy of wounds and burns.

b. Hyaluronan–Polysaccharide Composites.

Composites of chitosan–hyaluronan and alginate–hyaluronan composites based on physical interactions were prepared for use in drug delivery and tissue regeneration (60,108–110,117,118). This was described in detail above.

High-viscosity solutions of hyaluronan have been used in combination with basic fibroblast growth factor (FGF-2) as an injectable formulation for the treatment of bone fractures (150). The combination of hyaluronan and heparin has the potential to provide an even more efficient system for the sustained delivery of heparin-binding growth factors such as FGF-2. Therefore, polymeric gels of hyaluronan and heparin were synthesized, and the stability and activity of the released FGF-2 increased significantly (57).

Hyaluronan and carboxymethylcellulose (CMC), followed by carbodiimide-mediated "crosslinking" with lysine produces a bioresorbable membrane (Seprafilm®, Genzyme Corporation). The crosslinking chemistry most likely results in a material with *N*-acylurea linkages that confer electrostatic interactions as well as some covalent linkages. This membrane was approved by the FDA in 1996 for use in patients undergoing abdominal or pelvic laparotomy to reduce the incidence, extent and severity of postoperative adhesion (151). However, Seprafilm® is reported to suffer from handling difficulties—most likely due to the low percentage of covalent crosslinks—hampering its acceptance by surgeons for many applications. The same technology has been expanded to reduce the development of adhesions to synthetic non-absorbable mesh, such as hernia repair products. Sepramesh® is a composite of polypropylene mesh and a form of the Seprafilm® membrane. This composite reduces the development of adhesions to the surface of the hernia repair (152).

c. Hyaluronan–Collagen Composites.

Hyaluronan plays a prominent role in preventing scar formation during fetal wound healing. The *in vivo* evidence showed that application of hyaluronan reduced scar formation (153) and prevented joint contractures (154). The combination of hyaluronan and collagen may provide a suitable scaffold as a non-contractible biomaterial for skin regeneration. To prolong the persistence of hyaluronan in collagen fibrillar matrix (CFM), and simulate the lasting effects of hyaluronan in fetal wound healing, hyaluronan was crosslinked to collagen via periodate modification or CNBr-activation. The inhibition of CFM contraction by fibroblast was minimized for at least 7 days (155). Similar results were also observed in crosslinked CNBr-activated hyaluronan-collagen matrices (55). Biochemical and cytological analysis of these matrices suggested that hyaluronan strengthened the collagen fibrils and blocked direct communication between fibroblasts and the collagen fibrils.

Recently, a porous scaffold containing collagen and hyaluronan was fabricated by freezing–drying of the coagulates and subsequently crosslinking with EDCI to improve the mechanical stability of the composite matrix (156). The *in vivo* tests reported suggest that this matrix has promise for dermal tissue restoration (157).

Hyaluronan has been employed extensively in the treatment of osteoarthritis patients (158). However, controversial studies have been documented on the effects of hyaluronan on glycosaminoglycan synthesis by chondrocytes. Depending on molecular weight and biological source, hyaluronan has been reported to elicit inhibitory (159–161) as well as no effects (162,163) on cultured chondrocytes *in vitro*. However, in a 3D culture (collagen gel), hyaluronan enhanced chondrocyte proliferation as well as synthesis of both glycosaminoglycans and type II collagen, suggesting a benefit for repair of osteochondral defects (164).

A new osteoconductive collagen–hyaluronate matrix was synthesized for bone regeneration by crosslinking collagen fibers with peroxide-modified hyaluronate bearing active formyl groups (56). The resulting matrix was a 3D scaffold consisting of interconnected pores with an average size of 40 μm and a high pore volume/surface area ratio. The fraction of covalently bound hyaluronate in the matrix ranged from 5 to 25 wt%. When implanted in cranial defects in rats, the matrix demonstrated good biocompatibility and exhibited greater osteoconductive potential than matrices composed of either crosslinked collagen or crosslinked hyaluronate alone. The incorporation of hydroxyapatite may further improve the osteoconductivity in hyaluronan–collagen composites (165).

Composite materials consisting of hyaluronan and type I collagen also have been prepared by complexing both components into a coagulate followed by crosslinking with either glyoxal or starch dialdehyde (166). However, polyelectrolyte complex of hyaluronan and soluble type II collagen are readily formed and will precipitate from aqueous solutions. A new methodology was thus developed to make hyaluronan-type II collagen scaffold. In this method, 0.4 M NaCl was added to diminish the polymer-polymer electrostatic interactions. Next, EDCI was added to create the amide crosslinks between hyaluronan and collagen (167). This lyophilized and porous scaffold may be beneficial for cartilage regeneration, because hyaluronan and type II are the main components of cartilage.

Gelatin, the denatured form of collagen, is an excellent substrate for cell attachment, proliferation and differentitation. For instance, the physical incorporation of gelatin into esterified hyaluronan facilitated the osteochondral differentiation of culture-expanded, bone marrow-derived mesenchymal progenitor cell (168). The disadvantages of using gelatin as a scaffold material for tissue repair are its low biomechanical stiffness and rapid biodegradation. The conjugation of hyaluronan to gelatin matrices to prepare a covalent mimic of the ECM retarded degradation both *in vitro* and *in vivo*, and improved the biomechanical strength at the same time, which makes it useful as a cell delivery vehicle for tissue regeneration (34,35). Dipping an uncrosslinked hyaluronan-gelatin sponge into 90% (w/v) acetone–water containing EDCI resulted in a porous matrix, which could be used for either wound dressing or as a scaffold for tissue engineering (169). The incorporation of an antibiotic (silver sulfadiazine) in the sponge enhanced wound healing in a rat model (169).

IV. Conclusions

Hyaluronan possesses a combination of unique physical properties and biological functions that allow it to serve as an important starting material for the preparation of new biocompatible and biodegradable polymers. This review highlights the chemistries involved in the modification of hyaluronan, and the potential medical applications of these hyaluronan-based biomaterials. Important new products have already reached the marketplace, and the approval and introduction of an increasing number of medical devices and new drugs using hyaluronan-derived biomaterials can be anticiptated in the next decade.

References

1. Knudson CB, Knudson W. Cartilage proteoglycans. Semin Cell Dev Biol 2001; 12:69–78.
2. Fraser JR, Laurent TC, Laurent UB. Hyaluronan: its nature, distribution, functions and turnover. J Intern Med 1997; 242:27–33.
3. Band PA. In: Laurent TC, ed. The Chemistry, Biology and Medical Applications of Hyaluronan and Its Derivatives. London: Portland Press, 1998:33–42.
4. Vercruysse KP, Prestwich GD. Hyaluronate derivatives in drug delivery. Crit Rev Ther Drug Carrier Syst 1998; 15:513–555.
5. Prestwich GD, Marecak DM, Marecek JF, Vercruysse KP, Ziebell MR. Controlled chemical modification of hyaluronic acid: synthesis, applications, and biodegradation of hydrazide derivatives. J Control Release 1998; 53:93–103.

6. Luo Y, Kirker KR, Prestwich GD. In: Atala A, Lanza R, eds. Methods of Tissue Engineering. San Diego: Academic Press, 2001:539–553.
7. Barbucci R, Magnani A, Baszkin A, da Costa ML, Bauser H, Hellwig G, Martuscelli E, Cimmino S. Physico-chemical surface characterization of hyaluronic acid derivatives as a new class biomaterials. J Biomater Sci Polym Ed 1993; 4:245–273.
8. Campoccia D, Doherty P, Radice M, Brun P, Abatangelo G, Williams DF. Semisynthetic resorbable materials from hyaluronan esterification. Biomaterials 1998; 19:2101–2127.
9. Luo Y, Prestwich GD. Hyaluronic acid-*N*-hydroxysuccinimide: a useful intermediate for bioconjugation. Bioconjug Chem 2001; 12:1085–1088.
10. Barbucci R, Rappuoli R, Borzacchiello A, Ambrosio L. Synthesis, chemical and rheological characterization of new hyaluronic-based hydrogels. J Biomater Sci Polym Ed 2000; 11:383–399.
11. Barbucci R, Lamponi S, Borzacchiello A, Ambrosio L, Fini M, Torricelli P, Giardino R. Hyaluronic acid hydrogel in the treatment of osteoarthritis. Biomaterials 2002; 23:4503–4513.
12. Hermanson GT. Bioconjugate Chemistry. San Diego: Academic Press, 1996.
13. Kuo JW, Swann DA, Prestwich GD. Chemical modification of hyaluronic acid by carbodiimides. Bioconjug Chem 1991; 2:232–241.
14. Kuo JW, Swann DA, Prestwich GD. US patent 5356883, 1994.
15. Akima K, Ito H, Iwata Y, Matsuo K, Watari N, Yanagi M, Hagi H, Oshima K, Yagita A, Atomi Y, Tatekawa I. Evaluation of antitumor activities of hyaluronate binding antitumor drugs: synthesis, characterization and antitumor activity. J Drug Target 1996; 4:1–6.
16. Cera C, Terbojevich M, Cosani A, Palumbo M. Anthracycline antibiotics supported on water-soluble polysaccharides: synthesis and physicochemical characterization. Int J Biol Macromol 1988; 10:66–74.
17. Sakurai K, Miyazaki T, Kodera Y, Nishimura H, Shingu M, Inada Y. Anti-inflammatory activity of superoxide dismutase conjugated with sodium hyaluronate. Glycoconjug J 1997; 14:723–728.
18. Park YD, Tirelli N, Hubbell JA. Photopolymerized hyaluronic acid-based hydrogels and interpenetrating networks. Biomaterials 2003; 24:893–900.
19. Matsuda T, Magoshi T. Preparation of vinylated polysaccharides and photofabrication of tubular scaffolds as potential use in tissue engineering. Biomacromolecules 2002; 3:942–950.
20. Bulpitt P, Aeschlimann D. New strategy for chemical modification of hyaluronic acid: preparation of functionalized derivatives and their use in the formation of novel biocompatible hydrogels. J Biomed Mater Res 1999; 47: 152–169.
21. Crescenzi V, Francescangeli A, Taglienti A, Capitani D, Mannina L. Synthesis and partial characterization of hydrogels obtained via glutaraldehyde crosslinked of acetylated chitosan and of hyaluronan derivatives. Biomacromolecules 2003; 4: 1045–1054.
22. Crescenzi V, Francescangeli A, Taglienti A. New gelatin-based hydrogels via enzymatic networking. Biomacromolecules 2002; 3:1384–1391.
23. Pouyani T, Harbison GS, Prestwich GD. Novel hydrogels of hyaluronic acid: synthesis, surface morphology and solid-state NMR. J Am Chem Soc 1994; 116: 7515–7522.

24. Pouyani T, Prestwich GD. US patent 5,616,568, 1997.
25. Vercruysse KP, Marecak DM, Marecek JF, Prestwich GD. Synthesis and *in vitro* degradation of new polyvalent hydrazide cross-linked hydrogels of hyaluronic acid. Bioconjug Chem 1997; 8:686–694.
26. Moriyama K, Ooya T, Yui N. Hyaluronic acid grafted with poly(ethylene glycol) as a novel peptide formulation. J Control Release 1999; 59:77–86.
27. Prestwich GD, Vercruysse KP. Therapeutic applications of hyaluronic acid and hyaluronan derivatives. Pharm Sci Technol Today 1998; 1:42–43.
28. Prestwich GD, Luo Y, Ziebell MR, Vercruysse KP, Kirker KR, MacMaster JS. In: Abatangelo G, Weigel PH, eds. New Frontiers in Medical Sciences: Redefining Hyaluronan. Amsterdam: Elsevier, 2000:181–194.
29. Pouyani T, Prestwich GD. Functionalized derivatives of hyaluronan acid oligosaccharides: drug carriers and novel biomaterials. Bioconjug Chem 1994; 5:339–347.
30. Luo Y, Prestwich GD. Synthesis and selective cytotoxicity of a hyaluronic acid-antitumor bioconjugate. Bioconjug Chem 1999; 10:755–763.
31. Luo Y, Ziebell M, Prestwich GD. A hyaluronic acid–taxol antitumor bioconjugate targeted to cancer cells. Biomacromolecules 2000; 1:208–218.
32. Luo Y, Bernshaw NJ, Lu Z-R, Kopecek J, Prestwich GD. Targeted delivery of doxorubicin by HPMA copolymer–hyaluronan biocojugates. Pharm Res 2002; 19: 396–402.
33. Shu XZ, Liu Y, Luo Y, Robert MC, Prestwich GD. Disulfide crosslinked hyaluronan hydrogel. Biomacromolecules 2002; 3:1304–1311.
34. Shu XZ, Liu Y, Palumbo FS, Prestwich GD. Disulfide-crosslinked hyaluronan-gelatin hydrogel films: a covalent mimic of the extracellular materix for *in vitro* cell growth. Biomaterials 2003; 24:3825–3834.
35. Liu Y, Shu XZ, Gray SD, Prestwich GD. Disulfide-crosslinked hyaluronan-gelatin sponges: growth of fibrous tissue *in vivo*. J Biomed Mater Res 2004; 68A: 142–149.
36. Shu XZ, Liu Y, Palumbo SF, Luo Y, Prestwich GD. In-situ crosslinkable hyaluronan hydrogel for tissue engineering. Biomaterials 2004; 25:1139–1348.
37. Shu XZ, Ghosh K, Liu Y, Palumbo FS, Luo Y, Clark RA, Prestwich GD. Attachment and spreading of fibroblasts on an RGD peptide-modified injectable hyaluronan hydrogel. J Biomed Mater Res 2004; 68A:365–375.
38. Barbucci R, Magnani A, Casolaro M, Marchettini N, Rossi C, Bosco M. Modification of hyaluronic acid by sulfate group insertion to obtain heparin-like molecule: part I: characterisation and behaviour in aqueous solution towards H^+ and Cu^{2+} ions. Gaz Chim Ital 1995; 125:169–180.
39. Magnani A, Albanese A, Lamponi S, Barbucci R. Blood-interaction performance of different sulfated hyaluronic acids. Thromb Res 1996; 81:383–395.
40. Abatangelo G, Barbucci R, Brun P, Lamponi S. Biocompatibility and enzymatic degradation studies on sulphated hyaluronic acid and derivatives. Biomaterials 1997; 18:1411–1415.
41. Favia P, Palumbo F, D'Agostino R, Lamponi S, Magnani A, Barbucci R. Immobilization of heparin and highly-sulfated hyaluronic acid onto plasma-treated polyethylene. Plasma Polym 1998; 3:77–96.
42. Chen GP, Ito Y, Imanishi Y, Magnani A, Lamponi S, Barbucci R. Photo-immobilization of sulfated hyaluronic acid for antithrombogenicity. Bioconjug Chem 1997; 8:730–734.

43. Barbucci R, Magnani A, Lamponi S, Pasqui D, Bryant S. The use of hyaluronan and its sulphated derivative patterned with micrometric scale on glass substrate in melanocyte cell behaviour. Biomaterials 2003; 24:915–926.
44. Magnani A, Barbucci R, Montanaro L, Arciola CR, Lamponi S. *In vitro* study of blood-containing properties and effect on bacterial adhesion of polymeric surface with immobilized heparin and sulfate hyaluronic acid. J Biomater Sci Polym Ed 2000; 11:801–815.
45. Barbucci R, Magnani A, Lamponi S, Rappuoli R, Consumi M. In: Abatangelo G, Weigel PH, eds. New Frontiers in Medical Sciences: Redefining Hyaluronan. Amsterdam: Elsevier, 2000:203–212.
46. Coradini D, Pellizzaro C, Miglierini G, Daidone MG, Perbellini A. Hyaluronic acid as drug delivery for sodium butyrate: improvement of the anti-proliferative activity on a breast-cancer cell line. Int J Cancer 1999; 81:411–416.
47. Smeds KA, Pfister-Serres A, Hatchell DL, Grinstaff MW. Synthesis of novel polysaccharide hydrogel. JMS—Pure Appl Chem A 1999; 36:981–989.
48. Laurent TC, Hellsing K, Gelotte B. Crosslinked gels of hyaluronic acid. Acta Chem Scand 1964; 18:374–375.
49. Balazs EA, Leshchiner A, Band P. US patent 4,713,448, 1987.
50. Balazs EA, Leshchiner A. US patent 4,582,865, 1986.
51. Inukai M, Jin Y, Yomota C, Yonese M. Preparation and characterization of hyaluronate-hydroxyethyl acrylate blend hydrogel for controlled release device. Chem Pharm Bull 2000; 48:850–854.
52. Jin Y, Yamanaka J, Sato S, Miyata I, Yomota C, Yonese M. Recyclable characteristics of hyaluronate-polydroxyethyl acrylate blend hydrogel for controlled releases. J Control Release 2001; 73:173–181.
53. Leach JB, Bivens KA, Patrick CW Jr. Photocrosslinked hyaluronic acid hydrogels: natural, biodegradable tissue engineering scaffolds. Biotechnol Bioeng 2003; 82: 578–589.
54. Axen R, Porath J, Ernback S. Chemical coupling of peptides and proteins to polysaccharides by means of cyanogen halides. Nature 1967; 214:1302–1304.
55. Huang-Lee LLH, Nimni ME. Crosslinked CNBr-activated hyaluronan–collagen matrices: effects on fibroblast contraction. Matrix Biol 1994; 14:147–157.
56. Liu LS, Thompson AY, Heidaran MA, Poser JW, Spiro RC. An osteoconductive collagen/hyaluronate matrix for bone regeneration. Biomaterials 1999; 20: 1097–1108.
57. Liu LS, Ng C-K, Thompson AY, Poser JW, Spiro RC. Hyaluronate-heparin conjugate gels for the delivery of basic fibroblast growth factor (FGF-2). J Biomed Mater Res 2002; 62:128–135.
58. Asayama S, Nogawa M, Takei Y, Akaike T, Maruyama A. Synthesis of novel polyampholyte comb-type copolymers consisting of a poly(L-lysine) backbone and hyaluronic acid side chains for a DNA carrier. Bioconjug Chem 1998; 9:476–481.
59. Dahl LB, Laurent TC, Smedsrod B. Preparation of biological intact radioiodinated hyaluronan of high specific radioactivity: coupling of ^{125}I-tyramine-cellobiose to amino groups after partial *N*-deacetylation. Anal Biochem 1988; 175:397–407.
60. Oerther S, Maurin A-C, Payan E, Hubert P, Lapicque F, Presle N, Dexheimer J, Netter P, Lapicque F. High interaction alginate–hyaluronate association for the preparation of efficient biomaterials. Biopolymers 2000; 54:273–281.

61. Ohya S, Nakayama Y, Matsuda T. Thermoresponsive artificial extracellular matrix for tissue engineering: hyaluronic acid bioconjugated with poly(*N*-isopropylacrylamide) grafts. Biomacromolecules 2001; 2:856–863.
62. Benedetti L, Cortivo R, Berti T, Berti A, Pea F, Mazzo M, Moras M, Abatangelo G. Biocompatibility and biodegradation of different hyaluronan derivatives (Hyaff) implanted in rats. Biomaterials 1993; 14:1154–1160.
63. Zhong SP, Campoccia D, Doherty PJ, Williams RL, Benedetti L, William DF. Biodegradation of hyaluronic acid derivatives by hyaluronidase. Biomaterials 1994; 15:359–365.
64. Brun P, Abatangelo G, Radice M, Zacchi V, Guidolin D, Daga Gordini D, Cortivo R. Chondrocyte aggregation and reorganization into three-dimensional scaffolds. J Biomed Mater Res 1999; 46:337–346.
65. Benedetti LM, Topp EM, Stella VJ. Microspheres of hyaluronic acid esters-fabrication methods and *in vitro* hydrocortisone release. J Control Release 1990; 13:33–41.
66. Benedetti LM, Joshi HN, Goei L, Hunt JA, Callegaro L, Stella VJ, Topp EM. Dosage forms from polymeric prodrugs: hydrocortisone esters of hyaluronic acid. New Polym Mater 1991; 3:41–48.
67. Drobnik J. Hyaluronan in drug delivery. Adv Drug Deliv Rev 1991; 7: 295–308.
68. Hume LR, Lee HK, Benedetti LM, Sanzgiri YD, Topp EM, Stella VJ. Ocular sustained delivery of prednisolone using hyaluronic acid benzyl ester films. Int J Pharm 1994; 111:295–298.
69. Kyyronen K, Hume L, Benedetti LM, Urtti A, Topp E, Stella V. Methylprednisolone esters of hyaluronic acid in opthalmic drug delivery. Int J Pharm 1992; 80:161–169.
70. Nightlinger NS, Benedetti L, Soranzo C, Pettit DK, Pankey SC, Gombotz WR. *In vitro* release of huGM-CSF from esterified hyaluronic acid microspheres. Proc Int Symp Control Release Bioact Mater 1995; 22:738–739.
71. Ilium L, Farraj NF, Fisher AN, Gill I, Miglietta M, Benedetti LM. Hyaluronic acid ester microspheres as nasal delivery system for insulin. J Control Release 1994; 29:133–141.
72. Richardson JL, Ramire PA, Miglietta MR, Rochira M, Bacelle L, Callegaro L, Benedetti LM. Novel vaginal delivery systems for calcitonin 1. Evaluation of HYAFF calcitonin microspheres in rats. Int J Pharm 1995; 115:9–15.
73. Bonucci E, Ballanti P, Ramires PA, Richardson JL, Bacelle L. Prevention of ovariectomy osteopenia to rats after vaginal administration of HYAFF 11 microspheres containing calcitonin. Calcif Tissue Int 1995; 56:274–279.
74. Singh M, Briones M, O'Hagan DT. In: Abatangelo G, Weigel PH, eds. New Frontiers in Medical Sciences: Redefining Hyaluronan. Amsterdam: Elsevier, 2000:163–170.
75. Rooney P, Kumar S. Inverse relationship between hyaluronan and collagens in development and angiogenesis. Differentiation (Berlin) 1993; 54:1–9.
76. Wheatley SC, Isacke CM, Crossley PH. Restricted expression of the hyaluronan receptor, CD44, during postimplantation mouse embryogenesis suggests key roles in tissue formation and patterning. Development (Cambridge, UK) 1993; 119: 295–306.

77. Peterson PE, Pow CST, Wilson DB, Hendrickx AG. Distribution of extracellular matrix components during early embryonic development in the macaque. Acta Anat 1993; 146:3–13.
78. Shepard S, Becker H, Hartmann JX. Using hyaluronic acid to create a fetal-like environment *in vitro*. Ann Plast Surg 1996; 36:65–69.
79. Cortivo R, DeGalateo A, Castellani I, Burn P, Giro MG, Abatangelo G. Hyaluronic acid promotes chick embyro fibroblast and chondroblast expression. Cell Biol Int Rep 1990; 14:111–122.
80. Chen W, Abatangelo G. The functions of hyaluronan in wound repair. Wound Repair Regen 1999; 7:79–89.
81. Chen WYJ. In: Kennedy JF, Phillips GO, Williams PA, eds. Hyaluronan Volume 2 Biomedical, Medical and Clinical Aspects. Cambridge, UK: Woodhead Publishing, 2002:147–156.
82. Pianigiani E, Andreassi A, Taddeucci P, Alessandrini C, Fimiani M, Andreassi L. A new model for studying differentiation and growth of epidermal cultures on hyaluronan-based carrier. Biomaterials 1999; 20:1689–1694.
83. Burn P, Cortivo R, Zavan B, Vecchiato N, Abatangelo G. *In vitro* reconstructed tissues on hyaluronan-based temporary scaffolding. J Mater Sci Mater Med 1999; 10:683–688.
84. Andreassi L, Casini L, Trabucchi E, Diamantini S, Rastrelli A, Donati L. Human keratinocytes cultured on membranes composed of benzyl ester of hyaluronic acid suitable for grafting. Wounds 1991; 3:116–126.
85. Zacchi V, Soranzo C, Cortivo R, Radice M, Burn P, Abatangelo G. *In vitro* engineering of human skin-like tissue. J Biomed Mater Res 1998; 40:187–194.
86. Caravaggi C, Faglia E, Paola LD, Giglio RD, Caviani P, Mantero M, Gino M, Quarantiello A, Sommariva E, Pritelli C. In: Abatangelo G, Weigel PH, eds. New Frontiers in Medical Sciences: Redefining Hyaluronan. Amsterdam: Elsevier, 2000:313–320.
87. Harris PA, Francesco F, Barisono D, Leigh IM, Navsaria HA. Use of hyaluronic acid and cultured autologous keratinocytes and fibroblasts in extensive burns. Lancet 1999; 353:35–36.
88. Galassi G, Brun P, Radice M, Cortivo R, Zanon GF, Genovese P, Abatangelo G. *In vitro* reconstructed dernis implanted in human wounds: degradation studies of the HA-based supporting scaffold. Biomaterials 2000; 21:2183–2191.
89. Caravaggi C, Faglia E, Paola LD, Clerici G, Giglio RD, Sommariva E, Pritelli C, Manuela M, Caminiti M, Curci V, Fratino P In: Kennedy JF, Phillips GO, Williams PA, Hascall VC, eds. Hyaluronan Volume 2 Biomedical, Medical and Clinical Aspects. Cambridge, UK: Woodhead Publishing, 2002:79–86.
90. Tonello C, Zavan B, Cortivo R, Brun P, Panfilo S, Abatangelo G. *In vitro* reconstruction of human dermal equivalent enriched with endothelial cells. Biomaterials 2003; 24:1205–1211.
91. Catapano G, Bartolo LD, Vico V, Ambrosio L. Morphology and metabolism of hepatocytes cultured in Petri dishes on films and in non-woven fabrics of hyaluronan acid esters. Biomaterials 2001; 22:629–665.
92. von Heimburg D, Zachariah S, Low A, Pallua N. Influence of different biodegradable carriers on the *in vivo* behavior of human adipose precursor cells. Plast Reconstr Surg 2001; 108:411–420.

93. Halbleib M, Skurk T, de Luca C, von Heimburg D, Hauner H. Tissue engineering of white adipose tissue using hyaluronic acid-based scaffolds. I: *in vitro* differentiation of human adipocyte precursor cells on scaffolds. Biomaterials 2003; 24:3125–3132.
94. Aigner J, Tegeler J, Hutzler P, Campoccia D, Pavesio A, Kastenbauer CHE, Naumann A. Cartilage tissue engineering with novel nonwoven structured biomaterial based on hyaluronic acid benzyl ester. J Biomed Mater Res 1998; 42:172–181.
95. Huang L, Cheng YY, Koo PL, Lee KM, Qin L, Cheng JCY, Kumta SM. The effect of hyaluronan on osteoblast proliferation and differentiation in rat calvarial-derived cell cultures. J Biomed Mater Res 2003; 66A:880–884.
96. Awad HA, Erickson GR, Guilak F. In: Lewandrowski K-U, Wise DL, Trantolo DJ, Gresser JD, Yaszemski MJ, Altobelli DE, eds. Tissue Engineering and Biodegradable Equivalents: Scientific and Clinical Application. New York: Marcel Dekker, 2002:267–299.
97. Grigolo B, Lisignoli G, Piacentini A, Fiorini M, Gobbi P, Mazzotti G, Duca M, Pavesio A, Facchini A. Evidence for redifferentiation of human chondrocytes grown on a hyaluronan-based biomaterial (HYAFF11): molecular, immunohistochemical and ultrastructural analysis. Biomaterials 2002; 23: 1187–1195.
98. Girotto D, Urbani S, Brun P, Renier D, Barbucci R, Abatangelo G. Tissue-specific gene expression in chondrocytes grown on three-dimensional hyaluronic acid scaffolds. Biomaterials 2003; 24:3265–3275.
99. Solchaga LA, Dennis JE, Goldberg VM, Caplan AI. Hyaluronic acid-based polymers as cell carriers for tissue-engineering repair of bone and cartilage. J Orthop Res 1999; 17:205–213.
100. Radice M, Burn P, Cortivo R, Scapinelli R, Battaliard C, Abatangelo G. Hyaluronan-based biopolymers as delivery vehicles for bone-marrow-derived mesenchymal progenitors. J Biomed Mater Res 2000; 50:101–109.
101. Lisignoli G, Zini N, Remiddi G, Piacentini A, Puggioli A, Trimarchi C, Fini M, Maraldi NM, Facchini A. Basic fibroblast growth factor enhances *in vitro* mineralization of rat bone marrow stromal cells grown on non-woven hyaluronic acid based polymer scaffold. Biomaterials 2001; 22:2095–2105.
102. Lisignoli G, Fini M, Giavaresi G, Aldini NN, Toneguzzi S, Facchini A. Osteogensis of large segmental radius defects enhanced by basic fibroblast growth factor activated bone marrow stromal cells grown on non-woven hyaluronic acid-based polymer scaffold. Biomaterials 2002; 23:1043–1051.
103. Pelletier S, Hubert P, Payan P, Marchal P, Choplin L, Dellacherie E. Amphiphilic derivatives of sodium alginate and hyaluronate for cartilage repair: rheological properties. J Biomed Mater Res 2000; 54:102–108.
104. Johns DB, Rodgers KE, Donahue WD. Reduction of adhesion formation by postoperative administration of ionically crosslinked hyaluronic acid. Fertil Steril 1997; 68:37–42.
105. Lundorff P, Geldorp HV, Tronstad SE, Lalos O, Larsson B, Johns DD, diZerega GS. Reduction of post-surgical adhesions with ferric hyaluronate gel: a European study. Human Reprod 2001; 16:1982–1988.
106. Miyazaki T, Yomota C, Okada S. Development and release characterization of hyaluronan–doxycycline gels based on metal coordination. J Control Release 2001; 76:337–347.

107. Suh J-KF, Matthew HWT. Application of chitosan-based polysaccharide biomaterials in cartilage tissue engineering: a review. Biomaterials 2000; 21:2589–2598.
108. Denuziere A, Ferrier D, Domard A. Chitosan–chondroitin sulfate and chitosan–hyaluronate polyelectrolyte complexes: physico-chemical aspects. Carbohydr Polym 1996; 29:317–323.
109. Denuziere A, Ferrier D, Damour O, Domard A. Chitosan–chondroitin sulfate and chtiosan–hyaluronate polyelectrolyte complexes: biological properties. Biomaterials 1998; 19:1275–1285.
110. Lim ST, Martin GP, Berry DJ, Brown MB. Preparation and evaluation of the *in vitro* drug release properties and mucoadhesion of novel microspheres of hyaluronic acid and chitosan. J Control Release 2000; 66:281–292.
111. Picart C, Lavalle P, Cuisinier FJG, Schaaf P, Voegel JC. Buildup mechanism for poly(L-lysine)/hyaluronic acid films onto a solid surface. Langmuir 2001; 17: 7414–7424.
112. Picart C, Ladam G, Senger B, Voegel JC, Schaaf P, Cuisinier FJG, Gergely C. Determination of structural parameters characterizing thin films by optical methods: a comparison between scanning angle reflectometry and optical waveguide lightmode spectroscopy. J Chem Phys 2001; 115:1086–1094.
113. Picart C, Mutterer J, Richert L, Luo Y, Prestwich GD, Schaaf P, Voegel JC, Lavalle P. Molecular basis for the explanation of the exponential growth of polyelectrolyte multilayers. Proc Natl Acad Sci USA 2002; 99:12531–12535.
114. Schmidt CE, Shastri VR, Vacanti JP, Langer R. Stimulation of neurite outgrowth using an electrically conducting polymer. Proc Natl Acad Sci USA 1997; 94: 8948–8953.
115. Garner B, Georgevich B, Hodgson AJ, Liu L, Wallace GG. Polypyrrole–heparin composites as stimulus-responsive substrates for endothelial cell growth. J Biomed Mater Res 1999; 44:121–129.
116. Collier JH, Camp JP, Hudson TW, Schmidt CE. Synthesis and characterization of polypyrrole–hyaluronic acid composite biomaterials for tissue engineering applications. J Biomed Mater Res 2000; 50:574–584.
117. Miralles G, Baudoin R, Dumas D, Baptiste D, Hubert P, Stoltz JF, Dellacherie E, Mainard D, Netter P, Payan E. Sodium alginate sponges with or without sodium hyaluronate: *in vitro* engineering of cartilage. J Biomed Mater Res 2001; 57: 268–278.
118. Lindenhayn K, Perka C, Spitzer R-S, Heilmann H-H, Pommerening K, Mennicke J, Sittinger M. Retention of hyaluronic acid in alginate beads: aspects for *in vitro* cartilage engineering. J Biomed Mater Res 1999; 44:149–155.
119. Larsen NE, Balazs EA. Drug delivery systems using hyaluronan and its derivatives. Adv Drug Deliv Rev 1991; 7:279–293.
120. Larsen NE, Pollak CT, Reiner K, Leshchiner E, Balazs EA. Hylan gel biomaterial: dermal and immunologic compatibility. J Biomed Mater Res 1993; 27: 1129–1134.
121. Piacquadio DJ. In: Elson ML, ed. Evaluation and Treatment of the Aging Face. New York: Springer, 1994:304–308.
122. Hallen L, Johansson C, Laurent C. Crosslinked hyaluronab (hylan B gel): a new injectable remedy for treatment of vocal fold insufficiency—an animal study. Acta Otolaryngol 1999; 119:107–111.
123. Larsen NE, Lombard KM, Parent EG, Balazs EA. Effects of hylan on cartilage and chondrocyte cultures. J Orthop Res 1992; 10:23–32.

124. Pozo MA, Balazs EA, Belmonte C. Reduction of sensory responses to passive movements of inflamed knee joints by hylan, a hyaluronan derivative. Exp Brain Res 1997; 116:3–9.
125. Adams ME. An analysis of clinical studies of the use of crosslinked hyaluronan, hylan, in the treatment of osteoarthritis. J Rheumatol 1993; 20:16–18.
126. Kimmelman C, Edelstein C, Cheng H. Sepragel (hylan B) as a postsurgical dressing for endoscopic sinus surgery. Otolarynol Head Neck Surg 2002; 126:331–332.
127. Balazs EA, Eliezer-Pye IK, Dennebaum RA, Larsen NE, Whetstone JL. In: Kennedy JF, Phillips GO, Williams DF, Hascall VC, eds. Hyaluronan. Cambridge, UK: Woodhead Publishing, 2002; vol. 2:33–38.
128. Hu M, Sabelman EE, Tsai C, Tan J, Hentz VR. Improvement of schwann cell attachment and proliferation on modified hyaluronic acid strands by polylysine. Tissue Eng 2000; 6:585–593.
129. Ramamurthi A, Vesely I. Smooth muscle cell adhesion on crosslinked hyaluronan gels. J Biomed Mater Res 2002; 60:196–205.
130. Yui N, Nihira J, Okano T, Sakurai Y. Regulated release of drug microspheres from inflammation responsive degradable matrices of crosslinked hyaluronic acid. J Control Release 1993; 25:133–143.
131. Tomer R, Dimitrijevic D, Florence AT. Electrically controlled release of macromolecules from crosslinked hyaluronic acid hydrogel. J Control Release 1995; 33:405–413.
132. Tomihata K, Ikada Y. Preparation of crosslinked hyaluronic acid films of low water content. Biomaterials 1997; 18:189–195.
133. Glass JR, Dickerson KT, Stecker K, Polarek JW. Characterization of a hyaluronic acid Arg–Gly–Asp peptide cell attachment matrix. Biomaterials 1996; 17: 1101–1108.
134. Cooper ML, Hansbrough JF, Polarek JW. The effect of an arginine–glycine–aspartic acid peptide and hyaluronate synthetic matrix on epithelialization of meshed skin graft interstices. J Burn Care Rehabil 1996; 17:108–116.
135. De Iaco PA, Stefanetti M, Pressato D, Piana S, Dona M, Pavesio A, Bovicelli L. A novel hyaluronan-based gel in laparoscopic adhesion prevention: preclinical evaluation in an animal model. Fertil Steril 1998; 69:318–323.
136. De Iaco PA, Muzzupapa G, Pressato D, Bigon E, Rettore C, Dona M, Pavesio A, Bovicelli L In: Abatangelo G, Weigel PH, eds. New Frontiers in Medical Sciences: Redefining Hyaluronan. Amsterdam: Elsevier, 2000:345–349.
137. Solchaga LA, Gao J, Dennis JE, Awadallah A, Lundberg M, Caplan AI, Goldberg VM. Treatment of osteochondral defects with autologous bone marrow in a hyaluronan-based delivery vehicle. Tissue Eng 2002; 8:333–347.
138. Bernard GW, Pilloni A, Kang M, Sison J, Hunt D, Jovanovic S. In: Abatangelo G, Weigel PH, eds. New Frontiers in Medical Sciences: Redefining Hyaluronan. Amsterdam: Elsevier, 2000:215–231.
139. Solchaga LA, Goldberg VM, Caplan AI. In: Kennedy JF, Phillips GO, Williams DF, Hascall VC, eds. Hyaluronan. Cambridge, UK: Woodhead Publishing, 2002; vol. 2:63–66.
140. Tomihata K, Ikada Y. Crosslinking of hyaluronan acid with water-soluble carbodiimide. J Biomed Mater Res 1997; 37:243–251.
141. Jackson JK, Skinner KC, Burgess L, Sun T, Hunter WL, Burt HM. Paclitaxel-loaded crosslinked hyaluronan acid films for the prevention of postsurgical adhesions. Pharm Res 2002; 19:411–417.

142. Luo Y, Kirker KR, Prestwich GD. Cross-linked hyaluronic acid hydrogel films: new biomaterials for drug delivery. J Control Release 2000; 69:169–184.
143. Kirker KR, Luo Y, Nielson JH, Shelby J, Prestwich GD. Glycosaminoglycan hydrogel films as bio-interactive dressings for wound healing. Biomaterials 2002; 23:3661–3671.
144. Hu M, Sabelman EE, Lai S, Timek EK, Zhang F, Hentz VR, Lineaweaver WC. Polypeptide resurfacing method improves fibroblast's adhesion to hyaluronan strands. J Biomed Mater Res 1999; 47:79–84.
145. Ramire PA, E Milella. Biocompatibility of poly(vinyl alcohol)-hyaluronic acid and poly(vinyl alcohol)-gellan membranes crosslinked by glutaraldehyde vapors. J Mater Sci Meter Med 2002; 13:119–123.
146. Crescenzi V, Francescangeli A, Capitani D, Mannina L, Reiner D, Bellini D. Hyaluronan networking via Ugi's condensation using lysine as crosslinker diamine. Carbohydr Polym 2003; 53:311–316.
147. Yerushalmi N, Margalit R. Hyaluronic acid-modified bioadhesive liposomes as local drug depots: effects of cellular and fluid dynamics on liposome retention at target sites. Arch Biochem Biophys 1998; 349:21–26.
148. Yerushalmi N, Arad A, Margalit R. Molecular and cellular studies of hyaluronic acid-modified liposomes as bioadhesive carriers for topical drug delivery in wound healing. Arch Biochem Biophys 1994; 313:267–273.
149. Peer D, Margalit R. Physicochemical evaluation of stability-driven approach to drug encapment in regular and in surface-modified liposomes. Arch Biochem Biophys 2000; 383:185–190.
150. Radomsky ML, Thompson AY, Spiro RC, Poser JW. Potential role of fibroblast growth factor in enhancement of fracture healing. Clin Orthop Relat Res 1998; 355S:283–293.
151. Bowers D, Raybon B, Wheeless CR. Hyaluronic acid-carboxymethylcellulose films and perianastomotic adhesions in previously irradiated rats. Am J Obstet Gynecol 1999; 181:1335–1338.
152. Hooker GD, Taylor BM, Driman DK. Prevention of adhesion formation with use of sodium hyaluronate-based bioresorbable membrane in a rat model of ventral hernia repair with polypropylene mesh—a randomized, controlled study. Surgery 1999; 125:211–216.
153. Siebert JW, Burd AR, McCarthy JG, Weinzweig J, Ehrlich HP. Fetal wound healing: a biochemical study of scarless healing. Plast Reconstr Surg 1990; 85: 495–502.
154. Amiel D, Frey C, Woo SL, Harwood F, Akeson W. Value of hyaluronic acid in the prevention of contracture formation. Clin Orthop Relat Res 1985; 196:306–311.
155. Huang-Lee LLH, Nimni ME. Fibroblast contraction of collagen matrices with and without covalently bound hyaluronan. J Biomater Sci Polym Ed 1993; 5:99–109.
156. Park S-N, Park J-C, Kim HO, Song MJ, Suh H. Characterization of porous collagen/hyaluronic acid scaffold modified by 1-ethyl-3-(3-dimethylaminopropyl)carbodiimide crosslinking. Biomaterials 2002; 23:1205–1212.
157. Park S-N, Lee HJ, Lee KH, Suh H. Biological characterization of EDC-crosslinked collagen-hyaluronic acid matrix in dermal tissue restoration. Biomaterials 2003; 24:1631–1641.
158. Iwata H. Pharmacologic and clinical aspects of intraarticular injection of hyaluronate. Clin Orthop 1993; 289:285–291.

159. Wiebkin OW, Muir H. The inhibition of sulphate incorporation in isolated adult chondrocytes by hyaluronic acid. FEBS Lett 1973; 37:42–46.
160. Solursh M, Vaerewyck SA, Reiter S. Depression by hyaluronic acid of glycosaminoglycan synthesis by cultured chick embryo chondrocytes. Dev Biol 1974; 41:233–244.
161. Solursh M, Hardingham TE, Hascall VC, Kimura JH. Separated effects of exogenous hyaluronic acid on proteoglycan synthesis and deposition in pericellular matrix by cultured chick embryo limb chondrocytes. Dev Biol 1980; 75:121–129.
162. Larsen NE, Lombard KM, Balazs EA. Effect of hylan on cartilage and chondrocyte cultures. J Orthop Res 1991; 10:23–32.
163. Shimazu A, Jikko A, Iwamoto M, Koike T, Yan W, Okada Y, Shinmei M, Nakamura S, Kato Y. Effects of hyaluronic acid on the release of proteoglucan from the cell matrix in rabbit chondrocyte cultures in the presence and absence of cytokines. Arthritis Rheum 1993; 36:247–253.
164. Kawasaki K, Ochi M, Uchio Y, Adachi N, Matsusaki M. Hyaluronic acid enhances proliferation and chondroitin sulfate synthesis in cultured chondrocytes embedded in collagen gels. J Cellular Physiol 1999; 179:142–148.
165. Bakos D, Soldan M, Hernandez-Fuentes I. Hydroxyapatite–collagen–hyaluronic acid composite. Biomaterials 1999; 20:191–195.
166. Rehakova M, Bakos D, Vizarova K, Soldan M, Jurickova M. Properties of collagen and hyaluronic composite materials and their modification by chemical crosslinking. J Biomed Mater Res 1996; 30:369–372.
167. Taguchi T, Ikoma T, Tanaka J. An improved method to prepare hyaluronic acid and type II collagen composite matrices. J Biomed Mater Res 2002; 61:330–336.
168. Angele P, Kujat R, Nerlich M, Yoo J, Goldberg V, Johnstone B. Engineering of osteochondral tissue with bone marrow mesenchymal progenitor cells in a derivatized hyaluronan–gelatin composite sponge. Tissue Eng 1999; 5:545–553.
169. Choi YS, Hong SR, Lee YM, Song KW, Park MH, Nam YS. Studies on gelatin-containing artificial skin: II. Preparation and characterization of crosslinked gelatin-hyaluronate sponge. J Biomed Mater Res 1999; 48A:631–639.

Chemistry and Biology of Hyaluronan
H.G. Garg and C.A. Hales (editors)

Chapter 23

Medicinal Uses of Modified Hyaluronate

ROBERT J. MILLER and
LUIS Z. AVILA

Biomaterials Science and Engineering
Genzyme Corporation, 500 Kendall Street
Cambridge, MA 02142, USA

I. Introduction

Hyaluronic acid was first isolated from the vitreous humor by Karl Meyer and John Palmer 69 years ago (1). The proposed name, hyaluronic acid, was derived from one of its chemical constituents, uronic acid, and the hyaloid tissue that it was isolated from. Presently hyaluronic acid is called hyaluronate or hyaluronan because at physiological pH the polymer exists as a polyanion salt with bioavailable metal ions, such as sodium, potassium, calcium and magnesium. Hyaluronate exists as a simple repeating copolymer of D-*N*-acetylglucosamine and D-glucuronic acid, alternately linked β-1,3 and β-1,4 (see Figure 1 in Chapter 12) (2). Unlike heparin, chondroitin, and dermatan, which can have unique primary sequences by virtue of sulfation sites and specific epimeric centers, hyaluronate is essentially a simple non-branched polymer that differs from isolation source solely by its molecular weight. Hyaluronate is ubiquitous in nature and is produced by virtually every tissue in higher organisms and some bacteria. The only difference among commercial preparations of hyaluronate are any species-specific impurities that may be present and the molecular weight of the polymer. Biologically, the diversity in molecular weight has been shown to influence a variety of cellular processes such as inflammation (3–6), angiogenesis (7), fertility (8–10), and tissue lubrication (11). This influence in local cell biology has lead to the development of a hypothesis that hyaluronate homeostasis between anabolic biosynthesis and local catabolic degradation controls local cell reaction to inflammatory processes, infection and wound healing (3,12).

Since hyaluronate is a naturally occurring and ubiquitous material there has been significant commercial focus on the development of products either from hyaluronate or from chemically modified derivatives of hyaluronate. The current worldwide market for hyaluronate-derived products is substantial and in excess of one billion dollars annually. Most of this revenue is derived from the area of viscoelastics, such as Healon (Pharmacia) and Provisc (Alcon), that are used in the phacoemulsion of lens and intra-ocular lens insertion for the correction of cataracts (13). In this procedure, hyaluronate is inserted into the anterior chamber of the eye to afford a space filling role that helps protect the endothelial cells of the cornea perioperatively (14).

Recently, a variety of hyaluronate, and hyaluronate plus chemically cross-linked hyaluronate preparations have been approved as viscosupplements for patients with osteoarthritis of the knee (11,15–17) (see Chapter 20). These hyaluronate-based products include Artz® (Seikegaku) (18,19), Hyalgan® (Fidia) (20,21), and Synvisc® (Genzyme/Wyeth) (22–24). The exact mechanism for pain relief is unclear since some of the products provide relief for up to 1 year while the hyaluronate resides in the joint for less than 1 month (25–28).

Another recent area for the development of hyaluronate-based products is in reducing the incidence and severity of post-surgical adhesions. Hyaluronate's lubricious quality has been exploited to serve as a tissue protectant during surgery to reduce incidental tissue trauma. The hyaluronate-based product, Sepracoat®, has been shown in both animal models (29–31) and human clinical trials (32) to reduce the incidence and severity of tissue adhesion as a consequence of surgical trauma. This product is simply a 0.4% solution of hyaluronate that acts as a lubricant perioperatively to reduce incidental tissue damage due to desiccation and handling. The product does not function as a barrier because of its short residence time.

The first hyaluronate-based implantable device designed to reduce post-surgical adhesions by separating tissue planes was the Seprafilm® Bioresorbable Adhesion Barrier (33). This product was approved in 1997 for abdominal and gynecological surgery and is still the only product approved for both of these applications (34). Subsequent to this approval two more products have appeared on the market based on the same chemistry as found in Seprafilm®. The Sepramesh™ Biosurgical Composite is a prosthetic polypropylene mesh that is coated on one side with the same adhesion reduction material as in Seprafilm® (29). This device is used to repair a hernia defect in the abdominal wall while reducing the adherence of the underlying viscera to the mesh. A related product is the Seprapack Bioresorbable Nasal Packing, which is approved for separating tissue in the sinus and nasal cavities following surgery or trauma. In all of the above-mentioned devices the mechanism of action for adhesion reduction is similar in that the chemically modified hyaluronate slowly hydrates to form a gel that separates tissue planes for about 3–7 days until the initial inflammatory response abates and normal wound healing commences.

Of increasing interest is the potential ability of hyaluronate drug conjugates to target and control drug release to specific therapeutic areas. There are several receptors within the body that bind hyaluronate. The CD44 receptor has been implicated in helping orchestrate cellular attachment in the extracellular matrix (ECM) (35,36). This ubiquitous cell surface receptor is thought also to play a role in the local maintenance of homeostasis of hyaluronate in the ECM (37). A variety of other receptors such as RHAMM (CD168) (38–41), the HARE receptor on the hepatic sinusoidal endothelial cells (42,43), and the LYVE receptor in the lymph nodes (44–46) all present opportunities to bind hyaluronate in a tissue specific fashion. In addition, pathology can greatly affect the expression of hyaluronate receptors as described in some recent reviews of CD44 expression and hyaluronate binding in cancer and vascular disease (47). Because of this diverse and interesting biology hyaluronate affords a unique material to both deliver drugs and potentially target molecules to specific tissues in the body.

The medicinal uses of a series of products that have been generated from chemically modified hyaluronate have been discussed. This will span the development of adhesion reduction devices to hyaluronate drug conjugates that have the potential to deliver and/or target drugs in the body.

II. Surgical Adhesion Reduction

A. Etiology of Adhesion Formation

Surgically induced adhesions are a complication of invasive surgical procedures (48). Adhesions induced by either trauma, surgical or otherwise, and pathology (i.e., endometriosis) can lead to a variety of complications including infertility, bowel obstruction, pain, impaired joint mobility, and unintended complications during reoperation (49–53). The etiology of surgically induced adhesions is thought to be a result of incidental trauma caused by tissue manipulation (54), desiccation (54), tissue ischemia, and a foreign body reaction to particulates (55–57). This damage leads to a wound healing response that results in the production of a serous exudate that leads to fibrin clot production (58–60). If the fibrin clot is relatively short lived then it is resorbed with concomitant normal wound repair. If, however, the fibrin clot resides for a longer time period then cells of fibroblastic phenotype are recruited to the site. These cells begin to generate collagen leading to a scar-like tissue and adhesion formation. Over time remodeling of this tissue can lead to contraction and stricture, which if strategically placed around the viscera can lead to bowel obstruction. Therefore, reduction of adhesion formation can improve the outcome of a procedure and significantly reduce the complexity of reoperations (61).

B. Strategies to Reduce Post-surgical Adhesion Formation

There are essentially three ways to reduce adhesion formation: [1] reduce tissue damage during a procedure; [2] use a therapeutic that will affect fibrin resorbtion, such as a fibrinolytic or anti-inflammatory agent; or [3] use a barrier that will keep tissue planes separated until the fibrin clot has been resorbed and normal wound healing has taken place. Adhesion formation can be influenced by surgical technique (54,62–64) and the use of surgical irrigants such as Ringer's Lactate (65), Sepracoat™ (32), and solutions of icodextrins (66). Fibrinolytics, such as tPA (67,68), urokinase (67), streptokinase (67,69–71) and anti-inflammatory agents, such as naproxen (72), tolmetin (67,73), diphenhydramine (74), and aspirin (75), are examples of compounds that have been shown in animal models to reduce adhesions. Anti-proliferatives, such as paclitaxel, have also been shown to significantly reduce post-surgical adhesions in animal models (76). We will limit our discussion to the development of adhesion reduction barriers for the peritoneal cavity based on chemically modified hyaluronate.

C. Preparation of Modified Hyaluronates for Adhesion Reduction

Adhesion reduction using a barrier requires that the barrier reside long enough at the target site to keep the damaged tissue separated for about 3–7 days thereby allowing normal wound healing to occur (60). Unmodified hyaluronate will not remain at the target site long enough to keep tissue planes separated (26–28,77). Therefore, chemical modification of hyaluronate is required to extend the residence at the placement site long enough to affect adhesion formation.

A method to modify hyaluronate so that it can serve as a barrier is by interchain covalent cross-linking. There are four functional groups on hyaluronate that can react with cross-linking reagents. These groups are the hydroxyl, carboxyl, acetamido and the reducing end of the polymer (Fig. 1). The poor reactivity of the acetamido group in hyaluronate has resulted in a paucity of examples of derivatives prepared from the functional group. A novel hyaluronate

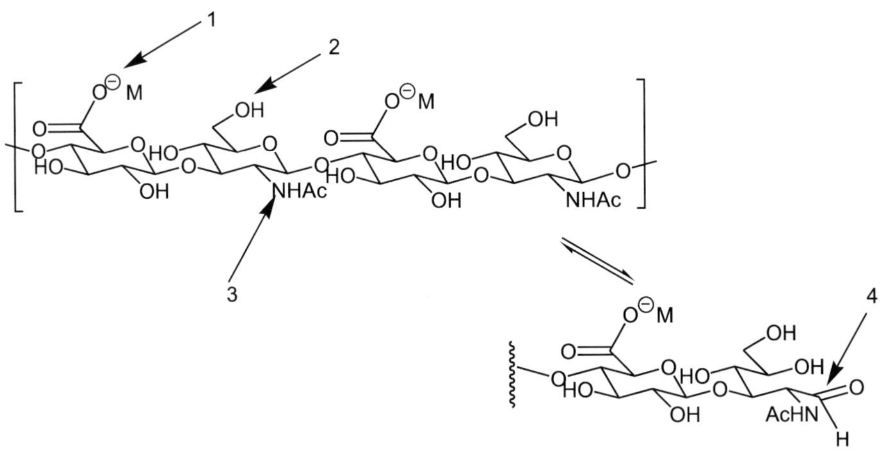

Figure 1 Functional groups on hyaluronate that can be chemically modified. Group 1, carboxyl; group 2, hydroxyl; group 3, acetamido; group 4, reducing end of polymer.

hydrogel has been prepared by first deacetylating hyaluronate with hydrazine then cross-linking the resulting amines with glutaraldehyde (78). The reducing end of hyaluronate does not lend itself to cross-linking hyaluronate because there is only one group per polymer. The most extensively published examples of hyaluronate cross-linking have been through the most prevalent functional group on the polymer, which are the hydroxyl or carboxyl groups.

Cross-linking through the hydroxyl groups, with bifunctional cross-linking reagents such as epichlorohydrin (79) and divinyl sulfone (80–83), has led to the development of some commercially available products for viscosupplementation (Synvisc®) (23,84–87) and soft tissue augmentation (Hylaform®) (88). These cross-links give functional groups that are hydrolytically quite stable under physiological conditions resulting in gels having residence times of months in the skin and years in the vitreous (80). Clearance of these derivatives from a placement site *in vivo* will depend almost entirely on the depolymerization of the hyaluronate chain.

Cross-linking via the hyaluronate carboxyl groups offers chemical diversity in that amide, acyl hydrazides or esters can be used thereby giving versatility in stability and degradation time. A very elegant method of cross-linking hyaluronate by an amide linkage has been done using either the Ugi or the Passerini reaction (Fig. 2). Reaction of hyaluronate, under Ugi conditions, with lysine, cyclohexylisocyanide, and formaldehyde yield an amide cross-linked hyaluronate (89).

Acyl hydrazide derivatives of hyaluronate form a derivative that possesses a nitrogen atom that is sufficiently nucleophilic at physiological pH to react with bifunctional cross-linking reagents (90,91) to give cross-linked hydrogels of hyaluronate. These molecules, despite their novelty and potential versatility, have not as of yet been developed into commercial products.

With regard to adhesions reduction one of the reasons for this lack of development might be that these materials by virtue of their high covalent cross-linking have residence time far in excess of the required 3–7 days postulated to affect adhesion formation. A promising variant on the technology is the introduction of bio-labile bond, such as a disulfide, in the cross-linking reagent into the cross-linked acyl hydrazide (92).

There have been some interesting examples of cross-linked hyaluronate based materials that reduce adhesions *in vivo*. A novel example of such a gel is the ferric ion complex with the carboxyl groups of hyaluronate (93). This iron-hyaluronate hydrogel successfully reduced adhesions in women undergoing peritoneal surgery under laparotomy for preservation of fertility in separate studies both in the United States (94,95) and Europe (96). This product, however, has been withdrawn from sale to the global market (97).

Another interesting and novel hydrogel for adhesion reduction has been produced by affecting internal, self-reaction of hydroxyl groups with the carboxyl moieties in hyaluronan (Fig. 3). This hydrogel made from this auto-crosslinked-polysaccharide (ACP) reduced surgical adhesions in rats (98) and rabbits (99). A recent human clinical trial in women undergoing hysteroscopic adhesiolysis

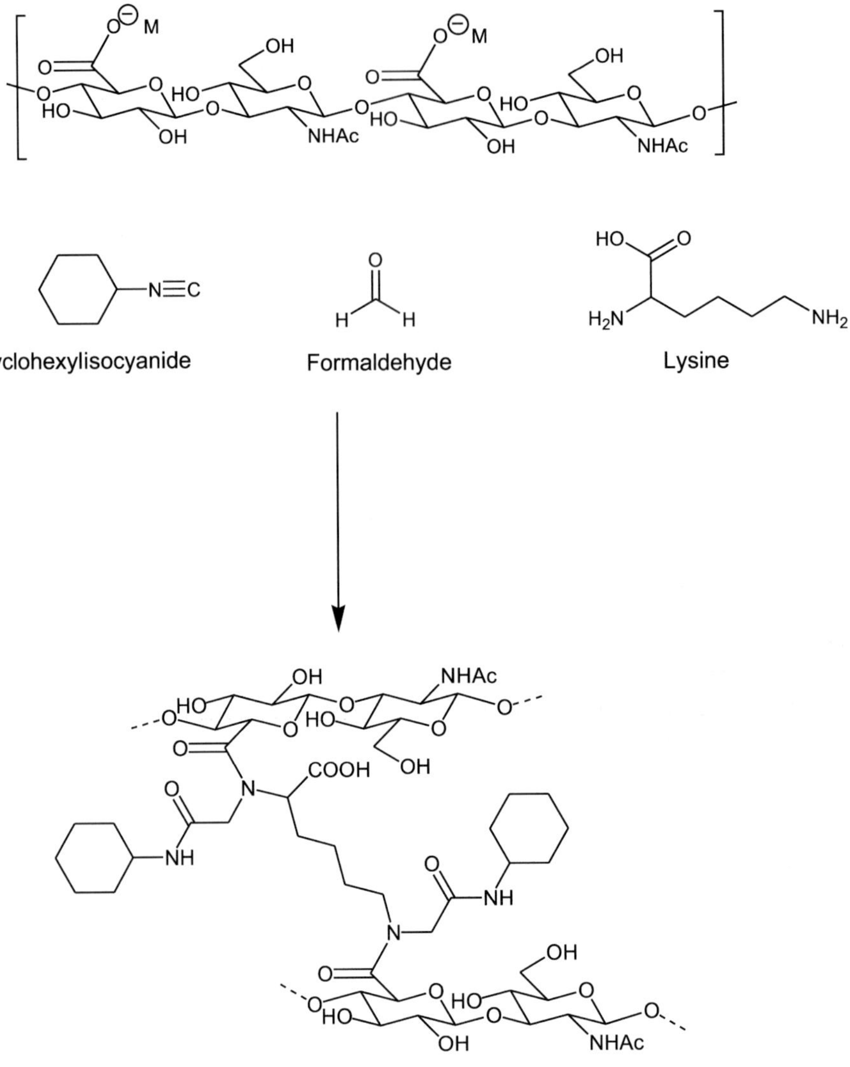

Figure 2 Putative reaction products for cross-linking of hyaluronate under the Ugi reaction conditions.

showed that patients that received the ACP gel at the time of surgery had significantly fewer adhesions at a 3 month follow-up (100).

The very first adhesion reduction device made from hyaluronate was approved in 1997 and is the Seprafilm® Bioresorbable Adhesion Barrier (34). This product is made from chemically derivatized sodium hyaluronate and sodium carboxymethyl cellulose (101). This device achieves the targeted residence time of 3–7 days, not by covalently cross-linking the polymer chains,

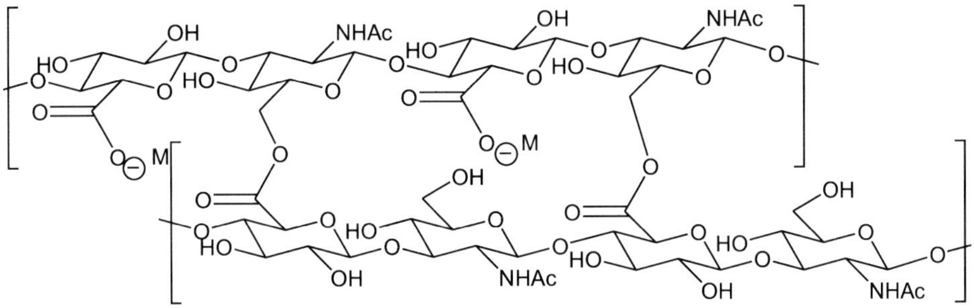

Figure 3 Putative structure of ACP hyaluronate.

but by chemically derivatizing the polymers so that there is a significant decrease in water solubility. This insolubilization is accomplished by derivatizing the carboxyl groups of both hyaluronate and carboxymethyl cellulose by the reaction of the water soluble carbodiimide, 1-ethyl-3-(3′-dimethylaminopropyl)carbodiimide hydrochloride (EDC) (Fig. 4). Under acidic conditions this reagent results in the modification of both hyaluronate and carboxymethyl cellulose to give an *N*-acylurea (102,103). The propylimido isomer has been shown, by isotopic labeling, to be the predominant of the two structural *N*-acylureas (104).

The partial modification of the carboxyl groups' results in a positive charge that can form a salt bridge with residual unreacted carboxyl on other polymer chains. This charge–charge interaction decreases water solubility and in turn extends the residence times *in vivo* of devices made from this material.

Seprafilm®, prepared from this process, has been shown to be very effective in reducing the incidence of post-surgical adhesions both in abdominal (105) and pelvic surgical procedures (106). The abdominal study was particularly compelling in that more than half (51%) of Seprafilm® recipients were adhesion-free, versus only 6% of untreated patients. Recently, it has been shown in a large multicenter study that there was no significant difference between Seprafilm® treated and control groups for abscess formation, peritonitis, and foreign body reaction (107). This safety with respect to peritonitis is consistent with results seen in a severe animal model for intra-abdominal sepsis where Seprafilm® did not adversely affect the progression of acute peritonitis or abscess formation (108). Seprafilm®, by virtue of its well demonstrated safety and efficacy in preclinical and human clinical studies, remains the only product approved for general surgical indications in the abdomen to this date.

Seprafilm® has been shown to reduce the incidence of adhesions to non-degradable, permanent tissue augmentation implants like polypropylene hernia repair meshes (109). This has led to the use of the diimide-modified hyaluronate and carboxymethyl cellulose materials in the Sepramesh™ Bioresorbable Composite. Sepramesh™ is a polypropylene mesh that has had laminated to one side of the mesh the same material that is in Seprafilm® (110). The placement of Sepramesh™ in the abdominal wall, with the adhesion reducing-side facing

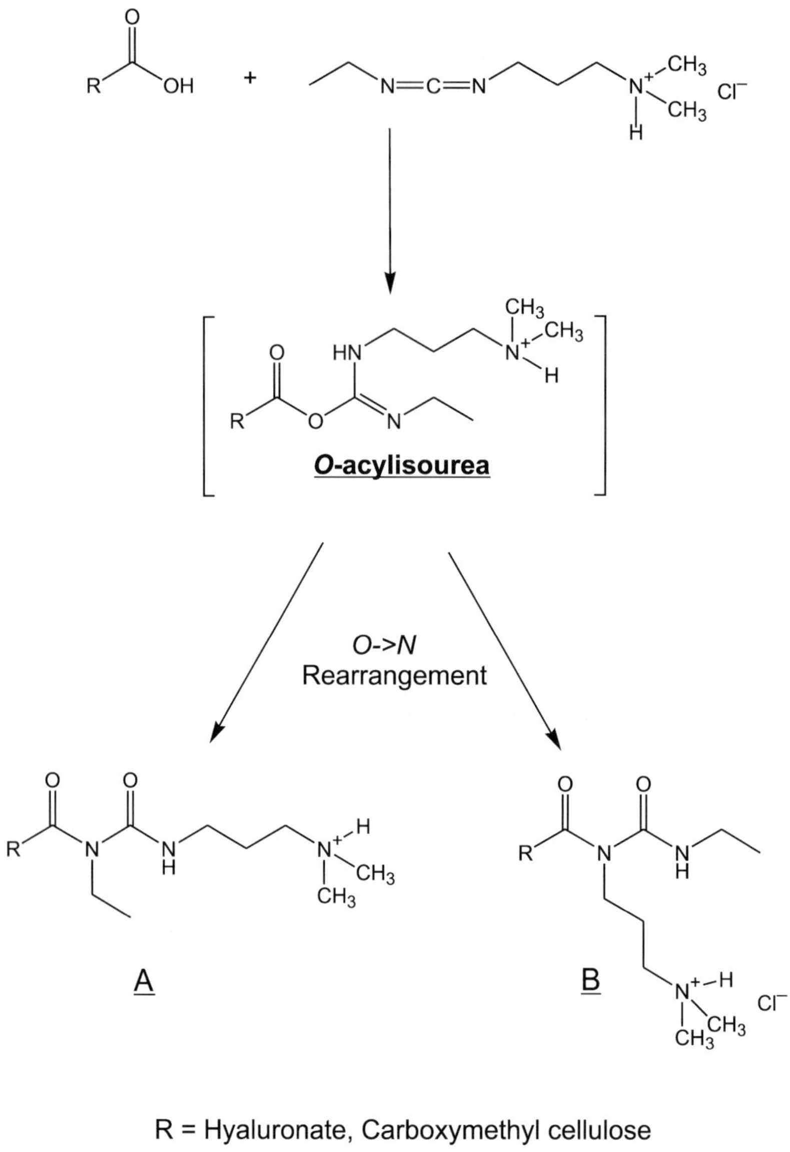

Figure 4 Scheme for the reaction of water soluble diimde, EDC with hyaluronate and carboxylmethyl cellulose under acidic reaction conditions (pH < 7). *N*-acylurea, B, is the predominant isomer under these reaction conditions.

the peritoneal cavity, has been shown to significantly reduce the incidence of adherence of the underlying viscera to the mesh (29,111). Reduction in the adherence of the viscera to a hernia repair mesh is thought to reduce the incidence of bowel fistula formation.

Additional derivatives have been introduced into the market to reduce sinus synechia (adhesions) following sinonasal surgery or trauma. Seprapak® Bioresorbable Nasal Packing is a foam that is made from the same material that is in Seprafilm® (101). This material functions to fill sinus cavities following surgery or trauma so that mucosal surfaces can remain separated during the healing process. A related product is Sepragel® Sinus which is prepared from divinyl sulfone cross-linked hylan (hylan B) (81). In a recently published study Sepragel® Sinus significantly improved the outcomes of patients that received this product following bilateral endoscopic ethmoidectomy (112). Lastly, an ethyl ester derivative of hyaluronate, Merogel®, is also commercially available. This material has been shown clinically to reduce the incidence of sinus adhesions following minimally invasive surgery to correct bilateral chronic sinusitis in patients (113).

III. Tissue Augmentation

A rapidly expanding area for the use of chemically modified hyaluronate has been in the use of these derivatives for soft tissue augmentation. The majority of applications for these types of products are in the aesthetics market to correct wrinkles and facial blemish such as acne scars. The hyaluronate-based products offer a significant advantage in that they can be used without a prerequisite skin test for sensitivity to bovine collagen products. Currently there are several hyaluronate-derived dermal fillers, such as Hylaform®, Restylane®, Perlane®, and Juvederm® that are sold in Europe. Hylaform is divinyl sulfone cross-linked hyaluronate-derived from chicken comb (hylan A) (Fig. 5) (80, 81).

Restylane®, Perlane® and Juvederm® are all prepared from bacterial-derived hyaluronate that is cross-linked with 1,4-butanediol diglycidyl ether under basic conditions (Fig. 6) (114). Both cross-linking reagents react with

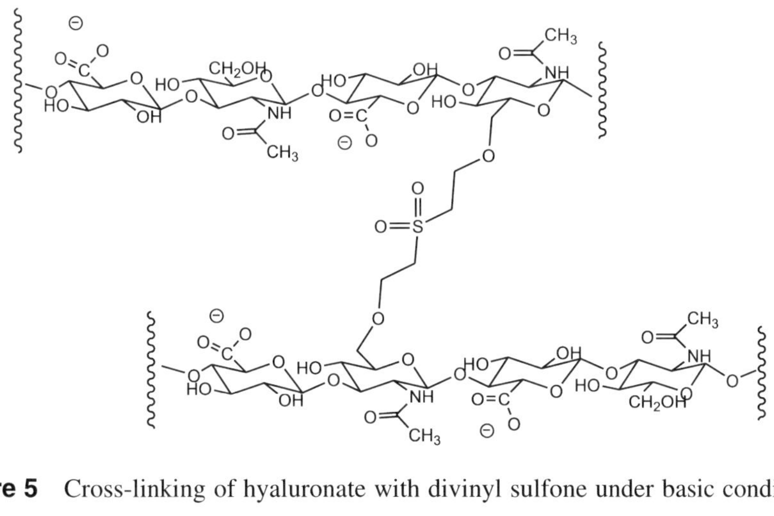

Figure 5 Cross-linking of hyaluronate with divinyl sulfone under basic conditions.

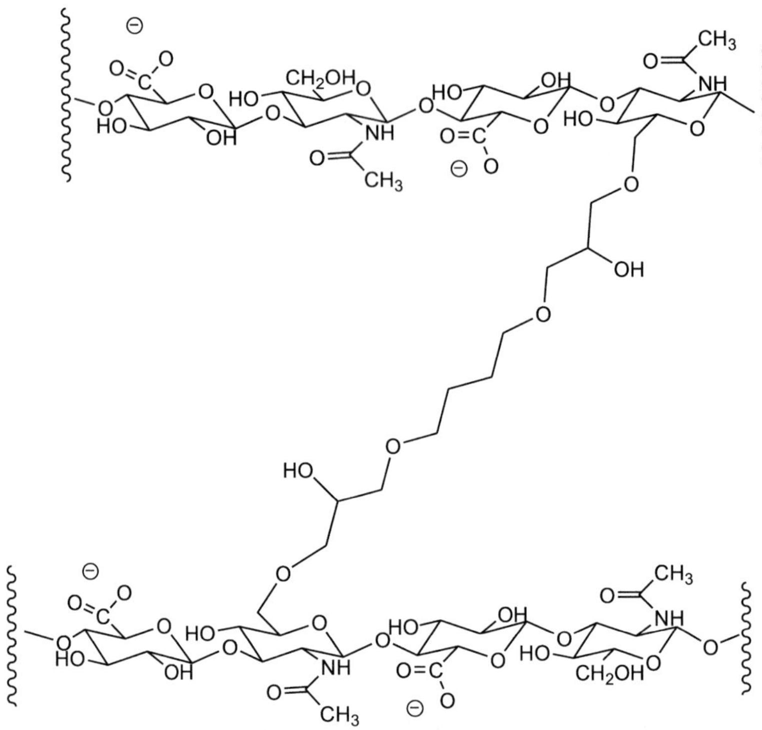

Figure 6 Cross-linking of hyaluronate with 1,4-butanediol diglycidyl ether under basic conditions.

the hydroxyl groups in hyaluronate to give chemically stable ether linkages that significantly extend the residence time at the placement site for several months. These products will afford aesthetic correction for about 3–6 months in humans (115–117). These products have the potential to provide mechanical correction of other soft tissues (e.g., vocal folds) in the body (88).

The regulatory approval of both cross-linked and chemically derivatized hyaluronates provides a unique platform to expand the application of these devices. One such expanded use would be to combine these devices with drugs in an attempt to alter the biology around the implant site or to serve as a drug delivery platform.

IV. Conjugation of Therapeutic Agents to Hyaluronate

In this section, we will discuss several strategies for coupling biologically active agents to hyaluronate and discuss the putative mechanisms for drug release both *in vitro* and *in vivo*. Specifically we will focus on the development of hyaluronate conjugates to anti-inflammatory agents and anti-proliferatives.

A. Conjugation of Taxol to Hyaluronate

Although Taxol has shown tremendous potential as an anti-cancer drug, its use is compromised by its poor aqueous solubility. Taxol conjugation to hyaluronate has been done to both increase water solubility and provide potential targeting to cells rich in hyaluronate cell surface receptors (CD44) (118). This hyaluronate derivative was prepared from the coupling of a Taxol succinylester to an acyl hydrazide of hyaluronate (Fig. 7). This approach was designed to preserve the structure of the active agent on eventual release from the matrix.

This approach arose as a consequence of the rearrangement of the putative *O*-acylurea intermediate in preference to reacting with amine nucleophiles under diimide reaction conditions at acidic pH (pH 4.5). In contrast, dihydrazides ($pKa = 2.0-3.0$) are substantially unprotonated and therefore remain nucleophilic under these conditions (119). The use of difunctional dihydrazides improves accessibility for conjugation and could be used to modify other chemical properties of the resulting conjugates (91,119).

The hyaluronate–Taxol conjugates showed selective toxicity toward human cancer cell lines that are known to over-express hyaluronate receptors while no toxicity was observed toward a mouse fibroblast cell line at the same concentration. The hyaluronate-ADH was shown to be completely non-toxic (118). Rapid uptake and selective cytotoxicity of hyaluronate–Taxol conjugates was blocked by either excess hyaluronate or by an anti-CD44 antibody but not by chondroitin sulfate. Release of Taxol from hyaluronate–Taxol in human plasma or in cell culture media revealed than the free drug was hydrolytically released by cleavage of the labile $2'$ ester linkage (118,120). A fluorescent–hyaluronate–Taxol conjugate showed cell-specific binding and uptake using flow cytometry and fluorescent confocal microscopy allowing direct correlation of uptake with selective toxicity (120).

B. Steroid Esters of Hyaluronate

Esters of hyaluronate have been prepared by reaction of quaternary ammonium salts of hyaluronate with alkyl halides in polar aprotic solvent such as methylsulfoxide (121). These esters hydrolyze under physiological conditions to give back the starting hyaluronate polymer and the corresponding alcohol (122). A hydrocortisone ester of hyaluronate has been shown to release intact hydrocortisone under physiological conditions (122) (Fig. 8).

The hydrolysis rate of the steroid esters and the ensuing steroid release were found to be independent of the degree of esterification. A 25% steroid esterified hyaluronate (HYC14) was found to release steroid at the same rate as a mixed ester (HYC13) containing 25% steroid and 25% ethyl ester. This suggested that the hydrolysis rate is primarily influenced by the local environment of the ester bond (122,123). The hydrolysis rates of partial hydrocortisone esters of HA was not affected by the presence or absence of esterases (124). This observation suggested that these esters are not good substrates of porcine liver esterase. In contrast, 21-acetate and -butyrate esters of hydrocortisone are rapidly cleaved by

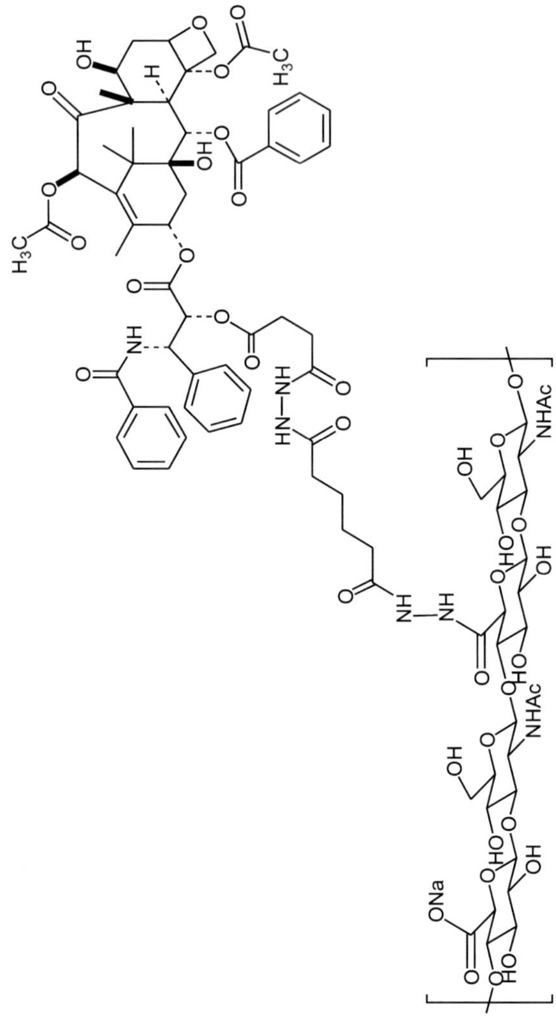

Figure 7 Structure of Hyaluronan–Taxol conjugate.

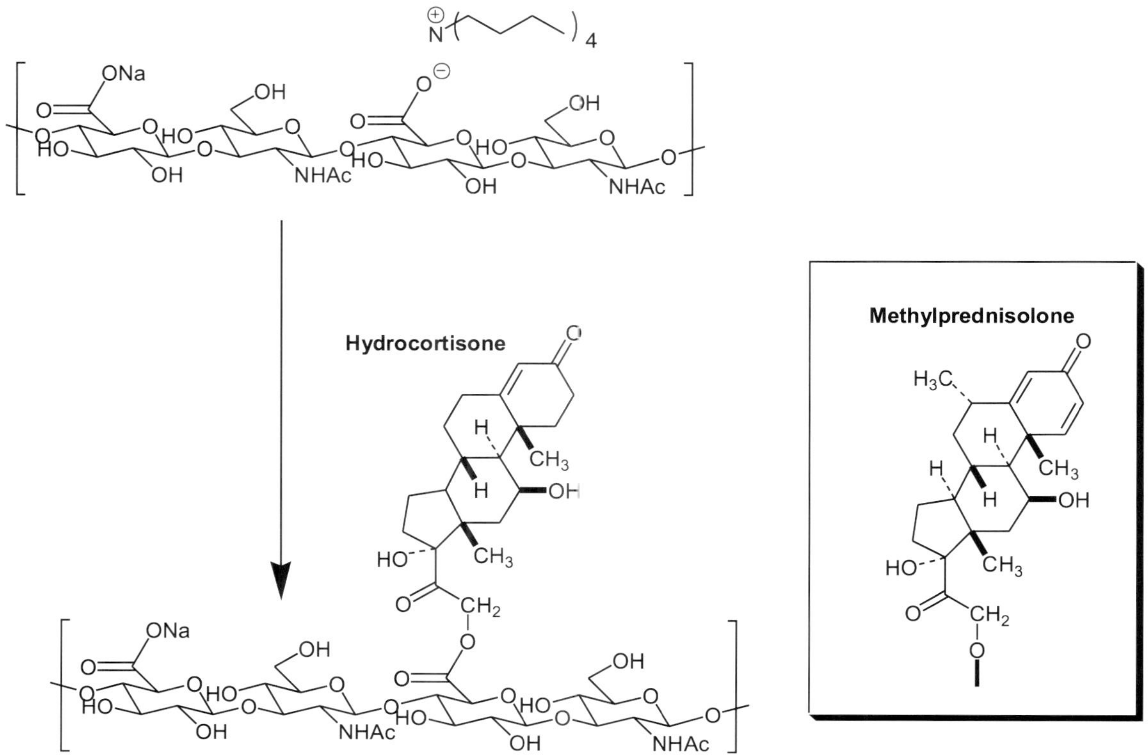

Figure 8 Preparation of ester conjugates of hyaluronan in organic phase using a tetrabutylammonium salt of hyaluronan.

this esterase (125). In a subsequent study, it was found that corticosteroid ester of hyaluronate (HCY141) was rapidly cleaved by synovial fluid extracts from patients with arthritic knees (126). This effect was not observed when the steroid ester was exposed to synovial fluid of healthy individuals suggesting that at least in arthritic knees there is a non-specific esterase that is capable of cleaving these steroid esters rapidly (126). The use of these steroid esters in arthritic knees are limited by this lack of predictability in release at least for this application.

A. Other Esters of Hyaluronate

Esters of hyaluronate have been shown to affect cell viability and handling for tissue engineering applications. One such example is HYAFF 11 which is a benzyl ester derivative of hyaluronate and has been shown to act as cell delivery vehicle *in vivo* (127,128). At 100% esterification, HYAFF 11 is a highly hydrophobic and water insoluble polymer (Fig. 9). HYAFF 11 is thought to degrade and clear *in vivo* by first hydrolyzing the benzyl esters to release hyaluronate and benzyl alcohol, which are products that have well known clearance pathways (129,130).

Because of the initial hydrophobicity these hyaluronate-based materials can function as tissue growth scaffolds for several weeks (128,129,131–133). The chemistry of HYAFF 11 easily renders it processable by conventional textile processes into woven and non-woven materials. A non-woven HYAFF 11-based matrix, prepared from fibers with a diameter $<20\ \mu m$ has been used as a tissue scaffold for the culturing of autologous cartilage cell (131,132,134,135). This effort has resulted in the development of an autologous cartilage replacement product that is currently being sold in some European countries (136). HYAFF 11 has also been shown to function as a tissue scaffold for fibroblasts and keratinocytes (128,133,137,138).

V. Summary and Conclusion

The biocompatibility and physiochemical properties of hyaluronate and its presence in virtually every tissue and fluid in the body make it an excellent material to develop medical devices from. The lubricity of hyaluronate at relatively low solution concentrations have been exploited to significantly lower the risk of complications in phacoemulsion and intra-ocular lens insertion and offers a new treatment for pain associated with early stage osteoarthritis.

Recently, several products derived from chemically modified hyaluronate have appeared on the market. Several hyaluronate-based adhesion reduction barriers now afford both general and gynecological surgeons products that can reduce the incidence of post-surgical adhesions while being able to bioresorb with time.

Conjugation of drugs to hyaluronate holds great promise for the generation of a new class of polymer based therapeutics. Drug delivery systems made from

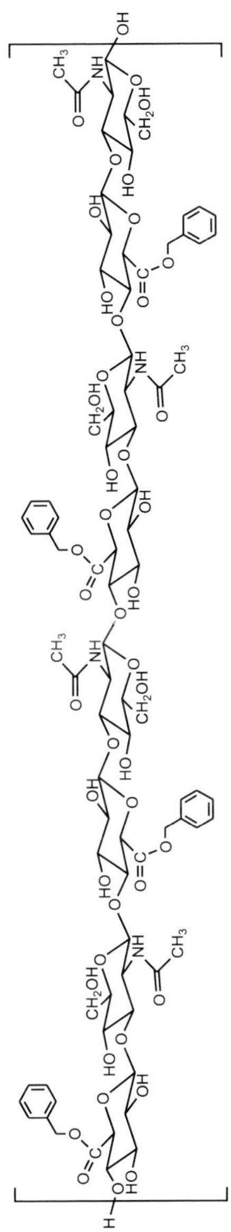

Figure 9 Benzyl ester of hyaluronate (HYAFF 11).

chemically modified hyaluronate have the ability to function as drug delivery platforms for the local delivery of therapeutics. Drugs conjugated to hyaluronate could also serve to target the drug to cells or tissues in the body that are rich in hyaluronate binding receptors.

Lastly, one of the most promising uses of chemically modified hyaluronate is in the area of tissue engineering. Hyaluronate is found in the ECM of virtually all tissues so its use as a cell delivery vehicle is quite logical because of its biocompatibility and ability to interact with cells. The ability of hyaluronate to serve as a cell delivery platform as well as have active agents conjugated uniquely positions this biomaterial as a versatile and powerful tool for the promising area of tissue engineering.

Acknowledgements

The authors would like to thank Dr James W Burns and Ms Alma E Weightman for their review and comments on this chapter.

References

1. Meyer K, Palmer JW. The polysaccharide of the vitreous humor. J Biol Chem 1934; 107:629–634.
2. Weissmann B, Meyer K. The structure of hyalobiuronic acid and of hyaluronic acid from umbilical cord. J Am Chem Soc 1954; 76:1753–1757.
3. Noble PW. Hyaluronan and its catabolic products in tissue injury and repair. Mater Biol 2002; 21:25–29.
4. Horton MR, McKee CM, Bao C, Liao F, Farber JM, Hodge-DuFour J, Pure E, Oliver BL, Wright TM, Noble PW. Hyaluronan fragments synergize with interferon-g to induce the C-X-C chemokines Mig and interferon-inducible protein-10 in mouse macrophages. J Biol Chem 1998; 273:35088–35094.
5. McKee CM, Penno MB, Cowman M, Burdick MD, Strieter RM, Bao C, Noble PW. Hyaluronan (HA) fragments induce chemokine gene expression in alveolar macrophages: the role of HA size and CD44. J Clin Investig 1996; 98:2403–2413.
6. Noble PW, McKee CM, Horton MR. Induction of inflammatory gene expression by low-molecular-weight hyaluronan fragments in macrophages. Wenner-Gren Int Ser 1998; 72:219–225.
7. Kumar S, Kumar P, Ponting JM, Sattar A, Rooney P, Pye D, Hunter RD. Hyaluronic acid promotes and inhibits angiogenesis. NATO ASI Series, Series A. Life Sci 1992; 227:253–263.
8. Ghosh I, Bharadwaj A, Datta K. Reduction in the level of hyaluronan binding protein 1 (HABP1) is associated with loss of sperm motility. J Reprod Immunol 2002; 53:45–54.
9. Salamonsen LA, Shuster S, Stern R. Distribution of hyaluronan in human endometrium across the menstrual cycle implications for implantation and menstruation. Cell Tissue Res 2001; 306:335–340.
10. Zhuo L, Yoneda M, Zhao M, Yingsung W, Yoshida N, Kitagawa Y, Kawamura K, Suzuki T, Kimata K. Defect in SHAP–hyaluronan complex causes severe female

infertility. A study by inactivation of the bikunin gene in mice. J Biol Chem 2001; 276:7693–7696.

11. Weiss C. Why viscoelasticity is important for the medical uses of hyaluronan and Hylans. Int Cong Ser 2000; 1196:89–103.
12. Turley EA, Noble PW, Bourguignon LYW. Signaling properties of hyaluronan receptors. J Biol Chem 2002; 277:4589–4592.
13. Lehmann R, Brint S, Stewart R, White GL Jr, McCarty G, Taylor R, Disbrow D, Defaller J. Clinical comparison of Provisc and Healon in cataract surgery. J Cataract Refract Surg 1995; 21:543–547.
14. Bresciani C, Lebuisson DA, Eveillard M, Grossiord JL, Drupt F, Montefiore G. Dynamic viscosity and corneal endothelial protection with Healonid, Healon GV, Provisc and Viscoat during phacoemulsification. J Fran D'Ophtal 1996; 19:39–50.
15. Uthman I, Raynauld JP, Haraoui B. Intra-articular therapy in osteoarthritis. Post Med J 2003; 79:449–453.
16. Prehm P. Hyaluronan in rheumatoid arthritis: a review. Int Cong Ser 2000; 1196: 129–135.
17. Tanaka N, Sakahashi H, Sato E, Hirose K, Ishima T, Ishii S. Intra-articular injection of high molecular weight hyaluronan after arthrocentesis as treatment for rheumatoid knees with joint effusion. Rheum Int 2002; 22:151–154.
18. Ghosh P, Holbert C, Read R, Armstrong S. Hyaluronic acid (hyaluronan) in experimental osteoarthritis. J Rheum Suppl 1995; 43:155–157.
19. Tateda C, Nagamine H, Iwadate K, Nakamura T. Effects of sodium hyaluronate (ME3710) in rabbit models of osteoarthritis (OA) and periarthritis of the shoulder (PS). Yak Chir 1995; 23:833–841.
20. Maheu E, Ayral X, Dougados M. A hyaluronan preparation (500–730 kDa) in the treatment of osteoarthritis: a review of clinical trials with Hyalgan. Int J Clin Pract 2002; 56:804–813.
21. Barrett JP, Siviero P. Retrospective study of outcomes in Hyalgan-treated patients with osteoarthritis of the knee. Clin Drug Investig 2002; 22:87–97.
22. Goorman SD, Watanabe TK, Miller EH, Perry C. Functional outcome in knee osteoarthritis after treatment with hylan G-F 20: a prospective study. Arch Phys Med Rehabil 2000; 81:479–483.
23. Adams ME, Atkinson MH, Lussier AJ, Schulz JI, Siminovitch KA, Wade JP, Zummer M. The role of viscosupplementation with hylan G-F 20 (Synvisc) in the treatment of osteoarthritis of the knee: a Canadian multicenter trial comparing hylan G-F 20 alone, hylan G-F 20 with non-steroidal anti-inflammatory drugs (NSAIDs) and NSAIDs alone. Osteoarthritis Cartilage 1995; 3:213–225.
24. Adams ME. An analysis of clinical studies of the use of crosslinked hyaluronan, hylan, in the treatment of osteoarthritis. J Rheum Suppl 1993; 39:16–18.
25. Laurent UB, Fraser JR, Engstrom-Laurent A, Reed RK, Dahl LB, Laurent TC. Catabolism of hyaluronan in the knee joint of the rabbit. Mater Biol 1992; 12: 130–136.
26. Lebel L, Gabrielsson J, Laurent TC, Gerdin B. Kinetics of circulating hyaluronan in humans. Eur J Clin Investig 1994; 24:621–626.
27. Fraser JRE, Laurent TC. Turnover and metabolism of hyaluronan. Ciba Found Symp 1989; 143:41–59.
28. Fraser JRE, Laurent TC, Engstroem-Laurent A, Laurent UGB. Elimination of hyaluronic acid from the blood stream in the human. Clin Exp Pharmacol Physiol 1984; 11:17–25.

29. van't Riet M, de Vos van Steenwijk Peggy J, Bonthuis F, Marquet Richard L, Steyerberg Ewout W, Jeekel J, Bonjer HJ. Prevention of adhesion to prosthetic mesh: comparison of different barriers using an incisional hernia model. Ann Surg 2003; 237:123–128.
30. Ozmen MM, Aslar AK, Terzi MC, Albayrak L, Berberoglu M. Prevention of adhesions by bioresorbable tissue barrier following laparoscopic intraabdominal mesh insertion. Surg Laparosc Endosc Percutan Tech 2002; 12:342–346.
31. Ballore L, Orru F, Nicolini F, Contini SA, Galletti G, Gherli T. Experimental results of the use of hyaluronic acid based materials (CV Seprafilm and CV Sepracoat) in postoperative pericardial adhesions. Acta Biomed Ateneo Parmense 2000; 71:159–166.
32. Diamond MP. Reduction of *de novo* postsurgical adhesions by intraoperative precoating with Sepracoat (HAL-C) solution: a prospective, randomized, blinded, placebo-controlled multicenter study. The Sepracoat Adhesion Study Group. Fertil Steril 1998; 69:1067–1074.
33. Alponat A, Lakshminarasappa SR, Yavuz N, Goh PM. Prevention of adhesions by Seprafilm, an absorbable adhesion barrier: an incisional hernia model in rats. Am Surg 1997; 63:818–819.
34. Kling J. Genzyme's Seprafilm gets FDA marketing nod. Nat Biotech 1996; 14:572.
35. Day AJ, Prestwich GD. Hyaluronan-binding proteins: tying up the giant. J Biol Chem 2002; 277:4585–4588.
36. Turley EA, Noble PW, Bourguignon LYW. Signaling properties of hyaluronan receptors. J Biol Chem 2002; 277:4589–4592.
37. Tammi MI, Day AJ, Turley EA. Hyaluronan and homeostasis: a balancing act. J Biol Chem 2002; 277:4581–4584.
38. Rein DT, Roehrig K, Schoendorf T, Lazar A, Fleisch M, Niederacher D, Bender HG, Dall P. Expression of the hyaluronan receptor RHAMM in endometrial carcinomas suggests a role in tumor progression and metastasis. J Can Res Clin Oncol 2003; 129:161–164.
39. Savani RC, Bagli DJ, Harrison RE, Turley EA. The role of hyaluronan-receptor interactions in wound repair. Bas Clin Derm 2000; 19:115–142.
40. Assmann V, Gillett CE, Poulsom R, Ryder K, Hart IR, Hanby AM. The pattern of expression of the microtubule-binding protein RHAMM/IHABP in mammary carcinoma suggests a role in the invasive behaviour of tumour cells. J Pathol 2001; 195:191–196.
41. Ziebell MR, Zhao Z-G, Luo B, Luo Y, Turley EA, Prestwich GD. Peptides that mimic glycosaminoglycans: high-affinity ligands for a hyaluronan binding domain. Chem Biol 2001; 8:1081–1094.
42. Weigel JA, Raymond RC, Weigel PH. The hyaluronan receptor for endocytosis (HARE) is not CD44 or CD54 (ICAM-1). Biochem Biophy Res Commun 2002; 294:918–922.
43. Zhou B, Weigel JA, Fauss L, Weigel PH. Identification of the hyaluronan receptor for endocytosis (HARE). J Biol Chem 2000; 275:37733–37741.
44. Banerji S, Ni J, Wang S-X, Clasper S, Su J, Tammi R, Jones M, Jackson DG. LYVE-1, a new homolog of the CD44 glycoprotein, is a lymph-specific receptor for hyaluronan. J Cell Biol 1999; 144:789–801.
45. Jackson DG. The lymphatics revisited. New perspectives from the hyaluronan receptor LYVE-1. Trends Cardiovasc Med 2003; 13:1–7.

46. Prevo R, Banerji S, Ferguson DJP, Clasper S, Jackson DG. Mouse LYVE-1 is an endocytic receptor for hyaluronan in lymphatic endothelium. J Biol Chem 2001; 276:19420–19430.
47. Toole BP, Wight TN, Tammi MI. Hyaluronan–cell interactions in cancer and vascular disease. J Biol Chem 2002; 277:4593–4596.
48. Ellis H. Medicolegal consequences of postoperative intra-abdominal adhesions. J R Soc Med 2001; 94:331–332.
49. Ellis H. The clinical significance of adhesions: focus on intestinal obstruction. Eur J Surg 1997:5–9.
50. Holmdahl L, Risberg B, Beck DE, Burns JW, Chegini N, diZerega GS, Ellis H. Adhesions: pathogenesis and prevention-panel discussion and summary. Eur J Surg 1997:56–62.
51. Ellis H, Moran BJ, Thompson JN, Parker MC, Wilson MS, Menzies D, McGuire A, Lower AM, Hawthorn RJ, O'Brien F, Buchan S, Crowe AM. Adhesion-related hospital readmissions after abdominal and pelvic surgery: a retrospective cohort study. Lancet 1999; 353:1476–1480.
52. Lower AM, Hawthorn RJ, Ellis H, O'Brien F, Buchan S, Crowe AM. The impact of adhesions on hospital readmissions over ten years after 8849 open gynaecological operations: an assessment from the Surgical and Clinical Adhesions Research Study. BJOG 2000; 107:855–862.
53. Parker MC, Ellis H, Moran BJ, Thompson JN, Wilson MS, Menzies D, McGuire A, Lower AM, Hawthorn RJ, O'Briena F, Buchan S, Crowe AM. Postoperative adhesions: ten-year follow-up of 12,584 patients undergoing lower abdominal surgery. Dis Colon Rectum 2001; 44:822–829.
54. Ellis H. The causes and prevention of intestinal adhesions. Br J Surg 1982; 69: 241–243.
55. Sjosten AC, Ellis H, Edelstam GA. Post-operative consequences of glove powder used pre-operatively in the vagina in the rabbit model. Hum Reprod 2000; 15: 1573–1577.
56. Ellis H. Experimental study of starch-induced intraperitoneal adhesions. Br J Surg 1991; 78:1020.
57. Kang N, Griffin D, Ellis H. The pathological effects of glove and condom dusting powders. J Appl Toxicol 1992; 12:443–449.
58. diZerega GS. Contemporary adhesion prevention. Fertil Steril 1994; 61:219–235.
59. diZerega GS. Biochemical events in peritoneal tissue repair. Eur J Surg 1997: 10–16.
60. DiZerega GS, Campeau JD. Peritoneal repair and post-surgical adhesion formation. Hum Reprod Update 2001; 7:547–555.
61. DeCherney AH, diZerega GS. Clinical problem of intraperitoneal postsurgical adhesion formation following general surgery and the use of adhesion prevention barriers. Surg Clin N Am 1997; 77:671–688.
62. Seinera P, Arisio R, Decko A, Farina C, Crana F. Laparoscopic myomectomy: indications, surgical technique and complications. Hum Reprod 1997; 12: 1927–1930.
63. Azziz R. Microsurgery alone or with INTERCEED Absorbable Adhesion Barrier for pelvic sidewall adhesion re-formation. The INTERCEED (TC7) Adhesion Barrier Study Group II. Surg Gyncol Obstet 1993; 177:135–139.
64. Khaitan E, Scholz S, Richards William O. Laparoscopic adhesiolysis and placement of Seprafilm: a new technique and novel approach to patients

with intractable abdominal pain. J Laparoendosc Adv Surg Tech A 2002; 12: 241–247.
65. Pagidas K, Tulandi T. Effects of Ringer's lactate, Interceed(TC7) and Gore-Tex Surgical Membrane on postsurgical adhesion formation. Fertil Steril 1992; 57: 199–201.
66. Verco SJ, Peers EM, Brown CB, Rodgers KE, Roda N, diZerega G. Development of a novel glucose polymer solution (icodextrin) for adhesion prevention: pre-clinical studies. Hum Reprod 2000; 15:1764–1772.
67. Buckenmaier CC 3rd, Pusateri AE, Harris RA, Hetz SP. Comparison of antiadhesive treatments using an objective rat model. Am Surg 1999; 65:274–282.
68. Buckenmaier CC 3rd, Summers MA, Hetz SP. Effect of the antiadhesive treatments, carboxymethylcellulose combined with recombinant tissue plasminogen activator and Seprafilm, on bowel anastomosis in the rat. Am Surg 2000; 66: 1041–1045.
69. James DC, Ellis H. The effect of streptokinase on the formation of peritoneal adhesions in rats and rabbits. Sangre 1964; 17:192–198 (a preliminary report).
70. James DCO, Ellis H, Hugh TB. The effect of streptokinase on experimental intraperitoneal adhesion formation. J Pathol Bacteriol 1965; 90:279–287.
71. Yagmurlu A, Barlas M, Gursel I, Gokcora IH. Reduction of surgery-induced peritoneal adhesions by continuous release of streptokinase from a drug delivery system. Eur Surg Res 2003; 35:46–49.
72. Miller JA, Ferguson RL, Powers DL, Burns JW, Shalaby SW. Efficacy of hyaluronic acid/nonsteroidal anti-inflammatory drug systems in preventing postsurgical tendon adhesions. J Biomed Mater Res 1997; 38:25–33.
73. Montz FJ, Monk BJ, Lacy SM, Fowler JM. Ketorolac tromethamine, a nonsteroidal anti-inflammatory drug: ability to inhibit post-radical pelvic surgery adhesions in a porcine model. Gynecol Oncol 1993; 48:76–79.
74. Avsar AF, Avsar FM, Sahin M, Topaloglu S, Vatansev H, Belviranli M. Diphenhydramine and hyaluronic acid derivatives reduce adnexal adhesions and prevent tubal obstructions in rats. Eur J Obstet Gynecol Reprod Biol 2003; 106: 50–54.
75. Golan A, Maymon R, Winograd I, Bukovsky I. Prevention of post-surgical adhesion formation using aspirin in a rodent model: a preliminary report. Hum Reprod 1995; 10:1797–1800.
76. Jackson JK, Skinner KC, Burgess L, Sun T, Hunter WL, Burt HM. Paclitaxel-loaded crosslinked hyaluronic acid films for the prevention of postsurgical adhesions. Pharm Res 2002; 19:411–417.
77. Lebel L. Clearance of hyaluronan from the circulation. Adv Drug Deliv Rev 1991; 7:221–235.
78. Crescenzi V, Francescangeli A, Segre AL, Capitani D, Mannina L, Renier D, Bellini D. NMR structural study of hydrogels based on partially deacetylated hyaluronan. Macro Bios 2002; 2:272–279.
79. Simkovic I, Hricovini M, Soltes L, Mendichi R, Cosentino C. Preparation of water-soluble/insoluble derivatives of hyaluronic acid by cross-linking with epichlorohydrin in aqueous NaOH/NH_4OH solution. Carbohydr Pol 1999; 41: 9–14.
80. Larsen NE, Pollak CT, Reiner K, Leshchiner E, Balazs EA. Hylan gel biomaterial: dermal and immunologic compatibility. J Biomed Mater Res 1993; 27:1129–1134.

81. Larsen NE, Pollak CT, Reiner K, Leshchiner E, Balazs EA. Hylan gel for soft tissue augmentation. Biotechnol Bioact Polym 1994:25–33 (Proc Am Chem Soc Symp).
82. Ramamurthi A, Vesely I. Smooth muscle cell adhesion on crosslinked hyaluronan gels. J Biomed Mater Res 2002; 60:196–205.
83. Ramamurthi A, Vesely I. Ultraviolet light-induced modification of crosslinked hyaluronan gels. J Biomed Mater Res 2003; 66A:317–329.
84. Hamburger MI, Lakhanpal S, Mooar PA, Oster D. Intra-articular hyaluronans: a review of product-specific safety profiles. Semin Arthritis Rheum 2003; 32: 296–309.
85. Hammesfahr JFR, Knopf Alan B, Stitik T. Safety of intra-articular hyaluronates for pain associated with osteoarthritis of the knee. Am J Ortho 2003; 32:277–283.
86. Kahan A, Lleu P-L, Salin L. Prospective randomized study comparing the medicoeconomic benefits of Hylan GF-20 vs. conventional treatment in knee osteoarthritis. Joint Bone Spine 2003; 70:276–281.
87. Karlsson J, Sjoegren LS, Lohmander LS. Comparison of two hyaluronan drugs and placebo in patients with knee osteoarthritis. A controlled, randomized, double-blind, parallel-design multicenter study. Rheumatology (Oxford, United Kingdom) 2002; 41:1240–1248.
88. Hallen L, Johansson C, Laurent C. Cross-linked hyaluronan (Hylan B gel): a new injectable remedy for treatment of vocal fold insufficiency—an animal study. Acta Otolaryngol 1999; 119:107–111.
89. Crescenzi V, Francescangeli A, Capitani D, Mannina L, Renier D, Bellini D. Hyaluronan networking via Ugi's condensation using lysine as cross-linker diamine. Carbohydr Polym 2003; 53:311–316.
90. Vercruysse KP, Marecak DM, Marecek JF, Prestwich GD. Synthesis and *in vitro* degradation of New polyvalent hydrazide cross-linked hydrogels of hyaluronic acid. Bioconj Chem 1997; 8:686–694.
91. Prestwich GD, Marecak DM, Marecek JF, Vercruysse KP, Ziebell MR. Controlled chemical modification of hyaluronic acid: synthesis, applications, and biodegradation of hydrazide derivatives. J Control Release 1998; 53:93–103.
92. Shu XZ, Liu Y, Palumbo F, Prestwich GD. Disulfide-crosslinked hyaluronan-gelatin hydrogel films: a covalent mimic of the extracellular matrix for *in vitro* cell growth. Biomaterials 2003; 24:3825–3834.
93. Huang WJ, Johns DB, Kronenthal RL. US patent #5532221.
94. Johns DB, Keyport GM, Hoehler F, diZerega GS. Reduction of postsurgical adhesions with Intergel adhesion prevention solution: a multicenter study of safety and efficacy after conservative gynecologic surgery. Fertil Steril 2001; 76: 595–604.
95. Thornton MH, Johns DB, Campeau JD, Hoehler F, Dizerega GS. Clinical evaluation of 0.5% ferric hyaluronate adhesion prevention gel for the reduction of adhesions following peritoneal cavity surgery: open-label pilot study. Hum Reprod 1998; 13:1480–1485.
96. Lundorff P, van Geldorp H, Tronstad SE, Lalos O, Larsson B, Johns DB, diZerega GS. Reduction of post-surgical adhesions with ferric hyaluronate gel: a European study. Hum Reprod 2001; 16:1982–1988.
97. http://www.lifecore.com/products/Ha_division.asp.
98. Kocak I, Unlu C, Akcan Y, Yakin K. Reduction of adhesion formation with cross-linked hyaluronic acid after peritoneal surgery in rats. Fertil Steril 1999; 72: 873–878.

99. Belluco C, Meggiolaro F, Pressato D, Pavesio A, Bigon E, Dona M, Forlin M, Nitti D, Lise M. Prevention of postsurgical adhesions with an autocrosslinked hyaluronan derivative gel. J Surg Res 2001; 100:217–221.
100. Acunzo G, Guida M, Pellicano M, Tommaselli GA, Sardo ADS, Bifulco G, Cirillo D, Taylor A, Nappi C. Effectiveness of auto-cross-linked hyaluronic acid gel in the prevention of intrauterine adhesions after hysteroscopic adhesiolysis: a prospective, randomized, controlled study. Hum Reprod 2003; 18:1918–1921.
101. Burns JW, Cox S, Walts AE. US patent #5017229.
102. Kuo JW, Swann DA, Prestwich GD. Chemical modification of hyaluronic acid by carbodiimides. Bioconj Chem 1991; 2:232–241.
103. Hamilton R, Fox EM, Acharya RA, Walts AE. US patent #4,937,270.
104. Pouyani T, Kuo JW, Harbison GS, Prestwich GD. Solid-state NMR of *N*-acylureas derived from the reaction of hyaluronic acid with isotopically-labeled carbodiimides. J Am Chem Soc 1992; 114:5972–5976.
105. Beck DE. The role of Seprafilm bioresorbable membrane in adhesion prevention. Eur J Surg Suppl 1997:49–55.
106. Diamond MP. Reduction of adhesions after uterine myomectomy by Seprafilm membrane (HAL-F): a blinded, prospective, randomized, multicenter clinical study. Seprafilm Adhesion Study Group. Fertil Steril 1996; 66:904–910.
107. Beck DE, Cohen Z, Fleshman JW, Kaufman HS, van Goor H, Wolff BG. A prospective, randomized, multicenter, controlled study of the safety of seprafilm adhesion barrier in abdominopelvic surgery of the intestine. Dis Colon Rectum 2003; 46:1310–1319.
108. Tzianabos AO, Cisneros RL, Gershkovich J, Johnson J, Miller RJ, Burns JW, Onderdonk AB. Effect of surgical adhesion reduction devices on the propagation of experimental intra-abdominal infection. Arch Surg 1999; 134:1254–1259.
109. Baptista ML, Bonsack ME, Delaney JP. Seprafilm reduces adhesions to polypropylene mesh. Surgery 2000; 128:86–92.
110. Greenawalt K, Gershkovich J. US patent #6,056,970.
111. Greenawalt KE, Butler TJ, Rowe EA, Finneral AC, Garlick DS, Burns JW. Evaluation of Sepramesh biosurgical composite in a rabbit hernia repair model. J Surg Res 2000; 94:92–98.
112. Kimmelman CP, Edelstein DR, Cheng HJ. Sepragel sinus (hylan B) as a postsurgical dressing for endoscopic sinus surgery. Otolaryngol Head Neck Surg 2001; 125:603–608.
113. Catalano PJ, Roffman EJ. Evaluation of middle meatal stenting after minimally invasive sinus techniques (MIST). Otolaryngol Head Neck Surg 2003; 128: 875–881.
114. ANG.gerup, US patent #5,827,937.
115. Narins Rhoda S, Brandt F, Leyden J, Lorenc ZP, Rubin M, Smith S. A randomized, double-blind, multicenter comparison of the efficacy and tolerability of Restylane versus Zyplast for the correction of nasolabial folds. Dermatol Surg 2003; 29:588–595.
116. Carruthers J, Carruthers A. A prospective, randomized, parallel group study analyzing the effect of BTX-A (Botox) and non-animal sourced hyaluronic acid (NASHA, Restylane) in combination compared with NASHA (Restylane) alone in severe glabellar rhytides in adult female subjects: treatment of severe glabellar rhytides with a hyaluronic acid derivative compared with the derivative and BTX-A. Dermatol Surg 2003; 29:802–809.

117. Duranti F, Salti G, Bovani B, Calandra M, Rosati ML. Injectable hyaluronic acid gel for soft tissue augmentation. A clinical and histological study. Dermatol Surg 1998; 24:1317–1325.
118. Luo Y, Prestwich GD. Synthesis and selective cytotoxicity of a hyaluronic acid–antitumor bioconjugate. Bioconj Chem 1999; 10:755–763.
119. Pouyani T, Prestwich GD. Functionalized derivatives of hyaluronic acid oligosaccharides: drug carriers and novel biomaterials. Bioconj Chem 1994; 5: 339–347.
120. Luo Y, Ziebell MR, Prestwich GD. A hyaluronic acid–taxol antitumor bioconjugate targeted to cancer cells. Biomacromolecules 2000; 1:208–218.
121. Della VF, Romeo A. Eur patent application, #216453.
122. Goei L, Topp E, Stella V, Benedetti L, Biviano F, Callegaro L. Drug release from hydrocortisone esters of hyaluronic acid: influence of ester hydrolysis rate on release rate. Polym Med 1992:85–92.
123. Goei L, Topp E, Stella V, Benedetti L, Biviano F, Callegaro L. Polymeric prodrugs: recent experience with devices formed from steroid esters of hyaluronic acid. Polym Med 1992:93–101.
124. Rajewski LG, Stinnett AA, Stella VJ, Topp EM. Enzymic and nonenzymic hydrolysis of a polymeric prodrug: hydrocortisone esters of hyaluronic acid. Int J Pharm 1992; 82:205–213.
125. Cheung YW, Po ALW, Irwin WJ. Resistance to enzymic hydrolysis as a parameter in drug potency. Int J Pharm 1985; 27:325–333.
126. Payan E, Jouzeau JY, Lapicque F, Bordji K, Simon G, Gillet P, O'Regan M, Netter P. *In vitro* drug release from HYC 141, a corticosteroid ester of high molecular weight hyaluronan. J Control Release 1995; 34:145–153.
127. Barbucci R, Magnani A, Baszkin A, Da Costa ML, Bauser H, Hellwig G, Martuscelli E, Cimmino S. Physicochemical surface characterization of hyaluronic acid derivatives as a new class of biomaterials. J Biomater Sci Polym Ed 1993; 4:245–273.
128. Milella E, Brescia E, Massaro C, Ramires PA, Miglietta MR, Fiori V, Aversa P. Physico-chemical properties and degradability of non-woven hyaluronan benzylic esters as tissue engineering scaffolds. Biomaterials 2001; 23: 1053–1063.
129. Campoccia D, Hunt JA, Doherty PJ, Zhong SP, O'Regan M, Benedetti L, Williams DF. Quantitative assessment of the tissue response to films of hyaluronan derivatives. Biomaterials 1996; 17:963–975.
130. Benedetti L, Cortivo R, Berti T, Berti A, Pea F, Mazzo M, Moras M, Abatangelo G. Biocompatibility and biodegradation of different hyaluronan derivatives (Hyaff) implanted in rats. Biomaterials 1993; 14:1154–1160.
131. Aigner J, Tegeler J, Hutzler P, Campoccia D, Pavesio A, Hammer C, Kastenbauer E, Naumann A. Cartilage tissue engineering with novel nonwoven structured biomaterial based on hyaluronic acid benzyl ester. J Biomed Mater Res 1998; 42: 172–181.
132. Brun P, Abatangelo G, Radice M, Zacchi V, Guidolin D, Daga Gordini D, Cortivo R. Chondrocyte aggregation and reorganization into three-dimensional scaffolds. J Biomed Mater Res 1999; 46:337–346.
133. Tonello C, Zavan B, Cortivo R, Brun P, Panfilo S, Abatangelo G. *In vitro* reconstruction of human dermal equivalent enriched with endothelial cells. Biomaterials 2003; 24:1205–1211.

134. Grigolo B, Roseti L, Fiorini M, Fini M, Giavaresi G, Nicoli Aldini N, Giardino R, Facchini A. Transplantation of chondrocytes seeded on a hyaluronan derivative (Hyaff-11) into cartilage defects in rabbits. Biomaterials 2001; 22:2417–2424.
135. Solchaga LA, Yoo JU, Lundberg M, Dennis JE, Huibregtse BA, Goldberg VM, Caplan AI. Hyaluronan-based polymers in the treatment of osteochondral defects. J Orthop Res 2000; 18:773–780.
136. Pavesio A, Abatangelo G, Borrione A, Brocchetta D, Hollander Anthony P, Kon E, Torasso F, Zanasi S, Marcacci M. Hyaluronan-based scaffolds (Hyalograft C) in the treatment of knee cartilage defects: preliminary clinical findings. Novartis Found Symp 2003; 249:203–217 (discussion 229–233, 234–208, 239–241).
137. Lisignoli G, Zini N, Remiddi G, Piacentini A, Puggioli A, Trimarchi C, Fini M, Maraldi NM, Facchini A. Basic fibroblast growth factor enhances *in vitro* mineralization of rat bone marrow stromal cells grown on nonwoven hyaluronic acid based polymer scaffold. Biomaterials 2001; 22:2095–2105.
138. Ambrosio L, Borzacchiello A, Netti PA, Nicolais L. Properties of new materials: rheological study on hyaluronic acid and its derivative solutions. J Macromol Sci Pure Appl Chem 1999; A36:991–1000.

Chemistry and Biology of Hyaluronan
H.G. Garg and C.A. Hales (editors)

Chapter 24

Hyaluronan in the Treatment of Ocular Surface Disorders

PASQUALE ARAGONA

Ocular Surface Unit, Department of Surgical Specialties, Section of Ophthalmology, University of Messina, Policlinico G. Martino, Viale Gazzi, I-98125 Messina, Italy

I. Introduction

The term ocular surface indicates those ocular structures that constitute a functional unit which is responsible for the protection of the eye from the external environment, so representing the interface between the eye and the outer world.

Constituents of this functional unit are the lacrimal apparatus, in charge of tear production and excretion, tears, lids, conjunctiva, limbus, corneal epithelium and the nervous structures which innervate the above areas (Table 1) (1,2).

The protection of the external part of the eye is of great importance since the transparency of the cornea, which represents the most anterior part of the visual system, must be perfect in order to guarantee the best quality of the refractive surface, so allowing the sharpest vision (3).

Several insults can damage the ocular surface. The shear forces of blinking across its surface, as well as several external factors such as wind, air conditioning, low air humidity, smoke, foreign bodies (including contact lenses) and microorganisms, like bacteria and viruses, can induce an ocular surface suffering which can be responsible for symptoms of ocular discomfort and can interfere with the quality of sight.

Among the components of the ocular surface functional unit, the tear film represents, by far, the most dynamic structure. It can be considered as an extracellular matrix since it plays an active role in regulating epithelial function

Table 1 Ocular Structures Contributing to the Ocular Surface Functional Unit

Muco-epidermal junction of the lid
Meibomian glands
Lacrimal apparatus (production and excretion of tears)
Tear film
Conjunctival epithelium
Limbal epithelium
Corneal epithelium
Sensory and motor nervous connections with the central nervous system

These structures are functionally connected by tear fluids, hormones, blood, nerves, cytokines, light, and lid movements.

and interaction with the surrounding tissues. In fact, it provides nutrients to the epithelium and acts as a communication pathway, distributing regulatory factors and transporting cells (e.g., inflammatory cells), which exert their activity onto the corneo-conjunctival epithelium (4–7).

The production and turnover of a proper tear film is essential for the health of the ocular surface. Tears nourish, cleanse, give an immunological protection and lubricate, through their physical properties, the surface of the eye. The interface between tears and air provides the initial and most powerful refractive surface of the eye (nearly 45 diopters) (8). In order to accomplish its functions, the tear film must be stable on the ocular surface where it is linked to the surface of the epithelial cells by means of the glycocalix, a mucinous structure produced by the corneo-conjunctival epithelium (9). The stability of the tear film is a function of its volume and composition. It was considered that a stable tear film resides on the ocular surface for at least 7–10 s. When changes in tear film quality occur, this will result in significant alterations of the optical quality of retinal images and, therefore, in a poor quality of vision; furthermore, ocular surface damage can occur, with stress to the corneo-conjunctival epithelium that will develop a squamous metaplasia (10).

A. The Tear Film Composition

Tears are a complex mixture composed of water, proteins, immunoglobulins, glycoproteins, lipids, products of epithelial metabolism, cellular debris and polymorphonuclear cells (2). However, it is impossible to exactly define the chemical composition of tear film at a particular moment due to its high variability depending upon the challenges with which the ocular surface has to deal. The tear film has been classically described as composed of three separate layers that, from the innermost to the outermost, are composed of mucin, water and lipids. These layers are thought to perform specific functions. The exact boundaries and thickness of each layer are still under discussion. Now, the tear film is considered to be composed of two main layers: under a thin lipid layer there would be an aqueous-mucin gel, in which the mucins have a decreasing gradient of concentration from the epithelium to the surface (11) (Fig. 1).

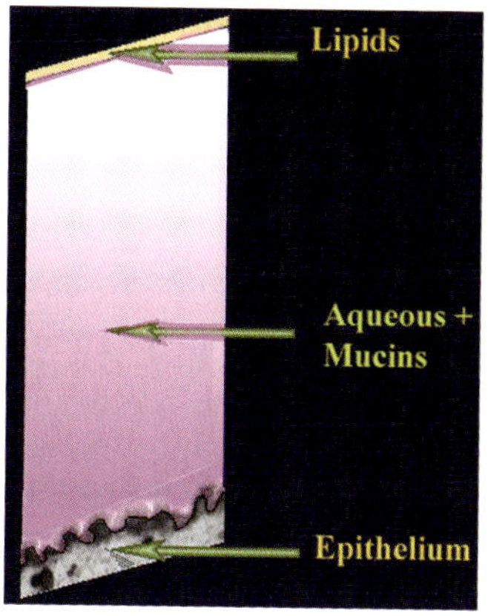

Figure 1 Schematic representation of the composition of the tear film.

1. The Mucous Component of the Tear Film

The tear film mucin is produced by conjunctival goblet cells and conjunctival and corneal epithelial cells. The corneal and conjunctival epithelia synthesize a mucin-like glycoprotein (MUC1) that constitutes the glycocalix (9). This mucin has a fundamental role for proper tear film spreading and wetting of the ocular surface. It renders the whole ocular surface hydrophilic and allows even spreading of the aqueous component over the eye. Without glycocalix, tears do not properly adhere to the eye and damage to the epithelium may occur, even in the presence of a normal aqueous tear production.

The bulk of the mucous content of the tear film is produced by conjunctival goblet cells. This comprises the mucins MUC4 and MUC5AC, which play an important role in tear film formation (12,13).

The mucins account for the non-Newtonian viscoelastic properties of the tear film, which determine a variable viscosity according to the shear rate of blinking. They maintain the optical quality of the tear film between blinks and protect the ocular surface during blinking, reducing the mechanical trauma to the surface (4,14,15).

2. The Aqueous Component of the Tear Film

The aqueous layer is produced by the main and accessory lacrimal glands under different stimuli (hormonal, parasympathetic, sympathetic). It is quantitatively the most important and creates the environment for the epithelium of the ocular surface, transporting nutrients and oxygen to the cornea, allowing cell movement over the ocular surface and washing away metabolic products, cellular debris and foreign bodies. The composition varies according to stimuli from the external environment or to changes either of hormonal or neuronal origin. These changes can influence the health of the epithelial cells. Many of the growth factors which are present in the tear film, such as EGF, TGF-α and HGF, derive from the lacrimal glands secretion (5,16,17). Furthermore, several proinflammatory substances such as HLA-DR, interleukin (IL)-1, IL-6, IL-8, produced locally or in adjacent structures, are transported by the aqueous phase and exert their action onto the ocular surface (6,18–21).

3. *The Lipid Component of the Tear Film*

The tear film lipids are produced by Meibomian glands, located within the tarsus in the lids. They are secreted on the lid margin from where they are distributed, at each blink, to constitute the outermost layer of the tear film. The role of the lipid layer is to prevent the evaporation of tears and to enhance the stability of the tear film (22–24). It also provides a smooth surface with excellent optical properties, allowing the formation of a sharp retinal image.

B. Tear Film Physical Properties

Most tear film functions are due to its peculiar physical properties. Tears flow from the glands and spread onto the ocular surface in their progress towards the lacrimal puncti, to be drained away, depending upon their viscous characteristics. Tear viscosity has been measured and it has been shown that they have a shear-thinning non-Newtonian fluid whose coefficient of viscosity is shear-rate dependent, as is commonly seen with dilute solutions of linear multipli-charged polymers (25–27). The portion of tear film thought to be responsible for this characteristic is the aqueous–mucous layer. The thickness and exact structure of this layer are still under discussion (28–30). However, the distribution of the tear film onto the ocular surface seems related to the surface and interfacial forces existing between its various layers, whereas gravitational forces seem to have little effect. The surface tension is also involved in aspects such as the respreading of tear film after each blink. It will influence the precorneal film thickness, the structure of upper and lower menisci, the interaction between the mucous–aqueous layer and lipid layer and, therefore, the tear film stability.

Although it is not exactly known how the mixing between mucins and water takes place after each blink, it was shown that in the aqueous phase of the tear film there are mucins of conjunctival origin such as the main gel-forming MUC5AC and the MUC1.

The viscosity and surface tension of tears resemble the characteristics obtained by the study of experimental models of dilute solutions of mucins or other linear multipli-charged polymers (31).

C. Tear Film Functions

As previously described, the tear film plays a fundamental role in maintaining the homeostasis of the ocular surface. Its activities depend on its physical and chemical properties; namely, the tear film plays the following functions:

Optic function, smoothing the irregularity of the corneal surface; protective filtering against ultraviolet and infrared rays; buffering activity for maintaining the pH against environmental pollutants; washing away cellular debris and metabolism products; antimicrobial activity for the presence of lysozyme, lactoferrin, secretory IgA; nourishing and reparative function for corneal epithelium due to its content of ions, vitamins and growth factors; transporting onto the ocular surface cells such as leukocytes, enzymes, electrolytes, products passing from the blood; lubricating activity, necessary to allow efficacious and non-traumatizing blinking. The tear film functions are summarized in Table 2.

D. Ocular Surface Disorders

When a modification of the tear film structure occurs, with consequent tear film instability, ocular surface stress will develop, resulting in a clinical condition known as dry eye. The classification of this disorder was carried out in 1995 by the National Eye Institute, dividing dry eye into two different types: aqueous layer disorders and tear evaporation disorders (32). This classification is very useful to focus on the main causative factors of the disorder, although the clinical presentation is often a mix of the two pathogenic pathways (i.e., a reduced aqueous production often results in an inadequate lipid layer spreading and in excessive tear film evaporation; meibomian gland disease is commonly associated with reduced aqueous secretion by the lacrimal gland). Aqueous layer deficiency is the most common cause of dry eye and is dependent on decreased secretion of the lacrimal glands, although increased evaporation of tears may also be involved. Main causes of tear aqueous deficiency are Sjögren's syndrome, senile hyposecretion, lacrimal gland traumatism, vitamin A deficiency, immune infiltration in the course of sarcoidosis or lymphoma, altered innervation of ocular surface (for example in diabetes, herpes infection or contact lens wearers) or lacrimal gland, conjunctival scarring following diseases like ocular cicatricial pemphigoid, tracoma or chemical burns (33).

Causes of lipid alterations are linked to Meibomian gland disfunction (34). An obstruction of the gland ducts can occur spontaneously or can be associated with skin diseases. A reduced lipid secretion will increase the evaporation of the tear film, reducing its stability. Conditions associated with Meibomian gland dysfunction are generalized sebaceous gland alteration such as those in acne

Table 2 Tear Film Functions

Optical (smoothing the irregularity of the corneal surface)
Protective (UV and IR rays, pH, O_2–CO_2 exchanges)
Cleaning (washing away cellular debris and metabolic products)
Antimicrobic (presence of lysozyme, lactoferrin, secretory IgA)
Corneal epithelium nourishing and reparation (due to its content of ions, vitamins and growth factors)
Vehicle (leukocytes, enzymes, electrolytes, products passing from the blood)
Lubricating (allowing an efficacious and non-traumatizing blinking)

rosacea or seborrhoeic dermatitis, blepharitis, atopic keratoconjunctivitis, or unbalanced androgen production (35).

An altered mucous production accompanies almost all forms of dry eye, since a modification of goblet cells production is present in all the irritating processes of the ocular surface. In fact, the evolution of dry eye is characterized by a squamous metaplasia of the conjunctiva, which will result in reduction and, in later stages, disappearance of conjunctival goblet cells. Furthermore, certain disorders have the conjunctiva as primary target. Among these are vitamin A deficiency, cicatrising conjunctival disorders, and irritation following chronic use of topical medications with preservatives.

Other causes of dry eye are those diseases characterized by an abnormal distribution of the tear film onto the ocular surface due to lid/globe inconsistency or to lid abnormality such as entropion, ectropion, lid margin irregularity, exophthalmos or ocular surface scarring.

In all these forms of dry eye, the therapy must be targeted to resolve the primary cause of the disorder and to recover a stable and efficient tear film. For the latter purpose, several tear substitutes have been proposed. Among these, those based on hyaluronan give encouraging results in replacing a stable tear film and ameliorating the healing of the ocular surface.

II. Hyaluronan and Ocular Surface

Hyaluronan (HA) is a polymer with molecular mass varying from $\sim 2 \times 10^5$ to $\sim 10 \times 10^6$, which is based on repeating disaccharide units of glucuronic acid and *N*-acetyl-D-glucosamine. It is present in virtually all biological fluids where it performs three functions: it expands the extracellular space by binding water and salt, interacts with extracellular molecules to form the extracellular matrix, and is recognized by cell surface receptors that activate intracellular signaling or HA internalization. HA is particularly prominent during embryonic development and at sites of wound healing.

A. Distribution of Hyaluronan in Normal Ocular Surface Tissues

Hyaluronan has been found in normal human tears, as demonstrated by a study carried out with ion exchange chromatography. This study was performed by collecting tears from healthy subjects by means of cellulose filter strips (Whatman no. 41) applied in the inferior conjunctival fornix. The results of the study showed the presence of esuronic acid and aminosugars in the quantity of about 1 mcg of esuronic acid obtained from each filter. As to the type of polysaccharide isolated, it was shown that 50% of the isolated material was HA, the other 50% being attributable mainly to glycoproteins (36). The presence of HA in tears was also confirmed by further studies that demonstrated the presence of HA both in tears and in lacrimal tissue where it was demonstrated along the basal membrane of acinar and ductal cells in close association with the CD44 receptor (37–39).

In the conjunctival epithelium, HA was found to be more pronounced in the deeper layers, although in some cases it appeared to be more diffuse in the superficial epithelium. The epithelial basement membrane region contained large amounts of HA, an observation similar to that obtained in the skin (40), suggesting that HA may mediate interactions between the epithelium and the basement membrane (41). HA was consistently demonstrated in the conjunctival stroma with a distribution which was quite different from that of corneal stroma (42). The intensity of HA stain was more evident than that for CD44 compared with the ratio of intensities of stain in other locations, suggesting that a different HA receptor might be present in the conjunctival stroma. The presence of a large amount of HA in the conjunctival stroma can mediate mesenchymal–epithelial interactions, important for continuous stem cell proliferation and differentiation, as demonstrated *in vitro* (43,44).

In the corneal epithelium, HA was co-localized with its receptor CD44 as it appears in other districts like skin (41). Staining was more evident in the basal layers, consistent with the possibility that HA might play a role in promoting epithelial migration and wound healing (45). Such localization was also consistent with the role that the interaction between HA and CD44 can play in the modulation of the adhesion of growing and migrating corneal epithelial cells, thus participating in the regulation of corneal epithelial regeneration.

Increased staining for HA was also observed on the surface of the superficial corneal cells. This HA can originate from tears and may play a role in stabilizing the preocular tear film. Furthermore, the great hydrophilic properties of HA suggest a possible role in the regulation of cornea hydration.

HA was seen in association with both cell membranes and cytoplasm, which implies possible intracellular metabolism and a role in signal transduction.

The peripheral cornea stained more intensely than the central cornea. Staining was absent in the corneal epithelial basement membrane; this feature differed from the conjunctiva and might reflect specific physiological functions unique to the cornea (42).

Variable amounts of HA were demonstrated in Bowman's layer and in the stroma of the peripheral cornea, gradually decreasing towards the central cornea, where it was undetectable. The presence of HA was also demonstrated in corneal stromal cells, suggesting that HA is synthesized in and secreted by these cells (46).

HA was absent in Descemet's membrane, while both HA and CD44 were present on endothelial cells with similar pattern in the central and peripheral cornea (42).

In addition, the CD44 receptor for HA was demonstrated in the ocular surface structure, always closely associated to HA. The only different HA/CD44 ratio of staining was in the conjunctival stroma where a higher amount of HA was demonstrated, suggesting the presence of a different HA receptor (44). Apart from this difference, the CD44 receptor for HA was demonstrated, with the same distribution of HA, in conjunctival epithelium, and in epithelial, stromal and endothelial cells of the cornea (46,47).

The distribution of HA in the normal ocular surface is summarized in Table 3.

B. Hyaluronan in Diseased Ocular Surface Tissues

The eyes of patients suffering from ocular surface disease such as dry eye are characterized by a deterioration of the corneal epithelium with a development of punctate erosions and increased permeability. Compared with normal eyes, the conjunctiva of such patients presents squamous metaplasia with decreased or abnormal goblet cells. In addition, HA and its receptor CD44 modify their pattern of distribution within the ocular surface structures.

It was shown that HA and its receptor CD44 were preferentially located in regions of active cell growth such as in the basal layers of the stratified epithelium (41). In an experimental model on rabbit of corneal epithelial wound healing after exposure to *n*-heptanol, it was shown that the HA concentration increased gradually in the epithelium starting from day 3, when the epithelium had completely covered the defect, until day 28 after wounding, when the thickness of the epithelium returned to the normal level. Furthermore, in the stroma, HA staining increased after 3 days and returned to normal levels after 56 days. The immunostaining for CD44 and fibronectin paralleled that of HA, suggesting that HA, CD44 and fibronectin cooperatively play a role in wound healing (48).

HA is also involved in the wound healing process which follows excimer laser treatment of the cornea for both refractive and therapeutic purposes. Its presence was found in the subepithelial layer of the stroma up to 44 months after the treatment, despite topical steroid treatment (49–51). HA was visualized in the anterior stroma of the excimer-treated eyes starting from 1 day after the procedure and lasting up to 60 days. Its amount was statistically significantly larger than baseline values starting from day 8 after treatment (52). Furthermore, HA in wound scrapings from human corneas with myopic regression after photorefractive keratectomy (PRK) has been associated with severe haze (53). Steroids have been claimed to reduce the HA content of corneas after PRK (54).

Table 3 Presence of HA in Normal Ocular Surface Structure

Preocular tear film	
Lacrimal tissue	Basal membrane of acinar cells
	Ductal cells
Conjunctiva	Epithelial basal cells
	Epithelial basement membrane
	Stroma
Cornea	Surface of the superficial epithelial cells
	Epithelial basal cells (peripheral cornea)
	Bowman's layer (peripheral cornea)
	Keratocytes
	Endothelial cells (both peripheral and central cornea)

In some late myopic regression, it was demonstrated that topical steroids were able to restore up to 3 diopters refraction within 1 week of treatment (55). The rapid hyperopic restitution suggested that a water displacement in the wound area could play a role in such a result. HA has been shown to have a strict association with subepithelial water accumulation in corneas after PRK (56). HA seems, therefore, to play an important role in wound healing modulation after excimer laser treatment of the cornea. The endogenous production of HA can also be considered a marker for wound healing reactivity (57).

Reverse transcription-polymerase chain reaction was used to evaluate the expression and distribution of CD44 in the healing corneal epithelium of rat. This study indicated that CD44 transcripts started to increase in the epithelial cells surrounding the wound margin 3 h after wounding and peaked at 18 h in the basal epithelial cell layer, at which time the epithelia were actively migrating. As the cells began proliferation after wounding, the density of CD44 mRNA label declined but was still significantly higher than that in control specimens. The label returned to basal level as epithelial cells reverted to their normal phenotype. Eighteen hours after wounding, CD44 immunoreactivity was elevated in the entire epithelium, from the leading edge to the limbal–corneal border. As happened for the mRNA, the cell surface CD44 declined as cells differentiated to re-establish the multilayered epithelium (58).

Another study, using immunohistochemical method with monoclonal antibodies against human CD44, showed that in normal corneas CD44 was expressed on the membranes of basal epithelial cells and on keratocytes. Enhanced expression was observed on the epithelium of corneas with inflammation and allograft rejection. The authors concluded that CD44, the HA receptor, may play an important role in corneal cell–cell and cell–matrix interactions. Its regulation is closely related to corneal inflammatory reaction (59).

According to the results of these studies, it appears that HA and its receptor CD44 are over-expressed by corneal epithelium in the course of wound healing, suggesting a role of these molecules in the reparative processes of the ocular surface.

III. Hyaluronan as a Tear Substitute

The therapy of dry eye was based for many years on the use of the so-called artificial tears, which are, to date, the mainstay for the treatment of dry eye syndrome. They were designed with a focus on physical properties relating to wetting of the ocular surface and usually were based on the presence of hydrophilic polymers, used with the intent to lubricate the eye during blinking.

The ideal tear replacement should have a composition that is compatible with the maintenance of a normal ocular surface epithelium. Furthermore, it should be able, when damage of the ocular surface exists, to provide an environment in which the epithelium could recover normal structure and function.

With this purpose, several artificial tears were designed using HA alone or in association with other molecules. Aiming to obtain an optimal solution for the treatment of ocular surface disorders, several molecular weights and several concentrations of HA were used in artificial tears formulation. HA used as a base for artificial tears can be produced from biological sources (the most frequent being rooster comb and certain strains of *Streptococcus* bacteria). The molecular weight used varies from $6-7 \times 10^5$ to 3×10^6 D. The concentration used varies from 0.015 to 0.4% according to the molecular weight used and, if in association with other mucomimetic molecules, in order to achieve a viscosity similar to that of tears. Furthermore, the osmolarity of the solution can be either isotonic or hypotonic compared with tears. The latter solution was designed for the treatment of severe dry eye, where tear hyperosmolarity was demonstrated (60).

In addition, hyalan, a family of cross-linked, high molecular weight derivatives from HA, has been used for artificial tears preparation at a concentration of 0.15%.

HA tear substitutes are available as both multidose or monodose formulations. The former has added preservatives such as benzalconium chloride, thiomersal or *N*-hydroxymethylglicinate + EDTA; the latter is preservative free.

IV. Hyaluronan Administration on the Ocular Surface

A. Effect on the Normal Ocular Surface

A solution of HA (0.75×10^{-6} D at a concentration of 0.2% in phosphate buffer) was administered as tear substitutes in normal conjunctival epithelium to determine whether a mid-term treatment is able to induce morphological changes. Morphological and morphometrical study on light and transmission electron microscopy (TEM) was carried out on CD1 male mice (Aragona P, unpublished data). By light microscopy, the investigators observed either sections stained with Alcian Blue pH 2.5/PAS or semithin sections stained with toluidine blue-pironine. Morphometric analysis included: goblet cell number/microscopic field and goblet cell area on light microscopy specimens; number and area of goblet cell mucin granules; and granular optical density studied on TEM sections. The mice were treated for 10 days with the application of 5 drops/day. An untreated group was also studied for control. On the 11th day, all the animals were sacrificed by intraperitoneal injection of 500 mg/kg of sodium pentobarbital and then decapitated.

The study gave the following results: in control animals, the goblet cells were mainly arranged in clusters; the AB/PAS stain demonstrated that AB-positive goblet cells were 10%, AB/PAS were 64% and PAS were 26% of the total. In the HA-treated animals, mainly isolated goblet cells showed the following staining characteristics: 0.5% AB-positive; 75% AB/PAS-positive and 24.5% only PAS-positive. The prevalence of the three staining characteristics was statistically significantly different in the two groups (Fig. 2).

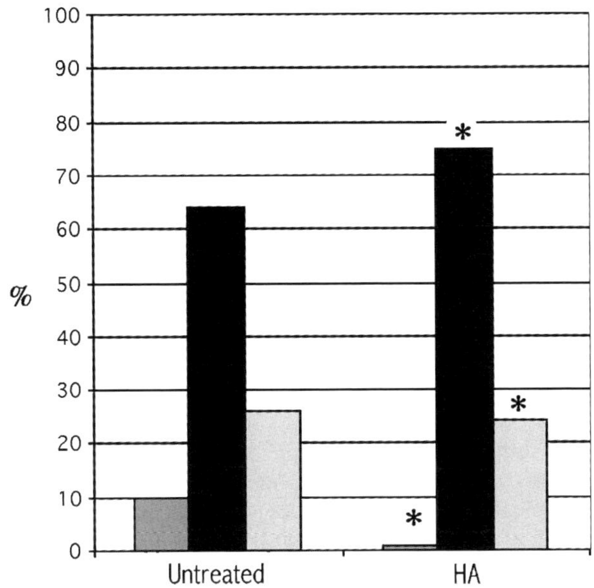

Figure 2 Results of the goblet cell staining with Alcian Blue/PAS stain. $*p < 0.01$ versus untreated animals.

The morphometrical study carried out on light microscopy specimens showed that no differences for the area and number/microscopic field of goblet cells were present between treated and untreated animals. On TEM sections (Fig. 3), the characteristics of the content of the mucin granuli present in goblet cell cytoplasm, i.e., the number of granules/Unit Area (UA) (4 μm^2), the area of granuli and their optical density, expressed in optical units (OU) of grey levels from 0 (black) to 255 (white), were studied. It was observed that the mucin granuli were reduced in number (12.9 ± 3.7 in the control group and 8.7 ± 3.2 in the HA-treated group), but with a bigger size (0.08 ± 0.06 μm^2 in the control group and 0.16 ± 0.08 in the HA-treated group). In addition, the optical density of mucin granules, studied at the TEM, demonstrated that the treatment with HA was able to induce a modification of the mucin content of granules that appeared to be less electron dense than that of untreated animals (88.4 ± 18.8 OU for the untreated animals and 160.4 ± 10.6 OU for HA-treated animals). The results of this study indicated that the treatment of the normal ocular surface with HA was able to induce specific modifications of the conjunctival epithelium with particular regard to the goblet cells structure (Table 4) (Aragona P, unpublished data).

The administration of artificial tears, based on a Hylan A solution, in healthy subjects working at a video display terminal (VDT) demonstrated a significant reduction of the blink frequency during the visual task. Furthermore, pretreatment with the Hyalan A solution reduced the ocular discomfort caused

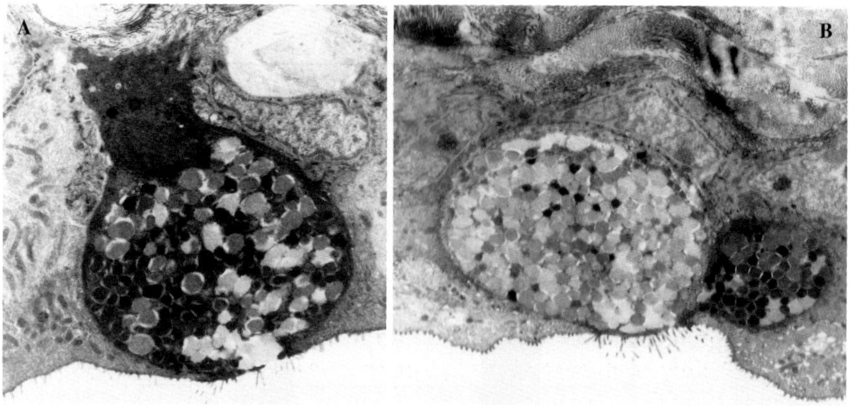

Figure 3 Transmission electron microscopy images of goblet cells. A: Goblet cells from untreated animals showed the cytoplasm filled with granules, which appeared mainly dark. On the apical surface, many scattered microvilli could be observed. B: Mice treated with 0.2%, 7.5×10^5 molecular weight hyaluronan. After 10 days of treatment, the goblet cells showed a cytoplasm filled with granules both electron-dense and electron-transparent. The apical surface showed many microvilli in cells that were not discharging their content on the conjunctival lumen.

by the work at a VDT in a significant manner and was more efficacious than a balanced salt solution (61).

Using a differential pachymetry map, the tear film thickness of healthy volunteers before and after the application of nonviscous, aqueous artificial tears and 0.3% HA was studied. The application of aqueous artificial tears disclosed a significant thickening of the superior part of the precorneal tear film compared with the inferior part. The 0.3% HA solution showed an even thickening of the precorneal tear film with no differences between the superior and the inferior regions (62).

B. Effect on the Diseased Ocular Surface

HA eye drops have been used in several trials for the treatment of dry eye (63–69).

It was shown that they increase precorneal tear film stability and corneal wettability, and reduce the tear evaporation rate and the healing time of corneal epithelium (66,67,70–73).

Table 4 Goblet Cell Modifications Following HA Treatment of Normal Conjunctiva

	N° goblet (cell/m.f.)	Goblet cells area (μm^2)	N° mucin gran./UA	Mucin granules area (μm^2)	Mucin granules optical density
Controls	14.4 ± 4.4	49.2 ± 10	12.9 ± 3.7	0.08 ± 0.06	88.4 ± 18.8
HA	14 ± 5.2	46.2 ± 9.7	8.7 ± 3.2*	0.16 ± 0.08*	160.4 ± 10.6*

*$p < 0.01$ versus control.

The eyes of patients suffering from severe dry eye syndrome are characterized by a deterioration of the corneal epithelium with development of punctate erosions and increased permeability. The conjunctival epithelium of such patients presents squamous metaplasia with decreased or abnormal goblet cells (74) and hence the need for a tear replacement able to provide an environment in which the epithelium can recover the normal structure and function. A topical application of HA was shown to confer both subjective and objective improvement in patients with dry eye syndrome or keratoconjunctivitis sicca (KCS) (68,75–78). Condon et al. (79) have reported a reduction in cell degeneration as assessed by rose bengal. Accordingly, Wysenbeek et al. (80) indicated that HA is able to protect the corneal epithelium.

In a controlled study aimed to explore the long-term effect of HA eye drops on the ocular surface of patients with moderate to severe dry eye, it was shown that, after 3 months of treatment, sodium hyaluronate improved impression cytology, studied with a grading system according to Nelson (69,74). At the same time, the difference with respect to a group of subjects treated with saline was statistically significant. The HA used, approximately 3×10^6 D of molecular weight, was manufactured by a process of continuous fermentation from streptococci and formulated to a 0.15% solution in phosphate-buffered saline at pH of 7.3. The study medication was well tolerated and no treatment-related adverse events occurred during the study. Impression cytology grades showed no significant differences between HA and saline groups after 1 and 2 months, whereas after 3 months of treatment patients treated with HA showed a statistically significantly lower alteration of impression cytology than those obtained in patients treated with saline ($p = 0.036$). These results may be explained as either a treatment-related effect or a detrimental long-term effect of saline on the ocular surface. In order to clarify this issue within a group, a comparison was also performed. Interestingly, the impression cytology grade remained unchanged over time in the saline group, whereas it improved in the HA-treated group. The positive effect over the baseline values became statistically significant after 3 months of treatment ($p = 0.024$). The cytological results obtained during the study are shown in Fig. 4.

Since an increased tear osmolarity has been shown to be responsible for the symptoms and epithelial changes of dry eye, and experimental studies on tissue cultures have proved a toxic effect of hyperosmolarity on epithelial cells (81), it was claimed that a good tear substitute for dry eye patients should be hypotonic (82). For this purpose, a controlled study to evaluate the efficacy of two unpreserved solutions of 0.4% HA, molecular weight of 7.5×10^5, with different osmolarity was carried out in Sjögren's syndrome patients with severe dry eye (83). The study solutions were HA-based eye drops and the two groups were treated as follows: group 1 with unpreserved hypotonic (150 mOsm/L) 0.4% HA eye drops; group 2 with 0.4% unpreserved isotonic sodium hyaluronate eye drops. In both eye drops, the HA had a molecular weight of 750,000 D. The eye drops were administered six times a day for 90 days and the subjects enrolled in

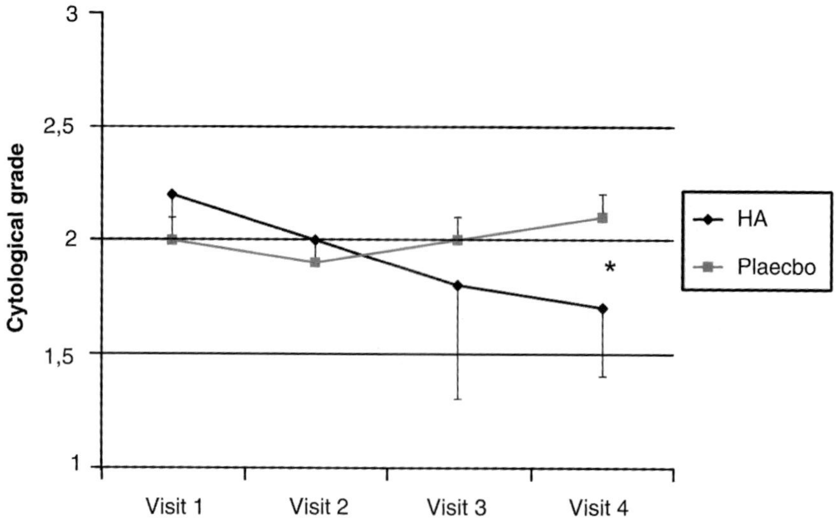

Figure 4 Results on conjunctival impression cytology specimens of the topical treatment with 0.15%, ~3×10^6 hyaluronan of dry eye patients.

the study underwent grading of subjective symptoms and clinical examination at time 0 and after 15, 30 and 90 days.

The results showed that symptoms were statistically significantly improved at day 15 in both groups but group 1 patients had a global score statistically significantly better than group 2 ($p = 0.02$).

At day 15, group 1 patients had an improvement from baseline values of B.U.T. ($p = 0.003$), fluorescein and rose bengal score ($p = 0.000001$ and $p = 0.0004$, respectively). Group 2 patients had, on day 15, an improvement of B.U.T. and fluorescein score compared with baseline values ($p = 0.05$ and $p = 0.0001$, respectively). A comparison between the two groups showed better results for group 1 patients at day 15 for rose bengal stain ($p = 0.01$) and for B.U.T. ($p = 0.05$) and fluorescein score ($p = 0.0003$) at day 90. The conjunctival impression cytology, evaluated by a scoring system according to Aragona (84–86), showed that group 1 had a statistically significantly better total score than group 2 starting from day 15 and lasting throughout the study ($p < 0.02$). In addition, group 2 patients showed an improvement from baseline values starting from day 30 ($p = 0.000005$). The results of the impression cytology study are shown in Figs. 5 and 6.

Therapies with hyposmotic agents in dry eye patients gave contradictory results in the past, since it was reported that dry eye patients prefer such treatment (87) but other data indicated no differences between the results with hypoosmolar or isoosmolar eye drops (68,88). In fact, in patients with a mild dry eye it was not possible to demonstrate any difference in dry eye symptoms relief between a hypotonic sodium hyaluronate solution of 215 mOsm/L compared with an

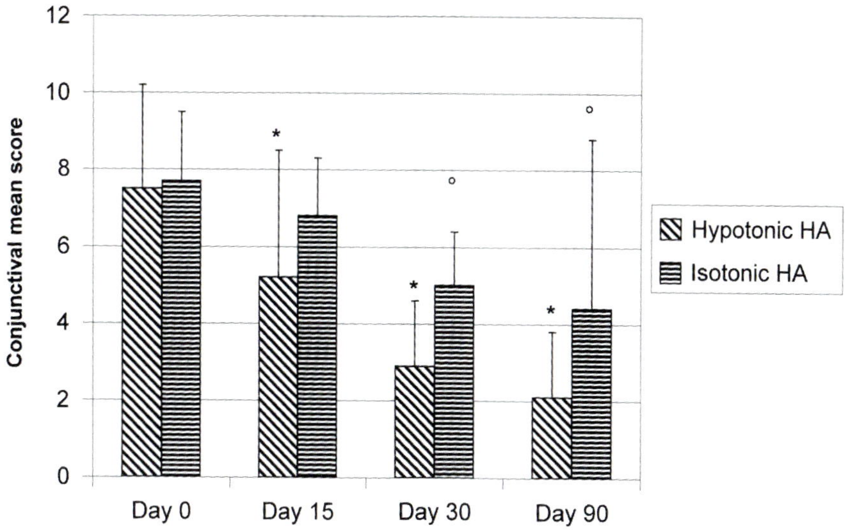

Figure 5 Conjunctival impression cytology score in patients treated with hypotonic and isotonic hyaluronate. $*p < 0.05$ versus isotonic HA; $^{\circ}p < 0.001$ versus baseline.

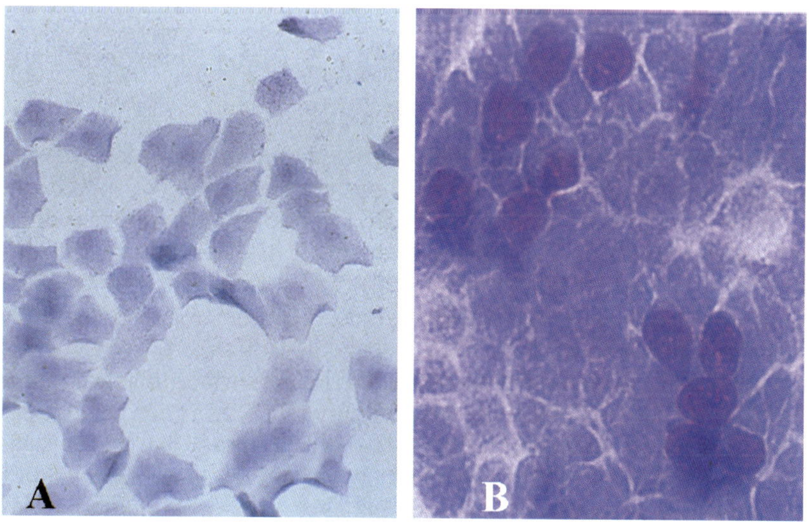

Figure 6 Impression cytology of the conjunctiva of patients with severe dry eye (Sjögren's syndrome) before (A) and 90 days after (B) treatment with hypotonic hyaluronan. At the end of the treatment, the squamous metaplasia was significantly improved with the appearance of goblet cells that, before treatment, were absent (PAS/Papanicolaou stain).

isotonic sodium hyaluronate solution (305 mOsm/L). Furthermore, both treatments determined an improvement in signs of corneo-conjunctival epithelial stress (68). However, the study carried out on Sjögren's syndrome patients with severe dry eye was able to demonstrate that a hypotonic solution of sodium hyaluronate with an osmolarity of 150 mOsm/L, lower than that previously tested, was able to induce a statistically significant improvement of ocular surface conditions, demonstrated either by impression cytology or by vital staining such as fluorescein and rose bengal. This improvement was better than that induced by an isoosmotic solution of HA of similar molecular weight. Therefore, it appears that tear substitutes with a lower osmolarity may be important to obtain a better therapeutic result in patients with highly compromised lachrymal gland secretion like Sjögren's syndrome patients with severe dry eye.

Other evidence of this therapeutic aspect for dry eye disorders arises from the results of the treatment of ocular discomfort in diabetic patients. In fact, diabetic patients often complain of ocular discomfort symptoms, which do not correspond to an evident alteration of the ocular surface. In such patients, an increased tear osmolarity has been demonstrated (89). This kind of patient has an advantage from treatment with hypotonic HA. In a comparative, prospective study carried out versus isotonic saline it was demonstrated that, unlike saline, hypotonic HA was able to statistically significantly reduce tear osmolarity and symptoms of ocular discomfort and to improve statistically significantly the corneal epithelium studied by fluorescein stain (90).

V. Other Uses of Hyaluronan in Ocular Surface Disorders

A. Hyaluronan as a Mask in Refractive Surgery

HA was used with success in assisting excimer laser surgery for the correction of irregular astigmatism resulting from previous corneal refractive surgery.

During corneal healing after excimer laser surgery, a variable degree of irregular astigmatism can occur. Most of the time, its effect on visual acuity and refraction is minimal but sometimes the corneal surface irregularity can cause light dispersion, which is responsible for the occurrence of decreased visual acuity and visual discomfort.

To obtain correction of the irregular corneal surface resulting from the previous surgery, photo-therapeutic keratectomy (PTK) has been used since 1994 (91). This procedure proved to be effective. Furthermore, the use of a viscous masking agent, such as 0.25% HA, was able to increase the efficiency of the procedure through protection of the valleys between the irregular corneal peaks, leaving these irregular peaks exposed to laser treatment (92). In fact, HA is able to cover the depressions of the corneal surface so effectively, masking them from the photoablation, that only protruding, irregular areas can be ablated. This characteristic is depending on HA physical properties, which are responsible for a photoablation rate of HA similar to that of the corneal tissue.

B. Hyaluronan as a Vehicle for Topical Treatments

The mucoadhesive properties of HA were used to increase the bioavailability of several drugs (93). It was tested for pilocarpine, gentamicin and tropicamide. Two studies were carried out on the use of pilocarpine associated to HA. The former showed that HA improved the efficacy of pilocarpine in terms of area under the miosis–time curve, showing no adverse effect (94). The latter demonstrated an increased pilocarpine retention in tear fluid and a 2-fold increase in drug concentration in the cornea and aqueous humor. A significant increase in miotic response was observed at a concentration just less than 0.1% HA. Pilocarpine solutions prepared from high molecular weight HA gave greater miotic response than those prepared from lower molecular weight samples (95).

The administration of a solution of gentamin sulfate in HA 0.25% demonstrated, in healthy volunteers, a higher concentration of gentamicin sulfate in tears than a solution in PBS. The difference was statistically significant at 5 and 10 min after the instillation. At 20 min, the concentration of gentamicin was still higher in the HA solution than in the PBS solution while, at 40 min, the concentration was comparable for both solutions. Thus, HA used as a vehicle demonstrated an increased availability of genatmicin sulfate for at least 10 min (96).

The addition of mucoadhesive polymers, such as HA, to aqueous solutions affects the ocular response of tropicamide (0.2%; w/v). The HA solution used was at the concentration of 0.1%. The mydriatic response of tropicamide was determined in adult male New Zealand rabbits, weighing 1.7–2 kg, by pupil diameter measurements at different times after instillation. The area under the mydriatic response–time curve (AUC 0–6 h) was interpreted as an indication of the bioavailability of tropicamide. The hyaluronic acid solution resulted in a higher AUC mydriasis/time value if compared with other mucoadhesive polymers such as 1% carboxymethylcellulose sodium salt, 0.2% polyacrylic acid and with a nonviscous formulation. The mydriatic effect remained up to 5.5 h for HA. If compared with the other solutions, HA enhanced the bioavailability of tropicamide, presenting a value of mucoadhesion similar to the reference mucin–mucin (97).

VI. Summary and Conclusions

In summary, existing studies about the presence and function of HA in the ocular surface and its role as a therapeutic tool show:

1. HA is found as a component of the normal ocular surface. Its presence was described in tears, lacrimal tissue, conjunctiva and cornea.
2. In the normal ocular surface, it has been demonstrated that the presence of the HA receptor CD44 is always closely associated with HA. HA plays a role in the regulation of corneal epithelial regeneration, promoting epithelial migration and wound healing through an interaction with the receptor CD44.

3. Evident immuno-staining for HA has also been observed on the surface of the superficial corneal cells. This HA can originate from tears and may play a role in stabilizing the preocular tear film. Furthermore, the great hydrophilic properties of HA suggest a possible role in the regulation of corneal hydration.

4. The structure of HA is that of a multi-charged molecule with non-Newtonian characteristics. In this respect, it is similar to tears so that it has been suggested as a tear substitute.

5. Where there is ocular surface disease, HA and its receptor CD44 modify their pattern of distribution within the ocular surface structures.

HA is involved in the reparative processes of the ocular surface following corneal surgery. In fact, it appears that HA and its receptor CD44 are over-expressed by corneal epithelium in the course of wound healing. HA has been shown in strict association with subepithelial water accumulation in the area of corneal wound healing.

6. A topical application of HA has been shown to confer both subjective and objective improvement in patients with dry eye syndrome or keratoconjunctivitis sicca.

7. The treatment of severe aqueous deficient dry eye and dry eye states related to increased tear osmolarity benefit from a hypotonic solution of HA.

HA was used with success acting as a mask in assisting excimer laser surgery for the correction of irregular astigmatism resulting from previous corneal refractive surgery.

8. The mucoadhesive properties of HA can be used with the purpose of increasing the residence time of drugs on the ocular surface.

HA is an essential component of tear film and the ocular surface. Its physical and chemical properties render it suitable for tear film replacement since it has been demonstrated that its administration proved to ameliorate the ocular surface in the course of dry eye disorders. Further studies are needed in order to achieve a better understanding about which physico-chemical characteristics, such as molecular weight and dilution, are more suitable to treat the various forms of ocular surface diseases.

HA's mucoadhesive characteristics also show a potential for a more diffuse use of HA as a vehicle to improve the residence time and the bioavailability of topical drugs used for the treatment of ocular disorders.

References

1. Stern ME, Beuerman RW, Fox RI, Gao J, Mircheff AK, Pflugfelder SC. The pathology of dry eye: the interaction between the ocular surface and lacrimal gland. Cornea 1998; 16(6):584–589.
2. Rolando M, Zierhut M. The ocular surface and tear film and their dysfunction in dry eye disease. Survey Ophthalmol 2001; 45(suppl 2):S203–S210.
3. Rieger G. The importance of precorneal tear film for the quality of optical imaging. Br J Ophthalmol 1992; 76:157–158.

4. Tiffany JM. Composition and biophysical properties of the tear film: knowledge and uncertainty. Adv Exp Med Biol 1994; 350:231–238.
5. van Setten GB, Macauley S, Humphreys-Beher M, Chegini N, Schultz G. Detection of transforming growth factor-alpha mRNA and protein in rat lacrimal glands and characterization of transforming growth factor-alpha in human tears. Investig Ophthalmol Vis Sci 1996; 37:166–173.
6. Thakur A, Willcox MD, Stapleton F. The proinflammatory cytokines and arachidonic acid metabolites in human overnight tears: homeostatic mechanisms. J Clin Immunol 1998; 18:61–70.
7. Li Q, Weng J, Mohan RR, Bennett GL, Schwall R, Wang ZF, Tabor K, Kim J, Hargrave S, Cuevas KH, Wilson SE. Hepatocyte growth factor and hepatocyte growth factor receptor in the lacrimal gland, tears, and cornea. Investig Ophthalmol Vis Sci 1996; 37:727–739.
8. Steiger A. Die Entstehung der spharishen Refraktionen des menschlichen Auges. Berlin, Karger: Doane Clinical Ophthalmology, 1913; vol. 1, Chapter 34.
9. Gipson IK, Inatomi T. Cellular origin of mucins of the ocular surface tear film. Adv Exp Med Biol 1998; 438:221–227.
10. Aragona P, Candela V, Scullica L. A new method for the evaluation of conjunctival alterations with impression cytology. In: Miglior M, van Bijsterveld OP, Spinelli D, eds. The Lacrimal System. Milano: Ghedini, 1993:77–82.
11. Dilly PN. Structure and function of the tear film. Adv Exp Med Biol 1994; 350: 239–247.
12. Inatomi T, Spurr-Michaud S, Tisdale AS, Zhan Q, Feldman ST, Gipson IK. Expression of secretory mucin genes by human conjunctival epithelia. Investig Ophthalmol Vis Sci 1996; 37:1684–1692.
13. Tei M, Moccia R, Gipson IK. Developmental expression of mucin genes ASGP (rMuc4) and rMuc5ac by the rat ocular surface epithelium. Investig Ophthalmol Vis Sci 1999; 40:1944–1951.
14. Dilly PN. Contribution of the epithelium to the stability of the tear film. Trans Ophthalmol Soc UK 1985; 104:381–389.
15. Danjo Y, Watanabe H, Tisdale AS, George M, Tsumura T, Abelson MB, Gipson IK. Alteration of mucin in human conjunctival epithelia in dry eye. Investig Ophthalmol Vis Sci 1998; 39:2602–2609.
16. Baldwin HC, Marshall J. Growth factors in corneal wound healing following refractive surgery: a review. Acta Ophthalmol Scand 2002; 80(3):238–247.
17. van Setten GB, Tervo K, Virtanen I. Immunohistochemical demonstration of epidermal growth factor in the lacrimal and submandibular glands of rats. Acta Ophthalmol (Copenh) 1990; 68:477–480.
18. Barabino S, Rolando M, Camicione P, Ravera G, Zanardi S, Giuffrida S, Calabria G. Systemic linoleic and gamma-linoleic acid therapy in dry eye syndrome with an inflammatory component. Cornea 2003; 22(2):97–101.
19. Brignole F, Pisella PJ, Goldshild M, De Saint Jean M, Goguel A, Baudouin C. Flow cytometric analysis of inflammatory markers in conjunctival epithelial cells of patients with dry eyes. Investig Ophthalmol Vis Sci 2000; 41:1356–1363.
20. Jones DT, Monroy D, Ji Z, Atherton SS, Pflugfelder SC. Sjögren's syndrome: cytokine and Epstein-Barr viral gene expression within the conjunctival epithelium. Investig Ophthalmol Vis Sci 1994; 35:3493–3504.
21. Tsubota K, Fujihara T, Saito K, Takeuchi T. Conjunctival epithelium expression of HLA-DR in dry eye patients. Ophthalmologica 1999; 213:16–19.

22. Mishima S, Maurice DM. The oily layer of the tear film and evaporation from the corneal surface. Exp Eye Res 1961; 1:39–45.
23. Mathers WD, Lane JA. Meibomian gland lipids, evaporation, and tear film stability. Adv Exp Med Biol 1998; 438:349–360.
24. Nagyova B, Tiffany JM. Components responsible for the surface tension of human tears. Curr Eye Res 1999; 19:4–11.
25. Hamano H, Mitsunaga S. Viscosity of rabbit tears. Jpn J Ophthalmol 1973; 17: 290–299.
26. Tiffany JM. The viscosity of human tears. Int Ophthalmol 1991; 15:371–376.
27. Tiffany JM. Viscoelastic properties of human tears and polymer solutions. Adv Exp Med Biol 1994; 384:267–270.
28. Danjo Y, Nakamura M, Hamano T. Measurement of the precorneal tear film thickness with a non-contact optical interferometry film thickness measurement system. Jpn J Ophthalmol 1994; 38:260–266.
29. Norn MS. The conjunctival fluid. Its height, volume, density of cells, and flow. Quantitative examinations and calculations on normal subjects. Acta Ophthalmol 1966; 44:212–222.
30. Prydal JL, Artal P, Woon H, Campbell FW. Study of human precorneal tear film thickness and structure using laser interferometry. Investig Ophthalmol Vis Sci 1992; 33:2006–2011.
31. Pandit JC, Nagyova B, Bron AJ, Tiffany JM. Physical properties of stimulated and unstimulated tears. Exp Eye Res 1999; 68:247–253.
32. Lemp MA. Report of the National Eye Institute/Industry worshop on clinical trials in dry eye. CLAO J 1995; 21:221–232.
33. Lemp MA. Dry eye syndromes: treatment and clinical trials. In: Sullivan DA, ed. Lacrimal Gland, Tear Film and Dry Eye Syndromes. New York: Plenum Press, 1994:553–560.
34. Mathers WD, Shields WJ, Sachdev MS, Petroll WM, Jester JV. Meibomian gland dysfunction in chronic blepharitis. Cornea 1991; 10:277–285.
35. Bron AJ, Benjamin L, Snibson GR. Meibomian gland disease. Classification and grading of lid changes. Eye 1991; 5:395–411.
36. Candela V, Aragona P, Calatroni A. Basi biochimiche della colorazione del muco lacrimale con alcian blu. Boll Ocul 1987; 66(1):113–117.
37. Oya T, Obata H, Miyata K, Tsuru T, Miyauchi S. Quantitative analyses of glycosaminoglycans in tear fluids in normal eyes and eye with corneal epithelial disorders. Nippon Ganka Gakkai Zasshi 1995; 99(3):302–307.
38. Miyauchi S, Morita M, Kuramoto K, Horie K. Hyaluronan and chondroitin sulfate in rabbit tears. Curr Eye Res 1996; 15(2):131–135.
39. Yoshida K, Nitatori Y, Uchiyama Y. Localization of glycosaminoglycans and CD44 in the human lacrimal gland. Arch Histol Cytol 1996; 59(5):505–513.
40. Meyer LJ, Stern R. Age-dependent changes of hyaluronan in normal skin. J Investig Dermatol 1994; 102(3):385–389.
41. Alho AM, Underhill CB. The hyaluronate receptor is preferentially expressed on proliferating epithelial cells. J Cell Biol 1989; 108(4):1557–1565.
42. Lerner LE, Schwartz DM, Hwang DG, Stern R. Hyaluronan and CD44 in the human cornea and limbal conjunctiva. Exp Eye Res 1998; 67:481–484.
43. Lutjen-Drecoll E, Schenholm M, Tamm E, Tengblad A. Visualization of hyaluronic acid in the anterior segment of rabbit and monkey eyes. Exp Eye Res 1990; 51(1): 55–63.

44. Tsai RJ, Ho YS, Chen JK. The effects of fibroblasts on the growth and differentiation of human bulbar conjunctival epithelial cells in an *in vitro* conjunctival equivalent. Investig Ophthalmol Vis Sci 1994; 35(6):2865–2875.
45. Inoue M, Katami C. The effect of hyaluronic acid on corneal epithelial cell proliferation. Investig Ophthalmol Vis Sci 1993; 34(7):2313–2315.
46. Asari A, Miyauchi S, Takahashi T, Kohno K, Uchiyama Y. Localization of hyaluronic acid, chondroitin sulfate, and CD44 in rabbit cornea. Arch Histol Cytol 1992; 55(5):503–511.
47. Eggli PS, Graber W. Ultrastructural distribution of Hyaluronan in rat cornea. Exp Eye Res 1993; 56:693–699.
48. Asari A, Morita M, Sekiguchi T, Okamura K, Horie K, Miyauchi S. Hyaluronan, CD44 and fibronectin in rabbit corneal epithelial wound healing. Jpn J Ophthalmol 1996; 40(1):18–25.
49. Weber BA, Gan L, Fagerholm PP. Short-term impact of corticosteroids on hyaluronan and epithelial hyperplasia in the rabbit cornea after photorefractive keratectomy. Cornea 2001; 20(3):321–324.
50. Weber BA, Gan L, Fagerholm PP. Wound healing response in the presence of stromal irregularities after excimer laser treatment. Acta Ophthalmol Scand 2001; 79:381–388.
51. Weber B, Fagerholm P. Presence and distribution of hyaluronan in human corneas after phototherapeutic keratectomy. Acta Ophthalmol Scand 1998; 76: 146–148.
52. Fitzsimmons TD, Fagerholm P, Harfstrand A, Schenholm M. Hyaluronic acid in the rabbit cornea after excimer laser superficial keratectomy. Investig Ophthalmol Vis Sci 1992; 33(11):3011–3016.
53. Fagerholm P, Hamberg-Nyström H, Tengroth B. Wound healing and myopic regression following photorefractive keratectomy. Acta Ophthalmol (Copenh) 1994; 72:229–234.
54. Fitzsimmons T, Fagerholm P, Harfstrand A, Schenholm M. Steroids after excimer laser decrease corneal hyaluronic acid content. Investig Ophthalmol Vis Sci 1992; 33(ARVO suppl):766.
55. Fitzsimmons T, Fagerholm P, Tengroth B. Steroid treatment of myopic regression: acute refractive and topographic changes in excimer photorefractive keratectomy patients. Cornea 1993; 12:358–361.
56. Weber B, Fagerholm P, Johanson B. Colocalization of hyaluronan and water in rabbit corneas after photorefractive keratectomy by specific staining for hyaluronan and by quantitative microradiography. Cornea 1997; 16:560–563.
57. Weber B, Fagerholm P. Plano and refractive keratectomy. Comparison of the wound healing response. Acta Ophthalmol Scand 1998; 76:537–540.
58. Yu FX, Guo J, Zhang Q. Expression and distribution of adhesion molecule CD44 in healing corneal epithelia. Investig Ophthalmol Vis Sci 1998; 39(5): 710–717.
59. Zhu SN, Nölle B, Duncker G. Expression of adhesion molecule CD44 on human corneas. Br J Ophthalmol 1997; 81:80–84.
60. Gilbard JP, Farris RL. Tear osmolarity and ocular surface disease in keratoconjunctivitis sicca. Arch Ophthalmol 1979; 97:1642–1646.
61. Acosta MC, Gallar J, Belmonte C. The influence of eye solutions on blinking and ocular comfort at rest and during work at video display terminals. Exp Eye Res 1999; 68:663–669.

62. Shimmura S, Goto E, Shimazaki J, Tsubota K. Viscosity-dependent fluid dynamics on the ocular surface. Am J Ophthalmol 1998; 125(3):386–388.
63. Sand BB, Marner K, Norn MS. Sodium hyaluronate in the treatment of keratoconjunctivitis sicca. A double masked clinical trial. Acta Ophthalmol (Copenh) 1989; 67(2):181–183.
64. Yokoi N, Komuro A, Nishida K, Kinoshita S. Effectiveness of hyluran on corneal epithelial barrier function in dry eye. Br J Ophthalmol 1997; 81(7):533–536.
65. Avisar R, Creter D, Levinsky H, Savir H. Comparative study of tear substitutes and their immediate effect on the precorneal tear film. Isr J Med Sci 1997; 33(3):194–197.
66. Hamano T, Horimoto K, Lee M, Komemushi S. Sodium hyaluronate eyedrops enhance tear film stability. Jpn J Ophthalmol 1996; 40(1):62–65.
67. Shimmura S, Ono M, Shinozaki K, Toda I, Takamura E, Mashima Y, Tsubota K. Sodium hyaluronate eyedrops in the treatment of dry eyes. Br J Ophthalmol 1995; 79(11):1007–1011.
68. Papa V, Aragona P, Russo S, Di Bella A, Russo P, Milazzo G. Comparison of ipotonic and isotonic solutions containing sodium hyaluronate on the symptomatic treatment of dry eye patients. Ophthalmologica 2001; 215:124–127.
69. Aragona P, Papa V, Micali A, Santocono M, Milazzo G. Long term treatment with sodium hyaluronate-containing artificial tears reduces ocular surface damage in patients with dry eye. Br J Ophthalmol 2002; 86:181–184.
70. Nakamura M, Hikida M, Nakano T, Ito S, Hamano T, Kinoshita S. Characterization of water retentive properties of hyaluronan. Cornea 1993; 12(5):433–436.
71. Tsubota K, Yamada M. Tear evaporation from the ocular surface. Investig Ophthalmol Vis Sci 1992; 33(10):2942–2950.
72. Snibson GR, Greaves JL, Soper ND, Prydal JI, Wilson CG, Bron AJ. Precorneal residence times of sodium hyaluronate solutions studied by quantitative gamma scintigraphy. Eye 1990; 4(Pt 4):594–602.
73. Chung JH, Fagerholm P, Lindstrom B. Hyaluronate in healing of corneal alkali wound in the rabbit. Exp Eye Res 1989; 48(4):569–576.
74. Nelson JD, Havener VR, Cameron JD. Cellulose acetate impression of the ocular surface. Arch Ophthalmol 1983; 101:1868–1872.
75. De Luise VP, Peterson SW. The use of topical Healon tears in the management of refractory dry-eye syndrome. Ann Ophthalmol 1984; 16:823–824.
76. Stuart JC, Linn JG. Dilute sodium hyaluronate (Healon) in the treatment of ocular surface disorders. Ann Ophthalmol 1985; 17:190–192.
77. Polack FM, McNiece MT. The treatment of dry-eyes with Na-hyaluronate (Healon). Cornea 1982; 1:133–136.
78. Gill GW, Frost JK, Miller KA. A new formula for half-oxidized hematoxylin solution that neither overstains nor requires differentiation. Acta Cytol 1974; 18: 300–311.
79. Condon PI, McEwen CG, Wright M, Mackintosh G, Prescott RJ, McDonald C. Double blind, randomised, piacebo controlied, crossover, multicentre study to determine the efficacy of a 0.1% (w/v) sodium hyaluronate solution (Fermavisc) in the treatment of dry eye syndrome. Br J Ophthalmol 1999; 83:1121–1124.
80. Wysenbeek YS, Loya N, Ben Sira I, Ophir I, Ben Shaul Y. The effect of sodium hyaluronate on corneal epithelium. Investig Ophthalmol Vis Sci 1988; 29:194–199.
81. Gilbard JP, Carter JB, Sang DN, Refojo MF, Hanninen LA, Kenyon KR. Morphology effect of hyperosmolarity on corneal epithelium. Ophthalmology 1984; 91(10):1205–1212.

82. Gilbard JP, Farris RL, Santamaria J II. Osmolarity of tears microvolumes in keratoconjunctivitis sicca. Arch Ophthalmol 1978; 96:677–681.
83. Aragona P, Di Stefano G, Ferreri F, Spinella R, Stilo A. Sodium hyaluronate eye drops of different osmolarity for the treatment of dry eye in Sjögren's syndrome patients. Br J Ophthalmol 2002; 86:879–884.
84. Aragona P, Candela V, Scullica L. A new method for the evaluation of conjunctival alterations with impression cytology. In: Miglior M, van Bijsterveld OP, Spinelli D, eds. The Lacrimal System. Milano: Ghedini, 1993:77–82.
85. Aragona P, Romeo GF, Puzzolo D, Micali A, Ferreri G. Impression Cytology of the conjunctival epithelium in patients with vernal conjunctivitis. Eye 1996; 10(6): 82–85.
86. Aragona P, Ferreri G, Micali A, Puzzolo D. Morphological changes of the conjunctival epithelium in contact lens wearers evaluated by impression cytology. Eye 1998; 12:461–464.
87. Gilbard JP, Farris RL. Tear osmolarity and ocular surface disease in keratoconjunctivitis sicca. Arch Ophthalmol 1979; 97:1642–1646.
88. Wright P, Cooper M, Gilvarry AM. Effect of osmolarity of artificial tear drops on relief of dry eye symptoms: BJ6 and beyond. Br J Ophthalmol 1987; 71:161–164.
89. Aragona P, Giuffrida S, Di Stefano G, Ferreri F, Di Benedetto A, Bucolo C, Cro M. Ocular surface changes in type 1 diabetic patients. Adv Exp Med Biol 2002; 506: 667–672.
90. Aragona P, Di Stefano G, Ferreri F, Ferreri G, Bucolo C, Cro M, Giuffrida S. Hypotonic hyaluronate treatment in type 1 insulin dependent diabetic patients. Investig Ophthalmol Vis Sci 2000; 40(4):B814 (Abstract no. 1439).
91. Gibralter R, Trokel SL. Correction of irregular astigmatism with the excimer laser. Ophthalmology 1994; 101:1310–1314.
92. Aliò JL, Belda JI, Shalaby AM. Correction of irregular astigmatism with excimer laser assisted by sodium hyaluronate. Ophthalmology 2001; 108(7):1246–1260.
93. Ludwig A, Van Ooteghem M. Evaluation of sodium hyaluronate as viscous vehicle for eye drops. J Pharm Belb 1989; 44(6):391–397.
94. Camber O, Edman P, Gurny R. Influence of sodium hyaluronate on the miotic effect in rabbits. Curr Eye Res 1987; 6(6):779–784.
95. Camber O, Edman P. Sodium hyaluronate as an ophthalmic vehicle: some factors governing its effect on the ocular absorpion of pilocarpine. Curr Eye Res 1989; 8(6): 563–567.
96. Bernatchez SF, Tabatay C, Gumy R. Sodium hyaluronate 0.25% used as a vehicle increases the bioavailability of topically administered gentamicin. Graefes Arch Clin Exp Ophthalmol 1993; 231(3):157–161.
97. Herrero-Vanrell R, Fernandez-Carballido A, Frutos G, Cadorniga R. Enhancement of the mydriatic response to tropicamide by bioadhesive polymers. J Ocul Pharmacol Ther 2000; 16(5):419–428.

Chemistry and Biology of Hyaluronan
H.G. Garg and C.A. Hales (editors)

Chapter 25

The Hyaluronan Synthases

PAUL H. WEIGEL

Department of Biochemistry and Molecular Biology and The Oklahoma Center for Medical Glycobiology, BMSB, Room 860, University of Oklahoma Health Sciences Center, Oklahoma City, OK 73104, USA

I. Introduction

The enzymes that make hyaluronan (HA), the hyaluronan synthases (HAS), were so difficult to solubilize and purify that 65 years elapsed between the identification of hyaluronic acid in 1934 and the first purification of an active synthase in 1999. Purification of HAS became possible after the first synthase gene was cloned in 1993 and the recombinant enzyme could be expressed and studied. Once the Group A *Streptococcus pyogenes hasA* gene encoding HAS was identified, the field developed rapidly, culminating with the cloning and characterization of three distinct isozymes for human HAS. The properties and characteristics of this novel family of polysaccharide synthases are summarized in this chapter.

After its discovery in 1934 by Meyer and Palmer (1), hyaluronan (hyaluronic acid or hyaluronate (HA) was found to be an ubiquitous, often abundant extracellular matrix (ECM) component in vertebrate tissues, particularly in cartilage, skin and vitreous humor (2). HA is also produced by some pathogenic bacteria. HA is a linear unbranched alternating polymer composed of two monosaccharides: β(1,4)-*N*-acetyl-D-glucosamine and β(1,3)-D-glucuronic acid. The molecular mass of native HA can approach 10^7 Da. Though simple in structure, HA is necessary for the normal development in vertebrates and plays important functions in normal health and in various diseases (e.g., as discussed in Chapters 11–18). HA is synthesized by the enzyme HAS.

Since our discovery in 1993 of the first HA synthase gene, from Group A *Streptococcus* (3,4), similar HA synthase genes (or cDNAs) have been identified in other bacteria (5–7), frogs, humans and many other mammals (8–16), and even in a virus that infects an algae, which in turn lives in a protozoan host (17). The GenBank database contains >20 sequence entries for HA synthases and more are reported every year as investigators continue to explore the role of these enzymes in biology. These various HA synthase enzymes comprise a large family of proteins with many common features and regions of amino acid sequence identity or similarity (for reviews, see Refs. 18–22). Based on their similarities and differences, the known HA synthase proteins have been divided into two categories, designated Class I and Class II (20). The HA synthase from *Pasteurella multocida* (pmHAS) is the only Class II member and this enzyme is quite different than all the other HA synthases. In this chapter, I will focus mainly on the characteristics of the large Class I HA synthase family.

II. Molecular Size and the Biology of Hyaluronan

Hyaluronan is continuously synthesized and secreted by fibroblasts, keratinocytes, chondrocytes and other specialized cells in ECMs throughout the body. The level of HA synthesis is very high in tissues wherein HA constitutes a large portion of the tissue mass, such as skin and cartilage. As described in Chapters 1 and 2, the mass of the HA in many tissues is very large, typically in the range of 2–5 MDa (2). Additionally, a hallmark characteristic of HA isolated from any natural source is its size heterogeneity. HA is very polydisperse with a range of masses present that usually spans more than two orders of magnitude. For example, a commercial preparation with an average mass of 1 MDa typically contains a distribution of sizes from <0.1 to >10 MDa. Previously, the physiological function of HA in the ECM was believed to be only structural or physical. Now, however, HA is also recognized as a pharmacologically active, signaling molecule.

Despite its simple structure, HA influences a wide range of cell functions and behaviors, including cell migration, differentiation and phagocytosis. It is especially important that some cells respond to very small fragments of HA (see Chapter 7). For example, angiogenesis is stimulated by small, but not large, HA (23) and activated macrophages are induced to express a large number of genes in response to small, but not large, HA (24). Small HA fragments stimulate signal cascades in target cells through specific cell surface receptors. Thus, large and small HA molecules have different physiological functions, and there is currently great interest in understanding the ability of HA synthases to produce HA of different sizes in different biological situations.

III. Hyaluronan Synthesis by Hyaluronan Synthases

A. Overview

Virtually all of the known enzymes utilize one, two or three substrates to produce one or two products. The HA synthase enzyme family is very unusual, and perhaps unique (18), in that these synthases have two different enzymatic activities (i.e., glycosyltransferases) within the same protein. Furthermore, the growing HA chain produced after each sugar addition becomes the substrate for the next sugar addition. HA synthases are also membrane-bound enzymes. In bacteria and eukaryotic cells, active HA synthase is localized to the plasma membrane. The overall reaction for the addition of one HA disaccharide unit to an HA chain (where n is the number of disaccharide units) is shown in Sch. 1. As discussed below, this is actually only one of two possible schemes.

$$\text{Scheme 1: UDP-GlcUA} + \text{UDP-GlcNAc} + (\text{HA})_n \xrightarrow{\text{HA synthase}} (\text{HA})_{n+1} + 2\text{UDP} \quad (1)$$

The two activated sugar precursors for HA biosynthesis are UDP-glucuronic acid (UDP-GlcUA) and UDP-*N*-acetylglucosamine (UDP-GlcNAc). In order to make a disaccharide unit and extend the growing HA chain, HA synthase possesses at least six distinct functions. These activities include two different sugar nucleotide binding sites, two different glycosyltransferase activities, one or more binding sites for the growing HA chain and the ability to transfer this HA chain within the enzyme, to set up the next round of saccharide addition. The UDP-sugar substrates are produced and used by the HA synthase inside the cell, and the HA chain is continuously transferred (translocated) across the membrane so that it is extruded to the cell exterior. In the case of bacteria, this HA forms a capsule. In the case of many eukaryotic cells, the result is either the formation of a pericellular HA coat surrounding the cell or the release and deposition of HA into the surrounding ECM.

B. Bacterial Hyaluronan Synthesis and the Discovery of Hyaluronan Synthases

Some bacteria can make the same HA polymer that mammals make, and because of this, these microorganisms have created a clever way to become pathogenic—by hiding behind a wall of HA inside the host they are infecting. These bacteria are not easily recognized by antibodies or attacked by phagocytes, because the HA capsule surrounding them has the same structure as the HA in the host tissues. Examples are: Group C *Streptococcus equisimilis*, which is a pathogen in animals and sometimes humans; *S. uberis*, which causes severe mastitis in cows; Group A *S. pyogenes*, which is a human pathogen; *P. multocida*, which is a fowl pathogen. Each pathogen, though infecting different animals, is able to hide from the host immune systems by surrounding itself with a thick HA layer. Although the HA capsule is a significant virulence factor contributing to the pathogenicity of these bacteria (25), it also provided a very good system in which biochemists could study HA synthesis. For example, Markovitz et al. (26) demonstrated in 1959 that membrane preparations from Group A *Streptococcus* could synthesize HA in the presence of the two UDP-sugars and Mg^{2+} ions, although this and other groups were subsequently unable to purify the HA synthase.

In order for streptococcal bacteria to synthesize an HA capsule, three different genes, which encode three different enzymes, must usually be present, arranged in an operon that is designated the *HA synthesis* (or *has*) operon (27). Two of the enzymes, UDP-glucose dehydrogenase (encoded by the *hasB* gene) and UDP-glucose pyrophosphorylase (*hasC*), are needed for the cell to produce large amounts of the two UDP-sugar precursors. The third enzyme is HA synthase, encoded by *hasA*. Three *Streptococcus* HA synthase genes have now been cloned; from *S. pyogenes*, *S. equisimilis* and *S. uberis*. These proteins are the smallest members of the HA synthase family at 417–419 amino acids, and their primary amino acid sequences are ~70% identical.

C. Discovery of the Eukaryotic Hyaluronan Synthases

After the Group A HA synthase was cloned, a family of related eukaryotic HA synthase genes was reported in other bacteria, mouse, human, frog, zebra fish and chicken (5–22). In human and mouse there are three different genes, encoding three distinct HA synthase isozymes. These genes are designated mouse *Has1*, *Has2* and *Has3* or human *Has1*, *Has2* and *Has3*. The three HA synthase genes are located on different chromosomes in both mouse and human (15). In humans the Has1 gene is on chromosome 19q13.3–13.4, Has2 is on chromosome 8q24.12 and Has3 is on chromosome 16q22.1. Each of the three mammalian HA synthases are active, i.e., able to synthesize HA, when expressed as recombinant proteins in a wide variety of cell types (28).

D. Expression of the Three Mammalian Hyaluronan Synthase Isoforms

Although the three gene products (the HAS proteins) catalyze the same biosynthetic reaction, the expression patterns for the three genes in different adult and embryonic tissues are quite distinct, as are their temporal patterns of expression during embryogenesis. The three HA synthase proteins may, therefore, be expressed in specific tissues at different times in order to perform different physiological functions.

E. Functions of the Three Mammalian Hyaluronan Synthase Isoforms

It is likely that the three HA synthase isozymes are subjected to distinct regulatory mechanisms that control factors such as the amount of each HA synthase isoform produced per cell, the activity of each enzyme, the size distribution of the HA product made and turnover of the HA synthase protein. However, few studies have been able to address these and other important questions about the biology of the HA synthases. The field presently lacks the necessary isoform-specific antibodies to perform localization or activity studies to determine which of the three HA synthases is expressed, and where, in the same cell and how their activities might be regulated by post-translational modifications.

The single or double knock-out mice, in which the *Has1* or *Has3* genes are not functional, are viable, fertile and, thus far, present no recognizable phenotype (19). The *Has2* knockout, however, is lethal during embryonic development because the cardiac cushion fails to develop normally, producing severe cardiac defects (29,30).

IV. Topological Organization of the Hyaluronan Synthase Proteins

The topology of the *S. pyogenes* HA synthase was determined experimentally (31) and the majority of the HA synthase protein, including both the amino and carboxyl termini, was found to be inside the cell (Fig. 1A). Four membrane domains pass through the membrane, giving only two small loops of the protein exposed to the extracellular side. Another two or more regions of the protein, including the large central domain, associate with the membrane as amphipathic helices or re-entrant loops that do not span the membrane. Thus, only about 5% of

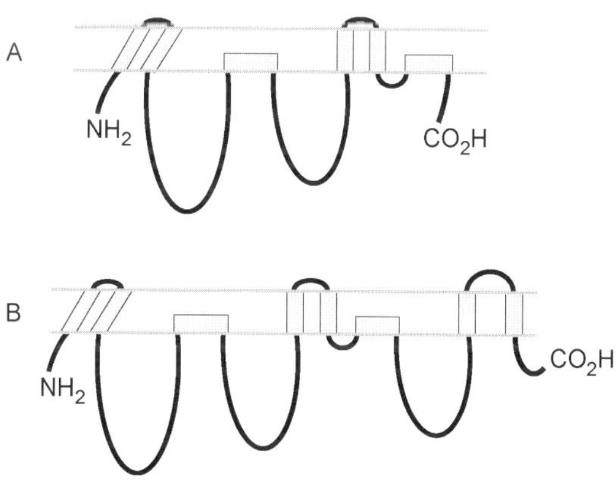

Figure 1 Topology of the streptococcal and eukaryotic HA synthases. The parallel lines indicate the cell membrane and the upper portions in A and B are outside the cell. The membrane topology of the recombinant *S. pyogenes* HA synthase expressed in *E. coli* (A) was determined experimentally (31). The topological organization of the other known streptococcal HA synthases are assumed to be the same because their amino acid sequences are >70% identical to spHAS. Although they have not been determined experimentally, the overall topologies of the eukaryotic HA synthases are also predicted to be the same within the NH_2-terminal two-thirds of these proteins that are homologous to the streptococcal enzyme (18–22).

the protein is exposed extracellularly. The topology of mammalian HA synthases has not yet been determined experimentally, but is predicted (Fig. 1B) to be essentially the same as the *S. pyogenes* enzyme, except that there are probably two additional transmembrane domains at the C-terminus (18).

V. Molecular Size of Functional Hyaluronan Synthases

Based on radiation inactivation analysis of two streptococcal HA synthases (32) and the *Xenopus* HAS1 (33), the active membrane-bound enzymes contain only a single HA synthase protein (rather than an oligomer), but the protein is associated with an additional mass of about 23 kDa. In the case of the streptococcal enzymes, this extra mass was identified as cardiolipin, a common phospholipid (32). Thus, the active streptococcal HA synthase is a complex of one HA synthase protein with about 14–18 molecules of cardiolipin. When purified to homogeneity in the absence of cardiolipin, both streptococcal HA synthases have very low activity (34). When this phospholipid is added back, however, enzyme activity increases up to 10-fold. The Group A enzyme was very specifically activated by cardiolipin, whereas the Group C enzyme was also activated by phosphatidylserine. Thus, within the Class I HA synthase family, there may be some differences in the particular phospholipids required for activity, but it is possible that all these HA synthases require one or more particular type of phospholipid in order to synthesize HA. We believe that the cardiolipin molecules help the enzyme to create a pore-like passage within the enzyme, through which the growing HA chain passes (34).

VI. Purification of Hyaluronan Synthases

A. Overview

HA synthases from any source have been extremely difficult to detergent-solubilize with retention of activity and then purify. The recombinant Group A and Group C streptococcal enzymes have been the easiest Class I HA synthases to purify (34) and have been studied more extensively. In an elegant study, Yoshida et al. (35) reported the purification and kinetic characterization of the recombinant mouse HAS1 enzyme. No other eukaryotic HA synthases have been purified. Based on the results from genetic (3,4), biochemical (36), and radiation inactivation studies (32,33) and studies with the purified enzymes (34,35), it is clear that the HA synthase is the only protein necessary for HA biosynthesis—when the UDP-sugars are provided. There is no evidence that HA synthases require a primer or any other proteins in order to synthesize HA. This finding was a surprise to the field, which believed for many years that multiple proteins would be required to perform the multiple functions needed to polymerize the HA polysaccharide.

B. Characteristics of the Hyaluronan Synthase Proteins

As noted above, the three streptococcal HA synthase proteins are ~70% identical. Amino acid sequence identities among the mammalian HA synthase family members range from ~50 to 95% (18). The streptococcal proteins are ~25% identical to the eukaryotic proteins, which are ~35% larger. All the HA synthase proteins characterized to date, except for the pmHAS enzyme, share numerous amino acid motifs and are membrane proteins with multiple predicted transmembrane domains (Fig. 1). The large group of related family members has been designated Class I HA synthases (20). The structurally and enzymatically distinct pmHAS is the only Class II HA synthase member. Mouse Has1, Has2 and Has3 are 583, 552 and 554 amino acids long, respectively. Human HAS1, HAS2 and HAS3 proteins contain 578, 552 and 553 amino acids, respectively.

The HA synthase proteins appear to be very compact and tightly folded. Even during electrophoresis in the presence of the strong detergent sodium dodecyl sulfate, the proteins behave as though they are ~15% smaller than their true size. Reducing agents do not change this behavior, which indicates that disulfide bonds are probably not present. The lack of disulfide bonds was confirmed biochemically for the streptococcal enzymes (37,38), which have only 4–6 Cys residues. These latter studies also demonstrated that Cys residues are not required for HA synthase activity. The Group A and Group C streptococcal Cys-null HA synthases retained from ~20 to 60% of the wild-type enzyme activity. Similar studies have not yet been performed with any of the eukaryotic HA synthases, probably because these larger enzymes have about three times as many Cys residues, making similar approaches more technically difficult.

C. Kinetic Characteristics of Hyaluronan Synthases

Although there are exceptions, the kinetic behaviors of HA synthases in membranes from various sources are generally very similar. The K_m values for UDP-GlcUA utilization are typically lower, e.g., 30–75 μM for the streptococcal and human HA synthases, than for utilization of UDP-GlcNAc, e.g., 250–1000 μM for the three human enzymes (28,39). The less efficient utilization of UDP-GlcNAc by HA synthases may reflect the fact that this precursor sugar is used for the biosynthesis of many more important structures than is UDP-GlcUA, e.g., cell wall biosynthesis in bacteria or the core chitobiosyl linkage region in the N-linked oligosaccharides of eukaryotic glycoproteins. If the K_m value for UDP-GlcNAc was lower, then HA synthesis might drain away this precursor so that cellular glycoconjugate biosynthesis was compromised.

It is more difficult to compare V_{max} values for various HA synthases because different groups use different methods to normalize the amount of HA synthesis to the amount of HAS protein. The HA polymerization rates for the streptococcal HA synthases in isolated membranes were estimated to be ~1200–2400 sugars per minute (36,39). At this elongation rate, it would take one active HA synthase molecule (or lipid complex) about 8–16 min to synthesize a single HA chain with a mass of 2 MDa. The rate of HA chain elongation in live cells has not been determined, but is likely to be faster than what has been measured *in vitro*.

VII. The Class I Hyaluronan Synthases Polymerize Hyaluronan by Addition to the Reducing End

A. Overview

Two previous studies using membrane preparations from eukaryotic cells (40,41) suggested that HA synthesis occurs at the reducing end. Since membranes contain other glycosyltransferase activities, however, these results might have alternate interpretations. Also, the contributions of the three mammalian HA synthase isoenzymes to these results was unknown. In order to confirm this important conclusion, we performed similar experiments using two different purified recombinant streptococcal HA synthases. Our results verify that addition of new saccharide units does, in fact, occur at the reducing end (42).

B. The Mechanism of Hyaluronan Chain Elongation at the Reducing End

The biochemical interactions and reactions involved in glycosidic linkage formation determine the nature of donor/acceptor relationships among the substrates. The mechanisms for biosynthesis of a polysaccharide are fundamentally different depending on whether the chain grows from the reducing or the nonreducing end. The mechanistic difference for addition of a disaccharide unit to a growing HA chain is illustrated between Sch. 1 (shown earlier) and Sch. 2 (below).

$$\text{Scheme 2: } \mathit{UDP}\text{-GlcUA} + \text{UDP-GlcNAc} + (\text{HA})_n\text{-}\mathbf{UDP} \\ \rightarrow (\text{HA})_{n+1}\text{-UDP} + \mathbf{UDP} + \mathit{UDP} \quad (2)$$

In Sch. 1, which is the situation for the Class II pmHAS (43), UDP is released from a precursor UDP-sugar (which is the donor) when this sugar is added to the nonreducing end of the HA polymer (which is the acceptor). Therefore, when one disaccharide unit is added, the two UDP groups that are released come from the two new sugars added.

However, for chain elongation at the reducing end (Sch. 2), the UDP-sugar added to the polymer chain is transferred intact, without cleavage of its UDP linkage. The UDP released during each transfer step comes from the HA-UDP intermediate formed by the addition of the previous sugar. Thus, of the two net UDP groups released when a disaccharide unit is assembled at the reducing end, only one UDP comes from the last two UDP-sugars added. The other UDP comes from the last sugar added prior to addition of the new disaccharide unit. The individual steps in this reaction mechanism are illustrated in Sch. 3.

Scheme 3: N-**UDP** + A-*UDP* → A-N-**UDP** + *UDP* (3)

A-N-**UDP** + A-*UDP* → A-N-A-*UDP* + **UDP**

A-N-A-*UDP* + N-**UDP** → A-N-A-N-**UDP** + *UDP*

In this example, GlcUA and GlcNAc are indicated by A and N, respectively, and the UDP groups for their UDP-sugars is indicated, respectively, in italic or boldface font.

C. Hyaluronan Chain Elongation at the Reducing End and Nomenclature

During chain elongation at the reducing end, the UDP-sugars are not the donors (as they are for additions that occur at the nonreducing end), instead they are the acceptors (22). Therefore, the Class I HA synthase transferase activity that utilizes UDP-GlcNAc actually creates the GlcUA(β1,3)GlcNAc linkage. In contrast, the Class II pmHAS activity that utilizes UDP-GlcNAc creates the GlcNAc(β1,4)GlcUA linkage (43). The multiple functions of Class I HA synthases, from the perspective of how these enzymes actually make HA, are summarized in Fig. 2.

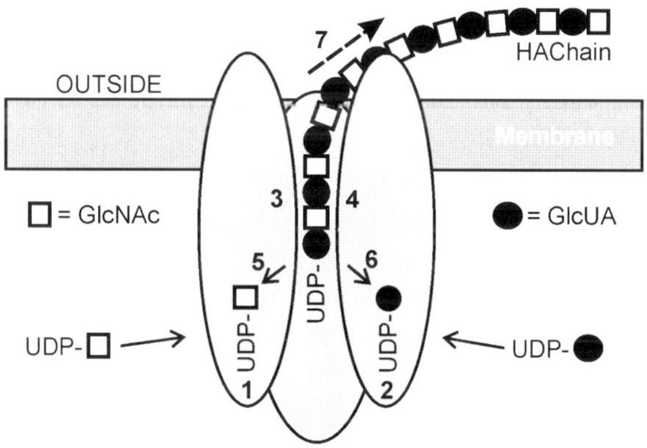

Multiple Functions of a Class I HA Synthase

1. UDP-GlcNAc Acceptor Binding
2. UDP-GlcUA Acceptor Binding
3. HA-GlcUA-UDP Donor Binding
4. HA-GlcNAc-UDP Donor Binding
5. HA-GlcUA-UDP: UDP-GlcNAc, beta-1,3(HA-GlcUA) transferase
6. HA-GlcNAc-UDP: UDP-GlcUA, beta-1,4(HA-GlcNAc) transferase
7. HA translocation through the membrane

Figure 2 Multiple functions of hyaluronan synthases. The model schematically illustrates the indicated functions of an active HA synthase, numbered one through seven. The UDP-sugar precursors, which are generated by enzymes that are not shown, are used in the cytoplasm as the synthase assembles an HA chain that extrudes into the exterior space outside the cell membrane.

In each cycle of monosaccharide addition at the reducing end, the released UDP is derived from the previously added monosaccharide (Schs. 2 and 3). The donor species transferred is actually the HA chain, which is still activated by its attachment to UDP. The systematic name for such a transferase activity, according to IUBMB nomenclature, would specify the *donor: acceptor, group transferred*. Thus, the activity that adds a GlcUA residue to a GlcNAc at the reducing end of the growing HA chain is a (HA)-GlcNAc-UDP: UDP-GlcUA, β(1,4)-hyaluronyl transferase. Similarly, a (HA)-GlcUA-UDP: UDP-GlcNAc, β(1,3)-hyaluronyl transferase is the activity that adds a GlcNAc to a HA-GlcUA-UDP chain.

The model in Fig. 2, in which the HA synthase itself is involved in HA transfer (translocation) across the membrane, is based on a variety of biochemical and genetic data (3,32,34,36). These studies showed that the streptococcal HA synthase is the only protein required for HA synthesis *in vitro*. Also, the *hasA* gene is the only gene required to transform a bacterial species that does not normally make HA, such as *Bacillus subtilis* (44) and *Enterococcus faecalis* (3) into a strain that makes HA. If the UDP-sugar precursors are present in such cells, then bacteria transformed with the *hasA* gene are able to make HA, using cytoplasmic precursors (31), and transfer it to the cell exterior as an HA capsule or a secreted product that accumulates in the medium. Based on these and other data, there is no indication that a polysaccharide transporter is required in order for HA to appear outside the cell.

D. Biological Implications of Synthesizing Hyaluronan at the Reducing End

When chain elongation occurs at the reducing end, the HA product will have UDP attached at the reducing end, which is from the last sugar unit added. Since either HA–UDP linkage is less stable under physiological conditions than either type of glycoside bond in HA, it will be susceptible to hydrolysis, even at near-neutral pH. Therefore, HA chains in commercial preparations that have been processed in various ways are unlikely to still contain UDP at their reducing ends. However, it is intriguing to postulate that the presence of a novel HA–UDP structural element at the reducing end of a newly released HA chain could provide a specific recognition site for a class of proteins designed to recognize this linkage. Although the implications of such interactions could be physiologically important, they have not yet been explored.

VIII. Are Hyaluronan Synthases Processive?

The Class I HA synthases appear to be processive, meaning that they may not release and rebind a growing HA chain (36). Rather, chain growth may cease when the enzyme releases an HA chain. This model is based, in part, on the inability of any Class I HA synthase tested to elongate short HA oligosaccharides.

For example, these enzymes do not recognize and extend HA tetrasaccharides obtained from testicular hyaluronidase digests. However, one problem with interpreting these negative data is that the substrate to test should be an HA–UDP fragment. Continued HA chain growth at the reducing end requires that the growing HA chain be attached to UDP (and thereby 'activated'). HA fragments are not expected to work, because they can only serve as acceptors, not donors as outlined in Sch. 3. Until UDP–HA intermediates can be isolated and tested for their ability to be extended by HA synthases, the question of whether these enzymes are truly processive cannot be answered.

IX. Size Distributions of Hyaluronan Products Made by Hyaluronan Synthases

A. Hyaluronan Synthases Do Not Produce Hyaluronan of a Unique Size

The products made by the HA synthases are not a particular, nor unique size. Rather, the HA products are a broad distribution of sizes that can span several orders of magnitude (e.g., 10^5-10^7 Da). Each enzyme is capable of starting and polymerizing one HA polymer after another. The final size of a given HA molecule may depend on how quickly the enzyme releases it. For example, an HA chain could grow to >40,000 monosaccharides, corresponding to a mass of >8 million Da, before it is released. No one knows what controls the HA release process by HA synthases. Presumably, one mechanism could be related to the UDP–HA linkage. Hydrolysis of the UDP–HA would terminate chain growth and favor HA release by the enzyme. Another possibility is that there are opposing HA-retention and HA-release forces within the HA–enzyme complex (45). Retentive forces such as H-bonding or hydrophobic interactions between HA and the synthase would work to hold the two together, whereas the release forces, e.g., created by Brownian motion, would tend to pull the HA chain away from the synthase. As the HA chain size increases, one might expect that the collisional release forces would increase and ultimately be greater than the retentive forces, resulting in HA chain release.

B. Different Hyaluronan Synthases can Produce Different Hyaluronan Size Distributions

Although all Class I HA synthases catalyze the same reaction, different enzymes may differ in the size distribution of HA they produce. This HA product size difference was demonstrated by Itano, Kimata and co-workers for the three human HAS isozymes expressed in transfected cells (21,28). HAS1 and HAS3 expression caused cells to synthesize and secrete HA of molecular mass ~0.2–2.0 MDa, whereas cells expressing HAS2 produced HA of molecular mass >2 MDa. A similar trend was observed for the size of HA products made by each of the three human HAS isozymes in isolated membranes. If these results are applicable to the situation in various human tissues and organs *in vivo*, then the actual HA size distribution made by HAS1, HAS2 or HAS3 could be different for specific biological purposes.

The biological implications of these differences in HA size are just beginning to be appreciated. The growing recognition that certain biological responses may be dependent on HA of a particular size range now meshes nicely with the discovery that HA synthase family members may be intrinsically different in the size range of HA they synthesize.

X. Summary and Conclusions

In 1993, we reported the first cloning of a glycosaminoglycan synthase, the Group A HA synthase. Within a decade, a large family of HA synthases has been identified, and many gene regulation studies, e.g., using RT-PCR, to follow HAS expression have been reported. Some, though far fewer, studies have used peptide-specific antibodies to study the expression of individual HAS protein isoforms. We are just beginning to explore the mechanisms by which the regulation of HA biosynthesis is achieved in bacteria, animal models and human disease. As more molecular tools are developed, particularly antibodies specific for only one of the three human isoforms, we will ultimately be able to understand the function and biology of HA in embryonic development, wound healing, angiogenesis, metastasis and many other important physiological processes.

Acknowledgements

The research described from the author's laboratory was supported by National Institutes of Health grant GM35978 from the National Institute of General Medical Sciences.

References

1. Meyer K, Palmer JW. The polysaccharide of the vitreous humor. J Biol Chem 1934; 107:629–634.
2. Abatangelo G, Weigel PH, eds. New Frontiers in Medical Sciences: Redefining Hyaluronan. Amsterdam: Elsevier, 2000.
3. DeAngelis PL, Papaconstantinou J, Weigel PH. Isolation of a *Streptococcus pyogenes* gene locus that directs hyaluronan biosynthesis in acapsular mutants and in heterologous bacteria. J Biol Chem 1993; 268:14568–14571.
4. DeAngelis PL, Papaconstantinou J, Weigel PH. Molecular cloning, identification, and sequence of the hyaluronan synthase gene from Group A *Streptococcus pyogenes*. J Biol Chem 1993; 268:19181–19184.
5. Kumari K, Weigel PH. Molecular cloning, expression, and characterization of the authentic hyaluronan synthase from Group C *Streptococcus equisimilis*. J Biol Chem 1997; 272:32539–32546.
6. DeAngelis PL, Jing W, Drake RR, Achyuthan AM. Identification and molecular cloning of a unique hyaluronan synthase from *Pasteurella multocida*. J Biol Chem 1998; 273:8454–8458.

7. Ward PN, Field TR, Ditcham WG, Maguin E, Leigh JA. Identification and disruption of two discrete loci encoding hyaluronic acid capsule biosynthesis genes *hasA*, *hasB* and *hasC* in *Streptococcus uberis*. Infect Immun 2001; 69: 392–399.
8. Itano N, Kimata K. Expression cloning and molecular characterization of hyaluronan synthase protein, a eukaryotic hyaluronan synthase. J Biol Chem 1996; 271:9875–9878.
9. Shyjan AM, Heldin P, Butcher EC, Yoshino T, Briskin MJ. Functional cloning of the cDNA for a human hyaluronan synthase. J Biol Chem 1996; 271:23395–23399.
10. Spicer AP, Augustine ML, McDonald JA. Molecular cloning and characterization of a putative mouse hyaluronan synthase. J Biol Chem 1996; 271:23400–23406.
11. Watanabe K, Yamaguchi Y. Molecular identification of a putative human hyaluronan synthase. J Biol Chem 1996; 271:22945–22948.
12. Fulop C, Salustri A, Hascall VC. Coding sequence of a hyaluronan synthase homologue expressed during expansion of the mouse cumulus–oocyte complex. Arch Biochem Biophys 1997; 37:261–266.
13. Spicer AP, Olson JS, McDonald JA. Molecular cloning and characterization of a cDNA encoding the third putative mammalian hyaluronan synthase. J Biol Chem 1997; 272:8957–8961.
14. Spicer AP, Seldin MF, Olsen AS, Brown N, Wells DE, Doggett NA, Itano N, Kimata K, Inazawa J, McDonald JA. Chromosomal localization of the human and mouse hyaluronan synthase genes. Genomics 1997; 41:493–497.
15. Spicer AP, McDonald JA. Characterization and molecular evolution of a vertebrate hyaluronan synthase gene family. J Biol Chem 1998; 273:1923–1932.
16. Yamada Y, Itano N, Zako M, Yoshida M, Lenas P, Niimi A, Ueda M, Kimata K. The gene structure and promoter sequence of mouse hyaluronan synthase 1 (mHAS1). Biochem J 1998; 330:1223–1227.
17. DeAngelis PL, Jing W, Graves MV, Burbank DE, Van Etten JL. Hyaluronan synthase of chlorella virus PBCV-1. Science 1997; 278:1800–1803.
18. Weigel PH, Hascall VC, Tammi M. Hyaluronan synthases. J Biol Chem 1997; 272: 13997–14000.
19. McDonald JA, Spicer A. In: Hascall VC, Yanagishita M, eds. Science of Hyaluronan Today 1998: Chap. 7, http://www.glycoforum.gr.ip.
20. DeAngelis PL. Hyaluronan synthases: fascinating glycosyltransferases from vertebrates, bacterial pathogens, and algal viruses. Cell Mol Life Sci 1999; 56: 670–682.
21. Itano N, Kimata K. Mammalian hyaluronan synthases. IUBMB Life 2002; 54: 195–199.
22. Weigel PH. Functional characteristics and catalytic mechanisms of the bacterial hyaluronan synthases. IUBMB Life 2002; 54:201–211.
23. Slevin M, Kumar S, Gaffney J. Angiogenic oligosaccharides of hyaluronan induce multiple signaling pathways affecting vascular endothelial cell mitogenic and wound healing responses. J Biol Chem 2002; 277:41046–41059.
24. Horton MR, Burdick MD, Strieter RM, Bao C, Noble PW. Regulation of hyaluronan-induced chemokine gene expression by IL-10 and IFN-gamma in mouse macrophages. J Immunol 1998; 160:3023–3030.
25. Wessels MR, Moses AE, Goldberg JB, Dicesare TJ. Hyaluronic acid capsule is a virulence factor for mucoid Group A streptococci. Proc Natl Acad Sci USA 1991; 88:8317–8321.

26. Markovitz M, Cifonelli JA, Dorfman A. The biosynthesis of hyaluronic acid by Group A *Streptococcus*. VI. Biosynthesis from uridine nucleotides in cell-free extracts. J Biol Chem 1959; 234:2343–2350.
27. Crater DL, van de Rijn I. Hyaluronic acid synthesis operon (has) expression in Group A *Streptococci*. J Biol Chem 1995; 270:18452–18458.
28. Itano N, Sawai T, Yoshida M, Lenas P, Yamada Y, Imagawa M, Shinomura T, Hamaguchi M, Yoshida Y, Ohnuki Y, Miyauchi S, Spicer AP, McDonald JA, Kimata K. Three isoforms of mammalian hyaluronan synthases have distinct enzymatic properties. J Biol Chem 1999; 274:25085–25092.
29. Camenisch TD, Spicer AP, Brehm-Gibson T, Biesterfeldt J, Augustine ML, Calabro A Jr, Kubalak S, Klewer SE, McDonald JA. Disruption of hyaluronan synthase-2 abrogates normal cardiac morphogenesis and hyaluronan-mediated transformation of epithelium to mesenchyme. 2000. J Clin Investig 2000; 106:335–336.
30. McDonald JA, Camenisch TD. Hyaluronan: genetic insights into the complex biology of a simple polysaccharide. Glycoconj J 2002; 19:331–339.
31. Heldermon CD, DeAngelis PL, Weigel PH. Topological organization of the hyaluronan synthase from *Streptococcus pyogenes*. J Biol Chem 2001; 276: 2037–2046.
32. Tlapak-Simmons VL, Kempner ES, Baggenstoss BA, Weigel PH. The active streptococcal hyaluronan synthases (HASs) contain a single HA synthase monomer and multiple cardiolipin molecules. J Biol Chem 1998; 273:26100–26109.
33. Pummill PE, Kempner ES, DeAngelis PL. Functional molecular mass of a vertebrate hyaluronan synthase as determined by radiation inactivation analysis. J Biol Chem 2001; 276:39832–39835.
34. Tlapak-Simmons VL, Baggenstoss BA, Clyne T, Weigel PH. Purification and lipid dependence of the recombinant hyaluronan synthases from *Streptococcus pyogenes* and *Streptococcus equisimilis*. J Biol Chem 1999; 274:4239–4245.
35. Yoshida M, Itano N, Yamada Y, Kimata K. *In vitro* synthesis of hyaluronan by a single protein derived from mouse *HAS1* gene and characterization of amino acid residues essential for the activity. J Biol Chem 2000; 275:497–506.
36. DeAngelis PL, Weigel PH. Immunochemical confirmation of the primary structure of streptococcal hyaluronan synthase and synthesis of high molecular weight product by the recombinant enzyme. Biochemistry 1994; 33:9033–9039.
37. Heldermon CD, Tlapkak-Simmons VL, Baggenstoss BA, Weigel PH. Site directed mutation of conserved cysteine residues does not inactivate the *Streptococcus pyogenes* hyaluronan synthase. Glycobiology 2001; 11:1–8.
38. Kumari K, Tlapkak-Simmons VL, Baggenstoss BA, Weigel PH. The streptococcal hyaluronan synthases are inhibited by sulfhydryl modifying reagents but conserved cysteine residues are not essential for enzyme function. J Biol Chem 2002; 277: 13943–13951.
39. Tlapak-Simmons VL, Baggenstoss BA, Kumari K, Heldermon C, Weigel PH. Kinetic characterization of the recombinant hyaluronan synthases from *Streptococcus pyogenes* and *Streptococcus equisimilis*. J Biol Chem 1999; 274: 4246–4253.
40. Prehm P. Synthesis of hyaluronate in differentiated teratocarcinoma cells: mechanism of chain growth. Biochem J 1983; 211:191–198.
41. Asplund T, Brinck J, Suzuki M, Briskin MJ, Heldin P. Characterization of hyaluronan synthase from a human glioma cell line. Biochim Biophys Acta 1998; 1380:377–388.

42. Tlapak-Simmons VL, Weigel PH. Hyaluronan synthesis by the purified Class I hyaluronan synthase from *S. pyogenes* and *S. equisimilis* occurs by addition to the reducing end. Glycobiology 2002; 2:708.
43. DeAngelis PL. Molecular directionality of polysaccharide polymerization by the *Pasteurella multocida* hyaluronan synthase. J Biol Chem 1999; 274:26557–26562.
44. Widner W, Behr R, Tang M, Hue T, Sternberg D, Von Dollen S, Sloma A, DeAngelis PL, Weigel PH, Brown SH. Hyaluronic acid production by recombinant *Bacillus subtilis* strains. 2003; Online proceedings of HA2003 meeting at Matrix Biology Institute website.
45. Heldermon C, Kumari K, Tlapak-Simmons V, Weigel PH. Streptococcal hyaluronan synthases and the synthesis of designer hyaluronan. In: Abatelangelo G, Weigel PH, eds. New Frontiers in Medical Sciences: Redefining Hyaluronan. Amsterdam: Elsevier, 2000:41–50.

Chemistry and Biology of Hyaluronan
H.G. Garg and C.A. Hales (editors)

Chapter 26

Molecular Genetic Dissection of Hyaluronan Function in the Mouse

ANDREW P. SPICER

Center for Extracellular Matrix Biology, Institute of Biosciences and Technology, Texas A&M University System Health Science Center, Houston, TX 77030, USA

JANET Y. LEE TIEN

Department of Human Anatomy and Cell Biology, University of California Davis, School of Medicine, Davis, CA 95616, USA

I. Introduction

It has become increasingly apparent that the high molecular mass glycosaminoglycan, hyaluronan (HA), is required for many morphogenetic processes during vertebrate development. This renewed understanding of the various developmental roles for HA has come about largely through the advent of gene targeting approaches in the mouse. To date, mutations have been engineered in the enzymes responsible for biosynthesis (HA synthases) and degradation (hyaluronidases) and for those proteins that bind to HA within the extracellular matrix and at the cell surface. Collectively, the phenotypes resulting from these mutations demonstrate that HA is critical for normal mammalian embryogenesis and for various processes in postnatal and adult life (Table 1) (1–7). In this chapter, our progress in understanding the biological functions for HA through targeted mutagenesis of the HA synthase 2 (*Has2*) and 3 (*Has3*) genes has been reviewed. More specifically, we have discussed the functional information obtained from investigation of Has2-deficient mouse embryos, provided an overview of the conditional gene targeting strategy being used to create tissue-specific deficiencies in Has2 function, and reviewed our progress in understanding the role of Has3-dependent HA biosynthesis.

Table 1 Phenotypes from Gene Knockout Mutations Relating to Hyaluronan Synthesis and Function

Gene name	Mutation	Phenotype	Conclusion	Reference
AGGRECAN (*Agc1*)	cmd (spontaneous)	Postnatal lethality due to severe chondrodysplasias	Aggrecan required for normal skeletogenesis	(1)
Link protein (*Crtl1*)	Targeted (null)	Postnatal lethality due to severe chondrodysplasias	Crtl1 required for normal skeletogenesis	(2)
CD44	Targeted (null)	Minor phenotypes related to lymphocyte trafficking and tumorigenecity	CD44 not absolutely required for normal development	(3)
RHAMM (*Hmmr*)	Targeted (null)	Viable and fertile; block in formation of desmoid tumors	RHAMM not required for normal development	(4)
VERSICAN (*Cspg2*)	Transgene insertion	E10.5 embryonic lethal	Versican required for normal endocardial cushion formation	(5)
Has2	Targeted (null)	E10.5 embryonic lethal	Has2 and HA required for endocardial cushion formation, epithelial-to-mesenchymal transformation, vasculogenesis and neural crest survival	(6)
Neurocan	Targeted (null)	Viable and fertile; mild defects in hippocampal LTP	Neurocan is not required for normal brain development	(7)
Has3	Targeted (null)	Viable and fertile tooth, fertility and skin phenotypes with Has2	Partial redundancy between Has2 and Has3	(a)

(a) Spicer et al. (unpublished data).

II. Hyaluronan Synthases

Hyaluronan is unique among the glycosaminoglycans, in that it is synthesized and released from the cell as a linear, unmodified polymer that is not covalently attached to a polypeptide backbone. HA is synthesized at the plasma membrane by any one of three HA synthases (see more details in Chapter 25) (8,9), integral plasma membrane proteins that draw off a cytosolic pool of activated UDP-sugar precursors as substrate. Heterologous expression of any one HA synthase in a given cell line, converts that line into a factory for HA biosynthesis (9,10). A preliminary expression screen indicated differential expression of *Has* genes during mouse embryonic development and in the adult, suggesting that promoters have diverged (9). In addition to differences in the regulation of gene expression, it is clear that each HA synthase has characteristic differences in enzymatic activity, primarily relating to V_{max} and relative binding affinities for the two UDP-sugar substrates (9,10). Thus, it is still formally possible that while each HA synthase catalyzes the biosynthesis of HA, the three enzymes may not be able to replace each other functionally during all aspects of development or adult life.

The three HA synthases are encoded for by three separate but related genes, which are located on separate autosomes (11). These genes are highly conserved in vertebrates, with conserved gene structures and amino acid sequence identities. For instance, human and mouse Has2 share 99% amino acid sequence identity with five amino acid substitutions over 552 amino acids. Of these five amino acid differences, four represent conservative substitutions. The Has2 and Has3 genes have the same gene structure, apart from differences in intron length and 5′ and 3′ untranslated regions (9,12).

III. Hyaluronan Synthase Gene Expression

Numerous studies have indicated that Has2 is the key gene with respect to the bulk of the HA biosynthesis that occurs in mammals (13–17). Has2 is expressed broadly during mouse embryogenesis, correlating with all previously described spatial and temporal localizations for HA ((18) and Tien and Spicer, manuscript in preparation). These include the endocardial cushions of the developing heart, craniofacial mesenchyme, precartilage areas, hypertrophic chondrocytes, neural crest cells and their derivatives (Fig. 1A). In contrast, Has1 is expressed only during gastrulation (E7.5–E8.5) in a more general pattern (data not shown), and Has3 is expressed only by the mesenchymal component of developing teeth and hair (Fig. 1B–F). Both teeth and hair develop through strong interactions between epithelial cells and mesenchymal cells (20). Within each developing hair follicle, Has3 expression is transient, correlating with the timing of condensation. In contrast, Has3 expression is detectable as early as E10.5 in tooth-forming cephalic crest and expression is maintained in the tooth-forming mesenchyme until at least the bell-stage of development.

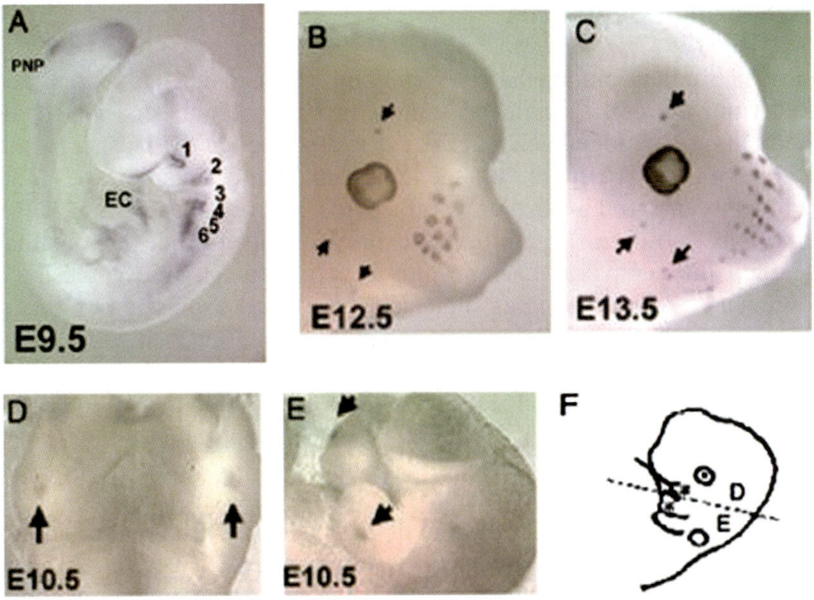

Figure 1 Expression of Has2 and Has3 in the developing mouse embryo. Whole-mount in situ hybridization was carried out essentially using published procedures (19) with digoxigenin-labeled antisense RNA probes for mouse Has2 or mouse Has3 or the equivalent sense probes. Colorimetric detection was employed. (A) Has2 expression was observed in neural crest cell populations and the endocardial cushions at E9.5. Branchial arches 1–6 are numbered. (B) and (C) Has3 is expressed within developing whisker and hair follicles in a transient fashion. (D)–(F) Has3 is expressed by tooth-forming cephalic neural crest cells within the first branchial arch at E10.5. Arrows indicate the areas of tooth-forming cephalic neural crest cells that are positive for Has3 expression. (F) Cartoon depiction of an E10.5 mouse embryo. The dashed line indicates the plane at which the embryos were cut to view the positive in situ signal in the maxillary (D) and mandibular (E) components of the first branchial arch. EC, endocardial cushions; PNP, posterior neuropore.

All *Has* genes are expressed to varying degrees in adult tissues (9). In contrast to the limited expression of Has1 in the embryo, this gene is the most widely expressed in adult tissues and is even expressed by various immune cells, including dendritic cells and activated T-cells (21). Has3 is also broadly expressed and appears to be the dominant HA synthase expressed in the epidermis (22). Has2 is expressed in a more restricted spatial pattern in the adult, but is the dominant HA synthase in the dermis (Spicer, unpublished data), cumulus cells of the ovary (23) and growth plate cartilage (16,24). Has2 is also the target for induction by many growth factors, proinflammatory cytokines and chemokines. These include members of the TGF-β/BMP superfamily, EGF, KGF, IL-1β and TNF-α (13–15,17,25–27). In addition, Has2 expression can be rapidly suppressed by glucocorticoids (13,28). In this regard, suppression of Has2 mRNAs and loss of HA biosynthesis may be part of the underlying pathology observed in long-term glucocorticoid treatment of humans. These patients often suffer from dermal thinning and bone loss/atrophy. Has3 can also be induced by various factors, including EGF, KGF, proinflammatory cytokines, interferon-gamma and mechanical stimulation (17,22,26,27,29). The Has3 gene has alternate exon 1s, with presumed alternate promoters (12,22). The respective expression domains for the alternate Has3 mRNAs are not understood at this time.

IV. Hyaluronan Synthase 2 Null Allele

To investigate the biological functions for the three HA synthases and their biosynthetic product, we have used conventional targeted mutagenesis. In each instance a similar strategy was used, with insertion of the selectable marker cassette leading to deletion of a substantial portion of the respective gene. While the targeted inactivation of Has1 (Itano, Spicer, Kimata and McDonald, unpublished data) and Has3 (Spicer and McDonald, unpublished data) produced viable homozygous null animals, mice homozygous for the Has2 null allele were not obtained; Has2 deficiency resulted in an embryonic lethal phenotype at E10.5 (6). Thus, HA is indispensable for normal mouse embryogenesis. It is perhaps not too surprising that loss of Has2 function resulted in embryonic lethality, given its normal pattern of expression. Has2 is expressed most highly in the mid-gestation mouse embryo where its expression domains do not overlap with those for Has1 or Has3.

Has2-deficient mouse embryos die in utero at E10.5 of multiple developmental abnormalities (6). The first apparent developmental defects are observed in the neural tube and adjacent somites at E8.5. The neural tube is often kinked and the somites are smaller in size and less organized. By E9.5, the neural tube is grossly abnormal and the flanking somites are significantly smaller and irregular in shape. At this time yolk sac abnormalities are also readily apparent (Fig. 2A and B); Has2-efficient embryos lack an organized yolk sac vasculature. This phenotype would be lethal in itself (30) and also confuses the interpretation of embryonic developmental defects as many structures/events within the developing mid-gestation embryo are dependent upon a functional vasculature. Indeed, the vasculature of the embryo proper is also grossly abnormal, with large disorganized networks of endothelium, rather than the normal complex and quite beautiful networks of vessels that are observed throughout the embryo by E9.5 (Fig. 2E and F) (6). The vascular defects have not been investigated in detail. It is important to note, however, that many studies have demonstrated a role for HA in angiogenesis (31,32). HA is a potent angiogenic stimulator in classic assays and can induce many pathways including those involved in cell proliferation and cell movement (33).

Perhaps, most progress has been made in understanding HA function during cardiac development (6,34,35). Has2-deficient mouse embryos die between E9.5 and E10.5. Upon visual inspection, the two most obvious defects relate to the yolk

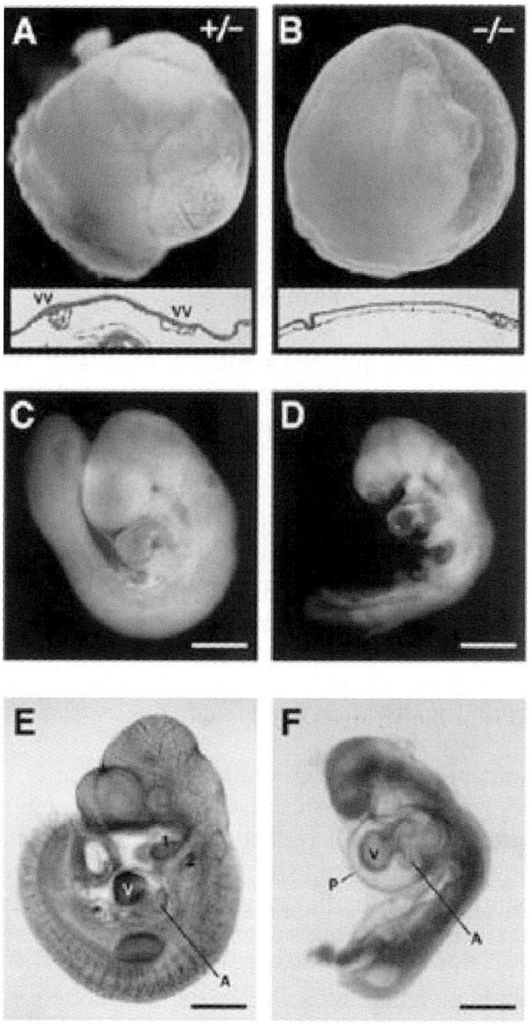

Figure 2 Multiple abnormalities in Has2-deficient embryos (30). A, B: Wild-type (A) and Has2−/− (B) yolk sacs at E9.5. Compared with wild type, Has2−/− embryos exhibit abnormalities in yolk sac vasculature including separation of the visceral endoderm from mesoderm obliterating the vitteline vessels (VV) and giving them a characteristic punctate appearance (B). C, D: E9.5 wild type (C) and Has2−/− (D) embryos. Grossly, the Has2−/− embryo (D) is smaller than its wild-type littermate and has smaller, disorganized somites. E, F: PECAM (CD31) staining of E9.5 wild type (E) and Has2−/− (F) embryos. The wild-type littermate exhibits a characteristic highly developed vasculature, whereas the Has2−/− embryo lacks an organized vasculature, and has an enlarged pericardial space (P). Branchial arches 1 and 2 are indicated by numerals. V, ventricle; A, atrium.

sac, which is observed immediately upon dissection, and the heart. Has2-deficient hearts can be most simply described as dilated, fluid-filled beating sacs with no apparent internal structures. The extent of the dilation can be extreme, with hearts being several times larger than the head in some instances. The outflow tract is present, but reduced in size and is usually not patent. Histological investigation revealed that the Has2-deficient hearts lacked endocardial cushions and trabeculations (foldings of the ventricular wall) and essentially consisted of juxtaposed myocardium and endocardium. This phenotype is strikingly similar to that previously observed in the heart defect (*hdf*) mouse, which resulted from the insertional inactivation of a transgene into the mouse versican (*Cspg2*) locus (5,36). Mice homozygous for the *hdf* allele fail to form endocardial cushions and lethality occurs at E10.5–E11.5.

The formation of endocardial cushions can be imagined most simply as a two-step process (35). First, ECM components, including HA and versican are produced by endocardial and myocardial cells as early as E8.0. The organization of a large, hydrated, HA-dependent ECM between the myocardium and endocardium acts to physically separate these two cell layers. This organized matrix is sometimes referred to as cardiac jelly. At approximately E9.5, a combination of signals emanating from myocardium and endocardium leads the endocardial cells to transform from an epithelial to mesenchymal state. The mesenchymal cells are proliferative and migratory and migrate into the HA-dependent ECM, seeding this matrix with mesenchyme. In contrast to the endocardial cushions of the atrioventricular canal, which orginate in situ through epithelial–mesenchymal transformation, within the outflow tract, the mesenchymal component of the endocardial cushions is neural crest in origin, often being referred to as the cardiac crest. Despite the difference in origin, both types of endocardial cushions fail to form in Has2-deficient mouse embryos (6).

Previously, when HA was thought of simply as 'a goo' (37), the supposition would have been that HA is required for cushion formation simply in its space-filling capacity. An organized, hydrated, open ECM is a necessary prerequisite for the epithelial–mesenchymal transformation. If no space is provided, cells cannot transform and invade. Strikingly, it is now clear that while HA is required for the establishment of an organized, hydrated ECM, it also acts in a signaling capacity and is required in this context for the epithelial–mesenchymal transformation (6,34,35).

It has been widely known that the transformation process is dependent upon growth factor signaling, with most work focusing upon members of the TGF-β superfamily (35,38). In these simple models, TGF-β is produced by the myocardium and signals to the endocardium. Investigation of other mouse knockouts suggested that members of the erbB receptor tyrosine kinases may also play a role (39–41), in addition to the Map kinase, Mekk3 (42). Through analysis of embryonic cardiac explant culture, a functional connection has been made between HA signaling and activation of the erbB receptor tyrosine kinase pathway (34). Only nanogram amounts of high molecular mass HA are required to rescue the epithelial–mesenchymal transformation of cardiac explants.

This rescue is dependent upon functional erbB receptor tyrosine kinases. Furthermore, HA signaling can be blocked by a soluble erbB antagonist (34). The precise mechanism of activation of erbB through HA is not yet known. It is possible that one or more HA receptors could be utilized, or that HA may activate erbB through an entirely novel mechanism. It is known that ras is a downstream mediator of the HA-dependent signaling cascade (6). The phenotype observed in Mekk3-deficient mouse embryos (42) is strikingly similar to that observed in Has2-deficient embryos. The yolk sac vasculature defects appear identical and endocardial cushions fail to form. In this instance, however, cardiac jelly is present and the failure of endocardial cushion formation occurs after the physical separation of the myocardium and endocardium has occurred. It is intriguing to speculate that Mekk3 may represent one of the major downstream effectors of HA signaling in different developmental contexts.

V. Hyaluronan Synthase 3 Null Allele

Based upon the observation that Has2 and Has3 are the most closely related HA synthases at the amino acid and gene structure levels (9,12), we investigated the phenotypes associated with compound loss of both Has3 and Has2 function. Animals doubly heterozygous for both Has2 and Has3 (Has2+/−, Has3+/−) were viable, healthy and fertile. However, a significant fraction of animals homozygous for Has3 and heterozygous for Has2 (Has2+/−, Has3−/−) presented with obvious malocclusions of the teeth before or shortly after the time of weaning (data not shown). These animals were obtained at a frequency of 14% on an outbred genetic background, suggesting that at least one other major gene is involved in modifying the tooth phenotype. As both Has2 and Has3 are expressed by the tooth-forming cephalic neural crest cells, it is highly likely that there is partial redundancy of function in this cell population, with respect to the source of HA. It is also likely that the malocclusions observed in these animals result from a defect that is intrinsic to the tooth-forming crest. This may be related to path finding of this cell population, or to proliferation or differentiation. All of these aspects are currently under investigation. The major modifier gene locus is also being sought. In addition to the tooth phenotype observed in compound Has2+/−, Has3−/− animals, we have observed reduced fertility and a variety of skin phenotypes, mostly related to excessive scratching. Has2 and Has3 are both expressed within the ovary (9,23,25) with Has2 being specifically induced during cumulus cell expansion (23,25). Litter sizes are reduced in *Ha*s2+/−, Has3−/− animals and numbers of litters per year are also reduced. Has3 is highly expressed within the epidermis (22), and we believe that the skin phenotypes may be simply explained in part by excessively dry skin, due to loss of HA-associated water. A marked reduction in the amount of HA within the epidermis, may affect terminal differentiation and loss of dead skin cells. Again, we are investigating the basis for these phenotypes.

VI. Hyaluronan Synthase 2 Conditional Gene Knockout

Overall, results from our conventional HA synthase gene knockouts have shown that Has2 is the primary HA synthase within the developing embryo and is necessary and sufficient for normal embryogenesis. In addition, Has2 and Has3 are partially redundant within certain cell populations. After E10.5, Has2 expression is maintained within the cushion tissue of the heart, but also becomes very obvious within the developing embryonic skeleton, where it is expressed by the precondensing cartilage, early cavitating joints and hypertrophic zones (Tien et al., manuscript in preparation). Thus, the ability to manipulate the Has2-dependent HA biosynthetic machinery in the embryo after E10.5 and in the adult would provide us with a wealth of new data regarding the various biological roles for HA. To achieve this goal, we have spent some time designing a vector system to apply allelogenic gene targeting (43) to the mouse Has2 gene locus (Fig. 3). Using this approach, recognition targets for two site-specific recombinases are employed and incorporated into the targeted allele, by homologous recombination. The first targeted allele is often hypomorphic, or leaky, resulting in a partial loss of function. This type of allele is extremely informative as it allows creation of lines with substantially reduced gene product function (43–46). Most conventional gene knockouts do not present with a phenotype in the heterozygous state. This would suggest that most developmental processes could proceed normally despite a significant reduction in the amount of any given gene product. We assume, for sake of argument, that most heterozygotes will produce roughly 50% of the wild-type level for a given gene knockout. The results obtained from the creation of hypomorphic alleles suggest that phenotypes are revealed when a given gene product is reduced to 5 or 10% of normal levels (43–46). Varying amounts of the gene product can be achieved by making the hypomorphic allele homozygous or by combining it with a null allele (Fig. 3). The strategy we have used to create our Has2 conditional allele is outlined in Fig. 3. A selectable marker cassette (PGKneo) has been inserted into intron 3, and is flanked by target sites for Flp recombinase. Recognition sites for Cre recombinase have been placed on either side of exon 3. Exon 3 is the smallest Has2 exon, encoding 34 amino acids of the enzyme, including a critical residue thought to play a role in coordinating the metal cation, which is required for enzyme activity (9,47). Deletion of this exon would thus result in the creation of a functionally null Has2 allele. We chose to flank exon 3 with loxP sites and to insert the neo cassette into intron 3, as we believe that this has the highest probability of generating a hypomorphic allele. In addition, exonic sequences for a gene that is transcribed from the opposite strand of Has2 are located within intron 1 of Has2, along with enhancer/regulatory sequences for Has2 and this new gene (Chao and Spicer, manuscript in preparation). Thus, we were hesitant to place a selectable marker cassette or loxP sites into or around exon 2. Correctly targeted clones, incorporating both loxP sites, were obtained at a frequency of 1 in 467 antibiotic resistant clones; four additional clones were obtained that only incorporated a single loxP site.

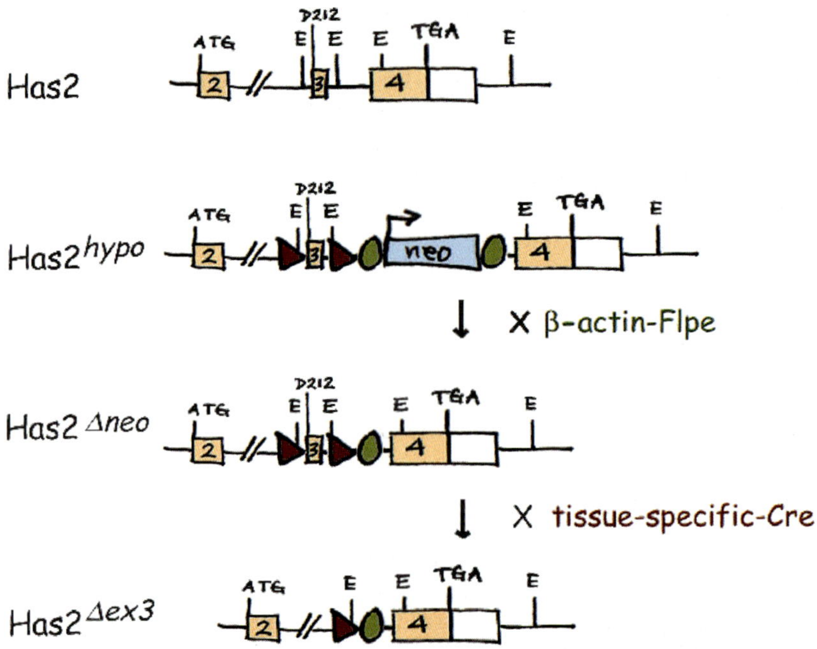

Figure 3 Strategy for allelogenic type conditional gene targeting of the Has2 Gene. Top, the previously described structure of the mouse Has2 gene. Homologous recombination should introduce two loxP sites and a FRT—flanked PGKneo cassette into the Has2 gene locus creating Has2^{hypo} as shown. This targeted Has2 allele can be manipulated in the ES cells or *in vivo* (in the mouse) by expression of Flp recombinase. In the presence of Flp, the PGKneo cassette is specifically deleted, leaving behind a single FRT site. This is expected to revert the hypomorphic Has2 allele to functionally wild-type allele, designated Has2$^{\Delta neo}$. The allele containing two loxP sites can also be manipulated *in vitro* (embryonic stem cells) or *in vivo*, this time using Cre recombinase. Cre recombinase will act to specifically delete exon 3, creating a functionally null Has2 allele as explained in the text. This allele is designated Has2$^{\Delta ex3}$. Cre recombinase is typically expressed under the control of a spatially and temporally regulated, or inducible promoter. Multiple transgenic mouse lines have now been reported in which Cre recombinase is regulated in this manner. Filled pink boxes, Has2 open-reading frame; open box, Has2 3′ untranslated region; filled blue boxes, PGKneo cassette; filled brown arrowheads, loxP sites; green ovals, FRT sites; numbers indicate exons for Has2; D212 indicates the relative position of the codon encoding the critical aspartate (D) residue at position 212 within the predicted polypeptide chain; (E) EcoRI sites. The start codon (ATG) and stop codon (TGA) are indicated.

The conditional Has2 allele will represent the most useful strategy for investigating HA function in many developmental processes and in normal adult physiology. First, the hypomorphic allele is generated. Next, the hypomorphic allele is reverted back to an essentially wild-type allele, using Flp recombinase *in vivo* (48). Once animals with the desired genotype have been derived, crosses are established in which offspring are predicted to have genotypes where Cre-mediated loss of Has2 function will result in cell populations or tissues with essentially no ability to synthesize HA. By performing all of these crosses on an Has3-null background, we will avoid the possibility of rescue/redundancy of Has2 through expression of Has3. Targeted clones and initial mouse lines have now been derived and we have demonstrated that these modified Has2 alleles can indeed be recombined *in vivo* by crossing with recombinase-expressing transgenic mouse lines.

VII. Summary and conclusion

HA has been previously thought of as 'a goo', quite simply a large, hydrated, slippery molecule. The results obtained from the molecular dissection of HA function in the mouse demonstrate that this simple polysaccharide has diverse functions during morphogenesis. HA can indeed act as a space filling, or perhaps more accurately, a space-creating molecule, acting as the nucleus around which an organized, hydrated HA-dependent ECM is assembled. This function is important in the outgrowth of the nasal processes (49), in the expansion of the cardiac jelly of the heart and in the formation of a large hydrated matrix around the neural tube prior to neural crest cell exit. A large organized HA-dependent extracellular matrix may effectively inhibit cell–cell adhesion. Accordingly, if such a matrix is endocytosed or degraded, cell–cell contact may be achieved. This is essentially the process underlying condensation, which occurs for instance during chondrogenesis and hair follicle development. HA may play a key role in the regulation of the timing of condensation and, hence, the number of cells that may be ultimately present within a given condensation. HA can also act as a signaling molecule in its own right, signaling directly through specific HA cell surface receptors and indirectly through other receptors to modify cell behavior. The best example of this function has come through investigation of the cardiac phenotype in Has2-deficient mouse embryos. HA can be made by any one of three HA synthases. Our data have demonstrated that Has2 and Has3 are partially redundant in function; tooth-restricted phenotypes were only observed in animals compound for Has2 and Has3 deficiency.

It is clear that we have just begun to appreciate the diverse biological roles for HA. A recent report, for instance, elegantly demonstrated a role for cell surface HA in antigen presentation and in T-cell activation (21). Through continued analysis of Has-deficient animals we expect to unlock many secrets and to be led through many twists and turns as we investigate this fascinating molecule.

Acknowledgements

This work has been partly supported through an R01 grant number AR47433 to APS and a Scientist Development Award 0030184N from the American Heart Association

National office to APS. This chapter is based on an invited review article published by us in the Glycoconjugate Journal with permission (50). Fig. 2 was printed with permission from the Journal of Clinical Investigation (6).

References

1. Watanabe H, Kimata K, Line S, Strong D, Gao LY, Kozak CA, Yamada Y. Mouse cartilage matrix deficiency (*cmd*) caused by a 7 bp deletion in the aggrecan gene. Nat Genet 1994; 7:154–157.
2. Watanabe H, Yamada Y. Mice lacking link protein develop dwarfism and craniofacial abnormalities. Nat Genet 1999; 21:225–229.
3. Schmits R, Filmus J, Gerwin N, Senaldi G, Kiefer F, Kundig T, Wakeham A, Shahinian A, Catzavelos C, Rak J, Furlonger C, Zakarian A, Simard JJ, Ohashi PS, Paige CJ, Gutierrez-Ramos JC, Mak TW. CD44 regulates hematopoietic progenitor distribution, granuloma formation, and tumorigenicity. Blood 1997; 90: 2217–2233.
4. Tolg C, Poon R, Fodde R, Turley EA, Alman BA. Genetic deletion of receptor for hyaluronan-mediated motility (Rhamm) attenuates the formation of aggressive fibromatosis (desmoid tumor). Oncogene 2003; 22:6873–6882.
5. Yamamura H, Zhang M, Markwald RR, Mjaatvedt CH. A heart segmental defect in the anterior–posterior axis of a transgenic mutant mouse. Dev Biol 1997; 186: 58–72.
6. Camenisch TD, Spicer AP, Brehm-Gibson T, Biesterfeldt J, Augustine ML, Calabro A Jr, Kubalak S, Klewer SE, McDonald JA. Disruption of hyaluronan synthase-2 abrogates normal cardiac morphogenesis and hyaluronan-mediated transformation of epithelium to mesenchyme. J Clin Investig 2000; 106:349–360.
7. Zhou XH, Brakebusch C, Matthies H, Oohashi T, Hirsch E, Moser M, Krug M, Seidenbecher CI, Boeckers TM, Rauch U, Buettner R, Gundelfinger ED, Fassler R. Neurocan is dispensable for brain development. Mol Cell Biol 2001; 21: 5970–5978.
8. Weigel PH, Hascall VC, Tammi M. Hyaluronan synthases. J Biol Chem 1997; 272: 13997–14000.
9. Spicer AP, McDonald JA. Characterization and molecular evolution of a vertebrate hyaluronan synthase (HAS) gene family. J Biol Chem 1998; 273:1923–1932.
10. Itano N, Sawai T, Yoshida M, Lenas P, Yamada Y, Imagawa M, Shinomura T, Hamaguchi M, Yoshida Y, Ohnuki Y, Miyauchi S, Spicer AP, McDonald JA, Kimata K. Three isoforms of mammalian hyaluronan synthases have distinct enzymatic properties. J Biol Chem 1999; 274:25085–25092.
11. Spicer AP, Seldin MF, Olsen AS, Brown N, Wells DE, Doggett NA, Itano N, Kimata K, Inazawa J, McDonald JA. Chromosomal localization of the human and mouse hyaluronan synthase genes. Genomics 1997; 41:493–497.
12. Monslow J, Williams JD, Norton N, Guy CA, Price IK, Coleman SL, Williams NM, Buckland PR, Spicer AP, Topley N, Davies M, Bowen T. The human hyaluronan synthase genes: genomic structures, proximal promoters and polymorphic microsatellite markers. Int J Biochem Cell Biol 2003; 35:1272–1283.
13. Jacobsen A, Brinck J, Briskin MJ, Spicer AP, Heldin P. Expression of human hyaluronan synthases in response to external stimuli. Biochem J 1999; 348:29–35.

14. Kaback LA, Smith TJ. Expression of hyaluronan synthase messenger ribonucleic acids and their induction by interleukin-1beta in human orbital fibroblasts: potential insight into the molecular pathogenesis of thyroid-associated ophthalmopathy. J Clin Endocrinol Metab 1999; 84:4079–4084.
15. Nishida Y, Knudson CB, Eger W, Kuettner KE, Knudson W. Osteogenic protein 1 stimulates cell-associated matrix assembly by normal human articular chondrocytes: up-regulation of hyaluronan synthase, CD44, and aggrecan. Arthritis Rheum 2000; 43:206–214.
16. Ohno S, Tanimoto K, Fujimoto K, Ijuin C, Honda K, Tanaka N, Doi T, Nakahara M, Tanne K. Molecular cloning of rabbit hyaluronic acid synthases and their expression patterns in synovial membrane and articular cartilage. Biochim Biophys Acta 2001; 1520:71–78.
17. Pasonen-Seppanen S, Karvinen S, Torronen K, Hyttinen JM, Jokela T, Lammi MJ, Tammi MI, Tammi R. EGF upregulates, whereas TGF-beta down regulates, the hyaluronan synthases Has2 and Has3 in organotypic keratinocyte cultures: correlations with epidermal proliferation and differentiation. J Investig Dermatol 2003; 120:1038–1044.
18. Fenderson BA, Stamenkovic I, Aruffo A. Localization of hyaluronan in mouse embryos during implantation, gastrulation and organogenesis. Differentiation 1993; 54:85–98.
19. Conlon RA, Rossant J. Exogenous retinoic acid rapidly induces anterior ectopic expression of murine Hox-2 genes *in vivo*. Development 1992; 116: 357–368.
20. Tucker AS, Sharpe PT. Molecular genetics of tooth morphogenesis and patterning: the right shape in the right place. J Dent Res 1999; 78:826–834.
21. Mummert ME, Mummert D, Edelbaum D, Hui F, Matsue H, Takashima A. Synthesis and surface expression of hyaluronan by dendritic cells and its potential role in antigen presentation. J Immunol 2002; 169:4322–4331.
22. Sayo T, Sugiyama Y, Takahashi Y, Ozawa N, Sakai S, Ishikawa O, Tamura M, Inoue S. Hyaluronan synthase 3 regulates hyaluronan synthesis in cultured human keratinocytes. J Investig Dermatol 2002; 118:43–48.
23. Fulop C, Salustri A, Hascall VC. Coding sequence of a hyaluronan synthase homologue expressed during expansion of the mouse cumulus–oocyte complex. Arch Biochem Biophys 1997; 337:261–266.
24. Mrosek E, Erggelet C, McDonald JA, Kurz H. Hyaluronan synthases in normal and regenerating joint cartilage. Cells Tissues Organs 2003; 173:93–104.
25. Elvin JA, Clark AT, Wang P, Wolfman NM, Matzuk MM. Paracrine actions of growth differentiation factor-9 in the mammalian ovary. Mol Endocrinol 1999; 13: 1035–1048.
26. Tanimoto K, Ohno S, Fujimoto K, Honda K, Ijuin C, Tanaka N, Doi T, Nakahara M, Tanne K. Proinflammatory cytokines regulate the gene expression of hyaluronic acid synthetase in cultured rabbit synovial membrane cells. Connect Tissue Res 2001; 42:187–195.
27. Karvinen S, Pasonen-Seppanen S, Hyttinen JM, Pienimaki JP, Torronen K, Jokela TA, Tammi MI, Tammi R. Keratinocyte growth factor stimulates migration and hyaluronan synthesis in the epidermis by activation of keratinocyte hyaluronan synthases 2 and 3. J Biol Chem 2003:00.
28. Zhang W, Watson CE, Liu C, Williams KJ, Werth VP. Glucocorticoids induce a near-total suppression of hyaluronan synthase mRNA in dermal fibroblasts and

in osteoblasts: a molecular mechanism contributing to organ atrophy. Biochem J 2000; 349:91–97.
29. Dowthwaite GP, Ward AC, Flannely J, Suswillo RF, Flannery CR, Archer CW, Pitsillides AA. The effect of mechanical strain on hyaluronan metabolism in embryonic fibrocartilage cells. Matrix Biol 1999; 18:523–532.
30. Copp AJ. Death before birth: clues from gene knockouts and mutations. Trends Genet 1995; 11:87–93.
31. West DC, Hampson IN, Arnold F, Kumar S. Angiogenesis induced by degradation products of hyaluronic acid. Science 1985; 228:1324–1326.
32. Lees VC, Fan TP, West DC. Angiogenesis in a delayed revascularization model is accelerated by angiogenic oligosaccharides of hyaluronan. Lab Investig 1995; 73:259–266.
33. Slevin M, Kumar S, Gaffney J. Angiogenic oligosaccharides of hyaluronan induce multiple signaling pathways affecting vascular endothelial cell mitogenic and wound healing responses. J Biol Chem 2002; 277:41046–41059.
34. Camenisch TD, Schroeder JA, Bradley J, Klewer SE, McDonald JA. Heart-valve mesenchyme formation is dependent on hyaluronan-augmented activation of ErbB2-ErbB3 receptors. Nat Med 2002; 8:850–855.
35. Schroeder JA, Jackson LF, Lee DC, Camenisch TD. Form and function of developing heart valves: coordination by extracellular matrix and growth factor signaling. J Mol Med 2003; 81:392–403.
36. Mjaatvedt C, Yamamura H, Capehart AA, Turner D, Markwald RR. The Cspg2 gene, disrupted in the hdf mutant, is required for right cardiac chamber and endocardial cushion formation. Dev Biol 1998; 202:56–66.
37. Toole BP. Hyaluronan is not just a goo!. J Clin Investig 2000; 106:335–336.
38. Camenisch TD, Molin DG, Person A, Runyan RB, Gittenberger-de Groot AC, McDonald JA, Klewer SE. Temporal and distinct TGFbeta ligand requirements during mouse and avian endocardial cushion morphogenesis. Dev Biol 2002; 248: 170–181.
39. Lee KF, Simon H, Chen H, Bates B, Hung MC, Hauser C. Requirement for neuregulin receptor erbB2 in neural and cardiac development. Nature 1995; 378: 394–398.
40. Kramer R, Bucay N, Kane DJ, Martin LE, Tarpley JE, Theill LE. Neuregulins with an Ig-like domain are essential for mouse myocardial and neuronal development. Proc Natl Acad Sci USA 1996; 93:4833–4838.
41. Erickson SL, O'Shea KS, Ghaboosi N, Loverro L, Frantz G, Bauer M, Lu LH, Moore MW. ErbB3 is required for normal cerebellar and cardiac development: a comparison with ErbB2-and heregulin-deficient mice. Development 1997; 124: 4999–5011.
42. Yang J, Boerm M, McCarty M, Bucana C, Fidler IJ, Zhuang Y, Su B. Mekk3 is essential for early embryonic cardiovascular development. Nat Genet 2000; 24: 309–313.
43. Meyers EN, Lewandoski M, Martin GR. An Fgf8 mutant allelic series generated by Cre- and Flp-mediated recombination. Nat Genet 1998; 18:136–141.
44. Gage PJ, Suh H, Camper SA. Dosage requirement of Pitx2 for development of multiple organs. Development 1999; 126:4643–4651.
45. Nagy A, Moens C, Ivanyi E, Pawling J, Gertsenstein M, Hadjantonakis AK, Pirity M, Rossant J. Dissecting the role of N-myc in development using a single targeting vector to generate a series of alleles. Curr Biol 1998; 8:661–664.

46. Raffai RL, Weisgraber KH. Hypomorphic apolipoprotein E mice: a new model of conditional gene repair to examine apolipoprotein E-mediated metabolism. J Biol Chem 2002; 277:11064–11068.
47. Charnock SJ, Davies GJ. Structure of the nucleotide-diphospho-sugar transferase, SpsA from *Bacillus subtilis*, in native and nucleotide-complexed forms. Biochemistry 1999; 38:6380–6385.
48. Rodriguez CI, Buchholz F, Galloway J, Sequerra R, Kasper J, Ayala R, Stewart AF, Dymecki SM. High-efficiency deleter mice showthat FLPe is an alternative to Cre-loxP. Nat Genet 2000; 25:139–140.
49. Burk DT. Morphological effects of Streptomyces hyaluronidase treatment on the outgrowth of the nasal processes in mouse embryos. J Craniofac Genet Dev Biol 1985; 5:385–398.
50. Spicer AP, Tien JL, Joo A, Bowling RA Jr. Investigation of hyaluronan function in the mouse through targeted mutagenesis. Glycoconj J 2003; 19:341–345.

Chemistry and Biology of Hyaluronan
H.G. Garg and C.A. Hales (editors)

Chapter 27

Functional, Structural and Biological Properties of Hyaluronidases

GÜNTER LEPPERDINGER

Institute for Biomedical Aging Research,
Austrian Academy of Sciences,
Rennweg 10, A-6020 Innsbruck, Austria

GÜNTHER KREIL

Institute of Molecular Biology,
Austrian Academy of Sciences,
Salzburg, Austria

I. Introduction

Hyaluronan (HA) is degraded by hyaluronidases, specific glycosyl hydrolases that are found in prokaryotes and eukaryotes. They show a wide range of properties, including differences in size, catalytic mechanisms, substrate specificities, affinities to inhibitors, and notably, varying pH dependency.

The action of a hyaluronidase was first observed some 75 years ago when Duran-Reynals co-injected antiviral vaccines together with extracts from testes (1). At that time, the compound from testes was termed a 'spreading factor'. Soon after, HA was isolated in pure form from the vitreous humor (2), and it could be shown that the spreading factor present in testes and other sources was a mucolytic enzyme (3). Hence, the term 'hyaluronidase' was introduced to denote that the enzyme degrades HA (4). The different types of hyaluronidases were classified into three distinct groups (5). The scheme is based on biochemical analyses of the enzymes and their reaction products: mammalian-type hyaluronidases, bacterial hyaluronidases and hyaluronidases from leeches, other parasites and crustaceans. This contribution only deals with the first two classes.

II. Mammalian-Type Hyaluronidase

Mammalian-type hyaluronidases have been classified in EC 3.2.1.35 (5,6). This group consists of endo-β-*N*-acetyl-hexosaminidases, which cleave the internal β1,4 glycosidic linkage between *N*-acetylglucosamine and glucuronate of HA producing oligomers (mainly hexasaccharides) as end products. Most of these enzymes also cleave 4-S and 6-S-chondroitin sulphate and catalyse transglycosylation (7,8).

In mammals, HA-degrading enzymes have been found in higher quantities in testes, liver lysosomes and serum. They are involved in controlling HA levels and are thus implicated in various diseases related to defects of HA metabolism, such as arthritis or tumor metastasis (9). At least six genes potentially coding for hyaluronidases are present in the human genome (10). These six genes define a novel paralogy group in the human genome. Three genes, HYAL1, 2, 3, are clustered on chromosome 3p21.3. The other three, HYAL4, PHYAL1—a pseudogene—and PH-20/SPAM1, are found on chromosome 7q31.3. It is noteworthy that each of these chromosomal regions has been found to be deleted in certain tumors (11–13). The tissue expression profile of the human enzymes is listed in Table 1 (14).

A. Protein Structure

The first structure of mammalian-type hyaluronidase could be deduced from a cDNA clone from venom glands of honey bees (15). This hyaluronidase, one of

Table 1 Expression Profile of Human Hyaluronidase mRNAs

	HYAL1	HYAL2	HYAL3	HYAL4	PH-20	HYALP1
Adrenal	+/−	++	+/−	−	−	++
Bone marrow	++	+/−	+++	+/−	−	+
Brain	−	−	+	−	−	+/−
Colon	+/−	+/−	+	−	−	+/−
Heart	++	++	+/−	−	−	+
Kidney	++	+	+/−	−	−	+/−
Liver	+++	+	+	−	−	+/−
Lung	+/−	++	−	−	−	+/−
Lymph node	+/−	+	+/−	+/−	−	+/−
Ovary	+/−	+	+/−	−	−	−
P.B. leucocytes	−	−	+/−	−	−	+/−
Pancreas	+/−	+	+/−	−	−	++
Placenta	−	+	−	++	−	+/−
Prostate	−	+	++	−	−	+/−
Skeletal muscle	+/−	+	−	+	−	++
Small intestine	+/−	+/−	−	−	−	+/−
Spinal cord	+	+	++	−	−	++
Spleen	++	+++	+/−	−	−	+/−
Stomach	−	+	+	−	−	+
Testis	−	++	+++	+/−	++++	+
Thymus	−	−	−	−	−	+/−
Thyroid	++	++	+	−	−	+++
Trachea	++	+	+/−	−	−	+

From Ref. 14.

the allergens of this venom, is a glycoprotein with a relative molecular weight of 41 kDa and a carbohydrate content of 7% (16). A search in the data banks then revealed that significant homology existed between the structure of the bee venom enzyme and PH-20, a protein isolated via a monoclonal antibody from guinea pig sperm (17). Subsequently, it could be demonstrated that PH-20 indeed has hyaluronidase activity (18,19). In subsequent years, the PH-20 enzyme has been investigated in greater detail (20). This enzyme is almost exclusively expressed in sperm, where it is located in the acrosome and the head bound to the membranes by a GPI-anchor. As opposed to other hyaluronidases, PH-20 is active in a broad pH range. Its biological role is thought to be the degradation of the HA-rich matrix of the cumulus layer, which surrounds the egg. In an attempt to identify the active site of PH-20, *in vitro* mutagenesis was performed. Two amino acids conserved in all hyaluronidases of this type, D111 and E113, were changed to N and Q, respectively (21). The E113Q mutation resulted in an inactive protein, while the D111N enzyme showed some residual activity. Taken together, this suggested that the pair of acidic residues conserved within the hyaluronidase family is an integral part of the enzyme's active centre.

In the search for further hyaluronidases, PH-20 served as a reference. Moreover, two genes originally termed LUCA 1 and 2, isolated from a chromosomal region frequently deleted in lung carcinomas, were clearly homologous to PH-20 (22–24). In subsequent studies, by a combination of protein purification and recombinant DNA techniques, it could be shown that these genes code for the hyaluronidases HYAL1 and 2, respectively.

The vertebrate enzymes exist in various isoforms, most likely resulting from post-translational proteolytic cleavage(s). For example, PH-20 can be liberated from its GPI-anchored state at the cell surface by proteolysis near the C-terminus, in some cases followed by a second proteolytic cleavage. These soluble forms, which are commercially available, are present in seminal fluid (25,26). The soluble fragments of bovine PH-20 also differ in their pH optima: the larger form is active around neutral pH while the 53 kDa form has a pH optimum of 4.0 (27). The conformational differences within the active centre, which cause such a shift, are unknown. In case of the HYAL1 enzyme, two smaller fragments have also been detected. In serum, only the 54 kDa isoform is present, while in urine a smaller 49 kDa form is found as well (28). The biological relevance of these variants is currently not known.

B. Tertiary Structure

The crystal structure of the bee venom hyaluronidase has been determined at 1.6 Å resolution for the enzyme alone and at 2.65 Å when co-crystallised together with an HA hexamer (29). The overall topology resembles that of a classical $(\beta/\alpha)_8$ TIM barrel which, however, is composed of only seven β-sheets. A long substrate-binding groove extends across the carboxy terminal end of the barrel (Fig. 1). Based on this structure, a detailed model for the reaction mechanism of this enzyme could be proposed (30). Hydrolysis proceeds via an acid–base

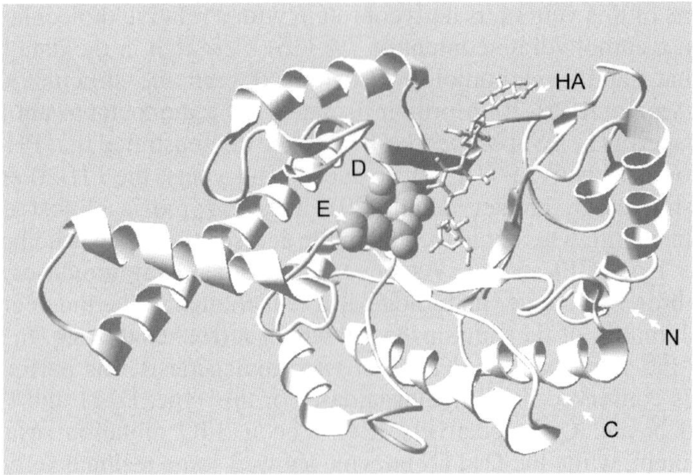

Figure 1 Three dimensional structure of bee venom hyaluronidase with a bound HA tetrasaccharide at 2.65 Å resolution. Ribbon presentation of the structure coordinates designated 1FCV (see Protein Data Bank depository of structures at www.pdb.org) is shown using Swiss pdb viewer (ver. 2.7). The single chain protein is composed of 350 amino acids and folds in one domain resembling a barrel composed of seven β strands, which is open between β1 and β2. The active site residues Glu113 (E) and Asp111(D) are shown as space-filled models. Asp111 keeps the side chain of Glu113 in the proper orientation for catalysis. N and C, the two ends of the polypeptide chain, are in close proximity at the rear of the enzyme (see arrows).

catalysis characteristic for retaining glycosidases. In this proposed mechanism, E113 acts as a proton donor, while the proximate D111 is required for the catalytically competent orientation of E113. Interestingly, in this acid–base catalysis, the *N*-acetyl group of the substrate acts as a nucleophile. The orientation of the *N*-acetyl group is influenced by D111. As shown in Fig. 2, hydrolysis proceeds with retention of the configuration at C1 of *N*-acetylglucosamine of the product. This type of mechanism has also been found in other hexosaminidases (glycosidase family 20) and chitinases (family 18) (31,32).

Compared to other mammalian hyaluronidases, the bee venom enzyme is folded in a single domain. Tertiary structures as well as the role of the additional carboxy terminal parts present in HYAL1 to 4 and PH-20 are unknown. These extra regions could possibly influence biochemical properties of these enzymes such as pH dependency, substrate binding and specificity or transglycosylation activity.

C. Biological Roles

1. PH-20

As already mentioned this enzyme is required for the sperm to penetrate the HA-rich cumulus layer (33). Indeed, it was thought that PH-20 was essential

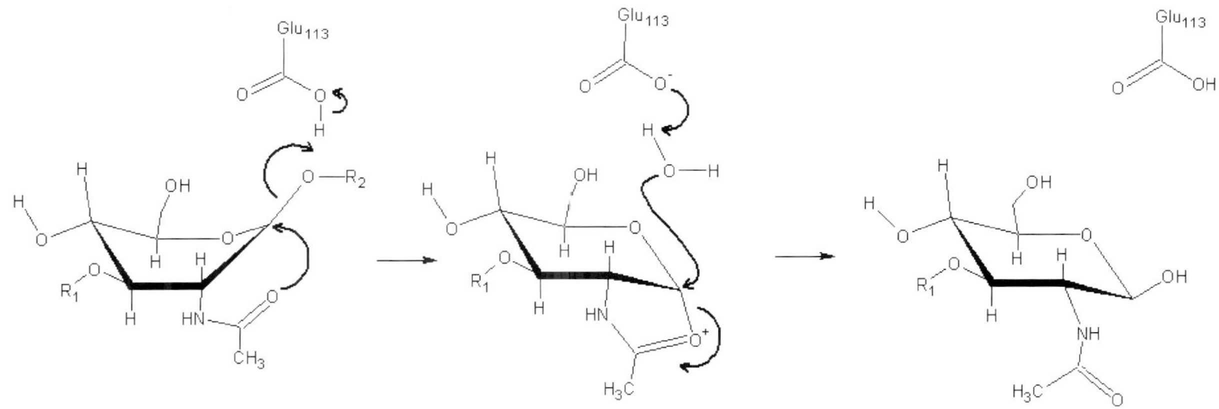

Figure 2 Reaction mechanism of mammalian-type hyaluronidases. Double-displacement substrate-assisted hydrolysis of HA proceeds after binding of *N*-acetylglucosamine in the boat conformation and formation of a covalent cyclic oxazolinium ion intermediate. Distortion of the sugar moiety facilitates the cleavage reaction by bringing the glycosidic oxygen nearer to Glu113 and the carbonyl oxygen of the acetamido group close to the anomeric C1 carbon. In this mechanism, the nucleophile for degradation of the polymer is supplied by the substrate.

for this process. However, sperm from mice deficient in PH-20 genes still degrades the cumulus layer and fertilizes eggs albeit with greatly reduced efficiency (34). Recent studies have shown that besides PH-20, HYAL3 also is expressed in the testis, which may explain why sperm of PH-20 − /− mice can still penetrate the HA-rich cumulus layer. Whether HYAL3 can also functionally replace PH-20 in the interaction with components of the zona pellucida (35) and in intracellular signalling (for a recent review see Ref. 20) remains to be investigated.

2. *HYAL1*

Human plasma and urine are known to contain HYAL1 (28,36). This enzyme is about 40% identical to PH-20. However, as it is only active at acidic pH, the function of the enzyme in the circulation remains elusive. The lysosomal storage disease mucopolysaccharidosis IX is caused by the lack of HYAL1 (37,38). Indeed, there is evidence that a 45 kDa processed form of HYAL1 resides in lysosomes, where it apparently plays an important role in the degradation of HA. Further roles have been assigned to HYAL1, which will be described in more detail in Chapter 13.

3. *HYAL2*

In humans and mice, this enzyme is present in many tissues with the notable exception of brain where it is expressed only during embryonic development and a short time after birth. This hyaluronidase has its optimum activity at pH 4 (24). Surprisingly, it cleaves high molecular weight HA only to a product of approximately 20 kDa, which corresponds to about 50 disaccharide units. Current evidence indicates that HYAL2 resides in lysosomes (24) as well as at the plasma membrane attached via a GPI anchor (39). The latter finding suggests a function in HYAL2 in tissue remodelling. Indeed, Hyal2 was shown to be involved in HA turnover during the early phase of lung injury and growth of astrocytomas in mice (40,41).

As mentioned previously, the gene for HYAL2 is located in a region of human chromosome 3p21.3 which is often deleted in lung carcinomas (11). Recent experiments have shown that HYAL2 serves as a receptor for a retrovirus (jaagsiekte sheep retrovirus, JSRV) (39). This virus replicates in the airways and causes epithelial cell tumors. The HYAL2 receptor protein was shown to bind to and thereby inactivates the RON receptor tyrosine kinase (42). The envelope protein of JSRV strongly binds to HYAL2. Yet, if this occurs, RON is liberated, and now activates Akt and mitogen-activated protein kinase pathways leading to oncogenic transformation. Contrastingly, HYAL2 was also found to counteract tumorigenesis. Overexpression of HYAL2 by means of a recombinant adenovirus suppressed the growth of tumor xenografts in mice (43).

Hyal2 from *Xenopus laevis* has also been investigated to some extent. As shown by overexpression in frog oocytes, in this species also, a form of Hyal2 exists which is bound to the membrane via a GPI anchor. Using highly sensitive detection methods (44), it could be shown that the enzyme is not only active at pH 4, but also to a small extent at neutral pH (45). Ectopic expression of this enzyme in *Xenopus* embryos leads to extensive degradation of HA in the extracellular matrix.

4. *HYAL3*

Only little is known about the product of this gene. It is present in many different tissues, particularly in testis, but no detailed biochemical studies have been performed. Expression of HYAL3 appears to be up-regulated, together with HYAL2, in differentiating chondrocytes (46). Both genes are activated by inflammatory cytokines such as IL-1 and TNF-α (47).

5. *HYAL4*

To date, no reports on the biological function or the biochemical properties of this protein have been published.

6. *Venom Hyaluronidases*

This enzyme is a component of venom in different animals such as bees, wasps, spiders, scorpions, fish, snakes and lizards (48). Only the structures of hyaluronidases from bee and wasps' venom have so far been elucidated. It is generally believed that these enzymes function as spreading factors for other constituents of the venom cocktail.

III. Bacterial Hyaluronidase

The function of the bacterial enzymes may be akin to that of 'spreading factors', i.e., to facilitate the diffusion of the microorganisms through the extracellular matrix. This could play a role in wound infections, pneumonia, meningitis and bacteremia in general (49,50). In contrast to the glycohydrolases found in the animal kingdom, the enzymes from bacteria are hyaluronate lyases. Enzymes of this type have been characterized from different bacterial strains of *Streptococcus*, *Staphylococcus*, *Clostridium*, etc., as well as from *Streptomyces hyalurolyticus* and streptococcal bacteriophages (51). The first complete gene of a hyaluronate lyase has been determined from *Streptococcus agalactiae* (52).

A. General Characteristics

HA lyases (EC 4.2.2.1) of bacteria and *Streptomyces* differ from animal hyaluronidases (hyaluronoglucosaminidases) by their mode of action (53). They catalyse a β-elimination reaction, rather than hydrolysis, of the β1,4-glycosidic linkage between *N*-acetyl-β-D-glucosamine and D-glucuronic acid residues of HA. This results in the formation of oligosaccharides with $\Delta^{4,5}$ unsaturated hexuronic acid residues at their non-reducing ends (2-acetamido-2-deoxy-3-*O*-(beta-D-gluco-4-enepyranosyluronic acid)-D-glucose). The enzyme from *Streptomyces hyalurolyticus* produces a mixture of unsaturated hexa and tetrasaccharides, while in the case of the lyase of *S. agalactiae*, disaccharides are the final product (54). These oligosaccharides stimulate pro-inflammatory or pathogenic processes, which are thought to facilitate the invasion of tissues or perturb the defence mechanisms of normal immune cells (9,55).

B. Structure

The molecular mechanism of the lyase action was unknown until the structures of the bacterial hyaluronate lyases from *S. pneumoniae* and *S. agalactiae* were elucidated by means of X-ray crystallography (53). Structural data of the native enzymes and their complexes with bound oligosaccharides of *S. pneumoniae* lyase, together with biochemical analyses, lead to the identification of the residues present in the catalytic centre (see later) (56). The enzyme structure contains two domains, with the catalytic domain built largely from helices resembling a horseshoe with an α5/α5-barrel structure (Fig. 3). The C-terminal domain consists almost entirely of β-sheets arranged in a four-layered sandwich. The two domains are connected by one flexible peptide linker; in addition, they interact through interface residues. The cleft present at one end of the α-domain is where the substrate binds and cleavage occurs. The internal surface of the cleft is largely positive with a few small strongly negative areas. Furthermore, the central part of the cleft has a patch of aromatic residues, Trp291, Trp292 and Phe343, that determine the size of the final product by interacting with two disaccharide units of HA (57). Hence, disaccharides are the smallest end products. The *S. agalactiae* enzyme structure has an additional β-sheet domain at its amino terminus comprised of 74 amino acids (58). This additional domain may be involved in substrate binding to the catalytic cleft. Furthermore, the structure of *S. agalactiae* HA lyase suggests that this might be a novel type of allosteric enzymes, where the substrate modulates the shape of the catalytic cleft.

C. Mechanism of Degradation

From the X-ray structure, the structure of the active site could be deduced (59). A catalytic mechanism known as ‘proton acceptance and donation’ was proposed (56,60). It involves five steps: [i] binding of the substrate within the catalytic cleft; [ii] interaction of the amide moiety of Asn349 with the carboxyl group of glucuronate; [iii] His399 then acts as a base and withdraws a hydrogen from C5 of the glucuronate, which leads to the formation of a double bond between C4 and C5; [iv] Tyr408 donates a proton to break the β1,4 glycosidic bond, which causes [v] cleavage of a disaccharide (Fig. 4). The catalytic residues His399 and Tyr408

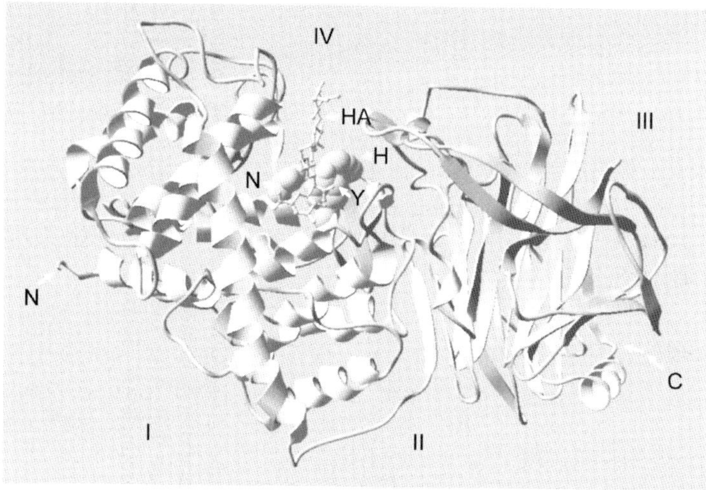

Figure 3 Three dimensional structure of *Streptococcus pneumoniae* hyaluronate lyase with a bound hexasaccharide at 1.7 Å resolution. The schematic illustration was drawn using Swiss pdb viewer (ver. 2.7) based on the structure coordinates designated 1c82 accessible from the Protein Data Bank. The structure is made up of 719 residues, The N-terminal (N) alpha helical domain (I) and the C-terminal (C) beta-sheet domain (III) are connected by a linker (II), and interact in addition through a surface interface of approximately 3550 $Å^2$. HA is bound and degraded in a cleft (IV) located in the alpha helical domain which is 30 Å in length and 10 Å deep and wide. The catalytic residues Asn349 (N), His399 (H) and Tyr408 (Y) are located in the cleft and depicted as space fill molecular structures.

are then reprotonated by exchange with water, regenerating the catalytic centre for a next round of degradation. From the structure of the *S. pneumoniae* enzyme complexed with a disaccharide unit, it emerged that the enzyme degrades the polymer from its reducing end.

IV. Summary and Conclusion

HA levels are determined by relative rates of synthesis and degradation. Abnormal degradation rates can be observed in a variety of diseases (tumor, shock, cancer, etc.) or bacterial infections. Hyaluronidases may be secreted and later on, taken up to be delivered to lysosomes. These enzymes may also be involved in receptor-mediated endocytosis of the substrate. More work remains to be done to fully characterize the biological roles of both bacterial and mammalian-type HA degrading enzymes. The structural information currently available and the structures of the complexes with the degradation product have opened new vistas about the reaction mechanisms of these enzymes.

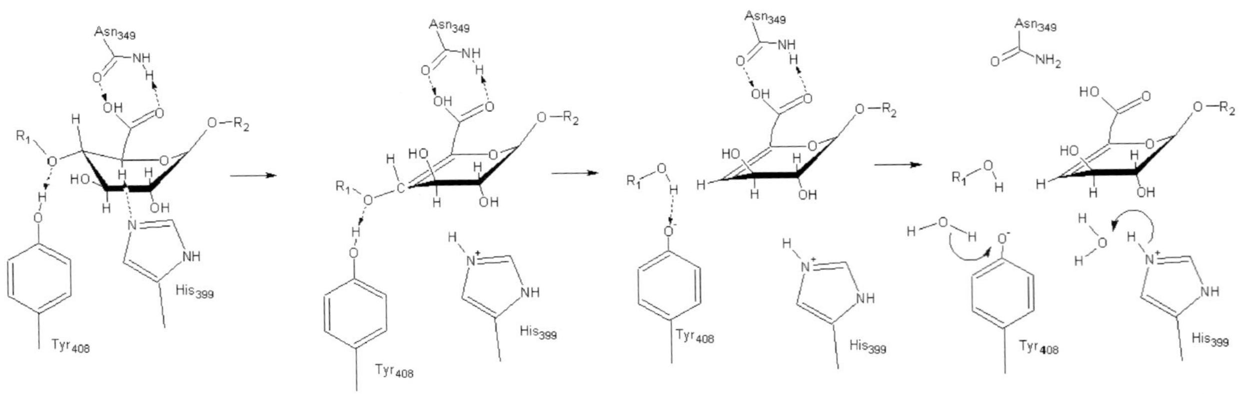

Figure 4 Enzymatic mechanism of the bacterial hyaluronate lyase: the catalytic residues, Asn349, His399 and Tyr408, catalyse cleavage of the β1,4 glycosidic bond by β-elimination in a five-step proton acceptance and donation mechanism. First the negatively charged substrate binds to the cleft, where the catalytic residues interact with a glucuronate unit. After cleavage of the β1,4 bond, and hydrogen exchange between His399 and Tyr408 and surrounding water molecules, the products are released or dislocated to enter the next cycle of degradation. The main end products are unsaturated disaccharide units.

Acknowledgements

GL is an APART fellow of the Austrian Academy of Sciences and supported by the Jubilee Fund of the Austrian National Bank (Project 10451). Work in the authors' institutes has been supported by the Austrian Science Foundation and the Jubilee Fund of the Austrian National Bank.

References

1. Duran-Reynals F. Exaltation de l'activité du virus vaccinal par les extraits de certains organes. C R Soc Biol 1928; 99:6.
2. Meyer K, Palmer JW. The polysaccharide of the vitreous humor. J Biol Chem 1934; 107:629–634.
3. Chain E, Duthie ES. A mucolytic enzyme in testis extracts. Nature 1939; 144: 977–978.
4. Chain E, Duthie ES. Identity of hyaluronidase and spreading factor. Br J Exp Pathol 1940; 21:324.
5. Meyer K. Hyaluronidases. In: Boyer PD, ed. Enzymes. New York: Academic Press, 1971; vol. V:307–320.
6. Kreil G. Hyaluronidases—a group of neglected enzymes. Protein Sci 1995; 4: 1666–1669.
7. Takagaki K, Nakamura T, Izumi J, Saitoh H, Endo M, Kojima K, Kato I, Majima M. Characterization of hydrolysis and transglycosylation by testicular hyaluronidase using ion-spray mass spectrometry. Biochemistry 1994; 33:6503–6507.
8. Saitoh H, Takagaki K, Majima M, Nakamura T, Matsuki A, Kasai M, Narita H, Endo M. Enzymic reconstruction of glycosaminoglycan oligosaccharide chains using the transglycosylation reaction of bovine testicular hyaluronidase. J Biol Chem 1995; 270:3741–3747.
9. Menzel EJ, Farr C. Hyaluronidase and its substrate hyaluronan: biochemistry, biological activities and therapeutic uses. Cancer Lett 1998; 131:3–11.
10. Csoka AB, Frost GI, Stern R. The six hyaluronidase-like genes in the human and mouse genomes. Matrix Biol 2001; 20:499–508.
11. Daly MC, Xiang RH, Buchhagen D, Hensel CH, Garcia DK, Killary AM, Minna JD, Naylor SL. A homozygous deletion on chromosome 3 in a small cell lung cancer cell line correlates with a region of tumor suppressor activity. Oncogene 1993; 8:1721–1729.
12. Hernandez JM, Mecucci C, Michaux L, Criel A, Stul M, Meeus P, Wlodarska I, Van Orshoven A, Cassiman JJ, De Wolf-Peeters C, Van den Berghe H. del(7q) in chronic B-cell lymphoid malignancies. Cancer Genet Cytogenet 1997; 93: 147–151.
13. Edelson MI, Scherer SW, Tsui LC, Welch WR, Bell DA, Berkowitz RS, Mok SC. Identification of a 1300 kilobase deletion unit on chromosome 7q31.3 in invasive epithelial ovarian carcinomas. Oncogene 1997; 14:2979–2984.
14. Csoka AB, Scherer SW, Stern R. Expression analysis of six paralogous human hyaluronidase genes clustered on chromosomes 3p21 and 7q31. Genomics 1999; 60:356–361.
15. Gmachl M, Kreil G. Bee venom hyaluronidase is homologous to a membrane protein of mammalian sperm. Proc Natl Acad Sci USA 1993; 90:3569–3573.

16. Kemeny DM, Dalton N, Lawrence AJ, Pearce FL, Vernon CA. The purification and characterisation of hyaluronidase from the venom of the honey bee, *Apis mellifera*. Eur J Biochem 1984; 139:217–223.
17. Lathrop WF, Carmichael EP, Myles DG, Primakoff P. cDNA cloning reveals the molecular structure of a sperm surface protein, Ph-20, involved in sperm–egg adhesion and the wide distribution of its gene among mammals. J Cell Biol 1990; 111:2939–2949.
18. Gmachl M, Sagan S, Ketter S, Kreil G. The human sperm protein PH-20 has hyaluronidase activity. FEBS Lett 1993; 336:545–548.
19. Lin Y, Kimmel LH, Myles DG, Primakoff P. Molecular cloning of the human and monkey sperm surface protein PH-20. Proc Natl Acad Sci USA 1993; 90: 10071–10075.
20. Cherr GN, Yudin AI, Overstreet JW. The dual functions of GPI-anchored PH-20: hyaluronidase and intracellular signaling. Matrix Biol 2001; 20:515–525.
21. Arming S, Strobl B, Wechselberger C, Kreil G. *In vitro* mutagenesis of PH-20 hyaluronidase from human sperm. Eur J Biochem 1997; 247:810–814.
22. Frost GI, Mohapatra G, Wong TM, Csoka AB, Gray JW, Stern R. HYAL1LUCA-1, a candidate tumor suppressor gene on chromosome 3p21.3, is inactivated in head and neck squamous cell carcinomas by aberrant splicing of pre-mRNA. Oncogene 2000; 19:870–877.
23. Sekido Y, Ahmadian M, Wistuba II, Latif F, Bader S, Wei MH, Duh FM, Gazdar AF, Lerman MI, Minna JD. Cloning of a breast cancer homozygous deletion junction narrows the region of search for a 3p21.3 tumor suppressor gene. Oncogene 1998; 16:3151–3157.
24. Lepperdinger G, Strobl B, Kreil G. HYAL2, a human gene expressed in many cells, encodes a lysosomal hyaluronidase with a novel type of specificity. J Biol Chem 1998; 273:22466–22470.
25. Cherr GN, Meyers SA, Yudin AI, VandeVoort CA, Myles DG, Primakoff P, Overstreet JW. The PH-20 protein in cynomolgus macaque spermatozoa: identification of two different forms exhibiting hyaluronidase activity. Dev Biol 1996; 175:142–153.
26. Meyer MF, Kreil G, Aschauer H. The soluble hyaluronidase from bull testes is a fragment of the membrane-bound PH-20 enzyme. FEBS Lett 1997; 413:385–388.
27. Oettl M, Hoechstetter J, Asen I, Bernhardt G, Buschauer A. Comparative characterization of bovine testicular hyaluronidase and a hyaluronate lyase from *Streptococcus agalactiae* in pharmaceutical preparations. Eur J Pharm Sci 2003; 18: 267–277.
28. Csoka TB, Frost GI, Wong T, Stern R. Purification and microsequencing of hyaluronidase isozymes from human urine. FEBS Lett 1997; 417:307–310.
29. Markovic-Housley Z, Miglierini G, Soldatova L, Rizkallah PJ, Müller U, Schirmer T. Crystal structure of hyaluronidase, a major allergen of bee venom. Struct Fold Des 2000; 8:1025–1035.
30. Markovic-Housley Z, Schirmer T. Structural evidence for substrate assisted catalytic mechanism of bee venom hyaluronidase, a major allergen of bee venom. In: Teeri TT, Svensson B, Gilbert HJ, Feizi T, eds. Carbohydrate Bioengineering: Interdisciplinary Approaches. Cambridge, UK: The Royal Society, 2002:19–27.
31. Drouillard S, Armand S, Davies GJ, Vorgias CE, Henrissat B. Serratia marcescens chitobiase is a retaining glycosidase utilizing substrate acetamido group participation. Biochem J 1997; 328:945–949.

32. Terwisscha van Scheltinga AC, Armand S, Kalk KH, Isogai A, Henrissat B, Dijkstra BW. Stereochemistry of chitin hydrolysis by a plant chitinase/lysozyme and X-ray structure of a complex with allosamidin: evidence for substrate assisted catalysis. Biochemistry 1995; 34:15619–15623.
33. Myles DG, Primakoff P. Why did the sperm cross the cumulus? To get to the oocyte. Functions of the sperm surface proteins PH-20 and fertilin in arriving at, and fusing with, the egg. Biol Reprod 1997; 56:320–327.
34. Baba D, Kashiwabara S, Honda A, Yamagata K, Wu Q, Ikawa M, Okabe M, Baba T. Mouse sperm lacking cell surface hyaluronidase PH-20 can pass through the layer of cumulus cells and fertilize the egg. J Biol Chem 2002; 277:30310–30314.
35. Vines CA, Li MW, Deng X, Yudin AI, Cherr GN, Overstreet JW. Identification of a hyaluronic acid (HA) binding domain in the PH-20 protein that may function in cell signaling. Mol Reprod Dev 2001; 60:542–552.
36. Frost GI, Csoka TB, Wong T, Stern R. Purification, cloning, and expression of human plasma hyaluronidase. Biochem Biophys Res Commun 1997; 236:10–15.
37. Natowicz MR, Short MP, Wang Y, Dickersin GR, Gebhardt MC, Rosenthal DI, Sims KB, Rosenberg AE. Clinical and biochemical manifestations of hyaluronidase deficiency. N Engl J Med 1996; 335:1029–1033.
38. Triggs-Raine B, Salo TJ, Zhang H, Wicklow BA, Natowicz MR. Mutations in HYAL1, a member of a tandemly distributed multigene family encoding disparate hyaluronidase activities, cause a newly described lysosomal disorder, mucopolysaccharidosis IX. Proc Natl Acad Sci USA 1999; 96:6296–6300.
39. Rai SK, Duh FM, Vigdorovich V, Danilkovitch-Miagkova A, Lerman MI, Miller AD. Candidate tumor suppressor HYAL2 is a glycosylphosphatidylinositol (GPI)-anchored cell-surface receptor for jaagsiekte sheep retrovirus, the envelope protein of which mediates oncogenic transformation. Proc Natl Acad Sci USA 2001; 98:4443–4448.
40. Li Y, Rahmanian M, Widstrom C, Lepperdinger G, Frost GI, Heldin P. Irradiation-induced expression of hyaluronan (HA) synthase 2 and hyaluronidase 2 genes in rat lung tissue accompanies active turnover of HA and induction of types I and III collagen gene expression. Am J Respir Cell Mol Biol 2000; 23:411–418.
41. Novak U, Stylli SS, Kaye AH, Lepperdinger G. Hyaluronidase-2 overexpression accelerates intracerebral but not subcutaneous tumor formation of murine astrocytoma cells. Cancer Res 1999; 59:6246–6250.
42. Danilkovitch-Miagkova A, Duh FM, Kuzmin I, Angeloni D, Liu SL, Miller AD, Lerman MI. Hyaluronidase 2 negatively regulates RON receptor tyrosine kinase and mediates transformation of epithelial cells by jaagsiekte sheep retrovirus. Proc Natl Acad Sci USA 2003; 100:4580–4585.
43. Ji L, Nishizaki M, Gao B, Burbee D, Kondo M, Kamibayashi C, Xu K, Yen N, Atkinson EN, Fang B, Lerman MI, Roth JA, Minna JD. Expression of several genes in the human chromosome 3p21.3 homozygous deletion region by an adenovirus vector results in tumor suppressor activities *in vitro* and *in vivo*. Cancer Res 2002; 62:2715–2720.
44. Müllegger J, Reitinger S, Lepperdinger G. Hapten-labeled hyaluronan, a substrate to monitor hyaluronidase activity by enhanced chemiluminescence-assisted detection on filter blots. Anal Biochem 2001; 293:291–293.
45. Müllegger J, Lepperdinger G. Degradation of hyaluronan by a Hyal2-type hyaluronidase affects pattern formation of vitelline vessels during embryogenesis of *Xenopus laevis*. Mech Dev 2002; 111:25–35.

46. Nicoll SB, Barak O, Csoka AB, Bhatnagar RS, Stern R. Hyaluronidases and CD44 undergo differential modulation during chondrogenesis. Biochem Biophys Res Commun 2002; 292:819–825.
47. Flannery CR, Little CB, Hughes CE, Caterson B. Expression and activity of articular cartilage hyaluronidases. Biochem Biophys Res Commun 1998; 251: 824–829.
48. King TP, Spangfort MD. Structure and biology of stinging insect venom allergens. Int Arch Allergy Immunol 2000; 123:99–106.
49. Spellerberg B. Pathogenesis of neonatal *Streptococcus agalactiae* infections. Microbes Infect 2000; 2:1733–1742.
50. Jedrzejas MJ. Pneumococcal virulence factors: structure and function. Microbiol Mol Biol Rev 2001; 65:187–207.
51. Sutherland IW. Polysaccharide lyases. FEMS Microbiol Rev 1995; 16:323–347.
52. Gase K, Ozegowski J, Malke H. The *Streptococcus agalactiae* hylB gene encoding hyaluronate lyase: completion of the sequence and expression analysis. Biochim Biophys Acta 1998; 1398:86–98.
53. Jedrzejas MJ. Structural and functional comparison of polysaccharide-degrading enzymes. Crit Rev Biochem Mol Biol 2000; 35:221–251.
54. Baker JR, Pritchard DG. Action pattern and substrate specificity of the hyaluronan lyase from group B streptococci. Biochem J 2000; 348:465–471.
55. Noble PW. Hyaluronan and its catabolic products in tissue injury and repair. Matrix Biol 2002; 21:25–29.
56. Jedrzejas MJ, Mello LV, de Groot BL, Li S. Mechanism of hyaluronan degradation by *Streptococcus pneumoniae* hyaluronate lyase. Structures of complexes with the substrate. J Biol Chem 2002; 277:28287–28297.
57. Li S, Jedrzejas MJ. Hyaluronan binding and degradation by *Streptococcus agalactiae* hyaluronate lyase. J Biol Chem 2001; 276:41407–41416.
58. Mello LV, de Groot BL, Li S, Jedrzejas MJ. Structure and flexibility of *Streptococcus agalactiae* hyaluronate lyase complex with its substrate. Insights into the mechanism of processive degradation of hyaluronan. J Biol Chem 2002; 277:36678–36688.
59. Jedrzejas MJ. The structure and function of novel proteins of Bacillus anthracis and other spore-forming bacteria: development of novel prophylactic and therapeutic agents. Crit Rev Biochem Mol Biol 2002; 37:339–373.
60. Ponnuraj K, Jedrzejas MJ. Mechanism of hyaluronan binding and degradation: structure of *Streptococcus pneumoniae* hyaluronate lyase in complex with hyaluronic acid disaccharide at 1.7 Å resolution. J Mol Biol 2000; 299:885–895.

Subject index

A

ACPTM 485
Acrosome 587
Acute lung injury (ALI) 153, 275, 279
Adaptor 91, 98, 103, 135–137, 157, 169, 170
Aggrecan 1, 2, 93, 99–101, 125, 130, 190–193, 206, 213, 223, 226, 232, 235, 251, 252, 396, 397, 466, 485, 570
Aggulatinatig factor 430–432, 437, 438
Aging 24, 71, 217, 231, 346, 351, 353, 357–361, 430, 433, 534, 585
Air-liquid interface (ALI) 324
Airway epithelium 324, 325, 327, 328, 331
Airway submucosal gland cells 324
Alcian-blue-PAS 324
Alginate 379, 487, 492
Allergen challenge 330, 331
Allosteric enzymes 592
Alltrans retinoic acid 399
Alveolar
  macrophages 257
  proteinosis 325
Alveoli 111, 247, 249, 255, 259, 278
2-Aminoacridone (AMAC) 31, 50
Analgesic effect 309
Angiogenesis 84, 98, 107, 126, 127, 129, 130, 137, 153, 154, 206, 225, 252, 253, 290, 293, 295, 296, 370, 374, 375, 408, 441, 505, 554, 564, 573
Antiproteases 326
Anti-RHAMM antibodies 130, 133, 136, 138, 140, 164, 328, 329
ARDS 271, 275, 278, 279
Arteries 249, 307, 308, 313, 314
Asthmatics 330, 331
Atherosclerosis 189, 307–310, 315
Aqueous humor 224, 225, 233, 373, 545
Autologous soft tissue-filler 484

B

Basal
  cells 396, 401, 403, 536
  lamina 211, 396, 398, 403
Bee venom 23, 587, 588
Bikunin 195, 207–212, 214, 217
  knockout mouse 214
Bioartificial liver devices 484
Biodegradation 71, 73, 378, 494
Biomaterials 346, 378–380, 475, 477, 480, 482, 483, 486, 487, 491, 494, 505
Biosynthesis 2, 22, 79, 92, 103, 111, 227, 274, 309, 351, 357, 358, 360, 361, 367, 506, 555, 558–561, 569, 571, 573
Biotin-labeled HA binding protein 53
Bleomycin 111, 249, 250, 254, 225
Body fluids 73, 286, 359, 361
Boltzmann's constant 12
Bovine serum albumin (BSA) 328
Bowman's layer 535, 536
Bradykinin 326, 332, 462
Brain link protein 1, 192
Brevican 93, 191–193, 223, 232
Bronchoalveolar lavage (BAL) 249, 324
Bronchoconstriction 330, 331
Bruch's membrane 224, 231
1, 4-butanediol diglycidyl 418, 513, 514

C

Cancer 77, 89, 90, 106–108, 125–127, 129, 132, 139, 142, 143, 153, 157, 159, 160, 162, 189, 213, 217, 253, 285–298, 400, 403, 406–408, 461, 478, 480, 507, 515, 590, 593
Carbodiimides 478
Carboxymethylcellulose 377, 428, 467, 492, 545
Catabolism 22, 71, 72, 75, 76, 102, 111, 227, 398, 440
Cataract 235, 422, 458, 459
Cationic proteins 332
Caveolae 398
CD44 23, 24, 73, 75, 83–111, 126, 129, 130, 136–139, 142, 154–166, 168, 170, 173, 174, 190, 191, 193, 194, 199, 200, 206, 211, 215, 216, 223, 224–227, 229, 231–234, 250, 251, 285, 287–292, 294–296, 298, 309–311, 314, 328, 352, 359, 373–375, 396, 400, 403–405, 408, 457, 460, 461, 466, 467, 478, 507, 515, 535–537, 545, 546, 570
Celiac disease 403

Cell
adhesion 156, 160, 166, 170, 190, 215, 216, 223, 289, 290, 292, 406, 487, 579
biological functions 457, 460, 461, 465–468
mesenchymal progenitor 485, 493
proliferation 127, 141, 142, 164, 224, 253, 295, 314, 329, 348, 358, 374, 375, 381–383, 399, 408, 440, 457, 535, 573
Cervix 403, 404
Chemoenzymatic synthesis 50, 55
Chitinases 588
Chitosan 486, 487, 492
Chondroitin sulfate 23, 29, 74, 76, 86, 87, 90, 92, 99, 208, 209, 211, 212, 224, 227, 228, 232–234, 252, 256, 261, 272, 273, 310, 327, 367, 422, 515
Choroid 224, 226, 231, 232, 234
Chronic
airways diseases 213
liver diseases 213
Ciliary
beat frequency 325
body 224, 227, 233
Ciliated cells 324, 328, 403
Clathrin coated pits 398
Clear cells 400, 530
Clinical studies 377, 381, 382, 425, 427, 437, 441, 488, 511
Cock's comb 2
Coil overlap 44–46, 423
Collagen-induced arthritis111, 136, 138, 214, 215
Colon cancer cells 157, 403, 480
Confocal-FRAP technique 5, 7
Conformation plot 64
Conjunctiva 224–226, 421, 529, 534, 536, 540, 543, 545
Cornea 225, 261, 439, 442, 459, 467, 506, 529, 531, 535- 537, 545
Corneal
epithelium 224–225, 467, 529, 530, 533, 535–537, 540, 541, 544, 546
wound healing 442, 467, 546
Cornified cells 396
Crohn's disease 217, 403
Crystal structure 23, 199, 327, 587
Cumulus
layer 75, 587, 588, 590
matrix stabilizing activity 210
oocyte complex (COC) 195, 210
Cutaneous wounds 379
Cyanogen bromide (CNBr) activation 481
Cytokines 73, 89, 98, 127, 168, 174, 253, 275, 309, 371, 372, 374, 379, 384, 404, 457, 530, 572, 573, 591
Cytosolic pool 571
Cytoplasmic vesicles 398
Cytoskeleton 84, 86, 90, 93–98, 103–105, 129, 131–132, 135, 137, 142, 156, 159, 160, 164–165, 168, 170, 173, 292, 431

D

Dendritic cells 24, 155, 168, 253, 257, 572
Depolymerization 29, 30, 72, 77, 78, 254, 330, 357, 360, 457, 468, 509
Dermis 71, 73, 102, 111, 126, 260, 285, 357, 368, 383, 396, 424, 466, 572
Descemet's membrane 535
Desmosomes 396
Dihydrazide 479, 480, 515
Disodium 8-amino-1, 3, 6-naphthalene trisulfonate (ANTS) 31, 51, 52
Disorders 231–235, 348, 356, 359, 361, 399, 488, 492, 529, 533, 534, 538, 544, 546
Divinyl sulfone 346, 481, 487, 509, 513
Drug delivery with hyaluronan 442
Dry eye syndrome 439, 537, 541, 546

E

Ectoderm 399
Elastase 209, 252, 258, 259, 261, 326
Elastic dyanamic modulus (G*) 419
Elastic LS
Quasi 56, 58, 59
Elastin 249–252, 258–260, 444
Electroblotting 53
Electroosmotic flow 29
Electrophoresis
agarose gel (AGE) 49, 53, 55, 276, 330, 331, 434, 435
capillary 29, 30, 43, 48, 52, 54, 100, 107, 247, 248, 484
polyacrylamide gel (PAGE) 31, 49, 50
β-Elimination 25, 591, 594
Emphysema 249, 258, 259, 261, 275
Enamel organ of teeth 399
Endocytosis 71, 74, 75, 84, 98, 107, 153, 154, 155, 159, 172, 174, 190, 194, 398, 591
Endothelial cell 73–75, 87, 100, 107, 108, 110, 127, 130, 137, 160, 161, 165, 1 66, 173, 174, 212, 226, 233, 248, 253, 289, 291–293, 307 309, 310, 374, 407, 421, 461, 480, 487, 488, 506, 507, 535, 536

Epican 87
Epidermis 73, 102, 324, 371, 383, 395, 396, 398–400, 405, 407, 466, 572, 576
Epithelial cell 73, 87, 100, 108, 127, 129, 141, 224, 226, 227, 233, 247–249, 255, 278, 291, 292, 324, 327–330, 395, 399, 400, 403–407, 467, 530, 531, 535, 536, 537, 541, 571, 590
Epithelium 171, 224, 225, 227, 229, 231–233, 324, 325, 327, 328, 331, 339, 340, 341, 380, 399, 400, 402–405, 408, 439, 466, 467, 529, 531, 533, 535–541, 544, 546
Esterification 378, 476, 481, 483, 515, 518
Etherification 481
Exoglucosidases 71
Extracellular matrix
  waterlogged 314

F

FACE 31, 32, 50, 51, 125, 198, 323
Fallopian tubes 368, 404, 408, 486
Familial hypercholesterolemia 310
Fertility 172, 505, 509, 570, 576
Fetal wounds 369, 370, 375
Fibrillin-1 252, 259
Fibrin polymerization 292, 310
Fibroblast growth factor 161, 168, 348, 372, 485, 492
Filaggrin 399

G

Gastric cancer cells 403
Gelatin 379, 490, 493, 494
Glands
  mammary 399, 400
  salivary 399, 400
  tear 399, 400
Glaucoma 232, 233
Glycosyltransferases 555
Goblet 328, 531, 534, 536, 538–541, 543
GPI anchor 587, 590, 591
Granual cell
Growth factors
  EGF 74, 89, 105, 161–163, 173, 194, 228, 333, 398, 399, 531, 573
  KGF 398, 399, 572, 573
GYNECARE INTERGEL 425, 486

H

Hair bulge 399
Half-life time 72, 73
HAPLN-190, 192, 193
HARE 74, 75, 154, 155, 172, 174, 194, 288, 461, 507
Has
  1 125, 309
  2 125, 129, 324
  3 125, 324
Healon 376, 421, 422, 433, 439, 442, 506
Heart defect 575
Heavy chains 198, 206–208, 210–212
Heparan sulfate proteoglycan 89, 105, 259
Heparinase 162, 327
Hepatitis 213
Hepatocellular carcinoma (HCC) 213
Hexosaminidases 25, 585, 588
High tidal volume 271, 277
Homeostasis 73, 138, 154, 174, 229, 272, 326, 421, 431, 505, 507, 532
Huggins equation 45–48
HYAFF 378–382, 442, 476, 478, 483–486, 518, 519
HYAL
  1 173, 287, 293, 586, 590
  2 173, 586, 590
  3, 586, 591
  4, 586, 591
Hyaladherins 93, 126, 129, 130, 136, 137, 154, 190, 191, 251, 373, 374
Hyalgan 433, 506
Hyalograft 3D 484
Hyaluronan
  binding domain 90, 228
  binding protein 224, 229, 231, 324, 330, 331
  butyrate 481
  cross linked network 5
  elastic properties 129, 3342, 415, 417, 419, 420, 426, 430, 434
  2-fold helix 4
  gels 49, 420, 424–426, 428, 441, 444, 489
  oligosaccharides 16, 24, 25, 31, 32, 35, 49, 51–53, 83, 89, 99, 100, 102, 104, 106, 109, 252, 314, 398, 407, 460
  analysis by NMR 35
  preparation 21, 35, 273, 416, 417, 419, 420, 422, 427, 431–438, 440
  random coil 3, 4, 23, 43, 415, 416
  receptors 23, 73, 74, 77, 92, 93, 98–108, 126, 127, 129, 136, 137, 153–155, 160–164, 168, 169, 171, 172, 174, 190, 194, 223, 226, 278, 285, 288, 289, 296–298, 309, 312, 327, 328, 352, 359, 373, 398, 406, 407, 430, 457, 461, 464,

466, 468, 469, 478, 502, 515, 520, 534, 576, 579
self diffusion 6–15
synthases 1, 125, 223, 250, 267, 287, 323, 398, 405, 553–564, 569, 571, 576, 579, 583
thiolated 480, 490
viscoelastic properties 4, 16, 60, 223, 344, 372, 415, 426, 531
Hyaluronate lyase
bacterial 25, 26, 168, 195, 326, 330, 334, 376, 377, 382, 416, 420, 443, 481, 513, 555, 591–594
lysosomal 71, 73, 76, 77, 174, 253, 324, 398, 590
mammalian-type 25, 585, 586, 589, 593
somatic 173
streptococcus 24, 538, 553–556, 591, 593
hyalurolyticus 26, 29, 32, 591, 592
pneumoniae 24, 592, 593
testicular, PH-20 75
venum 591
Hydrazinolysis 482
Hydrocortisone 399, 483, 515
Hydrodynamic volume 11, 14, 16, 26, 42–44, 48, 64
Hydrogen bonds 6, 11, 13–16, 200, 351–355, 360
Hylan
A 417–420, 423, 428, 433–435, 439, 487, 513, 539
B 346, 347, 417, 419, 424–428, 435, 442, 443, 487, 488, 513
Hylan B gel 346, 347, 417, 419, 424–428, 443, 444
Hylan G–F 20 431, 434, 435, 438, 488
Hylashield 439, 487
Hyperplasia 214, 215, 398, 399
Hypertrophy 399

I

Immune cells 73, 156, 249, 258, 572, 592
Immunomodular 257
Inflammation 22, 24, 53, 110, 111, 127, 136, 137, 142, 153, 154, 195, 196, 206, 213, 216, 233, 250, 257, 261, 274, 278, 279, 289, 326, 329, 347, 373, 374, 395, 400, 431, 432, 440, 442, 443, 461, 464, 465, 468, 505, 537
Inflammatory
bowl disease (IBD) 212, 215, 216
genes 24, 84, 90, 98, 99, 106, 107, 125–127, 129–134, 137, 153, 154, 161, 164, 167–169, 173, 174, 192, 194, 197–200, 206, 207, 212, 213, 216, 217, 223, 225, 229, 235, 249, 250, 252, 253, 257, 258, 289–291, 293, 295, 296, 314, 342, 359, 370, 371, 374, 375, 408, 441, 461, 465, 505, 553–558, 564, 569, 571–573, 577, 579, 586, 587, 590, 591
phase modulation 374
In situ crosslinkable hydrogel 450
Inter-a-inhibitor 191, 195
Interphotoreceptor matrix 224, 229
Interstitium 247, 249–252, 255, 258, 261, 273, 330, 400, 404
Intracellular 23, 71, 86–88, 94–96, 102–106, 110, 126, 130–132, 134–137, 156, 157, 160, 164, 166–168, 170, 172, 288, 289, 312, 351–353, 359, 373, 398, 460, 478, 534, 535, 590
Intrinsic viscosity 11, 43–48, 65, 250, 430
In vivo tissue regeneration 480
Iris 224, 226, 227, 233
Isourea coupling 481

J

Jaagsiekte sheep retrovirus 253, 590
Joint tissues 430, 431, 461

K

Kallikrein 326, 327, 332, 333
Keloids 371
Keratin 111, 399
Keratinocytes 24, 73, 102–104, 108, 111, 127, 163, 375, 381, 382, 396, 399, 405, 406, 466, 486, 518, 554
Kinase 24, 104, 106, 109, 129, 131, 133, 135, 139, 157, 159–163, 166–168, 170, 253, 292, 460, 575, 590
Kinin 330
Kininogens 326

L

Lacrimal system 232
Lacto
ferrin 326, 533
peroxidase 326, 332
Lamina propia
Laryngeal incompetence 346
Laser skin 381
Lens 224, 227, 233, 376, 421, 422, 439, 506, 518, 529, 533
Leukocyte protease inhibitor (LPI) 326

Life saving treatment 271
Light scattering 3, 9, 10, 41, 50, 55, 59
Link
  module 23, 75, 90, 94, 130, 154, 191–200, 206, 209, 327, 332
  protein 23, 86, 90, 93, 94, 106, 125, 154, 173, 191–193, 206, 211, 223, 225–228, 232, 235, 251, 252, 260, 351, 396, 397, 408, 570
Lipid rafts 90, 93, 94, 97
Lipoprotein-hyaluronan complexes 309
LUCA
  1 253, 587
  2 253, 587
Lung cell stretch 278, 279
Lymphocytes 73, 84, 104, 109, 110, 161, 164, 234, 308, 309, 374, 428, 461
Lymph nodes 72, 74, 79, 102, 108, 173, 174, 194, 212, 289, 478, 507
Lysosomal
  degradation 398
  storage disease 590
Lysosomes 71, 74, 76, 77, 79, 102, 174, 398, 586, 590, 593
Lysozyme 326, 533
Lysyl-bradykinin 326, 332
LYVE-1 75, 93, 126, 130, 154, 155, 170, 172–174, 190, 191, 194, 223, 225, 288

M

Macrophages 53, 73, 75, 94, 102, 155, 168, 169, 249, 250, 252, 254, 256, 257, 308–310, 340, 369, 374, 379, 428, 442, 554
MALDI 33–35
Map kinase 139, 296, 575
Mark-Houwink-Sakurada equation 43, 48
Martin equation 45–47
Maturation 24, 154, 155, 168, 171, 172, 253, 370, 400
Mechanical ventilation 271, 275
Medical application 347, 368, 376, 377, 457, 458, 468, 480, 486, 494
Melanoma 106, 107, 139, 253, 290, 291, 297, 298, 407
Metabolic labeling 396
Metaplasia 333, 403, 530, 534, 536, 541, 543
Metastasis 84, 87, 106–108, 110, 127, 139, 141, 153, 156, 160, 161, 165, 289–291, 293–296, 407, 564, 586
Microfibrils 259, 260
Mitogen-activated protein kinase 157, 590
Molecular weight
  distribution 42, 63
  standards 55
Motogenesis 167
Mucin 530–532, 538–540, 545
Mucociliary escalator 325
Mucopolysaccharidosis IX 590
Mucous blanket 325
Mucous cells 400
Myopia 235

N

Nasal mucosa 325
Neuraminidase 92
Neurocan 93, 191–193, 223, 228, 229, 232, 233, 570
NIF-NaHA 376, 421, 427, 430, 432, 433, 439
Nitric oxide synthase 254
NMR spectroscopy 35, 196, 354, 356
Non-clathrin-coated endosomes 73
Nonideal behavior 45

O

Ocular surface 529–541, 544–546
Ocular surgery 380
Opthalmic viscosurgical devices (OVDs) 422
Opthalmology 35, 223, 376, 529
Optic nerve 224, 232, 233
Organotypic keratinocyte cultures 398
Orthovisc 433, 434
Osteoconductive 493
Otology 379
Oviduct 404
Oxidized polypyrrole 487
Oxygen free radicals 398
Oxidative stress 330, 333

P

Passerini reaction 491, 509
Percutaneous embolization 443
Periciliary fluid 325, 326
Peritoneal
  dialysis 442
  fluid 442
Periodate oxidations 481
Peroxynitrite 48, 53, 78, 254, 257
Phagocytes 330, 355
Phagocytosis 127, 252, 256, 440, 554
Photopolymerized hydrogels 489
Photorefractive keratectomy 536

Physico-chemical properties 298, 348, 379, 489
Pinocytosis 73, 102, 398
Polyanion 9, 16, 49, 352, 505
Polydispersity 5, 41–44, 54, 55, 64, 430, 433–435
Post operative adhesions 377
Proton acceptance and donation 592, 594
Pseudoexfoliation syndrome 233
Psoriasis 399
Pulmonary alveolus 247, 254

R

Radius of gyration 9, 10, 58, 59, 64
Rayleigh factor 57
Reaction mechanism 561, 587, 589
Reactive
 nitrogen species (RNS) 329, 330, 332
 oxygen Species (ROS) 77–79, 127, 324, 329, 330, 332, 398
Receptors 23, 73, 74, 77, 92, 93, 98–107, 126, 127, 129, 136, 137, 142, 153–155, 160–164, 168, 169, 171, 172, 174, 190, 194, 223, 226, 278, 285, 288, 289, 296–298, 309, 312, 327, 328, 352, 359, 373, 398, 406, 407, 430, 457, 460, 461, 464, 466, 468, 478, 507, 515, 520, 534, 554, 576, 579
Receptor for hyaluronan-mediated motility (RHAMM) 108, 125, 126, 129–143, 154, 155, 164–168, 170, 285, 288–291, 309, 312, 328, 329, 332, 333, 352, 359, 373, 375, 457, 460, 461, 466, 507, 570
Reducing end
 modification 482
Regeneration 232, 234, 369, 461, 462, 464, 465, 541
Repair 103, 125–130, 136, 138, 139, 141–143, 153, 174, 189, 195, 225, 249, 258, 292, 341, 368–371, 374–376, 379, 380, 382, 384, 466, 484, 485, 487, 488, 492–494, 506, 507, 511, 512
Restenosis 307, 310, 314, 315
Restlane 488
Retina 224, 227–229, 231, 234, 421
Retinal pigment 224, 229, 231
Rheological characterization 343
Rheumatoid Arthritis 89, 153, 212–214, 216, 355, 462, 464, 488
RON receptor tyrosine kinases 590
RT-PCR 139, 328, 342, 564

S

Saliva 325, 400
Scar 138, 342, 345, 347, 348, 368–372, 374–376, 379, 382, 421, 428, 440, 486, 490, 492, 507
Sclera 224, 226, 231, 233, 235, 439
Sebaceous glands 399, 400
Sebocytes 398–400
γ-Secretase 106
Secretory leukoprotease inhibitor 326
Self-aggregation 250, 251
Seminal fluid 325, 404, 587
Seprafilm 429, 492, 506, 511, 513
Serous (dark) cells 400
Serum-derived hyaluronan-associated protein (SHAP) 207
Shear forces 6, 110, 529
Signal transduction 24, 84, 93, 98, 104, 105, 110, 135, 153, 155, 157–159, 165, 167–169, 171, 172, 296, 535
Sinusoidal endothelial cells 74, 212, 507, 461
Size exclusion chromatography (SEC) 42
Sjogrens syndrome 234, 235, 400, 541, 543, 544
Skin 21, 54, 72, 73, 84, 111, 127, 129, 138, 153, 189, 355, 357–360, 367, 368, 371, 373, 376, 381, 382, 395–400, 405, 406, 421, 425–427, 439, 440, 444, 465, 466, 484, 492, 509, 513, 533, 553, 554, 517, 576
Small intestine 54, 325, 586
Soft tissue augmentation 346, 377, 487, 509, 513
SPACR 223, 229, 231
Sperm
 flagella 328
 motility 329
Spinous cells 396, 399
Spreading factor 21, 585
Stabilin
 1 191, 194
 2 74, 191, 194, 198, 200
Stem cellSuperoxide
 dismutase 327, 398, 461, 478
Squamous metaplasia 403, 530, 534, 536, 541, 543
Streptococcus
 equisimilis 555
 pyogenes 553
 uberis 555, 556
Supramolecular aggregates 2
Surfactant
 lining layer 248
 phospholipids 248, 256

Synovial fluid 2, 21, 48, 53, 54, 71, 72, 78, 415, 420, 421, 429–437, 460–462, 464, 518
Synvisc 435, 488, 506, 509

T

Tear film 529–536, 540, 546–548, 550
Tendon Healing 376, 390
Terminal differentiation 396, 399, 407, 576
Therapeutic benefit 285, 293, 297
Thermoresponsive 482, 497
Thrombosis 308, 310, 311, 314
Thyroid-associated opthalmopathy 282
Tissue
  bulking agent 427, 487
  engineering 347, 377, 421, 424, 433, 443, 444, 465, 466, 475, 485, 487, 490, 494
  inhibitors of metalloproteases (TIMPs) 326
Trabecular meshwork 226, 233
Transcription factor 84, 89, 105, 157, 170, 297
Transglycosylation 25, 586, 588
TSG-6 93, 94, 191, 193–196, 199, 200, 311, 352, 374, 375, 408
Tumor necrosis factor stimulated gene 6 (TSG6) 209
Turnover 22, 71–74, 78, 102, 103, 111, 172, 173, 212, 226, 235, 293, 530, 556, 590

U

Ugi reaction 491
Ulcerative colitis 215, 309
Umbilical cord 2, 21, 71, 106, 256, 257, 416
Uptake 71–77, 103, 138, 173, 174, 212, 324, 373, 398, 478, 480, 515
Urine 74, 76, 77, 173, 207, 210, 259, 359, 404, 466, 587, 590
Urinary trypsin inhibitor 207
Urology 380
Uterotubal junction 404
Uterus 163, 404, 486

V

Van der waals association 353
Vascular diseases 324
Vehicle 298, 485, 488, 491, 494, 518, 520, 545, 546
Veins 249, 307, 314
Ventilator-induced lung injury 271, 275, 277
Versican 93, 100, 130, 191–193, 209, 223, 224, 226–228, 232, 233, 235, 252, 256, 259, 260, 277, 310, 311, 313, 406, 575
Vesnarinone 279
Viscoaugmentation 421, 424–427, 487
Viscoelastic properties 4, 16, 60, 223, 344, 345, 347, 348, 372, 415, 426, 531
Viscometry 42, 44, 48
Viscoprotection 421, 439
Viscoseparation 421, 427, 428
Viscosupplementation 421, 429–432, 434–438, 475, 486, 509
Viscosurgery 376, 421, 422
Visual cortex 232
Vitreous 1, 21, 48, 54, 189, 223, 225, 227, 228, 235, 250, 251, 261, 271, 273, 285, 355, 360, 376, 384, 415, 416, 421, 422, 424, 431, 432, 505, 509, 533, 585
Vocal
  cord augmentation 424
  folds 194, 339–349, 377, 425, 514

W

Wagner disease 235
Wound 405, 406, 408, 421, 440–442, 465–467, 484, 487, 488, 492–494, 506–508, 534–537, 545, 546, 564, 591

X

Xenopus 76, 558, 590, 591
X. laevis 76

Z

Zimm-plot 59
Zona pellucida 172, 590